The Play
of Po

The Play of Power

An Introduction to American Government

JAMES EISENSTEIN

The Pennsylvania State University

MARK KESSLER

Bates College

BRUCE A. WILLIAMS

University of Illinois at Urbana-Champaign

JACQUELINE VAUGHN SWITZER

Southern Oregon State College

St. Martin's Press
New York

Editors: Don Reisman, Beth A. Gillett
Senior development editor: Bob Nirkind
Managing editor: Patricia Mansfield Phelan
Senior project editor: Talvi Laev
Production supervisor: Dennis Para
Art director: Lucy Krikorian
Text design: Robin Hoffmann/Brand X Studios
Graphics: Academy Artworks
Photo research: Rose Corbett Gordon
Cover design: Lucy Krikorian
Cover art: Fred Otnes

Library of Congress Catalog Card Number: 95-70234

0 9 8 7 6
f e d c b a

For information, write to:
St. Martin's Press, Inc.
175 Fifth Avenue
New York, NY 10010

ISBN: 0-312-13662-5

Illustration Credits

Chapter 1: **Page 3** *(top):* Paul Conklin; *(bottom):* CNN, Inc. **7:** Greg Cranna/THE PICTURE CUBE; **9:** Bettmann; **12:** Wide World Photos; **14:** Bob Daemmrich/Stock, Boston; **18:** Bettmann; **22:** Paul Watson/Toronto Star/Sygma; **23:** Patrick Chauvel/Sygma

Chapter 2: **Page 29:** Bettmann; **32:** Bettmann; **34:** Pablo Bartholomew/Gamma Liaison; **51** *(left):* Charles Hires/Gamma Liaison; *(right):* Bob Fitch/Black Star; **58** *(left):* Daniel MacDonald/Stock, Boston; *(right):* Bob Daemmrich/Stock, Boston; **61:** Greg Nikas/THE PICTURE CUBE.

Chapter 3: **Page 69:** The National Archives; **71:** Bob Adelman/Magnum; **75:** The Metropolitan Museum of Art, Bequest of Charles Allen Munn, 1924. (24.90.1566a); **76:** Library of Congress; **77:** North Wind Picture Archives; **78:** Library of Congress; **83:** Library of Congress; **85:** Courtesy of The Council of Chiefs, The Onondaga Nation and The New York State Museum Albany; **87:** Bettmann; **90:** Bettmann; **95:** Tracy W. McGregor Library, Special Collections Department; University of Virginia Library; **100:** Wide World Photos; **104:** North Wind Picture Archives; **108:** Charles Gatewood/The Image Works

Chapter 4: **Page 117:** Bettmann; **121** *(left):* Joe Marquette; **121** *(right):* Paul Conklin; **123:** Remi Benali/Gamma Liaison; **129:** Wide World Photos; **130:** James Balog/Tony Stone Images; **140** *(left):* Frank Anderson/Lexington Herald-Leader; **140** *(right):* Wide World Photos; **143:** Rob Schoenbaum/Black Star; **154:** Wide World Photos

Chapter 5: **Page 163:** Wide World Photos; **166:** Wide World Photos; **171:** Bettmann; **172:** Wide World Photos; **175:** Bettmann; **179:** Rob Schoenbaum/Black Star; **183** *(left):* Bart Bartholomew/Black Star; **183** *(right):* Paul S. Howell/Gamma Liaison; **185:** Bettmann; **190:** Terrence McCarthy

Credits continue at the back of the book, on page R-25.

BRIEF

CONTENTS

CONTENTS

VIII

LIST OF BOXED FEATURES

PREFACE

Politics matters. Like it or not, it affects the lives of all readers of this book, as well as those of their families, friends, and fellow citizens. Therefore, learning how to understand politics makes good sense. And, since most students who enroll in an introductory American Government course will never take another political science class, a textbook such as *The Play of Power: An Introduction to American Government* can play a vital role in shaping their understanding of politics and government.

No book can achieve the goal of helping readers to understand American politics better without presenting an accurate, comprehensive, up-to-date description of the institutions, participants, and processes that produce decisions by government. The areas covered by our text are summarized in the figure appearing on the inside front cover of the text. The principal challenge that writers of an introductory American Government text face, however, is not that of providing an accurate description of government and politics, but rather the challenge of presenting this information in a way that helps students improve their ability to understand politics. Our answer to this challenge is to use a conceptual framework that is easily grasped and that serves the dual purpose of organizing information in each chapter and providing students with a set of concepts they can use to analyze current political developments.

THE CONCEPTUAL FRAMEWORK

Most students bring to the study of American Government a set of existing concepts and approaches, some sophisticated and useful, many incomplete and inadequate. This book presents a consistent, well-defined framework that students can use to improve their ability to make sense of politics. We summarize the dynamics of decision making in politics with the phrase "the play of power" and illustrate the application of this concept by drawing analogies to familiar games such as basketball and Monopoly. We use elements of the game metaphor to organize the content of each chapter, discussing the participants, their resources and strategies, the rules they follow, and the outcomes of their interaction. For easy reference, the inside back cover of the text provides a visual depiction of our conceptual framework.

To reduce the temptation to trivialize politics as "just another game," we stress the metaphor of politics as a "grand game" and take pains to underscore how profound the consequences of political decisions can be. Our testing of this approach has shown that students readily learn to transfer to politics their knowledge of how participants, rules, resources, and strategies combine to produce outcomes in games with which they are already familiar.

APPLYING CONCEPTS

People learn best by doing. Everyday skills such as cooking and gardening are not learned just by reading "how-to" books and watching videos but by cooking and gardening—that is, by applying knowledge. Consequently, throughout the book we encourage the active engagement of students in applying concepts and knowledge to the task of understanding political events. Each chapter ends with an "Applying Knowledge" feature that asks students to apply information and concepts from the chapter to a real-world scenario. In Chapter 11, for instance, students are asked to predict how various groups (e.g., men and women, blacks and whites, lower- and higher-income people) voted in the 1994 congressional election.

PROVIDING CURRENT INFORMATION AND EXAMPLES

Only by learning to interpret the political information delivered by media sources such as television, newspapers, and the information superhighway can students improve their grasp of political concepts and vocabulary. Throughout this text we provide up-to-date examples that rely on students' familiarity with current media sources. In addition, we use current examples to enhance student's awareness of the roles played by women and minorities in the political process. By the time many of our readers reach retirement age, 20 percent of the U.S. population will be Hispanic, 15 percent will be African American, and 10 percent will be Asian American. One cannot fully grasp how politics works without examining these groups' evolving role in the play of power.

PEDAGOGY

Central to our aim of helping students understand politics and American Government are eight special features intended to clarify and extend the text's conceptual framework:

• *Common Beliefs* features at the start of each chapter (e.g., "Rights as Immutable and Fair" in Chapter 5) encourage students to evaluate popular preconceptions about our government (as identified by current research and by data obtained from author surveys and national polls).

• *Participants* boxes (e.g., "Alexander Hamilton as Political Economist" in Chapter 2) focus on political leaders, institutions, and interest groups actively involved in the play of power.

• *Resources* boxes (e.g., "Intensity of Opinion as a Political Resource" in Chapter 7) emphasize how individuals and groups use their time, money, charisma, information, authority, and access to achieve their political goals.

• *Strategies* boxes (e.g., "Pornography, Feminism, and the First Amendment" in Chapter 5) focus on decisions that participants in the political process make about what goals to seek, when to seek them, what resources to mobilize, and how to mobilize them.

• *Rules* boxes (e.g., "The Motor Voter Bill—Battles over Rules in the Play of Power" in Chapter 11) illustrate how structural and procedural rules found in the Constitution, in legislation, and in informal practices shape the dynamics and outcomes of the play of power.

• *Outcomes* boxes (e.g., "Upholding Native American Fishing Rights" in Chapter 6) examine both the symbolic outcomes (e.g., emotional effects, status, and power) and material outcomes (tangible goods and services) inherent in politics.

• *Footnote* boxes (e.g., "The Nature and Content of Our Civil Religion" in Chapter 2) highlight topics that are not discussed in detail in the text but enrich the reader's grasp of the political process.

• *Applying Knowledge* features at the end of each chapter (e.g., "Identifying Elements of the Play of Power in Everyday Politics" in Chapter 1) tie chapter concepts to activities that invite students to develop and refine their critical thinking skills through active engagement with real-world issues and situations.

ANCILLARY PACKAGE

Our ancillary package has been designed to make the teaching of this text easier and more effective for the instructor as well as to stimulate students' interest in American Government and enhance their understanding of politics.

• *Instructor's Resource Manual*   This looseleaf binder, free to adopters, includes five of the ancillaries developed for this text:

- *Instructor's Manual*
- *Test-Item File*
- *Lecture Outlines and Lectures*
- *The St. Martin's Student Survey of Political Attitudes*
- *Documents*, a collection of primary-source materials

• *Instructor's Manual*   Written by co-author Jacqueline Vaughn Switzer, the *Instructor's Manual* includes the following components for each chapter: a detailed summary outline; group discussion questions; and class assignment projects that outline teaching goals and present in-class and take-home student activities. Also included is a list of multimedia sources and related transparencies and documents.

• *Test-Item File*   This test bank, developed by Cheryl D. Young and John J. Hindera of Texas Tech University, contains more than 2,000 questions in multiple-choice, true-false, fill-in, short-answer, and essay formats. This supplement is available in print, as well as on disk for DOS, MAC, and Windows.

• *Lecture Outlines and Lectures*   Produced by Jacqueline Vaughn Switzer, these materials provide lectures based on each text chapter, as well as skeletal outlines for the chapters. Also included are lists of additional readings on the lecture topics. This supplement will be provided in print as well as on disk.

• *The St. Martin's Student Survey of Political Attitudes*   Produced by Clyde Wilcox of Georgetown University, this questionnaire allows an instructor to administer, at the outset of the course, a survey of students' political issues and attitudes. When the results are sent to St. Martin's Press, they will be promptly tabulated and compared to national data. The survey will then be returned to the instructor in graph, table, or chart form for use throughout the course.

• *Documents*   This collection of key primary-source documents from American political history, compiled by Jacqueline Vaughn Switzer and available on disk, includes primary sources such as *The Articles of Confederation* and excerpts from *The Federalist Papers* as well as landmark Supreme Court decisions and memorable political speeches, both classic and contemporary.

• *Study Guide*   Prepared by Tamara Waggener and James Henson, both of the University of Texas at Austin, this supplement includes the following materials for each text chapter: an overview and learning objectives; a chapter outline; a list of key terms; a cross-chapter analysis; and a practice test containing approximately 50 questions.

• *The St. Martin's Guide to the Movies*   This supplement lists popular films that explore political themes and provides ideas for class discussion. One free movie of the instructor's choice will be provided with each adoption.

• *Transparencies*   A package of more than 60 full-color overhead transparencies is available for use in conjunction with this text. The transparencies highlight important graphs and charts from the main text.

• *Ralph Nader Presents: Practicing Democracy: A Guide to Student Action*   Written by Katherine Isaac of the Center for the Study of Responsive Law, this book presents students with the basic tools necessary for participation in government, detailing the ways in which students can lobby, employ the initiative and referendum processes, boycott, and educate the public in order to effect change.

• *Videos*   Also offered with this text are five videos developed by St. Martin's Press. These videos, each running approximately 25 minutes, include "Women and Politics," "Interest Groups in America," "Presidential Leadership," "The Selection and Confirmation of Supreme Court Justices," and "The Politics of Midterm Elections: The Case of the 104th Congress." Accompanying each of these tapes is a *Video Guide* that contains a summary of the text, a copy of the narrator's script, test and discussion questions, and a list of supplemental classroom activities.

It is difficult for anyone who has not written an introductory American government text to imagine how much hard work, vision, and cooperation this task requires, and on the part of how many people. The contributions of those who have helped make *The Play of Power* a reality deserve to be recognized here.

We wish first to acknowledge the efforts of the many professionals at St. Martin's Press who helped us to transform this book from an idea to a reality. Executive editor Don Reisman understood our vision for this text, convinced St. Martin's to publish it, and played a continuing role in its development. Acquisitions editor Beth Gillett came late to the project and contributed her guidance in seeing it through to publication. Senior development editor Bob Nirkind worked unflaggingly on all phases of manuscript preparation, coordinated multiple tasks and timetables, and provided myriad editorial and content suggestions and recommendations for our consideration. Director of development Barbara Heinssen oversaw the development process and played an increasingly important role as we neared completion of the task. Senior project editor Talvi Laev masterfully and patiently undertook the difficult chore of directing copy editing and the review and correction of galley and page proofs. Art director Lucy Krikorian furnished us with an excellent design for the book, as well as striving to ensure that the art program met often-difficult requirements. Production supervisor Dennis Para deserves credit for juggling numerous priorities and keeping the text on schedule, and associate editors Susan Cottenden and Anne Dempsey are owed a debt of gratitude for their efforts in coordinating and developing the ancillary package for this project. Photo researcher Rose Corbett Gordon merits special mention for giving us an outstanding photo program, as does copy editor Alice Vigliani for smoothing out all the rough edges and making our voices sound as one. Finally, Ed Stanford, president of the St. Martin's Press College Division, is owed our sincere thanks for providing timely and sage leadership at a critical point in the final months.

James Eisenstein would like to thank his wife, Virginia B. Eisenstein, for her ideas, encouragement, support, and criticisms, and for her outstanding editing of several chapters. He also thanks Lee Carpenter for her usual fine job of editing on a number of chapters. Finally, he acknowledges the invaluable efforts of all of the documents librarians at Penn State—especially Deb Cheney and Helen Sheehey, who repeatedly came up with hard-to-find statistics and information on short notice, and Joe Caracciolo for his help in producing computer-generated maps.

Mark Kessler would like to thank Bates College for providing funds for summer research assistants Karen Grady and Elizabeth Myrick, who did an outstanding job and contributed to the overall quality of this book. In addition, he would like to acknowledge the Department of Political Science at Bates, which provides a congenial and stimulating environment for projects that seek to integrate the experiences of historically neglected groups. He especially appreciates the many conversations that he had with Leslie Hill and with Bill Corlett, chair of the Department of Political Science, as well as the rearrangement of his responsibilities so that he could deliver his chapters in a timely fashion. Most of all, he would like to thank his wife,

Stephanie, and his children, Bob, Jen, and Hannah, for being patient, understanding, and encouraging throughout the writing of this book.

Bruce Williams would like to acknowledge several special debts. One is to his excellent research assistants, Peter Schirmer and Catherine Warren. Both cheerfully put up with numerous and probably unreasonable last-second demands for still more information from obscure sources. A second acknowledgment is to the truly wonderful library resources and personnel at the libraries of the University of Illinois. Most important, he owes a debt to Andrea Press and Jessie Press-Williams, who put up with a harried and distracted husband and father every time the UPS parcels from St. Martin's arrived at the front door.

Jacqueline Vaughn Switzer wishes to thank her husband, Steve, for his editorial assistance, patience, and love, and wishes to dedicate her contributions to this project to her students and teachers, from whom she has learned throughout the years.

We would also like to thank the members of our focus group, who provided us with insightful and constructive detailed reviews of our manuscript and gave up a weekend in order to meet with us to discuss this project. These individuals are:

Robert Delorme, California State–Long Beach
Harry D. Lawrence, Southwest Texas Junior College
Edward Platt, Indiana University of Pennsylvania
Lance Widman, El Camino College
Cheryl Young, Texas Tech University

Finally, we would like to express our sincere thanks and gratitude to the many instructors whose reviews of our manuscript were instrumental in the development of this text:

John R. Baker, Wittenberg University
Robert C. Bradley, Illinois State University
John Brigham, University of Massachusetts–Amherst
Kent M. Brudney, Cuesta College
Charles S. Bullock III, University of Georgia
Barbara C. Burrell, University of Wisconsin–Extension
Ronald J. Busch, Cleveland State University
Beverly A. Cigler, Pennsylvania State University–Harrisburg
Allan J. Cigler, University of Kansas
Phillip J. Cooper, University of Kansas
Paige Cubbison, Miami Dade Community College
Michael X. Delli Carpini, Barnard College
Maria de los Angeles Guido, Bentley College
Georgia Duerst-Lahti, Beloit College
Christopher M. Duncan, Mississippi State University
Larry Elowitz, Georgia College
Traci Price Fahimi, Irvine Valley College
Isao Fujimoto, University of California–Davis
F. Chris Garcia, University of New Mexico
Paul A. Gough, Northern State University
John C. Green, University of Akron

Kenneth P. Hagler, University of Maine
Meredith A. Heiser, Foothill College
Kathleen P. Iannello, Gettysburg College
W. Martin James, Henderson State University
Joseph Losco, Ball State University
Janet M. Martin, Bowdoin College
Michael W. McCann, University of Washington
Andrew Milnor, SUNY-Binghamton
Alan D. Monroe, Illinois State University
David Nice, Washington State University
Richard G. Niemi, The University of Rochester
Bernard O'Connor, Eastern Michigan University
Barbara L. Poole, Eastern Illinois University
Craig Ramsey, Ohio Wesleyan University
Mitchell F. Rice, Louisiana State University
Gerald E. Schenk, Marymount College
Diane E. Schmidt, Southern Illinois University of Carbondale
Larry Schwab, John Carroll University
Allan Spitz, University of Alabama in Huntsville
James A. Stever, University of Cincinnati
David L. Sudhalter, Bridgewater State University
Sharon A. Sykora, Slippery Rock University
Clifford J. Wirth, University of New Hampshire
Raymond B. Wrabley Jr., University of Pittsburgh at Johnstown

James Eisenstein
Mark Kessler
Bruce A. Williams
Jacqueline Vaughn Switzer

 **JAMES EISENSTEIN** (Ph.D., Yale University) is a professor in the Department of Political Science at The Pennsylvania State University. The author or co-author of numerous books on the operation of the legal process (including *Council for the United States, Felony Justice,* and *The Contours of Justice*), he now researches campaign finance. He has taught the introductory American government course since 1967 and has received a university-wide award for excellence in teaching.

 MARK KESSLER (Ph.D., The Pennsylvania State University) is a professor in and chair of the Department of Political Science at Bates College, where he specializes in public law, judicial politics, and administrative politics. He is the author of *Legal Services for the Poor: A Comparative and Contemporary Analysis of Interorganizational Politics,* as well as numerous articles in such journals as *Law & Society Review, Judicature, Law & Policy,* and *Administration & Society.*

 **BRUCE A. WILLIAMS** (Ph.D., University of Minnesota) is an associate professor at the University of Illinois at Urbana-Champaign. His teaching and research interests center on environmental policy making, grassroots citizen participation, and the influence of the mass media on democratic politics. He has recently co-authored *Democracy, Dialogue, and Environmental Disputes: The Contested Languages of Social Regulation* and *Public Opinion: Communication and Social Process.*

 **JACQUELINE VAUGHN SWITZER** (Ph.D., University of California, Berkeley) is associate professor and chair of the Department of Political Science at Southern Oregon State College. After teaching at the University of Redlands, she worked in the private sector for 10 years, developing expertise in the fields of campaigns and elections, the criminal justice system, and environmental policy. She is the author of *Environmental Politics* and frequently serves as a political analyst for the media.

The Play
of Power

CHAPTER 1

Understanding the Play of

Power in American Politics

POLITICS IS INESCAPABLE. Whether you are listening to the radio, watching CNN or the nightly news, reading a newspaper, or chatting with friends, you can hardly avoid hearing about politics and politicians. The Supreme Court announces a decision, a congressional committee debates health care legislation, the president's approval rating drops, a candidate attacks the record of an opponent—the list is endless.

How do we make sense of all this information? In many respects, understanding politics involves the same type of challenge that we encounter in many other aspects of our lives. Certain basic knowledge is required. For instance, without knowing that basketball pits two teams (each having five players) against each other, utilizes a ball that is passed or bounced on the floor, and awards points for field goals and free throws, a person watching or hearing a game can be utterly confused. Of course, most of us manage to indirectly acquire a rudimentary knowledge of things such as basketball and politics. Likewise, although we may not be able to describe in detail how Congress works or what specific powers the president has, we nonetheless have at least some knowledge that helps make sense of such things.

Understanding, however, requires more than facts. To be fully comprehended, facts have to be placed in context and related to one another through the use of *concepts* or *analytical tools*. That humans use such *mental tools* to understand the world is hardly surprising. After all, humans differ from other species in their extensive use of tools. For example, we

Americans obtain most of their knowledge about politics indirectly, through the mass media. Most information sources identify themselves with a familiar logo or image. How many of the sources pictured here do you recognize? Can you think of other categories of sources that you encounter regularly?

3

use such things as books, pliers, and computers, whereas baboons and dolphins do not. While concrete, physical tools enable us to manipulate the material world to awesome effect, but conceptual tools such as language, ideas, approaches, and metaphors also provide us with powerful means to understand the world. Indeed, humans' use of mental tools has led both to the invention of physical tools, machines, and medicines and to the techniques by which we use them effectively.

We all employ conceptual tools in our daily lives. Why did our basketball team lose a close game? Why did a promising evening turn into a disaster? What questions will the instructor ask on the next exam? Why does so much of every paycheck go to pay taxes? The answers people give to such questions draw heavily on conceptual frameworks. **Conceptual frameworks** consist of a set of concepts, beliefs, and predispositions along with notions about how to use the concepts to accomplish specific tasks. They provide a complex set of lenses through which people view and act in the world.

Conceptual frameworks are often implicit and incomplete rather than explicit and detailed. People may not be fully aware of their content or even that they possess them. Nevertheless, everyone uses a variety of conceptual frameworks all the time.

Suppose, for example, intelligent beings from outer space asked you to take them to your leader. You probably would not take them to a large tree, an automobile, or a movie star. But you might choose a minister, a relative, a gang leader, or a government official. Responses would differ because people use different conceptual frameworks in identifying a "leader."

Different conceptual frameworks lead to different courses of action. What might the residents of a quiet residential neighborhood decide to do to prevent an all-night convenience store from opening nearby? Depending on their conceptual framework(s), the residents might undertake any number of courses of action, including holding long periods of fervent prayer, drafting a letter of protest to the newspaper, hiring a lawyer to represent them before the zoning board, scheduling a meeting with the mayor, and/or sending a campaign contribution to their representative on the city council.

Conceptual frameworks differ in how accurate a view of the real world they provide and how effective they are in achieving specific purposes. This is expecially true of the conceptual frameworks people use to understand politics. This book assumes that the conceptual frameworks people bring to thinking about politics are not as fully developed or as helpful as they can be. A major objective, then, is to present new concepts and relations among them, a framework that is more accurate and does a better job of helping you understand politics. We hope you will benefit from considering a new conceptual framework that may strengthen your existing one.

Understanding politics poses a formidable challenge. After all, the structure of the American political system is extremely complex. It takes time and effort to learn the basic facts about this structure, the rules governing it, and the participants involved in making it work. As we will show in later chapters, many people are not interested enough in politics to make much of an effort to understand it. Others actively dislike politics and everything it represents. Finally, the quality of the factual knowledge and the conceptual frameworks employed by people often does not enable them to understand politics as well as they would like.

This textbook's fundamental goal is to improve your ability to understand American politics and the workings of government. We hope to achieve this goal by replacing those concepts and beliefs about politics that do not work well with others that work better. *In other words, we seek to offer factual knowledge and an accompanying conceptual framework that together will help you make sense of politics.* Moreover, the better you understand politics, the better are your chances of being able to participate effectively in it.

Why bother to improve your ability to understand American politics and government? Essentially, because *politics matters.* Whatever your ethnic background, gender, race, age, or class, you cannot escape the effects of politics any more than you can resolve to stop breathing. Indeed, the quality of the air you breathe, the money you pay to keep that air clean, and even how long you continue breathing at all depend on political decisions.

The rest of this chapter provides the foundation for the remainder of the book. We begin by examining the beliefs many readers bring to the study of politics. Next, we define two crucial terms: *politics* and *government*. We conclude by presenting the contents of a conceptual toolbox, based on our view of politics as "the play of power," that will enhance your ability to understand politics.

COMMON BELIEFS

POLITICS AND GOVERNMENT IN THE UNITED STATES

 Like Americans in general, readers of this text will differ in their knowledge of politics and level of political sophistication. In discussing some beliefs about politics typically held by students enrolled in an introductory course, we focus on those held by people with little political sophistication and background. In doing so, we are not implying that some readers are ignorant or wrong. In fact, some of these beliefs correspond in part to our own view of how politics really works. Rather, we list these preconceptions to help you recognize that you, like every other reader, come to this text with a conceptual framework already in place and in use.

Common beliefs about politics include the following:

1. The governmental system is ruled by law.
2. The United States Constitution is the prime embodiment of law.
3. The federal government consists of three branches—Congress, the president, and the Supreme Court—with a *separation of powers* where Congress makes the laws, the president executes them, and the Supreme Court interprets them according to the Constitution.
4. Governmental policy reflects the will of the public.
5. Elected officials make most of the important decisions, and the public controls and instructs these individuals through the power of public opinion and the outcome of elections.
6. Laws establish rights, define policy, and determine what happens in the real world.
7. Many politicians cannot be trusted and are dishonest; politics is therefore distasteful and unseemly.

From this book you will learn which of these beliefs reflect reality and which do not. Indeed, each of the following chapters begins with a brief discussion of some widely held beliefs about the topics to be covered in that chapter. Although a few of these initial beliefs provide helpful insights, most do not work well when they are used to try to understand politics and government in the United States.

Later in this chapter we introduce our own conceptual framework, captured by the phrase "the play of power." We believe it provides an effective framework for understanding each chapter's content. In addition, the diagram on the inside front cover of this text summarizes the patterns of interaction among the institutions and political participants discussed in later chapters. When we refer to the concept of "the play of power," we are referring to the general scheme presented in the diagram. ■

WHAT DO WE MEAN BY POLITICS?

Political scientists rely on two broad definitions of politics. The "classic" or traditional view defines **politics** as the *processes producing authoritative decisions of government that determine who gets what, when, and how.*[1] This definition captures the essence of politics: the combination of conflict and cooperation that people engage in collectively to determine in an official and approved way how to use the power possessed by governments. Much of the political process involves deciding how to distribute society's material goods and symbolic rewards when there are not enough of them to go around. Scarcity, which appears to be a fundamental condition of life, breeds conflict. Politics both generates and resolves conflict over who gets what.

Although it is a good initial definition of politics, the classic view is too narrow for our purposes. By focusing on decisions that allocate scarce resources, it diverts attention from the point that by acting together people can accomplish tasks and undertake projects that individuals, corporations, and interest groups cannot. In other words, politics should also be seen as a means of making possible, through cooperation, the provision of *collective benefits* to society that otherwise would not be available. This topic is discussed further in "Outcomes: The Nature of Collective Benefits."

Another problem with the classic definition of politics is that it emphasizes the decisions of government, thereby ignoring decisions made outside government that also profoundly affect people's lives. For example, when an industrial corporation closes a manufacturing plant or spends money to reduce the plant's emissions of toxic chemicals, the action affects people's lives as much as do most governmental actions. Of course, government can encourage, permit, discourage, or prohibit such actions, a fact that blurs the distinction between the "public" aspects of life shaped by politics and the "private" realm of life beyond the scope of government. Indeed, some political scientists find that few aspects of life escape the realm of politics; they argue that "the personal is political."[2] In this view, decisions made in families are as central to politics as those made in Congress or the White House.

These shortcomings cause us to favor the second, broader definition of politics, which is especially effective in focusing attention on the fate of groups that some political scientists (and their textbooks) have largely ignored, particularly women and ethnic and racial minorities. As political scientist Barbara J. Nelson concludes in her study of the development of political science as a discipline, "The message of the traditional study and practice of politics is that women do not belong."[3] Nelson notes that many of the founders of political science opposed giving women the right to vote and that several textbooks in use today pay little attention to women. One reason for this is that under the traditional definition of politics, women, African Americans, Hispanics, Asian Americans, and other groups that have been excluded from participation in governmental decision making are ignored.

We can formally state the second definition of politics as follows: *Politics consists of influence processes involving both conflict and cooperation, and occurring both within and outside of government, that authoritatively determine for a wide range of groups who gets what, when, and how.*

Although much of this book focuses on politics within government, we will use the broader definition to (1) discuss the role that previously ignored

THE NATURE OF COLLECTIVE BENEFITS

Most of what Americans read and think about politics reinforces the view that politics involves conflict and struggle. Insurance companies, hospitals, doctors, employers, and consumer groups argue about who should pay how much to provide what kind of health care to which people. Two candidates wage a vigorous campaign to determine who will win office and who will lose. Indeed, much of the terminology used to describe the play of power adopts language used to describe war and conflict: an election "campaign," a legislative "struggle," a debate in which one candidate was "on target" and "shot down" the "indefensible" claims of the other. In fact, many students of politics simply define it as conflict over the allocation of inherently scarce goods.

This view overlooks an important dimension of politics. The existence of politics in a social organization makes possible important outcomes that individuals acting alone or in families or small groups cannot produce by themselves. These are known as *collective benefits* or *public goods.*[1] Common examples include a strong national defense, clean air and water, and a system of public parks open to all.

The most distinctive feature of collective benefits is that they must be provided by government if they are to be provided at all. For example, no individual state, major corporation, labor union, or university could undertake the provision of armed forces to defend the United States. The task is too immense and the cost is too great. Another characteristic of collective goods is that everyone benefits from the outcomes they produce. If one person breathes clean air or enjoys the security of a strong defense, everyone else does too. In other words, no one can be excluded from consuming public goods. Finally, public goods cannot be financed voluntarily or by charging user fees. Because no one can be excluded from enjoying the goods, there is a strong temptation to avoid paying for

Even the largest private corporations would find it difficult to finance construction of a major highway interchange. Those who could probably would not, because it would not pay. Only government can provide some public goods.

what will be consumed anyway. Governments therefore require everyone to help provide public goods through taxation and other measures.

The ability of government to produce public goods makes it possible to provide new and important benefits such as better health care, cleaner air and water, and good roads that would not otherwise exist. An important aspect of the play of power involves deciding which of these cooperative possibilities will be realized. Of course, this does not mean that collective benefits are easy to achieve. Disputes arise over which public goods will be provided and who will pay for them. But these struggles differ in their essential nature and character from disputes about, for example, how high taxes on tobacco products will be or how much money will be appropriated to feed poor children.

[1] Political scientist Anthony Downs has written extensively on this topic. See, for example, "An Economic Theory of Political Action in a Democracy," *Journal of Political Economy* 65: 2 (1957), pp. 135–50.

groups have played, and (2) describe how the political system's decisions (including decisions to do nothing) have affected these groups. In doing so, we will present findings by political scientists who research how such groups affect and are affected by politics.[4]

WHAT DO WE MEAN BY GOVERNMENT?

Much of politics, and even more of what the news media or personal conversations convey about it, deals directly with the decisions of government. The term **government** refers to the formal institutions used by a society to make and enforce policy decisions about which collective benefits will be provided, in what quantities they will be provided, and who will pay for them. These institutions include the presidency, the United States Congress, state legislatures, and even the local traffic court that collects parking fines, along with hundreds of other federal, state, and local governmental bodies.

Several characteristics of the decisions made by these institutions qualify them as governmental. One is simply that they apply to everyone within the geographical boundaries of the society. More important, these decisions are *authoritative*—that is, they are backed by the possibility of the legitimate use of force.[5] For these decisions to be *legitimate*, members of society must believe that the decisionmaker has the right to decide, that he or she is competent to decide, and that the decision ought to be obeyed.

THE PLAY OF POWER AND THE GRAND GAME OF POLITICS

"The play of power" aptly describes the conceptual framework we bring to the task of understanding politics and government. This framework relies principally on one of the most frequently employed approaches to politics, an approach that views the "stuff" of politics as the exercise of *influence*. We will define this critical term in detail later. For now, your understanding of influence should be based on how you use the term in everyday language.

Political scientists have also used a concept closely linked to influence in studying politics: the concept of *power*. Although it is related to the notion of influence, power often involves physical coercion and force. It can and sometimes does replace the gentler mechanism of exerting influence to achieve goals. A neighborhood bully, a spouse abuser, an armed gang, and an attacking army all seek to impose their will by force.

In the normal course of politics in a stable society such as the United States, the part played by raw force in the public sphere of politics is small. However, governors and even the president sometimes dispatch troops to quell civil disturbances, and police often resort to force in making arrests. These infrequent instances of the use of raw power provide compelling evidence that the threat of force underlies the operation of politics. As we suggested when we defined *government*, the ability to employ physical coercion legitimately gives governmental actions an authoritative quality. However, another definition of power recognizes its possible use to accomplish shared goals and to empower people rather than merely coerce them.

Governments in stable societies rarely resort to raw physical power to exert influence. When they do, it is regarded as an extraordinary event that captures the attention of the entire nation. Urban riots in Detroit in 1967 caused President Lyndon Johnson to dispatch the federal troops pictured here. How many instances of raw coercion can you think of that have occurred in the United States in the past five years?

Just as definitions of politics that go beyond conflict broaden the understanding of politics, so do more expansive definitions of power.[6]

UNDERSTANDING THE PLAY OF POWER

In describing how the political process determines who gets what, political scientist Charles Lindblom uses the phrase *the play of power.* Lindblom writes that he uses the word *play*

> to evoke the same meaning as when one speaks of the state of play, say in a chess or football game or in a complex maneuvering in Congress to override a veto. The "play of power" suggests more complex and closer interconnections than are captured by the flat term "interactions" or by the general term "politics."[7]

The word *power* in the phrase captures the elemental fact that although influence normally animates political interaction, force lurks in the background. The term is so apt, in fact, that we have adopted it in the title of this textbook. The phrase *the play of power* is intended to remind you that power and influence lie at the heart of the political process.

How can we approach the task of describing and understanding the play of power? Our quotation from Lindblom suggests that knowledge about the playing of games is useful in analyzing the play of power. We therefore employ a game metaphor to better convey the complex interconnections among the elements of the political system depicted on this text's inside front cover.

The use of metaphors involves borrowing concepts from something familiar to make sense of something else. It is not unusual to use a metaphor to enhance understanding. All people use metaphors as mental tools. One book on this topic notes that "metaphor is pervasive in everyday life, not just in language but in thought and action. Our ordinary conceptual system, in terms of which we both think and act, is fundamentally metaphorical in nature."[8]

We are not the first to find the game metaphor useful. Many political scientists and journalists who write about politics use either the phrase *the game of politics* or specific images drawn from the notion of games. In fact, some political scientists employ concepts derived from the mathematically based theory of games to examine topics such as how nations negotiate with one another or how members of Congress interact with their constituents.[9] A leading journalist, Hedrick Smith, has even written a popular book about politics entitled *The Power Game* that uses a game metaphor to understand politics in Washington.[10]

USING THE "GRAND GAME" METAPHOR

However, there are dangers in thinking of politics as a game. The most serious danger is that use of the game metaphor will trivialize politics as "just another game." It is *not* just another game. As a prominent British politician aptly noted, "Politics is a blood sport."[11] Although a few other games (e.g., poker) have significant consequences beyond the psychological pleasure of "winning," none carries the importance of politics. The stakes of politics make it uniquely important and compelling. Politics matters. This point is so crucial that we elaborate on it in "Outcomes: Politics Ain't Beanbag."

Politics differs from most games in several other important ways. First, it is much more complex than most phenomena brought to mind by the word *game*. We use the term *grand game* to convey both the solemn significance and the complexity of politics. Politics is complex in part because the grand game consists of a multitude of "games within games." Just as sportscasters might call attention to the confrontation between two opposing star basketball players who are guarding each other, politics involves many games within games throughout the political process. Some players engage in a number of different games at the same time. As will be shown in Chapter 14, for example, the chief executive must simultaneously court public opinion, negotiate with foreign leaders, and work for the passage of legislation in Congress. Actions taken in any one of these spheres affect the outcome of efforts in the others. Complexity also characterizes the grand game of politics because of the large number and variety of players, the numerous and complicated rules that shape it, and the variety of strategies used by participants.

Second, the *authoritative* nature of the decisions reached through the grand political game distinguishes it from basketball and most other games. Raw physical force can be used legitimately by government to enforce political decisions. The nightstick and firearm that police officers carry, their

"POLITICS AIN'T BEANBAG"

The metaphor of politics as a grand game provides useful conceptual tools for understanding the play of power, but it also carries a great danger. By comparing politics to a game, we risk trivializing it.

One way to avoid allowing the metaphor of the grand game to diminish the significance of politics is to recall that "politics ain't beanbag." Beanbag is a form of dodgeball played in elementary school in which some players stand in a circle and others stand in the middle. Those in the middle try to avoid being hit by a small bag. Players can feel it when they get hit with the beanbag, but it doesn't really hurt.

Politics isn't a game like beanbag. It can hurt. In fact, political decisions carry profound consequences for human beings. Politics can lead to war, both foreign and civil. The history of the twentieth century is filled with horrifying examples of mass slaughter planned and carried out by political authorities to achieve political ends. The holocaust perpetrated by Hitler's Nazi Germany is perhaps best known. Stalin's regime killed many millions of Soviet citizens as part of its political program. Furthermore, such dire and terrifying instances of the extremes to which politics goes are not a thing of the past. Iraq's attacks against its Kurdish and other minorities, the "ethnic cleansing" in the former Yugoslavia, and the massacres of Tutsis by government troops and armed civilian

mobs in Rwanda provide chilling examples from the 1990s.

But one does not have to point to such profound and disturbingly frequent consequences of politics to understand its significance. Political decisions affect every aspect of people's lives. Throughout this book, we will describe the play of power surrounding decisions about who can vote, what restrictions are placed on free speech, which people will be enriched by government subsidies, who will pay how much in taxes, who may die for lack of adequate health care—the list goes on and on.

Much about politics turns people off. Politics is complicated and hard to understand. Negative campaign advertisements; news of politicians embroiled in scandal; bitterness over unemployment, arrest, or some other turn for the worse that is blamed on politics and politicians; and the negative picture of both painted by the news media make politics unappealing to many people. But there is another side to politics. It enables citizens to enjoy, through cooperation, collective benefits that otherwise would be unavailable. Regardless of people's attitudes toward politics, it is inevitable. Although it may be true that nothing in life is certain except death and taxes, it is also true that both death and taxes—and much else that is important in life—result from the play of power.

arrest of protesters, and other rare instances in which police or military coercion enforces decisions remind us that physical power lurks permanently in the background and guarantees that the grand game of politics is like no other game.

CONCEPTS FOR UNDERSTANDING
THE PLAY OF POWER

Conceiving of politics as a grand game works well because it suggests a number of useful conceptual tools drawn from games familiar to most people. For instance, basketball conjures up such notions as players, rules, strategy,

First elected to the Senate in 1958, Senator Robert Byrd (D-W.Va.) acquired detailed knowledge of the Senate's rules and developed unmatched skill in using them to exert influence. Such participants assume critical roles in the play of power. Byrd displayed his skill in early 1995 when he led the fight to defeat Senate passage of the balanced budget amendment to the Constitution.

conflict and cooperation, skill, the ebb and flow of strong emotion, winners and losers, and the like. Throughout this chapter we will rely on widespread familiarity with games such as basketball and Monopoly to introduce the principal conceptual tools that will help forge an understanding of American government and politics. We will apply these tools not only to familiar institutions such as Congress and well-known interest groups but also to groups such as women and minorities whose role in the play of power has traditionally been overlooked.

THE PARTICIPANTS

In basketball and Monopoly, the players are the principal participants. Of course, the banker in Monopoly and the referee in basketball play somewhat different roles. But both games involve only a handful of players, most of whom do similar things and participate at the same level of activity.

In the grand game of politics, tens of millions play—at one time or another, for brief periods, in a limited way (e.g., by voting). Further, their level of participation varies immensely, as do the activities in which they engage. The vast majority participate infrequently. Of course, a tiny percentage of the population engages in politics virtually full-time—the president and the cabinet, legislators, lobbyists, political reporters, the heads of governmental bureaucracies, and the rest of the cast of principal participants we will encounter in this textbook. These individuals are immersed in a network of games within the grand game, each game affecting the strategies they pursue in the others. Another small group devotes only a portion of its time to politics as a sideline to its principal occupation. This group includes labor union officials, corporate executives, local and state political party officials, members of local school boards or town councils, campaign managers, and unsuccessful candidates for public office. Finally, an even smaller collection of characters devotes its time to studying and writing about the other participants. These are, of course, the political scientists.

Participants in politics exhibit still other differences. Some come to the game with more resources than others. They vary in the *rate* at which they use resources (how hard they play) as well as the *skill* with which they use these resources. They also differ in their *value preferences*—that is, the goals they seek to attain by participating. *Personality traits* vary as well and can affect both how the game is played and its outcomes. Many people already use such concepts in analyzing the outcomes of basketball games or other sports or in explaining the outcome of a game of Monopoly. Thus, the concepts we use to think about players in commonplace games can also be used to think about participants in politics.

Two characteristics distinguish participants in the grand game of politics from those involved in other games. First, the list of participants is long. It includes not only citizens and leaders such as U.S. senators and the president, but a host of organizations such as legislative committees, interest groups, and newspaper editorial boards. Second, as we noted earlier, the most active participants in politics engage simultaneously in a number of smaller games within the grand game. Their activities and success in each of the component games shape their behavior in the others and in the grand game itself. To understand the significance of this, imagine playing in five never-ending Monopoly games at the same time, each with similar but not identical rules, some games with some of the same people, others with entirely new sets of players. Imagine further that your money, skill, and influence along with your debts and blunders transfer from one to another. This approximates the kaleidoscopic complexity of the situation in which active participants in American politics operate.

Most citizens participate in the game of politics only occasionally and show little interest except when they think it directly affects them. Voting constitutes the most common form of political activity aside from talking (and complaining) about politics and politicians. But elections occur several days a year at most, and, as Chapter 11 will show, fewer than half of those eligible actually vote in most elections. Even fewer people attend political meetings, give campaign contributions, write to their legislator, or volunteer in a political campaign. Despite myths that extol the crucial role played by the citizenry in governing the United States, most people remain spectators. You can easily confirm this by quizzing your friends and classmates (especially those not studying political science) about their level of interest in and concern about politics.

An analogy can be drawn between participation and roles in the grand game of politics and televised "big-time" college basketball. A few specialists such as players, coaches, sportswriters, and ticket-office personnel work nearly year-round at it. Basketball forms their principal occupation, with other activities being a sideline. Meanwhile, less skilled people sporadically play in intramural leagues or pick-up games. Finally, packed into arenas are the spectators, some watching only half-heartedly much of the time while engaging in other activities. But many people never go to games, are indifferent to the various teams' fate, and may even be hostile to the sport itself.

Political activists who call for mass rallies risk losing credibility if no one shows up or attendance is sparse. Most people remain spectators rather than participants in politics, however, and are reluctant to engage in even as undemanding an activity as going to a rally. Such was the response when Democrats in San Antonio, Texas, sponsored a rally to get out the Hispanic vote.

The basketball analogy evokes another theme that will recur in this textbook: the interaction between the principal participants in politics and the public. Some people read about a team's personnel before coming to the game. They are well informed, know each player's number, and appreciate the offensive and defensive strategies. However, many in the arena are relatively uninformed and uninterested. Yet there are occasions when even they become engrossed in the game, when they begin to lend their voices to cheer or howl, when they "get into the game" and become a "sixth player." It makes only a small difference in total noise volume whether any one fan screams or remains silent. Collectively, though, on such occasions the spectators can become a significant factor in influencing the game's outcome.

Politics, like basketball, is "played" at many levels and in a variety of settings. In collegiate basketball, the NCAA finals affect far more people than a game played before a small crowd at the local college—and yet the local game may be just as important to its participants. In many ways, the politics of school boards, sewer authorities, and town councils in the United States is the political equivalent of "small-time" basketball. In some places participation and interest in the local game of politics are consistently high, whereas in others they fluctuate or are consistently low.

Obviously, the analogy between college basketball and politics can be carried too far. (For example, basketball fans do not determine the principal players by majority vote.) Nevertheless, the analogy clarifies some important points about the variety of roles, divergence in levels of participation and skill, and other differences found among participants in the game of politics.

Rules perform two essential functions in games. First, they establish the basic structure of the game, the characteristics of the terrain on which it is played, and the identities of the principal participants. Second, they affect the way in which the game proceeds. To understand the role played by rules in politics, you need to know about both functions.

The rules established by the United States Constitution, supplemented by federal statutes and by state constitutions and laws, determine the major outlines of the grand game's *structure.* We will explore the origin, content, and implications of these constitutional rules in Chapter 3. Here, however, we introduce two crucial characteristics of the play of power in the United States created by the Constitution: (1) the existence of three **arenas of government** (the *executive, legislative,* and *judicial*) in which formal governmental decisions are made, and (2) the multiple **levels of government** (the *local, state,* and *national* levels). The multiple levels of government (called federalism) established by the Constitution fundamentally shape every aspect of American government and politics; we explore federalism in detail in Chapter 4. Finally, as we will discuss later in this section, rules also mold the resources brought by some principal participants to the play of the game.

In addition to structural rules, written rules of *procedure*—found in the Constitution, in statutes, and in decision-making bodies' internal rules—shape the play of the game of politics. For example, the Constitution declares that a two-thirds majority in the House and Senate must be reached in order to override a presidential veto of legislation (an action not easily accomplished). This rule permits the president to shape the content of some bills by threatening a veto. Furthermore, rules of procedure determining how cases come to court or how legislation wends its way through Congress convey advantages and disadvantages to the various participants. For example, procedural rules in Congress make it easier to block than to pass legislation.

But a study of written rules does not tell the whole story of how rules shape the play of power. Informal, unwritten rules fill gaps in the formal rules, establish additional procedures, and sometimes amend or repeal formal rules. In Monopoly, some people place in the middle of the board all fines levied by Chance and Community Chest cards, the income tax that falls four squares after Go, and get-out-of-jail fines. Anyone landing on Free Parking then gets the money. Some players adopt informal rules allowing borrowing from the bank. A person cannot always understand a Monopoly game merely by knowing the printed rules. Likewise, basketball officials establish informal rules concerning how closely they will call the game.

Even though informal rules are unwritten, in any given game they can be discovered through careful observation. The same is true in politics, where shared understandings, customs, and expectations expand on and sometimes override the formal rules. For example, attorneys and bankers compete according to formal rules for contracts to manage the sale of government bonds. However, an informal rule may require that they make campaign contributions to the officials who awarded them the contracts if they wish to obtain further work.

The control of games by rules is shaky. Virtually everyone can cite instances in which rules have been violated. After all, people cheat. Sometimes rules break down entirely, and the game degenerates into physical violence. Living room fights, stadium brawls, and even civil wars provide familiar examples. Sensitivity to the role of rules stimulates good questions. For any aspect of the play of power, one can ask about the written and unwritten rules. Which participants are helped and which are hurt by them? Who violates them, how, and how often? How do the rules affect the timing of events? How stable is the game, and what keeps it going? How would the game change if proposals for altering the rules were adopted?

THE ROLE OF INFLUENCE AND POWER

Written and unwritten rules determine who the principal participants are, the shape of the playing field, and the boundaries of accepted behavior. But the play of the game requires activity as well. In basketball, young women and men dribble and pass, push and shove, jump and shoot. In Monopoly, players roll dice; count as they move replicas of irons, thimbles, and hats around a square board; and pass play money back and forth. (Sometimes, when rules no longer control behavior and tempers flare, the players may resort to some of the physical behavior engaged in by basketball players!)

At the heart of activity in politics is *the exercise of influence*. Like many important and complex concepts, influence at its core is simple. In the physical world the displacement of objects results from the application of energy or force, which overcomes inertia and creates movement. In the world of social relations the equivalent process that moves the behavior of people involves the application of influence. **Influence** occurs when the behavior of one person is affected by that of another.[12] Analyze any aspect of your daily life and you will see that nearly everything you do in social situations results from a complex web of influence.

Throughout this book the discussions of how influence shapes the play of power rest on three assumptions about the nature of influence. One is the *reciprocal nature* of influence relationships. A second is the role played by *resources* in serving as the basis of influence. The third distinguishes three forms of influence: *exchange, persuasion,* and *authority.*

The reciprocal nature of influence Influence flows in many directions. In a two-person situation, one person may exert more influence than the other, but even people overtly and explicitly seeking to exert influence modify their behavior to accommodate the other person. Instructors may skip pop quizzes because of students' grumbles, dirty looks, and panic-stricken faces, not simply because grading quizzes is an abomination. When an elected official asks for a large campaign contribution, an intricate exchange often results. The politician takes care in framing the request and deciding how much to request. The contributor decides whether to donate and how much. Both usually acknowledge that the contributor will in the future have access—that is, the ability to talk face-to-face—and therefore an opportunity to persuade.

Even where some participants wield more influence than others, the weaker ones enjoy at least some leverage. In assessing the play of power, then, it is always useful to ask about the balance of influence—and, when one party dominates, about the influence exerted by the weaker person over the stronger.

How resources serve as the basis of influence   The term **resources** refers to things possessed that permit influence to be exerted. Think of any relationship that you have on a continuing basis. Why can employers influence employees' actions? Why did you accept your instructor's assignment to read this chapter? What is it that allows you to influence and be influenced by friends and relatives? The answers to these questions constitute a partial list of resources: money, personality, information, authority.

Where do resources come from? In part, the resources possessed by a participant depend on rules. Rules assign the resource of authority to those who hold specific positions. In 1993, under the constitutional rule that gives presidents the power to sign or veto legislation, President Bill Clinton signed a bill entitling employees to take time off to care for sick family members. President George Bush had used the same constitutionally based resource to veto a similar bill the year before. The rules that define the duties of specific positions also facilitate the acquisition of two other crucial resources: *information* and *access* to other key participants. Still other resources include *wealth, personality, reputation,* and *skill.* If one player exerts more influence than another in the same position with the same resources, it is often because the first is more skilled. Thus, consistent winners at chess succeed because they are better at manipulating forces of equal strength. This brief discussion of the nature of resources leads to the next question: How are resources used?

Types of influence: Exchange, persuasion, and authority   There are three ways in which resources are used to shape behavior. Much influence rests on **exchange**, a form of bartering. A student may share political science class notes in return for someone else's English notes. Legislators may win colleagues' votes by offering to support each other's favorite bills. Clearly, exchange can consist of positive inducements, as in these two examples. But it can also consist of threats or complex mixtures of explicitly and implicitly communicated potential rewards and punishments.

Persuasion, on the other hand, rests on argument and logic. In effective persuasion, party A convinces party B that what party A seeks does not differ from what party B desires. Thus, an instructor may explain that in order to really learn the marvelous mysteries contained in Chapter 2, students need the "incentive" of a quiz. Likewise, a candidate may assert that voting for him or her will produce the strong and healthy America we all want.

Influence based on **authority** is perhaps the subtlest and most complicated way to affect behavior. Authority can be exercised when the person influenced has *adopted a rule to obey* certain requests. Students accept their instructors' right to make assignments and give tests. Committee members accept the decisions made by committee chairs on when to meet and who will speak next. Authority depends on the existence of *legitimacy,* defined earlier as the belief that a decisionmaker has the right to issue commands, that he or she is competent to formulate them, and that such orders ought to be obeyed. Although many people associate politics with the exercise of authority, the play of power in politics involves the frequent use of resources through persuasion and exchange.

The president exercising his authority as chief executive enjoys widespread legitimacy. Here, in one of his first actions as president, Bill Clinton (with Vice-President Al Gore looking on) is signing an executive order removing restrictions on what counselors at federally funded clinics can say to women clients about abortion alternatives. The consequence of Clinton's affixing his signature to a piece of paper was a change in policies at clinics all over the country.

STRATEGY AND THE PLAY OF POWER

So far, we have represented politics as a solemn, grand, high-stakes game. Players or participants pursue their goals by seeking to influence others through authority, persuasion, or exchange. To exert influence, they must use their resources. The resources players possess, the ways in which these can be used, and the process for making decisions depend on written and unwritten rules defining the structure of the political game and its procedures. But the image conveyed by the phrase *the play of power* suggests that a dynamic element, activity, or "play" characterizes politics. Participants must make decisions about what goals to seek when, what resources to mobilize, and how to mobilize them. In other words, like participants in other games, they must adopt *strategies*.

The importance of timing   We can begin to convey the nature of political strategies by turning to the familiar game of Monopoly. Sophisticated observers understand the importance of *when* things happen in shaping the outcome. It is not just a matter of whether a player lands on Boardwalk when he or she is low on cash. Good players know when to mortgage certain properties to build houses on others, when to conserve cash to cover the costs brought on by bad rolls of the dice, and when to sit in jail for three turns. They also make good decisions about the *rate* at which they use resources and how many houses to buy at a time.

In politics, decisions about when to use which resources and at what rate to achieve goals are even more crucial. Of course, some events unfold outside of anyone's control or strategic plans. However, many episodes in the play of power reflect conscious decisions about when to use what portion of which resources. Whenever legislation is introduced, a press conference called, or a report on a governmental agency's failures released, one can appropriately ask, "Why now?" "Who decided on the timing?" and "What resources have been committed?" Thus, controlling the timing of actions and events and calculating how many resources to engage provides participants with crucial strategic options.

Changing the agenda   Sometimes Monopoly games take on added dimensions. The game may be transformed from a contest over who "wins" by getting all the property and money to one that involves feelings of self-worth and the nature of relationships that extend beyond the game. Brothers and sisters may come to think of it not as a mere board game but as a contest over fairness or a replay of a long history of sibling harassment. In other words, *what the game is about* changes.

In politics, this can be thought of in terms of a change in the **agenda**— the content of the issues being considered. For example, participants struggle over whether discussions involving abortion should use the term *freedom of choice* or the term *pro-life,* thereby shaping the emotions aroused by such terms. Likewise, opposing candidates strive to define whether the "real issue" in a campaign is the performance of the economy or the candidates' experience in foreign policy. Consequently, participants pursue strategies to shape the content of the agenda and the definition of each of the items on it in ways that best suit their own purposes.[13]

Agendas also determine the *order* in which items are considered and the *way* in which they are formulated and presented. Less controversial matters on which agreement can be reached may be placed first on a meeting agenda, while more divisive issues are left for the last few minutes. Because the sequence in which items are considered can affect outcomes, participants seek to affect the order of agendas as part of their strategy.

Forming a coalition   The dynamics of games can change when the players form **coalitions, temporary alliances in which people cooperate and pool their resources to achieve a common goal. In Mon**opoly, coalitions develop to prevent the victory of one player when other players forgo rent and lend money to coalition members. Coalitions take a central role in the play of power in American politics, too, as the discussion on political parties in Chapter 10 will emphasize.

The nasty deterioration that sometimes occurs in children's Monopoly games illustrates another component of strategy: shifting the *scope* of the conflict (i.e., the number and type of participants involved). Calling in mom or dad is a ploy that most young Monopoly players have either used or had used against them. Participants in politics pursue similar strategies when they try to expand the scope of the conflict by bringing in allies or mobilizing spectators. One way to do this is to raise the *visibility* of the dispute by attracting news media coverage through a leak to a reporter, a press conference, or a statement from a prominent politician. Another technique for expanding the scope of conflict is to redefine *what* the game is about—that is, to alter the agenda or the nature of the conflict at issue. Political scientist E. E. Schattschneider succinctly conveys the power and importance of this strategy when he concludes, "The substitution of conflicts is the most devastating kind of political strategy."[14] Clearly, the scope of the conflict is shaped both by changes in what the game is about and by the level of visibility.

Changing the rules Players in children's games sometimes use a different strategy when they are losing: They try to change the rules. The most active participants in the play of power usually understand the crucial importance of rules. The struggle over rules, unwritten as well as written, never ceases. Less active participants, including the general public, typically regard as unimportant struggles over issues such as whether citizens can register to vote on the day of an election or when they renew their driver's licenses. The long-standing battle over how candidates for Congress finance their campaigns has also left most people unmoved. But active participants pay special attention to battles over rules. To understand the play of power, one must recognize and assess strategies that seek to alter rules.

Finally, just as the home court advantage affects the outcome of sporting contests like basketball, *where* the play of power surrounding a decision occurs also confers advantages on some players. Participants in politics seek to determine both the *level* (local, state, or federal) and the *arena* (legislative, executive, or judicial) of government at which a decision is made. In Chapter 16 we will see how the National Association for the Advancement of Colored People (NAACP) chose to fight racial discrimination in the early 1950s by challenging school segregation at the federal level in the judicial arena rather than in Congress or in state courts.

The outcome of a variety of games, from mundane ones like Monopoly and basketball to the grand game of politics, depends on what the game is about (*agenda*), when things happen (*timing*), how many (and which) participants are involved (*scope*), how they cooperate (*coalitions*), the *rate* at which resources are used, the *visibility* of the game's play, the *rules* that control its conduct, and the *level* and *arena* where decisions are made. Much of the strategy involved in the play of power consists of the choices made by participants in trying to affect each of these factors. As we will see, established groups and participants with ample resources typically employ "conventional" strategies such as running for office, testifying, and making campaign contributions, whereas groups such as gay men and lesbians, African Americans, Hispanics, and the physically disabled, having fewer resources, have often had to resort to "unconventional" strategies including sit-ins, hunger strikes, and protest marches.

SYMBOLIC AND MATERIAL OUTCOMES

Avid fans enjoy talking about the outcomes of sporting events—who won and lost, why, and how. Lovers of politics focus on similar questions. Why did Bill Clinton defeat George Bush in the 1992 presidential election? How did Clinton pull off a one-vote victory in passing his budget in 1993 and win unexpected congressional approval of the North American Free Trade Agreement? Who is winning the battle for most influential White House adviser to the president? How did the resignation of Congressman Dan

Rostenkowski as chair of the House Ways and Means Committee affect the outcome of the battle over health care reform in 1994? The more one learns about politics, the more interesting anticipating these questions, seeking their answers, and discussing the key events in the play of power becomes.

But there is more to politics than the pleasure of watching how the game is played. As we have taken special care to emphasize, politics is not an ordinary or trivial game. The consequences of winning or losing in politics easily overshadow those in Monopoly and basketball. Ultimately, politics is about who gets what. It is the nature of the "what"—the outcomes—that makes politics a game like no other.

Throughout this book we distinguish between two fundamentally different (though intertwined) types of political outcomes: *material* and *symbolic*. The "what" of politics involves both the material and the symbolic, the tangible and the intangible.[15]

Material outcomes consist of tangible goods and services that people obtain or fail to obtain as the result of politics. Some recent college students understood very well the material outcomes resulting from decisions made during the Reagan era to reduce funding for college students. They received less money and sometimes had to leave school. Decisions determining the level of welfare benefits, changing the size of subsidies to farmers, or raising the tax on beer and wine affect the distribution of a familiar material good—money. Other decisions affect crucial aspects of material conditions of citizens' lives—their health, their physical freedom, their access to schools and jobs on an equal basis, even life itself in trials for crimes that carry the death penalty. All material outcomes share a common attribute: In one way or another they affect the conditions or distribution of physical objects (e.g., buildings, people, automobiles) or money.

Symbolic outcomes, on the other hand, involve feelings. They are psychological, internal, intangible. Most childhood games and athletic contests produce primarily symbolic rather than material outcomes. The winner in Monopoly simply "wins," often without bothering to reach over and actually take the play money, plastic houses and motels, and property deeds of the last surviving opponent. With the exception of real money exchanged after a bet is won or lost (and perhaps a hoarse voice), the basketball fan's outcomes are psychological or symbolic rather than material. Think about it. Why do many people feel good or bad depending on whether "their" team gets more points than another team?

Thinking about political symbolism Most people are not used to thinking of politics as involving symbolic as well as material outcomes. But this view is crucial to a clear understanding of politics. Symbolic politics is not inconsequential. People's feelings of self-worth, security or anxiety, and joy or despair are important. If this were not so, losers in Monopoly and supporters of defeated sports teams would not feel so bad. Thus, the "what" of politics includes the emotional and psychological reactions of both spectators and active participants.

In fact, the primary way in which most citizens consciously experience and react to politics is through emotional responses to information about distant events and people. With few exceptions (e.g., the day after income tax forms must be mailed), it is not obvious that people are affected materially by government. The same is not true of symbolic outcomes. People recognize and feel the impact of symbolic politics. For example, Supreme Court decisions regarding school prayer arouse strong emotions among advocates and opponents.

Of course, ordinary citizens are not the only people affected by symbolic politics. Full-time participants also respond psychologically to the threatening and reassuring symbols of politics. Consider carefully the following observation made by political scientist Murray Edelman. He states that for most people,

> most of the time politics is a series of pictures in the mind, placed there by television news, newspapers, magazines, and discussions. The pictures create a moving panorama taking place in a world the mass public never quite touches, yet one its members come to fear or cheer, often with passion and sometimes with action.[16]

The nature of political symbols   Because symbols play such an important role in the game of politics, we encounter them repeatedly. **Political symbols** include words, gestures, objects, and images that condense and convey emotion and meaning. The national anthem, the White House, the flag, and ideas such as "the United States of America," "freedom," "George Washington," "equality," "free enterprise," "Saddam Hussein," "terrorists," "ethnic cleansing," and dozens of other symbols evoke strong psychological responses in most Americans. Although we typically remain unaware of the

A single image can have great symbolic power. Photos such as this one of an American GI killed in 1993 while participating in the U.N.-sponsored peacekeeping mission in Somalia were broadcast widely in the news media. The photo of the soldier's corpse being dragged through the streets of Mogadishu aroused shock and horror both in the public and in Congress and helped swing opinion against continued American participation in the mission. In a matter of months, all U.S. troops were withdrawn from Somalia.

strong but subtle power of political symbols, occasionally some incident provides a sharp reminder. For instance, in 1993 the American public felt outrage at the sight of a dead American GI assigned to peacekeeping duties in Somalia being dragged through the streets.

If many political symbols arouse emotion, other symbols communicate status and power. Mere possession of these symbols impresses others by indicating that the owner is powerful. In Monopoly, the psychology of the game changes when a lucky player obtains the Boardwalk–Park Place monopoly. In politics, the president is surrounded by symbolic trappings that radiate a sense of power: the presidential helicopter, a coterie of Secret Service agents with firearms and earpieces, and (until recently) an individual with the infamous nuclear launch briefcase close at hand.

Because political symbols are so potent, it is little wonder that politicians pay particular attention to their use. The next time you have an opportunity to see the president speak on television, observe the setting carefully. The chief executive will probably be surrounded by carefully chosen symbols: the presidential seal, the flag, perhaps even a family photo placed in the background but clearly within camera range.

What difference does it make that political symbols stir such powerful responses? It makes a great deal of difference. One reason is that such feelings are important in their own right. Another is that the feelings evoked by symbols can and do influence people's behavior. Depending on how they are used, political symbols can be reassuring or threatening. When symbols

Political and religious symbols, skillfully employed, can rouse normally passive citizens to fervent political activity. Shown here is a rally led by the Ayatollah Khomeini, whose effective use of religious symbols mobilized millions of people in a mass movement that toppled the existing political regime in Iran in 1979. Most successful mass movements use symbols that evoke strong emotions in followers.

reassure, they lead to passivity, acceptance of the status quo, and support for decisions made. When symbols threaten, the resulting anxiety disposes people to take action, to protest, and to mobilize. Just as spectators at a sporting event can be moved to scream in unison, shout and clap, and engage in other emotional outbursts, so too do spectators and even some more active participants in politics allow the emotions evoked by symbols to affect their behavior. Much of the dynamics of interactions between leaders and followers can be understood by looking at how the symbols used threaten or reassure and lead to passivity or action.

The ability to distinguish between symbolic and material outcomes in politics constitutes a critical skill needed to understand American politics. It helps you comprehend your own feelings and emotions. By observing and analyzing the use of symbols, you can assess leaders' strategies and make predictions about how others will respond. Perhaps most important, the play of the political game is profoundly shaped by the flow of political symbols and the resulting responses of passivity or action. In the following chapters we will consider the concept of symbolic outcomes because it is so essential and useful in understanding politics.

SUMMARY

Politics and government are deadly serious, as the consequences of thermo-nuclear war and genocide show. Because "politics ain't beanbag," learning how to comprehend and engage in it makes sense for anyone who seeks to live an interesting and fulfilling life. Politics also presents a fascinating glimpse into the complexity, absurdity, and ingenious nature of human societies, and one can derive satisfaction and enjoyment from watching the play of power unfold. If this textbook succeeds in achieving its goal of helping to enhance your ability to understand American politics, it will pay lifelong dividends to you both as a citizen and as a thinking member of society.

Throughout this book we will employ concepts derived from the metaphor of politics as a grand game to understand the play of power. You will find that the metaphor works well as a conceptual framework for several reasons. First, many aspects of politics have much in common with less solemn and significant games. Second, nearly everyone has played games and is familiar with their principal elements (e.g., participants and rules) and with how these elements interact. Third, our metaphor helps avoid the pitfall of assuming that the American political system is static. Just as many everyday games evolve over time, so does the grand game of politics. Participants, resources, rules—indeed, all aspects—constantly change, sometimes slowly and sometimes quickly.

The text begins with a broad definition of politics as influence processes involving both conflict and cooperation and occurring both within and outside government. These processes authoritatively determine for a wide range of groups who gets what, when, and how. Government is seen as the formal institutions used by a society to make and enforce authoritative decisions—that is, decisions backed by the possibility of the legitimate use of force. Although much of the rest of the book deals with formal governmental institutions, our broad definition encompasses significant decisions made elsewhere, including those made by corporations and within families.

To help readers understand the play of power, we use concepts commonly associated with ordinary games. Participants, both individuals and organizations, take a variety of roles. A few engage actively in a number of "games within games," but most citizens remain spectators most of the time, entering the game only as voters. The structure of the game and the manner in which it evolves depend on both written and unwritten rules. Thus, structural rules create three arenas (executive, legislative, and judicial) and three levels (national, state, and local) where the play of power unfolds.

Participants pursue a variety of life goals through politics. They also possess resources, including money, knowledge, skill, access, and authority, that can be used to attain these goals through politics. They do so by trying to exert influence in three ways. Persuasion makes use of argument and logic, along with political symbols; exchange involves the trading of benefits and sanctions; authority relies on others' willingness to obey the person who holds authority. Participants in politics employ a variety of strategies in using these resources, varying the timing and rate of their use, choosing the level and arena where they are deployed, trying to control the definition of issues and the agenda, and forming coalitions.

The interaction of participants exerting influence by deploying their resources under existing rules through a variety of strategies produces outcomes, the authoritative decisions that form the heart of politics. These outcomes are both material and symbolic in nature. The entire process is called the play of power.

The content of this book thus hinges around the set of conceptual tools derived from the game metaphor and the notion of the play of power. We have placed a schematic summary of the contents of the conceptual toolbox on the inside back cover for easy reference.

The following chapters employ the concepts introduced here. After we describe the context of politics in the United States in Chapter 2, each of the remaining chapters in this text follows a similar format. We begin with a glimpse of a specific event or aspect of the play of power. We then present some of the common beliefs and concepts that many readers bring to the subject of the chapter. We proceed to consider participants, rules, resources, and strategies. Finally, we conclude each chapter by considering how the subject it addresses (e.g., political parties, Congress, the bureaucracy) affects outcomes and how it fits into the play of power in general.

APPLYING KNOWLEDGE

IDENTIFYING ELEMENTS OF THE PLAY OF POWER IN EVERYDAY POLITICS

 Excerpts from an article that appeared on the front page of the *New York Times* on October 26, 1989,[1] are reproduced here. These excerpts provide opportunities to apply what you have read in this chapter about the metaphor of politics as a grand game. Read the article carefully. Then see how many of the concepts summarized on the inside back cover of this book you can apply. Do the same for a current article in a daily national newspaper.

WASHINGTON, Oct. 15—President Bush is expected to announce proposals on Thursday to allow the Government to remove dangerous agricultural chemicals more quickly from the marketplace.

The measures come in response to growing public concern that current Government regulation of agricultural chemicals does not adequately protect consumers. But even as the President prepared to announce the proposals, environmental groups and some members of Congress criticized them as insufficient in some cases and a step backward in others.

The wide-ranging proposals, the substance of which was made available by the Natural Resources Defense Council and Congressional sources, include these: . . . [A summary of the proposals followed.]

The proposals will require legislation, said Janet S. Hathaway, a senior attorney at the Natural Resources Defense Council. She said she expected a "royal battle" in Congress.

William K. Reilly, the E.P.A. [Environmental Protection Agency] Administrator, would not comment on the proposals in a meeting with reporters this morning. But he said he was "fully behind" the President's measures.

Environmental groups gave the Administration some credit for proposing at least modest changes to speed up pesticide review. . . . The environmental groups tempered their praise of the President's plan by saying it would speed the cancellation process only slightly and have a small effect on how quickly the agency would act to suspend a pesticide.

The groups were particularly upset by the Administration's plan to revise the current "negligible risk" standard for a pesticide. . . . The environmental groups assert that this gives the Government far too much flexibility. They also oppose the change that would prohibit states from creating stronger pesticide laws than the Federal Government's. Although the states have rarely used their authority to make stricter standards, opponents of the President's plan say it gives the states a backup.

[1] Allan R. Gold, "Bush Would Cut Pesticide Rules; Proposals Stir Praise and Dismay," *New York Times*, October 26, 1989, p. A1. Copyright © 1989 by The New York Times Company. Reprinted by permission. ∎

Key Terms

conceptual framework (*p. 4*)

politics (*p. 6*)

government (*p. 8*)

arenas of government (*p. 15*)

levels of government (*p. 15*)

influence (*p. 16*)

resources (*p. 17*)

exchange (*p. 17*)

persuasion (*p. 17*)

authority (*p. 17*)

agenda (*p. 19*)

coalition (*p. 19*)

material outcomes (*p. 21*)

symbolic outcomes (*p. 21*)

political symbols (*p. 22*)

Recommended Readings

Edelman, Murray. *The Symbolic Uses of Politics*. Urbana: University of Illinois Press, 1964. A powerful description of the nature of symbolic politics and its relationship to material outcomes in the realm of economic regulation.

Lasswell, Harold. *Politics: Who Gets What When How*. New York: McGraw-Hill, 1936. A classic treatment of the nature of politics that continues to influence the study of political science.

Lindblom, Charles. *The Policy Making Process,* 2nd ed. Englewood Cliffs, N.J.: Prentice Hall, 1980. A cogent, concise treatment of politics and policy, with an especially good introduction to the concept of "the play of power."

Schattschneider, E. E. *The Semi-Sovereign People*. New York: Holt, 1960. Another classic approach to American politics that provides a few simple but powerful concepts (e.g., the scope and direction of politics) for understanding the play of power.

Smith, Hedrick. *The Power Game: How Washington Works*. New York: Random House, 1988. A well-done journalistic political history of Washington during the Reagan years that draws on the game metaphor to explain the workings of politics.

CHAPTER 2

The Economic, Social, and Ideological Context of Politics

DIFFERENCES IN HOW VARIOUS countries engage in the grand game of politics are so obviously a fundamental characteristic of the world that most people give little thought to them. In each country, the play of power unfolds differently. The news media report on foreign military coups, riots at religious shrines, the arrest and torture of political dissenters, the calling of new parliamentary elections. In the United States, Bill Clinton replaced George Bush as president in a peaceful ceremony; at the same time, forces seeking a change in governmental leadership were engaged in bloody civil wars in the Sudan, Somalia, Angola, and elsewhere.

The roots of political strife illustrate many of the characteristics that account for such differences. For example, the conflict that engulfed the former Yugoslavian states of Bosnia, Croatia, and Serbia in the early 1990s highlighted the role of ethnicity, religion, culture, and geography in shaping politics. But looking merely at political regimes in crisis misses the profound effect of these factors on the normal play of power in more stable societies such as ours.

Throughout this textbook we will describe a grand game of politics whose day-to-day functioning reflects the effects of the context or setting in which it evolves. However, before applying the metaphor to understand the play of power in the United States, we need to review (1) how the economy produces and distributes the material goods that sustain life, (2) the characteristics of the people who engage in the play of power, and (3) the political beliefs they hold.

Most Americans regard the events depicted in this photo, which shows outgoing president George Bush observing Bill Clinton's inauguration, as unremarkable. Indeed, pictures showing the peaceful transition of political power from one person to another appear routinely after every election. But such transfers are unheard of in some societies. The ceremonial trappings surrounding such rituals—the formal dress, the presidential seal, the chief justice in his robes of office—convey the high purpose and solemnity of the event.

COMMON BELIEFS

KEY ATTRIBUTES OF THE AMERICAN ECONOMY

 Understanding how politics and economics interact is difficult for many people because they accept a common belief that economics and politics are and should be distinct and unrelated. Anyone who reads newspaper opinion columns or the magazine advertisements for major corporations may have encountered the following argument: Government interference in the economy produces inefficiency, waste, and poor decisions because it violates the laws of economics. Fundamental to this argument is the belief that the economic and political realms are separate; that if only government would refrain from interfering with the free market, the economic realm would run by itself according to its own principles.

The view that economics and politics are separate rests on several other common beliefs. One is that free, competitive markets determine all prices—that is, we have a **free-market economy**. Another is that economic life revolves solely around transactions between private buyers and sellers. A third holds that, by and large, Americans enjoy relative equality in income and wealth compared to citizens of other countries.

Each of these common beliefs is partly or largely incorrect. Ironically, many people who think government should not interfere with the operation of the economy nonetheless praise or blame government for the state of the economy. Common beliefs about the economy need to be based on its true nature and how it relates to politics. No approach to understanding politics can ignore how the economy affects the play of power. ■

THE NATURE OF THE AMERICAN ECONOMY

Like people everywhere, Americans want a comfortable material life. However, they desire more than just that. They not only expect to avoid starving or having to live permanently in a hovel or college dorm room; they also wish to start or continue a family tradition of enjoying a higher material standard of living than their parents have. They look to government and politics to create the conditions that will allow them to do so.

Not everyone places the responsibility for a healthy economy on the political system. In many countries people call on such things as fate, karma, or the will of God to explain unemployment, disease, and poverty, never thinking to hold politicians or rulers accountable. In fact, for much of American history the seemingly inevitable ups and downs of the business cycle, not government action or inaction, were blamed for periodic economic crises. At least since the Great Depression, however, most Americans have held the political system responsible for the economy's performance and their personal well-being.

The logic used by many people in assigning blame or credit for the state of the economy lacks sophistication. If times are good, incumbents receive the credit; if times are bad, they get the blame. In Chapters 11 and 12 we will argue that the condition of the economy and voters' evaluation of it determine the outcome of presidential elections more than anything else does. This fact alone binds economics and politics together in an unbreakable embrace, but it hardly conveys the full extent or complexity of the relationship. As in most intimate embraces, the action flows both ways. Just as economic performance affects politics, so too do political decisions shape the economy.

Consequently it is impossible to understand the play of power without assigning a major role to the structure and dynamics of the economy and to the complex interactions between economics and politics. After all, a high proportion of governmental decisions deal with economic life. Furthermore, economic groups take a major role as participants in the play of power. The resources—especially wealth—that people bring to politics also reflect the operation of the economy. Politics and economics intertwine so closely that they can no more be isolated than reheated spaghetti can be separated from its sauce. This intimate connection between politics and economics dates from the earliest days of the Republic, as discussed in "Participants: Alexander Hamilton as Political Economist."

SOME CRUCIAL ATTRIBUTES OF THE ECONOMY

Our description of the American economy begins with three general features that profoundly shape the context in which the play of power unfolds.

The success of the economy Despite notable exceptions like the severe but short-lived recession that began in 1981, the American economy has been enormously productive since World War II. The huge quantities of goods and services produced provide most Americans with a comfortable life. Despite the fact that a significant minority—including millions of children—live in poverty, in both absolute and relative terms the economy has performed well.

The strong performance of the economy profoundly shapes American politics, helping to account for two characteristics that we will consider in later chapters: *low-tension politics* and *low levels of political activity*. Most people in the United States are not spurred to intense political activity by severe and widespread economic hardship. Instead, widespread prosperity has produced a generally high level of satisfaction. When the economy is sound, people feel they can ignore politics if they wish.

The capitalistic nature of the economy The American economy is basically *capitalistic* in structure, and this structure has profound consequences for politics. Most significant is the fact that ownership and control over most of the productive resources in the economy rest in private hands. This means that if economic growth is to continue, private organizations (and the individuals who run them), especially major business corporations, must continue to invest and produce as they strive for profit. They cannot be forced by government (or by any other body) to expand production, develop new products, or build new factories. They undertake these tasks voluntarily if they perceive conditions to be favorable enough to make it profitable to do so. Thus, our economy depends largely on decisions made by private entities instead of publicly or governmentally controlled organizations.

ALEXANDER HAMILTON AS POLITICAL ECONOMIST

Alexander Hamilton enjoys a well-deserved reputation as an advocate both of a strong central government and of manufacturing interests. It is less well known that the economic policies he adopted to manage a severe debt crisis generated powerful support for the new government created by the Constitution.[1]

To finance the American Revolution, the Continental Congress borrowed money and issued bills of credit, which circulated and became a form of currency. The individual states were supposed to raise the money to redeem them, with their respective shares being based on population. When the states failed to meet this obligation, the value of the bills of credit fell. The states had issued paper money themselves, and its value also fell rapidly.

Appointed secretary of the treasury in the new government established by the Constitution, Hamilton had to decide whether to pay full value to people who held bills of credit issued during the war or to pay only the amount that the speculators who bought them at a fraction of their original value had paid. After a bitter struggle, Hamilton's proposal to pay full value was adopted, as was his argument that the huge Revolutionary War debts incurred by the states should be assumed by the new national government and honored at their face value.

In a brilliant move, Hamilton raised the money needed to fund these policies by floating new notes (amounting to loans) backed by the revenues of the new national government. This was crucial: It meant that those who held the new notes had a significant financial stake in the ability of the new central government to raise revenue in the future through taxes. If the new government failed, they would lose money. The financial community consequently provided crucial and powerful support for Hamilton's proposal to establish a national bank and to increase taxes. The Whiskey Tax of 1791 was the first national tax to be imposed; when it led

Alexander Hamilton is better known for his death in a duel with Aaron Burr than as a master strategist whose policies strengthened the fledgling government of the new United States. But his policies outlining how Revolutionary War debts would be repaid did much to strengthen the new national government established by the Constitution in 1789.

to the Whiskey Rebellion, suppression of the rebellion helped establish the sovereignty of the new government.

Thus, Hamilton devised financial strategies that built the political institutions he wanted to see established. His solution to the problem of how to handle Revolutionary War debt fostered political support both for the new national government and for the principle of regular and substantial taxation to sustain its existence and growth.

---

[1] This description of Hamilton's strategy draws heavily from an unpublished paper by James Curtis, "The American Social Contract: Economic Foundations of Political Sovereignty," an earlier version of which was presented at the annual meeting of the American Political Science Association, September 3–6, 1987.

Consequently, the people who control the productive resources in the economy exert substantial influence in the play of power. We will explore this crucial fact, which is termed *the privileged position of business,* in greater detail in Chapter 9.

It is easy to overstate the consequences that flow from the concentration of economic power inherent in the capitalistic nature of the American economy. As you will see throughout this book, many other players compete with business in the play of power. Furthermore, deep disagreements and sharp conflicts divide business on many issues. We do not subscribe to the view that business essentially runs everything. On the other hand, the fact that the private sector does not always get everything it wants does not mean it is just another participant. It is not. The control that business exercises over investment and production decisions, which in turn affect the level of employment, personal income, and growth, gives government no alternative but to seek its cooperation. Business does not get everything it wants, but it gets much of what it needs, often without asking.

A related consequence of the capitalist economy in the United States is that the important decisions made in some countries by government—what goods will be produced, what production methods will be used, where factories will be built and where shut down—are made here by private institutions. Thus, some of the significant decisions about "who gets what" that could be made by public officials are made instead by nongovernmental participants in the United States. Looking merely at what governmental officials (whether elected or appointed) do misses much of what shapes who gets what, as suggested in our broad definition of politics in Chapter 1.

The globalization of the economy Increasingly, the fate of the American economy—its workers, its industrial firms, and its banking system—depends on the actions of both governmental and private organizations located elsewhere. In fact, enterprises that have no real home country (i.e., transnational corporations) play a growing role in the world economy. Thus, our description of the economic context must include the major features of the decline of U.S. economic sovereignty.

Decisions about the money supply and interest rate levels affect the performance of the domestic economy. When these decisions are being made, the reactions of foreign banks and markets must be taken into account. For example, in the summer of 1994 many people felt that domestic interest rates could not be increased without harming the economy. But others argued that stemming the fall in the value of the dollar caused by international currency traders necessitated increases nonetheless. Raising interest rates slows the domestic economy but increases the volume of foreign investments, strengthens the value of the dollar against other currencies, and improves the "balance of payments"—the relationship between how much money goes out of the country and how much comes in.

As transnational corporations' economic impact rises everywhere, their decisions produce significant effects on the American economy. Many corporations formerly based primarily in the United States now depend heavily on profits made elsewhere, while other American firms have been purchased by transnational corporations.

Finally, advances in transportation and communication technology mean that manufacturing processes can easily be moved almost anywhere

The globalization of the world economy has reached a point where most products can be manufactured virtually anywhere. American clothing manufacturers like Levi Strauss employ low-wage workers such as these people, who work in one of its factories in Bangladesh.

in the world, including places where wages are low and environmental regulations are less stringent. In 1994, for instance, the minimum wage was $1.80 a day in Indonesia, one of the more than 50 countries in which the Levi Strauss Corporation located its more than 700 plants worldwide producing Levis and Dockers jeans.[1] New trade agreements, such as the North American Free Trade Agreement (NAFTA) approved in 1993 and the General Agreement on Tariffs and Trade (GATT) approved in 1994, facilitate such movement. The emergence of a "global hiring hall" results in the loss of low-skill jobs in the United States; meanwhile, certain highly skilled jobs requiring an educated workforce may be gained. The overall impact of the increasing mobility of labor and capital is a matter of dispute. Many fear that wage levels everywhere will be depressed, particularly in high-wage places such as the United States. Whatever the precise effects of this and other aspects of globalization may be, the interconnections between the global and the American economy, and the importance of these interconnections have become a fact of political as well as economic life.

THE NATURE OF MARKETS IN THE UNITED STATES

The idea of a free, competitive market, as we argue later in this chapter, is central in the beliefs held by a strong majority of Americans. But for a market to be completely free of governmental control and fully competitive, several conditions must be met.

It is critical that the market consist of so many producers and consumers of each product that no one of them acting alone can affect prices. Thus, no one who grows and sells prunes can raise or lower the market price by putting more or fewer prunes on the market, and no prune purchaser can affect the price by buying more or fewer of them. A related condition is that new producers can easily enter the market when prices are high and good opportunities for profit exist.

Oligopolies   However, the nature of markets in the American economy departs radically from this ideal. In many important economic sectors, **oligopolies** emerge as industries mature and as larger and more efficient firms prosper—that is, a handful of producers control a significant proportion of the market. For example, a few firms dominate the domestic production of automobiles, aluminum, steel, and electrical generating equipment. In the 1970s the four largest firms accounted for 60 percent of market shipments in synthetic rubber, 82 percent of household refrigerators, 47 percent of metal office furniture, and 93 percent of chewing gum.[2] In the airline industry, eight airlines captured 90 percent of the domestic market in 1988.[3] Even among the very largest corporations, the biggest dominate. In 1986 the top 100 firms in *Fortune* magazine's list of the 500 largest firms held over 71 percent of the total assets and accounted for almost 70 percent of the total sales of all 500 firms. Furthermore, the relative share of both assets and sales of the top 100 firms has increased while the remaining 400 have lost ground in the past 20 years.[4]

The corporate and competitive sectors   The greater the degree of concentration and market domination, the more difficult it is for the classic market mechanism to operate. Outright collusion to fix prices, or more informal and subtle practices such as "price leadership" (where the largest producer sets prices and the other major producers follow its lead), result in substantial departures from a free market.

The growing concentration in manufacturing and financial services reduces the degree of competition in the **corporate sector** of the economy—that is, the part of the economy occupied by major private corporations. This does not mean, however, that prices never reflect the intersection of supply and demand beyond the control of producers that the ideal of a free market requires. For one thing, a handful of American firms in many economic sectors face increasing competition from foreign-based enterprises making the same product. Furthermore, the economy retains a fairly substantial **competitive sector** (composed of small businesses, partnerships, and small incorporated firms) in which the conditions found in the classic ideal of a true market are approximated: A number of firms enter and leave the market rapidly, there are a number of both new businesses and failures, and profit margins and wages are low.[5] Every year, hundreds of thousands of Americans who have managed to accumulate a little capital go into business for themselves, opening restaurants, laundries, repair shops, boutiques, and the like. As a careful monitoring of the business page of almost any newspaper will reveal, these businesses often fail. In 1991, for example, over 628,000 new businesses were incorporated but nearly 97,000 businesses failed.[6]

In short, enough competition remains that one can say that prices are shaped (though not absolutely determined) by the operation of market forces. However, the factors discussed here show that the market strays far from the classic ideal of the free market. Instead, we have a **modified market economy**, one that departs significantly and often from the classic ideal.

Related to the belief that the United States has a wholly free and competitive market system is the notion that markets can provide everything a society needs, with government playing a limited role. The problem is that markets fail in at least three ways, making governmental action inevitable. The concept of **market failure** is so essential that we discuss it separately in "Outcomes: When Markets Fail."

THE ROLE OF THE PUBLIC SECTOR IN THE ECONOMY

You may have heard politicians extol the virtues of "our system of free, private enterprise." Statements like this imply that most of the economy is indeed private and that it ought to be this way. Regardless of anyone's beliefs about the proper role of government, the reality is that government engages in massive economic activity. The locus of this activity is known as the **public sector** of the economy to distinguish it from the **private sector** of the economy, which is made up of partnerships, private businesses, corporations, and individuals who engage in economic exchange. Governments' revenues and expenditures are so huge that it is impossible to understand the American economy without considering the role of the public sector.

The **gross domestic product (GDP)** is the total value of all economic activity within the geographic boundaries of the nation in a given year. The federal government takes in approximately 19 percent of the total value of GDP in revenue derived from a variety of taxes.[7] It decides how approximately 24 percent of the GDP will be spent. The nearly 5 percent difference represents the annual federal deficit. The immortal words of Everett McKinley Dirksen, the silken-voiced Republican Senate leader of the 1960s, convey the significance of the federal government's impact on the economy: "A few billion dollars here, a few billion dollars there—pretty soon it adds up to real money." Here, we are talking about more than a few billion. For fiscal year 1994 (the 12-month period ending on September 30, 1994), the federal government's estimated revenues amounted to more than $1.25 trillion and its expenditures totaled almost $1.5 trillion, producing a deficit estimated at more than $203 billion. Another measure of the federal government's role in the economy is that it dwarfs every other organization in the number of people it hires, writing the paychecks for more than 3.1 million civilian employees.

These figures do not reflect the role played by state and local governments in the economy. Their revenues amount to almost as great a proportion of GDP as the federal government's. In fiscal year 1991, state and local governments took in more than $1 trillion, close to 19 percent of that year's GDP. Moreover, state and local governments employ far more people than does the federal government. In 1992, they had 15,700,000 employees, about five times the federal government's civilian workforce.[8]

Together, federal, state, and local governments take in and spend approximately 40 percent of the gross domestic product. They employ 18.5 million people, about one in every seven people working. Governments—not individuals, corporations, businesses, or other private entities—decide how the money will be raised and spent, and they inevitably have a major impact on the economy. These taxing and spending decisions are called **fiscal policy**.

WHEN MARKETS FAIL

Many economists, beginning with Adam Smith in the 1700s, have argued that when individuals pursue their own immediate self-interest in a free-market economy, good things happen and all of society benefits. The renowned British economist John Maynard Keynes wrote during the Great Depression that

> For at least another hundred years we must pretend to ourselves and to every one that fair is foul and foul is fair; for foul is useful and fair is not. Avarice and usury and precaution must be our goods for a little longer still. For only they can lead us out of the tunnel of economic necessity into daylight.[1]

But economists have long realized that the free market often fails to translate action motivated by greed into socially beneficial outcomes, leading to governmental regulation or intervention to rectify the problems produced by such failures.

Markets fail under three general conditions. First, they fail when economic transactions produce significant "externalities"—that is, substantial costs or benefits to individuals who are not involved in the buying or selling of the goods produced. Because these individuals are uninvolved, the costs and/or benefits they experience are ignored in the exchange of goods and services. For example, the production of industrial goods is often accompanied by environmental pollution, which affects those who live near the polluting factories. Because the "costs" of environmental pollution are not included in the price of the commodity, it is underpriced and sells in greater quantities than it should. Monopolies also produce externalities by imposing on consumers costs not determined by the actual costs of production. Another externality occurs, for example, when computer programs are copied illegally. Users receive value for which they have not paid, an externality that provides them with benefits rather than burdens. To deal with externalities, governments make the copying of computer programs illegal, regulate monopolies, and mandate pollution control.

Second, markets fail to produce public goods—that is, goods and services that, when provided, are not divisible so that those who do not pay for the goods cannot practically be excluded from benefiting. National defense is the classic case of a public good. If the state of Iowa were to stop paying its share of the cost of national defense, it would be impractical to exclude that state from enjoying the same protection received by other states. Anyone who will benefit anyway has every incentive to avoid paying and becomes what economists call a "free rider." Consequently, no rational person will pay, making it unlikely that private markets will produce public goods. Because some public goods are highly desirable (e.g., defense, clean air, or national parks), government must intervene in the private market either to produce the goods itself or to create incentives for their production by the private sector.

Finally, markets fail when consumers cannot get accurate information about the quality of a product. In these cases individual consumers will be unable or unlikely to acquire the information needed to make informed choices in the market. Thus, government may justifiably intervene to require the safety of all products.

---

[1] Quoted by E. F. Schumacher, *Small Is Beautiful: Economics As If People Mattered* (New York: Harper & Row, 1973), p. 24.

However, the national government's role in the economy goes beyond the effect of its fiscal policy. Among the other things it does to affect the economy in a major way are the following:

1. Providing a set of rules about the ownership and transfer of property, a minimum level of social peace, and a secure currency and banking system.
2. Setting **monetary policy**—that is, control of the supply of money through the setting of interest rates and other devices.
3. *Regulating* such economic activities as advertising, manufacturing processes, and marketing techniques. For example, the government regulates which farmers can plant how many acres of peanuts, the design of automobiles' safety features and the amount of pollutants that can be emitted from their tailpipes, the testing and introduction of new drugs, the design of nuclear power plants, the safety features of coal mines, the labels on processed food packages, and tens of thousands of other things.
4. Setting *tariffs* and quotas on imported goods.
5. Paying a variety of *subsidies* to individuals and businesses, including *transfer payments* or subsidies in the form of cash (e.g., welfare to the poor, disability payments to veterans, and crop subsidies to farmers); *tax subsidies* in the form of special deductions, such as investment credit to businesses and the deduction of interest payments on home mortgages; *benefits in kind*, such as the distribution of free milk and cheese to the elderly; and *credit subsidies*, whereby the government helps make money available at attractive interest rates or guarantees private loans (including student loans).
6. Directly producing goods and services that could be provided by private corporations, including the delivery of mail, the collection of trash, the sale (in some states) of alcoholic beverages, and the production of electricity.

THE DISTRIBUTION OF INCOME AND WEALTH

The cumulative impact of all these activities not only affects the operation of the economy but also shapes the distribution of income and wealth. The reason is obvious. Decisions about who will be taxed how and by how much, and who will be helped (and hurt) how much by the way in which these taxes are spent, unavoidably affect income and wealth.

The implications for politics are direct and profound. Money is a powerful resource, and the shape of its distribution among the population affects which people and groups exert how much influence in the play of power. Furthermore, intense conflict arises over proposals that seek to change who possesses how much wealth.

It is important to distinguish between income and wealth. *Income*, usually measured on a yearly basis, is the amount of new money made available for use from all sources: interest, dividends, salary, and the sale of assets. *Wealth* consists of the total value of assets owned. Thus, identical incomes may represent very different levels of wealth. Someone with no assets whatsoever (zero wealth) may have the same income—say, $30,000 a year—as someone who does not work but collects a little more than 5 percent interest on a nest egg of $600,000.

The basic facts regarding income and wealth in the United States can be summarized as follows:

- Income in the United States is distributed unequally.
- Wealth is distributed much more unequally than income.
- The degree of inequality in the distribution of income and wealth is not smaller than, and in most instances is greater than, that found in other major industrialized countries.
- The distribution of both income and wealth has changed very little over time; despite periodic minor changes, the current pattern resembles fairly closely that found at the turn of the century.
- There remain substantial differences among races, ethnic groups, and sexes in their share of the nation's wealth and income.
- The tax system, which is supposed to be **progressive**—that is, taking a larger proportion of the income of those who earn more by taxing them at a higher rate—in fact does not significantly affect the distribution of income or wealth toward either greater equality or greater inequality.

Regardless of when the distribution of income is assessed, a pattern of substantial inequality results. One student of this topic noted:

> From 1947 to 1969 income inequality across families declined modestly; from 1970 to 1979 it increased modestly; and from 1980 to 1984 it increased more sharply. But the most obvious feature of the postwar family income distribution was its stability.[9]

A study by the Ways and Means Committee of the U.S. House of Representatives examined the distribution of family income between 1977 and 1990 by using a common technique: ranking the income of all family units, dividing them into 10 equal groups (called *deciles*), and calculating what percentage of the total income of all families was received by each decile. If every family received exactly the same income, each decile would capture 10 percent of income. If the distribution of income were very unequal, the top decile would receive 99 percent of all income and the other 90 percent of families combined would receive just 1 percent.

Table 2.1 (on p. 40) presents the results. During the 13-year period studied, the distribution changed hardly at all, with the exception of the tenth (richest) decile. The degree of inequality in incomes is striking, as shown in Figure 2.1 (on p. 41). In 1990, the bottom 10 percent of families received just 1 percent of all income, whereas the top 10 percent received over 35 percent. The top half of our nation's families (deciles 6–10 combined) received 82 percent of all income. When the distribution of income among the highest-income families is examined, the pattern becomes even more distinct: The top 1 percent captured almost 12 percent of all income, as the bottom row of Table 2.1 shows.

How does the federal tax system, which at least in principle assesses higher rates of taxation on those who earn more, affect the distribution of income? The last column of Table 2.1 presents the results of an analysis of

TABLE 2.1
DISTRIBUTION OF FAMILY INCOME, SELECTED YEARS

DECILE	PERCENTAGE OF BEFORE-TAX INCOME RECEIVED				PERCENTAGE OF AFTER-TAX INCOME RECEIVED
	1977	1980	1985	1990	1990
First	1.1	1.0	0.9	0.9	1.0
Second	2.5	2.4	2.2	2.2	2.6
Third	3.9	3.8	3.6	3.6	4.0
Fourth	5.5	5.3	5.0	5.0	5.4
Fifth	7.1	6.9	6.4	6.4	6.8
Sixth	8.9	8.7	8.0	8.1	8.3
Seventh	10.9	10.6	10.2	10.1	10.2
Eighth	13.2	13.1	12.7	12.7	12.5
Ninth	16.6	16.6	16.5	16.4	15.9
Tenth	30.6	32.0	35.2	35.2	33.9
Top 5%	20.1	21.3	24.4	24.4	23.6
Top 1%	8.1	9.0	11.6	11.8	11.4

Source: U.S. House of Representatives, Committee on Ways and Means, *Background Material and Data on Programs within the Jurisdiction of the Committee on Ways and Means,* 101st Congress, 1st Session, March 15, 1989, Table 60, pp. 1043–45.

the effect on income shares after taxes for 1990. It shows that the federal tax system nudges the distribution of income toward greater equality, but not by very much.

Good information about the distribution of wealth is more difficult to gather than information about income, and the definitions used in measuring it are more varied.[10] Although the details of how wealth is distributed are murky, the general pattern is clear. Wealth is even more concentrated at the top than income. For example, one study estimated that in 1981 the top 0.8 percent of the U.S. population owned 20 percent of the wealth.[11] Another source claimed that the wealthiest 1 percent of households controlled 37.1 percent of all assets in 1989.[12] Since then, wealth has become even more concentrated.

THE RELATIONSHIP AMONG RACE, ETHNICITY, GENDER, AND INCOME

Because of the inequality in income, and especially in wealth, a small portion of the population holds a disproportionate share of this crucial political resource. But the pattern just described does not tell the full story, because it does not distinguish which groups are likely to be found among the rich and which among the poor.

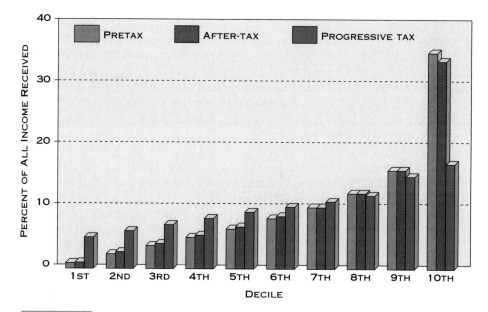

FIGURE 2.1

DISTRIBUTION OF FAMILY INCOME BEFORE AND AFTER TAXES, 1990 (BY FAMILY DECILE) If before-tax income were distributed equally, the olive bars for each family category would be the same height. The actual distribution is very unequal (see Table 2.1). A steeply progressive tax system would redistribute after-tax income as shown by the blue bars. The first few deciles would gain while the richest tenth would see its share plummet. The orange bars show actual after-tax income shares, demonstrating that the tax system hardly redistributes income at all. *Source:* U.S. House of Representatives, Committee on Ways and Means, *Background Material and Data on Programs within the Jurisdiction of the Committee on Ways and Means,* 101st Congress, 1st Session, March 15, 1989, Table 60, pp. 1043–45. Data for progressive tax supplied by author.

Women continue to earn much less on average than men, although the gap between earnings has narrowed slightly. In 1970 the median (i.e., the middle number in a sequence of numbers) yearly earnings for women who worked were just 33.5 percent of those of men; by 1980 the figure was 39 percent, and in 1991 it was 51 percent.[13] Average yearly earnings provide another measure of gender-related income inequality. In 1991 men averaged $35,850, women just two-thirds of that figure at $23,778. These differences persist even when education and age are taken into account. College-educated women in their prime earning years of ages 55 to 64 averaged $33,144 in 1991, compared to $59,089 for college-educated men of the same age group.[14]

Despite their lower earnings, the contributions made by women to the economy are enormous. In 1993, over 58 million women were in the civilian labor force, nearly 46 percent of the total.[15] Of course, these figures do not account for the substantial economic value of the unpaid work people do, especially in the home. Although the share of such work performed by men has increased somewhat, women account for the lion's share of uncompensated contributions to the economy.

Most of the gains made by women accrued to the better-educated and upper-income among them; the plight of poor women worsened. Between 1970 and 1987 the percentage of single-parent families (mostly female-headed) went from 13 percent to 27 percent of all families. Between 1980 and 1986 the income of married couples increased by 9 percent, but that of female-headed households grew by only 2 percent. The result is that despite the sizable income gains made by women, the total number of women living in poverty has also been growing.[16]

The income gap between women and men persists. In 1991, for example, the median income of married couples (which often had two wage earners) was over $41,000; female-headed families received less than half as much, about $18,000. Single women also earned less than single men ($14,321 versus $23,044 in median income).[17] The increasing numbers of women among the lowest-income earners have resulted in the *feminization of poverty.*

A similar pattern of income inequality can be found for African Americans and Hispanics. In 1987 the income of the average black family was only 56 percent of that of the average white family, the largest disparity since the 1960s. In 1992 white families' median income was over $39,000; for blacks it was $21,761 and for Hispanics, $24,900. Thus, African-American families still earned only about 55 percent of that earned by white families, and Hispanics about 63 percent.[18] By contrast, Asian and Pacific Islander families' median income exceeded that of whites by almost $4,000.[19] Furthermore, the effects of race and gender interact. In 1991, for instance, 68 percent of the children living in black (non-Hispanic) female-headed families lived in poverty; for female-headed Hispanic families, the figure was 70 percent.[20]

The implications of these figures for American politics over the coming decades are profound. The proportion of all children, but especially African-American and Hispanic children, who live in poverty is high and growing. In 1991 over 13 million children under age 16, more than one in five, lived in poverty. For Hispanics, the child poverty rate reached almost 29 percent; and for blacks, almost 33 percent.[21]

Both the magnitude and significance of these disparities in income by gender, race, and ethnicity can easily be glossed over. Figure 2.2 depicts visually just how great the income discrepancy is among various groups. If better information on inequality in total *wealth* among these groups existed, the differences would be even more striking. The persistence of such inequalities raises troubling questions about the exclusion of African Americans, Hispanics, and other groups from access to political power. The inequalities also contribute to these groups' difficulties in exerting influence because of the importance of money as a political resource. Finally, these inequalities have the potential to develop into significant political issues in their own right.

THE SOCIAL CHARACTERISTICS OF THE AMERICAN POPULATION

The population of the United States is not only large but extremely diverse. One need only look at the faces of the students in many public universities or read the last names carved into the Vietnam Veterans Memorial in

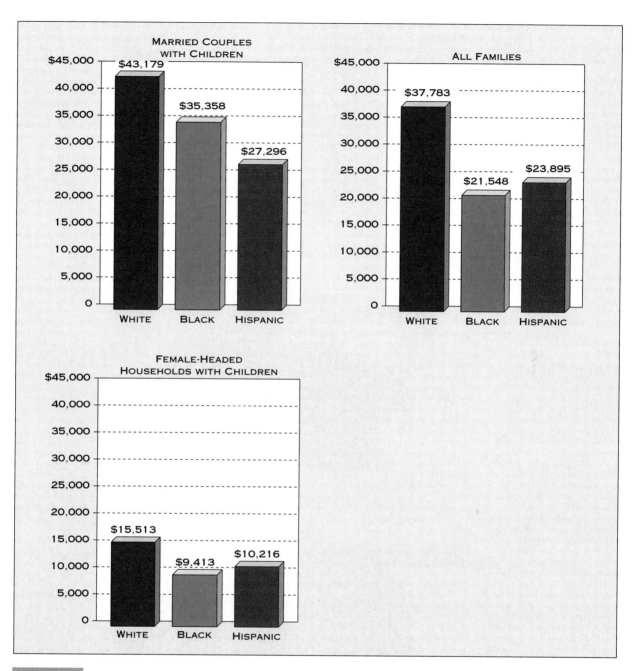

FIGURE 2.2

MEDIAN INCOME OF VARIOUS FAMILY GROUPS FOR WHITES, BLACKS, AND HISPANICS Striking differences in family median income are linked to family status, gender, race, and ethnicity. White, black, and Hispanic married couples with children earn more than all families in their group, and much more than families with children headed by females. In all three groups of families, whites receive a higher median income than do blacks or Hispanics. *Source:* U.S. Bureau of the Census, *Statistical Abstracts of the United States: 1993,* 113th ed. (Washington, D.C.: Government Printing Office, 1993), Tables 724 and 725.

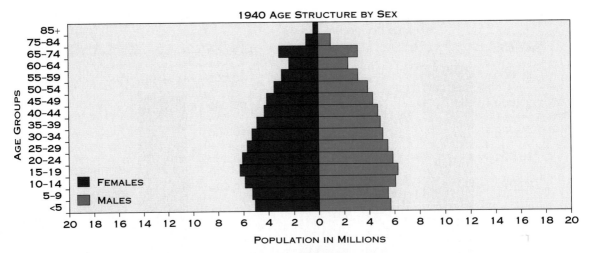

1940 Age Structure by Sex

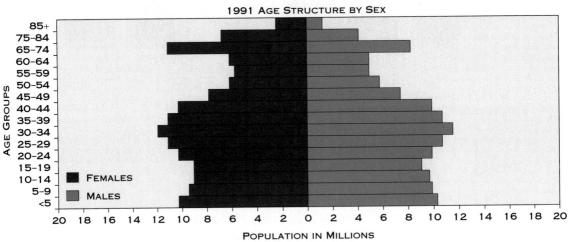

1991 Age Structure by Sex

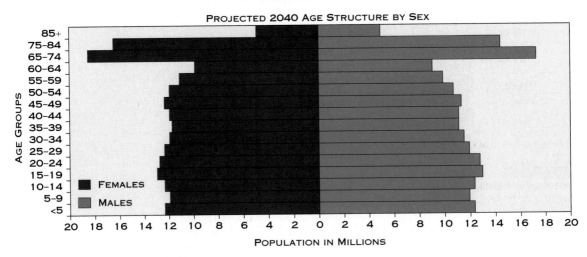

Projected 2040 Age Structure by Sex

FIGURE 2.3

THE SIZE AND AGE DISTRIBUTION OF THE U.S. POPULATION, 1940–2040 The length of the bars for 1940, 1991, and 2040 indicates that the population of the United States has grown dramatically in recent decades and will continue to do

(continues on p. 45)

Washington, D.C. Diversity and other social characteristics shape virtually every aspect of the play of power.

We begin our description by looking at changes in the number, age, and location of the U.S. population, characteristics whose political effects will be seen throughout this book. We then examine the diversity found in the races, religions, languages, and ethnicities of the American people, a diversity we call **social pluralism**. Because this social pluralism profoundly affects every facet of politics, we will describe it in some detail.

THE SIZE, AGE, AND DISTRIBUTION OF THE POPULATION

Dramatic changes have taken place in the composition of the population of the United States. Figure 2.3 summarizes two of the most significant changes. First, the number of people living in the United States has grown substantially. The 1940 census counted close to 132 million people. By 1991 the population had nearly doubled, to 252 million. In another 50 years the population may be over 364 million, according to one Census Bureau calculation. In Figure 2.3 the three breakdowns of the age of the population in each of these years, called *age pyramids,* are drawn on the same scale, so the increasing size of the bars depicts this growth. Second, as the shape of the age pyramids clearly shows, the proportions of older people grew dramatically in the 51 years between 1940 and 1991; by 2040 the two categories for people ages 65 to 84 will be the largest by far, and more of them will be women than men.

The population growth reflected in Figure 2.3 did not occur at the same rate everywhere, and projections suggest that regional differences will continue. In other words, as shown in Figure 2.4 (on p. 46), *where* people live is changing. As we will see later in this book, these differences have significant consequences for the composition of the House of Representatives and the role played by various states and regions in electing the president.

(*continued from p. 44*)
so. The 1940 population had fewer children as births declined during the Great Depression. Fewer than 7 percent of the population lived beyond age 65. By 1991 a bulge of people over 65, especially women, appears, totaling more than 20 percent of the population. Another bulge represents "baby boomers" between ages 30 and 45. By 2040, according to projections, there will be a huge increase in the number of people over 65, especially women, while the rest of the distribution will be almost square. The shape of the population age pyramid in 2040 has implications for the cost of health care, Social Security, and other programs that must be financed by the smaller numbers in their prime earning years (ages 30 to 60). With more aging women than men, issues involving the care of people over 65 will be linked to gender issues. *Sources:* (*top*) U.S. Bureau of the Census, *Census of the United States: 1940* (Washington, D.C.: Government Printing Office, 1941), United States Summary, p. 3; (*center*) U.S. Bureau of the Census, *Statistical Abstracts of the United States: 1993*, 113th ed. (Washington D.C.: Government Printing Office, 1993), Table 14; (*bottom*) U.S. Bureau of the Census, *Statistical Abstracts: 1993*, Table 17.

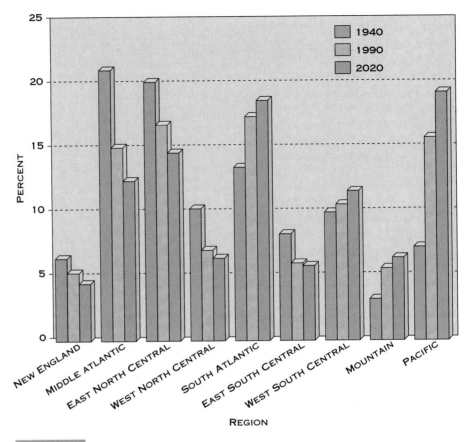

FIGURE 2.4

THE CHANGING DISTRIBUTION OF THE U.S. POPULATION, 1940–2020 Where people live makes a difference in politics. The number of electoral votes each region casts for president and the apportionment of members of the House of Representatives have changed and will continue to do so as the population shifts from the Northeast and Midwest to the South and West. For example, in 1940, states in the East held 27.3 percent of the population, whereas in 1990 they held 20.4 percent; during the same period the West jumped from a scant 10.6 percent to 21.3 percent. Based on these data, which areas of the country are likely to gain influence between 1990 and 2020, and which ones are likely to lose it? *Sources:* (1940) U.S. Bureau of the Census, *Census of the United States: 1940* (Washington, D.C.: Government Printing Office, 1941), No. 6, pp. 4–5; (1990 and 2020) U.S. Bureau of the Census, *Population Projections for States, by Age, Sex, Race, and Hispanic Origin: 1993 to 2020* (Washington, D.C.: Government Printing Office, 1993), Table 1.

SOCIAL PLURALISM

With the exception of the descendants of Native Americans, residents of the United States trace their origins to four continents. The majority, about 80 percent, came from Europe; about 12 percent from Africa; 3 percent from Asia; and most of the rest from Mexico and Central and South America. The ancestors of most immigrants came voluntarily, with the notable exception of those from sub-Saharan Africa, who were captured, enslaved, and loaded onto slave ships. Among these, almost unimaginably horrid conditions resulted in extremely high death rates en route.

The Census Bureau classifies approximately 9 percent of the population as "Hispanic," although the classification "Latino" is increasingly used in its place. Many, but by no means all, Hispanics speak Spanish; some even have ancestors from sub-Saharan Africa. The general classification "Hispanic" encompasses considerable diversity, as "Participants: Diversity within Diversity—The Case of Hispanics and Asians" (on p. 49) shows. Some are descendants of people who settled in areas that became part of the United States, including Florida, California, Texas, and other parts of the Southwest. These people did not come to the United States; rather, the United States literally came to them.

The diversity of descendants representing numerous continents further enriches social pluralism. The Spanish-speaking immigrants who joined those already living in what became the United States exhibited many combinations of racial origins: European Caucasians, African Negroes, and Native Amerindians, for example. Asia contributed significant numbers of Chinese, Filipinos, Japanese, Koreans, and Vietnamese. Virtually every ethnic group from Europe, many with a long tradition of mutual dislike, can be found in the population of the United States. The Census Bureau's 1990 breakdown of the total U.S. population of some 248,600,000 by race (which does not include a separate category of Hispanics) identified 12.1 percent of the population as black, 80.3 percent white, 2.9 percent as Asian or Pacific Islander, .8 percent as Native American, and the remainder (3.9 percent) as belonging to other races.[22]

Population projections indicate that the diversity of peoples residing in the United States will rise substantially, as is illustrated in Figure 2.5 (on p. 48). According to one Census Bureau estimate, by 2050 (within the expected lifetime of many readers), African Americans will constitute 15 percent of the population, Native Americans about 1 percent, Asians about 10 percent, and Hispanics 21 percent.[23]

Population diversity can be measured in many other ways, including by religion. Results from a 1990 survey reported in the *New York Times* found that 26.2 percent of Americans were Roman Catholics; 60 percent were Protestant (with the largest denomination, Baptists, accounting for 19.4 percent of the population); Jews, 2.2 percent; Muslims, .5 percent; and agnostics, atheists, and those with no religious beliefs, 8.2 percent.[24] Although these figures demonstrate considerable religious diversity, they do not fully convey the reality. Important differences affecting politics can be found among religious denominations within these broad categories. For example, among white Protestants, significantly more evangelicals voted for George Bush in the 1992 presidential election than did other white Protestants.[25]

CROSS-CUTTING MEMBERSHIPS AND REINFORCING CLEAVAGES

The U.S. population differs along virtually every dimension one can think of: race, religion, ethnicity, language, social class as determined by occupation and income, and even regional culture.[26] In some societies, if one characteristic of an individual is known, it is possible to predict many others with

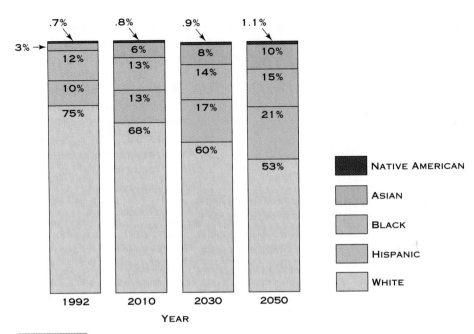

FIGURE 2.5

PROJECTED CHANGES IN THE COMPOSITION OF THE U.S. POPULATION, 1992–2050 The Census Bureau predicts small but steady increases in the proportion of Asian Americans, African Americans, and Hispanics in the population. Shown here are the Bureau's "middle series" projections, which, if correct, mean that by the year 2050 whites' share of the total population will drop from almost 75 percent to just over half (53%). Hispanics will account for over 21 percent, African Americans for 15 percent, and Asian Americans for 10 percent. The political resources of minority groups will grow, and issues involving race and ethnicity will continue to be among the most important in the play of power. *Source:* U.S. Bureau of the Census, *Statistical Abstracts of the United States: 1993*, 113th ed. (Washington, D.C.: Government Printing Office, 1993), Table 20 (middle series).

Because of rounding, percentages may not total 100.

a high degree of certainty. Thus, if one knows the caste to which someone in India belongs, one may also know much about their religious practices, their occupation and income, the hue of their skin, their language, and where they live. These attributes *reinforce* one another. And because individuals with distinct sets of reinforcing characteristics differ so much from each other, deep *cleavages* emerge between them. The term **reinforcing cleavages** refers to this pattern of social characteristics.

Successfully predicting Americans' social characteristics with a single piece of information is much harder. Church congregations can encompass people from a variety of occupations and backgrounds. There are wealthy African-American Roman Catholics, middle-class urban professional Hispanics, Irish Jews, and low-income labor union members who are white Anglo-Saxon Protestants. The people who gather in a common workplace, religious congregation, or educational institution can differ in gender, race, religion, and ethnicity. We call this diversity of combinations **cross-cutting memberships**. For example, religious affiliation "cuts across" every income

DIVERSITY WITHIN DIVERSITY—THE CASE OF HISPANICS AND ASIANS

Classifications such as "Hispanic" and "Asian" are useful for drawing broad generalizations. After all, some common interests and cultural ties bind people within each category together, as does the fact that they may encounter similar prejudicial attitudes from others. However, such categories also conceal considerable diversity among the individuals identified within them.

Hispanics come from many nations and cultures. Some have arrived recently; others are descended from those who came to what is now the United States hundreds of years ago.[1] Some speak Spanish almost exclusively, others hardly at all. They may be descendants of Europeans, Africans, or Amerindians, or a mixture. The Census Bureau currently refers to this diverse group as "Hispanics"; in the past, "Spanish," "Spanish-speaking," and "Latin American" have also been used.

The largest group of Hispanics came from Mexico, but significant numbers have immigrated from Puerto Rico as well. Even smaller numbers have come from Cuba, other parts of the Caribbean, and recently from Guatemala, Honduras, El Salvador, and other parts of Central America. In 1990 the largest group in the Hispanics population was Mexican American (5.4 percent of the U.S. population); 1.1 percent were Puerto Rican, .4 percent Cuban, and the rest from other Spanish-speaking countries.[2] The Cuban Americans who fled after Fidel Castro came to power in 1959 settled primarily in nearby Florida, especially Miami. Drawn from the more conservative and better-off segments of Cuban society, they quickly became politically active. For the most part they have supported the Republican Party. Mexican Americans and Puerto Ricans, by contrast, have favored the Democrats. However, their level of political activity was low until the 1960s.

Similar diversity characterizes the Census Bureau's "Asian and Pacific Islander" category, now 3 percent of the total population. Originally concentrated in the West, by 1980 over one-half lived elsewhere.[3] The Chinese were the first to immigrate in large numbers. By 1860 they numbered approximately 30,000, mostly settling in California; by the time further immigration was prohibited by the Chinese Exclusion Act in 1882, their numbers had grown to 125,000.[4] They were followed by Japanese, Filipinos, Indians, and Koreans, and most recently by Vietnamese and Laotians (Hmong). The 1992 census counted 1.65 million persons of Chinese origin. Filipinos made up the second largest number of Asians (1.41 million), followed by Japanese (850,000), Asian Indians (815,000), Koreans (799,000), and Vietnamese (615,000).

[1] For a thorough review, see F. Chris Garcia, ed., *Latinos and the Political System* (Notre Dame, Ind.: University of Notre Dame Press, 1988).
[2] U.S. Bureau of the Census, *Statistical Abstracts of the United States: 1993*, 113th ed. (Washington, D. C.: 1993), Table 18.
[3] Harry H. L. Kitano and Roger Daniels, *Asian Americans: Emerging Minorities* (Englewood Cliffs, N.J.: Prentice Hall, 1988), p. 5.
[4] Kitano and Daniels, *Asian Americans*, Chapter 3.

level. Many social scientists and social commentators have suggested that the American "melting pot" produced such a complex web of cross-cutting memberships that the country managed to avoid the intense conflicts that occur in countries where religious, racial, and economic cleavages reinforce one another.

Nevertheless, certain clusters of social characteristics occur more frequently than others. Social attributes are not distributed randomly. Many Roman Catholic congregations share an ethnic identity and similar occupations. Residential housing is typically segregated by race, religion, and economic status. As a result, public schools are often racially and economically homogeneous. For instance, many enclaves of white Anglo-Saxon Protestants live in the same suburban neighborhood, commute to work in high-status professions in the central city, and send their children to the same well-funded public schools. Likewise, many enclaves of inner-city, poor African-American families (often female-headed) have a difficult time finding work; their children attend some of the worst public schools in the land. Many inner-city schools are almost entirely black or Hispanic and are able to spend only a fraction per student of that spent for their virtually all-white suburban counterparts.[27] In this way social, economic, and racial cleavages reinforce one another to produce enduring patterns of inequality.

The ways in which racial and gender cleavages reinforce one another produce wide differences in income, wealth, and opportunity that clash with basic beliefs about equality held by Americans. Disparities in income, wealth, and poverty rates demonstrate the degree to which people of color and women have often failed to win material benefits in the play of power. These disparities call into question the degree to which the widely shared value of equality of opportunity is realized in practice. The picture is further complicated by differences within groups. Obviously, women disagree among themselves on a number of political issues, just as men do. Similar differences can be found within virtually any social group, no matter how narrowly defined.

THE POLITICAL CONSEQUENCES OF SOCIAL STRUCTURE

Some parents warn their children, "Never talk about politics and religion." Perhaps this is because both are touchy subjects and because social pluralism and cross-cutting memberships mean that it is often impossible to know a person's mindset. For example, a joke about another ethnic group loses its appeal when one of the listeners identifies with that group. Further, social characteristics limit the combinations of people found in the same group. Fewer memberships cut across race than across political party identification or religion. And although such characteristics as party identification or religion are not readily apparent, others such as race and gender are. All-male and all-white groupings occur frequently and thus are likely to be the site of sexist or racist humor.

The generation and moderation of conflict The combinations of social characteristics in the U.S. population have important political implications. Where they reinforce one another, they generate conflict. For example, when Roman Catholic Mexican-American migrant farm workers with little formal education work for low wages for well-educated, well-to-do white Protestant farm owners, the potential for sharp conflict across multiple social cleavages is high. Where patterns of cross-cutting memberships exist, they lead to the moderation of conflict. Many labor unions and veterans' organizations enroll members with different religious and ethnic backgrounds, and with different views on, say, abortion or the conflict between Protestants and Catholics in Northern Ireland. Groups lose their cohesion, and hence their political effectiveness, when they attempt to take action that divides their member-

Physical confrontations based on social differences, like the clash of Protestants and Catholics in Northern Ireland (*left*), occur relatively infrequently in the United States. But they do happen, as the photo (*right*) showing striking migrant workers in California demonstrates. The social differences between the impoverished strikers and their better-off employers—in wealth, education, social status, income, place of residence, language, and religion—reinforce one another; few cross-cutting memberships exist to moderate conflict.

ship. The social pluralism and prevalence of cross-cutting memberships in the U.S. population mean that many groups and organizations, although not all, contain a mixture of people that both moderates the views of members and restricts the range of issues on which the group can act.

These characteristics explain, at least in part, the relatively low level of conflict in American politics over issues involving religion, language, ethnicity, region, and even economic conditions. Of course, conflict over such issues occurs continuously in politics. But the *intensity* of that conflict pales when compared to that found in other societies. People kill each other in substantial numbers in many societies over differences in language, ethnicity, religion, and economic condition. The United States' **low-tension politics**, which has been a major characteristic of the post–World War II era, exists in part because of social pluralism and cross-cutting memberships. It helps to explain a major attribute of public opinion concerning politics that will be presented in Chapter 7—the fact that most people do not care about most current political issues most of the time.

One social characteristic, race, generates an important exception to low-tension politics. Indeed, the idealistic picture of an American melting pot in which cross-cutting memberships dampen political conflict can be sustained only by ignoring the enduring salience of race. Over the course of American history, race has been the most important source of political conflict, producing as much violence as any other issue, and perhaps more. Indeed, American political history appears quite different from the perspective of African Americans than it does from the perspective of whites. Racial divisions continue to be reflected in profound differences in the voting patterns and political attitudes of blacks and whites.

As we noted, issues deriving from other social divisions such as religion, language, and ethnicity sometimes arise, then recede. At any given time it is possible to find in the news an example of political conflict based on religion, language, or ethnicity. In 1989 a special election was held to fill the House seat left vacant on the death of Claude Pepper, a long-term congressman from the Miami area. The election pitted a Cuban-American woman against a white Jewish male. Despite their sharp differences regarding policy, ethnicity dominated the voting. For example, a *Miami Herald* poll found that 28 of every 29 Cuban voters planned to vote for their countrywoman and 24 of every 25 Jewish voters intended to support their co-religionist.[28]

When the performance of the economy creates widespread misery, as during the Great Depression of the 1930s, differences in economic status also become important sources of friction. When such issues arise, they can produce intense conflict. But they soon decline in importance and are replaced by some other issue, whereas race continues to generate intense opinions, conflict, and occasional violence.

The likelihood is high that, compared with most other issues, race will continue to generate sharp conflict. Anyone who is familiar with the political climate of today's college campuses can cite specific evidence for the staying power of race as a source of controversy. Even when campuses become integrated, there tends to be little development of the cross-cutting memberships that would lessen the intensity of racial conflict. Instead, black and white students often continue to go their separate ways. In this respect campuses sadly reflect the continuing patterns of racially segregated housing, social activities, and public schooling that characterize American society more generally. The connection between race and the distribution of economic rewards and opportunities, which we turn to next, further heightens political conflicts.

Determining the composition and resources of groups in American politics In Chapter 9 we examine the role played by interest groups in American politics. In order to exist, political interest groups require (1) shared interests or attributes, (2) an awareness of their common interests, and (3) a minimum frequency of interaction in a formal organization.

The social structure of American society defines the kinds of shared interests that may serve as the basis for the formation of an interest group. But it is not just a question of the numbers of people who share a characteristic. *Where* they live strongly influences the level of awareness of shared interests. If members are geographically concentrated, they are more likely to interact and recognize their common interests. Dispersion discourages organization. Furthermore, concentration enhances the value of political resources, including voting strength and consequently the likelihood that group members will win elective office. Figures 2.6 and 2.7 (on pp. 53 and 55) help clarify the significance of the fact that 12 percent of the population

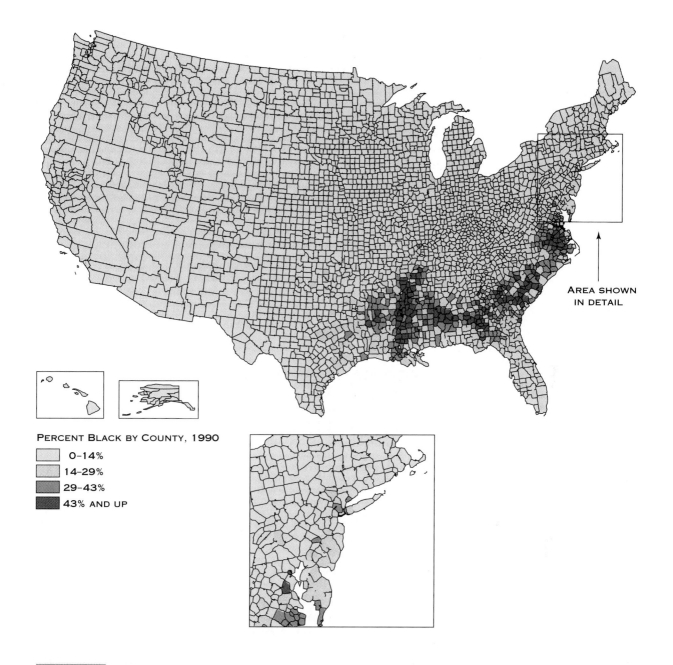

AREA SHOWN
IN DETAIL

PERCENT BLACK BY COUNTY, 1990

☐	0–14%
☐	14–29%
■	29–43%
■	43% AND UP

THE GEOGRAPHIC DISTRIBUTION OF AFRICAN AMERICANS IN THE UNITED STATES, 1990 (BY COUNTY) Where the increasing numbers of minorities projected in Figure 2.5 will live is bound to influence national politics. The concentration of blacks in the South makes possible the drawing of black majority congressional districts there. The inset shows more clearly the distribution of blacks in the urban centers of the Northeast, where their concentration is sufficient to elect black representatives to Congress. *Source:* U.S. Bureau of the Census, *Statistical Abstracts of the United States: 1993*, 113th ed. (Washington, D.C.: Government Printing Office, 1993). Data courtesy of Wessex, Inc. (Winnetka, Ill.). Graphic images created with ArcView® Version 2 software courtesy of Environmental Systems Research Institute, Inc.

is African American and about 9 percent Hispanic. The figures show that these groups are heavily concentrated geographically.

Other attributes of social characteristics affect the play of power as well. If a group's shared characteristic also serves as the basis for strong conflict, as it does for racial and many ethnic minorities, the likelihood increases that group members will recognize their common ties and form organizations that participate in the play of power. There is also increased likelihood that the organization's members will use their political resources at a higher rate, a function of the intensity of their commitment. Of course, social characteristics determine the amount of potential resources possessed. Groups enjoying high income, education, and social status more readily coalesce into organizations. This results both in overrepresentation of groups promoting the interests of the "better off" and in the success brought by such efforts to exert influence. On the other hand, lack of a common language can inhibit group formation. For example, it might be relatively hard to organize workers in a factory composed of one-third Vietnamese immigrants, one-third Hispanics, and one-third Appalachian whites. Thus, the number and type of interest groups in American politics and the influence wielded by them flow directly from the social structure of American society.

THE AMERICAN BELIEF STRUCTURE

College classrooms provide a familiar example of how every ongoing set of interactions among human beings depends on a minimal level of agreement among participants. You, your classmates, and your instructor agree on when and where to meet, who will decide what work will be required, and who will talk about what, when. Some high school and junior high school classrooms illustrate the consequences that flow from lack of agreement on such things: noise, inability to accomplish goals, even occasional violence.

Politics is no different. Without substantial agreement in the belief structures of citizens and political leaders, massive civil strife—even civil war—can break out with disastrous results. Even relatively stable and peaceful societies experience periodic episodes of violence and disorder arising from the clash of beliefs. Readers with a background in American history may be familiar with such episodes as the Civil War draft riots in New York City in 1863; race riots in East St. Louis at the end of World War I; riots in Harlem in 1943 and in Detroit, Newark, and Watts in 1967; and the Los Angeles riots of 1992.

Our examination of the context of politics concludes with a discussion of four areas in which basic agreement (we will use the term *consensus*) is needed in order for the play of power to proceed: (1) the *values* that government policies should promote; (2) the fundamental *rules and procedures* used to make decisions; (3) the *officials* who will make the decisions; and (4) the content of certain broad *policies* pursued by government.

SHARED VALUES IN THE AMERICAN POLITY

A society's principal **values** (i.e., its basic principles and beliefs from which attitudes flow) are expressed in a number of settings. In the United States,

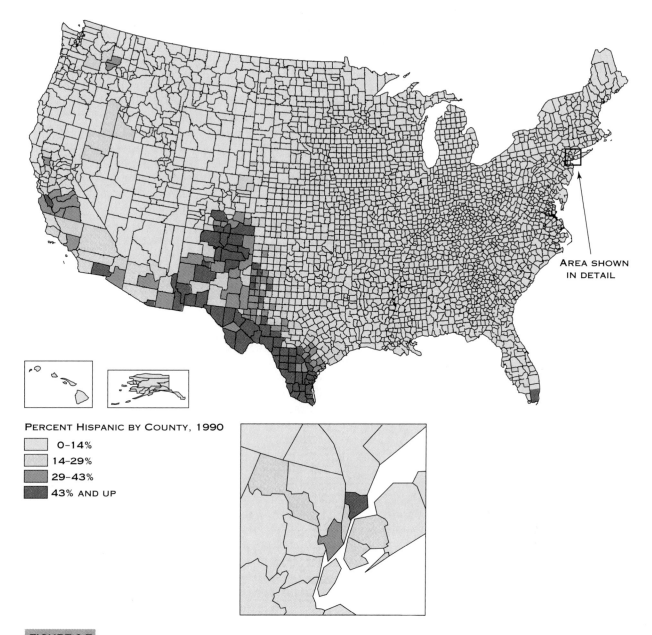

PERCENT HISPANIC BY COUNTY, 1990

☐ 0–14%
☐ 14–29%
▨ 29–43%
▨ 43% AND UP

AREA SHOWN
IN DETAIL

FIGURE 2.7

THE GEOGRAPHIC DISTRIBUTION OF HISPANICS IN THE UNITED STATES, 1990
(BY COUNTY) Hispanics, like African Americans, are not distributed uniformly
across the nation, and their location has similar implications for politics. Where
would you expect Hispanics to exert the most political influence? Recall that
Mexican and Cuban Americans and Puerto Ricans not only differ in their politics
in some respects but also live in different parts of the country. Which area(s) on
this map would you expect to have the highest concentrations of each group?
Which group is represented by the counties in the inset? What consequences do
the patterns shown in the inset have for politics? *Source:* U.S. Bureau of the
Census, *Statistical Abstracts of the United States: 1993,* 113th ed. (Washington,
D.C.: Government Printing Office, 1993). Data courtesy of Wessex, Inc.
(Winnetka, Ill.). Graphic images created with ArcView® Version 2 software
courtesy of Environmental Systems Research Institute, Inc.

presidential speeches routinely refer to core values in American society and politics. Consider, for example, this excerpt from a 1971 speech by President Richard Nixon:

> I would like you to join me in exploring one of the basic elements that gives character to a people and which make it possible for the American people to earn a generation of prosperity in peace. Central to that character is the competitive spirit. That is the inner drive that for two centuries has made the American workingman unique in the world, that has enabled him to make this land the citadel of individual freedom and of opportunity.
>
> The competitive spirit goes by many names. Most simply and directly, it is called "the work ethic." As that name implies, the work ethic holds that labor is good in itself; that a man or woman at work not only makes a contribution to his fellow man, but becomes a better person by the act of working.[29]

President Bill Clinton gave expression to other basic American values in his 1994 State of the Union address:

> [M]ay we, too, always remember who we are, where we come from, and who sent us here.
>
> If we do that, we will return over and over again to the principle that if we simply give ordinary people equal opportunity, quality education, and a fair shot at the American dream, they will do extraordinary things.[30]

Both speeches exalted some of the principal values held by Americans with respect to the operation of our capitalistic economy and democratic political system: competition, individual freedom and opportunity, hard work, and equality. Table 2.2 presents a more complete list of the principal values in American society, grouped under two overarching principles: **capitalism**, defined as an economic system characterized by the private ownership and control of productive resources; and **democracy**, defined as a system of government that seeks to place control of the authority of the state in the hands of the people to ensure that policies reflect their wishes.

The relationship between capitalistic and democratic values   The relationship between the values of democracy and those of capitalism is complex. Both esteem individual freedom and individual rights, limited government, and the guarantee of due process. However, they differ on some important details of the precise meaning of these terms. Other values central to democracy and capitalism clash so sharply that two political scientists, Herbert McClosky and John Zaller, have concluded that "the tension that exists between capitalist and democratic values is a definitive feature of American life that has helped to shape the ideological dimensions of the nation's politics."[31]

The conflict is especially deep with respect to the precise nature of "individualism" and "equality." Capitalism extols individualism as the right to pursue one's own interests, whereas democratic theory stresses the intrinsic worth of the individual. Human equality in democracy flows from a belief in the intrinsic worth of the individual, a sentiment nobly expressed

TABLE 2.2
PRINCIPAL VALUES ASSOCIATED WITH CAPITALISM AND DEMOCRACY

VALUES ASSOCIATED WITH CAPITALISM	VALUES ASSOCIATED WITH DEMOCRACY
• *Competition* • *Honest hard work* • *Self-reliance* and *individual initiative* • *Materialism,* the legitimacy of pursuing material possessions • The *pursuit of profit* as a legitimate and worthy activity • The sanctity of *private property* • *Private ownership* (rather than public ownership) of the means of production • *Free markets* rather than government decree as a means of determining prices and guiding economic activity • Individual *political freedom*	• *Individual rights* such as freedom of speech, freedom of religion, freedom of assembly, and the right to due process of law—that is, adherence to legal principles designed to protect individual rights • *Limited government,* the notion that individual rights limit what government can require of people and do to them • *Equality,* which can take the form of *equality of result* or *equality of opportunity* • Government based on *consent of the governed* and producing policies that reflect the will of the people

in the opening words of the Declaration of Independence: "We hold these truths to be self-evident: that all men are created equal." Human freedom in capitalism encourages individual initiatives that produce unequal outcomes in wealth and income.

The democratic belief in equality does not necessarily require *equality of result*.[32] But it does support the notion of *equality of opportunity*. Many people take this belief further to conclude that everyone is entitled at least to the basic necessities of life—that no one, for instance, should starve to death. But capitalistic values, especially the sanctity of private property, competition, and the individual pursuit of profit, readily come into conflict with equality. For example, democratic theory naturally concludes that if people are to be treated equally, they should have equal access to public accommodations such as restaurants and motels. When the Supreme Court upheld the public accommodations provisions of the Civil Rights Act of 1964, it did not end the objections on the part of some people that prohibiting restaurant owners from refusing to serve blacks or Asians infringed on their right to individual liberty.

How can the notion of democratic political equality be reconciled with the large disparities in the distribution of income and wealth described earlier in this chapter? One way to do so involves distinguishing between *political* and *economic* equality. In this view, equality extends only to inalienable political rights such as the right to vote and freedom of speech. The clash between the principle of equality and the inequality that actually prevails in wealth and income can be partially resolved by focusing on equality of opportunity rather than equality of result and by arguing that although many are poor, at least they have the opportunity to do better. But the degree to which disadvantaged racial and ethnic groups, as well as women, actually enjoy equality of opportunity is open to serious challenge.

EVOLVING CONCEPTIONS OF DEMOCRACY

Debate over the meaning of democracy has been waged in the United States since the eighteenth century. It is more than a debate over the meaning of a word. The term *democracy* is one of the most potent political symbols in our culture, one that advocates of a particular policy frequently apply to their preference as justification for its selection.

Although the debate has continued, the way in which it is conducted has changed, as have the areas of agreement and disagreement among its participants and the content of competing definitions. At the time of the American Revolution the term *democracy* was far from being a potent, positive symbol. Instead it meant intemperate rule by the many, to be avoided as much as rule by a king or aristocracy.[1] Only later did the term *liberal democracy* gain widespread acceptance.

Once the term became accepted, the debate shifted to whether the United States should be a "participatory" or a "representative" democracy. Even now, those advocating participatory democracy believe that true democracy requires a politically active citizenry that makes political decisions for itself. They celebrate New England town meetings and other forms of local and direct self-government; they argue that it is possible to perfect citizen self-government and

that "big government" far removed from the people is illegitimate and inherently anti-democratic. Advocates of representative democracy doubt whether average citizens are likely ever to have the intelligence or interest in politics necessary to make wise decisions. They believe that citizens ought to delegate decision-making authority, through elections or other means, to political leaders who will be able to act on their behalf. Advocates of representative democracy have been more successful in influencing the structure of government, but the values of participatory democracy remain a potent political symbol to those challenging the political status quo.

Another debate has involved how rule by the many can coexist with the principles of individual liberty and limited government. Decisions based on the consent of a majority require a process of discussion and decision making according to rules rather than the imposition of policies through the application of raw power. But what happens when majority decisions restrict the individual rights of minorities to liberty and freedom? Majority rule coexists uncomfortably with the principles of limited government and individual rights.

These debates have changed the meaning of democracy in America. Most African Ameri-

Conflict between democratic and capitalistic values will probably continue as long as American democracy itself. Indeed, the various solutions proposed to this conflict have contributed to the changing views of the very nature of democracy. The development of the continuing debate on the nature of democracy in the United States is summarized in "Footnote: Evolving Conceptions of Democracy."

Controversy over the conflicting values of equality and liberty continues. The debate over the provisions of health care reform in 1994, for instance, contained appeals to equality in seeking the goal of universal coverage and to liberty in seeking to protect the interests of small businesses and individuals' right to choose whether and how they would be covered by health insurance. But other value conflicts surface in American politics all the time. For example, people want to be safe from criminals but also to retain due process

Elements of both participatory and representative democracy characterize the American political system. Direct participation comes in many forms, including the New England town meeting pictured at left. For most people most of the time, voting for representatives who in turn will make governmental decisions on their behalf is the most active and direct form of participation in politics. Compared to participating in a town meeting, standing in line to vote and marking a ballot, as the college students shown at right are doing, is a passive activity.

cans lived in slavery until the end of the Civil War and did not enjoy full political rights (e.g., the right to vote) for another 100 years. A system that most people considered democratic excluded women from the most basic forms of political participation. Today the struggle for continued expansion of democracy involves the rights of gay men and lesbians and the physi-cally disabled. As the play of power unfolds during your lifetime, the meaning of democracy will surely continue its evolution.

[1] For a discussion of the early meaning of democracy, see Russell L. Hanson, *The Democratic Imagination: Conversations with Our Past* (Princeton: Princeton University Press, 1985), pp. 55–58.

rights prohibiting coerced confessions and illegal searches. Some argue that the state has a right to protect life by prohibiting abortions, whereas others invoke individual freedom and privacy rights to justify the legality of abortion. However, the presence of a variety of value conflicts in politics does not invalidate the basic point: Conflicts pitting political equality and capitalism against each other continue to play a central role.

According to political columnist Kevin Phillips, this tension helps to explain the cyclical character of Democratic and Republican electoral success and the public policies produced by their victories. Periods of Republican dominance have been characterized by an emphasis on capitalistic values and policies designed to provide the best possible environment for the unfettered pursuit of individual liberty and private gain. Although such periods (e.g., the Gilded Age of the Robber Barons at the end of the nineteenth century,

59

the Roaring 1920s, and the Reagan years) have been marked by dramatic increases in economic growth and the accumulation of vast wealth by some, they have resulted in increased inequality and neglect for those who are neither fortunate nor wise enough to succeed. Phillips argues that the excesses of such periods allow Democrats to gain electoral power by appealing to democratic values such as the need for equity that can be achieved by an activist government working to reduce economic and social inequality.[33]

The nature and role of consensus on values   The ability to maintain an overriding political consensus, despite such tensions, rests to a large degree on a widely shared belief that the United States Constitution provides the basis for an ideal government. The Constitution embodies the values of limited government, due process, and individual rights. It promotes the value of equality, however, only if one considers amendments abolishing slavery and extending the rights of citizenship, including the right to vote, to former slaves and to women. This caveat serves as a reminder that there have been and still are groups who do not yet fully enjoy such rights.

Of course, various groups in American society differ in their degree of adherence to these values. Furthermore, some of these values receive broader support than others. However, considerable evidence for the proposition that large majorities believe in the values listed in Table 2.2 (on p. 57).[34] These values find expression everywhere, not only in presidential speeches but in high school civics textbooks, candidates' campaign rhetoric, and Fourth of July speeches, to name just a few examples.

Like some religious beliefs, Americans' values regarding capitalism and democracy are widely shared, largely unexamined, and influential in a subtle, continuous, and incomplete way. They constitute, therefore, a **civil religion**, a set of beliefs and rituals held by a people about their collective political and economic (i.e., secular) life. In "Footnote: The Nature and Content of Our Civil Religion" we discuss the content of the American civil religion. Its wide acceptance explains in part the patriotic and nationalistic character of the American people.

Other shared values   The basic belief structure of Americans has other components besides those embodied in the notions of capitalism and democracy. Some beliefs have attained the status of conventional wisdom, finding expression in the popular press as assertions that need no evidence. In a discussion of what the Democrats had to do to capture the White House in 1992, the *Wall Street Journal* identified two such components of the belief structure:

> There clearly is a populist sentiment in the country. But it can't
> be backward-looking and it can't be defeatist. Two undeniable
> traits in the American political psyche are a "can-do" spirit and
> an eagerness to look ahead.[35]

Part of the "can-do" spirit holds that "problems can be solved." Therefore, once something (e.g., crime) is identified as a "problem," the task is to find the solution. If the problem continues, somebody must be at fault. In addition to being optimistic and having a "can-do" or "problems can be solved" attitude, Americans, like most people in the industrialized world, place

THE NATURE AND CONTENT OF OUR CIVIL RELIGION

The notion of a "civil religion" seems absurd at first. After all, religion deals with spiritual and sacred matters, whereas civil society encompasses secular concerns. But in many ways the practices and beliefs that accompany American politics share many characteristics with religious practices.

Like conventional religions, American politics has its revered and holy documents. The Constitution and Bill of Rights are enshrined in the National Archives, rising majestically from an underground vault precisely at 10:00 A.M. every morning to be viewed reverently by awed tourists. Our civil religion also has a number of revered symbols—the Statue of Liberty, the flag, the bald eagle—that appear in art, advertisements, and the backgrounds of campaign commercials on television. Many Americans display replicas of the bald eagle on their mailboxes and entryways.[1] In addition, the civil religion has its holy days, such as the Fourth of July and Presidents' Day (a joint celebration of Washington's and Lincoln's birthdays), and its rituals, such as the president's inauguration, the annual State of the Union address, and the hoopla surrounding presidential elections.

Most important, however, are the principal *values* and *tenets* of the civil religion summarized in Table 2.2 (on p. 57). The tenets of the American civil religion consist of statements about how things work; they are accepted by Americans on faith, without question, and without much hard evidence. These tenets usually have a normative component—a belief that they are good, that things *ought* to work the way the tenet says they do. We introduced seven of these principal tenets in Chapter 1, including the idea that our governmental system is ruled by law, that governmental policy reflects the will of the people, and that the structure of government exhibits a separation of powers, with Congress making the laws, the president executing them, and the Supreme Court interpreting them.

Other beliefs consistent with the values of the civil religion can be cited. For example,

Americans use patriotic symbols and colors in an amazing variety of ways. Symbols such as the national flag and images of the Statue of Liberty adorn lawns, houses, and T-shirts; are the subject of tattoos; and appear in advertisements and at political rallies. Another favorite patriotic icon is the eagle with shield and arrows, used here as a permanent door-top adornment.

many people believe they have a *civic duty* to vote. Further, many believe that voting makes a difference and that elections help to make government responsive to the wishes of the people.

The civil religion is important because it provides the conceptual framework used by many Americans when they think about and participate in politics. It also determines the symbols that stir them to action or lull them into complacency. It is, therefore, "an elaborate system of beliefs and ritual behaviors which defines for people what is right and what is wrong and why; what is possible and what is impossible; and the behavioral imperatives that follow from the beliefs."[2]

[1] For an excellent discussion of the civil religion and the history of its symbols and depictions in the United States, see Wilber Zelinsky, *Nation into State* (Chapel Hill: University of North Carolina Press, 1988), especially Chapter 6.
[2] This is the definition of *culture* provided by political scientists Francis Fox Piven and Richard Cloward in *Poor Peoples' Movements* (New York: Vintage, 1977), p. 1.

great faith in the ability of science and technology to solve problems. When technology itself produces problems, the response, in this view, is to apply more technology.

A final set of beliefs shared by many Americans and important for understanding the play of power is the longing for community. In the United States, the twentieth century has been marked by dramatic increases in geographic mobility as nuclear families move about in search of jobs and opportunities. How many of you grew up in the same town or city as your parents did? How many of you expect to return to your old neighborhoods when you graduate? In *Habits of the Heart*, Robert Bellah and his colleagues document the yearning of Americans they interviewed for the lost values that spring from attachment to political, religious, and social communities.[36]

Such yearnings make Americans wary of threats to their communities posed by the power exercised by "big government" and "big business." These institutions threaten both political and economic freedom and hence are viewed with great suspicion. Throughout American history, communal values have been expressed as a "democratic wish" for the possibility of perfecting an enlightened citizenry capable of ruling directly, without big government.[37] This is one of the reasons for continuing to celebrate New England town meetings and other forms of "direct democracy." It also helps explain why attacks on government and the call to "return government to the people" resonate with so many Americans.

BELIEFS ABOUT RULES AND PROCEDURES

Continuation of the play of power in American politics rests on widespread acceptance of the basic rules that define the structure of government and the procedures used to run them. Indeed, Americans express very high levels of support for many of the rules and procedures used in the play of power. When asked in the abstract about such things as freedom of speech, due process, equality of treatment, and the legitimacy of elections as a means to choose leaders, overwhelming majorities express support.

At the same time, however, support for the practical application of these abstract principles is much lower. For example, 97 percent of a sample of the general public said they believed in freedom of speech. But when these people were asked, "Do you believe in it to the extent of allowing fascists and communists to hold meetings and express their views in the community?" only 22 percent said "yes."[38] When asked whether the United States was meant to be "a Christian nation" or "a country made up of many races, religions, and nationalities," 20 percent chose the first response and 73 percent chose the second (7 percent declined to choose); but when asked whether a community's civic auditorium should be used by "atheists who want to preach against God and religion," only 17 percent said "yes," whereas 68 percent thought Protestant groups wanting to hold a revival meeting should be able to use it.[39] When it comes to specific *attitudes* on a variety of issues, sharp disagreement exists despite widespread acceptance of broad principles such as freedom of speech and religion.

In the United States the fairness of the procedures used to arrive at a decision sometimes takes on as much importance as the content. Decisions reached using fair methods are more likely to be seen as good. Often, critics of a decision focus on criticizing the procedures used to make it ("We weren't notified of the meeting." "The discussion was not full and free." "There was no environmental impact statement."). Thus, if values such as equality or freedom to speak and be heard are violated, they become grounds for questioning the procedures employed.

CONSENSUS ON OFFICIAL DECISIONMAKERS

Although disagreement exists on the precise role that the president, Congress, and the courts should play in the political system, there is broad consensus on what their general responsibilities are. The legitimacy and power of the Constitution, which establishes these positions and identifies their basic functions, gives them legitimacy as well. This is a straightforward and unsurprising fact, and we do not need to elaborate further on it except to note that where a population fails to agree on who should make decisions, civil war can result, with tremendous loss of life and destruction of property. This lesson is taught all too well by the catastrophic recent history of Lebanon, the Sudan, Afghanistan, Romania, Liberia, Somalia, Rwanda, Yemen, Bosnia, and other countries rent by civil war in part because too little agreement existed on who should make what decisions.

POLICY CONSENSUS IN THE UNITED STATES

Some agreement on basic policies to be pursued by an organization must exist among its members. For example, how long could an association continue if it were made up of people all concerned about abortion but including equal numbers of ardent pro-lifers and pro-choicers?

The degree of consensus on the policies government ought to pursue is lower than that on basic values, procedures, and officials. After all, policies require decisions on specifics; and attitudes on abortion, health care, and taxes clash. Much of politics consists of conflicts over which legitimate competing values should prevail. Achieving sufficient consensus on controversial issues to enact significant change can be extraordinarily difficult, as demonstrated by the struggle over health care reform in the early 1990s. When politicians succeed in building consensus on policy, as President George Bush did in building domestic and foreign support for the Gulf War in 1989, the praise they receive is well deserved. More often, they receive criticism for failing the challenge of forging consensus, a fate that Bill Clinton suffered after the defeat of health care reform in 1994.

However, the prevalence of conflict over many policies should not obscure the fact that in several crucial broad areas of policy, widespread support exists. For example, nearly everyone supports a "strong national defense." Although disagreements arise over the details of what is necessary to acquire it, few in either the political elite or the mass public question its desirability. Furthermore, most people accept the notion of the *positive state,* the belief that government should act to improve the conditions of life. Foremost among the policies supported are Social Security and Medicare: One needs to search long and hard to find elected politicians who advocate abolishing either. Even the threat of small cuts in Social Security benefits rouses the ire of a broad segment of the population, especially the elderly. Widespread approval also exists for governmental intervention in the economy to ensure that inflation and unemployment remain low and that business prospers. Similarly, most Americans approve of at least some assistance to needy children, the unemployed, the disabled, and the destitute.

Finally, despite the reality of a "mixed" public-private economy described earlier in this chapter, most people agree that economic production should remain in private hands. Capitalism is regarded as a good economic system, **socialism** (defined as an economic system characterized by ownership and control of productive resources by government) as a bad one. In the late 1950s, 89 percent of the general public disagreed with the proposition that some form of socialism would be better than the existing system, and it is unlikely that opinions have changed much since.[40]

THE SIGNIFICANCE OF CONSENSUS

Without minimal consensus on values, procedures and rules, officials, and policies, there would be no political system or play of power. But beyond this obvious point, what difference does the content of such consensus make for politics?

First, consensus subtly but powerfully structures the entire game of politics. Existing rules and procedures favor some interests and hurt others, granting better access to decision making for some and increasing the potency of some resources over others. Consensus about policy excludes consideration of some proposals that routinely occupy an important part of the agenda in other political systems. Thus, virtually no one proposes to nationalize (i.e., place under governmental control) basic industries such as steel, banking, and chemical manufacturing. Existing consensus simply makes such proposals unrealistic. When politicians make a mistake and propose something that runs against this consensus, they soon learn better. In the early 1980s, tentative proposals discussed within the Reagan administration to scale back the growth in Social Security payments met with a firestorm of protest. Whether lack of support for proposals running counter to existing consensus on policy stems from indoctrination and manipulation or represents genuine and free choice makes no difference for the fate of such proposals. However, it is important to examine which groups are considered in reaching consensus. Sometimes consensus results from failure to consider the preferences of groups excluded from the political process who oppose the policy.

Second, the existing consensus defines which symbols will evoke powerful responses from the citizenry. Politicians who invoke images of the flag, the Constitution, George Washington, free enterprise, and the concept of social security arouse powerful reactions in their audiences because of the content of this consensus. Images such as "José Martí" (the Cuban hero who led a revolution against Spanish rule), "guaranteed annual income," and "equality in the distribution of income," which arouse people elsewhere, leave Americans largely unmoved.

Third, broad consensus, common positive responses to symbols such as the Constitution and the flag, and positive feelings about the American system enhance the *legitimacy* of the political system, its decisionmakers, and its outcomes. This means that people accept the authority of political officials, thereby enabling the officials to exert influence through authority. It also prevents some abuses of existing rules, both because agreement on procedures and officials produces "internal restraints" and because it provides a rallying point around which to organize opposition to perceived violations of fundamental beliefs.

SUMMARY

Every political system reflects the structure of its economy, the social characteristics of its population, and the beliefs of its people. To understand the play of power in the United States, it is crucial to begin with a description of the size and shape of the field on which it unfolds and the nature of the players.

A popular theme in American political culture states that politics and economics are and ought to be distinct and separate. It assumes that the United States has a competitive free-market economy unaffected by government. However, markets are not entirely free. Because of market failure, government intervenes to control prices where monopolies exist, to produce public goods, and to impose regulations. Furthermore, oligopolies result in a substantial concentration of economic power in the corporate sector, which undermines the degree of competition in the domestic economy. Nevertheless, competition exists due to the increasing globalization of the economy and to the existence of its competitive sector. Although a purely free market does not exist, many of its elements do, producing a modified market economy in the United States. At the same time, the size of the public sector (including state and local governments) guarantees government a major role in the shaping of economic life. The federal government is the single largest employer; it takes 19 percent of the GDP in taxes and spends approximately 24 percent on a variety of programs and activities that shape the goods and services produced. Furthermore, the performance of the economy depends heavily on both fiscal and monetary policy. Regulations, trade restrictions, government enterprises (e.g., the Postal Service, the Tennessee Valley Authority), loan guarantees, and a variety of subsidies and transfer payments also affect overall economic activity.

These activities, combined with the operation of the tax system, affect the distribution of income and wealth. The distribution of income is relatively unequal compared to that in other Western democracies, and the distribution of wealth is even more unequal. The distribution of both has remained largely unchanged for decades partly because in practice the tax system is not progressive. Furthermore, wealth and income differ substantially by gender, race, and ethnicity. Women, racial minorities, and disadvantaged ethnic groups have much less wealth and income.

The economic structure profoundly shapes American politics. It determines who possesses one of the most important political resources—money. Those in the private sector who control production and finance wield substantial influence in American politics, which leads to the privileged position of business. Finally, many (but not all) of the major interest groups active in American politics seek to further their economic self-interest, and many of the principal issues fought in the play of power relate to the economy.

The social structure in the United States displays considerable social pluralism, with diversity in religion, race, ethnicity, language, and regional identity. The complex distribution of these characteristics, producing many cross-cutting memberships, moderates conflict; but some reinforcing cleavages promote it. Where the lines of social cleavage are not sharp, low-tension politics results. Reinforcing cleavages, like those involving race, can produce intense and even violent political conflict. The diversity of the U.S. population finds expression in the variety of issues on the political agenda and the wide range of political interest groups engaged in the play of power.

The belief structure in the United States, which can be thought of as a civil religion, exhibits a broad consensus on four aspects of politics. First, basic values associated with the notions of capitalism and democracy receive especially strong support, although democratic norms of equality sometimes clash with capitalistic norms of freedom. Second, agreement exists on the rules and procedures (e.g., the use of free elections to select some decisionmakers). Third, Americans largely agree on who should make decisions, accepting the role of Congress (and other legislative bodies), courts, and chief executives. Finally, some consensus exists on the broad outlines of many policies—the continuation of a capitalistic (rather than socialistic) economic system, a strong national defense, and a positive state that bears some responsibility for seeing that the economy performs well and ensuring that the unemployed, the destitute, the elderly, and the sick receive help (especially in the form of Social Security and Medicare).

This consensus shapes the agenda of politics (ruling out some proposals, such as nationalization of major industries, and favoring others), defines which symbols will be powerful, and enhances the legitimacy and authority of government, thereby contributing to the stability of the political system.

APPLYING KNOWLEDGE

IDENTIFYING HOW CONTEXT SHAPES THE PLAY OF POWER

In this chapter we argue that the social, economic, and belief structures of political systems profoundly shape the play of power. Because such characteristics are both well known and change slowly, it is easy to overlook their significance, especially in societies that are politically stable.

To get into the habit of thinking about how these factors shape the play of power, you will find it useful to examine the experience of strife or civil war in unstable political systems. Choose a country whose current political strife is described in the daily newspapers and read carefully articles that describe the characteristics of the conflict. Then write an essay identifying the particular attributes of the social, economic, and belief structures in that country that help explain why the conflict arose, who the participants are, what their goals are, and what resources they possess.

Key Terms

free-market economy (*p. 30*)

oligopoly (*p. 35*)

corporate sector (*p. 35*)

competitive sector (*p. 35*)

modified market economy (*p. 35*)

market failure (*p. 36*)

public sector (*p. 36*)

private sector (*p. 36*)

gross domestic product (GDP) (*p. 36*)

fiscal policy (*p. 36*)

monetary policy (*p. 38*)

progressive tax (*p. 39*)

social pluralism (*p. 45*)

reinforcing cleavages (*p. 48*)

cross-cutting memberships (*p. 48*)

low-tension politics (*p. 51*)

values (*p. 54*)

capitalism (*p. 56*)

democracy (*p. 56*)

civil religion (*p. 60*)

socialism (*p. 64*)

Recommended Readings

Garcia, F. Chris, ed. *Latinos and the Political System*. Notre Dame: University of Notre Dame Press, 1988. A collection of articles exploring the little-studied politics of Hispanics in the United States.

Garreau, Joel. *The Nine Nations of North America*. Boston: Houghton Mifflin, 1981. A readable analysis of the distinctive regional cultures in the United States.

Kozol, Jonathan. *Savage Inequalities*. New York: Crown, 1991. A gripping and impassioned description of the woeful conditions that severe underfunding has imposed on inner-city schools.

McClosky, Herbert, and John Zaller. *The American Ethos*. Cambridge, Mass.: Harvard University Press, 1984. An in-depth analysis of a survey of Americans' attitudes toward capitalism and democracy.

Phillips, Kevin. *The Politics of Rich and Poor*. New York: Random House, 1990. A leading conservative's examination of politics, policy, and inequality in American society.

CHAPTER 3

The Constitution:

Rules of the Game

RULES SHAPE THE CONTOURS and outcomes of the simplest games. Among other things, rules define the participants, establish boundaries within which participants act, outline the timing of important events, and specify who shall make decisions.

The rules are closely linked to the game's legitimacy. They provide a set of ideals that may be compared with a perceived reality of how the game is actually played. If the rules are seen as fair and the actual game is perceived to operate according to the rules, then the rules confer legitimacy on the game and its outcomes are likely to be accepted by all participants, even those who lose. If, on the other hand, the rules are perceived to be unfair and the game's actual operation is thought to diverge from the rules, the game will lack legitimacy. Consequently, some of the participants—especially those who lose—will not accept the game's outcome and may refuse to play again.

Rules established in the Constitution shape the play of power in American politics in similar ways and with similar consequences. For instance, constitutional rules define the most direct participants in the play of power—a president, a House of Representatives, a Senate, and a Supreme Court—and specify the minimum age for officeholders.

The Constitution establishes boundaries for all participants, specifying powers that may be exercised as well as limitations on power. For example, Congress is given the power to pass legislation, but it may not administer programs that it passes. The president and executive agencies administer programs but may not bypass Congress and pass their own legislation. The Constitution also shapes the timing of some events, such as elections, and specifies the terms of office for participants. Constitutional rules define who makes significant decisions. Only the House of Representatives, for instance, may initially appropriate money for

The solemn and beautiful presentation of the original Constitution on display at the National Archives, the guard who continuously stands watch next to it, and the quiet and respectful demeanor of most visitors reinforce the view that the Constitution is a sacred document. Its value is also reflected in the precautions taken to preserve it, such as lowering it at night into a vault some 20 feet below the floor.

public programs; the president is the only participant who may nominate a Supreme Court justice; and the Senate is the only participant that may confirm or reject the president's nominee.

Constitutional rules in American politics have been accepted by nearly all participants and, therefore, have shaped the play of power strongly and continuously throughout our nation's history. And those excluded from participation in American politics at the time of the Constitution's adoption—women, African slaves, and Native Americans—have demanded inclusion in a political structure that they deemed legitimate.

We begin our exploration of rules created by the Constitution with an examination of some common beliefs about it. We focus on the American public's reverence for the document adopted some 200 years ago—a document considered to be the "supreme law of the land." We then discuss the Constitution's framing, tracing its history from colonial times through the American Revolution and the revision of the nation's first constitution, the Articles of Confederation. We elaborate on the positions and strategies of principal participants in debates over the Constitution and give some attention to those excluded from participation. We also examine strategies used in seeking the Constitution's ratification by the states. Finally, we assess the governmental structure established by the Constitution, the rules governing relations between citizens and government contained in the Bill of Rights, and rules contained in other amendments to the document. The chapter concludes with a discussion of the Constitution's role in the play of power.

COMMON BELIEFS

OUR REVERENCE FOR THE CONSTITUTION

 As our discussion of belief structures in Chapter 2 showed, Americans revere the notion of "law." The Constitution, as the chief expression of law in the American political system, receives nearly universal adoration. Many Americans believe that the Constitution, a document reflecting the widely shared values of limited government, individual rights, equality, and due process, provides the foundation for a fair and just government.

Americans have such tremendous respect for the Constitution that discussions of it often use descriptive terms usually reserved for sacred texts.[1] Our Constitution is "glorious," "superior," "hallowed," "honored," and "radiant." Indeed, on hearing such descriptions by political officials and ordinary citizens, one often has difficulty distinguishing between constitutional provisions and biblical pronouncements such as the Ten Commandments. This phenomenon reflects a long-standing view that the Constitution expresses universal and timeless principles rooted in nature.[2] It also explains why beliefs about the Constitution are a principal component of the American civil religion.

Reverence for the Constitution is reflected in our survey of a large introductory American politics class. Nearly three-quarters of the students agreed with the statement "The Constitution is a very special, remarkable document, one that Americans should view with pride." Only 7 percent of the students disagreed with this statement. Similarly, nearly three-quarters of these students agreed with the statement "When I think about the Constitution, I experience feelings of admiration, respect, and even awe."

These strong, positive feelings encourage Americans to hold political officials and institutions, rather than the rules established by the Constitution, responsible for perceived governmental failure. Although political officials are frequently denounced and vilified, their actions are not perceived to emanate from a defective constitution. Instead, unacceptable action and inaction are seen to be the product of personal deficiencies. For example, an African-American gang leader interviewed on ABC's *Nightline* program in the aftermath of the Los Angeles riots in the spring of 1992 characterized the Constitution as "the greatest document ever written." The riots, from his perspective, were not a response to racial bias in the Constitution; rather, they reflected the African-American community's belief that prejudiced political officials and the police routinely ignore constitutional mandates.

This example suggests that veneration of the Constitution may extend across racial, ethnic, gender, and class lines.[3] Although this phenomenon has not been studied systematically, we do know that constitutional ideals played an important role in progressive social movements such as those battling for women's rights, the rights of the disabled, and civil rights for African Americans. Martin Luther King Jr. mixed religious and constitutional ideals in many speeches that expressed his aspirations for a more just, more free, more egalitarian America.[4] In his famous "I Have a Dream" speech, delivered during the 1963 March on Washington, King

Because the Constitution expresses fundamental American political values, it is sometimes used to point out discrepancies between American ideals and political reality. Here Martin Luther King Jr. is shown delivering his famous "I Have a Dream" speech, in which he urges America to live up to the ideals set forth in the Constitution by treating all citizens equally.

referred to the Constitution as a "promissory note to which every American was to fall heir." The note, said King, promised that both blacks and whites "would be guaranteed the unalienable rights of life, liberty, and the pursuit of happiness." King argued that "America has defaulted on this promissory note insofar as her citizens of color are concerned" and urged that it make good on its commitments.[5]

Consistent with the American public's reverence for the document, those who wrote the Constitution also are viewed in an extraordinarily favorable light. Our civics books and other popular writings portray the Framers as brilliant, selfless, public-spirited individuals with the wisdom to discover the timeless truths inscribed in the Constitution.[6] Acceptance of this portrayal leads some to believe that the Framers knew how the blueprint they created would work. Indeed, some believe that an ideal government may be achieved merely by identifying and applying the Framers' original intentions in drafting various provisions in the document.

Unfortunately, Americans' reverence for the Constitution and its framers is typically not grounded in a deeper knowledge of the document's history or provisions. What is the "necessary and proper" clause, and why was it included in the Constitution? Why does the Constitution prohibit states from "impairing the obligation of contracts"? What did the Framers intend by constitutional provisions related to slavery? What historical circumstances and social interests led to the adoption of a provision giving Congress the power to tax? If you are like many Americans, you will not be able to provide a very detailed answer to any of these questions.

In this chapter we seek to provide a context for understanding the Constitution. Although the Constitution has played a significant role in establishing a foundation for over two centuries of democratic government, we view it not as a divinely inspired document expressing universal and timeless truths but as a series of political compromises that benefit some and disadvantage others. We view the process of constitution making as emerging from the play of power, with identifiable participants using their resources to secure rules beneficial to them in the larger play of power shaped by the document.

---

[1]See Michael Kammen, *A Machine That Would Go of Itself: The Constitution in American Culture* (New York: Knopf, 1987).

[2]Edward S. Corwin, *The "Higher Law" Background of American Constitutional Law* (Ithaca: Cornell University Press, 1955).

[3]However, see Stuart Scheingold, *The Politics of Rights: Lawyers, Public Policy, and Political Change* (New Haven: Yale University Press, 1974).

[4]Randall Kennedy, "Martin Luther King's Constitution: A Legal History of the Montgomery Bus Boycott," *Yale Law Journal* 98 (1989), pp. 999–1067. Not everyone, of course, views the Constitution in such favorable light. For a series of essays that are critical of the document, see Bertell Ollman and Jonathan Birnbaum, eds., *The United States Constitution: 200 Years of Anti-Federalist, Abolitionist, Feminist, Muckraking, Progressive, and Especially Socialist Criticism* (New York: New York University Press, 1990).

[5]This speech may be found in James M. Washington, ed., *A Testament of Hope: The Essential Writings of Martin Luther King, Jr.* (New York: Harper & Row, 1986), pp. 217–20.

[6]Stanley M. Elkins and Eric McKitrick, "Youth and the Continental Vision," in Leonard W. Levy, ed., *Essays on the Making of the Constitution,* 2nd ed. (New York: Oxford University Press, 1987), p. 215. ∎

THE CONTEXT OF THE
CONSTITUTION'S FRAMING

In the sections that follow, we examine some of the salient conditions leading to the convening of a constitutional convention in Philadelphia in 1787. We focus first on important participants in colonial America, the social and economic interests they represented, political struggles among them, and conflicts between the colonies and England. We also discuss significant events preceding the American Revolution and describe and analyze important political documents of the times: the Declaration of Independence and the Articles of Confederation. We conclude this section with an exploration of a debtor's rebellion in western Massachusetts that had a significant impact on the Constitution's writing.

THE SOCIAL STRUCTURE OF COLONIAL AMERICA

Understanding the ideals and principles expressed by the Constitution requires that we examine the society from which it emerged. Colonial politics consisted of competition, conflict, and alliance among seven distinct segments of society, each with unique interests: northern commercial interests (merchants); southern planters (large landholders); small farmers; laborers; shopowners; artisans; and royal officeholders (representatives of the Crown).[1] Despite the participation of these seven segments, the play of power in colonial politics failed to include as participants a majority of the population. Even though slaves, free blacks, Native Americans, and women played important roles in society, they were excluded from political participation.

Although conflicts over issues of trade and taxes occasionally divided them throughout the 1700s, an alliance among northern merchants, southern planters, and royal officeholders formed a political and economic elite that, by and large, suppressed more radical demands for change from small farmers, laborers, shopowners, and artisans. Until the middle of the eighteenth century, this colonial aristocracy was content to remain within the British Empire. But with increasing British interference in the colonies' political and economic affairs, northern merchants and southern planters split with royal officeholders and sought the support of the other sectors of colonial society to oppose British policy. Although merchants and planters simply desired reform of British action and policy, radical forces included in the new alliance (and drawn from small farmers, laborers, shopowners, and artisans) used this opportunity to expand their influence and press for independence. The radical forces, such as those agitating for independence as part of direct action organizations like the Sons of Liberty, helped to set in motion a series of events precipitating the armed conflict that came to be known as the American Revolution.[2]

BRITISH POLICY AND COLONIAL RESPONSE

Tensions between England and the American colonies began to build in response to major changes in British policy following the French and Indian War in 1763. With its victory in the war, England won possession of Canada and all French territory east of the Mississippi River, acquisitions creating a significant financial burden and new administrative challenges. British troops arrived in the colonies to protect them from possible attack by Native American tribes whose lands had been taken by the colonists and to help in administering the expanded Empire. To help ease the debt incurred during the war and finance the colonies' future defense, Britain levied direct taxes on them for the first time.

Even in the 1760s, taxes roused the ire of Americans. A series of acts from 1763 to 1774 threatened to drain the pocketbooks of northern merchants and southern planters (see Table 3.1).[3] Since Parliament, rather than colonial assemblies, passed these measures, they were seen to constitute "taxation without representation," a phrase that became a rallying cry for independence.

Throughout the 1760s, northern merchants and their southern planter allies mobilized support among more radical sectors and, together, organized mass protests and demonstrations. Colonists harassed, intimidated, and occasionally assaulted British revenue collectors, many of whom resigned their posts as a result. Merchants, planters, shopowners, small farmers, laborers, and artisans cooperated to organize nonimportation agreements against British products.

TABLE 3.1
BRITISH INTERFERENCE IN COLONIAL AFFAIRS, 1763–1774

ACTION	DESCRIPTION
Proclamation of 1763	Closed frontier west of Alleghenies to further settlement, land purchases, or patents.
Sugar Act (1764)	Levied duties on imports of sugar, molasses, indigo, coffee, wine, linens.
Stamp Act (1765)	Levied duties to be paid by affixing revenue stamps on legal documents, bills of sale, liquor licenses, playing cards, newspapers.
Quartering Act (1765)	Required colonies to supply British troops with provisions.
Declaratory Act (1766)	Stated that Parliament had authority to enact laws binding colonies "in all cases whatsoever."
Townshend Act (1767)	Levied duties on imported glass, red and white lead, painters' colors, tea, paper.
Restraining Act (1767)	Suspended power of New York assembly until it complied with Quartering Act of 1765.
Tea Act (1773)	Granted monopoly in imported tea to East India Company.
Boston Port Act (1774)	Closed port of Boston pending restitution to East India Company for tea dumped in Boston Tea Party.
Massachusetts Government Act (1774)	Altered charter of Massachusetts to facilitate more effective British control.
Quebec Act (1774)	Extended boundaries of Quebec Province; appeared to violate colonial charters by annexing western lands.

The Stamp Act produced political protest throughout the colonies. This woodcut shows a group of protesters carrying the effigy of the New Hampshire stamp distributor through the streets of Portsmouth in 1765. Note that the parade is led by men carrying a coffin to symbolize the death and burial of the Stamp Act. Protesters accused Parliament of passing laws that constituted "taxation without representation," a phrase that became an important symbol for political mobilization against England.

Colonial women played an important role in implementing these agreements by refusing to buy British commodities, persuading others to join boycotts, harassing those who violated the nonimportation agreements, organizing "spinning associations" to produce clothing and other products that previously came from England, and raising funds for a colonial army.[4] For a more detailed discussion of their role, see "Participants: Women in Revolutionary America (on pp. 78–79)."

By 1769 the British had stationed some 2,000 soldiers in Boston, a center for colonial resistance, to put down protests and restore order. On March 5, a group of protesters led by a runaway slave, Crispus Attucks, gathered to demonstrate outside the Boston Custom House. A large, boisterous crowd gathered and tensions began to mount. Some members of the crowd taunted the guards on duty and threw objects, knocking one soldier to the ground. He then fired into the crowd, with other soldiers following his example. After the smoke cleared, five colonists lay dead and six others were wounded. Colonial pamphleteers, seeking to use this event to ignite rebellion, labeled it the "Boston Massacre." The first to be shot and killed was the former slave, Crispus Attucks. The irony of a former slave becoming the first casualty of the American Revolution is underscored by historian Lerone Bennett, who writes: "By a paradoxical act of poetic justice, it was this American—an oppressed American, born in slavery with, it is said, African and Indian genes—who carried the American standard in the prologue that laid the foundation of American freedom."[5]

Paul Revere's engraving of the conflict in 1769 at the Boston Custom House helped to rally others for rebellion. At right, the British officer appears to order his troops to fire on an unarmed, peaceful assembly. The Custom House is labeled "Butcher's Hall," and smoke escapes from a gun barrel pointing out of the window. Colonists committed to independence called this tragedy the "Boston Massacre," a name that helped incite others to action against England.

Suffering financially from the effects of the nonimportation movement in the colonies, British merchants lobbied successfully to repeal many of the most objectionable measures. By and large, northern merchants and southern planters in the colonies viewed repeal of these measures as the end of the matter and a complete victory in the struggle with England. However, radical elements representing laborers, shopowners, small farmers, and artisans sought to go further, continuing their agitation for independence from Britain. Led by individuals such as Samuel Adams, Christopher Gadsden, and Patrick Henry, radicals argued that British rule was fundamentally unjust and, therefore, had to end.

Parliament presented those agitating for independence with a golden opportunity to transform colonial dissatisfaction with British rule into revolution when it passed the Tea Act in 1773. Compounding the financial injury to colonial commerce, the Act permitted the East India Company to sell

Radicals, pushing for independence from England, staged the Boston Tea Party to express resistance to British policy and goad England into taking repressive action. Having succeeded, they then used England's punitive response to further mobilize colonial resistance.

tea through its own agents, effectively bypassing colonial merchants. Once again, northern merchants and southern planters allied themselves with their more radical adversaries to oppose British policy. The radicals, led by Samuel Adams, were determined to use this turn of events to force Britain into taking action that would alienate its moderate supporters in the colonies. Showing tremendous skills in the play of power, they accomplished their goals brilliantly in staging a major symbolic political act, the Boston Tea Party, dumping some 20,000 pounds of East India tea into Boston Harbor.

As the radicals had hoped, England's response to the Tea Party was swift and severe. In 1774, Parliament passed the Boston Port Act, closing Boston Harbor; the Massachusetts Government Act, altering the state's charter by giving the Crown more local political appointments as a way of bringing the colony under closer control of the British government; and the Quebec Act, restricting settlement in western lands. Demonstrating their skill in defining issues, the radicals labeled these measures the "Intolerable Acts," a name that came to be widely accepted in the colonies. Ironically, rather than solidifying British rule in the colonies, the Intolerable Acts served to win more general public support for the cause of independence.

Many colonists involved in the politics of the day believed that a **republican form of government** was the most desirable. Such a government was designed to protect individual liberty from intrusions by kings (the one), nobles and aristocrats (the few), and even the majority of the people (the many). Republicanism's basic tenets include a government based on popular

WOMEN IN REVOLUTIONARY AMERICA

The role of "free" white women in colonial America was limited by rigid gender roles and widespread beliefs in their inferiority. Influenced by liberal social theories that distinguished between public and private spheres of life, colonial society viewed public life as the exclusive province of men. Women were relegated to the private sphere, working in the household and caring for the family. Thomas Jefferson, whose views on women were quite typical of the times, argued that women lacked the capacity to engage in the rough-and-tumble of politics. Instead, women's role in politics was "to soothe and calm the minds of their husbands returning ruffled from political debate."

Few political pamphlets or tracts written during the revolutionary period addressed the status of women. One exception is a pamphlet penned by James Otis, who asked in 1764, amid arguments that England was violating the liberty of American colonists, "Are not women born as free as men?" A few women, such as Mercy Otis Warren, wrote political pamphlets that helped to increase popular support for independence. Moreover, in a now-famous letter to her husband, John, who was engaged in debate over the Constitution, Abigail Adams urged that the Constitution's framers "remember the ladies" and protect them against the "naturally tyrannical" tendencies of their spouses. Tongue in cheek, Abigail appropriated the Revolution's rhetoric to threaten John: "If particu-lar care and attention is not paid to the ladies, we are determined to foment a Rebellion, and will not hold ourselves bound by any laws in which we have not voice, or Representation." John Adams dismissed this plea rather casually: "I cannot but laugh" at "your extraordinary Code of Laws.... Our masculine systems ... are little more than Theory.... In practice you know we are the subjects. We have only the names of Masters."

Within the limitations of colonial society, women played a significant role in the movement for independence. Spinning associations organized by the Daughters of Liberty assisted the nonimportation movement, producing goods subject to the boycotts. Women refused to buy English products and policed local merchants suspected of hoarding scarce commodities. During the Revolutionary War some traveled with the troops, filling traditional women's roles as cooks and nurses. Others served in less traditional roles as spies and messengers for the Continental Army. A few, like Deborah Sampson, disguised themselves as men and joined the fighting. Groups of women patriots in Philadelphia, Trenton, Maryland, and Virginia collected funds for George Washington's troops.

After the war, women's legal status remained virtually unchanged. The common-law principle of *coverture*, a restriction giving husbands exclusive control over their spouse's property, remained in place. Only one state,

consent, very limited in its powers, and shielded from the sometimes misguided and irrational judgments of the majority. Even though many in the colonies had perceived England as a superb example of republican government, the growing dissatisfaction with British policy toward the colonies led many pamphleteers during this prerevolutionary period to begin painting a very different picture of British politics.

Pamphleteers described England as "one mass of corruption," "tottering on the brink of destruction" and suffering "internal decay."[6] John Adams argued that the Tea Act exposed British intentions and "threw off the mask of legality."[7] In her satirical poem "A Political Reverie," Mercy Otis Warren predicted a revolution because freedom no longer existed in Britain.[8] And Thomas Jefferson, in a clever use of symbolism to define the issue, interpreted

Philip Dawes's *A Society of Patriotic Ladies* depicts a group of colonial women meeting to sign an agreement to boycott British goods. Produced in 1775, this mezzotint portrays common stereotypes of women engaged in politics as physically unattractive and neglectful of family responsibilities. Observe the exaggerated physical features of the two women in the right foreground and the unattended child sitting under the table.

New Jersey, permitted women to vote. Its voting law, passed in 1790, referred to voters as "he or she" and applied equally to white and black women. New Jersey repealed the law in 1807. It would be another century before women became full-fledged voting members of the American political community.

Sources: Linda K. Kerber, *Women of the Republic: Intellect and Ideology in Revolutionary America* (Chapel Hill: University of North Carolina Press, 1980); Mary Beth Norton, *Liberty's Daughters: The Revolutionary Experience of American Women* (Boston: Little, Brown, 1980).

British policy as "a deliberate and systematical plan of reducing us to slavery."[9] Of course, it is ironic, and illustrative of a major contradiction in early American life, that a society in which slavery played such a crucial role was stirred by words accusing the British of trying to reduce the free segments of society to "slavery."

The qualities attributed in these writings to British society and people—corruption, decay, and illegality, for example—were juxtaposed with those required for a truly republican society. In particular, colonial writers emphasized the significance of personal virtue, sacrificing self-interest for the general welfare of society. As "Resources: Republican Virtue and American Diversity" (on p. 81) indicates, the concept of virtue was significant in defining the types of people who initially were perceived to possess the potential to become good republican citizens. These people—white, male, and Christian—increasingly viewed themselves as different from the British, as Americans rather than as British subjects. These Americans came to view their country, a country built on what was increasingly viewed as the ruins of a corrupt England, as destined to play a special role in history. As John Adams expressed it, "America was designed by Providence for the theatre on which man was to make his true figure, on which science, virtue, liberty, happiness, and glory were to exist in peace."[10]

Thus, events in 1773 and 1774 pushed the colonies beyond the point at which reconciliation with England seemed plausible. In 1774 delegates from all the colonies convened in the First Continental Congress to decide how to proceed. Shortly thereafter, the Congress requested a total boycott of all British goods. Radicals pushed for consideration of complete independence from British rule. Two years later, the Declaration of Independence was drafted and signed.

THE DECLARATION OF INDEPENDENCE

By 1775 a revolutionary movement had begun. Colonists in Massachusetts were fighting British troops at Lexington and Concord. In June 1776, the Second Continental Congress appointed a committee of five—John Adams of Massachusetts, Benjamin Franklin of Pennsylvania, Thomas Jefferson of Virginia, Robert Livingston of New York, and Roger Sherman of Connecticut—to draft a proclamation declaring American independence from British rule.

The Declaration is perhaps the most extraordinary document written in American history.[11] The task of writing it was delegated to Jefferson, who was respected by his colleagues for his "peculiar felicity of expression." Philosophically, the Declaration expressed clearly and forcefully the views of liberal philosophers, most notably John Locke, that individuals have natural, divinely given rights that are inalienable and that may not be abridged or taken away by government. Governments, according to Locke, are established among people entering into a *social contract*, the exclusive purpose of which is to preserve the *natural rights* of all individuals. If governments violate the social contract's terms, individuals have the right to remove rulers or disband the government. These principles are easily identified in the eloquent writing of Jefferson:

> We hold these truths to be self-evident, that all men are created equal, that they are endowed by their Creator with certain unalienable rights, that among these are life, liberty, and the pursuit of happiness. That to secure these rights, governments are instituted among men, deriving their just powers from the consent of the governed. That whenever any form of government becomes destructive of these ends, it is the right of the people to alter or to abolish it, and to institute new governments. . . .

In political terms, the Declaration sought to unify the various interests dividing the colonies along social, economic, and geographic lines by cataloguing the widely shared grievances against Britain. Jefferson's extensive list of unacceptable British practices—22 paragraphs of specific charges

REPUBLICAN VIRTUE AND AMERICAN DIVERSITY

We often think of political resources as exclusively material in nature, consisting of such tangible things as money and votes. But the images evoked in the symbols we use to communicate often have crucial effects on our politics as well. The images evoked by the word *virtue* in republican ideology during the revolutionary period illustrate how language may subtly define legitimate participants in the play of power, as well as exclude entire groups.

Many colonial leaders viewed the American Revolution as an attempt to gain independence from a corrupt and decaying British Empire. Revolutionary leaders hoped to construct a republican society built on a foundation of virtuous citizens. Rooted in Enlightenment thought and Protestant theology, republican ideology viewed virtue—sacrificing self-interest for the general welfare of society—as an important prerequisite for a truly republican nation. Republican thought conceived of the virtuous citizen as rational, self-governing, and self-restrained. Virtuous citizens dedicated their lives to hard work, were frugal in their personal lives, and mastered their passions and instinctual needs.[1]

Virtue was contrasted with its opposites, vice and corruption. Corrupt societies, in John Adams's words, were populated by people with "vicious and luxurious and effeminate Appetites, Passions, and Habits."[2] In republican ideology, "luxury"—a "dull animal enjoyment" leaving "minds stupefied and bodies enervated"—constituted the crucial attribute to be avoided. Luxury appealed to the passions and senses, draining individuals of energy and making them "effeminate" and "weak."[3]

Colonial leaders associated women, African slaves, and Native Americans, each in their own way, with qualities of vice and corruption.[4] Many revolutionary leaders considered these groups to lack sufficient control over their passions and instinctual needs. Native Americans and African slaves, for example, were perceived as uncivilized creatures of passion, wild and primitive. Women were delicate, emotional, and passionate, rather than rational and self-restrained.[5] Even white women lacked qualities appropriate for participation in public life. But republican thought also emphasized a private, feminine version of virtue. Encapsulated in the notion of "republican motherhood," feminine virtue manifested itself in domestic duties, especially in the raising of male children. Virtuous (white upper- and middle-class) republican women "would stay in their homes and, from that vantage point, shape the characters of their sons and husbands in the direction of benevolence, self-restraint, and responsible independence."[6]

Thus, republican ideology helped to construct a national identity based on an image of virtuous citizens who were white, male, and Christian.[7] This ideology served to rationalize the appropriation of Native American lands, the exploitation of African slave labor, and the subordination of women. These ideas were to have a lasting impact on the new nation, serving for two centuries as ideological battlegrounds in political struggles for egalitarian reform.

[1] Bernard Bailyn, *The Ideological Origins of the American Revolution* (Cambridge, Mass.: Harvard University Press, 1967); Ronald Takaki, *Iron Cages: Race and Culture in 19th Century America* (New York: Oxford University Press, 1990).

[2] Takaki, *Iron Cages*, p. 5.

[3] Takaki, *Iron Cages*, p. 5.

[4] Winthrop Jordan, *White over Black: American Attitudes toward the Negro, 1550–1812* (Baltimore: Penguin, 1969); Linda Kerber, *Women of the Republic: Intellect and Ideology in Revolutionary America* (Chapel Hill: University of North Carolina Press, 1980); Takaki, *Iron Cages*; Ruth H. Bloch, "The Gendered Meanings of Virtue in Revolutionary America," *Signs* 13 (1987), pp. 37–58.

[5] Mary Beth Norton, *Liberty's Daughters: The Revolutionary Experience of American Women, 1750–1800* (Boston: Little, Brown, 1980).

[6] Kerber, *Women of the Republic*, p. 231.

[7] Takaki, *Iron Cages*, p. 12.

THE DECLARATION OF SENTIMENTS, 1848

As part of a strategy to further the cause of women's equality, the women who wrote the Declaration of Sentiments in 1848 borrowed the style, form, and rhetoric of the Declaration of Independence. Running through this important document is the conviction that the American Revolution had made implicit promises to women that had not been kept. Excerpts from the Declaration of Sentiments follow.

We hold these truths to be self-evident: that all men and women are created equal; that they are endowed by their Creator with certain inalienable rights; that among these are life, liberty, and the pursuit of happiness; that to secure these rights governments are instituted, deriving their just powers from the consent of the governed. Whenever any form of government becomes destructive of these ends, it is the right of those who suffer from it to refuse allegiance to it, and to insist upon the institution of a new government, laying its foundation on such principles, and organizing its powers in such form, as to them shall seem most likely to effect their safety and happiness. . . . When a long train of abuses and usurpations, pursuing invariably the same object, evinces a design to reduce them under

The women who wrote the Declaration of Sentiments deliberately used the style and language of the Declaration of Independence to underscore their conviction that its words did not apply to their lives. Shown here are Susan B. Anthony (*left*) and Elizabeth Cady Stanton.

absolute despotism, it is their duty to throw off such government, and to provide new guards for their future security. Such has been the patient sufferance of the women under this government, and such is now the necessity which constrains them to demand the equal station to which they are entitled.

against the King—appealed to the various constituencies composing colonial America. This "propaganda technique," as one writer calls it, assisted in mobilizing colonists for the Revolutionary War.[12]

On July 2, 1776, the Second Continental Congress voted unanimously for independence. Two days later, the Declaration of Independence was adopted. With the important exception of a provision in Jefferson's original draft indicting the King for continuing the slave trade—a provision deleted to gain the support of representatives from Georgia and South Carolina—the Declaration was adopted almost precisely as it had been originally penned by Jefferson.

Throughout our nation's history, the Declaration has served as an inspiration for those with grievances against American government and society.

The history of mankind is a history of repeated injuries and usurpations on the part of man toward woman, having in direct object the establishment of an absolute tyranny over her. To prove this, let facts be submitted to a candid world. . . .

[Seventeen paragraphs with specific complaints follow. Here are a few of them.]

He has never permitted her to exercise her inalienable right to the elective franchise.

He has compelled her to submit to laws, in the formation of which she has no voice.

He has withheld from her rights which are given to the most ignorant and degraded men—both native and foreigners.

He has taken from her all right in property, even to the wages she earns.

In the covenant of marriage, she is compelled to promise obedience to her husband, he becoming, to all intents and purposes, her master—the law giving him power to deprive her of her liberty, and to administer chastisement.

He has denied her the facilities for obtaining a thorough education, all colleges being closed against her.

He has created a false public sentiment by giving to the world a different code of morals for men and women, by which moral delinquencies which exclude women from society, are not only tolerated, but deemed of little account in man.

He has endeavored, in every way that he could, to destroy her confidence in her own powers, to lessen her self-respect, and to make her willing to lead a dependent and abject life. . . .

Now, in view of this entire disfranchisement of one-half the people of this country, their social and religious degradation—in view of the unjust laws above mentioned, and because women do feel themselves aggrieved, oppressed, and fraudulently deprived of their most sacred rights—we insist that they have immediate admission to all the rights and privileges which belong to them as citizens of the United States.

Source: Elizabeth Cady Stanton, Susan B. Anthony, and Matilda Joslyn Gage, eds., *History of Woman Suffrage,* Vol. 1 (New York: Fowler & Wells, 1881), pp. 70–71.

Groups excluded from the play of power have often employed as an explicit political strategy a comparison of their unequal status to the ideals of the Declaration. For example, in 1848 a group of women organized by Elizabeth Cady Stanton and Lucretia Mott met at Seneca Falls, New York, to draw up a "Declaration of Sentiments and Resolutions." Modeled closely on Jefferson's Declaration, which they believed had left them out, this declaration emphasized that the power of men over women was no more derived from the consent of the governed than it earlier had been from the King of England (see "Strategies: The Declaration of Sentiments, 1848"). This document, with its echoes of Jefferson's Declaration, became an important device seeking to unite women with diverse interests in the struggle for gender equality.

THE GREAT LAW OF PEACE

The text of the Haudenosaunee constitution, The Great Law of Peace, begins with the planting of the Tree of the Great Peace. This tree, a white pine, serves as a metaphor for unity among nations. The actual tree and meeting site for the confederation are located at the center of the confederacy, on the land of the Onondaga Nation near Syracuse, New York. The Great Law of Peace enunciated democratic ideals such as equal suffrage, recall of officials, and the possible use of initiative and referendum measures for voting on important issues. It provided for constitutional change through an amending process and prohibited infringement on free expression and unauthorized entry into dwelling places.

Each of the five nations—the Senecas, Oneidas, Mohawks, Cayugas, and Onondagas (a sixth, the Tuscaroras, later joined after moving north from the Carolinas)—maintained its own council governed by *sachems*, or nation leaders. Women played a significant role in Haudenosaunee society. These women, called clan mothers, chose the leaders and had the power to remove them. The Haudenosaunee had 50 leaders, 9 each from the Mohawk and Oneida nations, 14 from the Onondagas, 10 from the Cayugas, and 8 from the Senecas. When the nation leaders meet at Onondaga, the meeting is called a Grand Council. Only members of the original five nations can be chosen as leaders.

The Grand Council meets to discuss issues of common concern to the member nations as well as to other peoples.

Members of the Grand Council debated issues while sitting close to a fire. Separate deliberations were conducted by two nations, the Senecas and Mohawks, designated as "older brothers," and two nations, the Cayugas and Oneidas, referred to as "younger brothers." If the two nation pairs announced different decisions, the Onondagas, designated as the "keepers of the flame," would decide the issue. If the four nations reached consensus, the Onondagas had the authority to disagree and send the issue back for further deliberation. If the four nations could not agree on an issue, the Onondagas could make the final decision. In this way, The Great Law of Peace established a decision-making structure that included the equivalent of a two-house legislature (the older and younger brother nations) with an executive or presidential (Onondaga) veto that could be overridden.

Thus, there are clear similarities between The Great Law of Peace and both the Articles of Confederation and the United States Constitution. Most historians agree that some of the Constitution's framers interacted frequently with Haudenosaunee leaders and were familiar with the nations' Great Law of Peace. However, scholars disagree about the extent to which the Haudenosaunee influenced the framing of the

THE ARTICLES OF CONFEDERATION

After declaring independence, the American colonies needed to establish a government. In November 1777 the Continental Congress adopted the Articles of Confederation and Perpetual Union, the first set of rules governing relations among the several former British colonies. Ratified by all the states by 1781, the Articles were in force for some 12 years, then were replaced by the United States Constitution in 1789.

Although the Articles of Confederation constitute the first set of written rules applying to the former colonies, they are not the first constitution drafted in the geographical area now known as the United States. The first written constitution in North America was drafted long before Columbus

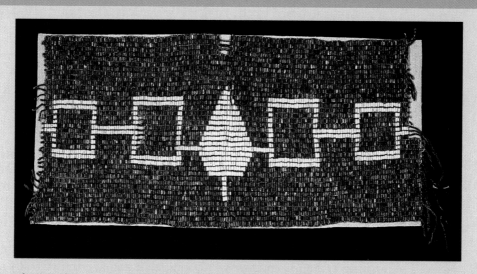

The Haudenosaunee Confederacy was established and organized on the basis of rules recorded in wampum belts. The Great Law of Peace—the remaining portion of which is shown here—included many provisions similar to those found much later in the United States Constitution, leading some scholars to argue that the Haudenosaunee influenced the writers of the more recent document.

American constitutions adopted by European Americans. There is greater agreement with the statement that "the Iroquois [Haudenosaunee] Constitution provided a written preview of some of the governmental values later to be adopted by the whites in America."

Sources: Chief Irving Powless, Onondaga nation; Bruce E. Johansen, *Forgotten Founders: How the American Indian*

Helped Shape Democracy (Boston: Harvard Commons Press, 1982); Vine Deloria Jr. and Clifford M. Lytle, *American Indians, American Justice* (Austin: University of Texas Press, 1983); Elizabeth Tooker, "The United States Constitution and the Iroquois League," *Ethnohistory* 35 (Fall 1988), pp. 305–36; Donald A. Grinde Jr. and Bruce E. Johansen, *Exemplar of Liberty: Native Americans and the Evolution of Democracy* (Los Angeles: American Indian Studies Center, UCLA, 1991).

began his famous voyage. Five Native American nations—the Senecas, Oneidas, Mohawks, Cayugas, and Onondagas—in what is now the northeastern United States formed an alliance some four hundred years ago to resolve problems of blood feuds and provide for their common defense. This group of nations, known to the French as the Iroquois, adopted a constitution, referred to as "The Great Law of Peace," recorded in wampum belts.[13] As we relate in "Participants: The Great Law of Peace," the influence of The Great Law of Peace on the United States Constitution has been hotly debated.

Reflecting fears that a strong national government would replace the British monarchy, the Articles of Confederation sought to limit the central government's power. This document established a **confederation**, a loose association of independent states agreeing to cooperate in certain specified areas. The Articles conferred powers on the central government only over foreign affairs and military policy and were careful to preserve the power of states by retaining ultimate authority within their borders. The political system constructed by the Articles of Confederation carried a clear message that a strong central government constituted the greatest threat to liberty.

Weaknesses of the Articles in practice   In many important respects, the confederation created by the Articles—with its limitations on the central government and its preservation of state sovereignty—expressed the democratic aspirations of many who fought for independence.[14] But the Articles also made it virtually impossible for the states to act in concert to resolve common problems.

The first signs that the form of government created by the Articles might be seriously flawed came during the Revolutionary War. Because Congress lacked the power to tax the states, and the states failed to heed the national government's call to contribute funds for the war effort, the Continental Army often went unpaid.[15]

After the war, the Confederation found it impossible to negotiate commercial treaties with other nations. States could not be compelled to comply with agreements, and foreign nations hesitated to enter into any agreements with a nation that could not meet its obligations. Within the Confederation, Congress could do nothing to prevent states from engaging in trade wars with one another.

These deficiencies led many from all walks of life to believe that constitutional reform was necessary. But growing opposition to the Articles also reflected the social, political, and economic cleavages evident in prerevolutionary America. The colonies' victory in the war and the new government established thereafter changed, in dramatic fashion, the configuration of power in the states.[16] Royal officeholders representing the Crown were stripped of their power and privileges, forced to flee with their allies to Canada or England. Although some members of the prerevolutionary elite— northern merchants and southern planters—supported the Articles of Confederation as a means of creating autonomy for their governing groups, more radical forces solidified their power within several states and moved to democratize state governments.

State legislatures representing the interests of small farmers, laborers, artisans, and shopowners enacted laws that filled the prerevolutionary political establishment with horror. The 1780s were a time of economic depression, a result of the war and the loss of many British markets. In response, some states passed laws authorizing the circulation of paper money and declared it legal tender; suspended actions on debts; permitted debts to be paid in kind, rather than in cash; and protected citizens from mortgage foreclosures. All these measures aided an increasing number of debtors while damaging the interests of creditors.

The Annapolis Convention   In the midst of the fear and panic created by these policies among businesspeople and property owners, five states accepted an invitation from the Virginia legislature to meet in Annapolis, Maryland, to consider a common policy on interstate commerce. The Annapolis Convention met in September 1786. Alexander Hamilton and James

Madison used the occasion to persuade the gathering to adopt a resolution calling for a convention of all the states in Philadelphia the following May to revise the constitution to meet the needs of the nation.

Shays's Rebellion Shortly after the Annapolis Convention, a debtor's rebellion in Massachusetts convinced many American leaders that the upcoming conference of states needed to go beyond merely amending the Articles. The rebellion came in the context of the economic depression gripping the new nation and the collapse of agricultural prices.[17] Unlike some other states, the Massachusetts state legislature—dominated by financial and merchant interests—refused to provide relief measures for struggling farmers. To make matters worse, it sought to pay back its war debts by levying heavy taxes that fell disproportionately on farmers.

When farmers in the western Berkshire hills failed to obtain the debt relief they demanded from the state legislature, they found themselves faced with foreclosures on their debt-ridden lands. Led by a former captain in the Continental Army, Daniel Shays, they sought to prevent the county courts of western Massachusetts from operating until after the next election. For several days the angry and desperate farmers struck fear into the hearts of state leaders by forcing local courts to close, attempting to capture the federal arsenal at Springfield, and threatening to lay siege to Boston. These events provoked a call by the Massachusetts government to Congress for help in putting down the rebellion and restoring order. Because Congress could not act, the disturbance had to be suppressed by the state militia. In 1787 a newly

Farmers led by Daniel Shays closed a state courthouse to prevent farm foreclosures and dramatize their need for tax relief. The action, known as Shays's Rebellion, struck fear into creditors and other American leaders, showing the Confederation's weakness and the need to design a stronger central government. After the next election, when Shaysites took control of the state legislature and granted some relief for debtors, many American leaders concluded that the general public could not be trusted with the vote.

elected legislature in Massachusetts, with a large contingent of members sympathetic with the goals of the rebellion, granted some of the farmers' demands.

Shays's Rebellion, as it came to be known, confirmed the worst fears of a social and economic elite that felt itself and its world threatened by postrevolutionary America.[18] The rebellion underscored for them the dangers of democracy and the need for a central government that could act quickly and effectively to ensure domestic order. James Madison argued that the new nation no longer had to fear the tyranny of the few over the many. Rather, he continued, "it is much more to be dreaded that the few will be unnecessarily sacrificed to the many."[19]

REVISING THE RULES OF THE GAME

By 1787 most of America's social, political, and economic leaders believed that the new nation was in grave danger of collapsing. The government created by the Articles of Confederation was unable to pay its debts, further the nation's interest in the world, establish a national economy, or deal with civil disobedience. These were perceived as serious structural flaws in need of correction. These concerns convinced the states to send delegates to the Constitutional Convention in Philadelphia. As we describe in the sections that follow, the delegates engaged in an intricate, rough-and-tumble, and protracted play of power that established the fundamental rules of American politics, most of which still apply today.

PARTICIPANTS AT THE CONSTITUTIONAL CONVENTION

The convention that drafted the new American constitution was organized primarily by northern merchants and southern planters—those interests that had been weakened after the Revolution and under the Articles of Confederation. Fifty-five delegates selected by seven states convened at the State House in Philadelphia in May 1787. In the four months that it took to draft the new document, five additional states sent representatives. Convinced that the Articles of Confederation were fundamentally flawed, and with the terror of Shays's Rebellion firmly planted in their minds, the delegates quickly abandoned their mandate to simply amend the Articles and began deliberations that would produce a dramatically revised constitution.

The delegates came from various regions of the country, from states that varied in size and major economic activity, and they engaged in differing occupations. Although the delegates were young, averaging 42 years of age, most had obtained considerable rank and standing and possessed wealth well above the national average.[20] The Constitution's framers were primarily lawyers, politicians, merchants, financiers, and planters. Those interests and social classes that had seized control of many state governments after the Revolution—small property-owning farmers, artisans, and laborers—had virtually no representation. Some of those who could have represented these interests, such as those associated with the radical forces that fought for independence, were so wary of the Convention that they refused to participate. For example, Patrick Henry of Virginia remarked that he "smelt a rat

Few paintings or drawings of the Constitution's framers depict the conflicts and disagreements that divided them during the Constitutional Convention in 1787. Most convey the impression that the Framers agreed on most issues, thus masking the intense debates and many compromises from which the Constitution emerged.

in Philadelphia" and chose to stay home. Likewise, Rhode Island refused to send a delegation, fearing that an amended constitution would impede state power and establish a tyrannical central government. Having no choice in the matter, women, Native Americans, free blacks, and African slaves were also excluded from participation.

The Convention has been described as a "fractious body," with coalitions of state delegations shifting from issue to issue.[21] However, on a few central propositions the Framers agreed. First, they agreed that the Confederation's current political and economic situation was grave, making change imperative. As representatives of states, they could not establish a government that abolished states altogether. Imagine the reaction a representative to the Convention would receive on returning to the state legislature to report its abolition! While retaining a role for the states, a revised constitution had to establish a stronger central government that could deal effectively with the domestic economic problems of credit, commerce, and interstate rivalries, elevate the nation's international standing, and curb what was perceived as the democratic "excess" of state governments established after the Revolution.

In an important sense, the Framers hoped to change the balance of political power among social interests and economic classes in the new nation. Some, for example, complained that "gentlemen of property" frequently lost state electoral contests to candidates from the "lower classes."[22] John Jay was perhaps the most candid and blunt, proclaiming that "the people who own the country ought to govern it."[23] As we will see, these preferences are reflected in the Constitution, leading one historian to conclude that "the Constitution was intrinsically an aristocratic document designed to check the democratic tendencies of the period."[24]

THE MOTIVATION OF THE FRAMERS

For nearly a century, scholars have argued about the Framers' motives. In 1913, historian Charles Beard argued that the moral principles purportedly reflected in the Constitution were merely a means to secure the general population's consent for a government that furthered the Framers' narrow economic interests.[25]

In the 1950s an alternative explanation emerged that portrayed the Framers as motivated primarily by ethical and philosophical principles.[26] n fact, the Framers did seek to fashion a constitution reflecting some of he major philosophical principles of the time. However, contemporary riters suggest that the two explanations are not mutually exclusive—that the moral and philosophical principles animating the Constitution also served the long-term political and economic interests of the mercantile and property-holding classes so dominant at the Constitutional Convention.[27] The constitution produced by the Framers created a government that, in Alexander Hamilton's words, "fit the commercial character of America," a government capable of promoting commerce, protecting property, and curbing the influence of radical state legislatures bent on providing relief to debtors. The new constitution shielded the commercial and propertied classes from the "democratic excess" in states that had proven hostile to their interests.

PROGRAMS OF ACTION AND THE GREAT COMPROMISE

The Philadelphia Convention began its work in late May 1787, meeting continuously until mid-September. Delegates met formally at the State House, or what is now called Independence Hall, and informally at several local "watering holes" such as City Tavern, the Indian Queen, and the Black Horse. The delegates decided at the outset of the Convention that voting on specific measures would be by states and that a majority of states present would decide any question.

During the first weeks of debate, Edmund Randolph of Virginia presented a plan for constitutional revision, suggested by fellow Virginian James Madison, that offered a centralist solution to the problems perceived in the Articles of Confederation. It became clear that the delegates had more in mind than simply amending the Articles when they agreed to debate what became known as the Virginia Plan.

The **Virginia Plan**, as summarized in Table 3.2, proposed a substantially stronger national government and a system of representation in the national legislature based on the free population of each state or the proportion of each state's revenue contribution. Randolph also proposed a second legislative chamber, with members elected by representatives in the first chamber. Because the states varied radically in size, population, and wealth, the rules to be adopted regarding representation in the national legislature had important implications for the resources available to the various participants in the play of power. Thus, some perceived that the Virginia Plan unfairly

TABLE 3.2
THE VIRGINIA PLAN, THE NEW JERSEY PLAN, AND THE CONSTITUTION OF 1787

PROPOSED CONSTITUTION	PROVISIONS REGARDING BRANCHES OF GOVERNMENT		
	LEGISLATURE	JUDICIARY	EXECUTIVE
Virginia Plan	Two chambers, with representation by population. Popularly elected lower house elects uppers house. Broad powers, absolute veto over laws passed by states.	National judiciary chosen by legislature.	Single-term executive elected by legislature.
New Jersey Plan	One chamber, with equal state representation. Same powers as under Articles, plus power to tax and regulate commerce.	Appointed by president.	Plural executive, removable by legislature upon petition by majority of state executives.
Constitution of 1787	Two chambers, a House based on population, and a state-elected Senate based on equal state representation. Broad powers, including taxing and regulating commerce.	Appointed by president, with Senate approval.	Single executive chosen by electoral college.

allocated more resources to larger states (through greater representation) than to smaller states.

The Virginia Plan also defined relations between levels of government. Under this proposal, Congress obtained broad and expansive powers. The Plan permitted Congress to veto state legislation, empowering it "to negative all laws passed by the several states, contravening in the opinion of the National Legislature the articles of Union." The Virginia Plan envisioned a congress with authority to compel state compliance if necessary, permitting it "to call forth the force of the Union against any member of the Union failing to fulfill its duties under the Articles thereof." The Plan gave Congress broad and indefinite powers in all cases where the states proved "incompetent." Finally, the Virginia Plan conferred broad appointment powers on the legislature, permitting it to select a single-term executive and a life-term judiciary.

The resolutions offered by Randolph and Madison constituted an astounding victory for those seeking the Confederation's total demise. They completely obliterated the notion of state sovereignty, a concept at the heart of the Articles of Confederation. But while the Convention debated the

Virginia Plan, additional delegates arrived at the State House and began to organize opposition to it. Their opposition took the form of an alternative program, introduced by William Patterson of New Jersey, known as the New Jersey Plan (see Table 3.2 on p. 91).

Patterson offered the **New Jersey Plan** as a modification, rather than a drastic revision, of the Articles of Confederation. It dealt with the Articles' major weaknesses, proposing to expand Congress's authority by adding the powers to tax and regulate commerce. The Plan attracted support from smaller states by retaining state equality of representation in a one-chamber legislature and creating an executive directly subject to state control. Unlike in the Virginia Plan, Congress would not appoint the national judiciary. Judges would be appointed by the president, who would be chosen by Congress.

Given that the rules proposed in these plans would shape the play of power in such fundamental ways, it is hardly surprising that debate on these proposals was so intense that a decision could not be reached. When the Convention threatened to disband in hopeless deadlock over the two plans, delegates from the medium-size state of Connecticut offered a third plan. Known as the **Great Compromise**, this plan called for a two-house legislature, or **bicameral legislature**, allowing for equal state representation in the Senate (two senators for each state) and representation based on population in the House of Representatives. This scheme granted a veto power to both large and small states, since all legislation would require the approval of both legislative chambers. Because the large states would bear a disproportionate share of the tax burden, the compromise plan provided that all revenue bills would originate in the House, where population determined voting strength. Although this compromise did not garner an enthusiastic response from many delegates, and a few delegates were so upset that they packed up and went home, most of the delegates preferred compromise to either a breakup of the union or a continuation of the Confederation. The delegates thus accepted the Great Compromise and then proceeded to negotiate further compromises, providing for a president selected by an "electoral college" of citizens chosen by states and for a supreme court appointed by the president with the Senate's approval.

SLAVERY IN THE CONSTITUTION: THE INSIDIOUS COMPROMISES

A more fundamental issue, touching the core of the new nation's ideals, had to be resolved before negotiations leading to the Great Compromise could be concluded. If national supremacy was to be the guiding principle of a new union of states, disagreements between northern and southern states on the potentially explosive issue of slavery had to be worked out somehow. Nearly 90 percent of the more than half-million African slaves resided in five southern states—Georgia, Maryland, North Carolina, South Carolina, and Virginia—where they constituted one-third of the total population.[28] Thus, the question of representation in the House of Representatives brought the slavery issue directly into the debates leading to the Great Compromise. Given the large number of slaves in the South, the question was quickly defined: Would slaves be counted in the population formula used in apportioning representatives?

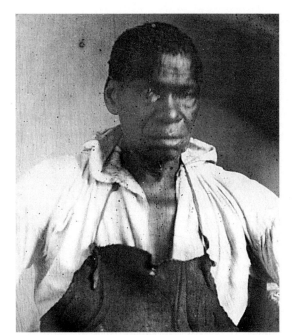

Illustrating the contradictions in revolutionary American society, this photo (taken c. 1845) shows Isaac Jefferson, a former slave from the plantation of Thomas Jefferson, author of the Declaration of Independence. Although some strongly opposed slavery, it was widely tolerated and even supported during the revolutionary period.

Northern and southern delegates ultimately reached an agreement in what became known as the **Three-Fifths Compromise**. Representatives in the House would be apportioned "according to their respective numbers, which shall be determined by adding the number of free persons . . . and . . . three-fifths of all other persons." These "other persons," of course, were slaves. Slaves would not be permitted to vote or otherwise participate in politics. But for purposes of determining the number of representatives due each state in the lower house, five slaves equaled three whites. Nowhere is the bargaining characterizing the Convention better illustrated than in the compromise struck to count some human beings as three-fifths of a person. The utter devaluation of African slaves in this formulation is underscored by historian Winthrop Jordan: "Framing a national constitution forced men to say it outright: the Negro as a slave was but three-fifths a man."[29]

Questions regarding slavery nearly destroyed the prospects for union. However, the debates over slavery focused not on its morality or ethics but, rather, on whether or how much slaves could contribute to calculations on representation. Delegates understood that rules defining slaves for purposes of determining the number of representatives in each state would contribute significantly to the relative resources enjoyed by northern and southern states. The southern states, of course, demanded that slaves count. Northern delegates, not all of whom opposed slavery in principle, argued that property should not be counted or that other forms of property owned by citizens of northern states should also count. Northern delegates expressed deep concerns that the Three-Fifths Compromise would unfairly permit southern states to increase their national political clout. Indeed, the South's share of the national population and, therefore, its share of representation increased from 41 percent when only free white persons were counted to more than 46

TABLE 3.3
PROVISIONS RELATING TO SLAVERY IN THE CONSTITUTION

PROVISION	CONTENT
Article I, Section 2	Representation in House apportioned on basis of population, computed by counting all free persons and three-fifths of slaves
Article I, Section 2; Article I, Section 9	Two clauses requiring that direct taxes be apportioned based on population including three-fifths of slaves; purpose of preventing Congress from levying head tax on slaves to encourage emancipation
Article I, Section 9	Congress prohibited from abolishing international slave trade to the United States until 1808
Article IV, Section 2	States prohibited from emancipating fugitive slaves; runaway slaves required to be returned to masters
Article I, Section 8	Congress authorized to call up states' militias to suppress insurrection, including slave uprisings
Article IV, Section 4	National government obliged to protect states against domestic violence, including slave uprisings

Source: Derrick Bell, *And We Are Not Saved: The Elusive Quest for Racial Justice* (New York: Basic, 1987), p. 34. Copyright © 1987 by BasicBooks, Inc. Reprinted by permission of BasicBooks, a division of HarperCollins Publishers, Inc.

percent under the three-fifths formula.[30] Southern influence in presidential politics was equally enhanced, because a state's presidential electors equaled the total number of its senators and representatives.

In the debates that ensued over representation, southern delegates made it clear that they would never agree to join a union if Northerners refused to compromise. North Carolina's William Davie, for example, explained that he was certain that his state would refuse to participate in a union if slaves were not counted, threatening that if northern states "meant . . . to exclude them altogether the business was at an end."[31] Northern delegates reluctantly accepted the compromise as the price of constitutional reform.

As the Convention proceeded, northern delegates bowed to southern demands on other matters relating to slavery (see Table 3.3). Although some northern delegates hoped to end the slave trade in America, a compromise was struck prohibiting Congress from interfering in this practice for 20 years. In return for northern concessions on the slave trade, South Carolina provided crucial support for deletion of a provision requiring a two-thirds majority in Congress to pass commercial regulations, a provision vehemently opposed by northern merchants.[32] Northern delegates also agreed to a fugitive slave clause requiring the return of runaway slaves.

So it was that a nation that had recently fought a war for liberty, and adopted a Declaration of Independence proclaiming the equality of all people, institutionalized slavery in its Constitution. This grand irony is captured in the words of political scientist Donald Robinson: "Never was a people more thoroughly or knowledgeably committed to liberty, or more acutely aware of the contaminating power of slavery. Yet rarely, if ever in history, has the institution of slavery formed so fundamental and so pervasive a part of a political community."[33] The enslavement of Africans would continue for some 80 years until North and South could no longer reconcile their differences, which culminated in a divisive and bloody civil war.

THE CONSTITUTION: NEW RULES OF THE GAME

It is difficult for us today to comprehend how radically the Constitution changed the structure and rules of the play of power from the way they had been framed in the Articles of Confederation. In the compromises hammered out in Philadelphia, the Framers succeeded in changing the relative power of varying arenas and levels in the play of power. The new rules of the game significantly enhanced the prestige and power of the national government. This power could be used to promote commerce and protect property from radical state legislatures that sought to provide debtor relief. A strengthened Congress gained a mandate to tax. National courts and a potentially strong presidency were established. But the Framers were careful to limit the new government's powers and to protect against concentration of power in any individual or institution. To limit the power of national institutions and protect against the "excesses" of democracy, the Constitution provided for a bicameral legislature, checks and balances, indirect election of representatives (selection of the president by an electoral college and of senators by state legislatures), and staggered terms of office. Structural features such as federalism and separation of powers further protected citizens' liberty and property from powerful governmental institutions and democratic abuse. The Framers, being extremely skillful politicians, also developed rules that would help secure popular support and ratification.[34] Thus, the Constitution included a provision for the direct popular election of representatives in the House and, subsequently, a bill of rights appended to the basic document.

The Framers' political achievements were substantial. The Great Compromise and the Three-Fifths Compromise solidified and unified the merchant and planter groups while weakening those interests dominant under the Articles. The Great Compromise, agreed to in the context of centralizing power, reassured those fearing that their locality, state, or region would lose influence in the new nation. The compromises on slavery convinced merchants and southern planters that they could work together in the future.

Somehow, amid all the bargaining and compromises, the Framers managed to portray themselves as a unified body. Indeed, the words *Framers* and *Founders* continue to constitute a potent political symbol, as suggested by the fact that Americans customarily capitalize the words, as they do *God* and *the Constitution*. Further, when the Framers are viewed as a group, their differences and the compromises they hammered out are obscured. Consequently, the terms *Framers* and *Founders* promote the belief that "*the* intentions of *the* Framers" exist and can—indeed, should—be used to decide complex legal issues and policy matters.

KEY STRUCTURAL FEATURES

In the following sections we take a closer look at some of the major structural features and rules of the game created by the Constitution: federalism, the separation of powers, checks and balances, and representative government. We focus attention on how these features further the Framers' goals of expanding national power, preventing the concentration of power, curbing state legislative power and democratic "excess," protecting property, and helping to secure popular support. The significance of these structural features and rules will be evident throughout the rest of the text.

Federalism Compared to the structure defined in the Articles of Confederation, federalism was a movement in the direction of a strengthened national government. **Federalism**, a system of government providing for a division of power between national and state governments, increased national power while not completely impeding the states' authority. National and state governments could exercise authority over persons and property within their own spheres. Competition between the different governmental levels, it was hoped, would prevent power from becoming concentrated at either level. Federalism, then, furthered the Framers' goal of enhancing national power while simultaneously providing a countervailing force in states to guard against a tyrannical national government. The multiple levels created by federalism also increased the options available to participants seeking to devise strategies to be employed in the play of power.

Separation of powers The Framers sought to protect against power's being concentrated within any one national governmental institution by creating a **separation of powers** that clearly distinguished and specified the authority and resources of institutions operating in different arenas—legislative, executive, and judicial. Congress was given the authority to make laws, the executive branch was to implement and enforce laws, and the judicial branch had responsibility for interpreting laws. The three branches used different methods for selecting personnel, linking each arena to a different constituency. In practice, the strict separation of powers among the different arenas has evolved into a system where "separate institutions share power."[35]

Checks and balances The Constitution not only conferred specific powers on each branch but also created a system of **checks and balances** between branches, giving each branch authority and resources that could be employed to block or modify actions of the other branches. Checks and balances provided another mechanism for preventing the concentration of power in any one institution or arena (see Figure 3.1). Congress, for example, has the authority to pass laws, but the president may veto and the courts may interpret congressional legislation. Congress may exercise authority over the president by overriding a veto by an unusually large (two-thirds) majority or by instituting impeachment proceedings. It may also propose new legislation or constitutional amendments to counter Supreme Court interpretations of congressional legislation. Checks and balances promote outcomes based on negotiation and compromise, and influence through exchange and persuasion, rather than through the pure exercise of authority.

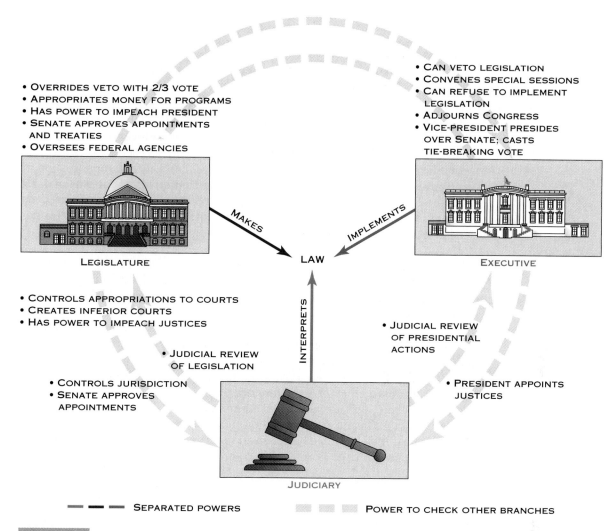

- OVERRIDES VETO WITH 2/3 VOTE
- APPROPRIATES MONEY FOR PROGRAMS
- HAS POWER TO IMPEACH PRESIDENT
- SENATE APPROVES APPOINTMENTS AND TREATIES
- OVERSEES FEDERAL AGENCIES

- CAN VETO LEGISLATION
- CONVENES SPECIAL SESSIONS
- CAN REFUSE TO IMPLEMENT LEGISLATION
- ADJOURNS CONGRESS
- VICE-PRESIDENT PRESIDES OVER SENATE; CASTS TIE-BREAKING VOTE

LEGISLATURE

MAKES

LAW

IMPLEMENTS

EXECUTIVE

- CONTROLS APPROPRIATIONS TO COURTS
- CREATES INFERIOR COURTS
- HAS POWER TO IMPEACH JUSTICES

- JUDICIAL REVIEW OF LEGISLATION

INTERPRETS

- JUDICIAL REVIEW OF PRESIDENTIAL ACTIONS

- CONTROLS JURISDICTION
- SENATE APPROVES APPOINTMENTS

- PRESIDENT APPOINTS JUSTICES

JUDICIARY

— — — SEPARATED POWERS   ▩ ▩ ▩ POWER TO CHECK OTHER BRANCHES

FIGURE 3.1

SEPARATION OF POWERS AND CHECKS AND BALANCES The political system created by the Constitution has three separate branches, each with its own specified powers and functions. Each branch may check the power of the other two. Most of the checks depicted in this figure are written explicitly into the Constitution, although some, such as the power of judicial review, have evolved through court interpretations and political practice. The separation of powers, combined with checks and balances, has produced a government in which separate institutions share power, often arriving at outcomes after negotiation and compromise.

Representative government   Electoral rules and structures created by the Framers provided for a form of limited democracy. Although the Constitution created a political structure that was much more democratic than almost anything else then existing in the world, the Framers were careful to limit the direct decision-making authority of American citizens. In a number of important respects, the Constitution merged democratic and antidemocratic elements: the popular election of members of the House, the indirect election of senators by state legislatures and of the president by the Electoral College, and the appointment of the national judiciary by the president with senatorial approval. The Constitution is silent on the issue of who shall be permitted to vote, leaving this decision to the states, most of which imposed stringent racial, gender, religious, and property restrictions on the franchise. The electorate may vote only for governing officials, not for or against laws or constitutional amendments. And political officials, after election, are in no way formally bound to cast votes that reflect majority sentiment among those who elected them. These arrangements accommodated the Framers' fears of democracy, based on their experiences with state governments under the Articles, and their recognition that popular support and ratification depended on rules ensuring that government was at least minimally responsible to the people. Therefore, the Framers devised an innovative approach, a **representative government** whereby power resides in the people but is exercised by elected or appointed representatives.

The contours of the new play of power   The major structural features and rules created by the Constitution—federalism, separation of powers, checks and balances, and representative government—produced a highly complex play of power in which power is extraordinarily dispersed and fragmented. The fears of governmental tyranny, democratic excess, and infringement on property rights are reflected in a governmental blueprint that seeks to prevent any individual, official, or institution from acquiring too much power relative to others. It also provides citizens with only indirect means of altering the outcomes of governmental activity. Consequently, one of the enduring challenges for the government created by the Constitution has been to act boldly and affirmatively in response to changing circumstances. For citizens, the challenge has been to identify the participants and governmental arenas and levels responsible for both positive and negative policy outcomes.

THE CONSTITUTION'S SEVEN ARTICLES

The Constitution includes seven distinct articles. The first three outline the new government's structure, the powers of the separate branches, and their rules and internal operations. The remaining articles declare the supremacy of federal law, define relationships among states, and identify procedures for ratifying and amending the Constitution. In the following sections we provide a brief introduction to the content of the seven articles. We will discuss many of these constitutional rules in greater depth in later chapters on Congress, the presidency, and federal courts.

Article I: The legislative branch   The importance that the Framers assigned to the legislature is shown by the fact that it is defined in the Constitution's first article. Composed of 10 sections and some 50 subsections, by far the longest and most detailed article, Article I includes provisions that further all of the Framers' major goals.

   Article I defines the composition, method of selection, and terms of office for a two-chamber, or bicameral, legislature. A bicameral legislature had the advantage of providing a check on concentrated power within the legislative branch—particularly a check by the Senate on the popularly elected House. Directly elected members of the House of Representatives would serve relatively short two-year terms, thereby encouraging popular support for the new government. The Senate, on the other hand, would be elected indirectly by state legislatures for longer six-year terms, with such terms staggered so that the terms of only one-third of the members expired every two years. The scheme created a further buffer between the Senate and the electorate, as staggered terms protected the institution against rapid changes in popular preferences expressed by state legislatures.

The legislative branch created by Article I greatly expands congressional power as compared to the power of the legislature existing under the Articles of Confederation. Article I, Section 8, for example, consists of 18 subsections that expressly grant Congress a variety of powers, including the power to tax, borrow, regulate commerce, coin and regulate the value of money, declare war, and maintain a military force. In a certain sense these provisions may be viewed as a constraint on congressional power, as the Constitution implies that any powers not explicitly stated are not granted. This is known as the doctrine of **enumerated powers**. But the Framers sought to enhance the national government's power, so they included the **necessary and proper clause**, also referred to as the **elastic clause**, in subsection 18. This clause grants Congress the power "to make all laws which shall be necessary and proper" to carry out its enumerated powers. Such an expansive, open-ended grant of authority signaled that the Framers conceived of the congressional powers enumerated in the Constitution as sources of strength rather than constraint.

Article I, Section 9 includes explicit limitations on the powers of Congress. Some, such as one prohibiting Congress from suspending the right of **habeas corpus**—the right of individuals to appear in court before being jailed—were designed to reassure the public that the Constitution did not create a potentially tyrannical government. Other limitations were meant to protect important mercantile, creditor, and property interests. For example, provisions in Section 9 prohibit Congress from privileging the port of one state over another and from levying duties on articles exported from any state. Further, reflecting the Framers' deep concerns about Shays's Rebellion and the possibility of slave uprisings, the Constitution includes an exception to the prohibition against suspension of habeas corpus, "in cases of Rebellion" when "the public safety may require it."

In Article I the Framers also sought to protect commercial, creditor, and property interests from radical state legislatures. Section 10 prohibits states from taxing imports and exports or placing regulations on commerce outside their own borders. States are also proscribed from issuing paper money—a common practice in states under the Articles of Confederation—and paying debts in any currency other than gold and silver coin. Perhaps most significant for creditors, a **contract clause** prohibits states from "impairing the obligation of contracts." This meant that states could no longer relieve debtors of their contractual obligations, prevent creditors from foreclosing on mortgages, or declare moratoriums on debt.

Article II: The executive branch Article II is composed of four sections that define the president's duties, powers, and qualifications as well as the procedures for election of the president through an electoral college. The simple act of establishing an executive branch with a president at its head

Presidents can check the power of Congress by vetoing legislation. President Clinton, during his 1994 State of the Union address, waved a pen that he threatened to use to veto any health care legislation that did not include universal coverage.

strengthened the national government more than had the Articles of Confederation, which had created no chief executive office or administration.

Consistent with the Framers' desire to create an office that could provide decisive action and direction, a number of powers are expressly conferred on the president in Article II. The president acts as commander-in-chief and negotiates treaties, subject to Senate approval. The president also has the power to appoint major executive branch and judicial personnel, convene Congress in special sessions, veto congressional legislation, and grant pardons and reprieves, except in cases of impeachment. In addition, the president is granted the relatively open-ended power to "take care that the laws be faithfully executed." Article II requires that the president inform Congress periodically of the "State of the Union," a practice that has evolved into an annual address to all branches of the national government and the American people.

Powers granted to the president in Article II reflect the Framers' desire that the president and executive branch be important participants in the play of power. But the Framers also sought to shield the president from popular pressures by making the office subject to indirect election by a college of electors who were not bound to cast votes consistent with popular preferences in their states.

Article III: The judicial branch   The Constitution's third article creates a federal judiciary, promoting the Framers' goals of strengthening the national government and curbing democratic influences. The national judiciary was also expected to be a bulwark against abridgements of liberty and property by the national executive and legislative branches.

Article III establishes a Supreme Court as the highest court in the nation, a court that is expressly given the power to resolve conflicts between state and national laws. Further, Article III gives federal courts jurisdiction over disputes between citizens of different states, disputes that arose frequently in the context of interstate trade rivalries and that had to be resolved if a national economy was to be created.

The Framers hoped to insulate federal courts from popular majorities. The Constitution provides for the appointment of judges by a president, who is not popularly elected,* with the advice and consent of the Senate, whose members, at the time of the Constitution's adoption, were not directly elected. Federal judges serve life terms "during good Behavior," with a salary that cannot be lowered during their time on the bench by Congress or the president.

The Constitution makes no explicit mention of what would become the Court's most significant resource. **Judicial review**, the power to invalidate state laws and acts of Congress and the executive, was declared in 1803 by Chief Justice John Marshall in *Marbury v. Madison*.

Articles IV through VII Other topics are covered in the remainder of the Constitution, Articles IV through VII. Article IV, among other things, seeks to provide an even playing field among states in politics and economics. One provision, the **full faith and credit clause**, requires states to honor and enforce the governmental actions of all other states. This means, for instance, that a person cannot escape the judgment of a state court by moving to another state. Another provision, the **privileges and immunities clause**, compels states to guarantee citizens of other states all citizenship rights in their states. Both clauses promote uniformity among states, an important prerequisite in creating national political and economic systems. Article IV also "guarantee[s] to every state in this Union a Republican Form of Government" and vows that the federal government will intervene when invited by authorities to restore order in states experiencing "domestic Violence." This provision was a direct response to the Confederation's inability to intervene in Shays's Rebellion when requested to do so by the Massachusetts government. Finally, Article IV provides for the admission of new states to the union and guarantees to existing states that no territory will be taken from them without their consent.

Article V specifies procedures for amending the Constitution. These procedures are cumbersome and difficult, requiring that Congress either propose an amendment by a two-thirds vote in each chamber or call a national constitutional convention in response to a request by two-thirds of the states. Of these two methods, only the first has been used. Since the Bill of Rights was adopted in 1791, thousands of amendments have been introduced in Congress, but only 24 received enough votes to be sent to the states. Only 17 of the 24 received the three-fourths majority vote necessary for ratification by the states. In short, changing the rules governing the play of power established in the Constitution of 1787 has proven to be nearly impossible.

---

* The president is elected by the Electoral College.

Article VI contains the heart and soul of the Constitution—the **suprem-acy clause**, which declares that the Constitution, federal laws, and federal treaties "shall be the supreme law of the land," superseding state laws and constitutions. Article VI also requires both national and state officials to take an oath pledging to support and uphold the national constitution. Finally, this article contains a major gift for creditors at the time of ratification. It declares that all debts of state and nation accumulated prior to the Constitution's adoption will be honored by the new nation. This provision pleased and reassured members of the commercial classes to whom most of the debt accumulated during the Revolutionary War was owed. It went a long way as well toward convincing foreign nations that the new union could be trusted to honor its obligations.

Rules and procedures for ratifying the Constitution are outlined in the final article, Article VII. Rather than requiring unanimity, as did the Articles of Confederation, the Constitution mandated that only 9 of 13 states ratify. Ratification would occur in special state conventions called for this specific purpose, rather than in state legislatures. It is to the ratification process that we now turn.

THE SELLING OF THE CONSTITUTION

After the delegates signed the Constitution in September 1787, they had to persuade the states to accept it. This was no easy task, taking some two and a half years to complete (see Table 3.4). Less than three months after the Framers departed from Philadelphia, Delaware became the first state to ratify the Constitution, doing so unanimously in December 1787.

By early 1788, five states had ratified, with New Jersey and Georgia also providing unanimous support. The Constitution's margin of victory in

TABLE 3.4
THE ROAD TO RATIFICATION BY THE STATES

DATE	STATE	VOTE IN STATE LEGISLATURE
December 7, 1787	Delaware	Unanimous
December 12, 1787	Pennsylvania	46–23
December 18, 1787	New Jersey	Unanimous
January 2, 1788	Georgia	Unanimous
January 9, 1788	Connecticut	128–40
February 6, 1788	Massachusetts	187–168
April 28, 1788	Maryland	63–11
May 23, 1788	South Carolina	149–73
June 21, 1788	New Hampshire	57–46
June 25, 1788	Virginia	89–79
July 26, 1788	New York	30–27
November 21, 1789	North Carolina	195–77
May 29, 1790	Rhode Island	34–32

Connecticut and Pennsylvania, the other two states ratifying early in the process, was narrower, achieved only after vigorous and contentious debate. In June 1788 New Hampshire became the ninth state to ratify. Although New Hampshire constituted the last state required for ratification, a strong union depended on all states' approving. Virginia and New York, crucial states with large populations, followed New Hampshire, ratifying the Constitution by the slimmest of majorities after some opponents were reassured that a bill of rights would be amended to the Constitution. North Carolina joined the union in 1789, after the Bill of Rights had been submitted by Congress to the states. Rhode Island, the state that refused to send delegates to Philadelphia, held out until 1790, when it became the thirteenth state to approve the new government.

Throughout the process of writing and ratifying the Constitution, supporters of constitutional change demonstrated great political skill and sophistication. Those who supported a strong central government and a national commercial economy, people who might have been called "Centralists" or "Nationalists," appropriated the name **Federalists** to suggest a commitment to decentralization and state sovereignty, as well as a link to revolutionary principle. The Constitution's opponents, those genuinely committed to state sovereignty and decentralization, were labeled by the Constitution's supporters as **Antifederalists**, a label that stuck.[36] In the battle over names, a battle holding great symbolic significance, the Constitution's supporters clearly won.

In the struggle for ratification, the Federalists made excellent use of their substantial resources to persuade state conventions and the general public. The Federalists benefited from superior organization and a leadership that included some of the most talented, energetic, and prominent people of the time. Three of their leaders—James Madison, Alexander Hamilton, and John Jay—published a series of 85 eloquent articles, known as *The Federalist*, in the Constitution's defense.[37] Referred to by one scholar as "inspired propagandists,"[38] these writers, publishing under the pen name "Publius," provided trenchant critiques of Antifederalist positions and extraordinarily forceful arguments for ratification. The support of George Washington, Revolutionary War hero and national statesman, greatly enhanced the legitimacy of the Federalist cause. Although a few Antifederalist leaders, such as George Clinton, Patrick Henry, Richard Henry Lee, and Mercy Otis Warren, had achieved some national prominence, most were local people without broad influence or connections.

Despite their disadvantages in the contest over ratification, the Antifederalists stubbornly resisted defeat in several states. To persuade the Constitution's opponents to support the new union, Federalists reluctantly promised to adopt a bill of rights subsequent to the Constitution's ratification. Most historians point to this Federalist concession as a crucial prerequisite for the Constitution's ratification in several states.

THE BILL OF RIGHTS

Amid Antifederalist demands for a second constitutional convention to reconsider the entire framework of government, and mindful that wary delegates to state ratifying conventions had been promised a bill of rights,

Few individuals were more dedicated to the idea of independence from England than Mercy Otis Warren. Her writings played an important role in mobilizing support for the American Revolution, and later she was a significant force in debates over the Constitution's ratification. An Antifederalist, Warren was concerned that the new Constitution would create an aristocracy and that it did not include a bill of rights.

Federalist leaders in the first Congress began work on a bill of rights. James Madison, initially opposed to the idea, led the effort to draft amendments. According to Madison, a bill of rights "will kill the opposition everywhere, and by putting an end to disaffection to the Government itself, enable the administration to venture on measures not otherwise safe."[39]

The Bill of Rights places specific boundaries on each of the three branches (see Table 3.5). The First Amendment prohibits Congress from enacting legislation that would establish religion, inhibit its free exercise, or regulate speech, the press, the right of people to assemble together, or their right to petition government. The Second, Third, and Fourth Amendments outline restrictions on the executive branch, while Amendments Five through Eight set limits on courts. Amendments Nine and Ten declare that rights not enumerated in the Bill of Rights are retained by the people or by the states.

OTHER CONSTITUTIONAL AMENDMENTS

Seventeen additional amendments have been added to the Constitution since the Bill of Rights was ratified in 1791 (see Table 3.6 on p. 106). Six of these expand the numbers and types of persons eligible to vote. Four more alter governmental structure and procedures. Five additional amendments expand or limit governmental powers. And one amendment, the Eighteenth, sought to legislate substantive policy by prohibiting the manufacture and sale of liquor. This amendment survived for 14 years, after which it was repealed by the Twenty-First Amendment.

For constitutional rules to be changed or modified through amendment, a two-stage process must be followed: proposal and ratification. The Consti-

TABLE 3.5
THE BILL OF RIGHTS

BRANCH RESTRICTED	AMENDMENT	CONTENT
Congress	1	No laws establishing religion; regulating religion, speech, press, assembly; preventing the petitioning of government
Executive	2, 3, 4	No infringing on right to bear arms (2); prohibition on arbitrarily taking homes for militia (3); not engaging in search and seizure of evidence without search warrant (4)
Courts	5, 6, 7, 8	No trials on serious offenses without grand jury; immunity from being tried twice for same offense; no property taken without just compensation (5); guarantee of speedy trial; presentation of charges; confrontation of hostile witnesses; immunity from testimony against oneself (6); guarantee of trial by jury for serious offenses (7); bail and punishment may not be excessive (8)
Federal government	9, 10	All rights not enumerated in Constitution reserved to states or people

tution specifies two different methods for completing each stage; the resulting four avenues for constitutional amendment are depicted in Figure 3.2 (on p. 107). Amendments may be proposed either by a two-thirds vote in the House of Representatives and Senate or by a national convention called by Congress at the request of two-thirds of state legislatures. In reality, all proposed amendments have resulted from a congressional vote; none has emerged from a national convention.

In the 1980s, however, a growing movement in favor of an amendment requiring a balanced budget came close to gaining the votes in state legislatures needed to compel Congress to summon a national convention. From the early 1970s to 1988, 32 of the required 38 states had voted to convene a national convention to consider the balanced budget issue. But in 1988 the Alabama legislature rescinded its earlier vote in favor of a convention, fearing that a convention might overstep its authority and propose other constitutional changes. In recent years Congress has debated a constitutional amendment requiring a balanced budget. In 1995, a few weeks after the Republicans took control of Congress, more than two-thirds of the House of Representatives voted for such an amendment. In the Senate, however, the amendment fell one vote short of a two-thirds majority.[40]

After amendments are proposed, Congress directs that they be ratified either by three-fourths of state legislatures or by constitutional conventions held in three-fourths of the states. The state convention method has been used only once, for the Twenty-First Amendment repealing Prohibition (the Eighteenth Amendment).

TABLE 3.6
CONSTITUTIONAL AMENDMENTS RATIFIED SINCE THE BILL OF RIGHTS

AMENDMENT NO.	PROPOSED	RATIFIED	CONTENT
AMENDMENTS EXPANDING ELECTORATE			
15	1869	1870	Gives right to vote to all males regardless of race.
17	1912	1913	Permits direct election of senators.
19	1919	1920	Gives right to vote to women.
23	1960	1961	Gives right to vote for president to residents of Washington, D.C.
24	1962	1964	Gives right to vote to all classes; abolishes poll tax.
26	1971	1971	Lowers voting age to 18.
AMENDMENTS ALTERING STRUCTURE, PROCEDURES			
12	1803	1804	Requires Electoral College to vote separately for president and vice-president.
20	1932	1933	Changes date of presidential inauguration; eliminates "lame duck" session of Congress.
22	1947	1951	Limits president to two terms.
25	1965	1967	Describes succession of president and vice-president in case of death, resignation, removal from office, disability.
AMENDMENTS EXPANDING OR LIMITING GOVERNMENTAL POWER			
11	1794	1795	Restricts federal court jurisdiction over cases in which individual sues state.
13	1865	1865	Eliminates slavery; declares that states may not allow persons to be kept as property.
14	1866	1868	Declares that states may not deprive any person of life, liberty, or property without due process; guarantees citizens equal protection of law.
16	1909	1913	Gives Congress power to collect income tax.
27	1789	1992	Prohibits Congress from passing pay raise to take effect before next House election.
AMENDMENTS ENACTING SUBSTANTIVE PUBLIC POLICY			
18	1917	1919	Prohibits manufacture and sale of liquor.
21	1933	1933	Repeals Eighteenth Amendment.

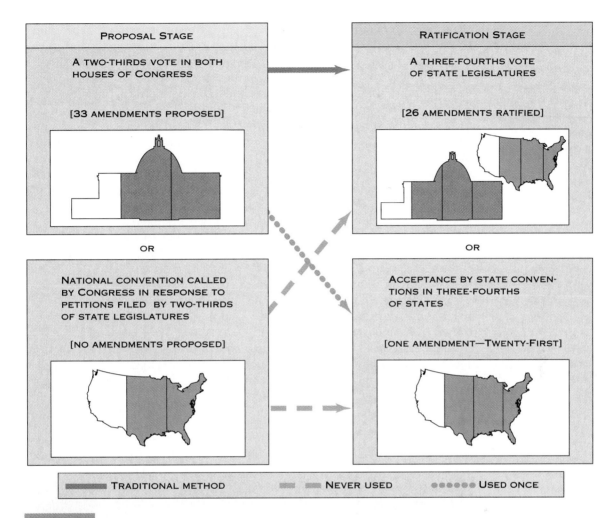

PROPOSAL STAGE	RATIFICATION STAGE

PROPOSAL STAGE

A TWO-THIRDS VOTE IN BOTH HOUSES OF CONGRESS

[33 AMENDMENTS PROPOSED]

RATIFICATION STAGE

A THREE-FOURTHS VOTE OF STATE LEGISLATURES

[26 AMENDMENTS RATIFIED]

OR

OR

NATIONAL CONVENTION CALLED BY CONGRESS IN RESPONSE TO PETITIONS FILED BY TWO-THIRDS OF STATE LEGISLATURES

[NO AMENDMENTS PROPOSED]

ACCEPTANCE BY STATE CONVENTIONS IN THREE-FOURTHS OF STATES

[ONE AMENDMENT—TWENTY-FIRST]

▬▬▬ TRADITIONAL METHOD   ▬ ▬ NEVER USED   ●●●●●● USED ONCE

FIGURE 3.2

THE FOUR AVENUES FOR CONSTITUTIONAL AMENDMENT There are two stages in amending the Constitution: proposal and ratification. Although Congress cannot control the proposal stage, it chooses the method of ratification and establishes time limits for state consideration. Both stages require extraordinary majorities for passage and therefore constitute difficult hurdles for those seeking constitutional change. Consequently, constitutional change most often occurs as a result of informal processes, such as court interpretation and other political practices.

As Table 3.6 indicates, most amendments achieved ratification within a few years of being proposed. However, the most recent addition to the Constitution, its Twenty-Seventh Amendment, constitutes a dramatic exception to this rule. This amendment, which prohibits Congress from giving its members a pay increase that takes effect before the next House election, was originally proposed with the Bill of Rights in 1789. Only six states had ratified the amendment by 1792. It took another 81 years for a seventh state to ratify, in 1873. Finally, in 1992, 203 years after it was proposed, 38 of 50 states had approved the amendment, meeting the three-fourths requirement. At first some members of Congress seemed ready to challenge the validity of ratification occurring so many years after the amendment had been proposed. But owing to fear that opposition could be interpreted by voters as an attempt to block an important spending reform, no such challenge materialized. Congress now routinely places limitations on the number of years that amendments may circulate in states before ratification.

Since 1787 Congress has introduced and debated thousands of amendments. Only a small percentage have actually been proposed to the states. However, once proposed, amendments have a great chance of achieving ratification. Only seven amendments submitted to the states ultimately failed to be ratified.

The Equal Rights Amendment (ERA), which would guarantee equal rights to women, is an important example of an amendment that failed to be ratified. The culmination of nearly a half-century of struggle by women's groups, the ERA was approved by two-thirds majorities in the House and Senate in late 1971 and early 1972. The amendment's language was simple and direct. Section 1 declared that "equality of rights under the law shall not be denied or abridged by the United States or by any State on account of sex." Section 2 conferred power on Congress to pass laws to enforce equal rights for women. A third section declared that the amendment would take effect two years after its ratification.[41]

Supporters had seven years, until March 1979, to secure ratification. The ERA appeared headed for certain victory after whizzing through 35 of the required 38 states, 22 in 1972 alone. But opposition formed rapidly and a "Stop ERA" movement was organized. In 1974 three states that had ratified in 1973—Idaho, Nebraska, and Tennessee—rescinded their approval, actions raising novel and important legal questions about whether a state could

The Equal Rights Amendment was one of the rare constitutional amendments that failed to be ratified after being sent to the states. After some 10 years of struggle, in 1982 the amendment fell six states short of ratification.

rescind a vote for ratification. The Supreme Court dodged the issue, arguing that challenges to the three states' actions raised "political questions" better decided by Congress. By the end of 1978 the ERA had stalled, unable to win ratification in more than 32 states. In an unprecedented action, Congress granted ERA supporters an extension of the ratification deadline until June 1982. Despite the extension, the amendment failed to become part of the Constitution.

INFORMAL METHODS FOR CHANGING THE CONSTITUTION

Although the Framers hoped to discourage frequent changes in the Constitution by requiring extraordinarily large majorities to propose and ratify amendments, it is impressive that only 27 changes have passed. Despite dramatic changes in the participants and other elements of American politics, the formal rules established by the Constitution have remained much the same.

However, the formal amendment process is only one way to alter the rules governing the play of power. Constitutional rules create a very general framework that does not provide unambiguous answers to all possible questions. Thus, other institutions have contributed to defining the Constitution's meanings, and such meanings often have evolved over time. In fact, courts throughout our nation's history have infused meaning into the Constitution through their interpretations of specific provisions. Ambiguous phrases such as *interstate commerce, necessary and proper,* and *privileges and immunities* gain substantive content through court opinions that resolve concrete problems and disputes.

Congress, too, has played a role in filling in the details of constitutional provisions. For example, Congress passed the Judiciary Act of 1789, legislation creating the lower federal court structure that we have today. The federal bureaucracy emerged from Congress's authorization of executive branch organizations, which were not described at all in the Constitution.

Presidential practice also has given meaning to the Constitution. Although the president is given few formal powers in the Constitution, the powers of the presidency in relation to Congress have grown steadily as political practice responded to events such as war and domestic economic crises. Moreover, institutions receiving no mention in the Constitution, most notably political parties, have developed, grown, and asssumed important roles in the play of power.

THE CONSTITUTION
IN THE PLAY OF POWER

In an important sense, the Constitution is a rule book that guides the play of power examined in this text. It provides the fundamental framework for American politics, creating its structure and shaping relations among institutions and other participants. Constitutional rules define who participates and what many of their attributes are. The Constitution itself, as a symbol of a just government, is used by participants as a resource in the play of power.

Although at times it is vague in its mandates and open to varying inter-pretation, the Constitution provides the general framework in which par-ticipants pursue their goals in the play of power. The Constitution may be viewed as a grand blueprint or map of the play of power, specifying dif-ferent routes by which participants may secure both material and symbolic benefits.

The constitutional map creates and highlights multiple points of access to political authority in different levels of government—local, state, and national—and in varying national institutional arenas, the presidency, courts, and two houses of Congress. This structure discourages swift and decisive action by dispersing power among many participants at different levels and in different arenas. In addition, rules regarding elections and the right to vote encourage the formation of multiple groups. By defining the powers and duties of various institutions, governmental levels and arenas, and specific participants, the Constitution shapes the resources and, in turn, the political strategies employed in the play of power. For example, a presi-dent trying to curb governmental spending can not simply command Con-gress to cut appropriations. Instead, the Constitution provides the president with the power to veto unacceptable appropriations bills.

Constitutional rules also help to specify how the play of power proceeds. For example, those seeking increased governmental appropriations for spe-cial projects must begin their quest in the House of Representatives, where, the Constitution tells us, money bills originate. Similarly, treaties with foreign nations must be negotiated by the executive branch and then gain approval by the Senate. You will see the many effects of the constitutional blueprint throughout this text's description of the play of power and its outcomes.

THE CONSTITUTION AS A RESOURCE IN THE PLAY OF POWER

In addition to providing the framework in which the play of power is con-ducted, the Constitution has itself become a resource used by participants. We have already acknowledged the extraordinary respect for the Constitu-tion exhibited by Americans. This respect is a source of symbolic capital that may be used by participants as they pursue their goals in the play of power.

Throughout our nation's history, political leaders have appropriated con-stitutional language and provisions to justify their decisions, actions, and failure to act. During the Watergate scandals of the early 1970s, for example, then-president Richard Nixon refused to release tape recordings that might incriminate him in a coverup of illegal activity, claiming that forcing him to turn over the tapes violated the power of "executive privilege" implied in Article II of the Constitution. Nixon relinquished the tapes only when he was compelled to do so by a federal judge whose interpretation of the Constitution differed from Nixon's. Political officials also frequently use the Constitution's symbolic capital to avoid stating positions on controversial issues. Representatives facing re-election in districts where opinion is di-vided on abortion policy, for instance, may sidestep this issue by simply arguing that they follow the Constitution's mandate, as interpreted by the United States Supreme Court.

Generations of individuals and groups seeking to change their status in American society have placed their demands in the language of "rights" guaranteed in the Constitution's amendments.[42] Women, African Americans, the disabled, gay men and lesbians, and other historically disadvantaged groups have used a political discourse of rights to argue for the application of constitutional provisions to their groups. The probability that these claims will be examined seriously and responded to by political leaders is enhanced by their grounding in the Constitution's familiar and symbolically powerful language.

DEFINING THE PARTICIPANTS IN THE PLAY OF POWER

Finally, the Constitution plays a crucial role in defining who shall be permitted to participate in the play of power at all. Although the Constitution created a political system that was as inclusive as any existing in the world at that time, its exclusion of the vast majority of the population makes it much less democratic than we expect today.

The Constitution of 1787 defined each African slave as three-fifths of a person and conferred on them no right or authority to participate in politics. Native Americans were referred to as "foreign nations" in the Constitution and given no rights of citizenship. In addition, the Constitution failed to mention women, one-half of the nation's population. By granting the right to vote only to those who, in 1787, could vote in state lower house elections, the Constitution had the effect of excluding women from political participation, because no state permitted women to vote at that time. Moreover, when the Constitution employed a pronoun, it invariably used the masculine forms, *he* and *him*.

The Constitution's exclusionary nature reflected and helped to promote efforts by American leaders to define a national identity during the post-revolutionary period. One of the challenges facing America at that time was to distinguish itself from England, from which it had gained independence. America in the 1780s was a diverse nation composed of people from a variety of ethnic and racial backgrounds, including Irish, Scotch, French, German, Swedish, Native American, and African peoples. In 1790 only approximately 65 percent of the American population was of English origin.[43] However, the need to encourage the feeling of American distinctiveness, coupled with the need to promote a strong and unified nation and government, led American leaders to emphasize the unity among America's population rather than its diversity. Describing the new nation and its peoples, John Jay in *The Federalist*, No. 2, wrote: "Providence has been pleased to give this one connected country, to one united people, a people descended from the same ancestors, speaking the same language, professing the same religion, attached to the same principles of government, very similar in their manners and customs."[44]

Such a view of who constitutes an American excludes Native Americans and those of African descent. In one of its initial actions, the first Congress codified this view of the American citizen when, in the Naturalization Act of 1790, it specified that only "white foreigners" would be eligible for

American citizenship. Only after the Civil War and more than 150 years of political and constitutional struggle were all the diverse peoples residing in the United States granted formal political equality.

SUMMARY

Disputes between England and the colonies, and among competing social, political, and economic interests within the colonies, led to the American Revolution and the founding of a new, independent nation. The Articles of Confederation, the nation's first constitution (adopted in 1777), conferred sovereignty on states and created a weak national government with few formal powers. The national government's weaknesses under the Articles led to their replacement by the Constitution of 1787.

The Constitution's framers sought to achieve several goals. They hoped to create a stronger national government that would promote commerce and a national economy. They sought to curb the power of state legislatures, many of which they believed to be overly influenced by the public. The Framers also hoped to design a government with sufficient internal checks to prevent infringement on liberty and property. Finally, the Framers included provisions that would help to secure popular support.

The Constitution included several crucial structural features. It created a federal structure, dividing powers between national and state governmental levels. It created separate national governmental arenas—legislative, executive, and judicial—and gave each specific powers, functions, and duties. The Constitution also created checks and balances, by which each branch of the national government could curb the power of other branches. Finally, the Constitution established electoral rules and structures for a limited democracy. A bill of rights added soon after the Constitution's ratification contains provisions protecting individuals from governmental tyranny and abridgement of rights.

The major structural features are contained in the Constitution's seven articles. Article I describes the powers and duties of the legislative branch. Article II defines the president's powers, duties, qualifications, and term of office, as well as procedures for electing the president. Article III creates a national judiciary. Articles IV through VII range broadly over many topics, including provisions outlining the obligations of states to citizens of other states, restoring order in states that experience civil disobedience, describing procedures for ratifying and amending the Constitution, and announcing the national government's supremacy over states.

The Constitution provides the general framework for the play of power in American politics. It also helps to define who participates in the play of power and the resources that may be used, and it shapes the strategies employed in pursuing outcomes. Because it is so revered by the American public, the Constitution itself is used as a resource by participants in the play of power. The specific ways in which the Constitution has shaped and otherwise influenced the play of power will be examined further throughout the remainder of this book.

SYMBOLS AND THE CONSTITUTION

 The year 1987 marked the Constitution's two-hundredth anniversary. The following excerpts are taken from a bicentennial speech delivered by President Ronald Reagan on September 17, 1987, in front of Independence Hall in Philadelphia. As you read, try to identify and interpret the symbols used by Reagan in referring to the Constitution and the Framers. What do these remarks teach us regarding American beliefs about the Constitution?

[I]t wasn't the absence of problems that won the day in 1787. It wasn't the absence of division and difficulty; it was the presence of something higher—the vision of democratic government founded upon those self-evident truths that still resounded in Independence Hall. It was that ideal, proclaimed so proudly in this hall a decade earlier, that enabled them to rise above politics and self-interest, to transcend their differences and together create this document, this Constitution that would profoundly and forever alter not just these United States but the world. In a very real sense, it was then, in 1787, that the Revolution truly began. For it was with the writing of our Constitution, setting down the architecture of democratic government, that the noble sentiments and brave rhetoric of 1776 took on substance, that the hopes and dreams of the revolutionists could become a living, enduring reality. . . .

If our Constitution has endured, through times perilous as well as prosperous, it has not been simply as a plan of government, no matter how ingenious or inspired that might be. This document that we honor today has always been something more to us, filled with a deeper feeling than one of simple admiration—a feeling, one might say, more of reverence. One scholar described our Constitution as a kind of covenant. It is a covenant we've made not only with ourselves but with all of mankind. As John Quincy Adams promised, "Whenever the standard of freedom and independence has been or shall be unfurled, there will be America's heart, her benedictions and her prayers." It's a human covenant; yes, and beyond that, a covenant with a Supreme Being to whom our Founding Fathers did constantly appeal for assistance.

It is an oath of allegiance to that in man that is truly universal, that core of being that exists before and beyond distinctions of class, race, or national origin. It is a dedication of faith to the humanity that we all share, that part of each man and woman that most closely touches on the divine. And it was perhaps from that divine source that the men who came together in this hall 200 years ago drew their inspiration and strength to face the crisis of their great hopes and overcome their many divisions. After all, both Madison and Washington were to refer to the outcome of the Constitutional Convention as a miracle; and miracles, of course, have only one origin. . . . [T]he guiding hand of providence did not create this new nation of America for ourselves alone, but for a higher cause: the preservation and extension of the sacred fire of human liberty. This is America's solemn duty.

Source: Office of the Federal Register, National Archives and Records Administration, *Public Papers of the Presidents of the United States, Ronald Reagan,* Book II (Washington, D.C.: Government Printing Office, 1989), pp. 1040–43.

Key Terms

republican form of government (*p. 77*)

confederation (*p. 85*)

Virginia Plan (*p. 90*)

New Jersey Plan (*p. 92*)

Great Compromise (*p. 92*)

bicameral legislature (*p. 92*)

Three-Fifths Compromise (*p. 93*)

federalism (*p. 96*)

separation of powers (*p. 96*)

checks and balances (*p. 96*)

representative government (*p. 98*)

enumerated powers (*p. 99*)

necessary and proper clause/ elastic clause (*p. 99*)

habeas corpus (*p. 99*)

contract clause (*p. 99*)

judicial review (*p. 101*)

full faith and credit clause (*p. 101*)

privileges and immunities clause (*p. 101*)

supremacy clause (*p. 102*)

Federalists (*p. 103*)

Antifederalists (*p. 103*)

Recommended Readings

Beard, Charles. *An Economic Interpretation of the Constitution of the United States.* New York: Macmillan, 1913. The classic study that provoked decades of debate among historians over the motivations of the Framers. Beard argues that the Framers were motivated by financial self-interest.

Becker, Carl L. *The Declaration of Independence: A Study in the History of Political Ideas.* New York: Vintage, 1922. The classic study of the political philosophy embedded in this important document.

Brown, Robert. *Charles Beard and the Constitution.* Princeton: Princeton University Press, 1956. An influential refutation of the Beardian view of the Framers' motivations in writing the Constitution.

Jordan, Winthrop. *White over Black: American Attitudes toward the Negro, 1550–1812.* Baltimore: Penguin, 1969. A study of the attitudes of white Americans toward African slaves and Native Americans in colonial and revolutionary America.

Kammen, Michael. *A Machine That Would Go of Itself: The Constitution in American Culture.* New York: Knopf, 1987. A fascinating study of American beliefs about the Constitution and the prominent place this document occupies in our national culture.

Kerber, Linda K. *Women of the Republic: Intellect and Ideology in Revolutionary America.* Chapel Hill: University of North Carolina Press, 1980. An examination of attitudes about and the roles played by women in revolutionary America.

Main, Jackson Turner. *The Antifederalists: Critics of the Constitution, 1781–1788.* Chapel Hill: University of North Carolina Press, 1961. Recounts the political struggle for the Constitution's ratification from the perspective of its most important opponents.

Ollman, Bertell, and Jonathan Birnbaum, eds. *The United States Constitution: 200 Years of Anti-Federalist, Abolitionist, Feminist, Muckraking, Progressive, and Especially Socialist Criticism.* New York: New York University Press, 1990. A collection of essays that criticize the Constitution from a variety of perspectives.

Wood, Gordon S. *The Creation of the American Republic, 1776–1787.* New York: Norton, 1969. A leading study of the transformation of American political thought and practice from the beginning of the Revolution to the writing of the Constitution.

Federalism and Intergovernmental Relations

IN DECEMBER 1993 President Bill Clinton signed into law the Brady Bill establishing a five-day waiting period before the purchase of a handgun can take place. The bill, named for James Brady, who as President Ronald Reagan's press secretary had been grievously wounded in a thwarted assassination attempt on the president, was vigorously contested by groups opposing gun control and strongly supported by those favoring more careful screening of handgun owners. A key provision of the bill required that local law enforcement officials conduct background searches of those applying to purchase a handgun.[1] The law was challenged in federal court, and in May 1994 a district court judge ruled the provision unconstitutional. On what grounds did the judge base his ruling? If you've read Chapter 3 and remember what's in the Bill of Rights, you probably have an answer—the judge based his ruling on the Second Amendment: "A well regulated militia, being necessary to the security of a free State, the right of the people to bear and keep arms shall not be infringed." Although this may be an obvious answer, it is incorrect.

In fact, the judge based his ruling on the Tenth Amendment: "The powers not delegated to the United States by the Constitution, nor prohibited by it to the States, are reserved to the States respectively, or to the people." He ruled the Brady Bill unconstitutional because, even though it requires state and local governments to carry out its provisions, Congress allocated no money for this purpose. Hence, local police would be diverted from their current assignments to fulfill the law. In effect, by requiring states and localities to carry out its bidding, Congress was establishing an **unfunded mandate**—a task that states and localities are required by the federal government to carry out but for which the federal government provides no funds. The issue of unfunded mandates, as we shall see, is especially troubling in an era of substantial federal budget deficits and fiscally

As Jim and Sarah Brady look on, President Clinton signs the Brady Bill mandating a waiting period before purchase of a handgun. Federal officials get the credit, but who foots the bill? Whether the federal government has an obligation to pay for the policies it requires states and localities to enact is part of current debates over unfunded mandates, as well as broader and more enduring arguments about the relationship among state, local, and national governments.

strained states and localities. Under the Brady Bill's unfunded mandate, Congress was diverting—unconstitutionally, in the judge's opinion—state and local public funds from tasks assigned by state and local government. In short, the Brady Bill's provision was struck down (although, of course, the federal district court ruling is being appealed and may not stand) not because it violated the right to bear arms but because it violated the division of labor among national, state, and local governments.

This division of labor, called *federalism,* is the subject of this chapter. Here we will address a variety of questions: Why is the responsibility for making political decisions in the United States divided among three different levels of government (local, state, and national)? How has the balance of responsibility shifted among these three levels over the course of American political history? More important, what are the consequences of these arrangements for the play of power in American politics? Which participants are advantaged and which are disadvantaged by American federalism? Understanding the play of power in American politics requires an appreciation of the differences and similarities between politics at the national, state, and local levels.

In this chapter we first examine some common, often unstated beliefs about the role of federalism in American politics. We then provide a more careful definition of federalism and discuss its evolution. We consider the ways in which the idea of federalism influenced the framers of the Constitution and how it has influenced Americans' struggle with the diversity of a large, multicultural democracy. We also explore the ways in which approaching politics as a play of power helps us to understand the impact of federalism on American politics. In particular, we show how each level of government constitutes a separate play of power with its own rules, and we characterize strategies working to various participants' advantage and disadvantage. Understanding American politics requires an appreciation for the ways in which participants strategize to shift political struggles to the levels of government that will provide the most favorable arena for pursuing their interests.

COMMON BELIEFS

THE ROLE OF FEDERALISM

 What should the role of the federal government be in our lives? How much can we depend on state and local government to meet our needs? We don't often discuss federalism when debating politics in the United States. Many of us are unsure what federalism is or why it was established in the first place. It turns out, though, that whether we know it or not, preconceptions about the best way of dividing power among national, state, and local government—the essence of federalism—underlie many of our thoughts about politics as well as the rhetoric of political leaders. Paramount among common beliefs about federalism are beliefs about what role the federal government should play in the political life of the nation.

On the one hand, many of us believe that the federal government is, almost by definition, a meddlesome presence in our lives. Castigating the inability of Congress to address our specific problems, criticizing a supposedly bloated federal bureaucracy and its constant intrusions into our lives—these are the routine stuff of political discourse. As we debate issues such as health care reform, welfare reform, and crime, these sorts of arguments are constantly raised. Underlying them is the notion that the federal government is too remote, has too much power, and ought to be gotten off our backs. Polls reveal that the number of Americans who hold such negative opinions about the federal government is increasing. In 1989, only 33 percent of Americans said they got the most for their money from the national government as opposed to other levels of government. This percentage was down to 26 percent in 1992. A related belief is that, inasmuch as government is necessary in our lives, it ought to be responsive and close to the people and that, hence, the greatest possible reliance ought to be placed on state and local government. In 1992, only 42 percent of people polled expressed trust and confidence in the federal government, but 51 percent and 60 percent expressed trust and confidence in state and local governments respectively.[1]

On the other hand, although faith in the federal government seems to be declining, many Americans also appear to want a more active and vigorous national government. This is especially evident when people are asked about specific policies. In 1989, the American Council on Intergovernmental Relations conducted a poll revealing that 75 percent of the public supported federal food labeling standards, 71 percent supported federal water pollution standards, and 61 percent supported federal regulation of Medicaid requirements.[2] These findings suggest that many people, despite being skeptical about government, regard the national government as the only authority that has the resources and ability to truly act on behalf of all the American people. This view is most commonly expressed by leaders and citizens arguing for policies that will benefit minorities or the poor. Politicians holding this view decry the limitations and parochialism of state and local government and cite programs such as the New Deal, Medicare and Medicaid, and civil and voting rights legislation as triumphs of a strong federal government.

Reconciling these two conflicting preconceptions requires a careful analysis of how politics differs at the state, local, and federal levels. Although many people associate the most common belief about federalism—that the federal government is too big and should be limited—with political conservatives, and the belief that only the national government is capable of solving many of the nation's problems with liberals, we intend to show that such an outlook vastly oversimplifies the significance of federalism for the play of power. We will argue that participants in the play of power, regardless of political ideology, try to locate political authority at the level of government that maximizes their chances of winning. These chances vary with the particular policy at stake and the particular participants involved. It is not at all clear that liberals always win when the arena is the national level, or that conservatives always win when the play of power occurs at the state or local level. Rather, the implications of federalism must be considered within the boundaries of each specific case.

[1]Poll figures are from Timothy J. Conlan, "Federal, State, or Local? Trends in the Public's Judgment," *Public Perspective* 4: 2 (January/February 1993), pp. 3–5.
[2]Conlan, "Federal, State, or Local?" p. 5.

FEDERALISM, CONFEDERATIONS, AND UNITARY SYSTEMS

Let us begin by defining what we mean by federalism and distinguishing it from other sorts of political arrangements.[2] *Federalism* is a system of government in which a central or national government shares sovereignty with lower units of government, such as states, towns, or cities. By the term **sovereignty** we mean that a government has established an unchallenged claim to rule over a particular geographic area and people. Within that area, citizens generally agree to abide by the laws of the government and accept the government's claim to the use of coercion in order to enforce those laws. The federalist arrangement is quite different from a **unitary system of government**, such as Great Britain's, in which sovereignty is vested entirely in the national government. Under federalism, lower-level governments are guaranteed sovereignty not simply by specific laws passed by the central government but also by the national constitution, which stands above both the central government and the lower units of government. This is significant because in a federal system the autonomy of the lower units cannot be revoked by the central government. Instead, the central and lower levels of government exercise independent and limited authority within certain policy arenas and within the geographic boundaries of their own jurisdictions.

An important advantage of federalism is that it allows a group of political elites at the national level to rule over vast amounts of territory by allowing local elites to retain some autonomy. This is how the Romans were able to control their vast empire. Lacking the power or resources to control the culture of the lands they conquered, the Romans were interested primarily in the power to tax, which was reserved for the central government. Local rulers were allowed to maintain a great deal of power and thus had less reason to resist Roman dominance.[3]

Although the concept has ancient roots, the United States was the first political system to formalize federalism in a written constitution. The United States Constitution specifies that political authority is shared between the states and the national government. It is precisely this sharing of authority, along with the independence of state governments, that is guaranteed by the Tenth Amendment and that provides the grounds for the judge's ruling on the Brady Bill. His ruling affirms the federal principle that Congress, a national legislative body, in exercising its authority cannot dictate what policies states pursue in their own areas of authority. In essence, this means that American citizens hold a dual citizenship: They are citizens both of the nation and of the state in which they live. As with the separation of powers among the three branches of government, the counterbalancing of state and national authorities was another constitutional solution to the problem of how to establish a strong central government while at the same time preventing it from gaining too much power and authority.

The limited autonomy of subnational governments produces an extraordinarily rich and complex political structure in the United States. In addition to the 50 states with their substantial political powers, there are approximately 83,000 subnational jurisdictions. These include 38,000 cities and towns, 3,100 counties, and 15,000 school districts. Most of these units have the power to tax, spend, and make public policy.[4]

One of the characteristics of a federal system is that it provides many different sites for political decision making, such as the U.S. Congress (*left*) and Maryland's state legislature (*right*). Whether this is an advantage or a disadvantage depends on who you are and the issue you are concerned about.

If we examine other political systems, or our own over time, we find there are many ways of dividing power between national and local governments. Federalism is a type of arrangement that falls between two other kinds of arrangements. At one end are more unitary systems like Great Britain's, in which Parliament is the national legislative body that can pass laws without considering the preferences of local governments. Moreover, unlike Americans, who are citizens of both their state and the nation, the British people are subjects only of the Crown, or national government.

At the other end are *confederations*—systems dominated by regional governments that have joined together and that grant little or no independent sovereignty to the central government. In such a system there is great variation in the policies and characteristics of regional governments. In the United States, the Articles of Confederation created a federal system that was very close to this end of the continuum. Remember that under the Articles the central government had very little power and no real sovereignty, existing instead at the pleasure of the member state governments.

Even after distinguishing among unitary systems, federalism, and confederations, it is important to recognize that there are no hard-and-fast boundaries between these power-sharing arrangements. The United States, even while remaining a federal system, has seemed more like a unitary system at some points in its history and more like a confederation at other times. Especially before the Civil War, the country was much closer to being a confederation than it is today. In fact, the term *confederacy* was chosen by the seceding southern states to emphasize their view of the federalist system established by the Framers. Moreover, the very term *United States of America* was treated as a plural noun prior to the Civil War (e.g., "the United States *are* committed to the principle of federalism"), reflecting the idea that state autonomy and differences among regions lay at the center of the American political system. Only after the Civil War did the term come to be treated as a singular noun (e.g., "The United States *is* committed to the principle of federalism"). As we shall see, the modern definition of federalism, incorporating an extremely powerful central government and much more limited state autonomy, is much closer to a unitary system than to a confederation. This definition emerged only after the Great Depression.

However, even though the federal government is far more powerful today, movement toward a more unitary system of federalism has not been constant. Especially during the Republican administrations of Presidents Richard Nixon and Ronald Reagan, there were attempts to increase the autonomy of the states and limit the authority of the federal government. Reagan's often-cited appeal to "get government off our backs" was mainly directed at the federal government; it reflected the belief that the nation had moved too close to a unitary model and that it needed to restore what he saw as the principles of federalism. Hence he called his policies "the New Federalism." These arguments are again being made by Republicans in the 104th Congress. However, as the Reagan administration supported policies that reduced the authority of federal policymakers, this meant increasing the powers of states and localities. Indeed, we will find that neither Reagan nor any other president has consistently supported either the delegation or the centralization of political power.

Finally, it is important to understand that the distinction between more unitary systems and more purely federal systems is not equivalent to any distinctions we might make between totalitarian and democratic states. That is, federal systems are not necessarily more democratic, nor are unitary systems necessarily more authoritarian. The former Soviet Union, for example, was a federal system, whereas Great Britain is a much more unitary system.

THE ONGOING STRUGGLE
BETWEEN STATE AND NATION

Although we don't often recognize it, debates over federalism in the United States—whether political power should be more or less decentralized—involve some of the most basic issues that can be raised in a democratic system. Are we a single nation with a single people?[5] Some argue that the differences among groups, regions, races, and so forth are subordinate to our common interest as Americans. This view implies strong powers for the central government. Others argue, however, that we are many peoples whose differences are enduring and that political arrangements must recognize such differences. From this perspective, there needs to be autonomy for state and local governments to reflect the unique needs and aspirations of our nation's many peoples.

The debate over whether we are one people or many peoples has been at the heart of American democracy from its beginning, and it remains alive today. It was central to the drafting of the Constitution and continues to underlie current arguments over the role the federal government should play in our lives. Current debates over multiculturalism can be seen as the latest manifestation of this controversy. Are we all Americans, a single people, who, despite differences among groups, share a common culture, moral system, political history, and so forth? This argument is made by many who support laws that would make English the official language of their state or locality and who contest the use of hyphenated terms of self-identification such as *Asian-American, African-American, Polish-American,* or *Italian-American.* In this view, the role of education is to socialize all Americans to a single set of values and a single history that we all can share. To do otherwise—to focus only on our differences—risks hyperfragmentation and degeneration into separate ethnic and racial enclaves. Historian Arthur Schlesinger Jr. makes this case when he writes that there is currently an "ethnic upsurge" in the United States that "today threatens to become a counterrevolution against the original theory of America as . . . a common culture, a single nation."[6]

Others would argue that we should consider America a land of many cultures and peoples, each with its own distinct history and values. In this view, the experience of each group has been different, and attempts to construct a single story of America risk obscuring the unique struggles and triumphs of these different peoples.[7] Consequently, education must be tailored at least partially to the specific experiences of different groups of students. In his multicultural history of the United States, Ronald Takaki says that "By allowing us to see events from the viewpoints of different groups, a multicultural curriculum enables us to reach toward a more comprehensive understanding of American history."[8] Or, as Donna Shalala, secretary of health and human services in the Clinton administration, says, "Every student needs to know much more about the origins and history of the particular cultures which, as Americans, we will encounter during our lives."[9]

The debate over multiculturalism has far-reaching consequences for American society in general and American politics in particular. Samuel Beer, a political scientist, suggests that this debate underlies the development of federalism in the United States. It leads to two very different theories

Are we one people, or are we many different peoples? This question is at the heart of ongoing, often heated debates over multiculturalism, diversity, and federalism.

of democratic government: national and compact theories (see Table 4.1).[10] Beer argues that these two theories have been used throughout American history to justify rearranging the balance of power between state and national governments.

NATIONAL THEORIES OF GOVERNMENT

The **national theory of government** holds that we are a single people. It locates sovereignty in the entire citizenry of the United States. This approach suggests that the Constitution (whose Preamble begins with the phrase *We the People*), the ultimate authority of the government, comes from the national constituency. Such authority is prior to and operates over and above

TABLE 4.1
NATIONAL AND COMPACT THEORIES OF GOVERNMENT

	NATIONAL THEORY	COMPACT THEORY
Source of sovereignty	The entire national citizenry ("We the People . . .")	The states whose representatives came together to write the Constitution ("The United States *are* committed to freedom of speech")
Constitutional faction closest to the theory	Federalists (James Madison, Alexander Hamilton)	Antifederalists (James Wilson)
Level of government assumed to be best able to respond to the will of the people	National	State or local
Preferred model of democracy	Representative	Participatory
Characteristic statements	"The Union is older than any of the states and, in fact, it created them as States. . . . The Union and not the states separately produced their independence and their liberty. . . . The Union gave each of them whatever of independence and liberty it has."—*Abraham Lincoln* ". . . [O]ne people. That is what we are and that is what we need to remain."—*David Dinkins*	"The Federal government did not create the states; the states created the Federal government."—*Ronald Reagan* ". . . the very idea of an *American People*, as constituting a single community, is a mere chimera. Such a community never for a single moment existed—neither before nor since the Declaration of Independence."—*John C. Calhoun*

the states, whose representatives came together to write the document. This view identifies the national government as the level most capable of acting in the overall interests of all the people. The role of the central government is to help forge this national identity. Centralized political power, with appropriate checks and balances, has significant advantages in overcoming the parochial interests of those who may oppose national purposes.

To see why this is so, consider the case of governmental attempts to regulate products such as tobacco. In states such as Kentucky or North Carolina that produce these products, there is considerable opposition to controlling tobacco use, because to do so would reduce the incomes of tobacco farmers and cigarette companies. Such locally important economic interests are powerful political forces in these states, even though they are much less significant at the national level. (Few states have tobacco farms, and most economic analysis reveals that cigarette companies produce few, if any, jobs in non-tobacco-growing states.) These interests struggle to keep policy questions about tobacco under the control of state legislators whom the interests stand a good chance of influencing. Let's assume for a moment that it is in the overall interest of Americans to discourage smoking (indeed, public opinion polls reveal that an overwhelming majority of Americans support limiting the use of cigarettes). Only by vesting power in a central government can a national majority overcome the opposition of strong state powers to pursue policies, such as eliminating tobacco subsidies, that would (1) reduce the production of a commodity, and (2) serve a national purpose at the expense of local interests that would otherwise resist such policies. This was the argument made at the Constitutional Convention by those (e.g., Alexander Hamilton and James Madison) who feared that narrow sectional and state interests might stand in the way of overriding national economic interests, as had been the case under the Articles of Confederation.

Even though economic issues continue to be of decisive importance in debates over federalism, they are not the only issues at stake. A strong national government was also necessary to define and protect the political rights of minorities who had been excluded from fair access to political life. As we will discuss, in the 1960s the federal government overrode white majorities in many southern states who favored the continued segregation of blacks and whites and the exclusion of blacks from virtually all forms of political participation.

Presidents and other political figures who support a larger role for the federal government usually speak in terms that assume this national theory. In his first inaugural address, in 1933 in the midst of the Great Depression, Franklin Delano Roosevelt defended the expansion of the federal government that would emerge under the New Deal by speaking of the primary need to restore "a sound national economy" and "a rounded and permanent national life."[11] This approach to government appeals to the basic rights and privileges accorded to all citizens of the United States. Only the federal government is in a position to guarantee such rights.

COMPACT THEORIES OF GOVERNMENT

In opposition to the national theory, the **compact theory of government** argues that the states existed prior to the framing of the Constitution and

STATEN ISLAND PROPOSES TO SECEDE FROM NEW YORK CITY

In November 1993, the residents of Staten Island voted overwhelmingly to secede from New York City. This referendum began an extended legislative struggle, still unresolved, over whether the state of New York would actually allow Staten Island to become the state's third largest city (thirty-ninth largest in the nation). Even with a population of 380,000, Staten Island is New York's smallest borough; it is generally suburban with a predominantly white population—quite different from the other four larger, more urban, and more racially diverse boroughs. Especially interesting is the degree to which the debate over Staten Island's proposed secession was cast in terms of compact versus national theories of government. (Of course, in this particular controversy the unit of government reflecting the national idea of "all one people" is the city of New York, not the entire United States.) Clearly, although the competing theories of government have their origins in the creation of American federalism, they are still alive and continue to animate political debate today. Indeed, if we did not have a federal system that grants autonomy to local governments, there would be no point to secessionist movements in the first place.

Staten Island's secessionist movement began in 1981 after federal courts ruled that the New York City Board of Estimate's scheme of representation was unconstitutional. The Board was the most powerful legislative body in city government; like the United States Senate, which gives each state equal representation regardless of population, each New York borough had an equal voice. However, the courts ruled that this policy violated the Fourteenth Amendment's guarantee of one person, one vote. (The details of the new governing arrangements that emerged are not important here, except in that, because they were based on population, they greatly diluted the influence of Staten Island.)

The Staten Island secessionists argued for a compact theory of government:

> With only 3 of 51 votes in the City Council, Staten Islanders were convinced they had no real say in local government, not that "local" was the best word for a City Hall with a budget bigger than that of any state in the union except New York and California. What did Staten Island need with equipment to battle skyscraper fires in Manhattan? Or cops to babysit United Nations diplomats? Why should suburban Staten Islanders have to comply with urban building code[s]? . . . Why shouldn't they run their own schools, patch their own potholes, develop their own waterfront, tend their parks, and control their dump?[1]

that because their representatives came together to write the document, the actual authority for the national government is ultimately rooted in the states. The Constitution is a "compact" between them. Such arguments were made by southern states when they seceded from the Union. Southerners, like John C. Calhoun of South Carolina, argued that they were simply dissolving the compact because the more numerous and powerful northern states were pursuing policies increasingly disadvantageous to the interests of the South. Although these arguments have been discredited because they were used to support slavery, it is important to see that even the cataclysm of the Civil War did not settle them.

One reason that compact theory survives in a variety of forms is that it advocates a model of democracy that is quite different from the democratic

The secessionists echoed the lament of contemporary opponents of strong central governments that these governments are far removed from and unresponsive to the wishes of the people. As compact theory suggests, the boroughs had come together to form the modern city of New York in 1894 and still existed as separate entities that could dissolve that compact when it was no longer in their interest.

In contrast, opponents of secession drew on the national theory of government, claiming that the concept of a single people is paramount if democratic government is to survive:

> Islanders were the envoys of an increasingly fragmented and dangerous age in which politics were organized around the shibboleths of racial identity. The rhetoric of self-governance put a nice face on white flight, a wholesale attempt to escape the headaches of presumably inner-city problems like drugs, crime, homelessness. Moreover, secession's threat to pluralism—the social contract at the heart of democracy—wasn't limited to Staten Island. Breakaway fevers were firing up parts of California, Idaho, Florida, and Illinois; citizens who might once have tried to resolve differences with the beleaguered arts of negotiation and compromise were now bent on redrawing maps and building walls.[2]

The arguments made by opponents of secession echo those made by critics of secession in the 1860s and raise again the idea that what is really at the heart of secessionist movements is race.

The debates reflect deeply felt attitudes about democracy and self-governance and what constitutes a political community. Indeed, it is clear that secessionists and their opponents are not motivated by political expedience or narrow economic self-interest. Despite overwhelming support for independence among its residents, Staten Island receives far more in city services than it pays in city taxes. David Dinkins, mayor of New York until he was defeated by Rudolph Giuliani in 1993, was an outspoken opponent of secession who applied the theory of a single people to events in New York: "One city—one people. That is what we are and that is what we need to remain."[3] Yet, as a liberal and an African American, he was clearly disadvantaged by the continued presence of Staten Island in city politics. In fact, because the island votes overwhelmingly Republican and Dinkins lost by a narrow margin, he would still be mayor if Staten Island had seceded.

[1]Chip Brown, "Escape from New York," *New York Times Magazine*, January 30, 1994, p. 22.
[2]Brown, "Escape from New York," p. 22.
[3]Brown, "Escape from New York," p. 44.

model assumed by the national theory. Underlying compact theory is the idea that because state governments are closer to the people they govern, they are better able to reflect the differences in preferences that endure between groups and regions. When governments are unable or unwilling to adopt policies that mirror the preferences of specific groups of citizens, the most basic values of democracy are violated and reform is needed. Thus, although no one seriously argues that any state will secede from the Union, the question of whether local governments will combine into larger units or maintain their autonomy is a contemporary variant of arguments based on compact theory. New York City, for example, is made up of five boroughs. As "Strategies: Staten Island Proposes to Secede from New York City" describes, in recent years the borough of Staten Island has threatened to secede from the city and form a separate municipality because its demands for public services are not being met.

At the national level, political figures who support increasing the autonomy of the states tend to assume a compact theory of sovereignty in their arguments. In contrast to Roosevelt's use of the national theory, Ronald Reagan, a president who worked to reduce the size and authority of the federal government, appealed to compact theory in his first inaugural address in 1981: "The Federal government did not create the states; the states created the Federal government."[12]

NATIONAL THEORIES, COMPACT THEORIES, AND DEMOCRACY

Although the debates over compact and national theories may seem arcane, it is important to understand that much is at stake when we decide whether to assign a specific responsibility to state or national government. On the one hand, there is much to be said for placing government closer to the groups of citizens who are affected by policies. Indeed, there is a powerful strain of political thought in the United States that views democracy as being participatory and maintains that for this participation to occur, governmental decisions must be made close to the people. Recall that **participatory**, or **communal, democracy** assumes that citizens will not delegate decision-making power to elected representatives, but will make most governmental decisions directly for themselves in local forums.

This vision of participatory, or communal, democracy was held by several of the Framers. It was used as the basis for their criticisms of the great powers the Constitution gave to the central government. Those who opposed the Constitution, the Antifederalists, believed that democracy in the United States ought to be participatory. James Wilson, an Antifederalist supporter of participatory democracy, argued that citizens must be provided with direct experience at governing on their own behalf.[13] Thomas Jefferson, assuming such a vision of democracy, argued that the country should remain a largely agrarian nation so that most citizens would own their own land and thus develop a stake in their government. He contended that the maintenance of such an economic system would preserve the rough economic equality required for true democracy to function. In this vision, the citizens themselves would make most of the decisions of government. Such a model of democracy would require that government be easily accessible and close to the people; the New England town meeting is the model of this form of democracy. Although the United States did not develop along the lines envisioned by either Wilson or Jefferson, these arguments still carry great weight in American politics. Moreover, they help to account for the great suspicion that many Americans have of government—especially big government that is far removed from the people on whom its decisions are imposed.

On the other hand, in contrast to the vision of participatory democracy was the suspicion of many of the Framers that the average citizen was unable to make wise political choices. Remember that for many at the Constitutional Convention, the term *democracy* was equivalent to what we might term *mob rule* today. In this view, **representative democracy**, in which citizens delegate authority to make decisions to others, was the more appropriate form of government. Further, many of the Framers believed that only at the national level

Chicago's Hull House, established in 1889 by social worker Jane Addams, provided a primary settlement house for underprivileged immigrants. Addams's activities typified the political role that women played at the local level as well as outside electoral politics. As long as we focus exclusively on national politics or on the activities of governmental players, we will tend to underestimate the political participation of women.

would it be possible to overcome the narrow interests of specific localities and make decisions in the interest of the entire nation.

Obviously, both arguments have a great deal of validity. Many people support citizens' making decisions for themselves and having a direct role in politics, whereas others are much more skeptical about the ability and willingness of average citizens to make wise decisions, especially in today's complex world.[14] This is one reason why the nation has moved back and forth along the continuum of federalism between centralized and decentralized arrangements.

It is important to note that the debates between advocates of participatory and representative democracy and compact and national models of federalism are not simply arguments between conservatives and liberals. Rather, the implications of centralization and decentralization are complex and differ from issue to issue. Even though conservative political figures, such as former president Reagan, argued for decentralization of governmental authority to the state and local levels, such decentralization is often favored by progressives as well because it furthers the interests of local minorities or disadvantaged groups that might be influential in certain regions but not at a national level.

This seems to be the case historically for the role of women in politics. It has been argued that, long before women received the right to vote in 1920, they were active in politics. However, since the activities of women were largely at the local level and took place outside the confines of electoral politics, their work is invisible if we focus only on representative politics and/or the national level:

> [F]rom the time of the Revolution women used, and sometimes pioneered, methods for influencing government from outside electoral channels. They participated in crowd actions in colonial

COAL MINING ON BLACK MESA

Balancing an overriding national interest against specific local interests is at the heart of federalism—it is one reason why we grant autonomy to all three levels of government. Federalism, though, assumes that all groups will be represented in this division of labor, an assumption that is particularly hard to sustain when it comes to Native Americans. The degree to which Native peoples and their culture fail to find an avenue to express their interests is apparent in the controversies surrounding the mining of coal on the lands of the Navajo and Hopi.[1]

Throughout the 1980s and 1990s, federal energy policy placed a premium on mining low-sulphur coals. This priority was driven by several developments: a concern with air pollution that made clean-burning, low-sulphur coal an attractive source of energy; projections that there would be a dramatic increase in the demand for coal and coal technology in the expanding Pacific Rim nations; and the Persian Gulf crisis, which reinforced the need to develop our own domestic energy sources rather than relying on more vulnerable Middle Eastern oil. All these factors made the decision to mine low-sulphur coals, a relatively clean source of energy, justifiable in the national interest.

One of the richest sources of "steam" coal, a particularly low-sulphur variety, is in the Black Mesa of Colorado. However, this land is part of the Hopi and Navajo reservations (the former located within the boundaries of the latter). The tribal councils of both tribes have accepted large payments from the Peabody Coal Company in return for the rights to mine in the Black Mesa. With respect to our usual ideas about federal-

Questions regarding whether reservations are sovereign nations and whether tribal councils adequately represent the interests of Native peoples affect the health and safety of Native American miners. These questions have resulted in much uncertainty as to whether these miners are protected by federal regulations.

ism, this would be enough to decide the issue. But is it in the interests of the local peoples to allow their land to be mined? What forms of government in our federal system have been established to allow potential conflicts to be resolved and for traditional peoples to participate in this resolution?

Remember, though, that Indian reservations are considered sovereign nations. Can the

America and filled quasi-governmental positions in the nineteenth century; they circulated and presented petitions, founded reform organizations, and lobbied legislatures. Aiming their efforts at matters connected with the well-being of women, children, the home, and the community, women fashioned significant public roles by working from the private sphere.[15]

national interest of the United States be considered the same as the national interests of the Hopi and the Navajo? Can tribal councils, which have been imposed on Native peoples by the U.S. government, be considered adequate for registering the desires of all the Native Americans they represent? It turns out that many of the Hopi and Navajo people do not feel they have been allowed to participate in the decision to mine the Black Mesa. In fact, even as the U.S. government recognizes only the tribal councils it created, more traditional Hopi political structures continue to operate. These traditional structures reject the legitimacy of the tribal councils and the agreements they have reached for mining coal:

> Many of the traditional Hopi and Navajo people who currently live in the Black Mesa region firmly believe that the earth cannot be owned. They view themselves as the caretakers of the earth, believing that they must live in harmony with the laws of the Creator in order to maintain the balance for the entire planet. To both the traditional Hopi and the traditional Navajo, this means living a simple life without electricity or running water, completing cycles of ceremonies and offering prayers each day.[2]

These people reject the agreement between Peabody and the tribal councils, arguing that it will devastate their sacred grounds. Indeed, mining operations will substantially alter the natural ecology of the region, and part of the agreement is to relocate Hopi and Navajo who have used the Black Mesa for sheepherding. Moreover, in some ways the relationship between tribal councils on the one hand and Peabody Coal and the federal government on the other is more like the relationship between industrialized and less developed countries than like the relationship between private businesses and ordinary local and state governments. As is often the case with multinational investment in poor countries, "Many [Navajo and Hopi] have grown poorer as the projects consume land and destroy resources that were once available for subsistence living. Others have grown richer, accumulating more capital, assets, and access to land rights."[3]

While traditional Hopi and Navajo continue to fight Peabody Coal, their struggle, and many others like it, raise troubling questions about the degree to which the distribution of political authority in American federalism continues to ignore and suppress the interests of Native Americans. When we decide on policies that are in the national interest, we must always ask who was allowed to participate in the formulation of an overriding common interest and, in the case of Native peoples, whose nation we are talking about anyway.

[1]Our discussion is based on Kathy Hall, "The Impacts of the Energy Industry on Navajo and Hopi," in Robert D. Bullard, ed., *Unequal Protection: Environmental Justice and Communities of Color* (San Francisco: Sierra Club Books, 1994).
[2]Hall, "The Impacts of the Energy Industry on Navajo and Hopi," p. 131.
[3]Hall, "The Impacts of the Energy Industry on Navajo and Hopi," p. 136.

It is important to note that Native Americans, too, have been left out of most debates over compact and national theories of sovereignty. The development of a reservation system in the nineteenth century created tribal councils that relate poorly to and are largely ignored in most discussions of federalism. Prior to 1934, reservation lands were shrinking as the federal government sought to turn them over to private owners, including whites who purchased the land at rock-bottom prices from individual Native Americans to whom title had passed. In 1934 the Indian Reorganization Act was signed into law, reversing the policy of privatizing reservation lands and white ownership. The government provided funds for the repurchase of tribal lands and the creation of local self-government. Even then, many tribes refused to participate in this policy, rejecting the idea that the government in Washington could tell them what to do and how to govern themselves.[16]

Reservations, and the tribal councils that govern them, are considered sovereign nations. Although this unique status is often ignored by most Americans, it was recognized in an "unprecedented" meeting in 1994 between President Clinton and 542 tribal leaders.[17] The economic implications of this status were driven home by a 1987 Supreme Court ruling that reservations, being sovereign nations, are not subject to the gambling regulations of state governments. This ruling has freed tribes to develop lucrative gambling operations that are free of state taxes and regulation. In fact, 81 tribes run gambling operations, from which they collectively received a net income of $567 million and net profits of $142 million in 1992 alone.[18]

Although some tribal leaders argue that the exemption from state laws and regulations provides rich development opportunities for tribal lands, others see a much darker side to the ambiguous position of Native American reservations in the federal system. "Participants: Coal Mining on Black Mesa" (on pp. 130–31) describes how this conflict has played out in the American Southwest. Indeed, sociologist Robert Bullard argues:

> Because of more stringent state and federal environmental regulations, Native American reservations, from New York to California, have become prime targets for risky technologies. Native American nations are quasi-sovereign and do not fall under state jurisdictions. Similarly, reservations are "lands the feds forgot," and their inhabitants must contend with some of America's worst pollution. . . .
>
> Few reservations have infrastructures to handle the risky technologies that are being proposed for their communities, and more than 100 waste disposal facilities have been proposed for Native American lands.[19]

THE SHIFTING BALANCE OF POWER

As the United States has grown and the demands placed on government have changed, popular opinion has sometimes favored the national, at other times the compact theory of federalism. The result has been constant change in the balance of power between national, state, and local levels of government. As a consequence, the federal system today bears little resemblance to the system envisioned by the Framers.

LAYER-CAKE FEDERALISM

Even though the contours of federalism have become more complex over time, the design of federalism outlined in the Constitution seems straightforward. In trying to balance the issues raised by debates over various theories of democracy and questions about whether we are one or many peoples, the authors of the Constitution arrived at an elegant solution. They provided for two distinct and separate spheres of authority for the states and national government.[20] This model is sometimes called **layer-cake federalism**, because in it the powers of the two levels of government are as separate and distinct as the layers in a cake.[21]

The national government was generally restricted to economic tasks that promote commerce. This was especially important to those like Alexander Hamilton who, unlike Jefferson, saw the need for establishing a political system capable of supporting a commercial republic. A major source of controversy at the Constitutional Convention was, in fact, the balance of economic policy-making power between state and national government.

As in American politics today, this economic debate swirled around the government's debt. Two such questions were debated at the Constitutional Convention: Who would be responsible for the sizable debts run up during the Revolutionary War? and Which levels of government would be allowed to borrow money (i.e., run a deficit) in the future? The Federalists, who supported a strong national government, took the position that only the national government should be allowed to borrow money guaranteed by the "full faith and credit" of the United States. The Antifederalists, fearing accurately that this would limit the economic autonomy of the states that had been allowed to issue such "bills of credit" under the Articles of Confederation, opposed the provision. In the end, the Federalists triumphed and Article I, Section 8 of the Constitution gave the national government the right to borrow money on the credit of the United States, whereas Article I, Section 10 prohibited the states from issuing such bills of credit. This provision, along with the interstate commerce clause and other measures (e.g., centralization of the responsibility for national defense in the federal government), all severely undercut the economic policy-making powers of the states.

Throughout American history, the struggle over where to locate the authority to assume debt in a federal system has been at least as much about political power as it has been about the economic health and well-being of the nation.[22] As the Federalists and Antifederalists saw, the power to borrow money—and the resulting responsibility to raise taxes to pay off that governmental debt—is an extremely significant power. It not only allows the national government to direct the course of the nation's economic growth (by allowing only this level of government to spend more than it raises in taxes), it also greatly expands the power of the central government in all other public policy arenas. Similarly, attempts in the 104th Congress to pass a balanced budget amendment are more than just efforts to put the country on a sound economic footing. In fact, there is disagreement among economists about whether the current deficit poses a threat to the nation's fiscal well-being. Moreover, virtually all agree that there are times when running a deficit is part of sound economic policy—in times of economic depression, for example, when deficits help to stimulate the economy. Rather, most Republicans and many Democrats in the 104th Congress supported the balanced budget amendment as a way to limit the power and autonomy of the national government in particular, and all of government more generally. Restricting the federal government's power to borrow money, coupled with the prohibition against deficits at the state and local levels, would inevitably restrict the scope of government in our society.[23]

In contrast to the powers expressly granted to the national government, the Tenth Amendment reserves all remaining powers of government for the states. This includes public education, transportation, and public safety. In general, the Constitution assigns direct coercive powers over citizens to the states. Even though the Framers were especially concerned with the federal role in creating a strong commercial republic, they apparently felt that other, more directly coercive powers were best left to levels of government closer to the people on whom the laws would be enforced. As a result, states are the level of government that develops and enforces criminal justice codes, regulates the usage of private property, promulgates divorce laws, and monitors entry into various professions.

Reflecting the considerable diversity among states, there is a great deal of difference in how crimes are defined and punished from state to state. Thus, if you are arrested for using illegal drugs, what happens to you depends on the state in which you are arrested. Some states have much more lenient drug laws than others. If you are arrested for selling drugs in Michigan, regardless of the amount you sold or your previous criminal record, you will be given a mandatory life sentence with no possibility of parole. If you commit the same crime in Wisconsin, your sentence may be far less severe.

Although the layer-cake model of federalism is easy to understand, it does not address the dramatic changes that were to occur in the American political economy, especially during the twentieth century. Even in the nineteenth century, the line between state and national governments could blur. For example, although education is a state and local responsibility, the national government provided funding in the 1860s for the development of land-grant universities to provide assistance to farmers.

MARBLE-CAKE AND PICKET-FENCE FEDERALISM

The crisis of the Great Depression during the 1930s effectively ended the era of layer-cake federalism. As the unemployment rate soared, reaching 25 percent at its worst, businesses went bankrupt, banks collapsed, and people lost their jobs, homes, and life savings. The level of need went far beyond the limited capabilities of states and localities to provide even the most basic services for their citizens. As a result, the federal government became intimately involved in areas of public policy that had previously been thought to reside outside its sphere of authority. Federal legislation was passed providing unemployment compensation, Aid to Families with Dependent Children, Social Security, and a host of other programs that we take for granted today. In addition, the powers of the national government to regulate the economy were dramatically expanded.

The New Deal of the 1930s ushered in a new stage of federalism, sometimes called **cooperative federalism** or **marble-cake federalism**. In this form of federalism, the responsibilities of state and national governments are not neatly layered, but rather—as in a marble cake—there is a complex intermingling of responsibilities. In cooperative federalism, the national government developed a system of grants-in-aid that provided funds to help states and localities provide services specified by the central government. Usually this money is provided on a matching basis: For every dollar the state spends on the program, the federal government provides a certain number of dollars. So, while the federal government provides a financial

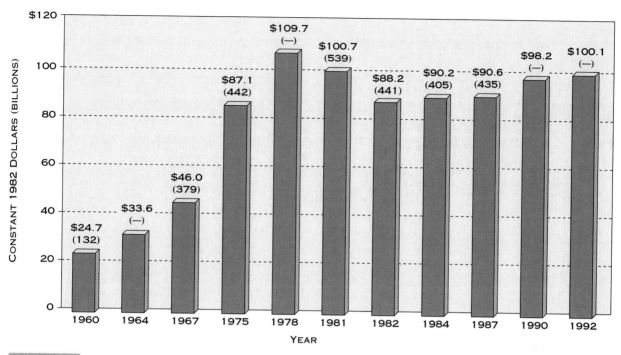

FIGURE 4.1

FEDERAL AID TO STATE AND LOCAL GOVERNMENTS, SELECTED YEARS Note the dramatic rise and then leveling off of federal aid (in terms of constant dollars as well as the number of grants). What are the implications of these trends for government services at all three levels of our federal system? *Source:* Data from U.S. Advisory Commission on Intergovernmental Relations, *The Federal Role in the Federal System: The Dynamics of Growth—A Crisis of Confidence and Competence* (Washington, D.C.: ACIR, July 1980), pp. 120–21; *Revenues and Expenditures,* Vol. 2 of *Significant Features of Fiscal Federalism, 1990* (Washington, D.C.: ACIR, August 1990), p. 42. Data quoted in Laurence J. O'Toole Jr., "American Intergovernmental Relations: An Overview," in Laurence J. O'Toole Jr., ed., *American Intergovernmental Relations,* 2nd ed. (Washington, D.C.: Congressional Quarterly Press, 1993), p. 9.

Number of grants for each year is shown in parentheses. Dashes indicate that data for those years are not available.

incentive, the state is free to decide the level at which specific services will be provided. The increasing role that the federal government has come to play in lower levels of government is hard to overestimate. In 1902 federal aid to state and local governments was approximately $28 million in five grant programs. By 1992 there were over 400 grant programs providing slightly more than $100 billion to lower levels of government[24] (see Figure 4.1).

The federal highway program enacted during the Eisenhower administration is one of the most widely cited examples of a successful federal matching program. Started in the 1950s, this program financed the building of a modern interstate highway system that greatly aided postwar economic development, increased our reliance on the automobile, and lessened our support for public transportation. The program was based on a 90 percent–10 percent match: For every dollar that a state spent on building portions of the system within its boundaries, the federal government paid nine dollars.

In cooperative federalism, the national government has an extensive role in specifying the services and aid that all citizens will receive from their government. Further, as suggested by the metaphor of the marble cake, in any given area of public policy there is an intermingling of state and federally mandated policies. As the number of grants-in-aid grew, a new wrinkle in federalism emerged. Categorical grants created strong alliances between governmental bureaucrats at the national, state, and local levels who administered these grants. So, for example, national officials at the Environmental Protection Agency develop strong linkages with state and local officials who administer environmental programs, such as the Clean Water Act and the Clean Air Act. These highly technical programs become familiar to the officials who work for environmental agencies but are very difficult for elected officials at the national, state, or local level (who must deal with a wide range of policy concerns) to understand or oversee. The result is what has been called **picket-fence federalism** "because of the narrow ties that run between higher and lower levels of government, like the crosspieces in the fence."[25] The impact of picket-fence federalism is that public policy becomes fragmented as each picket becomes isolated and insulated from the others. "At the state and local levels, this state of affairs causes great consternation among elected officials, who desire control over local administrators."[26] All three models of federalism are illustrated in Figure 4.2.

PAST THE MARBLE CAKE TO PICKING AT THE CRUMBS: UNFUNDED MANDATES

The reason why marble-cake federalism could be called cooperative federalism is that although the national government initiated a wide range of policies, it also used the national taxing powers to pay for these programs, thereby providing a sizable financial incentive to states and localities. However, as concern over limiting the federal deficit has restricted the ability of the federal government to fully finance the programs it mandates, states are often forced to pay for such programs. (This reflects the connection between limitations on governmental borrowing and other aspects of governmental power.) As we saw at the outset of this chapter, such unfunded mandates are a growing source of conflict between national, state, and local governments.

Unfunded mandates are the reason why state and local officials' old theme song "Give Me Money!"—a favorite throughout the lean years of the Reagan and Bush administrations—has finally given way to a new tune, "Leave Me Alone!"[27] Consider the 1994 revisions to the 1974 Safe Drinking Water Act. In order to guarantee the safety of America's drinking water, the act requires local governments to build state-of-the-art water treatment facilities and to monitor drinking water for a wide variety of contaminants. As originally passed, the act seemed to be a program entirely consistent with the model of picket-fence federalism. It linked environmental officials in Washington, states, and localities responsible for drinking water by establishing technical requirements that had to be met in all localities. Further, given the widespread concern of most Americans with drinking-water safety, the goals of the act had wide public support. However, in the 20 years between the act's passage and its revision in 1994, much had changed in

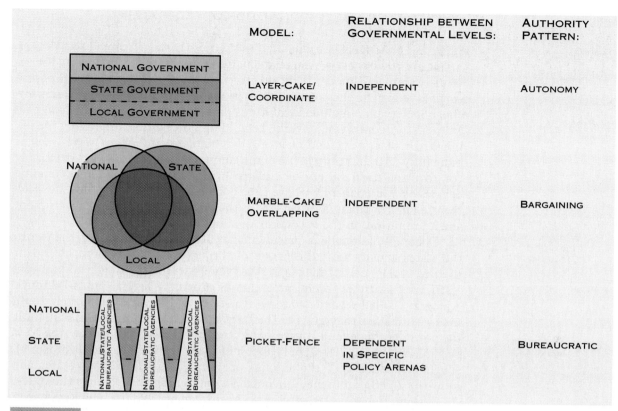

	MODEL:	RELATIONSHIP BETWEEN GOVERNMENTAL LEVELS:	AUTHORITY PATTERN:
NATIONAL GOVERNMENT / STATE GOVERNMENT / LOCAL GOVERNMENT	LAYER-CAKE/ COORDINATE	INDEPENDENT	AUTONOMY
NATIONAL / STATE / LOCAL	MARBLE-CAKE/ OVERLAPPING	INDEPENDENT	BARGAINING
NATIONAL / STATE / LOCAL (NATIONAL/STATE/LOCAL BUREAUCRATIC AGENCIES)	PICKET-FENCE	DEPENDENT IN SPECIFIC POLICY ARENAS	BUREAUCRATIC

FIGURE 4.2

THREE MODELS OF FEDERALISM Shown here are the layer-cake, marble-cake, and picket-fence models of federalism. Note that the three models assume very different authority relationships among levels of government in a federal system. *Source:* This figure borrows from Deil S. Wright, "Models of National, State, and Local Relationships," in Laurence J. O'Toole Jr., ed., *American Intergovernmental Relations,* 2nd ed. (Washington, D.C.: Congressional Quarterly Press, 1993), p. 76.

the dynamics of federalism. Concern over rising federal deficits made it increasingly difficult for Congress to provide the money for localities to carry out its mandates.

Yet public support for environmental protection remained strong. The result was that the cost burden of complying with federally mandated standards fell on state and local governments. This brought howls of protest from organizations representing local governments. Consequently, revision of the Safe Drinking Water Act became a highly charged and controversial topic as conservatives in Congress, already opposed to stringent federal environmental standards, joined forces with state and local officials who opposed additional federal burdens without additional federal funds to pay for them. Mayor Sharpe James of Newark, New Jersey, a city required to build a $41-million water treatment facility that would be paid for out of an already severely stretched city budget, explained: "If I only have $1 and I'm trying to pay for police, fire, education and cleaning up the street, and the federal government says take 50 cents of that to build this water treatment plant . . . we can't function."[28]

Since the New Deal, various administrations have attempted to alter the balance of power between state and national government. Following Roosevelt's example, Democratic presidents have often sought to enact new programs that expand the power of the national government. Lyndon Johnson's Great Society plan included ambitious programs, funded by the federal government, for combating poverty, rebuilding inner cities, improving the nation's school system, and ensuring that African Americans were brought more fully into the political life of the country. Such programs are extremely controversial for a variety of reasons. In the context of the long-standing debates over federalism we have been considering, these programs are often seen as an unwarranted expansion of the powers of the federal government and a limitation on the autonomy of states and localities.

Generally, Republican presidents have sought to grant greater power to the states and to limit the power of the national government. President Richard Nixon, calling his plans "the New Federalism," sought to limit the restrictive nature of federal programs by providing federal funds—through a program called **revenue sharing**—to states and localities that they could use essentially as they saw fit. The Reagan administration also tried to limit the power of the federal government. Reagan's vision, also called "the New Federalism" (though it might be better termed "the New New Federalism"!), sought to reduce both the amount of federal funds flowing to states and localities and the influence of federal government in the affairs of the states. As a tribute to his efforts, on Reagan's birthday in 1995 the new Republican majority in the U.S. House of Representatives specifically chose to pass legislation outlawing unfunded mandates.

Since the New Deal, this moving back and forth between greater federal power and greater state autonomy has resulted in an extremely complex federal system. Along with grants-in-aid and the now defunct revenue-sharing program, there are many different types of financial aid provided by the federal government. **Block grants** provide money to states and localities for very broadly defined policies (e.g., those relating to criminal justice). Reflecting the decentralizing thrust of the "Contract with America," House Republicans in the 104th Congress support converting federal grants under the 1994 Omnibus Crime Bill and a variety of educational programs (e.g., Head Start and school lunch programs) from grants-in-aid, which have specific spending rules, to more general block grants that would allow greater leeway to state and local governments. Indeed, many scholars argue that we can no longer characterize the entire system by a single metaphor—like a layer cake or a picket fence. Rather, what has developed is a system of intergovernmental relations that differs dramatically from policy area to policy area.

The emergence of, first, cooperative federalism (in its marble-cake and picket-fence varieties) and then a more complex system of intergovernmental relations (with its blurring of the neat lines between levels of government and increasing conflict over who will pay the costs of national programs) raises anew the issue of whether we are one people or many peoples. There is no simple way to analyze this complex system; rather, the implications of federalism must be considered within the confines of specific policy arenas. In order to provide you with the tools for doing your own analysis of the implications of centralizing and decentralizing political power in the complex system of intergovernmental relations, we now discuss the role of federalism in the play of power.

FEDERALISM IN THE PLAY OF POWER

To understand the role of federalism in the play of power in the United States, it is important to remember the distinction between symbolic and material politics. At the symbolic level, placing greater powers in the hands of state government appeals to the values of direct democracy and positioning government closer to the people. Implicit in these appeals are the old themes that we as a nation are many peoples and that too much power in Washington is dangerous to our liberty. These themes have great resonance in American politics and appeal to our suspicions of big government and our desire for direct or communal democracy. Ronald Reagan in particular was continuously effective at making such appeals. In his 1982 State of the Union address he said:

> Our citizens feel they've lost control of even the most basic decisions made about the essential services of government. . . . And they're right. . . . The main reason for this is the overpowering growth of Federal grants-in-aid programs during the past few decades.

Reagan's solution was to provide greater policy-making authority to state and local governments:

> Some will say our States and local communities are not up to the challenge of a new and creative partnership. . . . This administration has faith in State and local governments and the constitutional balance envisioned by the Founding Fathers. We also believe in the integrity, decency, and sound good sense of grassroots Americans.

At a more material level, allowing states and localities to exercise autonomy provides greater variation among them and more experimentation about what works and what doesn't work in public policy. This is an especially attractive idea at a time when the federal government's resources are limited and the public problems faced by the country seem to grow more complex every day. In the 1970s, for instance, New Jersey's policies to pay for the cleanup of leaking and unsafe hazardous waste sites became the model for the federal Superfund.[29]

More recently, the role of state experimentation has become prominent in debates over reform of the welfare system. The state of Wisconsin, under the direction of a conservative Republican governor, has experimented with "workfare." In this state, welfare recipients are required to either work or enter a job-training program; in any event, the length of time any individual can remain on welfare is limited. This approach heavily influenced the Clinton administration and the president's promise to "end welfare as we know it." The program Clinton presented to Congress in 1994 included a

State experimentation in policy areas serves as a testing ground for other states and the nation. Here a local school council meets under Kentucky's Educational Reform Act (*left*), and welfare recipients participate in Wisconsin's workfare program (*right*).

provision that all welfare recipients be enrolled in some form of job-training program and that they be limited to two years of governmental support. Many governors and the 104th Congress have included workfare provisions in their plans for reforming the welfare system.

Education is also an area that has seen much experimentation by the states. The state of Kentucky has adopted a far-reaching public school reform program—the Kentucky Educational Reform Act—that places considerable power in the hands of local parents and teachers and removes power from bureaucratically entrenched superintendents and district school boards.[30] Thus, Kentucky's reforms reflect the view of those who favor participatory democracy and believe that the only way to improve the educational system in that state is through the direct involvement of parents. Other states and the federal government are carefully monitoring the Kentucky experience to see whether its innovative program is suitable for their own jurisdictions. Obviously, using the states as "laboratories" for public policy would be impossible if welfare and educational policies were set by the federal government and enforced uniformly among all 50 states.

Although there are good arguments for the decentralization of power, opponents of decentralization appeal to another set of symbols to argue for greater power being placed in the hands of the national government. These arguments use the symbolic appeal of the national theory of government and its underlying notion that we are a single people and that it is the task of the federal government to forge a national identity. These arguments suggest that individual states are often unwilling or unable to provide certain basic goods and services to their citizens. Only the federal government, for reasons that we will soon explore, is capable of raising the money to provide such services to all Americans. Further, there are certain rights that all Americans ought to enjoy, regardless of where they live. The nature of these specific rights is open to debate, but they are usually defined to include, among other things, a decent education, decent housing, and the right to exercise one's political rights. President Clinton's arguments about the need to reform the health care system were often couched in this rhetoric when he argued that every American is entitled to decent medical care, regardless of how well off he or she is or where he or she lives. We cannot decide such issues at an abstract or symbolic level. In considering the role of federalism in the play of power, it is important to be very specific about what policies and what players we are discussing.

When we examine the role of federalism in the play of political power, we find that there is much more than symbolic politics going on in debates over the appropriate balance of power among levels of government. Politics at the material level is greatly influenced by the distribution of power among levels of government. When we consider the ways in which governmental policies affect the distribution of tangible goods and services, especially their impact on economic policies, we understand that national, state, and local governments are quite different. Further, these differences affect different players differently. For example, state, local, and national politics vary in the ease with which majorities can be constructed for any one policy. The sorts of interest groups, and their relative strengths, vary among levels of government and with the specific policy being analyzed. As a result, the issue under discussion and your own job and income level will strongly affect your perspective on whether policy responsibility should be vested in federal, state, or local government.

FEDERALISM, EXIT COSTS, AND PUBLIC CHOICE

To see how federal, state, and local levels of government differ, let us return to our idea of politics as a play of power. The politics of federalism is driven by the resources that various actors possess and the ease with which they are able to move, or threaten to move, between different jurisdictions. One of the greatest advantages of federalism is that the existence of various sites for different plays of power provides a variety of opportunities for most citizens. Just as if you don't like baseball you can choose to play or follow basketball, darts, or chess, so, too, the myriad state and local governments provide a wide variety of choices. Thus, one of the advantages of federalism is that the autonomy of state and local governments allows for different mixtures of goods and services, providing many choices for citizens. Rather than sharing a single educational system, a single welfare system, a single governmental policy for support of the arts, and so forth, as would be the case in a more unitary system, citizens in a federal system can choose from a variety of public goods and services. However, it must be noted that not all citizens are able to take advantage of those choices, both because of the problems of moving between different types of governments and because of the differing resources that citizens have. We will see why this is so when we examine the fierce competition over economic development that has arisen between states and localities.

When we play a game, one of our assumptions is that everyone who plays will obey the rules and that everyone will remain in the game. If some players can, as the phrase goes, "take their ball and go home," then obviously the game will not last very long. In fact, if you are the person who has the only ball, or if your departure makes it impossible for the game to continue, then you might be able to convince the other players to change the way the game is played to suit you better. Your power will be especially convincing if you have the only ball and there are several groups in the neighborhood who want to play ball.

The concept of "exit costs" was developed by economists to explain why this is so. In general terms, exit costs help us understand why the person with the ball has such power. **Exit costs** are the burdens associated with leaving an economic transaction or market. If there are many games nearby, the costs of leaving any one game and playing in another are quite low. If, on the other hand, there are no other games nearby and you would need to get a ride to another game, then the exit costs of leaving the local game are much higher. Also, in general, exit costs affect a government's ability to make and enforce its laws. In general, the exit costs at the local level are much lower than those at the state level, which are in turn much lower than those at the federal level.

For example, if you don't like the schools in a particular town (and schools are one of the policy areas that are most closely controlled and financed at the local level), it is relatively easy to move to an adjoining town with schools that are more to your liking. In such a move, you may be able to keep your current job, you'll be able to keep seeing your friends, and so forth. In contrast, if you don't like the statewide laws governing inheritance taxes, for example, you'll have to move to a new state. Although this is possible (indeed, such considerations often affect well-off retirees' decisions about where to live), it is harder than simply moving from one town to the next. Finally, if you think that the federal government is taking too big a bite out of your salary each year, then, assuming you aren't willing to violate the law, the only way to escape such policies is to move to a different country. This is a considerable task that not many people are willing to undertake. In short, the costs of exiting from the national government are higher than the costs of exiting from a state or locality. For instance, when Colorado repealed and Cincinnati rejected ordinances that would explicitly prohibit discrimination on the basis of sexual preference, gay and lesbian groups tried to organize an economic boycott of both areas. Such actions are possible because of the relatively low exit costs of not doing business in those jurisdictions. On the other hand, if anti-gay policies were enacted by the federal government, it would be considerably more difficult to protest through the boycott strategy.

The wide variety of choices for citizens has been noted as a distinct advantage of a federal system by political scientists who have developed the **public choice perspective**—an attempt to apply the insights and assumptions of microeconomics to the study of politics. From this perspective, the multiple, autonomous governments in a federal system have the advantage of allowing citizens to choose from among the combinations of public goods and services offered by state and local governments. Providing a wide range of choices is especially effective at the local level, where exit costs are relatively low. Government ceases to be a singular and systemwide monopoly supplier; as a result, citizens can choose among the various public policy offerings of competing jurisdictions. The possibility that citizens will "vote with their feet" forces governments to be efficient and responsive to citizens' demands, because they can move from one jurisdiction to another. In short, from the public choice perspective, the advantage of federalism is that it can create "quasi-marketplaces" among competing jurisdictions, thereby providing many of the advantages assumed to flow from the operation of private markets.[31] Public choice theorists have been vigorous defenders of multiple, small, and overlapping jurisdictions and opponents of attempts to create larger units of government pursuing more uniform public policy goals.[32] "Applying Knowledge: Public Choice and the Public School System" at the end of this chapter describes a widely debated proposal to break the monopoly of public schools by allowing students and parents to choose which school to attend.

The public choice focus on the differences between small, overlapping jurisdictions and large, unitary governments helps illuminate another characteristic of federalism. Multiple, smaller jurisdictions have important advantages for formerly disenfranchised groups who try to gain political power, because it is much easier for members of such groups to win state and local elections than national elections. Thus, for example, while the numbers of women gaining elective office in the United States remains alarmingly low, the numbers are considerably higher at the state and local than at the national level. In 1990, 16 percent of mayors, 14 percent of other municipal officials, and 17 percent of state legislators were women. In 1994, only 13 percent of Congress members (48 representatives and 8 senators) were women.[33]

The relative autonomy of the numerous governmental jurisdictions in a federal system allows many forms of political and cultural diversity to flourish in the United States. It is here that federalism connects to issues of multiculturalism, as we discussed earlier in this chapter. Consider the case of gay men and lesbians. Some jurisdictions—Cincinnati and the state of Colorado, for example—have rejected ordinances that prohibit discrimination on the basis of sexual preference. In contrast, New York City and

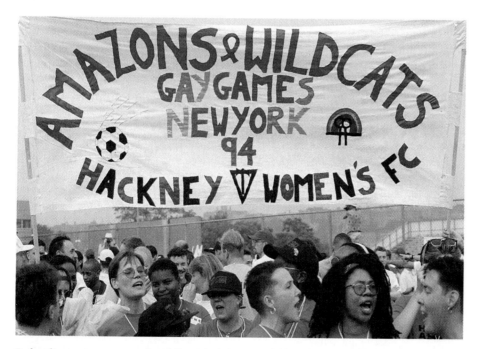

Federalism may protect the rights of minorities, but it may also preserve discriminatory practices. In 1994 New York City hosted the Gay Games, which gave gay men and lesbians an opportunity to participate in a public event that received wide media coverage and reinforced their claim to protection from discriminatory practices. On the other hand, some localities use the autonomy provided by federalism to resist extending civil rights protection to gay men and lesbians.

San Francisco have large and vocal gay and lesbian communities and, consequently, governments that have been relatively sympathetic to the rights of homosexuals. For instance, the city of New York lent its support to promoting the Gay Games, an athletic event that took place in June 1994. The Gay Games were covered prominently, and favorably, in the local media and were loudly supported by many local politicians, including Mayor Rudolph Giuliani. People sympathetic to granting rights to gay men and lesbians have an incentive to move to places like New York or San Francisco. In contrast, those who object to granting such rights have the option of moving to places like Cincinnati or Colorado. Public choice theorists would point to the existence of smaller, autonomous governments as the reason for this diversity. If we had a more unitary system, the national government would set the policy toward gay men and lesbians. If there were a national majority in favor of granting such rights, all jurisdictions would be forced to follow this path. On the other hand, if there were a national majority against granting such rights, all jurisdictions might be prevented from extending such protection.

THE PLAY OF FEDERALISM

Why do you need to know about exit costs and public choice? Well, owing to differing exit costs, different levels of government have varying abilities to pursue certain types of policies. On the one hand, the possibility that citizens can exit helps to hold local governments responsive to their demands and is consistent with democratic principles. On the other hand, not all citizens and groups are equally able to exercise the threat of exit. Most significantly, state and local governments are much less able to adopt **redistributive policies**—policies that benefit poorer citizens and are paid for by wealthier citizens—than is the federal government. The reason has to do with exit costs. The ability of citizens, businesses, and investment capital to move easily between local governments, less easily between state governments, and with most difficulty between nations creates a type of politics we'll call *the play of federalism.* The play involves creating, at each level of government, the mixture of goods and services that will satisfy the demands of constituents while at the same time attracting, or at least not inciting to exit, those individuals and organizations that pay taxes to support such programs. Another feature of this play of power is that groups that support specific policies will try to locate the authority for making those policies at the level of government best able to respond to their demands. It is important to remember that this dynamic is created by the existence of a federal system. If all state and local governments were simply carrying out uniform policies established by the central government, then incentives to move between states and localities to gain advantages would not exist.

To see how this play of power operates, let's start with a simple example. Suppose that the mayor and city council of one town—let's call it Bleeding Heart—decide that they must do something about the poor and homeless within its borders. So they decide to enact a generous program of welfare benefits, homeless shelters, free meals, and so forth. The money for these programs must come from somewhere, so the town raises taxes—which, of course, are not paid by those who benefit from these policies, since they earn little and pay little in taxes. Instead, the programs are paid for by the better-off taxpayers of Bleeding Heart, who do not benefit directly from the programs because they have homes, jobs, and food. In this way, the policies that guide the programs are redistributive because they take income from the better off to benefit the less fortunate.

Right next door to Bleeding Heart is a second town—let's call it Social Darwinville—whose leaders are not quite as concerned about the poor. In fact, Social Darwinville has no welfare benefits, no homeless shelters, and no free meals. Instead, the mayor and city council keep taxes extremely low and spend a good proportion of the money they do raise on a local opera company, a local horse stable, a local symphony orchestra, and a police force to make sure that the homeless are made to feel unwelcome.

What is going to happen over time? Given the low exit costs (remember, the two communities are right next door to each other), we can expect that wealthier citizens will tend to move to Social Darwinville, where the taxes are lower and the services more to their liking. On the other hand, the less well-off citizens will tend to find Social Darwinville a relatively inhospitable place and will want to move to Bleeding Heart, where they can take advantage of the redistributive policies. As time goes on, the tax base of Social Darwinville will expand and that of Bleeding Heart will fall, making it ever more difficult to support the generous redistributive policies. In the end, we would expect that the citizens of Bleeding Heart, especially those who are better off, will vote the mayor and council out of office and support politicians who adopt a set of policies more like that in Social Darwinville.

The point of this example is that the lower the exit costs of moving from one jurisdiction to the next, the more difficult it is for governments to enact policies that will benefit the less fortunate. To the extent that redistributive policies are to be pursued, they must be adopted at a level of government where exit costs are high enough to make it difficult for those taxpayers who will bear the burden to escape from the policies. For this reason, redistributive policies are likely to be more effective when pursued by the national government. To return to our example, if the policies pursued by Bleeding Heart had been federal policies required by all localities, then Social Darwinville would also have adopted those policies. Because wealthy taxpayers would have had to leave the country to escape paying for the policies and few would have done so, there would have been little movement, by either wealthy or poor citizens, between the two communities.

This dynamic also helps explain why different levels of government rely on different types of taxation. Governments use three kinds of taxes to raise revenue: property, sales, and income taxes. The income tax has the most potential for redistributing wealth; even a constant tax rate will take more from those with higher incomes than from those with lower incomes. Because of this redistributive impact, the federal government (whose high exit costs limit the number who will leave the country to avoid the tax) is best able to rely on the income tax to raise revenue. States and localities are much less able to tax income, given the more real possibility that the wealthy will move to avoid higher taxes. Indeed, some states use the absence of an income tax as a device to attract wealthy taxpayers. Local governments face the most difficult task in raising monies, since it is easiest for individuals to move to escape high taxes. For this reason local governments raise the overwhelming bulk of their revenues from the property tax, because, unlike income or capital, land cannot be moved across jurisdictional lines. States, in contrast, rely on the sales tax as their largest single source of revenue.

This analysis gives a somewhat different perspective on arguments about federalism that are couched in the symbolic language of placing government closer to the people. When politicians advocate increasing the responsibility assigned to states and localities and cutting back on the responsibilities of the federal government, they are supporting something more than just the placing of governmental decisions closer to the people. Such decentralization also makes it much less likely that government will adopt policies that try to benefit the less well off through tax revenues paid primarily by those who are better off. Thus, to the extent that states and localities are granted greater policy-making authority, we can expect fewer redistributive policies to be passed and implemented. Within the play of federalism, this is one reason why more conservative politicians who oppose redistributive policies tend to support greater state and local autonomy. On the other hand, liberal politicians, whose constituents tend to support redistribution, usually favor an expanded role for the federal government.

It is therefore not surprising that Republican administrations, which garner most of their votes from the middle and upper classes, are inclined to advocate greater reliance on states and localities for policy making. Similarly, it is not surprising that the "Contract with America" contains many provisions that would increase the autonomy of state and local governments. On the other hand, it is equally unsurprising that Democratic administrations, which depend on support from poorer voters (and did so especially while the New Deal Coalition was still intact), or Democrats resisting Republican proposals in the 104th Congress, have tended to support greater reliance on the federal government.

Debates within the rhetoric of federalism go beyond whether government will redistribute income from the better off to the poor. The debates also consider the role government will play in fostering economic growth—or **developmental policies**—and which groups will influence such policies. Just as states and localities are constrained from taxing the wealthy to benefit the less fortunate, they have great incentive to attract taxpayers—either individuals or businesses—that will expand the local economy and increase the tax base. If a state's economy grows, it can increase its revenues without having to increase tax rates. This is a highly desirable outcome because increasing tax rates are likely to lead to the exit of well-off taxpayers as they seek advantages in states with lower tax rates. To see how this competition works in a specific policy arena, we'll now explore the play of federalism in the competition among states to attract automobile manufacturers interested in building factories in the United States.

INTERSTATE COMPETITION IN THE PLAY OF FEDERALISM

Between 1980 and 1986, Japanese automobile manufacturers located five factories in five U.S. states: Illinois, Indiana, Kentucky, Michigan, and Tennessee. These factories employed approximately 14,000 workers and produced 1 million automobiles per year. To challenge Japanese dominance in the small-car market, General Motors announced in 1985 the construction of a Saturn plant in Spring Hill, Tennessee, that would employ 3,000 workers and produce 250,000 cars per year. A large number of new firms, producing parts and related goods and services, were expected to spring up around these factories. These massive investments, all made within a short period of time and in geographically contiguous states, led some observers to dub this region of the country the Japanese Auto Alley.[34]

The bidding war that erupted over these plants illustrates many of the issues we have discussed about exit costs and the competition between states and localities for developmental policies. The automobile companies had valuable resources—jobs and investment—and, much as happened with the player who owned the only ball in town, states competed with one another to attract the automobile manufacturers to "play" in their state. As Table 4.2 (on p. 148) indicates, the five states had good reason to actively court these firms: All five had unemployment rates above the national average during 1980–86 (although in 1986 Indiana did fall below this level). Further, these states were particularly hard hit by the decline in manufacturing during the recession of the late 1970s and did not participate fully in the bicoastal economic expansion of the early 1980s. Indeed, we would expect an intensive and escalating bidding war to erupt over the location of these plants. Economically depressed states are the most likely to enter into costly bidding for investment because the pressure on state politicians to create jobs at any cost is severe. Moreover, the decisions were all made within a relatively short period of time; and each state, with the exception of Tennessee, competed unsuccessfully for a plant before finally landing another. Thus, for most political leaders the memory of losing out in the competition was quite fresh.

Figure 4.3 (on p. 149) illustrates the costs per worker to the state for each plant. Evidently the bidding for these plants escalated rapidly throughout the six-year period. All the successful state packages included state funds for worker training (indeed, this one factor accounted for between 33% and 48% of the cost of each package), land purchases, site renovations, and transportation improvements. It is clear that the incentives offered by the winning state in each case provided the baseline for the next round of bidding. In 1980, for example, Tennessee provided $11 million in incentives and tax abatements to Nissan for a plant that would employ approximately 3,000 workers. By 1986, the state of Kentucky provided $65 million in worker training for a plant that would also employ 3,000 workers.

To understand the role that federalism plays in this bidding process, we should first note that the economic logic of such large subsidies is questionable if we adopt a national, rather than state, perspective. If we assume that Japanese auto manufacturers were going to locate their factories somewhere in the United States (the reasons for making this assumption are spelled out below), then large state subsidies did not affect the total number of jobs created, but simply their location. That is, worker training subsidies and the like simply lowered the cost of doing business for firms such as Toyota, Mazda, and Nissan, while depleting already hard-pressed state treasuries, without any overall increase in the national economy. The economic logic of such large-scale subsidies is further brought into question when one considers that there was no guarantee to the states involved about how

TABLE 4.2
CHARACTERISTICS OF SELECTED AUTOMOBILE PLANT LOCATION DECISIONS

LOCATION AND ANNOUNCEMENT DATE	PLANT	AUTO COMPANIES INVOLVED	COMPANY INVESTMENT	STATE INVESTMENT (IN MILLIONS)	EMPLOYEES	STATE UNEMPLOYMENT RATE IN ANNOUNCEMENT YEAR	% STATE UNEMPLOYMENT RATE ABOVE NATIONAL AVG.
Smyrna, Tennessee 1980	Nissan Motor Manufacturing Company	Nissan	$745–848 million	$ 22.0 Road access 11.0 Worker training $ 33.0 Total	3,000	7.3%	2.82%
Flat Rock, Michigan 1984	Mazda Motor Manufacturing Corporation	Mazda (Ford will buy 50% of cars)	$745–750 million	$ 19.0 Worker training 5.0 Road improvement 3.0 On-site railroad improvement 21.0 Economic development grant loan .5 Water improvement $ 48.5 Total	3,500	11.2	49.3
Spring Hill, Tennessee July 1985	Saturn Corpo-ration	General Motors	$3.5–4.79 billion	$ 30.0 Worker training 50.0 Road improvement $ 80.0 Total	3,000	8.0	11.1
Bloomington/ Normal, Illinois October 1985	Diamond-Star Motors Corporation	Chrysler, Mitsubishi	$500–700 million	$ 17.8 Road improvement 11.0 Site acquisition 14.5 Water improvement 40.0 Worker training $ 83.3 Total	2,500–2,900	9.0	25.0
Georgetown, Kentucky December 1985	Toyota Motor Manufacturing USA	Toyota	$823.9 million	$ 12.5 Land purchase 20.0 Site preparation 47.0 Road improvement 65.0 Worker training 5.2 Toyota families education $149.7 Total	3,000	9.5	31.9
Lafayette, Indiana December 1986	Fuji Isuzu	Fuji Heavy Industries, Isuzu Motors Ltd.	$480–500 million	$ 55.0 State funds 26.0 Federal subsidies 3.0 Water improvement 2.0 Road improvement $ 86.0 Total	1,700	6.7	Below national avg. in 1986

Source: Adapted from H. Brinton Milward and Heidi Hosbach Newman, "State Incentive Packages and the Industrial Location Decision," in Ernest J. Yanarella and William C. Green, eds., *The Politics of Industrial Recruitment* (New York: Greenwood Press, 1989), Tables 4–5.

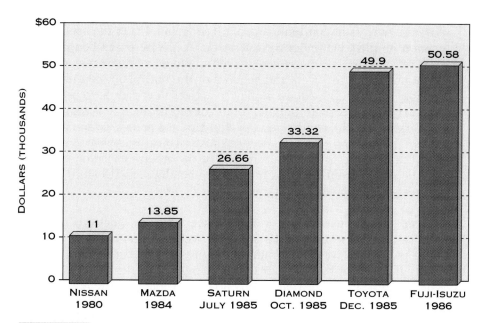

FIGURE 4.3

STATE COSTS PER PLANT EMPLOYEE FOR SIX AUTOMOBILE FACTORIES The level of subsidies provided by states to attract large manufacturing operations such as automobile factories has increased dramatically in recent years. Are the expenditures worthwhile for states? How you answer this question will depend on the time frame you adopt and your assumptions about the relationship between a private market economy and a federal system. *Source:* Adapted from data in H. Brinton Milward and Heidi Hosbach Newman, "State Incentive Packages and the Industrial Location Decision," in Ernest J. Yanarella and William C. Green, eds., *The Politics of Industrial Recruitment* (New York: Greenwood Press, 1989), p. 45.

long such plants would actually remain open. Indeed, a heavily subsidized Volkswagen Rabbit plant in Pennsylvania closed down in 1988 after only 10 years of operation.[35]

Although the bidding may be difficult to rationalize from a national economic perspective, the pressures placed on state political leaders by the interstate economic competition resulting from the play of federalism make the entire process quite understandable.

As arguments made for maintaining or increasing the autonomy of state governments become more successful, state officials are increasingly held responsible for state economic growth. This accountability seems to have played a role in the decision of state officials to enter into bidding for the automobile plants. Media coverage of plant location decisions plays an important role in emphasizing this accountability. Consider the following summary of newspaper coverage of two such location decisions: the Subaru-Isuzu factory in Indiana and the Toyota factory in Kentucky:

> [N]ewspapers presented accounts of the new auto plants that stressed the views of political officials and business leaders, and provided much less coverage of persons from other community sectors. Moreover, topic content and coverage stressed the

positive aspects of community growth, in general, and the specific economic initiative under consideration. . . . [Newspapers] do not take the initiative in presenting to their readers a broad range of questions about the costs and benefits of the new plant, or the wisdom of providing public money to attract new plants, to provide tax abatements, to provide funds for recruitment and training of workers, and to acquire and develop land for private corporations. The actions of newspapers in these two cases appear to support the hypothesis that they are part of the growth coalition and see their own profit-oriented interests to be aligned with similar interests in the business community.[36]

Given this sort of coverage, it is easy to see why the long-term costs of incentive packages would not be carefully weighed against their benefits and why politicians would be under great pressure to land an auto plant, especially if the state had lost out in a previous competition.

You should remember that, even though national politicians are also held responsible for economic performance, the dynamics of competition for investment is quite different at the federal and state levels. This explains the bargaining power of automobile firms in their dealings with state governments and their ability to extract increasingly lucrative incentive packages.

Consider the case of the Japanese automobile firms. For quite some time, these firms had been contemplating locating plants in the United States in order to be closer to their largest foreign market. Growing protectionist sentiment in Congress during the 1980s made it even more imperative to do so. Thus, Japanese firms were faced with the choice of either building plants in the United States or risking being closed off from a large portion of their market. However, the strong bargaining position of the United States as a national entity, which would have avoided the need for incentives, was eliminated when Japanese firms were allowed to deal directly with state governments. The dynamics of federalism left the plants' location to negotiations between state governments and Japanese corporations. As more and more states entered the competition for auto plants,[37] the bidding intensified and less and less consideration was paid to whether the costs to the winner would be balanced by the benefits flowing from the factories. Interestingly enough, an American firm seeking to locate in Japan would never be placed in such an advantageous position in relation to local Japanese communities: The terms of foreign investment in Japan are carefully controlled by the national government.

Finally, the bargaining position of state governments was further eroded by a more general feature of a private market economy. The decisions and calculations of the firms being courted remained private. Public officials could not know whether features of their state (e.g., characteristics of the labor force, proximity to transportation networks) would lead firms to locate there with much lower incentives than other states were offering.[38] All such public officials can know is that the personal costs of failing to land a plant could be enormous at the polls and in the press.

We have seen that states are involved in an increasingly fierce competition for investment and growth. We have also discovered that this competition may be driven by political forces emanating from the play of federalism, rather than from any economic rationale. From the perspective of the national economy, many incentive packages may be entirely irrational. However, the political realities facing state officials make such careful economic calculations impossible.

THE DEMOCRATIC RHETORIC AND STRUCTURAL REALITIES OF FEDERALISM

The ways in which economic competition and democratic rhetoric interact in debates over federalism make it extremely complicated to analyze the policy implications of assigning authority to various levels of government. Decentralization, although appealing on democratic grounds, can have different results when we consider the limits of state and local governments. For example, both the Reagan and Bush administrations supported a general decentralization of responsibility for implementing many environmental regulations. In 1995, the Republicans in Congress pursued a variety of regulatory reforms that would support decentralization again. Supporters argue for these policies on familiar democratic grounds. They claim that allowing states and localities to determine their own levels of enforcement better reflects the differences among regions. They go on to argue that such decentralization, taking advantage of the possibilities for diversity inherent in federalism, would be more democratic and more efficient.

However, as we have seen, the symbolic language of greater democracy is not all that is operating in the play of federalism. Economic competition, which is also created by federalism, means that many states have found it difficult to pay for increased enforcement of environmental regulations. That is, the funding of effective enforcement mechanisms would have required raising taxes or imposing costs on industry that would run counter to the goal of economic development. The result has been that many states that were hard pressed economically have been unable to enforce federal environmental regulations.[39]

When we consider federalism we need to keep in mind the symbolic arguments about democracy, which are important. In fact, economic development is not the only goal that local governments pursue. "Outcomes: Gay and Lesbian Rights, the Religious Right, and Economic Development" chronicles the story of a small town in Texas that almost turned down a multimillion-dollar computer operation because of the company's policy of providing benefits to the partners of gay and lesbian employees. However, even though questions of democratic theory are emphasized by political leaders, they are not the only thing at stake in the arguments over whether to centralize or decentralize political authority. The economic competition created by the divided responsibility of federalism is also a key factor. The limited options available to state and local governments, shaped by exit costs and the competition for economic development, impose structural constraints that are part of the play of federalism.

In fact, interests struggle over the level of government that will be the site for the plays of power that most concern them. This is especially so when businesses consider the governmental regulations that will shape the way they operate. As we have seen, corporations can choose the state in

GAY AND LESBIAN RIGHTS, THE RELIGIOUS RIGHT, AND ECONOMIC DEVELOPMENT

In early 1993 the Apple Computer Corporation came to Williamson County, Texas—a small county not far from Austin—and made a proposal that almost any local government would love. The computer giant proposed to build an $80-million office complex that would create almost 1,500 new jobs and pour up to $300 million into the local economy. In return, the corporation wanted $750,000 in tax abatements over seven years from the county. Given what we have said about the pressures on local governments in a federal system to compete for economic development, we would expect that County Commission approval would be a mere formality. However, that did not happen. What actually occurred tells us much about what happens in a federal system when local values clash with those more generally accepted by the nation as a whole, or at least by the mainstream mass media, and about the pressure that private corporations can exert on local officials.

It turns out that Apple Computer extends health benefits not only to spouses of employees but also to live-in partners, regardless of sex. It also turns out that there are many conservative Christians in Williamson and that they are well organized and politically active. The month before the Apple vote, the school board, controlled by a newly elected conservative Christian majority, had fired the school superintendent for recommending that prayers before football games be discontinued. When it became public knowledge that Apple extended benefits to homosexual couples, conservative Christians vigorously opposed granting the tax abatements, saying they didn't want the company in their community. In December 1993, the County Commission voted 3–2 to refuse Apple's request. The key vote came from Commissioner David Hays, who, although he had initially supported the plan, changed his vote after being lobbied by the Christian right. "If I had voted yes," Hays said later, "I would have had to walk into my church with people saying, 'There's the man who brought homosexuality to Williamson County.' "[1] Aside from the likelihood that there are already gay men and lesbians in Williamson, the Commission's rejection is interesting because it raises many of the issues we have been discussing in this chapter. Under federalism, local autonomy is a way of preserving values that may be at odds with those of the wider society. However, whatever we think about the virtues of preserving local cultural values, it is extremely difficult to do so when economic pressures are brought to bear.

The reaction to the Williamson decision also illustrates these dynamics of federalism. First, the Commission's decision made it into the national media. The network news, the *New York Times*, and *Time* magazine all ran stories portraying the county, in the words of the *Houston Chronicle*, as "homophobes and economic rubes."[2] Second, other localities, both in Texas and outside the state, immediately offered to put together incentive packages to attract Apple. Third, local businesses in Williamson banded together to bring pressure on the Commission to reverse itself and endorse the Apple deal. Thus, in this case the general consensus in the mainstream media, opposing discrimination on the basis of sexual preference, came together with the competition between local governments to attract private corporations. As a result the County Commission reconvened, one commissioner switched his vote, and the Apple deal was approved.

[1]The quote and the information in this paragraph are from Richard Lacayo, "Take This Job and Shove It," *Time*, December 13, 1993, p. 44.
[2]Roy Bragg, "Williamson County Panel Makes U-turn," *Houston Chronicle*, December 8, 1993, p. A25.

which their headquarters will be located. This allows them to select the state government that will be their primary regulator. In some cases an industry can seek to be subject to federal regulation rather than to a plethora of individual states' regulations.

Federalism's effect on regulation is well illustrated by the scandals surrounding failed savings and loan institutions (S&Ls). It is important to understand the role of federalism in the S&L crisis. The thrift institutions had the option of operating under either federal or state charters; they could even shift their charters from one level of government to another.[40] State-chartered S&Ls charged that many states' regulators were overly sensitive to the wishes of consumer groups and that they attempted to artificially depress the interest rates on home mortgages. In order to escape such restrictions, many S&Ls shifted—or threatened to shift—to the federal charter system. Given the pro-business climate of the Reagan and Bush administrations and the corresponding weakness of consumer groups in Washington, D.C., the S&Ls believed that federal regulators would grant them greater latitude to set adjustable interest rates; such rates, the bankers argued, are necessary to ensure adequate levels of profit. In effect, S&Ls threatened to take their ball and move to another game of regulation.

What was the nature of this ball? Why would states be concerned about such an exit from their regulatory control? Well, when a financial institution shifts its charter to the federal system, the state loses more than the power to regulate; it also loses substantial tax revenues. The mere threat of such a shift created tremendous pressures on the states to align their regulations with federal standards.

The pressures on state regulators, however, arose from more than intergovernmental competition for regulatory powers and revenues. Interstate competition for these resources also become increasingly fierce. Several pioneering states—notably, Texas and Virginia—attracted banks to their state systems by eliminating many restrictions on home mortgage interest rates. This overall weakening of the regulations on S&Ls is blamed by many analysts for the disastrous collapse of many of these institutions in the 1980s.[41]

STATES' RIGHTS VERSUS CIVIL RIGHTS

As we have seen, the play of federalism adds a great deal of complexity and ambiguity to American politics. Nowhere are these issues more starkly posed than in struggles over the federal role in the civil rights movement. On the one hand, the struggles illustrate the degree to which local autonomy has been used to exclude specific groups from having a fair opportunity to participate in the political, economic, and social life of their communities. On the other hand, the growing and changing role of the federal government in the electoral politics of states and localities raises troubling issues about how to guarantee fair opportunities for all citizens while still allowing communities to make their own political decisions.

Although we don't usually think of the civil rights movement in these terms, struggles by African Americans to gain access to the political system cannot be understood without considering the significance of federalism.

Policies enacted by the federal government, and the validation of these policies in the courts, illustrate how far the United States has come in moving along the continuum of federalism to a much stronger role for the central government. At the same time, as these policies have evolved, they have raised important questions about what form this intervention can take while still preserving the state and local autonomy at the heart of federalism.

The struggle that began in the 1950s and 1960s in cities such as Montgomery, Alabama, where African Americans organized a boycott of the public transit system, is heroic testimony to the courage and power of ordinary people (such as Rosa Parks, who precipitated the boycott by refusing to sit in the back of a public bus) to change their lives.[42] However, it is also important to see that a central aspect of the civil rights struggle was the gradual but growing support for the movement that developed at a national level. Events like the landmark *Brown v. Board of Education* Supreme Court case in 1954 (which ruled that segregated public schools were unconstitutional), the murders of civil rights workers, and the horrifying pictures and films of the brutal police tactics used to disperse black protesters in Alabama

Understanding the civil rights struggle requires a comprehension of the complex interactions among local, state, and federal governments. Was southern resistance to integration an attempt to preserve a different way of life—one of the supposed benefits of federalism—or was it something else? In December 1956, the Reverend Ralph D. Abernathy (*first row*) and the Reverend Martin Luther King Jr. (*second row, left*) participated in the Montgomery bus boycott in an attempt to break segregation in public transportation in the South.

led to the growing involvement of the federal government in the civil rights movement during the Eisenhower, Kennedy, and Johnson administrations. Even though this intervention was at times grudging and marked by cynical political calculation, without the intervention of the federal government it is unlikely that the movement could have overcome the staunch resistance of white Southerners who had at their disposal the full panoply of forces of government.

Southern political leaders employed the rhetoric of states' rights to argue that the national government had no authority to interfere in the affairs of state governments, even when those governments supported a system of segregation that prevented African Americans from participating fully in the public life of these states. The fact that these arguments strike many of us as strange today is, in part, due to the dramatic changes that have occurred in our understanding of federalism. What difference does a changing definition of federalism make? Throughout the nineteenth and early twentieth centuries, the dominant model of layer-cake federalism maintained a strict separation of state and national governmental powers. The Supreme Court, for example, ruled in a number of cases that the Bill of Rights limited the national government and not the actions of state governments. In effect, the Court resolved that we Americans are many peoples and that an individual's rights depend on the capacity in which those rights are asserted—as a citizen of either a state or the national government.

Following this line of argument, white Southerners maintained that states were entitled to withhold rights from their citizens even though those very same citizens might be entitled to those very same rights in their status as citizens of the national government. However, remember that in layer-cake federalism, state and federal citizenship rights are two different things. This meant that segregation and the denial of basic rights to African Americans could be maintained in the South.

Breaking down the barriers to participation imposed by segregation policies in the South required not just the passage of landmark federal programs, such as the Civil Rights Act of 1964 and the Voting Rights Act of 1965. It also required a fundamental shift in the way we view federalism and the balance of powers between state and national governments. That this change had occurred by 1964 is clear when we examine the reasoning of the Supreme Court in the case of *Katzenbach v. McClung* (1964).[43] The case dealt with a local restaurant in Birmingham, Alabama. Even though the Civil Rights Act of 1964 prohibited discrimination in public accommodations, Ollie's Bar-B-Que refused service to African Americans. Ollie's lawyers argued that since the restaurant was not involved in interstate commerce, legislation passed by the national government, including the Civil Rights Act, did not apply. The Court ruled against this argument, saying that although Ollie's was a local restaurant, food and supplies were bought from companies outside the state of Alabama. This made a sufficient connection to interstate commerce, and therefore the Civil Rights Act did apply. The broad interpretation of the connection to interstate commerce of even the smallest businesses brought all individuals, in their role as citizens of the nation and regardless of their state citizenship, under the umbrella of nationally guaranteed rights. This sea change in the Court's understanding of federalism was a necessary part of the success of the civil rights movement.

The expansion of the power of the national government transformed the political systems of many southern states by dramatically increasing the political participation of African Americans. At the same time, it raises questions about the degree to which such dramatically increased federal power can be made compatible with the positive features of state autonomy—especially the notion articulated by the Antifederalists that placing government closer to the people is more democratic because it allows citizens to make decisions for themselves. Federal efforts to enforce the Voting Rights Act of 1965 are relevant here. This legislation was a response to the recognition that the structure of state and local governments in the South made them unable or unwilling to provide open access to the ballot box for black citizens. Federal efforts focused initially on mandating open access to the ballot and, thus, a fair electoral process. Therefore, efforts were aimed at ensuring that blacks were able to participate on a fair and equal footing. Once fair participation was achieved, the specific outcome of elections was left to citizens themselves.

Of course, fair and equal access is difficult to define.[44] Beyond simply providing access to the polls, federal officials and the courts soon had to decide what constituted an equal vote as states and localities devised election rules and drew jurisdictional boundaries that diluted the impact of newly enfranchised black voters. Such determinations changed over time as blacks became an ever more powerful force in many states and as the specific contours of southern politics changed.

However, federal efforts did not stop there. Instead, enforcement of the Voting Rights Act by the federal government began to focus on whether black officials were actually getting elected, rather than simply whether black citizens had access to the ballot box. Here we see a shift from attempts to guarantee open access to the electoral process to a focus on achieving specific outcomes. Federal officials began to define open access to the ballot in terms of the number of blacks or minority officeholders who were actually elected. Although this was an easier goal to implement and enforce than "one person, one vote," it shifted attention away from the more significant question of how to define effective political participation by minority citizens.

This shift in goals brought the federal government into conflict with the legitimate issue of local self-determination. It raised disturbing questions about whether a strong national government can exercise its power in the pursuit of national goals without trampling on the abilities of citizens in diverse states and localities to make decisions for themselves. Indeed, the campaigns of Republican presidents Nixon, Reagan, and Bush all made pointed use of these issues in their calls for scaling back the power of the central government. There are no easy answers to these questions. What is clear is that definitions of federalism are central to many current political debates.

In fact, political scientist William Riker argues that the issue of race is central to any evaluation of federalism. He suggests that it was only the passage of national civil rights legislation that allowed the realization of the full advantages of a federal system:

> The civil rights reforms of the 1960s . . . eliminat[ed] the protection for local repression. With the racial dimension of judgment thus removed it became possible for the first time in American

history to value federalism unambiguously as a deterrent to statism, a deterrent that liberals could readily support and that statists might believe restrictive and unpleasant.[45]

157

CHAPTER 4

FEDERALISM AND

INTERGOVERNMENTAL

RELATIONS

SUMMARY

The political economy of American federalism dramatically affects the prospects of various players in American politics at different levels of government. In this chapter, we have explored the changing relationship between levels of government by first examining the different approaches to assigning responsibility to national and regional governments: federalism, confederations, and unitary systems. While the United States is characterized as a federal system, relationships among levels of government have at some points in history moved us closer to being a unitary system, while at other times we have moved closer to being a confederation. Such changes are a result of continued conflict between two theories of political legitimacy: national and compact theories. Conflict between those who want to view the United States as a single people (national theorists) and those who see America as many peoples (compact theorists) continues to inform debates over how much authority ought to be vested in state, national, and local governments.

As the relationship among national, state, and local governments has shifted, so too has the way public policies are made and the way funds are raised to finance them. The national crisis of the Great Depression moved the United States from a strict separation of policy-making authority (captured by the layer-cake model of federalism) to a more complex and intertwined system (captured by the marble-cake model). Struggles over specific public policies from the 1960s through the 1990s have resulted in a complex sharing of power that differs from policy arena to policy arena and is captured by the picket-fence model of intergovernmental relations.

In fact, one of the most basic ways in which federalism affects the play of political power is through its complexity. Unlike the situation in more unitary systems, as the American system of federalism has evolved it has become increasingly difficult to determine which level or levels of government are responsible for specific public policies. This means that federalism disproportionately disadvantages those players with less detailed knowledge about the ins and outs of intergovernmental relations, especially those who do not enter into the play of power very often—as is the case for most citizens. For such players, the sheer frustration of determining whether state, national, or local government is the place to get their problems solved may act as a barrier to fuller political participation.

The complexity introduced by the evolution of federalism has a profound impact on the play of power. Indeed, confusion over the dynamics of federalism can be used to manipulate the system and the ire of citizens. Often elites use arguments that appeal to the desire for democracy: They assail distant government and support placing responsibility for decisions closer to the people, when in fact the governments to which policy-making responsibility is assigned are woefully limited in their ability to respond to the wishes of local citizens. We suggest that arguments over economic power, while couched in a variety of rhetoric, have been at the center of debates over federalism since the framing of the Constitution. They remain central to debates over assigning responsibilities to various levels of government. The location of policy-making responsibility at the state or national level has important implications for various categories of players, for "recurring philosophical disagreements about the division of power between Washington and the states almost always mask very practical disputes about how people live, which people will enjoy the greatest benefits of our federal bargain, and which are going to bear the heaviest burdens."[46]

APPLYING KNOWLEDGE

PUBLIC CHOICE AND THE PUBLIC SCHOOL SYSTEM

Public education in the United States has been under attack for many years. As Scholastic Aptitude Test scores decline and American students seem to do worse academically than their peers in other countries, one wonders whether the schools are capable of preparing students for jobs in the twenty-first century—to say nothing of educating capable citizens.

A solution to the decline of America's public schools comes from the public choice theorists who have championed the idea of providing a wider range of school alternatives and, in particular, have supported developing voucher systems for public education. Currently, if children wish to attend public school they must attend the schools to which they are assigned by their local Board of Education. (There are some exceptions, as in the case of magnet schools.) Because schools have a captive audience (so the public choice argument goes), they have little incentive to respond to the demands of parents or children. Nor do they have much incentive to provide a good education at a reasonable cost. As a result, schools have become overly bureaucratic, inefficient, and dominated by the self-interested demands of teachers' unions.[1] To remedy this problem, some analysts suggest that parents should be able to choose their children's schools. In this way, schools would compete with one another to attract students. Faced with a quasi-market in education, educators would have to pay attention to the demands of their "customers" (i.e., parents and students). Those schools that do a good job in responding to customer demands would prosper and grow, whereas those that do not would shrink and, perhaps, close down. There would be an incentive for schools to learn about what works best in other schools and to adopt relevant methods. In short, the public school system would become flexible, competitive, and responsive.

One approach to establishing choice would allow parents to send their children to any public school that had room. In 1987, the state of Minnesota adopted such a choice system, which allows parents to send their children to any public school, regardless of district, as long as (1) there is room, (2) the move does not hurt efforts to desegregate the schools, and (3) the parent provides transportation. The public monies allocated for each student follow him or her from school to school. The system is regarded as so successful that in 1991 it was expanded to allow groups of teachers to establish their own new schools, thereby simulating the establishment of new and innovative firms in the private market.[2]

An even more dramatic proposal for school choice is the voucher system. Under a voucher system, parents would be allowed to apply a set amount of public money (pegged to the amount of state and local funds provided) for their children to attend any school—public or private, secular or parochial. Voucher proposals differ in how much they address equity issues: Some allow parents to add to the voucher value in order to pay for private school tuition (thus benefiting wealthier families), while other proposals limit school tuition to the value of the voucher (thus providing equal access for all students).

This issue of equity has raised the ire of critics of greater school choice. One of the most vocal and eloquent voices criticizing choice and voucher plans and insisting that equity be the paramount concern in public education is Jonathan Kozol. He argues that public education has an obligation to *all* students and that there will be great disparity in the ability of poorer families to take advantage of voucher or other choice systems. Therefore, although the idea of free choice in public schools seems a good way to allow everyone to seek out the best possible education, Kozol warns us,

> this is not the situation that exists. In the present situation, which is less a field of education options than a battlefield on which a class and racial war is being played out, the better schools function effectively, as siphons which draw off not only the most high-achieving and the best-connected students but their parents too; and this, in turn, leads to a rather cruel, if easily predictable scenario. . . . Having obtained what they desire, they secede, to a degree, from the political arena. The political effectiveness of those who have been left behind is thus depleted. Soon enough, the failure . . . of the schools that they attend confirms the wisdom of those families who have fled to the selective schools. This is, of course, exactly what a private school makes possible; but public schools in a democracy should not be allowed this role.[3]

What do you think?

Questions

1. What is the role of public education in a democracy? Is there an overriding national interest it ought to serve, or is education essentially a local matter? How would advocates of national and compact theories of government answer this question?
2. Would a voucher system make for better schools? What about a choice system like Minnesota's?
3. Would all students benefit from a voucher system? From a choice system? Which students would be more likely to benefit, and which would be less likely? Which localities, if any, would be better off and which would be worse off under each system?

4. Is the private market an appropriate model for providing public services like education?
5. Should a voucher system allow students to use public funds to attend private schools, or only public schools? What about religious schools?

[1]For examples of this sort of argument, see David Osborne and Ted Gaebler, *Reinventing Government: How the Entrepreneurial Spirit Is Transforming the Public Sector* (New York: Plume, 1992), pp. 93–104; and John Chubb and Terry Moe, *Politics, Markets, and America's Schools* (Washington, D.C.: Brookings Institution, 1988).
[2]Osborne and Gaebler, *Reinventing Government*, p. 100.
[3]Jonathan Kozol, *Savage Inequalities: Children in America's Schools* (New York: Crown, 1991), p. 110.　■

Key Terms

unfunded mandate (*p. 117*)

sovereignty (*p. 120*)

unitary system of government (*p. 120*)

national theory of government (*p. 124*)

compact theory of government (*p. 125*)

participatory democracy/communal democracy (*p. 128*)

representative democracy (*p. 128*)

layer-cake federalism (*p. 133*)

cooperative federalism/marble-cake federalism (*p. 134*)

picket-fence federalism (*p. 136*)

revenue sharing (*p. 138*)

block grant (*p. 138*)

exit costs (*p. 142*)

public choice perspective (*p. 142*)

redistributive policies (*p. 144*)

developmental policies (*p. 146*)

Recommended Readings

Beer, Samuel H. *To Make a Nation: The Rediscovery of American Federalism.* Cambridge, Mass.: Harvard University Press, 1993. A thorough historical analysis of the origins of federalism and its advantages and disadvantages by one of the most respected writers on the subject.

Elazar, Daniel. *American Federalism: A View from the States*, 3rd ed. New York: Harper & Row, 1984. A classic study of the federal system from the perspective of the states. The author provides the best discussion available of the impact of state political culture on politics.

Orren, Karen. *Corporate Power and Social Change: The Politics of the Life Insurance Industry.* Baltimore: Johns Hopkins University Press, 1974. Orren provides an excellent history of the development of the insurance industry in America. More important, she analyzes the connection between the industry's economic power and its political clout flowing from its ability to use federalism to its advantage, especially with respect to attempts at governmental regulation.

Peterson, Paul E. *City Limits.* Chicago: University of Chicago Press, 1981. An original and influential analysis of the structural limitations on politics created by the existence of capitalism and federalism.

Peterson, Paul E., Barry G. Rabe, and Kenneth K. Wong. *When Federalism Works.* Washington, D.C.: Brookings Institution, 1986. An insightful series of case studies exploring how various types of public policies operate in a federal system.

Szasz, Andrew. *Ecopopulism: Toxic Waste and the Movement for Environmental Justice.* Minneapolis: University of Minnesota Press, 1994. This book explores the ways in which the shift of focus in the environmental movement to the state and local levels has helped to radicalize and reinvigorate the movement. The author is especially strong in demonstrating the connection between local environmental activism and broader issues of political participation and democracy.

Constitutional Rights as Rules

FOR MANY YEARS African-American children in Davis Station, South Carolina, walked nearly seven miles to attend the segregated public school in Summerton.[1] Then, some 50 years ago, a group of parents pooled their resources and bought a dilapidated bus to transport their children to school. Concerned about the condition of the bus, a local minister, Joseph DeLaine, convinced the parents to petition the county's school board for a newer and safer bus. The parents asked for a bus in the same condition as those used to transport white children to all-white schools. Their request reflected the principle, announced in 1896 by the United States Supreme Court in *Plessy v. Ferguson*,[2] that providing separate facilities for blacks and whites does not violate the Constitution if those facilities are equal. The school board refused.

Hoping to sue the school over this issue, Reverend DeLaine recruited a parent—one whose children were directly harmed by this decision—to bring the action to court. Levi Pearson, a local farmer with no formal schooling who believed that education would provide new and better op-

portunities for his children, served as one of the plaintiffs. Although the case was initially dismissed on technical grounds, it was resubmitted by Thurgood Marshall on behalf of the National Association for the Advancement of Colored People's (NAACP) Legal Defense Fund. Marshall would use this case to do more than ask for a new bus. The lawsuit invoked the Fourteenth Amendment's guarantee of equal protection of law to challenge the entire system of school segregation and the "separate but equal" doctrine established in *Plessy*. Over time, the case became bundled together with four other lawsuits—including the Kansas case of *Brown v. Board of Education of Topeka*—and arrived on the Supreme Court's docket.[3]

The combined cases, now simply called *Brown v. Board of Education*, were argued twice before the Supreme Court.[4] In 1952, Edward T. McGranahan, attorney general for President Harry Truman, filed a brief asking the Court to accept the NAACP's arguments and declare school segregation unconstitutional. Thirty social scientists provided an appendix to the

As an attorney for the NAACP's Legal Defense Fund, Thurgood Marshall (*center*) led a legal team that argued *Brown v. Board of Education* before the Supreme Court. The case was contested as part of a well-designed legal strategy begun in the 1930s to overturn the "separate but equal" doctrine established in *Plessy v. Ferguson*. Flanking Marshall here are the team's other members, George E. C. Hayes (*at Marshall's right*) and James M. Nabrit. Marshall went on to become the first African-American associate justice of the Supreme Court.

NAACP's brief, arguing that segregation damaged both white and black children psychologically. John W. Davis, Democratic presidential candidate in 1924, argued for the school boards in defense of segregation. After this first hearing, the Court asked both lawyer teams to answer a number of questions regarding the intent of the Fourteenth Amendment's framers and scheduled the case for reargument during the Court's next term. In December 1953, *Brown v. Board of Education* was reargued.

During this second hearing, the NAACP's argument to overturn *Plessy* enjoyed support from a brief filed by Herbert Brownell, President Dwight Eisenhower's attorney general. Although the Eisenhower administration supported desegregation, the brief also expressed the hope that the Court would allow a transition period before it occurred—a much more moderate position than that expressed previously by the Truman administration.

On May 17, 1954, Chief Justice Earl Warren announced the decision of a unanimous Court. Segregation of public schools "solely on the basis of race" denied African-American children "equal educational opportunity" even though "physical facilities and other tangible factors" may be equal. This interpretation of the concept "equality" in the Constitution's Fourteenth Amendment created a new rule that explicitly overturned the old "separate but equal" rule announced in *Plessy*. In *Brown,* the Court decided that "separate educational facilities are inherently unequal" because of segregation's psychologically damaging effects on black children's personal development.

Almost 20 years after *Brown,* the Supreme Court expanded the scope of its decision by ruling in *Keyes v. School District No. 1, Denver, Colorado* that Mexican Americans, who along with other Hispanics were often treated by local school districts as white, should be considered an identifiable minority group for purposes of desegregation.[5] In subsequent decisions, federal courts applied this rule to Puerto Ricans in New York and Boston.[6]

In Chapter 3 we examined the Constitution as rules defining the general framework of government in the United States. In this chapter we elaborate on the view, introduced in Chapter 3, that the Bill of Rights, and rights outlined in later constitutional amendments, serve as rules that both limit and expand the scope of government created in the Constitution. Our emphasis is on the Supreme Court's interpretation of **constitutional rights**—provisions in the Constitution's amendments that outline relationships between government and citizens—and the ways in which such interpretations evolve historically.

The role of constitutional rights as rules, and the evolution of the rules through court interpretation, is clearly illustrated by the case of school desegregation. In this case the Supreme Court was asked to answer a crucial question: Do laws and policies that establish separate schools for blacks and whites violate the Fourteenth Amendment's guarantee of equal protection of law even if the facilities are approximately equal in value? The Court gave two very different answers to this question in *Plessy v. Ferguson* and *Brown v. Board of Education.* In *Brown* the Supreme Court overturned the "separate but equal" rule established almost 60 years earlier in *Plessy.* The Court's new interpretation of "equality" in the Fourteenth Amendment's equal protection clause sought to create a dramatically different framework in which local school districts were to operate. In subsequent decisions the Supreme Court and other federal courts were asked to establish a variety of rules regarding how desegregation would be implemented, such as defining which racial and ethnic groups would count as minorities for purposes of desegregation. The rules established in *Brown* and later cases created new governmental responsibilities to act affirmatively in providing equal access to educational facilities.

We begin our examination by discussing some common beliefs about interpretations of rights by courts, focusing on the notion that such interpretations are fair and constant over time. We then focus on several of the most significant Supreme Court decisions that infuse meaning into constitutional rights in areas such as free expression, assembly, and religion; the procedural rights of criminal defendants; privacy and reproductive freedom; and equality rights, including affirmative action. Finally, we assess the role of constitutional rights as rules in the broader play of power.

RIGHTS AS IMMUTABLE AND FAIR

 In Chapter 3 we described the reverence that Americans have for the United States Constitution. Our admiration for this document applies with even greater force to constitutional provisions and amendments that express widely shared values such as equality, due process, and other individual rights. Constitutional rights, embroidering a document that creates a limited government, are thought to provide a framework for a political and legal system that is just, fair, and sensitive to the needs of individuals.

In a summary of research on American beliefs about constitutional rights, political scientist Stuart Scheingold suggests that such views take root early in life.[1] Children associate the rules established by constitutional rights with social order, as protections against violence and bodily harm. In addition, survey research suggests that constitutional rights are perceived by children as both unchanging and fair.[2] As children grow older they come to see legal rules in more complex ways. For instance, the belief that rights and their interpretation by courts never change "is softened by perceptions of movement and adaptation."[3] Over time, children increasingly value individual freedoms and connect constitutional rights with their protection.

Survey research also suggests that approval for the Supreme Court—the institution most responsible for giving meaning to constitutional rights—increases with age and comes to be linked with such attributes as "infallibility," "knowledgeability," and "power."[4] Although the propensity to criticize specific Supreme Court decisions increases with age, there is apparently "a reservoir of good will" in public perceptions of the Supreme Court.[5] By and large, the general public views the Supreme Court as a "visible but distant governmental institution closely associated with legal and constitutional norms," such as freedom, equality, and fairness.[6]

In this chapter we explore the rules produced by Supreme Court interpretations of constitutional rights. We examine the substantive content of these rules and assess, whenever appropriate, how the Supreme Court's interpretation of fundamental rights has evolved. Constitutional rights as interpreted by the Supreme Court are politically significant because they define relationships between the government and the citizenry. These relationships, in turn, may affect how participants operate in the play of power. Consequently, the rules developed by the Supreme Court are often highly contested by opposing parties with an important stake in the outcome.

[1] Stuart A. Scheingold, *The Politics of Rights: Lawyers, Public Policy, and Political Change* (New Haven: Yale University Press, 1974), pp. 62–71.
[2] Scheingold, *The Politics of Rights*, p. 64, citing Robert D. Hess and Judith V. Torney, *The Development of Political Attitudes in Children* (Chicago: Aldine, 1967), pp. 52–54.
[3] Scheingold, *The Politics of Rights*, p. 64
[4] Scheingold, *The Politics of Rights*, p. 65, citing David Easton and Jack Dennis, *Children in the Political System: Origins of Political Legitimacy* (New York: McGraw-Hill, 1969), pp. 278–79.
[5] Walter F. Murphy and Joseph Tanenhaus, *The Study of Public Law* (New York: Random House, 1972), p. 43; discussed in Scheingold, *The Politics of Rights*, p. 70.
[6] Scheingold, *The Politics of Rights*, p. 70.

INTERPRETING CONSTITUTIONAL RIGHTS

All the rights examined in this chapter are associated with the Constitution. All are listed explicitly in amendments to the Constitution or have been interpreted by the Supreme Court as being implied in explicit constitutional language. Many of these rights are included in constitutional provisions intended to protect citizens against improper governmental action, protections of what are commonly referred to as **civil liberties**. Civil liberties embedded in constitutional rights limit governmental action; they include prohibitions against laws that infringe on free speech, assembly, and the exercise of religion, as well as protections against unreasonable searches and seizures, cruel and unusual punishment, and deprivations of life, liberty, and property without due process of law.

Other constitutional rights are included in provisions that require the government to take positive steps to (1) protect citizens from illegal activity engaged in by other citizens or other governmental units and levels, and (2) provide certain benefits and opportunities. These governmental obligations are often referred to as **civil rights**. For example, the Fourteenth Amendment featured in our chapter-opening case study contains a provision that requires the government to guarantee equal protection under the law.

Because many of the key terms in constitutional provisions outlining civil rights and civil liberties—terms such as *due process, equal protection,* and *speech*—are vague and ambiguous, the federal courts and, especially,

In cases involving civil liberties, the Supreme Court draws lines between permissible and impermissible actions by individuals and the government. The Court permits some types of expression mixed with conduct, as in political demonstrations. But during the Vietnam War it upheld a law prohibiting the burning of draft cards, ruling that such activity constituted "conduct," not "speech."

the United States Supreme Court play a crucial role in determining their precise meaning. Rules established in the process of interpreting the ambiguous terms in constitutional provisions are political in the sense that some people will benefit from them and others will lose out. In our example of school segregation, the Court's interpretation of the Fourteenth Amendment was a grand political victory for integrationists, whereas segregationists suffered a dramatic defeat—the reversal of the "separate but equal" rule. Thus, it should come as no surprise that the Court's interpretation of constitutional rights is often a site of political struggle, with opposing interests advocating the interpretations and rules most beneficial to them and their cause. In the sections that follow, we discuss some of the most significant examples of how the Supreme Court has defined the rules embedded in constitutional rights—both those that directly protect citizens from government (civil liberties) and those that obligate the government to guarantee protection from the illegal actions of others (civil rights).

THE FIRST AMENDMENT: EXPRESSION, ASSEMBLY, AND RELIGION

The Constitution's First Amendment is written as follows:

> Congress shall make no law respecting an establishment of religion, or prohibiting the free exercise thereof; or abridging the freedom of speech, or of the press; or the right of the people peaceably to assemble, and to petition the government for a redress of grievances.

As you can see, the First Amendment establishes rules defining inappropriate congressional actions that infringe on important rights. Congress appears to be prohibited from passing legislation establishing an official religion, hindering the free exercise of religion, or limiting free speech, press, or the right to assemble and petition the government.

It seems as if the First Amendment would be relatively easy to apply to concrete situations if it is interpreted literally. Any congressional law limiting the rights mentioned above would simply be struck down as unconstitutional. But in reality only one Supreme Court justice in our history, Justice Hugo Black, argued consistently that the words *no law* mean that absolutely no restrictions on First Amendment rights are permissible. Instead, the Court has struggled to draw lines between constitutional and unconstitutional actions that touch on expression, assembly, and religion.

Free expression and assembly   Free expression includes the freedom of both speech and press. In these areas the Court has drawn lines creating exceptions to absolute protections against governmental infringement (see Table 5.1 on p. 168). For example, the Court has sought to establish a rule to distinguish protected and unprotected expression espoused by political dissenters. These rules have changed over time. During World War I, when fear of political radicalism spread rapidly throughout the country, the Court affirmed convictions of socialists such as Eugene Debs (see "Participants: Eugene V. Debs and Political Dissent" on pp. 170–71), communists, anarchists, and others

TABLE 5.1
SELECTED SUPREME COURT DECISIONS CREATING EXCEPTIONS TO PROTECTION OF FREE EXPRESSION

Subversive political expression

Schenck v. United States (1919) Upheld a conviction for distributing pamphlets opposing the military draft. Ruled that the government may restrict speech that constitutes a "clear and present danger" of producing an evil that the government has the right to prevent.

Gitlow v. New York (1925) Upheld a conviction for writing and distributing pamphlets advocating the overthrow of the government. Ruled that the government may restrict writings that have a "bad tendency" to result in evil.

Brandenburg v. Ohio (1969) Overturned the conviction of a Ku Klux Klan leader for remarks made at a rally. Ruled that to restrict political speech, the government must show that the speech incites "imminent lawless action."

Obscenity

Roth v. United States (1957) Ruled that obscene materials are not protected by the First Amendment. To determine if material is obscene, one must ask "whether to the average person, applying contemporary community standards, the dominant theme of the material taken as a whole appeals to prurient interest." To restrict, one must show that material is "utterly without redeeming social importance."

Miller v. California (1973) Defined obscenity as materials "which, taken as a whole, appeal to the prurient interest in sex, which portray sexual conduct in a patently offensive way and which, taken as a whole, do not have serious literary, artistic, political, or scientific value." Community standards used shall be local, not national.

Fighting words

Chaplinsky v. New Hampshire (1942) Upheld the conviction of a person publicly addressing someone as "damned fascist" and "goddamned racketeer." Established the rule that "fighting words," words that could result in injury to the listener and/or the speaker, are not protected by the First Amendment.

Terminiello v. Chicago (1949) Protected anti-Semitic speech by an extremist Catholic priest, ruling that he could not be punished solely on the grounds that such expression invites disputes or stirs people to anger. To be restricted, such expression must be "shown likely to produce a clear and present danger of serious substantive evil that rises far above public inconvenience, annoyance, or unrest."

Libel

New York Times v. Sullivan (1964) Reversed a lower court's judgment awarding Sullivan, a city commissioner of Montgomery, Alabama, $500,000 in damages from the *New York Times* for publishing a civil rights group's advertisement criticizing unnamed Alabama officials. The Court ruled that public officials suing for libel must prove "actual malice" or "reckless disregard for the truth."

opposed to the war because their anti-war expression was seen to constitute a "clear and present danger" to the republic.[7]

The **clear and present danger rule** was first announced by Justice Oliver Wendell Holmes in 1919 in *Schenck v. United States.*[8] In words that became famous, Holmes argued that the "most stringent" reading of the First Amendment "would not protect a man in falsely shouting 'fire' in a theater and creating a panic." According to Holmes, "the question in every case is whether the words used are used in such circumstances and are of such a nature as to create a clear and present danger that they will bring about the substantive evils that Congress has a right to prevent."

In *Schenck*, anti-war leaflets distributed by Charles Schenck and several associates were deemed to pose a "clear and present danger" to the war effort. Holmes suggested that the Court's rule should not be interpreted to mean that the government could suppress dissent under any and all circumstances. He explained the Court's rationale for this rule as follows:

> When a nation is at war, many things that might be said in time
> of peace are such a hindrance to its effort that their utterance will
> not be endured so long as men fight and that no Court could re-
> gard them as protected by any constitutional right.

The Court's interpretation of the right to free speech in the First Amendment, and its development of the clear and present danger rule, suggested that the relation between the government and its citizens—in particular, the government's ability to limit the right of citizens to disagree with its policies—would vary according to circumstances. Critics of this rule argue that it renders political dissent meaningless, because dissent may only be expressed when it has little chance to persuade or is irrelevant to current governmental policies and practices.[9]

The Court did not consistently apply the clear and present danger rule to all cases involving political speech. It established and applied another rule even more restrictive of expression, deciding that anti-war speeches or writings advocating the government's overthrow could be prohibited because they had a "bad tendency"—a possibility, even if remote—to produce an evil result, such as war resistance or armed revolution.[10] Most recently the Court has relied on a more liberal rule, protecting subversive political expression unless it produces or incites "imminent lawless action."[11] Although in an important sense this rule reinforces the idea that there are exceptions to absolute First Amendment protections, the requirement to demonstrate a close and direct connection between speech and action provides stronger protection against governmental suppression of political speech that challenges prevailing policies.

Exceptions to First Amendment protection have been created in other areas as well. For example, in *Chaplinsky v. New Hampshire* the Court held that words that "inflict injury or tend to incite an immediate breach of the peace"—expressions that the Court referred to as "fighting words"—do not convey ideas and, therefore, are not protected by the First Amendment.[12] Consequently, it upheld the conviction of a Jehovah's Witness for calling a city marshal a "goddamned racketeer" and a "damned fascist" in a public place.

The Court has been careful to define "fighting words" narrowly so that its rule does not suppress protected speech. For example, in *Terminiello v.*

EUGENE V. DEBS AND POLITICAL DISSENT

Eugene V. Debs was born in Terre Haute, Indiana, in 1855 and began working for the railroad at age 15. In time he became the national secretary for the Brotherhood of Locomotive Firemen, a national union. Debs helped organize the American Railway Union and became its president in 1893. As president he supported the Pullman Strike of 1894, a national strike that turned violent, and was ordered by a federal court to help put an end to the action. When he disobeyed the order, he was held in contempt of court and sent to prison. Debs was a founding member of both the Social Democratic Party and the American Socialist Party. He ran for president of the United States five times and received nearly a million votes in the 1912 and 1920 elections. In 1920, Debs ran for president while serving a 10-year prison sentence for seditious speech.

Debs's imprisonment for seditious speech resulted from an anti-war talk he gave in Canton, Ohio, in 1918. Speaking to a Socialist Party convention, he clearly recognized the danger of expressing disagreement with the government's involvement in World War I. Debs began his speech by saying, "I realize that . . . there are certain limitations placed on the right of free speech. I must be exceedingly careful, prudent, as to what I say and even more careful and prudent as to how I say it." A few sentences later he seemed to throw caution to the wind, stating that he "would a thousand times rather be a free soul in jail than to be a sycophant and coward in the streets." Even if he were to be jailed, he said, "they cannot put the Socialist movement in jail."

Almost the entire speech that Debs delivered that day focused on a broad-scale critique of capitalism and recent actions of the government in arresting other political and labor leaders. Only once in the entire speech did Debs mention war, declaring that "the master class has always declared the wars, the subject class has always fought the battles. The master class has all to gain and nothing to lose, while the subject class has had nothing to gain and all to lose—especially their lives." Debs went on to urge "continuous, active, and public opposition to the war, through demonstrations, mass petitions, and all other means within our power." As a result, Debs was indicted and convicted of violating the Espionage Act of 1917, a congressional law prohibiting expression that sought to interfere with the nation's military efforts. He appealed his conviction, which was upheld in 1919 by a unanimous Supreme Court.

Writing for the Court, Justice Oliver Wendell Holmes argued that although the main theme of Debs's speech, socialism, was protected by the First Amendment, the speech's

Chicago the Court struck down the conviction for disturbing the peace of an openly anti-Semitic Catholic priest for words expressed in an address to a right-wing extremist group, Christian Veterans of America.[13] As some 1,500 angry protesters outside the building threw rocks, bottles, and bricks, Terminiello denounced "Communistic, Zionistic Jews" and referred to the protesters as "slimy scum." His audience responded to the tone and content of his address by chanting "Dirty kikes!" and "Kill the Jews!" The Court overturned this conviction, ruling that the most provocative speech, even if it incites people to anger, is constitutionally protected. In *Terminiello* the Court narrowed its definition of "fighting words" by ruling that free expression may not be suppressed "unless shown likely to produce a clear and present danger of serious substantive evil that rises far above public inconvenience, annoyance, or unrest."

intent "to obstruct the recruiting service" could be constitutionally restricted by the Espionage Act. The Court concluded that Debs's opposition to the war was expressed in such a way that its "natural and intended effect" would be to obstruct military recruiting. The day after the decision, Debs issued a statement to the press:

> The decision is perfectly consistent with the character of the Supreme Court as a ruling class tribunal. It could not have been otherwise. So far as I am concerned, the decision is of small consequence. . . . Great issues are not decided by courts, but by the people. I have no concern in what the coterie of begowned corporation lawyers in Washington may decide in my case. The court of final resort is the people, and that court will be heard from in due time.

Ultimately, Eugene Debs went to prison for expressing opposition to American involvement in World War I, but he was released in 1921 by order of President Warren Harding.

Sources: Ray Ginger, *Eugene V. Debs: The Making of an American Radical* (New York: Collier, 1962); John C. Domino, *Civil Rights and Civil Liberties: Toward the 21st Century* (New York: HarperCollins, 1994), pp. 14–15.

Throughout our nation's history the Supreme Court has been reluctant to protect the expression rights of political dissidents during wartime. Eugene Debs was sentenced to 10 years in prison for giving a speech criticizing American involvement in World War I. The Court affirmed the conviction, arguing that Debs's words would obstruct military recruitment.

Expression defined as "obscene" also does not enjoy constitutional protection. Rules defining obscenity seem to have been more difficult for the Court to determine than rules for any other concept in its history. Indeed, at one point Justice Potter Stewart threw up his hands on this question, commenting that he could not define obscenity but could identify it when he saw it.[14]

As Table 5.1 (on p. 168) indicates, the Court under the leadership of Chief Justice Earl Warren first established the rule regarding obscenity in *Roth v. United States.*[15] Justice William J. Brennan, writing for the Court, defined obscenity in this case as material appealing to "prurient interest" or inciting lustful thoughts. The Court, and lower courts using this rule, would determine "whether to the average person, applying contemporary community standards, the dominant theme of the material taken as a whole

PORNOGRAPHY, FEMINISM, AND THE FIRST AMENDMENT

More clearly than most issues, obscenity and pornography illustrate how rules developed by the Supreme Court in the course of interpreting constitutional rights are politically significant and highly contested by opposing parties. Indeed, critics of the Court's approach, including radical feminists and religious fundamentalists, joined forces in a strategy to write legislation providing alternative rules that would be much more restrictive.

The regulation of sexually explicit material has divided feminists. Some feminists have been very critical of the Court's approach for permitting the circulation of material that can be considered damaging to women. For instance, feminist legal scholar Catharine MacKinnon argues that the Court's rules regarding obscenity in the *Roth* and *Miller* cases reflect a male understanding of this issue. She contends (1) that the perspective of the "average person" referred to in the rule will be interpreted by judges, who are predominantly men, (2) that "contemporary community standards" are determined primarily by men, and (3) that the "prurient interest" that obscene material must appeal to relates to male, not female, reactions to such material. According to MacKinnon, the Supreme Court's rule on obscenity

Catharine MacKinnon has been at the forefront of feminist attempts to regulate or ban pornography. This issue has split the feminist community, with many criticizing MacKinnon's approach for permitting the censorship of constitutionally protected expression.

means simply that sexually explicit material that may be banned or regulated is material that makes a *man* blush.

MacKinnon and other like-minded feminists refuse to use the term *obscenity* to refer to material that may be proscribed. Instead, to distinguish their views from the Court's opinions, they use the term *pornography* to mean sexually

appeals to the prurient interest." In an effort to provide the greatest possible protection to expression, Brennan's rule required that a work be deemed obscene only if it was "utterly without redeeming social importance."

The *Roth* decision shows that rules constructed by the Court can themselves be ambiguous. Think of the many difficult questions lower courts now confront. Does the material in question appeal to "prurient interest"? What is the "dominant theme" of the material, "taken as a whole"? How would the "average person" interpret the material? What are "contemporary community standards"? Is the material "utterly without redeeming social importance"? In time, the Warren Court lost interest in determining whether particular works were obscene and found almost all material to have some redeeming social value.

A more conservative Court, led by Chief Justice Warren Burger, sought to establish a more restrictive rule. In its 1973 decision in *Miller v. Califor-*

explicit material that, they argue, discriminates against women by degrading and humiliating them—by portraying women as willing victims of sexual violence. Further, they believe pornography perpetuates the subordination of women by socializing its readers and viewers to accept its depictions of sexual relations. Together with political activist Andrea Dworkin, MacKinnon helped write an antipornography ordinance, passed in 1984 in Indianapolis, that defined pornography as "the graphic, sexually explicit subordination of women through pictures and words." The ordinance permitted women to sue in order to ban such material and even obtain monetary damages if, among other things, it portrayed women as sexual objects or commodities, as enjoying pain or humiliation, or as experiencing pleasure in rape.

Less than 90 minutes after the ordinance was signed into law, a coalition of book and magazine publishers, distributors, and sellers filed a lawsuit challenging the ordinance as violating the First Amendment. They were supported in their efforts by another group of feminists, who called themselves the Feminist Anti-Censorship Taskforce (FACT). In a brief filed with the federal court hearing the case, FACT argued that the ordinance's rules regarding pornography were so vague that almost anything could be restricted. In addition, FACT argued that the notion that sex degrades women has been a part of many efforts in U.S. history to restrict women's equality. According to the brief, this notion was invoked in passing laws that reinforce gender stereotypes, such as the ideas that women have no will of their own, require special protection from sexual activity, and are always the "victim" in sexual activity. The brief expressed concern that enforcement of the Indianapolis ordinance would "reinvigorate those discriminatory moral standards which have limited women's equality in the past." The federal court, in *American Booksellers Association v. Hudnut*, accepted these arguments and struck down the ordinance. The Supreme Court affirmed the decision.

Sources: Nan D. Hunter and Sylvia A. Law, "Brief Amici Curiae of Feminist Anti-Censorship Taskforce et al., in *American Booksellers Association v. Hudnut*," *University of Michigan Journal of Law Reform* 21 (Fall 1987–Winter 1988), pp. 69–135; Catharine A. MacKinnon, *Toward a Feminist Theory of the State* (Cambridge, Mass.: Harvard University Press, 1989), pp. 195–214; Steven H. Shiffrin and Jesse H. Choper, *The First Amendment* (St. Paul, Minn.: West, 1991), pp. 190–204; Nadine Strossen, *Defending Pornography: Free Speech, Sex, and the Fight for Women's Rights* (New York: Scribner, 1995).

nia, the Burger Court made two significant changes in the obscenity rule.[16] First, rather than examining material for any minimal social value, the new rule required "serious literary, artistic, political, or scientific value." Second, the community standards employed in judging works would be local rather than national. In reality, the new approach did little to curb the circulation of pornographic materials. Consequently, the rules established by the Court have been intensely criticized both by conservative moralists and by feminists. We feature the criticism of these rules by anti-pornography feminists in "Strategies: Pornography, Feminism, and the First Amendment."

In the area of free press, the Court has ruled on many occasions that **prior restraint**, the censoring of a story before its publication or broadcast, is unconstitutional.[17] But the Court also has created an exception to First Amendment protection of the press for **libel**, the publication of a story that falsely damages someone's character.[18] The Court's rule requires that public officials suing for damages must prove "actual malice"—an intent to falsely discredit a person—or "reckless disregard for the truth" on the part of the journalist or publication, as is noted in Table 5.1 (on p. 168).

Constitutional rules derived from the First Amendment—especially its provisions guaranteeing free speech, free assembly, and the right to petition government—affect the use by social movements of unconventional political strategies, such as protests and demonstrations (see Table 5.2). For one thing, the Court distinguishes between "speech" and "conduct," protecting the former while permitting regulation or prohibition of the latter. The bound-

---

TABLE 5.2
SELECTED SUPREME COURT DECISIONS AFFECTING UNCONVENTIONAL POLITICAL BEHAVIOR

Cox v. Louisiana (1965)   Struck down convictions of civil rights protesters arrested under local ordinances prohibiting breach of peace. Ruled that local governments may not use such vaguely worded ordinances to arbitrarily prevent some protest activity.

Adderly v. Florida (1966)   Upheld the conviction of civil rights protesters for demonstrating on the grounds of a county jail in violation of a local "trespass" ordinance. The Court found the ordinance precisely drawn and unrelated to the content of any political message of protest. The Court found this action to be primarily "conduct," not "speech."

United States v. O'Brien (1968)   Upheld a conviction for burning a draft card during the Vietnam War. The Court ruled that burning a draft card constitutes "conduct," unprotected by the free speech provision in the First Amendment.

Lloyd Corporation v. Tanner (1972)   Did not uphold the right of anti-war protesters to distribute handbills at a privately owned shopping mall. The Court ruled that private property may not be used at will for communication of political expression; "this court has never held that a trespasser or an uninvited guest may exercise general rights of free speech on private property used non-discriminatorily for private purposes only."

Clark v. Community for Creative Non-Violence (1984)   Upheld a Park Service rule prohibiting camping in District of Columbia parks, even if the act was a symbolic protest of homelessness. Ruled that the government has a legitimate interest in regulating the "time, place, and manner" of park use.

Frisby v. Schultz (1988)   Upheld a local Wisconsin ordinance prohibiting picketing in private neighborhoods. Anti-abortion protesters were legitimately prevented from picketing at the private residence of a doctor who performs abortions.

Texas v. Johnson (1989)   Overturned a conviction for burning an American flag as part of a political protest. The Court ruled that the action of burning the flag is "symbolic speech" protected by the First Amendment.

aries between "speech" and "conduct" are extremely difficult to draw; consequently, the Court has rendered decisions that are seemingly irreconcilable. For instance, in *United States v. O'Brien* the Court ruled that burning a draft card to protest American involvement in the Vietnam War constituted "conduct" unprotected by the First Amendment, whereas in *Texas v. Johnson* it decided that burning an American flag as part of a political protest was **symbolic speech**—conduct with a purpose to express a political view—and, therefore, constitutionally protected.[19]

The Court has also decided that the right to assemble for purposes of political protest may be regulated at certain times and locations if the rules do not discriminate against particular viewpoints and if they exist only to provide authorities with advance notice of actual crowd control problems.[20] In addition, the Court has ruled that, under many circumstances, political demonstrations may be prohibited on private property, such as a shopping mall.[21]

Freedom of religion   The First Amendment protects religious practices in two ways. It prohibits the government from establishing an official, government-supported religion; and it guarantees the free exercise of religious beliefs. The First Amendment's **establishment clause** erects a wall between church and state, and its **free exercise clause** prohibits government from interfering with the practices of religious groups. As with the free speech guarantee, neither of these clauses has been interpreted by the Supreme Court as absolute. Table 5.3 (on p. 176) identifies several significant rulings in these areas.

In 1978, American Nazi Party members threatened to march through Skokie, Illinois, a community with many residents who had survived the Holocaust. Courts have regularly been asked to determine whether such demonstrations are constitutionally protected speech or are so threatening and potentially damaging psychologically to local residents that they should be prohibited.

TABLE 5.3
SELECTED SUPREME COURT DECISIONS ON RELIGION

Establishment of religion

Engel v. Vitale (1962)   Struck down a New York law requiring the daily reading in public schools of a 22-word nondenominational prayer.

Abington School District v. Schempp (1963)   Found that reading from the Bible and reciting the Lord's Prayer in public schools violated the establishment clause.

Lemon v. Kurtzman (1971)   Struck down state laws providing state financial aid to parochial schools for "secular educational services" and a salary supplement for teachers of secular subjects. The Court developed a three-part standard: A law is valid if (1) it has a secular purpose, (2) its principal or primary effect neither advances nor inhibits religion, and (3) it does not foster excessive governmental entanglement with religion.

Lynch v. Donnelly (1984)   Permitted the display of a crèche on property owned by the city of Pawtucket, Rhode Island. Although the crèche has obvious religious connotations, the Court ruled that this display passed the *Lemon* test because the city had a "legitimate secular purpose" to recognize and celebrate a holiday (Christmas) that touches the lives of everyone.

Wallace v. Jaffree (1985)   Struck down an Alabama law requiring public school teachers to observe a moment of silence for meditation or voluntary prayer. Found that the law violated the *Lemon* test because it had a religious purpose.

Edwards v. Aguillard (1987)   Struck down the 1982 Louisiana "Creationism Act" mandating that evolution can be taught in public schools only if creation science is taught as well. Applying the *Lemon* test, the Court ruled that the law did not have a secular purpose.

Free exercise of religion

West Virginia State Board of Education v. Barnette (1943)   Struck down a compulsory flag salute law. Ruled that the law interfered with the free exercise of religion for Jehovah's Witnesses, who are prohibited by their faith from worshiping graven images. Overruled a previous decision upholding the law.

Wisconsin v. Yoder (1972)   Held that the Amish were not subject to the compulsory school attendance law because of their belief that education beyond the eighth grade teaches values that clash with their religious creed.

Goldman v. Weinberger (1986)   Ruled that the military could impose a dress code on an Orthodox Jewish rabbi and Air Force captain, prohibiting him from wearing a yarmulke (skullcap).

Employment Division v. Smith (1990)   Ruled that Native Americans fired from their jobs for using an illegal drug, peyote, as part of an off-duty religious ceremony could be denied state unemployment compensation benefits.

Many of the landmark establishment clause cases raise issues of possible church and state entanglement in public schools. The Supreme Court has been asked on numerous occasions to determine the constitutionality of such practices as prayer in the public schools, state financial aid to parochial schools, teaching a biblical version of creation in place of or in addition to evolution, and the use of public school facilities for religious activities. In addition, the Court has ruled on whether displaying religious symbols—such as the crèche and the menorah—on public property constitutes the government's endorsement of a particular religion. Although the Supreme Court has been somewhat inconsistent on these issues,[22] it developed a set of rules in the case of *Lemon v. Kurtzman* that it employs frequently to determine whether governmental policies violate the establishment clause.[23] The law (1) must not have a religious purpose, (2) must not promote or inhibit religion, and (3) must not cause "excessive government entanglement" with religion. Of course, each of these rules is in need of further interpretation. Table 5.3 shows how these rules have been applied in certain areas.

In one area, school prayer, the Court has consistently interpreted the establishment clause to prohibit state involvement. In 1962, in *Engel v. Vitale*, the Court struck down a New York law requiring the daily reading of a brief nondenominational prayer: "Almighty God, we acknowledge our dependence upon Thee, and we beg Thy blessings upon us, our parents, our teachers, and our country."[24] According to the Court, this prayer constituted a state-sponsored religious activity "wholly inconsistent" with the establishment clause.

One year after *Engel*, the Court struck down a Pennsylvania law requiring daily Bible reading and recitation of the Lord's Prayer in public schools in the case of *Abington School District v. Schempp*.[25] The Court ruled that the state's activity in this area violated a constitutional requirement for governmental neutrality in matters of religion. The Court has returned to the issue of school prayer more recently in connection with state-mandated moments of silence. In 1985 it ruled, in *Wallace v. Jaffree*, that an Alabama law requiring a moment of silence in public schools for meditation or voluntary prayer violated the establishment clause.[26] Applying rules it had developed in the *Lemon* decision, the Court suggested that this law was invalid because it did not have a secular purpose—although it hinted that a more carefully drawn law calling only for a moment of silence, and not "voluntary prayer," might be upheld.

In the area of free exercise, the Supreme Court has interpreted the word *religion* quite broadly to include almost any set of beliefs guiding choices about how to live. And it has distinguished religious beliefs from actions based on those beliefs. The government may not interfere with beliefs, but it may regulate or prohibit antisocial or illegal actions unprotected by the First Amendment. Thus, although the Court struck down a law compelling schoolchildren to salute the flag[27]—a law challenged by Jehovah's Witnesses because it forced them to worship graven images, contrary to their faith—it also ruled that Native American employees fired for using the illegal drug peyote during an off-duty religious ceremony could legitimately be denied state unemployment compensation. The Court ruled that "generally applicable, religion-neutral laws that have the effect of burdening a particular religious practice" need only be reasonable to be valid.[28]

Four amendments to the Constitution describe how the government should behave in criminal justice proceedings. Provisions in these amendments are meant to protect those suspected of or arrested for criminal activity from misconduct by the police and other agents of government. The amendments read as follows:

> FOURTH AMENDMENT The right of the people to be secure in their persons, houses, papers, and effects, against unreasonable searches and seizures, shall not be violated, and no Warrants shall issue, but upon probable cause, supported by Oath or affirmation, and particularly describing the place to be searched, and the persons or things to be seized.

> FIFTH AMENDMENT No person shall be held to answer for a capital or otherwise infamous crime, unless on a presentment or indictment of a Grand Jury, except in cases arising in the land or naval forces, or in the Militia, when in actual service in time of War or public danger; nor shall any person be subject for the same offence to be twice put in jeopardy of life or limb; nor shall be compelled in any criminal case to be a witness against himself, nor be deprived of life, liberty, or property, without due process of law; nor shall private property be taken for public use, without just compensation.

> SIXTH AMENDMENT In all criminal prosecutions, the accused shall enjoy the right to a speedy and public trial, by an impartial jury of the State and district wherein the crime shall have been committed, which district shall have been previously ascertained by law, and to be informed of the nature and cause of the accusation; to be confronted with the witnesses against him; to have compulsory process for obtaining witnesses in his favor, and to have the Assistance of Counsel for his defence.

> EIGHTH AMENDMENT Excessive bail shall not be required, nor excessive fines imposed, nor cruel and unusual punishments inflicted.

These amendments touch on the actions of nearly every participant at each stage of the criminal justice system. As you can see, many of the key phrases in these amendments, such as "unreasonable searches and seizures," "compelled to be a witness against himself," "assistance of counsel," "impartial jury," "speedy trial," "due process," and "cruel and unusual punishment," are so vague that the Supreme Court has been called on to define them more precisely and apply them to concrete circumstances. Table 5.4 (on pp. 180–181) describes a few of the Court's attempts to define these important procedural rules.

The interpretation of many of these crucial terms, and the application of the Fourth, Fifth, Sixth, and Eighth Amendments to criminal justice systems at the state and local levels, is a relatively recent phenomenon. The Warren Court—the most liberal Court in our nation's history—established most of the central precedents defining the rules of criminal procedure in

The Supreme Court's interpretations of the Constitution's Fourth, Fifth, Sixth, and Eighth Amendments are meant, among other things, to guide police conduct toward criminal suspects. These decisions regulate acts such as searches for evidence, arrests, pretrial procedures, trials, jury selection, and sentencing.

the 1960s. For example, in *Mapp v. Ohio* the Warren Court decided that material seized in an illegal search could not be introduced as evidence in a state court—a rule that became known as the **exclusionary rule**.[29] Extending a rule developed many years earlier for the federal courts, the Warren Court based its judgment on the Fourth Amendment's protection against unreasonable searches and seizures.

The Warren Court continued to expand protections for criminal defendants in *Miranda v. Arizona*, ruling that the Fifth Amendment's protection against self-incrimination required police officers to inform suspects of their rights before they underwent interrogation.[30] Immediately after arrest, police officers were to offer suspects what came to be known as the **Miranda warnings**: that (1) suspects had the right to remain silent, (2) any information they divulged could be used against them in court, (3) they had the right to speak to an attorney before police questioning and have the attorney present at any interrogation, and (4) if they could not afford an attorney, one would be appointed before any interrogation took place. In the case of *Gideon v. Wainwright*, the Warren Court interpreted the Sixth Amendment to mean that anyone accused of a serious crime had the right to legal counsel. If a defendant could not afford an attorney, the government had an obligation to appoint one.[31]

The rules established by the Warren Court that increased protections for criminal defendants proved to be controversial. The Supreme Court was accused by candidates for public office of tying the hands of the police, permitting criminals to avoid punishment by invoking new legal loopholes, and generally contributing to crime. In the 1970s and 1980s, Republican presidents Richard Nixon, Gerald Ford, Ronald Reagan, and George Bush implemented campaign promises by appointing federal court judges and Supreme Court justices who were less sympathetic to the rights of criminal

TABLE 5.4
SELECTED SUPREME COURT DECISIONS ON THE RIGHTS
OF CRIMINAL DEFENDANTS

Unreasonable searches and seizures

Mapp v. Ohio (1961)   Found that evidence obtained in violation of the Fourth Amendment is inadmissable in state court. The rule developed is referred to as the "exclusionary rule."

United States v. Leon (1984)   Established the "good-faith" exception to the exclusionary rule. The exclusionary rule may not apply where evidence is seized by police in good-faith reliance on what they wrongly believe is a validly issued search warrant.

Self-incrimination

Miranda v. Arizona (1966)   Found that to protect a defendant against compulsory self-incrimination, the following rights and warnings must be communicated by police officers to the defendant at the time of the arrest: (1) the right to remain silent, (2) that anything said can be used against the defendant in court, (3) the right to discuss the case with a lawyer before being questioned, and (4) that a lawyer will be appointed by the court if the defendant cannot afford one.

New York v. Quarles (1984)   Found that the Miranda warnings are not required when overriding considerations of public safety are present.

Assistance of counsel

Gideon v. Wainwright (1963)   Established the right to legal counsel in serious criminal cases. If the defendant cannot afford an attorney, the government must supply one.

(continued on facing page)

defendants. Consequently, the Supreme Court under the chief justiceships of Warren Burger (from 1969 to 1986) and William Rehnquist (from 1986 to the present) did not extend procedural protections in many areas and introduced many exceptions to the general rules established during the 1960s.

For example, the Court in 1984, in the case of *United States v. Leon,* established a **good-faith exception** to the exclusionary rule.[32] In this case the police obtained a search warrant from a judge based on a tip from an unreliable informant. Although "probable cause" had not been firmly established, the Court permitted the evidence to be used at trial because the police had acted in "good faith." In *New York v. Quarles* the Court ruled that "overriding considerations of public safety"—such as potential danger for the arresting officer—may justify the immediate questioning of a suspect without first giving the Miranda warnings against self-incrimination.[33] Also, in 1990,

TABLE 5.4 (CONTINUED)
SELECTED SUPREME COURT DECISIONS ON THE RIGHTS
OF CRIMINAL DEFENDANTS

Assistance of counsel (continued)

Argersinger v. Hamlin (1972)   Extended the right to legal counsel to
cases involving less serious offenses (misdemeanors), for which a
short period of incarceration is a possible penalty.

Cruel and unusual punishment

Furman v. Georgia (1972)   Found that although the death penalty per se
is not cruel and unusual punishment, racially discriminatory imposition of the death penalty in Georgia does constitute cruel and
unusual punishment. In the years following the decision, some 35
states enacted new death penalty statutes.

Gregg v. Georgia (1976)   Upheld a new two-stage sentencing procedure
developed in Georgia after the *Furman* decision: A trial is divided into
verdict and sentencing stages, both taking place before the same jury.
At the sentencing, the jury must find the defendant guilty beyond a
reasonable doubt of at least 1 of 10 statutorily aggravated circumstances before applying the death penalty. The State Supreme Court is
required to review every death sentence to determine whether it was
imposed as a result of prejudice and whether it was disproportionate
to penalties in similar cases. The Court ruled that this statute
protects against the death penalty's being imposed in an arbitrary and
capricious manner.

McCleskey v. Kemp (1987)   Found that statistical evidence of racial
disparity in imposition of the death penalty does not prove racial bias
or that the death penalty is cruel and unusual punishment. The Court
ruled that "apparent disparities in sentencing are an inevitable part of
our criminal justice system" and that state legislatures are appropriate bodies to evaluate studies and change procedures.

the Court in *Illinois v. Perkins* ruled that a law enforcement officer could
pose as a prison inmate and use a confession from an unwitting actual
inmate without giving the Miranda warnings.[34] Writing for the Court, Justice
Kennedy argued that the Fifth Amendment "does not confer a right to
boast."[35]

The Supreme Court did not tackle the most significant and difficult
issue in criminal justice, the imposition of the death penalty, until 1972.
In that year the Court, in *Furman v. Georgia*, struck down a death penalty
law, arguing that the random, arbitrary, and perhaps racially discriminatory
manner in which it was administered violated the Eighth Amendment's
prohibition against cruel and unusual punishment.[36] In response to this
decision, many state legislatures enacted new criminal codes with explicit
sentencing guidelines. The Court approved these new criminal statutes in
Gregg v. Georgia, and "death rows" in prisons reopened throughout the

country.[37] Most recently the Court, in 1987, in *McCleskey v. Kemp,* ruled that statistical evidence showing that the death penalty was four times more likely to be imposed on murderers of whites than on murderers of blacks was insufficient to prove unconstitutional racial bias.[38]

PRIVACY, REPRODUCTIVE FREEDOM, AND SEXUAL ORIENTATION

Most of the Court's decisions on speech, religion, and defendants' rights have involved interpretations of terms and phrases written explicitly into constitutional amendments. By and large, the Court in these decisions did not identify or create new rights but, rather, anchored its rulings in words appearing in the Constitution's text.

In the 1960s and 1970s the Court took more dramatic steps to protect a right of privacy not appearing in any of the Constitution's amendments (see Table 5.5). In 1965, in its decision in *Griswold v. Connecticut,* the Court struck down a law prohibiting the use of contraceptives, arguing that it violated a right of privacy applying to married couples.[39] Writing for the Court's majority, Justice William O. Douglas suggested that the "specific guarantees in the Bill of Rights have penumbras," or areas of partial illumination, that help give them "life and substance." Specific rights in the First, Third, Fourth, and Fifth Amendments created a "zone of privacy" that, among other things, protected the marital relationship from governmental intrusion. In addition, said the Court, the zone of privacy was protected by the general language of the Ninth Amendment, guaranteeing that rights enumerated in the Constitution "shall not be construed to deny or disparage others retained by the people."

In 1973 the Court extended the right to privacy to a woman's decision to abort a fetus in *Roe v. Wade.*[40] In *Roe,* the Court devised a set of rules regarding the government's ability to regulate abortion at different periods

Norma McCorvey (*left*), a 21-year-old divorcée, could not get an abortion in Texas in 1969 because a state law prohibited abortions except when necessary to save the life of the mother. McCorvey challenged the law in court and was represented by Sara Weddington (*right*), a graduate of the University of Texas Law School. To spare McCorvey personal stigma from bringing this controversial case to court, her name was changed in court documents to Jane Roe.

TABLE 5.5
SELECTED SUPREME COURT DECISIONS ON PRIVACY

Contraception

Griswold v. Connecticut (1965)   Struck down a law prohibiting use of contraceptives. The Court found that the law affected the intimacy of the marital relationship and represented governmental intrusion into the protected zone of privacy. According to the Court, the "very idea" of allowing police to search bedrooms for contraceptives "is repulsive to the notions of privacy surrounding the marriage relationship."

Abortion

Roe v. Wade (1973)   Invoked the right to privacy to strike down a Texas law making it a felony to have an abortion except to save the mother's life. The Court found that the state's interest in protecting prenatal life must be balanced with a woman's right to privacy. It announced different rules to be applied to different trimesters of pregnancy. Prior to the end of the first trimester, the abortion decision is left to a woman and her doctor. During the second trimester, the government may regulate in ways "reasonably related to maternal health." During the final trimester, the government may regulate or prohibit in the interest of protecting "potentiality of human life."

Maher v. Roe (1977)   Held that the government is not required by the Constitution to pay for abortions for women who are poor. Upheld Connecticut's refusal to reimburse Medicaid recipients for elective abortion expenses.

Harris v. McRae (1980)   Upheld the Hyde Amendment, a federal law banning the use of Medicaid funds for abortions performed on women who are poor.

Akron v. Akron Center for Reproductive Health (1983)   Struck down a law requiring women seeking abortions to wait at least 24 hours after receiving counseling including the statement that "the unborn child is a human life from the moment of conception."

Thornburgh v. American College of Obstetricians and Gynecologists (1986)   Struck down a requirement in Pennsylvania law that women seeking abortions must be given specific "informed consent" information intended to discourage abortion.

Webster v. Reproductive Health Services (1989)   Upheld a Missouri law banning the use of public funds, hospitals, and clinics for abortions.

Planned Parenthood v. Casey (1992)   Upheld a 24-hour waiting period and teenage parental consent restrictions on abortion. Adopted an "undue burden" standard to evaluate restrictions.

Sexual orientation

Bowers v. Hardwick (1986)   Upheld a Georgia sodomy law, refusing to extend the right to privacy to homosexual conduct.

in a woman's pregnancy. Dividing the term of pregnancy into trimesters, the Court determined that a woman had the most autonomy during the first trimester, when government could not regulate the abortion decision. The government's authority to regulate abortion increased during the second trimester and was greatest during the third, or final, trimester.

Since *Roe*, the Court has generally reaffirmed a woman's right to privacy. Although it has ruled that the government is not required by the Constitution to pay for abortions obtained by women who are poor,[41] it has struck down laws containing a variety of restrictions on abortion, such as those including requirements for spousal notification, a waiting period after exposure to anti-abortion literature, and dissemination of "informed consent" information intended to discourage abortions.[42]

The Court nearly changed the rules announced in *Roe* in 1989 in *Webster v. Reproductive Health Services*.[43] In a 5–4 decision, the Court upheld a Missouri law barring the use of public funds, hospitals, and clinics for abortions not needed to save the life of the mother. The personal papers of the late Justice Thurgood Marshall, released shortly after his death, suggest that for a time during the *Webster* deliberations five justices—Rehnquist, Kennedy, Scalia, White, and O'Connor—appeared ready to sign an opinion written by Chief Justice Rehnquist that overturned *Roe v. Wade*.[44] Three justices—Marshall, Blackmun, and Brennan—proposed a dissent, saying, "*Roe* no longer survives." But, illustrating how small-group dynamics can affect Court opinions, Justice Sandra Day O'Connor ultimately could not accept such a sweeping decision, forcing the opinion to be drastically modified to attract majority support.

The Court appears to have altered its standard for evaluating restrictions on abortion in the 1992 decision of *Planned Parenthood v. Casey*.[45] Upholding portions of a state law requiring (1) women to delay abortions for 24 hours after viewing an anti-abortion presentation, and (2) teenagers to obtain the consent of one parent or a judge before undergoing an abortion, the Court adopted a standard proposed by Justice O'Connor that asks whether the restriction imposes an "undue burden," a "substantial obstacle in the path of a woman seeking an abortion before the fetus attains viability."

The right to privacy, to this point, has been applied by the Court exclusively to heterosexual couples and women seeking abortions. The Court refused to apply this right to gay men in the 1986 case of *Bowers v. Hardwick*.[46] In this case the Court upheld a Georgia law prohibiting sodomy, arguing that privacy interests related to marriage, contraception, and abortion had little in common with any right to engage in homosexual conduct.

THE POST-CIVIL WAR AMENDMENTS AND EQUALITY

To this point, we have described constitutional rules emerging from provisions in the Bill of Rights, the Constitution's first 10 amendments, which were ratified in 1791. We now turn our attention to rules guaranteeing equality that are contained in three amendments—the Thirteenth, Fourteenth, and Fifteenth—adopted after the Civil War.

The Thirteenth Amendment, ratified in 1865, outlaws slavery with the words "Neither slavery nor involuntary servitude . . . shall exist within the United States, or any place subject to their jurisdiction." The Fifteenth Amendment, ratified five years later, sought to guarantee the right to vote to African-American males. According to this amendment, "the rights of citizens of the United States to vote shall not be denied or abridged by the United States or by any state on account of race, color, or previous condition of servitude."

Ratified in 1868, the Fourteenth Amendment contains several important provisions. It provides that former slaves are citizens: "All persons born or naturalized in the United States, and subject to the jurisdiction thereof, are citizens of the United States and the State wherein they reside." It also includes three significant clauses, proclaiming that "no state shall" (1) "abridge the privileges and immunities of citizens of the United States," (2) "deprive any person of life, liberty, or property without due process of law," or (3) "deny to any person within its jurisdiction the equal protection of the laws."

Shortly after the Fourteenth Amendment's passage, the Supreme Court interpreted its privileges and immunities clause narrowly.[47] Consequently, the Fourteenth Amendment's other two major clauses, the due process clause and the equal protection clause, have played more significant roles in the legal and political struggle for equality. Among other things, the **due process clause** has been used by the Court to apply provisions in the Bill of Rights to the actions of state governments. The **equal protection clause** has been a crucial resource for reformers seeking to guarantee equality to all citizens.

Applying the Bill of Rights to the states   The Bill of Rights was adopted in large measure to satisfy the fears of Antifederalists who believed that the federal government created in the new constitution would be too powerful. Given its historical origins and the fact that its First Amendment begins with the phrase "Congress shall make no law," the Bill of Rights appeared to apply only to the actions of the federal government. Indeed, although several litigants made the claim that the Bill of Rights also applied against action by state government, the Court ruled prior to the Civil War that the substantive protections contained in this document (found primarily in its first eight amendments) restricted only the federal government.[48]

This interpretation changed dramatically after the adoption of the Fourteenth Amendment, as the due process clause became the vehicle through which provisions in the Bill of Rights were applied to the states. The Court's

Denied legal counsel because he could not afford a lawyer, Clarence Earl Gideon in 1962 petitioned the Supreme Court to overturn his conviction. The resulting landmark case, *Gideon v. Wainwright*, supported the right to legal representation for all individuals even if the cost must be borne by the federal government.

argument centered on two important ideas. First, the Fourteenth Amendment, which includes the important phrase "No state shall," prohibits states from engaging in a variety of actions. Second, if provisions in the Bill of Rights are part of the nebulous "due process" concept that also appears in the Fourteenth Amendment, then such rights may be protected against state infringement (because "no state shall" violate "due process"). In a series of cases decided between 1897 and 1970, the Court ruled that almost every provision in the Bill of Rights included "fundamental" rights that were part of the "due process" concept. These provisions were then absorbed or incorporated by the term *due process* in the Fourteenth Amendment and were applied against the states. Table 5.6 describes some significant decisions applying these rights to the states.

TABLE 5.6
SELECTED SUPREME COURT DECISIONS APPLYING THE BILL OF RIGHTS TO THE STATES

PROVISION	CASE
First Amendment	
Establishment of religion	*Everson v. Board of Education* (1947)
Free exercise of religion	*Cantwell v. Connecticut* (1940)
Free speech	*Gitlow v. New York* (1925)
Free press	*Near v. Minnesota* (1931)
Free assembly, petitioning of government	*Dejonge v. Oregon* (1937)
Fourth Amendment	
No unreasonable search and seizure	*Mapp v. Ohio* (1961)
Fifth Amendment	
No double jeopardy (person may not be tried twice for same crime)	*Benton v. Maryland* (1969)
No self-incrimination	*Malloy v. Hogan* (1964)
Just compensation	*Chicago B. & Q. Railroad v. Chicago* (1897)
Sixth Amendment	
Speedy trial	*Klopfer v. North Carolina* (1967)
Public trial	*In re Oliver* (1948)
Jury trial	*Duncan v. Louisiana* (1968)
Confrontation by adverse witness	*Pointer v. Texas* (1965)
Counsel	*Gideon v. Wainwright* (1963)
Eighth Amendment	
No cruel and unusual punishment	*Robinson v. California* (1962)

The constitutional rights applied to the states did not reach *all* people living in the United States. Prior to 1968, Native Americans subject to the jurisdiction of tribal governments were unaffected by the Court's decisions. Native American tribes are not deemed to be states, instead being categorized by the Supreme Court as "domestic dependent nations" and "distinct, independent political communities."[49] In 1896 the Court ruled that the Constitution placed no limits on tribal self-government.[50] These decisions gave tribal governments the authority to resolve intratribal disputes and prohibited individuals from using the federal courts in many cases.

Many years later, after holding hearings that produced testimony regarding abuse of power by some tribal officials, Congress passed the Indian Civil Rights Act of 1968.[51] This legislation applied to tribal governments many provisions from the Bill of Rights, including protections of free speech, free exercise of religion, free press, assembly, and petition. It also included many rights of criminal defendants, such as protections against unreasonable searches and seizures, self-incrimination, and cruel and unusual punishment. Other provisions of the Bill of Rights are not included. For example, because some tribal governments are theocratic (governed by religious principles and laws), the First Amendment's prohibition of religious establishment is not part of the legislation. Some rights included in the legislation are modified versions of those in the Bill of Rights. For instance, requiring that lawyers be provided to all indigent criminal defendants had the potential to bankrupt certain tribes. Consequently, Congress required a right to legal counsel only if a defendant could afford to hire an attorney.

Many Native Americans were critical and resentful of the act, viewing it as the imposition of a foreign value system on tribal life. In particular, as we discuss in Chapter 16, tribal courts operate on the premise that conflict resolution should benefit the entire tribal community rather than punish an individual offender whose rights must be scrupulously protected. According to Vine Deloria and Clifford M. Lytle,

> Tribes are reluctant to abandon their past traditions by placing too much reliance on the whites' legal procedures and practices. While borrowing some Anglo-American notions about the system of justice, tribal courts are struggling to preserve much of the wisdom of their past experiences. Many tribal judges continue to operate as the head of a family might in solving problems. The desired resolution of an intratribal dispute is one that benefits the whole Indian community (family) and not one designed to chastise an individual offender.[52]

Thus, one source of Native American apprehension about the legislation was that it "would go a long way toward transforming them [tribal courts] into dark-skinned replicas of the non-Indian courts."[53]

Equal protection of law As we mentioned earlier, many key terms in the Constitution and its amendments are vague and in need of interpretation. This is clearly the case for the term *equal protection of law*. The Supreme Court's interpretation of the equal protection clause has been complex. The Court, over time, has developed three different levels of review—strict, intermediate, and minimal—to examine laws that allegedly discriminate (see Table 5.7). The Court has also created a hierarchy of groups, ruling that laws affecting only a few of them would receive the Court's most serious level of review. The origin of its general approach may be traced to a 1944 decision, *Korematsu v. United States*,[54] a case involving a challenge to the federal government's policy of placing Japanese Americans in detention camps during World War II (see "Outcomes: *Korematsu v. United States*" on pp. 190–91). In upholding the government's policy, the Court promulgated the following rules:

> [All] legal restrictions which curtail the civil rights of a single racial group are immediately *suspect.* This is not to say that all such restrictions are unconstitutional. It is to say that courts must subject them to the most *rigid scrutiny. Pressing public necessity* may sometimes justify the existence of such restrictions; racial antagonism never can.[55] (italics added)

These rules, of course, required further interpretation and elaboration. They came to mean that some categories in law—such as those of race and national origin—would be considered by the Court as "suspect," presumed to be based on unreasonable group antagonisms or prejudice. As such, these "suspect categories" would be subject to the Court's highest level of scrutiny, "strict" or "rigid" scrutiny. Laws examined in this way could only be justified by a "pressing public necessity," or what the Court later called a "compelling" or "overriding" governmental need.

For example, in the *Brown* decision the Court viewed race-based distinctions in laws affecting public schools as involving a "suspect category" (African Americans) and, therefore, being worthy of the Court's "strict" scrutiny. Seeing no "compelling governmental need" for the laws producing

TABLE 5.7
LEVELS OF REVIEW AND RULES USED BY THE SUPREME COURT TO DETERMINE VIOLATIONS OF THE EQUAL PROTECTION CLAUSE

CLASSIFICATION	LEVEL OF SCRUTINY	QUESTION ASKED BY COURT
Race, national origin	Strict	Does the government have a "compelling need" for classification?
Gender	Intermediate	Does the government have an "important objective"? Is the distinction "substantially related to the objective"?
Other (age, sexual orientation, wealth)	Minimal	Does the government have a "rational basis" for distinction?

segregation in this case, the Court struck them down as violations of the equal protection clause. Since then the Court has struck down many other laws and practices that discriminate on the basis of race, such as those creating segregated recreational facilities, prohibiting interracial marriages, and providing tax exemptions for private schools that discriminate. In addition, the Court, in the case of *Swann v. Charlotte-Mecklenburg Board of Education*, interpreted the equal protection clause—in combination with federal laws—to require that affirmative steps such as busing be employed to achieve racial balance in schools.[56]

In a series of complex, sometimes contradictory, and highly controversial decisions, the Court has permitted colleges, universities, and other organizations to take affirmative action in remedying past discrimination through programs promoting the recruitment and retention of minorities (and, in some cases, women).[57] In *Regents of the University of California v. Bakke*, for example, the Supreme Court ruled that it was legitimate for the University of California Medical School at Davis to take race into consideration in admissions decisions as a way of enhancing diversity in the student population, but that fixed quotas for minority student positions were unacceptable. Although affirmative action programs are permissible under certain conditions, such as when past intentional discrimination can be demonstrated, the Court also has ruled that racial classifications of any sort—even those privileging historically disadvantaged groups—are inherently suspect and call for strict scrutiny. Hence in some situations white males may bring successful "reverse discrimination" lawsuits.

Gender distinctions in law, unlike race, have not been treated by the Court as a "suspect category" requiring "strict" scrutiny. Nor does the Court require the government to show a "compelling need" to uphold the law. Instead, as Table 5.7 shows, the Court uses a second or intermediate level of review, asking whether the law serves an "important governmental objective" and is "substantially related to those ends."[58] The Court, then, requires the government to show less of a need (not "compelling" or "overriding," only "important") to sustain legislation that creates gender distinctions. For example, the Court ruled in *Rostker v. Goldberg* that restricting women from military service served "important governmental objectives" and so upheld discriminatory portions of the Military Selective Service Act.[59] But the Court has also ruled that some gender distinctions, while perhaps serving important objectives, are not "substantially related to those ends." For instance, in *Craig v. Boren* it struck down a law allowing teenage women, but not teenage men, to drink alcoholic beverages, arguing that this distinction was not substantially related to an important governmental objective.[60]

A third, minimal level of Supreme Court review is applied to laws that make other types of distinctions, such as those based on age, wealth, and sexual orientation. Groups falling into such categories—such as the elderly, the poor, and gay men and lesbians—may successfully challenge laws perceived to discriminate against them only if the government cannot show any "rational basis" for the distinction made in law. When the Court employs this minimal level of review, it typically upholds the law or practice in question. In the case of *Bowers v. Hardwick* the Court ruled that Georgia had a "rational basis" to prohibit consensual homosexual behavior because of social taboos against it.

KOREMATSU V. UNITED STATES

During World War II, Japan's bombing of the U.S. naval base at Pearl Harbor on December 7, 1941, incited an intense campaign for the detention of Japanese Americans living on the West Coast. California's leading newspaper, the *Los Angeles Times*, warned on the day after Pearl Harbor was bombed that California was "a zone of danger" and urged "alert, keen-eyed civilians" to collaborate with military personnel "against spies, saboteurs, and fifth columnists [secret sympathizers]." The *Times* raised special concerns regarding Japanese Americans: "We have thousands of Japanese here. . . . Some, perhaps many, are good Americans. What the rest may be we do not know, nor can we take a chance in the light of yesterday's demonstration that treachery and double-dealing are major Japanese weapons." The campaign peaked in February 1942, when President Franklin D. Roosevelt signed an executive order authorizing military commanders to designate military areas from which Japanese Americans might be excluded. General John DeWitt, commanding general of the Western Defense Command, issued orders excluding Japanese Americans from broad areas in the western United States and began evacuating some 120,000 residents.

The evacuees were moved to detention camps, euphemistically referred to as "relocation centers." A new federal administrative agency called the War Relocation Authority (WRA) located 10 of these camps in isolated, semi-desert areas of the country. The camps, with capacities ranging from 8,000 to 20,000 residents, consisted of rows of army barracks surrounded by barbed-wire fences and guard towers staffed by armed military personnel. In less than a year the WRA concluded that deten-

Survivors of the Japanese-American detention camps established during World War II held services in 1992 to mark the fiftieth anniversary of their internment. Shown here is Fred T. Korematsu, the litigant in the Supreme Court decision upholding the use of the detention camps. Korematsu lights candles that represent the various camps located throughout the United States.

tion was harmful for the evacuees and sought to institute a program of loyalty oaths followed by resettlement in nonrestricted areas. Some Japanese Americans refused to sign the loyalty oaths. About 18,000 were isolated in the Tule Lake Segregation Center, where they were labeled "the Disloyal."

One California-born and -educated Japanese American, Fred T. Korematsu, refused to report for "relocation." In contesting his conviction for this offense, he challenged the constitutionality of Roosevelt's policy, charging that it violated

RIGHTS AS RULES IN THE PLAY OF POWER

Many Americans believed at the time of the Constitution's ratification that it required the addition of a bill of rights to protect individual freedoms and restrict governmental power. The first 10 amendments, along with the

the Fourteenth Amendment's equal protection clause. On December 18, 1944, the Supreme Court, in a 6–3 vote, upheld Roosevelt's executive order, concluding that it was valid because "military necessity" justified the action. In a stinging dissent, Justice Frank Murphy wrote that Roosevelt's executive order went "over the brink of constitutional power" into "the ugly abyss of racism." Arguing against the majority's reliance on "military necessity" to justify the policy, Murphy called for racial tolerance and made a plea for cultural pluralism:

> A military judgment based upon such racial and sociological considerations is not entitled to the great weight ordinarily given to judgments based upon strictly military considerations. . . . I dissent, therefore, from this legalization of racism. . . . All residents of this nation are kin in some way by blood and culture to a foreign land. Yet they are primarily and necessarily a part of this new and distinct civilization of the United States. They must accordingly be treated at all times as the heirs of the American experiment and as entitled to all the rights and freedoms granted by the Constitution.

In February 1976, some 34 years after its issuance, the exclusion order was formally revoked by President Gerald Ford. Placing the wartime actions in the context of the constitutional bicentennial celebrations occurring in 1976, Ford insisted that "an honest reckoning" must take account "of our national mistakes as well as our national achievements." Apologizing for the nation, Ford continued:

> We now know what we should have known then—not only was the evacuation wrong, but Japanese Americans were and are loyal Americans. On the battlefield and at home, Japanese Americans—names like Hamada, Mitsumori, Marimoto, Noguchi, Yamasaki, Kido, Munemori, and Miyamura—have been and continue to be written into our history for the sacrifices and contributions they have made to the well-being and security of this, our common Nation.

In 1983, almost 40 years after the Supreme Court decided the case, Fred Korematsu's conviction was finally reversed. In the Civil Liberties Act of 1988, Congress authorized the payment of $20,000 to each surviving detainee, along with formal letters of apology. The payments and letters were meant to compensate for what the Federal Commission on Wartime Relocation and Internment of Civilians called the "grave injustice" motivated by "racial prejudice and war hysteria."

Sources: Peter Irons, *Justice at War: The Story of the Japanese American Internment Cases* (New York: Oxford University Press, 1983); S. Frank Miyamoto, "Japanese in the United States," in Hyung-Chan Kim, ed., *Dictionary of Asian-American History* (New York: Greenwood Press, 1986); Roger Daniels, *Asian America: Chinese and Japanese in the United States since 1850* (Seattle: University of Washington Press, 1988); Katherine Bishop, "Treating the Pain of 1942 Internment," *New York Times*, February 19, 1992, p. A15.

post–Civil War amendments, created a framework that included limits on governmental power over individuals (civil liberties) and individual entitlements to particular benefits and opportunities (civil rights). In its interpretation of constitutional rights, the Supreme Court seeks to specify the relationship between government and citizens, identifying the boundaries of governmental authority as well as its responsibilities to citizens. Thus, constitutional rights play a significant role in American politics because they help identify the contours of governmental power, authority, and responsibility.

Equally important, constitutional rights as interpreted by federal courts offer an important measuring device for citizens who are concerned about the actual operations and actions of government. Does the government operate properly in passing legislation restricting individual liberties? What obligations, if any, does the government have to provide a minimal level of benefits to its citizens? The specification of constitutional rights in constitutional provisions and court interpretations moves rights from the level of abstraction to something tangible that may be invoked by individuals and groups who perceive some form of injustice.

Thus, constitutional rights are meaningful not only because they map the boundaries of governmental power but also because they may be used as resources by citizens seeking to further their cause. In Chapter 6 we assess the impact of rights by exploring the many ways in which they are employed as resources by individuals and groups in the broader play of power.

SUMMARY

Constitutional rights—provisions in constitutional amendments outlining relations between government and citizens—can be seen as rules that create an important part of the context for the play of power in our country. These rules and their interpretation by federal courts provide protections (1) to civil liberties by specifying constraints on government, and (2) to civil rights by outlining the government's responsibility to protect individuals and offer opportunities.

Young children view rights as unchanging and fair to all. As they get older, children gain an appreciation for individual freedoms and come to believe that constitutional rights are essential for their protection. Americans of all ages tend to view the Supreme Court favorably as the nation's repository for fundamental legal and constitutional norms.

Throughout the years the Supreme Court has worked to infuse meaning into constitutional rights in a number of areas. It has provided interpretations of the First Amendment's guarantee of free speech, especially the many exceptions to absolute freedom in this area. Specifically, the Court has created exceptions for political speech constituting a "clear and present danger," obscenity, fighting words, and libel. But the Court has also protected freedom of the press by prohibiting any prior restraint on publication. It has interpreted rules affecting unconventional political behavior, such as protests and demonstrations, and it has provided protection for symbolic speech under many circumstances.

In considering the relationship between governmental activity and religion, it is useful to focus on the Supreme Court's interpretation of the First Amendment's establishment clause and its free exercise clause. Also important are rules contained in the Fourth, Fifth, Sixth, and Eighth Amendments defining the rights of criminal defendants. For example, the police are required to issue "Miranda warnings" to defendants, and an exclusionary rule bars the admission in court of evidence obtained illegally by the police. Important differences can be identified in the interpretation of criminal procedural rights between the liberal Warren Court and the more conservative Burger and Rehnquist Courts; an example is the recent creation of a good-faith exception to the exclusionary rule.

The Court's discovery of a privacy right and its application in the area of reproductive freedom are noteworthy. However, this right has not been applied to the private sexual activities of gay men.

There is broad relevance in the purpose, use, and interpretation of the post–Civil War amendments to the Constitution. In our discussion we have examined the use of the Fourteenth Amendment's due process clause in applying provisions of the first eight amendments to the actions of state governments. We have compared the extension of rights to the states with legislation that applies part of the Bill of Rights to Native American tribal governments. Moreover, we have examined how the Supreme Court's analysis of the Fourteenth Amendment's equal protection clause has evolved, and we have compared its approach to laws distinguishing among people on the basis of race, national origin, gender, and other attributes.

Constitutional rights and their interpretation do indeed provide a context for the play of power. Rights as rules permit citizens to judge the operations of government. As Chapter 6 indicates, rights may also be used as resources in the broader play of power.

A P P L Y I N G   K N O W L E D G E

INTERPRETING CONSTITUTIONAL RIGHTS

In this chapter we have examined how constitutional rights, as interpreted by courts, define the relationship between government and citizens. Here we reproduce portions of an article that appeared in the *New York Times* in 1991. The article describes the arguments of lawyers before the Supreme Court in a case involving the constitutionality of including a prayer at public school graduation ceremonies. After reading the excerpt, analyze how the different participants view the relationship between government, religion, and the citizenry.

WASHINGTON—The question before the Supreme Court today was whether public school graduation ceremonies can include benedictions that refer to God. . . . At issue is an appeal by school officials in Providence, R.I., of a ruling that the inclusion of prayers in a middle-school graduation had the unconstitutional effect of advancing religion.

The [Bush] Administration and the school officials are asking the Court to permit government to sponsor expressions of religious belief as long as no one is coerced into participating. . . . Several justices . . . appeared particularly startled when Charles J. Cooper, representing the school officials, said the Constitution would permit a state to designate an official religion as long as no one was forced to practice the designated faith. . . .

[Bush Administration Solicitor General] Kenneth Starr said he was not urging the Court to overturn the classroom prayer decisions "because of concerns about coercion that are at the heart of religious liberty." Even if a child was free not to join a classroom prayer, he said, "there is a powerful, subtle, indirect pressure in the classroom" that makes organized prayer in that setting unconstitutional.

By contrast, Mr. Starr said, a commencement exercise "is much more in the nature of a celebration." Prayer at graduation, he added, is as constitutionally inoffensive as the blessings offered at Presidential inaugurations. Prayers at public ceremonial occasions . . . are simply "acknowledgements of God and of the role of God in our life as a nation."

[Earlier in the argument,] Justice Anthony Kennedy . . . disputed Mr. Cooper's assertion that because attendance at a graduation ceremony was voluntary, students who were offended by prayers there were free not to attend.

"In our culture, graduation is a key event in a young person's life," Justice Kennedy said. "It is a very substantial burden to say that he or she can elect not to go."

Source: Linda Greenhouse, "Court Appears Skeptical of Argument for Prayer," *New York Times,* November 7, 1991, p. A22. Copyright © 1991 by The New York Times Company. Reprinted by permission. ■

Key Terms

constitutional rights *(p. 164)*	establishment clause *(p. 175)*
civil liberties *(p. 166)*	free exercise clause *(p. 175)*
civil rights *(p. 166)*	exclusionary rule *(p. 179)*
clear and present danger rule *(p. 169)*	Miranda warnings *(p. 179)*
prior restraint *(p. 173)*	good-faith exception *(p. 180)*
libel *(p. 173)*	due process clause *(p. 185)*
symbolic speech *(p. 175)*	equal protection clause *(p. 185)*

Recommended Readings

Abraham, Henry J., and Barbara A. Perry. *Freedom and the Court,* 6th ed. New York: Oxford University Press, 1994. A systematic survey of rules developed by the United States Supreme Court in interpreting constitutional rights.

Cortner, Richard C. *The Supreme Court and the Second Bill of Rights.* Madison: University of Wisconsin Press, 1981. An in-depth study of the process by which the Supreme Court applied the Bill of Rights to the actions of states through the Fourteenth Amendment's due process clause.

Domino, John C. *Civil Rights and Liberties.* New York: HarperCollins, 1994. An excellent survey of the facts, legal issues, and constitutional questions surrounding major Supreme Court rulings, from the Warren Court to the present.

Goldstein, Leslie Friedman. *The Constitutional Rights of Women*, rev. ed. Madison: University of Wisconsin Press, 1988. A casebook in constitutional rights focusing on major decisions of the Supreme Court affecting women.

Irons, Peter. *The Courage of Their Convictions.* New York: Penguin, 1990. Tells the stories of 16 individuals who took their claims of injustice all the way to the Supreme Court.

MacKinnon, Catharine A. *Toward a Feminist Theory of the State.* Cambridge, Mass.: Harvard University Press, 1989. A provocative analysis of constitutional rights from a leading feminist legal scholar. Chapters focus on such issues as pornography, abortion, rape, and sexual equality.

Strossen, Nadine. *Defending Pornography: Free Speech, Sex, and the Fight for Women's Rights.* New York: Scribner, 1995. Strossen, a feminist and president of the American Civil Liberties Union, critiques the feminist anti-pornography movement and argues that the cause of women's equality is best promoted through the abolition of any laws restricting sexually explicit material.

Tushnet, Mark V. *The NAACP's Legal Strategy against Segregated Education, 1925–1950.* Chapel Hill: University of North Carolina Press, 1987. A comprehensive account of the personalities, issues, and strategies in the NAACP's campaign to end school segregation.

CHAPTER 6

Constitutional Rights as Resources

Rights as Political Resources
Participants in the Politics of Rights
Limitations of Rights as Resources
The Politics of Rights

Constitutional Rights as Resources in the Play of Power

WE HAVE EXAMINED how constitutional rights serve as rules that seek to limit and expand the scope of the government created by the Constitution. However, understanding the significance of constitutional rights requires that we look beyond the rules announced in federal court decisions to the ways in which they are used as resources by participants in the broader play of power. Although the historic decision in *Brown v. Board of Education*, the lawsuit discussed in Chapter 5, may have seemed like the end of a long road for Thurgood Marshall and the National Association for the Advancement of Colored People (NAACP), in reality it was only the beginning. Even before the Court announced its decision, some of the problems involved in actually desegregating schools after *Brown* were signaled in the explosive reactions of southern elites to the Eisenhower administration's relatively moderate position in support of slow and gradual desegregation. For example, Arkansas representative E. C. Gathings accused

the administration and its attorney general of "trying to subvert the will of the people." Georgia's Governor Eugene Talmadge went even further, arguing that "radical elements are vying with each other to see who can plunge the dagger deepest into the heart of the South."[1]

Consistent with the position of the Eisenhower administration, the Supreme Court failed to order immediate desegregation of schools. The Court waited another year to address the question of timing; it then issued an ambiguous directive that desegregation should be carried out "with all deliberate speed."[2]

Although *Brown II*, as the directive came to be known, was intended to reassure southern whites that their world would not change overnight, the Court's desegregation decisions shook southern opinion, resulting in swift, open, and absolute defiance.[3] Litigants and other participants in such cases were often harassed and punished. In Davis Station, South Carolina, the local bank terminated litigant Levi Pearson's credit so he

In 1963, Alabama governor George Wallace stood in the doorway of the University of Alabama to block implementation of a desegregation order. Court decisions merely begin the process of guaranteeing rights to citizens. In many cases implementation is a highly contested event.

could not buy fertilizer for his farm, and his white neighbors refused to loan him a harvesting machine as they had in the past, causing his crops to rot in the field. Reverend DeLaine, who had helped instigate Pearson's lawsuit, was forced to leave the county after angry whites burned down his home.[4]

Politicians, community leaders, and newspapers throughout the South damned the Supreme Court, charging it with flagrant abuse of power, usurpation of state legislative powers, and abandonment of settled constitutional principle. In 1956 a group of 101 southern senators and representatives signed and presented to Congress a "Declaration of Constitutional Principles," also known as the "Southern Manifesto," containing a scathing indictment of Brown and the Warren Court. This document portrayed the Brown decision as an exercise of "naked power" with "no legal basis." State political leaders were even more strident in their opposition to Brown. Alabama's Governor George Wallace proclaimed: "I draw the line in the dust and toss the gauntlet before the feet of tyranny and I say segregation now, segregation tomorrow, segregation forever." Mississippi's governor expressed his strong opposition to desegregation by promising that "Ross Barnett will rot in a federal jail before he will let one nigra [sic] cross the sacred threshold of our white schools."[5]

"Massive resistance," as many referred to it, led some southern states to develop legislative strategies to avoid compliance. Louisiana amended its constitution to require that "all elementary and secondary schools . . . shall be operated separately for white and colored children" and mandated that public funds be withheld from any schools violating the state's segregation policy. Virginia passed a resolution of "Interposition and Nullification," calling the Supreme Court's interpretation of the Fourteenth Amendment "a clear and present threat to the several states."[6] Alabama, Arkansas, Florida, Georgia, and Mississippi gave local school districts the power to close the states' public schools and make tuition grants from public funds to white children attending newly created "private" schools.[7]

In 1958 the Court and the federal executive branch struck back against these measures in a case involving desegregation in Little Rock, Arkansas. At issue were the actions of Arkansas governor Orval Faubus in preventing nine black children from entering Little Rock's Central High School. Under the governor's direction, laws enacted to prevent desegregation were to be implemented by the National Guard, called out by Faubus to prevent the nine children from walking through the school's doors. As tensions rose, President Dwight Eisenhower— after a lengthy delay—ordered Army troops to Little Rock to guarantee the nine children's safety and to maintain order.

After effective federal intervention in Little Rock, open defiance to Brown began to diminish. Nonetheless, in 1963 Alabama governor George Wallace activated the National Guard to prevent the implementation of desegregation orders in Mobile, Birmingham, and Macon County. President John Kennedy countered by federalizing the National Guard and declaring Wallace's actions an "obstruction of justice." Two weeks later, five federal district judges in Alabama issued an injunction preventing Wallace from interfering with these schools' operations.

Ten years after Brown, advocates of school desegregation had little to celebrate. In 1964 slightly more than 1 percent of black students in the 11 states of the former Confederacy were attending school with whites. In three states there was absolutely no racial integration in public schools.[8] Consequently, those pursuing racial justice began to lose faith in law and legal processes. Increasingly, blacks made their demands through protests, picketing, demonstrations, and other forms of direct political action. In some cases protesters called for the implementation of court decisions and the realization of constitutional ideals. Martin Luther King Jr. frequently made reference in his speeches to constitutional provisions and Supreme Court decisions that had not been realized. When King urged blacks in Montgomery, Alabama, to boycott city buses to protest racial discrimination, he invoked constitutional rights to inspire mass action. "We are not wrong," said King in December 1955, because "if we are wrong, the Supreme Court of this nation is wrong. If we are wrong, the Constitution of the United States is wrong."[9]

Brown v. Board of Education ruled that segregated public schools were inherently unequal. *Brown II* called for desegregation "with all deliberate speed." But some 40 years after *Brown* the schools in Summerton, South Carolina, remain highly segregated. These photos show students in the virtually all-black Scotts Branch High School and students leaving Clarendon Hall, a whites-only Christian academy established in response to *Brown*.

Impatience with the slow rate of desegregation contributed to demands for federal incentives and sanctions. These demands—expressed in both violent and nonviolent ways—helped spur congressional passage of the Civil Rights Act of 1964. Title IV of the act authorized the federal government to produce studies and report to Congress and the president on the progress of school desegregation, give technical and financial assistance to local school systems seeking to desegregate, and file lawsuits against recalcitrant school districts on behalf of those unable to initiate legal action themselves. Further, Title IV authorized the federal government to terminate financial assistance to any recipient of federal aid engaged in discriminatory practices.[10]

As a consequence of the federal government's clear commitment to desegregation and civil rights for African Americans, and as the Supreme Court continually prodded recalcitrant school districts and encouraged measures such as busing[11] to implement earlier decisions, the pace of desegregation quickened. Yet, even as an increasing number of African-American children in the South attended desegregated schools, Hispanics in the South and Southwest remained largely segregated. For many years the federal government's major school desegregation enforcement agency—the Office of Civil Rights—considered them white for purposes of desegregation.[12] Consequently, Hispanics could remain segregated without arousing federal concern. In addition, local school districts could achieve some measure of "desegregation" by sending black and Hispanic students to the same schools, leaving all-white schools undisturbed.

The Office of Civil Rights changed this policy in 1970, declaring that it would now be concerned with discrimination and segregation on the basis of national origin. This was partly the result of certain Court rulings that Mexican Americans and Puerto Ricans should be considered minority groups for purposes of desegregation.[13] Thus, segregated school districts could no longer legitimately treat Hispanics as whites in order to satisfy desegregation requirements.

At present, southern school districts are the most integrated in the nation, although in most places desegregation efforts have slowed considerably and resegregation appears to be occurring in some districts.[14] The Northeast now has the most segregated schools; this is especially evident in inner-city areas of urban districts. No region of the country has fully desegregated its schools. Nationwide, nearly 70 percent of African-American and 75 percent of Hispanic students continue to attend segregated schools.[15] Even in the South, where governmental pressure to desegregate was strongest, some districts never achieved even a semblance of racial balance. As southern school districts began to desegregate, some whites chose to establish and send their children to all-white private schools.

Throughout this chapter we illustrate the role of constitutional rights in the broader play of power by returning to the story of *Brown* and school desegregation. We emphasize that constitutional rights have a strong symbolic life but often fail to be realized materially. We begin by discussing some common beliefs about rights, focusing on the notion that all citizens possess rights that may be effectively invoked for redress when abridged by government, private institutions, or other citizens.

COMMON BELIEFS

THE MYTH OF RIGHTS

 Imagine the following scenario: In response to a study showing that a majority of American youth hold liberal political values and are likely to vote for national Democratic candidates, a Republican president and Republican-controlled Congress pass legislation prohibiting anyone under the age of 23 from voting. This law, of course, appears to contradict the Constitution's Twenty-Sixth Amendment guaranteeing that federal and state governments will not deny the right to vote to any citizen who is 18 years of age or older.

Many of you who are reading this book would be directly affected by this law. And those of you who are older than 23 can imagine how you would feel if you were among those excluded from voting. How would you react to this law? What would you do? Your reaction might be as follows: "You can't do that! It's unconstitutional! It's against the law!" This expression of outrage and the course of action you would likely follow emerge from a set of common beliefs about rights, law, courts, and litigation that has been labeled the **myth of rights**.[1]

The myth of rights suggests that constitutional rights are possessed by all citizens, regardless of such attributes as class, race, gender, or religion. These rights simply cannot be abridged or denied. If they are violated, victims may hire attorneys, mobilize the law by filing litigation, and obtain a declaration from a court that a right has been denied. It is assumed that the court's declaration will provide an effective remedy for the victim because it is self-implementing. Therefore, in the case of the hypothetical law denying voting rights, a court's decision that the law violates the Twenty-Sixth Amendment would result in the immediate restoration of voting rights to all citizens ages 18 to 23.

Beliefs based on the myth of rights view rights as significant political resources. Although studies show some differences based on class and race in Americans' acceptance of these perceptions,[2] many believe that constitutional rights can be mobilized and employed in litigation to bring about desired change under almost any circumstances. In our own survey of a large introductory American politics class, two-thirds of the students agreed with the statement "When the rights of American citizens are violated, going to court provides an effective way to see that the denied right is restored."

In this chapter we provide an alternative understanding of the role of constitutional rights in the play of power, based on Stuart Scheingold's notion of a **politics of rights**. This understanding sees constitutional rights as political resources of unknown value that are often employed in political struggles. Thus, we do not assume that constitutional rights are always invoked when violations occur, that courts always declare the right and provide an effective remedy, or that rights are self-implementing. Instead, we believe that the utility of rights depends on the circumstances. Rights are sometimes, but not always, mobilized by victimized parties. Court decisions announcing rights sometimes, but not always, provide effective and immediate remedies. And rights are rarely self-implementing; rather, they are employed as part of broader strategies to force implementation. Thus, we contend that, depending on the circumstances, rights may be helpful, of little value, or even harmful to individuals and groups seeking remedies for perceived wrongs.

[1] Stuart A. Scheingold, *The Politics of Rights: Lawyers, Public Policy, and Political Change* (New Haven: Yale University Press, 1974).
[2] After reviewing the literature, Scheingold concludes that African Americans and other historically disadvantaged groups are less likely to accept the central tenets of the myth of rights. Scheingold, *The Politics of Rights*, Chapter 5. ■

RIGHTS AS POLITICAL RESOURCES

In Chapter 5 we analyzed how rights and their interpretation by courts provide rules affecting the play of power. We now turn to the ways in which constitutional rights serve as resources for participants seeking to advance their goals and interests. We look first at the many participants in the politics of rights and then examine limitations on the use of rights as resources. Finally we assess important indirect contributions of rights as resources in the play of power.

PARTICIPANTS IN THE POLITICS OF RIGHTS

Participants who play important roles in the politics of rights will vary depending on the issues involved in particular cases. The police, for example, are likely to become crucial participants in most issues involving criminal justice procedures. Newspaper organizations will likely play an important role in issues touching on freedom of the press. Although the great diversity of rights struggles makes it difficult to generalize about the specific participants involved, the *Brown* school desegregation case shows that contests over legal rights can involve enormous numbers of participants.

The school desegregation case shows that participants in all three governmental arenas (judicial, executive, and legislative) and at all levels (federal, state, and local) are engaged, at times intensely, in the politics of rights. To fully understand the dynamics of this politics, we must look beyond participants who are obviously important, such as the United States Supreme Court or even all federal courts.

When Americans believe their rights have been denied, they often take political action. And when some assert that their rights have been violated, others may feel so threatened that they mobilize themselves politically. Here citizens in Lewiston, Maine, express their support for a referendum that would repeal a gay rights ordinance passed by the city council.

Other important participants in the school desegregation case included state courts, which implemented *Brown* or found ways to circumvent it. Litigants such as Levi Pearson and lawyers such as Thurgood Marshall, who worked for a national interest group, the NAACP, were crucial in bringing the issues to court. (See "Participants: The Mexican American Legal Defense and Educational Fund" for an example of another group that brought civil rights litigation to courts.) Other interested parties—such as the government through the office of the attorney general, and a group of social scientists—contributed legal briefs containing important arguments.[16] The federal executive branch's Office of Civil Rights defined which groups would be considered racial minorities for purposes of desegregation. Members of Congress played varied roles, from the Southerners who criticized *Brown* and incited defiance in the "Southern Manifesto" to those who helped pass the Civil Rights Act of 1964. Presidents Eisenhower and Kennedy issued orders to enforce desegregation mandates, using Army troops and the National Guard.

State and local politicians, community leaders, and newspapers in the South helped spur resistance to desegregation. Local white residents in some communities reacted to *Brown* by building private all-white schools as alternatives to desegregation. Local school boards devised a variety of strategies to either comply with or defy the Supreme Court. Finally, civil rights activists and leaders such as Martin Luther King Jr. engaged in direct political action to press for social change.

THE MEXICAN AMERICAN LEGAL DEFENSE AND EDUCATIONAL FUND

Throughout our nation's history, the Mexican-American community has experienced great difficulty in achieving its goals through conventional political strategies in legislative and executive arenas. Lacking organization and substantial financial resources, Mexican Americans existed on the margins of the play of power.

Although interest group organizations such as the League of United Latin American Citizens (LULAC) had represented the Mexican-American community for years, by the 1960s some community leaders saw a need for an explicit legal strategy such as that used by the NAACP's Legal Defense Fund in its assault on racial segregation in schools. The importance of legal tactics became clear to Pete Tijerina, a lawyer and leader of LULAC, when he brought a tort claim before an all-Anglo jury on behalf of a Mexican-American client.

Learning that his client could not afford to challenge what Tijerina perceived as a discriminatory practice, Tijerina sent another LULAC member to attend a 1967 conference of the NAACP's Legal Defense Fund. Within a year, the Mexican American Legal Defense and Educational Fund (MALDEF) was incorporated, assisted by a $2.2-million Ford Foundation start-up grant.

In its early years MALDEF had numerous problems. Because so few Mexican Americans were attorneys, it had trouble recruiting experienced litigators who understood community issues. When it failed to define its mission clearly, MALDEF was inundated with legal cases, most involving fairly routine issues that could be quickly settled out of court rather than important constitutional issues affecting the entire Mexican American community. In addition, MALDEF had no legal strategy to guide its attorneys' actions and, consequently, lost a number of early cases. The most significant defeat was at the Supreme Court level in the case *San Antonio School District v. Rodriguez* (1973), where MALDEF failed to convince the Court that education was a fundamental right protected by the Fourteenth Amendment. The

Considered the "father" of MALDEF, Pete Tijerina played an integral role in the creation of the defense fund in 1968. Tijerina saw that a legal organization was needed to protect the civil rights of Mexican Americans, and he took the lead in convincing Hispanic leaders that such a group would provide a vital service to their community.

Court ruled that the state of Texas could not be required to subsidize poorer school districts.

In response to these problems, MALDEF changed leadership, reorganized, and focused on bringing a few important cases to appellate courts. In the 1970s it spent much of its time, effort, and financial resources on cases presenting educational issues affecting the Mexican-American community. These changes produced a number of important legal victories, including MALDEF's campaign to challenge school districts' practice of excluding the children of undocumented immigrants unless they paid tuition. After several years of litigation, MALDEF won a major victory in *Plyler v. Doe* (1982). In this case the Supreme Court ruled that children of undocumented immigrants were protected by the Fourteenth Amendment's equal protection clause and that, thus, school districts could not exclude them from enrollment.

Source: Karen O'Connor and Lee Epstein, "A Legal Voice for the Chicano Community: The Activities of the Mexican American Legal Defense and Educational Fund, 1968–82," *Social Science Quarterly* 65 (June 1984), pp. 245–56.

The myth of rights suggests that mobilizing rights in law and obtaining a declaration from a court vindicating a rights claim are effective remedies to individuals and groups experiencing some violation. The Bill of Rights, post–Civil War amendments, and other constitutional amendments are portrayed as crucial resources employed by historically disadvantaged groups seeking to enhance their status and influence in the play of power. Groups lacking access to institutions in other political arenas and lacking appropriate resources, such as money and organizational strength, that translate into influence in such institutions use rights defined in law to make their demands on more accessible judicial institutions.

Judicial institutions are perceived by many as operating under different rules than more explicitly political institutions. Many Americans believe that judicial decisions are not the products of bargaining and compromise among competing interests, as are decisions made in the political arena. Judges are perceived to make more objective, more neutral decisions than do political decisionmakers. Federal judges with life tenure are believed to be relatively independent and insulated from political pressure. Consequently, judicial policies are expected to be influenced less by the relative strength of parties than are policies made in legislative bodies and executive agencies. Discussing judicial policy regarding the poor, one commentator expresses this view: "Courts are designedly insulated from the usual levers of political influence and thus are particularly charged with ensuring that the benefits of the rule of law reach the nation's poor."[17] A myth-of-rights perspective identifies decisions such as *Brown* as evidence of the progressive potential of constitutional rights, law, and litigation.

Many scholars find that the myth-of-rights perspective is simplistic and ignores limitations on effective use of constitutional rights as resources in the play of power. We now consider these limitations, including the restrictive language and practice of rights in law, problems in mobilizing constitutional rights, the political nature of legal institutions and judges, and the limited capacity of courts and judges to implement their decisions. This body of work challenges the view that established legal institutions can be employed fruitfully to produce social change without significant support from other participants in the broader play of power.

The restrictive language and practice of rights in law   Of enormous importance to Americans are liberal values such as individualism, private property, and limited government. These values are clearly reflected in the language of legal claims based on rights and in court opinions declaring rights. How are conflicts involving rights claims framed in litigation and discussed in court opinions announcing violations of rights? They are stated as contests between individuals, even though specific cases may be part of broader political conflicts between groups or social classes. The political implications of framing conflict in this way are significant; it has been argued that "success depends on establishing a personal entitlement and often turns on distinguishing one's cause from others with similar claims."[18] Thus, using strategies that involve making rights claims in courts may have the unintended consequence of "driving a wedge between potential allies."[19]

The language we use to discuss rights may also defuse or depoliticize conflict. Rights claims often become a crucial part of arguments over issues

Despite laws against segregation, in some cities minority students still attend school in separate facilities. Here Hispanic and African-American students share a classroom in a public school in Newark, New Jersey.

that are fundamentally political. Does segregation in public schools violate the rights of Hispanic and African-American schoolchildren? Do laws restricting abortion violate a woman's right to choose? Do court decisions striking down abortion restrictions violate a fetus's right to life? These questions have clear political dimensions. However, when they are framed in terms of rights, the underlying political issues may be obscured. Imagine that you are part of a student coalition protesting your college's or university's decision to raise tuition by 25 percent without consulting student organizations. As you march in front of the main administration building demanding to see the university's financial records, one of the students is arrested and charged with trespass. The rest of the crowd disperses. What will you and other students talk about the next day? How will you frame the issue emerging from the previous day's experience? Quite possibly, your attention will shift away from the general issues affecting all students—how tuition decisions are made, why student input is not solicited, whether the tuition increase is justified—to the issue involving the individual student's "right of free speech."

Some feminist scholars argue that the language of rights is "gendered," reflecting the viewpoints, experiences, and interests of men. Legal scholar Rhonda Copelon traces this phenomenon to the adoption of the Constitution and the Bill of Rights, arguing that both were "written by and for white propertied men." According to Copelon, the gendered nature of rights

> was reflected not only by the fact that only white propertied men were deemed entitled to the rights of citizenship, but also by the nature of the rights themselves. Women were not excluded by,

but rather invisible to, the Constitution makers. . . . The rights chosen to be included in the Bill of Rights protected what was exclusively male activity in the realm of male authority.[20]

In particular, many provisions in the Bill of Rights applied to activity in the public sphere, a sphere of social life closed to women at the time. Freedom of speech, press, and assembly, for example, failed to protect women's activity, which took place, by and large, in the home. Likewise, constitutional provisions protecting private property, such as the Fifth Amendment's just compensation clause ("nor shall private property be taken for public use, without just compensation"), disproportionately benefited men because the husbands of white women had exclusive control over their spouses' property. Consequently, both the language used in talking about rights and the way in which rights are used by and in courts ignore the activities and experiences of women.

Analyzing the language of rights as it applies to equality, Michael McCann suggests that the Supreme Court has confined the Fourteenth Amendment's reach to *individuals* deprived of their rights by *state action*.[21] In other words, racial discrimination is defined as an action inflicted on an individual victim—rather than an entire group or class—by an official "perpetrator."[22] The focus on individual perpetrators and victims diverts attention from the actions of individuals not identified as perpetrators, of institutions, and of society at large. Further, it fails to investigate the possibility that a history of past discrimination continues to adversely affect members of victimized groups. Moreover, this way of thinking about inequality and discrimination leaves the impression that victims may gain redress for wrongs committed against them, so that individuals who are struggling for economic survival have only themselves to blame. In other words, the language we use to discuss equal rights places the root causes of inequality beyond the reach of litigation. McCann writes:

> The result has been not only that the Constitution remains indifferent to the deprivations and needs of the American underclass generally, but that constitutional remedies for past discriminatory acts contributing to racial, ethnic, and other forms of social injustice also have been limited by their refusal to address the structural character of class inequality.[23]

Mobilizing rights in law   A crucial assumption of the myth of rights is that people whose rights have been violated will use the law and legal system to protect themselves—that people who experience an abridgment of their constitutional rights will mobilize those rights by filing a lawsuit or otherwise making a claim. Research on civil litigation and legal mobilization raises questions about this view.

As part of a major study of civil litigation, 1,000 households were surveyed in each of five federal judicial districts: South Carolina, eastern Pennsylvania, eastern Wisconsin, New Mexico, and central California.[24] Participants were asked whether they had experienced any of a variety of civil legal problems in the past three years. If they had, they were asked to give detailed information on how they handled the situation(s). The findings suggested that individuals are likely to take action when they experience a

Booklets like the ones shown here seek to educate citizens about their rights so that they can use the legal system to protect themselves. However, besides a lack of knowledge, there are many other obstacles to asserting one's rights in courts and other arenas.

problem involving such things as landlord-tenant relations, debt, child custody or support, and personal injury caused by an accident. People are much less likely to take remedial action when the problem involves discrimination. Creditors are the most likely to take action, making some kind of claim in nearly 95 percent of the reported disputes. And nearly 600 of 1,000 post-divorce grievances, involving such things as child custody and support, led to the hiring of a lawyer, with about 450 going all the way to court. On the other hand, discrimination victims rarely took any action. Only 29 of 1,000 discrimination grievances resulted in a lawyer's being retained, and only 8 went to court.[25] These findings suggest that rights in law and the Constitution protecting individuals against discrimination are rarely used by victims.

Why aren't rights in law mobilized by victims of discrimination? Why aren't lawyers hired and lawsuits filed in most cases? For one thing, some discrimination victims lack knowledge of their rights and how to make claims when their rights are violated. Some of those who do have the needed knowledge lack the financial resources required to retain an attorney and press their claims in court. Moreover, those who are less well off financially are often less organized, so that fewer organizations exist to provide resources and attorneys to individuals without sufficient personal resources. "Resources: Attorneys and Job Discrimination Cases" (on p. 208) examines the difficulties experienced by working people in retaining private attorneys to litigate job discrimination claims.

In an effort to equalize matters for the poor, the federal government in the mid-1960s established and funded programs to provide them with civil legal assistance.[26] However, research suggests that a variety of conditions

ATTORNEYS AND JOB DISCRIMINATION CASES

Imagine you are a Hispanic woman or an African-American man who has been bypassed for promotion because of your race and/or gender. What would you do? The myth of rights advises you to retain an attorney and gain redress in court. After all, some job discrimination cases are brought by lawyers in the federal Equal Employment Opportunity Commission. But the limited number of attorneys in the agency can file only about 500 cases per year, and they usually have a huge backlog of cases that have not been investigated. In addition, in recent years private attorneys have been rejecting employment discrimination cases as never before. A committee of lawyers and judges appointed by Congress to study the federal court system reported in 1990 that large numbers of "claimants ... find it difficult to litigate in federal court because they cannot find counsel to take their cases." In fact, a survey conducted in 1991 by the National Employment Lawyers Association, a group of 1,000 attorneys representing employees in disputes with employers, found that nearly half of its members rejected more than 90 percent of the job discrimination cases brought to them.

Why are private attorneys reluctant to take many of these cases? First, job discrimination cases are much less lucrative for attorneys than are other types of civil litigation—for example, personal injury cases, which permit courts to award punitive damages. Job discrimination litigation is also very expensive, requiring funds before trial for such things as the hiring of expert witnesses. Moreover, these cases are ex-

tremely time-consuming, often taking years to reach the courts. And the cases are difficult to win. According to many lawyers, after 12 years of Republican administrations that appointed conservative attorneys to the federal bench, federal courts have become increasingly hostile to discrimination claims.

The one exception to attorneys' reluctance to bring such cases to court is in the area of age discrimination. These cases are more lucrative because, under federal law, juries may award monetary damages equal to twice the amount of back pay lost on account of the discrimination. In addition, those experiencing discrimination in this area tend to be white male executives who have lost their jobs in management cutbacks. Some lawyers find these executives to be more "attractive" clients than women or people of color. According to one Washington lawyer,

> Age discrimination is still the white male's preserve. Typically, the plaintiff is a middle- or upper-management employee who has been laid off. They are much more sympathetic figures to juries. More importantly, they make bigger salaries so they can pay the up-front costs of litigation. And if you win and get a back pay award, the amount the lawyer gets is even bigger.

Source: Steven A. Holmes, "Workers Find It Tough Going Filing Lawsuits over Job Bias," *New York Times,* July 24, 1991, p. A1.

prevent poverty attorneys in many programs from mobilizing law against powerful local organizations or filing litigation in a form that may have a broad policy impact benefiting poor people as a class.[27] Many of these programs depend on established individuals, groups, and organizations—such as local bar groups, judges, and social service agencies—for funding, political support, assistance in case processing, and final decisions favorable to their clients, and may be reluctant to mobilize legal rights in ways that might disturb or alienate those providing crucial resources to the program.

In a study of individual responses to racial and gender discrimination, Kristin Bumiller emphasizes the psychological costs that constrain legal mobilization.[28] She sought to discover why victims of racial and gender discrimination rarely file lawsuits or otherwise make claims seeking redress.[29] Table 6.1 summarizes the reasons given by those included in Bumiller's survey for choosing not to complain about perceived discrimination.

In general, Bumiller argues that people's reluctance to invoke the law after experiencing discrimination results from the fact that anti-discrimination law creates a category, "victim," that people don't like to apply to themselves. Many discrimination victims interviewed by Bumiller sought to avoid taking on this identity. Some wanted to avoid feelings of powerlessness associated with the "victim" identity. Some feared that others in the community would consider them "troublemakers" or "crazy zealots." Discrimination victims were also constrained from mobilizing rights for other, more practical reasons, such as fear of being fired from a job or a desire to maintain decent personal relations at work. Consequently, Bumiller found that many

TABLE 6.1
REASONS FOR NOT COMPLAINING ABOUT PERCEIVED DISCRIMINATION

RESPONSE TO "WHY DIDN'T YOU COMPLAIN?"	NO. OF RESPONSES	% OF RESPONDENTS
Would do no good/Final result the same/Way of life	35	19.9%
Not worth it/Not a major problem	18	10.2
Need immediate resolution	17	9.7
Gave up/Don't know/I'm easygoing	17	9.7
Fear of retaliation, image of "troublemaker"	16	9.1
Don't know whom to complain to	14	8.0
Could not prove/No evidence	13	7.4
Avoid dealing with them in future	12	6.8
Don't want to cause trouble	9	5.1
Excessive time or cost	9	5.1
Rules cannot be changed	8	4.5
Other	8	4.5

Source: Adapted from Kristin Bumiller, *The Civil Rights Society: The Social Construction of Victims* (Baltimore: Johns Hopkins University Press, 1988), p. 27.

Women in predominantly male workplaces may face discrimination or harassment. However, many pressures on the victims of such behavior keep them from filing legal claims or complaining publicly about their experience.

discrimination victims justified the actions of those who discriminated against them. Perpetrators "didn't mean it" or had "simply made an administrative decision." Some were fatalistic, saying that "discrimination is part of life." Bumiller's research challenges the view in the myth of rights that an even playing field exists for victims and perpetrators of discrimination. Although the designation of rights in anti-discrimination law supports the view that anyone may use the law to protect and promote his or her interests, differences in power relations and social status limit the mobilization of rights by those who are most dependent on them.

The political nature of courts and judges   The myth of rights suggests that courts play the role of protecting individual rights against infringement by tyrannical majorities. Courts are well equipped to play this role because judicial decisions are thought to be more objective and neutral than decisions made by political officials in political institutions. Federal judges are electorally unaccountable, being insulated by their life tenure from political pressures.

These views are modified significantly by scholars who investigate relations between federal courts, especially the Supreme Court, and other political branches. In an influential study, Robert Dahl suggested that the Supreme Court, far from being the guarantor of minority rights, served primarily to legitimize policies developed by Congress and the president.[30] Examining the Court's use of judicial review through the early 1950s, Dahl concluded that the Court, for most of its history, simply joined what he called the "law-making majority" and helped it to govern. Although the Court might occasionally issue a decision that challenged prevailing political opinion[31] and thereby delay the will of political leaders and institutions, "the policy views dominant on the Court will never be out of line for very long with the policy views dominant among law-making majorities of the United States."[32]

In a more recent study, Gerald Rosenberg provides further evidence of the Court's propensity to be influenced by prevailing majority sentiment.[33] He shows that during periods when Congress expressed its opposition to the Court by introducing legislation to reverse or alter decisions, the Court typically has backed away from its commitments. For example, Rosenberg argues that the political pressures emanating from white southern congressmen after *Brown* led the Court to distance itself from the desegregation issue. After the "Southern Manifesto" signaled the beginning of "massive resistance" in the South, "the Court heeded these attacks by avoiding major civil rights decisions until well into the 1960s."[34] In terms of education, according to Rosenberg, the Court "did not issue a full opinion after *Brown* until the Little Rock crisis in 1958. After the crisis passed, there was silence again until 1963."[35] Rosenberg concludes:

> At times of congressional ire with the Court, the Court succumbs to congressional preferences. Depending on the Court, then, to defend unpopular minorities or opinion against political hostility is misplaced. The very American notion of the courts as the "ultimate guardians of our fundamental rights" may be dangerously wrong.[36]

Implementing rights in legal decisions   A large body of research challenges the view in the myth of rights that court decisions that declare rights are self-implementing or that one may expect decisions to take effect routinely and without controversy.[37] In fact, legal decisions are not self-executing and almost always require others to act. Courts lack powerful tools to enforce their policies, and their decisions are often ineffective in the face of opposition. In other words, court decisions enter a play of power where those whose interests are harmed by judicial policies are likely to resist.

The *Brown* decision and its aftermath provide what is perhaps the best illustration of the problems that courts experience in policy implementation. As we described earlier in this chapter, the "Southern Manifesto" expressed the views and intentions of many political officials in the South:

> The unwarranted decision of the Supreme Court in the public school cases is now bearing the fruit always produced when men substitute naked power for established law. . . . We pledge ourselves to use all lawful means to bring about a reversal of this decision which is contrary to the Constitution and to prevent the use of force in its implementation.[38]

Although southern segregationists did not succeed in reversing the *Brown* decision in the courts, they managed to ensure that, for more than 10 years, very few African-American children attended public schools with whites. In "Outcomes: Upholding Native American Fishing Rights" (on p. 212) we describe how political opposition to a court decision outlining fishing rights for Native American tribes presented a significant obstacle for social and economic change benefiting another historically disadvantaged group.

Limitations on rights as political resources—limitations in the language and practice of rights, problems in legal mobilization, the political nature of legal institutions and judges, and the inability of courts to implement

UPHOLDING NATIVE AMERICAN FISHING RIGHTS

The politics of rights suggests that legal victories do not necessarily produce large tangible benefits, especially for disadvantaged groups. This principle is illustrated in the case of Native American fishing rights in the Northwest.

Native Americans in Washington State have been in court for decades seeking to uphold fishing provisions in their treaties. In 1974, Federal District Judge George H. Boldt handed down one of the most controversial court decisions in Washington State history, recognizing Native American fishing rights guaranteed in treaties negotiated between the United States and several tribes in 1854 and 1855.

Judge Boldt interpreted one crucial sentence—"The right of taking fish, at all usual and accustomed grounds and stations, is further secured to said Indians, in common with all citizens of the territory"—to mean that Native American tribes could harvest up to one-half of all the salmon in Washington's waters. The decision was affirmed by the United States Supreme Court.

Reaction to Judge Boldt's decision was swift and loud. A group opposed to Native American treaty rights organized quickly and initiated a campaign to impeach Judge Boldt. Bumper stickers began appearing around the state with the message "Let's give 50 percent of Judge Boldt to the Indians." A bomb exploded in the Federal Building in Tacoma. Native Americans attempting to fish for salmon faced harassment, threats of bodily harm, and guns fired into the air.

Siding with the opposition forces, the state of Washington filed litigation, arguing that courts should not order specific allocations of the salmon resource, leaving this decision to the free market. State administrative agencies wrote regulations that sought to restrict Native American fishing. In Congress, a number of treaty modifications, or abrogation bills, were introduced by Washington State senators and representatives with ties to non–Native American fishing interests. One such bill passed. According to a study conducted several years after Judge Boldt's decision, resistance to the court's declaration of

Disagreements over whether Native Americans have the right to hunt and fish on their reservations without following state wildlife conservation rules have fueled many protests. In 1966, Native American fishermen asserted their tribal rights by staging a "fish-in" on the Nisqually River in Olympia, Washington.

tribal fishing rights resulted in an increase in the tribe's share of the salmon catch from 2 percent of the total to only about 17 percent.

Courts in the Northwest continue to issue judgments that appear to protect Native American treaty rights. In a recent decision, a federal court extended Judge Boldt's ruling on salmon to cover shellfish. As with the earlier decision, reaction from opponents was swift and explosive. Owners of property in Washington, concerned that Native Americans would trespass on their property, threatened legal action. The shellfish industry appealed the decision and suggested that it would try to overturn Judge Boldt's original decision. Opponents have openly threatened to harass any Native American who harvests shellfish. According to a representative of the shellfish industry, "If they go ahead and hit the beaches, there's probably going to be more people shooting over their heads and other problems."

Sources: Rita Bruun, "The Boldt Decision: Legal Victory, Political Defeat," *Law and Policy* 4 (July 1982), pp. 271–98; Timothy Egan, "Indians of Puget Sound Get Rights to Shellfish," *New York Times,* January 27, 1995, p. A12.

their decisions—lead some commentators to argue that litigation invoking constitutional rights should rarely, if ever, be considered as a strategy in the broader play of power. To them, the potential costs of litigation in financial resources, time, energy, and talent outweigh the meager benefits that can be expected from winning in court. Activists' resources are diverted to institutions that can do very little to further their cause and may ultimately weaken efforts at using other, more effective strategies, such as lobbying and strategic voting.[39]

THE POLITICS OF RIGHTS

Because compelling evidence challenges the notion that judicially declared rights directly produce significant social change, a subtler understanding of the role of rights as resources has evolved from Scheingold's notion of the politics of rights. Rather than treating judicially created rights as accomplished social facts and assuming that these declarations translate directly into policy and behavior, this perspective views rights as "authoritatively articulated policy goals" that have an uncertain connection to change. Although both the myth of rights and the politics of rights see rights as resources, the politics of rights views rights "as political resources of unknown value in the hands of those who want to alter the course of public policy."[40]

From this alternative perspective, rights may have a variety of indirect effects on social movements, movement activists, and members of historically disadvantaged groups. Specifically, rights may have beneficial or harmful psychological effects on members of social movements, other interested parties, and disadvantaged groups; may help to set and clarify an agenda for the social movement and the nation; may contribute to political mobilization; and may be employed as political leverage in bargaining situations.

Psychological effects   We have acknowledged the respect and reverence that Americans have for law, courts, and the Constitution. Because Americans attach such symbolic significance to the law and courts, the successful assertion of rights can have a variety of beneficial psychological effects, especially for members of disadvantaged groups. Asserting one's rights and winning a declaration of rights from a court can give one the feeling of belonging to the community or of being a full-fledged citizen of the country. Successful assertion of rights may also empower individuals, who now come to believe that they are capable of taking effective action on their own behalf.[41] Moreover, assertion of rights that results in judicial confirmation of rights violations may enhance self-esteem for members of disadvantaged groups, a phenomenon suggested in the following discussion by African-American legal scholar Patricia Williams:

> For the historically disempowered, the conferring of rights is symbolic of all the denied aspects of humanity: Rights imply a respect which places one within the referential range of self and others, which elevates one's status from human body to social being. For blacks, then, the attainment of rights signifies the due, the respectful behavior, the collective responsibility properly owed by a society to one of its own.[42]

Asserting rights can increase feelings of solidarity among members of diverse social movements and disadvantaged groups. In this protest a variety of interest groups, including organized labor, civil rights groups, and women, come together for a common purpose.

The successful assertion of rights may also contribute to feelings of collective identity for members of social movements and disadvantaged groups. Highly visible lawsuits filed on behalf of women, African Americans, Native Americans, Hispanics, gay men and lesbians, and the disabled may encourage individuals to consider how their values and interests are shared by broader groups and movements. One writer about the impact of rights on people of color suggests that "rights serve as a rallying point and bring us closer together."[43]

The beneficial psychological effects of successfully declared rights may be seen in the comments of those who participated in the civil rights movement. An attorney for the NAACP's Legal Defense Fund assessed the psychological impact of *Brown* as follows: "I think the greatest impact of the *Brown* decision was on the black community itself. It was a statement to the black community that they had a friend, so to speak, the Supreme Court. And so it emboldened the communities of blacks around the country to move forward to secure their own rights."[44] The reflections of an activist with the Albany Movement of the Freedom Riders emphasize the hopes created by *Brown:* "I remember when the 1954 Supreme Court decision came, my father saying, 'Now that's the supreme law of the land.' Like, the Supreme Court, that's it. I remember him reading from it in the house, and it being a really high time."[45] A member of the Women's Political Council of Montgomery, Alabama—a group that participated in a boycott of the city's segregated buses—commented that she and other African Americans felt as if they had won self-respect, were no longer "aliens" in their own city, and began to believe that "America was our country too," after the Supreme Court declared segregation in Montgomery buses unconstitutional.[46]

Of course, court decisions declaring rights may also have detrimental psychological effects on losing parties and those who identify with them.

When litigation involves group conflict—as it often does when constitutional rights are invoked—the good feelings and reassurance produced in members of winning groups are matched by the threat felt by those on the losing side. The quotations in the opening pages of this chapter from Alabama's governor George Wallace, Mississippi's Ross Barnett, and the "Southern Manifesto" convey the enormous psychological threat that desegregation constituted for many white Southerners.

Publicity and agenda setting   Rights and litigation may be used by individuals, groups, and lawyers to publicize their cause. In some instances, increasing a case's or cause's public visibility is meant to "expand the scope of conflict" in an effort to attract allies and coalition partners.[47] Regardless of the eventual outcome in court, news stories about the case have the potential to increase public support for the cause and financial contributions to the group, movement, organization, or law firm responsible for bringing the action.[48] (See "Strategies: Creation Science in the Courts" on p. 216.)

Publicity also may be used to help force a defendant to make concessions. For example, one group of public interest lawyers brought an employment discrimination lawsuit against Pacific Telephone and Telegraph Company, demanding 10,000 additional jobs for Hispanics.[49] Seeking publicity for their cause, the lawyers challenged the company's license renewals before the Federal Communications Commission and filed for a rate reduction before the state public utilities commission, rather than taking the conventional route of filing a complaint with the Fair Employment Commission or the Equal Employment Opportunity Commission. The lawyers expected the route they took to provide the maximum amount of publicity for the issues involved in the case. As the media focused attention on the case, the lawyers conducted a massive publicity campaign against the company. Pacific Telephone and Telegraph responded quickly, entering into an affirmative action agreement specifying goals and timetables for hiring additional Hispanic workers.

Even if media attention does not result in immediate concessions, the publicity surrounding legal action may place new issues on the public's agenda. One study of **comparable worth** litigation—legal action demanding wages for women equal to those earned by men in comparable jobs—highlights the importance of court decisions for placing wage equity on the national agenda.[50] According to the study, media coverage of the comparable worth issue increased dramatically after visible federal court decisions in the early 1980s.[51] About one-third of the articles featured in three national newspapers and several weekly newsmagazines studied focused exclusively on lawsuits and court decisions, and over two-thirds spent considerable time discussing legal cases. Many activists interviewed for the study believed that court decisions had important effects on public debate. One activist, for example, believed that the Supreme Court's treatment of comparable worth in *County of Washington v. Gunther*—ruling that female prison guards could not be paid less than male prison guards, even if they did not do exactly the same work—changed people's reaction toward the general idea substantially. The activist argued:

Suddenly an obscure concept, understood by few and belittled by many, became an issue with content and weight. Women and

CREATION SCIENCE IN THE COURTS

A variety of legal and political strategies are available to groups and movements using constitutional rights as resources in the play of power. The specific strategy used depends on such variables as the group's purposes and resources. Groups with enough financial resources, highly skilled lawyers, and support from a network of other sympathetic groups and interests may effectively use a test case strategy, carefully selecting issues, cases, and litigants to develop constitutional precedents supporting their cause. Organizations lacking such resources may inject their views into disputes in progress by using a strategy of writing briefs to accompany cases in federal courts. In the initial stages, when the movement or group desperately needs members and financial resources, it may deliberately bring attention to itself by filing litigation that generates publicity. The case of creation science groups—organized primarily in the 1960s and 1970s—shows how strategies may evolve as groups accumulate resources and mature.

Ever since the turn of the century, Darwin's theory of evolution (which asserts that humans evolved from animals) has been the subject of concern among some religious groups. Groups such as the World's Christian Fundamental Association formed in the early twentieth century to seek legislation banning the teaching of evolution in public school classrooms. Such groups successfully pressured for restrictive legislation in several southern "Bible belt" states. Tennessee's law banning the teaching of evolution was challenged by the American Civil Liberties Union in the famous "Scopes monkey trial" case of 1927, which involved the prosecution of a biology teacher, John Scopes, for teaching evolution. A state supreme court upheld the statute. As time went on, the movement faded, although the laws it helped pass often remained on the books.

The issue resurfaced in the 1960s and 1970s as biology texts increasingly incorporated evolution as a central idea. Consequently, several organizations—such as the Creation Research Society, the Institute for Creation Science, the Creation Science Research Center, and the Bible Science Association—were formed for the purpose of elevating the biblical version of creation to the status of a science so that it might be taught along with evolution. Lawyers for the groups, many of them volunteers, began to argue that teaching evolution but not creation science violated the First Amendment. During the 1970s these groups launched an intense legal and political campaign to promote a policy of "balanced treatment."

In their infancy, when they lacked organizational and financial resources, the groups' major strategy was to file litigation primarily for the sake of publicity. Unlike in the litigation campaigns of more established groups, the cases, issues, and litigants were not chosen strategically. Indeed, most of these cases were dismissed at early stages. However, the groups used the cases to address publicly issues of concern to them, portraying their struggle as one between "good" and "evil." The strategy worked well, because the litigation and surrounding publicity attracted many new members and financial contributions. The groups' increased strength allowed them to effectively pressure local school boards, textbook committees, and state education bureaucracies to enact "balanced treatment" policies. Although the movement ultimately lost a major Supreme Court battle in 1987 (in *Edwards v. Aguillard*), it is now powerful enough to carry out multiple strategies in a variety of political arenas.

Source: Wayne V. McIntosh, "Litigating Scientific Creationism, or 'Scopes' II, III . . . ," *Law and Policy* 7 (July 1985), pp. 375–94.

unions awakened to its possibilities. . . . The Supreme Court's decision in *Gunther* strengthened the significance of comparable worth. . . . The issue could no longer be disparagingly dismissed.[52]

Thus, the idea of comparable worth reached the public agenda and became a part of the national conversation only after legal reformers filed litigation and courts rendered opinions.

The assertion of rights, and court decisions announcing rights, may also assist the development of social and political movements by clarifying and focusing the movements' agendas. One writer suggests that agenda building around particular rights claims may emerge in various ways—by emphasizing conflicts between judicially declared rights and practices violating those rights, by exploiting contradictions between settled understandings of different rights, or by extending the logic and applicability of court-announced rights.[53]

The case of *State v. Wanrow*[54] is a good illustration of the relationship between rights and movement goals. In *Wanrow* a jury convicted a Native American woman of second-degree murder for shooting a white man who she believed had tried to molest her child. Although the woman claimed to know that the man had a history of child molestation and had previously attempted to molest her child, the trial court instructed the jury to consider only the circumstances "at or immediately before the killing" and to apply the "equal force" rule for self-defense—a rule prohibiting the use of force greater than the assailant used—to the woman's claim that she acted in self-defense. After reading the trial transcript, the woman's attorney team realized that the judge's instructions prevented the jury from considering the woman's state of mind "as shaped by her experiences and perspective as a Native American woman when she confronted this man."[55] The jury had no information on the lack of police protection in such circumstances, the woman's knowledge of the man's history of child molestation, or the woman's belief that she could only defend her child and herself with a weapon. The "equal force" rule, moreover, prevented the jury from evaluating her claim that she acted in self-defense.

In court, the lawyers challenged the law of self-defense as being biased by gender. They argued for women's "equal right to trial," a right requiring a new self-defense rule. The Washington Supreme Court agreed that the rule was gender biased, and it announced a new rule based on the individual defendant's "perceptions." According to this court, "The respondent was entitled to have the jury consider her actions in the light of her own perceptions of the situation, including those perceptions which were the product of our nation's 'long and unfortunate history of sex discrimination.' "[56]

Other lawyers in other courts with different cases began to make arguments similar to that used by the Washington court. Lawyers making rights claims and courts listening to such arguments began to consider how women's experience shaped their actions in circumstances such as women murdering men following a sexual assault.[57] In this way the judicial determination in *Wanrow* clarified and focused the legal and political agendas of women's groups concerned with violence against women and with the treatment of women by the criminal justice system.

Political mobilization   A crucial element of the politics of rights concerns the relationship between rights and political mobilization. According to the politics of rights, rights announced by courts may persuade members of a normally quiescent population of their cause's merit and move them to join with others in political activity. Movement leaders and organizations may use the symbolism of rights—manifested in our reverence for rights and law—to mobilize political support for implementation and for broader political goals.

Comparable worth litigation, for example, was followed by political organization around the issue by unions, feminist groups, and state agencies.[58] According to Michael McCann,

> strategically timed and highly publicized, such actions effectively generated widespread attention to the issues, raised expectations among women workers, and became the initial focus of organizing efforts within unions, feminist organizations, and other allied groups. This has included activating many thousands of previously organized supporters around the wage equity issue as well as providing a catalyst to new union local affiliates and formation of women's groups.[59]

Political mobilization followed wage equity litigation, despite the fact that legal victories were few and far between and none conferred judicial legitimacy on the concept of comparable worth, as many activists desired.

In recent years, decisions on the siting of hazardous waste facilities have been framed as civil rights issues. Using statistics showing that such facilities are placed disproportionately in minority communities, the Environmental Protection Agency's Office of Civil Rights has used the Civil Rights Act of 1964 to challenge specific siting decisions. In this photo citizens protest a plan to place a hazardous waste recycling plant in St. Gabriel, Louisiana, a minority community.

Some evidence exists that the *Brown* decision had similar effects on political mobilization.[60] Looking back at this period, Bayard Rustin, leader of the Congress of Racial Equality, remarked:

> When the Supreme Court came out with the *Brown* decision in '54 things began rapidly to move. Some of us had been sitting down in the front of these buses for years, but nothing happened. What made '54 so unusual was that the Supreme Court in the *Brown* decision established black people as being citizens with all the rights of all other citizens. Once that happened, it was very easy for militance, which had been building up, to express itself in the Montgomery bus boycott of '55–'56.[61]

Historian Aldon Morris suggests that *Brown* had a tremendous impact in helping the NAACP to organize the African-American community. This was significant, because local NAACP leaders were responsible for articulating black resistance to segregation. According to Morris, "The winning of the 1954 decision was the kind of victory the organization needed to rally the black masses behind its program; by appealing to blacks' widespread desire to enroll their children in the better equipped white schools, it reached into black homes and had meaning for people's personal lives."[62] *Brown*'s symbolic importance and its utility for political mobilization is also evidenced in the repeated timing of marches and demonstrations to coincide with the anniversary of the Supreme Court's landmark decision.[63] Although *Brown* was not implemented for over 10 years, the gap between the decision's promise and the reality of continued segregation encouraged increased political mobilization and agitation.[64]

Bargaining leverage　The politics of rights suggests that rights may be used by individuals, groups, and social movements as bargaining resources in political negotiations. "Voluntary" settlements often result from successful litigation or anticipated reactions accompanying threats of litigation. The wage equity activists interviewed by McCann ranked legal action as one of the two most effective resources for negotiating policy change. Leverage was often produced simply by filing litigation. According to McCann, "what is striking is that the overwhelming percentage of legal leveraging actions never reached the trial stage. In most cases, simply filing formal charges achieved the desired result; in a few cases, mere verbal threats of a lawsuit broke negotiation deadlocks."[65]

Drawbacks to a politics of rights　Although rights at times contribute indirectly to social change, there are drawbacks for movements and groups pursuing a politics of rights.[66] For instance, feelings of empowerment and effective political mobilization depend on the acceptance of at least a portion of the myth of rights. In particular, people must accept the notion that judicially declared rights can be redeemed. But acceptance of the myth of rights varies by social class.[67] Middle-class and working-class Americans accept the myth of rights to a greater extent than do the poor. Therefore, the potential for rights to have positive psychological effects and spark political mobilization is greatest for such middle-class groups as consumers, environmentalists, right-to-life and pro-choice advocates, the disabled, and feminists, rather than groups such as the homeless. It may be the case that in the area of civil rights *Brown* and other court victories have had the greatest impact on those African Americans who, being upwardly mobile, have been most empowered by them.[68]

The outcomes of a politics of rights are uncertain as well, because rights may be used as resources by anyone interested in the issues raised by the litigation. In civil rights, *Brown* and other court decisions provided so serious a threat to many white Southerners that organizations formed to resist desegregation through lobbying and outright defiance. Gerald Rosenberg found tremendous growth after *Brown* in organizations such as the White Citizens Councils and the Ku Klux Klan.[69] In a similar way, the right-to-life movement expanded rapidly after *Roe* declared that a woman's right to privacy prevented many restrictions on abortions. Pro-life groups were able to lobby many state legislatures effectively for restrictions on *Roe*, leading to an eventual retreat from *Roe* by a more conservative Supreme Court. Conversely, following the Court's decision in *Webster v. Reproductive Health Services*—a decision signaling the Court's retreat from *Roe*—pro-choice forces seemed to become reinvigorated.

CONSTITUTIONAL RIGHTS AS RESOURCES IN THE PLAY OF POWER

Because Americans revere the Constitution, rights emanating from it play an important symbolic role in the play of power. American politics is expected to be conducted according to rules outlined in the Constitution. Many of these rules appear in the form of individual rights. When participants in the play of power violate the rules, these rights may be mobilized, litigation may be filed, and adjustments may be made that are consistent with court mandates. The view that rights eventually determine how the play of power operates constitutes a major source of the game's legitimacy—the belief on the part of many Americans that rights guarantee a political system that is fair and just. As Stuart Scheingold argues, "the American mainstream has increasingly come to think about the rights guaranteed by the Constitution as a repository of values that—either as they stand or as they may be perfected in response to changing conditions—provide the moral basis and the political resources to build a just social order."[70]

In addition to serving as rules that lend legitimacy to the game, rights serve as resources for participants pursuing substantive political goals. Rights are used in political struggle because of their important place in American culture. Consequently, rights as interpreted and declared by courts become an arena of conflict among competing groups. Judicially declared rights are one of many different types of resources and constraints that shape the terms of political struggles. Although all citizens may use rights as resources, differently situated individuals and groups have grossly unequal capabilities to use legal resources to further their interests. Like patterns of inequality we describe in other elements of the play of power, inequalities in the use of rights as resources fall along lines of class, race, and gender. Those with an abundance of other resources, such as money, can more easily mobilize rights, hire attorneys to press their claims in courts, or use rights in making demands in legislative and executive arenas. And for those seeking to change dominant views and perspectives, rights—steeped as they are in mainstream American values—are likely to be a constraint.

Few people, however, view rights as constraints. Indeed, rights increasingly define the terms of political debate in the United States. Think about how we discuss abortion policy in the United States. One side asserts a woman's right to privacy and choice, and the other emphasizes the unborn child's right to life. The translation of political and policy questions into a language of rights makes it difficult, if not impossible, to define, discuss, and rationally analyze significant social problems.[71] Moreover, political debates conducted in rights language restrict opportunities for the kind of ongoing political dialogue that is necessary in a democratic society. Once the abortion debate is defined in terms of rights that are in conflict, there is little space for the debate to proceed. Such debates present obstacles to political compromise, the identification of common ground, and the recognition that responsibilities to the community exist simultaneously with individual rights.

SUMMARY

Constitutional rights can be used as resources in the play of power. Throughout this chapter we have compared preconceived notions about the role of rights in American politics—the myth of rights—with an alternative perspective, the politics of rights.

The myth of rights suggests that all citizens have rights that may be invoked and used to gain redress when abridged by the government, private institutions, or other citizens. Rights are viewed as self-implementing, becoming immediately effective after a declaration from a court.

The alternative, based on the notion of a politics of rights, suggests that rights are resources of unknown value that are used by opposing sides in political battles. To understand the unpredictability of using rights to further political causes, one must consider limitations in the language used to discuss rights, constraints on political mobilization, and the politics surrounding court decisions and implementation. The usefulness of rights as political resources seems to depend in large part on the distribution of political power and the configuration of political interests that exist in specific cases.

Rights may also be used as effective resources in political struggles, even if their value cannot be specified with any degree of certainty. Rights may be used to set the national agenda, help clarify a social movement's agenda, assist in political mobilization, or provide some leverage in political bargaining situations, as in the case of the comparable worth issue. Judicially declared rights may provide important psychological benefits to members of social movements and historically disadvantaged groups who win in court.

Thus, constitutional rights occupy an important place in the play of power, possessing symbolic capital desired by contending groups pursuing different policy goals. The role of rights in conferring legitimacy on the political system and as resources that can be used indirectly to change the terms of debate, build social movements, and bring about the distribution of political resources makes their definition, interpretation, and use highly contested aspects of the play of power.

APPLYING KNOWLEDGE

CONSTITUTIONAL RIGHTS AND POLITICAL MOBILIZATION

 Here we reproduce an advertisement by the American Civil Liberties Union that appeared in the *New York Times* on January 19, 1992. What links between constitutional rights and political mobilization are illustrated by the advertisement?

ROE V. WADE
1973-1992?

This is not a death announcement.
It is a birth announcement.

The birth of a bold new initiative to guarantee the right of all women to decide for themselves whether or when to bear a child – without the interference of government.

Today, instead of mourning the imminent demise of *Roe v. Wade*, the landmark Supreme Court decision that affirmed this right 19 years ago this week, we call on all Americans to join us in our new and vital struggle.

A hostile Court

For the past few years, an increasingly hostile Supreme Court has steadily chipped away at reproductive rights. Doctors working at federally-assisted family planning clinics have been barred from even discussing abortion. In many states, severe restrictions have been imposed, and some states have even banned certain forms of birth control. At this moment, there are four major cases pending in the courts, any of which could turn out to be the vehicle for overturning *Roe v. Wade* entirely. One case is before the Court right now.

But even though the odds appear to be against us in the courts, we can still win.

A new strategy

Today, we launch a national campaign of historic importance. We hereby call for an Act of Congress – whether by federal statute or if necessary a Constitutional amendment – to explicitly guarantee the right to reproductive choice, leaving no room for restrictive interpretation by a hostile Court.

When women finally won the right to vote in 1920, it was not because the Supreme Court gave it to them. They fought for it – and won it – in the streets, in state legislatures, in the Congress, and finally through a Constitutional amendment. Reproductive rights must now be won the same way – by mobilizing public support and forcing political change.

We begin by calling for the swift passage of the Freedom of Choice Act, now pending in Congress, without any restrictions or weakening amendments. This Act would guarantee the rights that have been protected by *Roe v. Wade*. And we will not end our fight until the right to choose is secure once and for all.

How you can help

First, write to your Representative and Senators urging their unqualified support of the Freedom of Choice Act.

Second, please join our campaign. Check the box below and we'll send you a campaign kit containing detailed information about how you can actively participate.

And we urgently need your financial support. For 20 years, the ACLU has handled and won most of the major court cases involving the right to choose. We have also been a powerful and effective lobbying force, not only in Washington but in every state in the country. All of this is tremendously expensive, and we urgently need the help of every American who believes in this fundamental human right.

Congress will not listen unless they hear your voice – loudly and clearly. Over and over again, if necessary. As the women's suffrage movement did not rest, neither will we.

The Supreme Court will act at any moment. It's up to you to act first.

American Civil Liberties Union
132 West 43 Street, Dept. RW, New York, NY 10036
❏ I want to help. Enclosed is my tax-deductible contribution of:
❏ $25 ❏ $100 ❏ $1000 ❏ Other
❏ Please send me a campaign kit with information about how I can participate.
Please make check payable to: "ACLU Foundation" to help support our litigation and public education activities.
Name
Address
City, State, Zip
Telephone
ACLU Nadine Strossen, President; Ira Glasser, Executive Director

Key Terms

myth of rights (*p. 200*)

politics of rights (*p. 201*)

comparable worth (*p. 215*)

Recommended Readings

Bumiller, Kristin. *The Civil Rights Society: The Social Construction of Victims.* Baltimore: Johns Hopkins University Press, 1988. An examination of why so many people who experience discrimination are reluctant to complain about it.

Glendon, Mary Ann. *Rights Talk: The Impoverishment of Political Discourse.* New York: Free Press, 1991. A provocative analysis of the many ways in which rights consciousness adversely affects political debate in the United States.

Handler, Joel F. *Social Movements and the Legal System.* New York: Academic Press, 1978. Based on an examination of several cases, this study outlines a theory of the conditions under which court decisions produce changes benefiting social movements.

Horowitz, Donald L. *The Courts and Social Policy.* Washington, D.C.: Brookings Institution, 1977. Based on several case studies, this work argues that courts lack the capacity to render decisions that produce intended social changes.

Johnson, Charles A., and Bradley C. Canon. *Judicial Policies: Implementation and Impact.* Washington, D.C.: Congressional Quarterly Press, 1984. A survey of theories and empirical research on the impact and implementation of court decisions.

McCann, Michael W. *Rights at Work: Pay Equity Reform and the Politics of Legal Mobilization.* Chicago: University of Chicago Press, 1994. An important study of the pay equity issue that examines the significance of litigation's indirect effects on the goals of social movements.

Rosenberg, Gerald N. *The Hollow Hope: Can Courts Bring about Social Change?* Chicago: University of Chicago Press, 1991. An empirical study arguing that litigation intended to produce significant reforms in civil rights, women's rights, abortion, the environment, and criminal law has had little, if any, effect.

Scheingold, Stuart A. *The Politics of Rights: Lawyers, Public Policy, and Political Change.* New Haven: Yale University Press, 1974. An important book arguing that constitutional rights are not self-implementing but, rather, are resources that may be used as part of broader political struggles.

The Nature and Role
of Public Opinion

IN EARLY 1989 CONGRESS dearly wanted a pay raise. At first things looked promising. A federal commission charged with recommending pay increases had proposed a 51 percent raise to $135,000 per member. Outgoing president Ronald Reagan approved, and president-elect George Bush expressed his support for the measure. Unless both houses of Congress voted against the raise, bigger paychecks would soon become reality. Congressional Democratic and Republican leaders informally agreed that the House would not vote, ensuring that the raise would go through.

There was only one problem: Congress failed to anticipate the public's reaction. The proposal ignited a firestorm of protest. Radio talk show hosts derided it and urged listeners to send tea bags to their representatives in a modern-day version of the Boston Tea Party. The resulting flood of tea bags and howls of protest provided tangible evidence of what opinion polls showed—that 85 percent of the public opposed the increase. The carefully negotiated agreement between the House and Senate fell apart when House Speaker Jim Wright bowed to the inevitable after House members defeated a motion to adjourn, demanding he promise a vote on the pay raise. When the roll was called, House members, like senators several weeks earlier, voted overwhelmingly against the increase.

This incident illustrates what Barry Sussman, the chief of polling for the *Washington Post*, calls the "irresistible force" of public opinion.[1] President Bill Clinton experienced it early in his administration when public opinion turned against his first two nominees for attorney general after it became known that they had failed to pay Social Security taxes for immigrants caring for their children. The president, sensing defeat in the Senate, withdrew both nominations. Sussman colorfully assesses the role played by public opinion in such incidents:

To put it directly, politicians as a rule tend to regard the people with terror. Public opinion may erupt at any moment. The citizenry is often condemned as selfish, ignorant, fickle, demanding, easily duped. It is all of those things, packaged in sticks of TNT. From time to time in recent years, surging public opinion—spurred by indignation, outrage, or a sense of urgency—has stopped government leaders in their tracks, pushing them into unexpected, sometimes gradual, sometimes abrupt, changes in policy.[2]

Politicians' statements sometimes reflect this view of the power of public opinion. After many months of growing popular reaction against President Richard Nixon as the Watergate scandal unfolded, he resigned when it appeared that the House of Representatives would vote to impeach him. In a short speech following his dramatic swearing-in ceremony, President Gerald Ford acknowledged the role of public opinion with the simple words "Here, the people rule."

If such incidents show the power of public opinion, other evidence points to its weakness. The refusal of Congress to enact strong gun control legislation between 1968 and 1993 despite solid public support and strong majorities provides the most dramatic example. Many other examples exist as well. For instance, in mid-September 1994, opinion polls found that approximately two-thirds of the public continued to oppose an invasion of Haiti even after President Clinton gave a nationally broadcast address seeking to build support for it. Nevertheless, Clinton proceeded with invasion plans. He even ordered planes carrying an initial wave of paratroopers to take off while the diplomatic mission in Haiti led by former president Jimmy Carter was negotiating a peaceful entry for U.S. troops.

Clearly, the power of public opinion varies. Sometimes public opinion determines policy; at other times it is ignored; and often its effects fall somewhere in between. Accordingly, the nature and influence of citizens' opinions in the play of power are explored in this chapter. After all, public opinion is a principal mechanism through which the general public affects the play of power.

THE INFLUENCE OF PUBLIC OPINION IN POLITICS

 Americans' beliefs about democracy support the proposition that governmental actions should reflect the wishes of the people. As a result, many Americans approach the topic of public opinion with established beliefs about its nature and role. For example, in our survey of students on the first day of an American government class, 85 percent agreed that it is important to remember that "the public" exists when looking at what the government does. Almost as many (79%) agreed that governmental decisions ought to reflect what the public wants. But only about half said they believed that the public plays a role in determining governmental policies. Furthermore, these students did not believe the public had a sound idea of the policies it wanted government to adopt: Only 28 percent agreed it did, whereas 56 percent disagreed.

These beliefs reflect the complex relationship between public opinion and governmental policy. Most people understand that the public participates in the play of power, but many recognize that it does not always know what it wants or succeed in getting it. This chapter will help you gain a deeper understanding of the role of public opinion in American politics. ■

THE NATURE OF POLITICAL OPINIONS

We begin by examining the nature of "opinion." Just what is an opinion? What makes an opinion *political*? What kinds of things do political opinions judge? What attributes do political opinions have that are relevant to understanding their impact? Finally, how do an individual's opinions relate to one another? Throughout this chapter we will refer to the results of polls that provide much of the information available about public opinion. ("Participants: Who Measures Public Opinion?" on page 245 describes the various organizations that conduct these polls.)

VALUES, ATTITUDES, AND OPINIONS

In Chapter 2 we identified the basic *values*—or underlying beliefs such as liberty, freedom of speech, free enterprise, and equality—found in the American civil religion. Such values or basic beliefs often serve as the basis of **attitudes**, defined as predispositions to favor or oppose, like or dislike, or approve or disapprove of that which one encounters.[3] Since they are predispositions, people are not always fully aware of them. Because attitudes usually derive from fundamental beliefs, they tend to be unchanging and to cover a broad range of topics.

Opinions are expressions of support or opposition regarding specific policies or people based on underlying attitudes. They are held more consciously, are narrower in subject, and are more likely to change than attitudes. Opinions can be tapped by asking very specific questions (e.g., "Do you

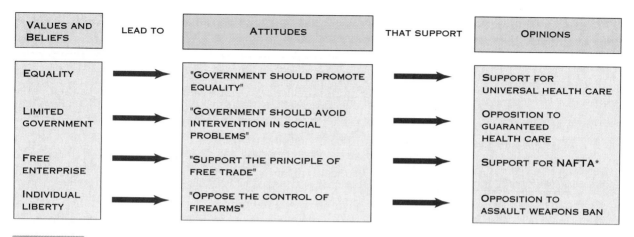

VALUES AND BELIEFS	LEAD TO	ATTITUDES	THAT SUPPORT	OPINIONS
EQUALITY	➡	"GOVERNMENT SHOULD PROMOTE EQUALITY"	➡	SUPPORT FOR UNIVERSAL HEALTH CARE
LIMITED GOVERNMENT	➡	"GOVERNMENT SHOULD AVOID INTERVENTION IN SOCIAL PROBLEMS"	➡	OPPOSITION TO GUARANTEED HEALTH CARE
FREE ENTERPRISE	➡	"SUPPORT THE PRINCIPLE OF FREE TRADE"	➡	SUPPORT FOR NAFTA*
INDIVIDUAL LIBERTY	➡	"OPPOSE THE CONTROL OF FIREARMS"	➡	OPPOSITION TO ASSAULT WEAPONS BAN

FIGURE 7.1

THE RELATIONSHIP AMONG VALUES, ATTITUDES, AND OPINIONS Basic values, such as a belief in free enterprise, can lead to attitudes (such as belief in free trade) that in turn can produce opinions on specific policies. Values are basic to individuals and change little, while attitudes and especially opinions are less deeply rooted and more likely to change.

*North American Free Trade Agreement

favor or oppose a five-day waiting period before the purchase of a handgun?"), whereas attitudes (e.g., generalized opposition to the control of firearms) can only be inferred from behavior or from a number of related opinions.

Figure 7.1 illustrates the relationship among values, attitudes, and opinions. For example, your belief in individual liberty might lead to an attitude opposing the control of firearms, which might cause you to express an opinion opposing a ban on private ownership of assault weapons. Of course, in practice the connections are not always logical or consistent. Some attitudes are not tied closely to fundamental beliefs, and some opinions do not rest on underlying attitudes. Furthermore, people may hold conflicting values and attitudes, thus making it difficult to predict their opinions.

People form opinions about everything. Many of these opinions relate to private rather than public matters. Our concern here is with **political opinions**, opinions about *public matters* such as proposed legislation or the president's job performance or how society deals with spousal abuse. Why should you be concerned about opinions and the attitudes and values that mold them? The answer is obvious but significant: People act on their opinions. The play of power unfolds through human action. Only by studying the opinions that shape action can you gain a full understanding of the play of power.

THE OBJECTS OF PUBLIC OPINION

Consider four typical questions used by pollsters to assess public opinion:

1. "Would you say the government is pretty much run by a few interests looking out for themselves, or that it is run for the benefit of all the people?"

2. "Do you approve or disapprove of the job Bill Clinton is doing as president?"
3. "Generally, do you think students today are as willing to work for the good of the country as other groups, or not?"
4. "Would you favor or oppose a law that would require a person to obtain a police permit before he or she could buy a gun?"

These questions ask about four distinct categories or objects of political opinion (see Figure 7.2). The first pertains to the *political regime* or the system generally. Questions about the regime measure such things as trust in government, whether people think society is "on the right path" or not, and whether the economy is getting better or worse. The second category examines views of the role of *decisionmakers* such as political leaders and institutions. How well is a president, senator, or mayor doing? Should nominees for attorney general be withdrawn because they fail to pay Social Security taxes for child-care workers? Specific *groups* and people's attitudes toward them provide a third focus for opinions. How much confidence do people have in religious leaders, in educators? What do they think about people on welfare? Do they think the Democrats or the Republicans can do a better job of handling the economy? Finally, opinions on specific *issues of policy,* such as questions about U.S. intervention in Bosnia or support for health care reform, make up the fourth category.

CRUCIAL ATTRIBUTES OF POLITICAL OPINIONS

Much of the information about public opinion presented by the news media reports only a single attribute of an opinion, its **direction**. Direction refers to whether the opinion is favorable or unfavorable. Questions measure direction by posing choices between "approve" and "disapprove," "like" and "dislike," and "support" and "oppose." Thus, opinion about the proposed congressional pay increase showed a strong negative direction because 85

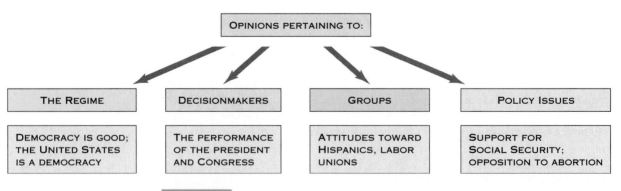

FIGURE 7.2

THE OBJECTS OF PUBLIC OPINION Public opinions deal with four general topics: the regime, decisionmakers, groups, and policy issues. The bottom row of boxes provides examples of specific opinions that might be held for each of the four objects of public opinion.

percent of respondents disapproved. Although the direction of a political opinion is important, other less familiar attributes are crucial as well.

One of the other aspects is the **salience** of the opinion—that is, its significance, prominence, or degree of importance to an individual. Most Americans obligingly answer questions about all sorts of issues that are unimportant to them. If you asked people, "Do you approve or disapprove of the job Newt Gingrich is doing as House Speaker?" many would give an answer (indicating *direction*) even if they found the topic uninteresting and irrelevant. If you asked, "How important is it to you whether Speaker Gingrich is doing a good or poor job?" many would answer, "Not very important," revealing the issue's low salience for them.

Opinions on gun control, abortion, and other "hot" topics differ on another crucial dimension—intensity. **Intensity** refers to how strongly an opinion is held. It can be measured by asking how strongly one likes or dislikes, agrees or disagrees. Often, salience and intensity are similar. The issues that are most important to people generate the strongest opinions.

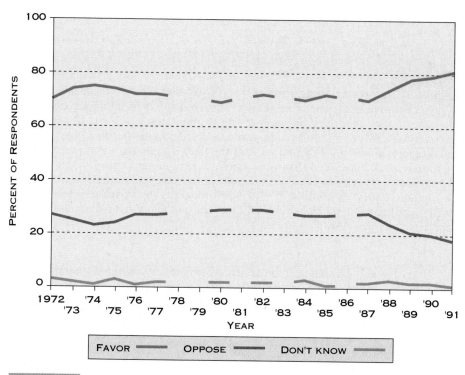

FIGURE 7.3

ATTITUDES ON GUN CONTROL, 1972–1991 The question asked was, "Would you favor or oppose a law that would require a person to obtain a police permit before he or she could buy a gun?" Stable and strong majorities persisted for 20 years, with approval rising steadily from 1987 on. However, the question (and the graph) reveal nothing about the intensity of these opinions. *Source:* Harold W. Stanley and Richard G. Niemi, *Vital Statistics on American Politics,* 4th ed. (Washington, D.C.: Congressional Quarterly Press, 1994), p. 36.

Gaps are for years with no data.

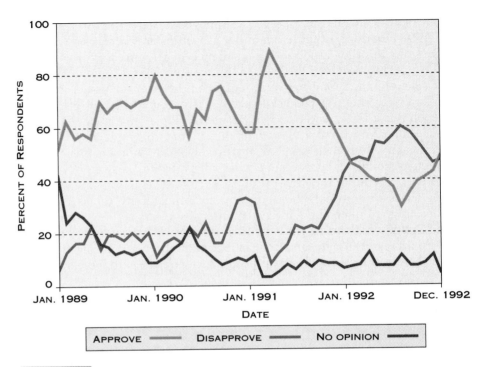

FIGURE 7.4

APPROVAL RATINGS FOR GEORGE BUSH, 1989–1992 The question asked was, "Do you approve or disapprove of the way George Bush is handling his job as president?" Note that the proportion of "no opinion" responses was 40 percent at the start of Bush's term but soon dropped rapidly. Approval remained fairly high and disapproval low until the outbreak of the Gulf War in early 1991, when approval soared. It declined sharply to a point where more people disapproved than approved, then rebounded after the election. *Source:* George Gallup Jr., *The Gallup Poll: Public Opinion, 1989–1992* (Wilmington, Del.: Scholarly Resources, 1993). Copyright © 1993 by Scholarly Resources Inc. Reprinted by permission of Scholarly Resources Inc.

But intensity and salience can be analyzed separately. When someone recognizes the importance of an issue but cannot make up his or her mind about it, the opinion is salient but unlikely to be intense. You may regard the question of whom to support for president as important and may think about it a lot, but if you are undecided or uncertain, your tentative opinion on whom to vote for may lack intensity.

Significantly, the **stability**—that is, the extent to which there is change over time—of the first three opinion attributes differs. The march of events produces a kaleidoscopic pattern, with the direction, salience, and intensity of opinion remaining stable on some issues, shifting gradually on others, and changing rapidly on still others. The *direction* of opinion on gun control, for instance, showed remarkable stability for many years, as is shown in Figure 7.3 (on p. 229). In contrast, approval ratings for George Bush as president swung widely over several years, as Figure 7.4 illustrates. However, both *salience* and *intensity* can vary over time. In an economic recession, for example, the salience and intensity of opinions about unemployment and the state of the economy rise. As conditions improve, other issues—

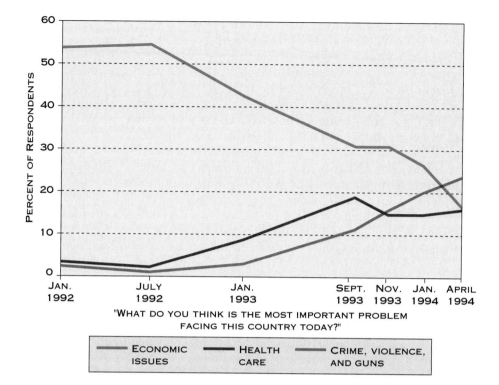

FIGURE 7.5

CHANGES IN THE SALIENCE OF NATIONAL ISSUES, JANUARY 1992 TO APRIL 1994 The graph plots "spontaneous mentions" (a good way to measure salience) to the question "What do you think is the most important problem facing this country today?" During the recession, economic issues were most salient. As economic conditions improved, other issues such as crime and health care rose from near invisibility to significant levels. *Source: New York Times,* May 10, 1994, p. A19. Copyright © 1994 by The New York Times Company. Reprinted by permission.

Based on polls conducted by the *New York Times* and CBS News, the latest April 21–23 with 1,215 adults.

such as crime, drugs, or health care—become more salient. As Figure 7.5 indicates, this is precisely what happened as the economy improved after Bill Clinton's election in 1992.

THE IMPORTANCE OF INTENSITY

People act on intense opinions. Those who feel particularly strongly will risk arrest, injury, and sometimes death. Even those with few resources can do remarkable things when intense opinions impel them to action. In the early 1960s, African-American college students risked physical attack and arrest when they sat in at lunch counters, used bus terminal waiting areas, and drank from water fountains marked "whites only." Their actions inspired others and energized the civil rights movement, which won passage of major civil rights legislation in 1964 and 1965.[4] Intensity also explains the success enjoyed by gun control opponents for decades. Their intense opposition to gun control has caused members of the National Rifle Association (NRA) to act swiftly and angrily when informed about pending proposals. Their phone calls and letters to legislators have drowned out the opinions of the majority, which, although favoring such measures, did so with little intensity. (See "Resources: Intensity of Opinion as a Political Resource.")

Because intensity determines the rate at which political resources are used, knowing the elements that determine its level helps us understand the play of power. Opinions tend to be more intense and salient when individuals perceive an effect on their material *self-interest.* Issues affecting jobs, income, and taxes usually produce strong opinions, as do those regarded as life-threatening: war, a nearby toxic chemical dump, AIDS.

But symbolic issues linked to deeply held values and beliefs about such matters as race, religion, and ethnicity can generate equally intense opinions. Examples abound. Many Hispanics hold intense opinions about establishing English as the nation's official language, and about bilingual education. U.S. policy toward Poland prompts great concern among Polish Americans. Almost any frank classroom discussion about abortion or race-related issues will quickly demonstrate the intensity of opinions based on deeply held beliefs.

Issues that affect both symbolic and material outcomes often produce especially strong opinions. Abortion is one such issue. War is another. In the late 1960s and 1970s, opposition to the Vietnam War by many college-age men and women arose from their moral opposition to U.S. involvement and their fear that they or their friends would be killed. Personality characteristics also affect intensity. Some people, regardless of the degree of perceived self-interest, become more easily excited and concerned about politics than others.

RELATIONSHIPS AMONG ATTITUDES: CONSISTENCY AND IDEOLOGY

Political opinions do not exist in isolation. People hold opinions about many aspects of the same general topic. The play of power that produces political decisions on topics such as abortion, gun control, health care, and national

DOONESBURY   **by Garry Trudeau**

Mike Doonesbury's initial opinion lacks intensity and stability. Confronted with contrary information, he quickly changes his mind. The pollster's classification of him as "undecided" illustrates some pitfalls of polling and the importance of looking carefully at "no opinion" figures.

INTENSITY OF OPINION AS A POLITICAL RESOURCE

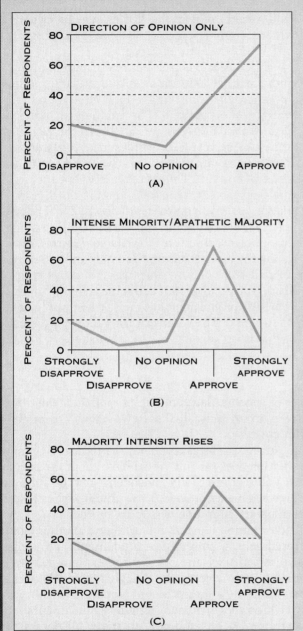

The shape of the distributions of the *direction of opinion* that result when individual opinions are aggregated says nothing about the crucial attribute: the *intensity* of opinion. Allowing respondents to indicate whether their approval or disapproval is "strong" provides a measure of intensity. The breakdown between "strong" and "not so strong" approval or disapproval can produce very different patterns. Most interesting are situations in which graphs of direction alone suggest that a strong majority favors one position, but the minority holds very intense views and the majority is apathetic. In such cases the rate at which citizens mobilize their political resources to influence the outcome of the issue is likely to differ considerably from what is suggested by looking simply at direction.

For example, assume that a poll reveals the following distribution of approval and disapproval on the issue of gun control: 75 percent approve, 20 percent disapprove, and 5 percent have no opinion. (This is the pattern that has appeared in polls over several decades.) The distribution of opinion illustrated in the figure "Gun Control and Intensity of Opinion" is known as a "J" curve. But this information reveals nothing about the distribution of intensity. For decades, those favoring gun control did not feel strongly; those opposing it did, and, guided by the National Rifle Association (NRA), they acted on their convictions. This pattern, with an apathetic majority and an intense minority, enabled the latter to dominate citizen efforts on the issue and resulted in the defeat of nearly every gun control bill. In the early 1990s, the majority opinion became somewhat more intense and legislation began to pass.

GUN CONTROL AND INTENSITY OF OPINION Section A shows *direction of opinion* only and indicates a large majority of respondents (roughly 75%) favoring gun control. But Section B, showing *intensity of opinion* (measured by asking respondents whether they "strongly" approve or disapprove), reveals that a minority was intensely against gun control (18% strongly disapproved) whereas the majority was more apathetic (5% strongly approved). Later, as Section C shows, even though the total percentage of respondents favoring gun control remained around 75 percent, a larger proportion of them felt strongly about the issue, so that the number of intense opponents of gun control was matched by the number of intense proponents. Such a shift occurred in the early 1990s, explaining why Congress began to enact gun control measures.

HOW ACCURATE ARE POLLS?

Have you ever been asked your views in an opinion survey? Do you know anyone who has? If you are like most students, you are probably skeptical about how a poll of 1,500 people can reveal the opinions of 100 million.

Some skepticism makes sense. Polls that use biased or unclear questions, poor methods for selecting a sample of respondents, and improper techniques for analyzing responses *do* produce misleading results. For instance, a widely cited poll reportedly found that one person in five thought it was possible that the Nazi extermination of Jews during World War II had never happened. The question, however, was confusing: "Does it seem possible or does it seem impossible to you that the Nazi extermination of the Jews never happened?" A second poll used a clearer question: "Does it seem possible to you that the Nazi extermination of the Jews never happened, or do you feel certain that it happened?" It found 91 percent of those polled certain that the killing had happened, whereas only 1 percent said it was possible that the killing had never happened.[1]

In contrast, correctly conducted polls can provide reliable information about a large group by employing *sampling*, a technique used in a variety of situations. For example, a severe drought and hot weather in 1988 in the American Midwest resulted in widespread contamination of corn with a potent cancer-causing fungus called aflatoxin.[2] This fungus grows in corn kernels and can enter the food chain in milk, cornflakes, and other processed foods. Inspecting each kernel of corn in each load of corn would have been tedious and expensive. Instead, one company inserted nine hollow tube probes, each to a different depth, to withdraw *samples* from each truckload of corn delivered for processing. The tiny proportion of kernels so sampled were then tested. The company also randomly tested a small portion of its boxes of finished product.

defense involves a number of specific issues on which opinions exist. Students of public opinion refer to an individual's views about aspects of a general topic as an **opinion cluster**.

Some fairly obvious questions can be asked about the nature of opinion clusters.[5] How much information does the individual have about the general issue? Are opinions within a cluster internally consistent? Do the opinions in the cluster predict the person's behavior well? How salient is the cluster? The answers to each question vary from one person to another.

People differ in how their opinion clusters relate to one another. Some people have highly structured beliefs, grounded in fundamental principles, that produce highly consistent attitudes and opinions to explain politics—that is, they have an **ideology**. Well-developed ideologies provide a person with a worldview that can be used to interpret and explain a broad range of events in a consistent fashion. For instance, religious fundamentalists, orthodox Marxists, and extreme liberals and conservatives all possess an ideology. Less well developed belief systems, like those of some people who think of themselves as "moderate" liberals or conservatives, produce less encompassing explanations and only moderately consistent opinions. And many Americans adhere to no particular ideology at all. Their beliefs are not systematically organized, and their opinion clusters display little structure or consistency. It is not surprising, therefore, that people's opinions are often inconsistent. For example, many support both increased governmental

Public opinion surveys rely on the same techniques. As long as the sample accurately represents the total population and the questions are unbiased, good estimates of the number of people in the total population holding each view can be obtained. Of course, even well-conducted polls provide results that differ from the "real" number. But statisticians can estimate the likelihood that survey results will "deviate," or differ, from the actual number—the "sampling error." Samples of 1,500 to 2,000 give results with a sampling error of only 2 or 3 percent most of the time, regardless of the size of the population studied.

The accuracy of polls can be tested by, for example, comparing election predictions to actual outcomes. Between 1950 and 1988 the average difference between the Gallup poll's final pre-election survey and the actual result in 18 national elections was just 1.4 percent.[3] In 1988 six major polls predicted that Bush would win,

and all but one of them came within three percentage points of the actual difference. In 1992 the final Gallup poll estimated that Clinton would get 49 percent, Bush 37 percent, and Ross Perot 14 percent. Clinton actually got 43.3 percent, Bush 37.7 percent, and Perot 19 percent. Gallup hit Bush's percentage on the nose and correctly predicted the winner but overestimated Clinton's vote and underestimated Perot's by about 6 percent each.

[1] Michael R. Kagay, "Poll on Doubt of Holocaust Is Corrected," *New York Times*, July 8, 1994, p. A10.
[2] The information on aflatoxin is from Scott Kilman, "Fungus in Corn Crop: A Potent Carcinogen Invades Food Supplies," *Wall Street Journal*, February 23, 1989, p. A1.
[3] Robert Erikson, Norman R. Luttbeg, and Kent L. Tedin, *American Public Opinion*, 4th ed. (New York: Macmillan, 1991), p. 29.

spending on social programs and substantial cuts in taxes, or cuts in governmental spending without any reduction in the programs government supports.

THE "PUBLIC" IN PUBLIC OPINION: GENERAL AND ATTENTIVE PUBLICS

When politicians and the news media use the term *public opinion*, they usually mean the opinions of the *general public*. But who qualifies for membership in "the public"? When epidemiologists talk about public health, they include infants, and the elderly, and the infirm. When people speak of public opinion, they exclude babies and nursing home residents. But what about 15-year-olds, nonresident aliens, or the millions of people who never vote or participate in politics?

The seeming precision of the word *public* in discussions of public opinion conceals great ambiguity. The point is obvious but not trivial. It has important implications for interpreting statements that purport to reveal what "public opinion" is. Just as one can ask who the Framers had in mind when they began the Preamble to the Constitution with the words "We the People," so too can one consider just what part of the total population is meant by the term *the public*.

Political scientist V. O. Key defines **public opinion** as "those opinions held by private persons which governments find it prudent to heed."[6] By "those opinions" we mean *political* opinions (which we have defined as opinions about public matters) held by some defined *group* of people. But a crucial question remains: *Whose* political opinions (i.e., which groups' views) is it prudent to heed?

On the major issues of politics, active participants in the play of power consider the political opinions of most adults, that is, the **general public**. Polls conducted by the news media, for example, seek the views of a random sample of all adults—or, just before elections, of all likely voters—and report the results as "public opinion."

Obviously, some people care much more about any given issue than does most of the general public. Whether it is gun control, abortion, support for college scholarships, or how commercial real estate is depreciated for tax purposes, the issue will be more salient to some people than to others and will generate better-informed and more intense opinions. We use the terms **attentive public** and **special public** to refer to those people who know and care about a particular issue. Governments—indeed, many active participants—find it prudent to heed the opinions of attentive publics for the simple reason that their intensity and salience make such publics likely to use their resources to affect outcomes in the play of power.

One can find an attentive public for virtually every issue in politics. On most issues the general public's opinions have little salience and intensity, are quite unstable, and lead to little or no action. Attentive publics, however, care very much about the issue, and their intense feelings often prompt them to participate actively. In assessing the effects of "public opinion" in the play of power, one should ask: Who is the attentive public, and what does it think? What does the general public think?

THE ORIGINS OF POLITICAL VALUES, ATTITUDES, AND OPINIONS

People acquire language, religion, and culture from their environment. The same holds true for political values, attitudes, and opinions. After all, no part of the genetic code produces Democrats or proponents of gun control.

The term **political socialization** refers to the process by which people form political values, attitudes, and opinions. Like language skills and religious beliefs, many fundamental political beliefs develop during childhood. But political learning continues as the circumstances and events encountered in life change. We therefore need to look at both childhood and adult socialization.

CHILDHOOD POLITICAL SOCIALIZATION

Studies of how children learn about politics, politicians, and policies produce few surprises. The **agents of socialization**—that is, the people and institutions imparting this knowledge—include family members (especially parents), elementary and secondary schools, religious leaders and teachings, peers, the larger community, and the mass media.

Two-parent families in which politics is regarded as important and the adults hold the same views exert the greatest influence on offspring. Children in such families learn they are "Republicans" or "Democrats" before they know what these terms mean. They absorb their parents' attitudes toward the political regime, groups, politicians, and policies. Disagreement among parents weakens the impact of the immediate family, as does their lack of interest in politics. Families headed by single parents with little time, energy, or regard for politics also shape attitudes less. Thus, some children's family life predisposes them to regard politics positively and to develop well-formed views that match those of the adults. Others learn to see politics as something harmful and to be avoided, and still others see it as uninteresting and irrelevant.

The impact of schools rivals that of the home, especially where the family regards politics as unimportant. Most schools teach the Pledge of Allegiance, how to salute the flag, the story of Thanksgiving, and the benevolent role of authority figures such as the president and the police. Grade schools promote the fundamental values of the dominant civil religion, giving potency to a variety of political symbols such as the flag, Abraham Lincoln, "free enterprise," and "equality." Later schooling introduces deeper knowledge about governmental structure. Although some high school classes teach critical thinking about politics, most emphasize acquiescence to authority and idealized depictions of how well our system follows democratic ideals. Indeed, many of the common beliefs about politics presented throughout this book are acquired in secondary school.

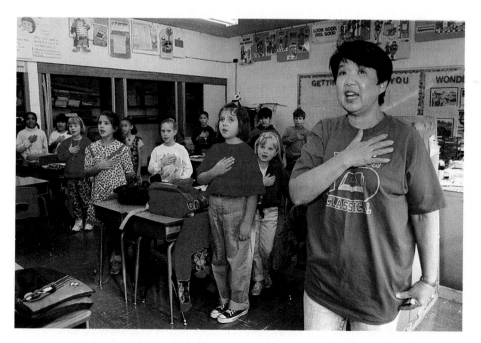

Virtually all grade schools in the United States begin the day with the Pledge of Allegiance. This remarkable uniformity of practice reveals the importance our society attaches to teaching basic political values and loyalty. Based on your own schooling, list some other acts of political socialization performed by schools.

You can probably identify other agents of socialization from your own childhood, including classmates, playmates, family friends, neighbors, and the mass media. The media warrant a closer look. The average American watches 20,000 hours of television by the time she or he reaches college age. Spending the equivalent of 2,500 eight-hour days (more than six years!) in front of the tube exposes the viewer to much that shapes political attitudes. While the precise effects of sustained viewing are difficult to determine, some scholars argue that heavy viewers are more likely to view the real world as resembling the world of television: violent, dangerous, yet mostly middle-class.[7]

For many people, childhood socialization produces deeply held values and fundamental beliefs consistent with the American civil religion that last throughout their lives. Some children also acquire important stable attitudes, such as a psychological attachment to a political party. Thus, many of the intense, salient, and stable opinions and attitudes held by adults about the political regime and groups (and some about basic policies) stem from childhood socialization.

ADULT POLITICAL SOCIALIZATION AND OPINION ACQUISITION

Political learning continues during adulthood, but through different agents. For one thing, the mass media assume greater importance. After all, few citizens meet politicians face-to-face, attend legislative sessions, or witness political events firsthand. Information from the media, especially television, profoundly structures the play of power, as is shown in Chapter 8 (where we discuss how the media shape adults' opinions).

Adults' political opinions also reflect the impact of personal experiences. Interaction with co-workers, neighbors, and friends continuously shapes opinions. Simply ask a physician or pharmacist about health care, a real estate agent about federal tax policy, or textile and auto workers about tariffs and free trade to observe the effects of occupation on opinions. Occupation makes some issues salient. Furthermore, in most informal social groups, some people serve as "opinion leaders." These individuals are more interested in politics and better informed, and they help mold the attitudes of others who look to them for guidance.

Other life circumstances and events also shape opinions. When young couples buy a home and have children, their attention turns to property tax rates and the quality of schools. The death of a child because of a drunk driver or a stray bullet is likely to produce strong opinions on policies regarding alcohol consumption or gun control.

The organizations that adults belong to and identify with serve as "reference groups" on issues affecting the organizations. For example, NRA members rely on the association for cues on issues affecting gun ownership, and many churchgoers learn about current issues (e.g., abortion bills pending in the state legislature, or famine relief efforts abroad) from their clergy.

Although the diversity of life experiences in the population means that different people will have different political attitudes and opinions, some forces affect everyone. Dramatic and disruptive political events such as the Great Depression, World War II, the assassination of President John Kennedy, and the Vietnam War leave their mark on the entire population, adults and children alike.

Finally, popular opinion evolves in response both to the effects of public policies and to the actions and words of political leaders and interest groups (see "Strategies: Influencing Public Opinion on Health Care Reform" on pp. 240–41). Popular support for Social Security and Medicare, for instance, rose as these programs took effect and people liked what they achieved. Support for a U.S.-led invasion to expel Iraqi forces from Kuwait in 1991 soared after Operation Desert Storm achieved a rapid victory. These examples illustrate a crucial point: Even though people normally think that public opinion influences policy and politicians, influence can flow the other way as well. Policy and politicians also help determine adults' political opinions.

THE CONTENT OF THE GENERAL PUBLIC'S OPINIONS

We now turn to some specific attitudes and opinions held by the general public about each of the four objects of political opinion: the political regime, the principal participants and institutions, major social groups, and some important policy issues. In doing so, we will look at patterns of the opinions of individuals who form the general public. We conclude our discussion by examining levels of political knowledge and the intensity and stability of American political opinion.

DECLINING TRUST IN GOVERNMENT

The general public expresses inconsistent views about the political regime. Americans believe that democracy is the best form of government and that the United States is a democracy. But although they still generally like their system of government, they express a growing lack of *trust in government* as it actually works.[8] A "trust in government" scale can be derived by combining the answers to five questions. Figure 7.6 (on p. 242) plots changes in responses to two of these questions. The responses show increasing distrust. Scholars debate the significance of these changes. Some argue that low levels of trust breed cynicism, low participation, and general hostility toward the political regime, all of which threaten its stability and survival. Others doubt the changes are significant.[9] Regardless, the changes shown in Figure 7.6 are substantial.

DECLINING CONFIDENCE IN POLITICIANS AND INSTITUTIONS

As trust in government has declined, so has the degree of confidence expressed by citizens in most of the principal participants and institutions in American politics. When poll respondents were asked, "As far as the people running [various institutions] are concerned, would you say you have a great

INFLUENCING PUBLIC OPINION ON HEALTH CARE REFORM

President Clinton's efforts to reform health care triggered massive lobbying campaigns by opponents and proponents to influence the opinions of both the general public and special publics.

The best-known effort to shape general opinion came in a series of television ads produced for the Health Insurance Association of America (HIAA). They featured a fictional white, middle-class couple, "Harry and Louise," and Louise's friend "Libby," who all professed support for the principle of health care reform but predicted dire consequences if Clinton's plan were passed. The HIAA spent $10.5 million on advertising in 1993 and another $3.5 million in 1994. The impact of the ads extended far beyond their initial broadcast. Portions were rebroadcast when the ads themselves became news. A highly publicized attack on the ads by Hillary Clinton, who accused the health insurance industry of lying, only increased their visibility.

Other groups joined the fray. The Democratic National Committee waged a "National Health Care Campaign" that included television advertising. In February 1992 the American Federation of Labor and Congress of Industrial Organizations (AFL-CIO) announced it would spend $10 million to support the Clinton plan. A group favoring the plan, Families USA, gathered statistics and hard-luck stories, which it publicized effectively. For example, it called a press conference the day before a man was jailed for lending his health insurance card to an injured, uninsured friend so the friend could get emergency care.[1] Various television news broadcasts and talk shows carried his story.

These efforts went beyond television. The American Medical Association printed brochures to be distributed in member doctors' waiting rooms. The Catholic Church and the Christian Coalition passed out postcards to churchgoers. Several groups sent what critics termed "fright mail" to the elderly, claiming Clinton's bill would cut Medicare benefits. Their efforts produced "hundreds of thousands of letters to Congress" opposing Clinton's plan, according to officials of one organization.[2] Many groups took out full-page newspaper ads.

Shown here is one of the infamous "Harry and Louise" advertisements, paid for by the Health Insurance Association of America, attacking provisions of the 1994 Clinton health care plan. Both the ads and Hillary Clinton's attacks on them became news themselves.

Some groups tried to shape the opinions of special publics. The 600,000-member National Federation of Independent Businessmen (NFIB) was perhaps the most active. One report noted that the NFIB "stoked its members' anger with mailings and faxes," organized public forums in key legislators' districts, and hired a telemarketing firm to call members and connect them to their member of Congress if they said they were willing to speak against to Clinton's plan.[3]

When President Clinton first introduced his plan to reform health care, it won widespread public support. By the time it was defeated in late 1994, opponents' efforts, combined with the criticisms of congressional opponents, had eroded this support significantly.

[1] Tamar Lewis, "Hybrid Organization Serves as a Conductor for the Health Care Orchestra," New York Times, July 28, 1994, p. A20.
[2] Robert Pear, "'Liars' Try to Frighten Elderly on Health Care, Groups Say," New York Times, May 27, 1994, p. A18.
[3] Neil A. Lewis, "Lobby for Small-Business Owners Puts Big Dent in Health Care Bill," New York Times, July 6, 1994, p. A1.

In addition to sponsoring television advertisements, a number of interest groups tried to affect public opinion on the Clinton health care plan through paid newspaper advertising. This ad, placed by Single Payer Action, appeared in the first section of the *New York Times* on August 5, 1994.

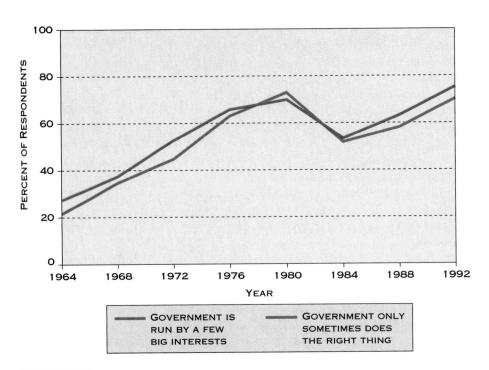

FIGURE 7.6

DISTRUST OF GOVERNMENT, 1964–1992 This graph plots answers to two questions: "Would you say that the government is pretty much run by a few big interests looking out for themselves or that it is run for the benefit of the people?" and "How much of the time do you think you can trust the government in Washington to do what is right—just about always, most of the time, or only some of the time?" In 1964 fewer than 30 percent of respondents expressed distrust. The level more than doubled by 1980, dipped briefly during Ronald Reagan's first term, then returned to its previous high level. *Source:* Stephen C. Craig, *The Malevolent Leaders* (Boulder: Westview, 1993), p. 11, Table 1.1. Reprinted by permission of Westview Press, Boulder, Colorado.

deal of confidence, only some confidence, or hardly any confidence at all in them?" beginning in the early 1970s the number who answered, "A great deal," for the Supreme Court, the executive branch or White House, and Congress declined, and it has remained low.[10] Only President Reagan's popularity in the mid-1980s temporarily broke the pattern of decline, which continued into the 1990s. In September 1994, for example, a *New York Times*/CBS News poll found that 63 percent of respondents disapproved of the way Congress was handling its job, while just 25 percent approved.[11]

PUBLIC OPINION TOWARD AFRICAN AMERICANS, WOMEN, AND OTHER GROUPS

Racial divisions in the U.S. population produce the most persistent and deep divisions in politics. The content of and changes in opinions about race explain much about the play of power at any given time.

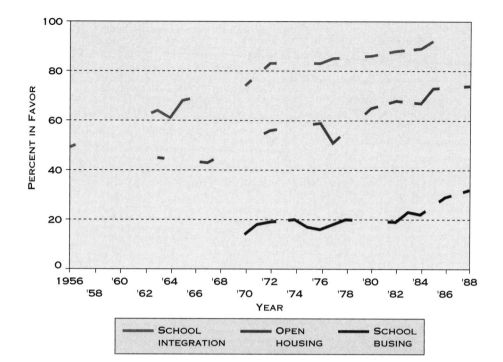

FIGURE 7.7

TRENDS IN OPINION ON RACIAL ISSUES, 1956–1988 The school busing
question was, "In general, do you favor or oppose the busing of black and white
school children from one district to another?" The open housing question asked
respondents to agree or disagree with the statement "White people have a right to
keep blacks out of their neighborhoods if they want to, and blacks should respect
that right." (The graph shows the percentage of whites disagreeing slightly or
strongly.) The school integration question was, "Do you think white students and
black students should go to the same schools or to separate schools?" Support for
all three policies rose fairly steadily, if slowly. Many people still oppose these
policies, especially school busing. *Source:* Richard G. Niemi, John Mueller, and
Tom W. Smith, *Trends in Public Opinion: A Compendium of Survey Data* (New
York: Greenwood Press, 1989), pp. 174, 182, and 184, Tables 8.8, 8.17, and 8.19.

Gaps are for years with no data.

Recent decades have seen dramatic shifts in white Americans' attitudes
toward African Americans. As late as 1944, 52 percent of whites thought
that "white people should have the first chance at any kind of job." By 1972
the number supporting such blatant racial discrimination plunged to just 3
percent.[12] In 1958, 53 percent of all respondents said they would not vote
for a generally well-qualified black candidate for president; by 1987 the figure
dropped to 13 percent.[13] Surveys measuring approval of policies furthering
the rights of African Americans, such as school integration, integration of
residential neighborhoods, and school busing also found increasing support,
as Figure 7.7 shows. Nevertheless, substantial opposition to such policies
still exists and has shown signs of increasing in the early 1990s. For instance,
the number of whites who felt that "we have gone too far in pushing equal
rights in this country" rose between 1988 and 1994 from 47 to 51 percent.[14]

When asked whether they support "preferential treatment" for women and minorities to make up for past discrimination, whites express opposition by huge margins. Similarly, public support of busing to promote school integration, although growing slowly, still finds little favor. In 1972, 20 percent approved; by 1988, that figure had increased to 32 percent, but in 1993 it still hovered just above 31 percent.[15]

Dramatic changes in opinions about women in politics have occurred as well. In 1937, only one-third of respondents indicated they would "vote for a woman for president if she qualified in every other respect." In 1988, 85 percent said yes when asked: "If your party nominated a woman for president, would you vote for her if she were qualified for the job?"[16] In 1975, a poll found close to half agreeing that "most men are better suited emotionally for politics than are most women."[17] By 1988, those agreeing declined to one in three, still a substantial bloc.

Attitudes toward other groups who are disadvantaged in the play of power confirm that the general public continues to hold discriminatory views. For example, a survey found that people stereotype Hispanics almost as negatively as they do blacks, seeing them as prone to violence, dependent on welfare, unpatriotic, and lazy.[18] A 1993 Gallup poll found that 30 percent of respondents thought Hispanics were more likely to commit crimes than other people in society; 37 percent thought this of blacks; 15 percent thought it of Asians; and just 6 percent thought it of whites. The same survey found that whereas 75 percent thought that Irish immigrants benefited the country, 29 percent thought that Mexican immigrants did, and only 19 percent thought that Haitians did.[19]

Public opinion toward disadvantaged groups reflects differences in support of *equality of opportunity* and *equality of result.* There is strong approval for guaranteeing equality of opportunity—for instance, policies upholding the right to attend a nonsegregated school, to compete for a job on equal terms, to seek elective office, and to buy a home anywhere. But support evaporates for governmental actions intended to achieve equality of result, such as preferential hiring of women and minorities and mandatory school busing to achieve racial integration.[20] For instance, a 1987 Gallup poll found that fully 63 percent of respondents disapproved of hiring and promoting women and minorities over better-qualified men and women to achieve better racial and gender balance.[21]

As the play of power unfolds, different groups receive attention, causing opinions about them to develop. In the early 1990s, for example, attitudes toward illegal immigrants in California and Florida gained salience and intensity, as did public opinion about issues involving homosexuals in many places. Support for these groups' *equality of opportunity* appeared to be weakening.

ATTITUDES AND OPINIONS ABOUT MAJOR PUBLIC POLICIES

Americans' beliefs about the economy and the government's role in it contain important contradictions. A large majority likes private ownership of property and free enterprise, and dislikes socialism; this suggests that opposition to government's interference in the economy should be strong. But an equally large majority agrees that a strong central government must act to

WHO MEASURES PUBLIC OPINION?

During the last six months of the 1988 presidential campaign, more than 100 public opinion polls published the results of their measures of support for candidates George Bush and Michael Dukakis.[1] Even more polls accompanied the 1992 election. Non-election-period polls of public opinion appear with increasing frequency in the news media. Increased attention to the public's views is elevating its status as a participant in the play of power. It also raises a question: Who conducts these polls?

At one time, most polls were conducted by commercial polling organizations such as Gallup, Roper, and Harris. Although the organizations lost money in polling public opinion, the publicity they garnered enhanced their prestige and visibility and made them attractive to potential paying clients. Academic studies (e.g., by the Survey Research Center at the University of Michigan and the National Opinion Research Center at the University of Chicago) also measured public opinion. However, their results appeared years after the poll had been taken, and then only in academic publications.

Today, a number of other organizations conduct polls.[2] You are probably most familiar with those undertaken jointly by newspapers and television networks. CBS teams up with the *New York Times*, ABC with the *Washington Post*, and NBC with the *Wall Street Jour-*nal. CNN, the Associated Press, and major newspapers such as the *Los Angeles Times* and the *Minneapolis Star-Tribune* also conduct polls. *Newsweek* employs the Gallup organization; *Time* employs the Yankelovich organization. Less familiar are the extensive polls conducted by political consulting firms for the candidates and political party organizations that hire them. Results of such polls usually remain confidential, although some are selectively leaked to bolster their clients' prospects. Policy research organizations conduct polls sponsored by governmental agencies. Some agencies, such as the U.S. Census Bureau, conduct their own polls.

Although the quality varies, an increasing flood of fairly reliable information about at least some aspects of the public's opinions is now available. As a result, efforts to exert influence through persuasion in the play of power increasingly employ the symbol of "public opinion" by mentioning poll results.

[1] Frank W. McBride, "Media Use of Preelection Polls," Chapter 8 in Paul J. Lavrakas and Jack K. Holley, eds., *Polling and Presidential Election Coverage* (Newbury Park, Calif.: Sage, 1991), p. 184.

[2] For an overview of organizations that conduct polls, see Norman M. Bradburn and Seymour Sudman, *Polls and Surveys: Understanding What They Tell Us* (San Francisco: Jossey-Bass, 1988), Chapter 4.

solve economic problems. Furthermore, when asked whether governmental regulation of business is needed to keep industry from becoming too powerful, or whether such regulation does more harm than good, more respondents support regulation (45%) than see it as doing harm (28%).[22] In addition, most Americans think government should help those who cannot make ends meet. Even Ronald Reagan, the most conservative of our recent presidents, expressed support for such help when he vowed that his cuts in social services would leave untouched the "social safety net" required to guarantee everyone the basic necessities of life.

Despite distaste for "handouts" and welfare, few people are willing to watch children die of starvation or disease caused by malnutrition, or to allow the elderly to freeze to death because of lack of heat. Strong support exists for governmental programs to assist the elderly and the poor, including

Social Security, Medicare and Medicaid, day care, and college loans. A poll taken during the middle of the Reagan era found that seven of eight Americans felt the government "definitely" or "probably" had a responsibility to provide a decent standard of living for the elderly.[23] In fact, substantial majorities in 1988 wanted to increase spending on Social Security and to protect and improve health care, day care, and "assistance to the poor." Only when people were asked about "welfare" instead of assistance to the poor and food stamps did support drop significantly.[24]

Opinions on foreign policy issues change quickly as crises come and go. Just how quickly new crises emerge and old ones fade is shown by the number of foreign countries that dominated the foreign-policy news in the late 1980s and early 1990s—Lebanon, Grenada, Panama, Kuwait, the Soviet Union, Bosnia, Somalia, and Haiti are some more memorable examples. However, a few foreign policy issues endure. Figure 7.8 plots responses to a question about whether the United States should take an active part in or stay out of world affairs. The figure shows some modest but inconsistent shifts over time within a general pattern of stability. In contrast, views

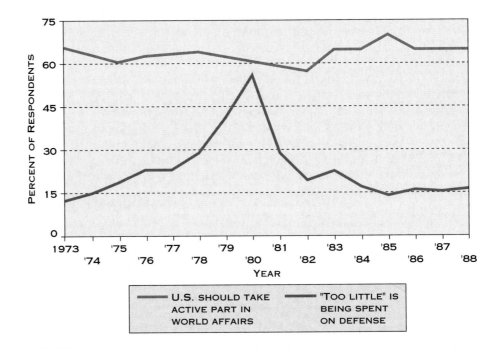

FIGURE 7.8

STABILITY OF ATTITUDES REGARDING TWO FOREIGN POLICY QUESTIONS, 1973-1988 The stability of the direction of opinion on foreign policy issues can vary. When people were asked, "Do you think it would be best for the future of this country if we took an active part in world affairs, or if we stayed out of world affairs?" the responses barely changed over 15 years. But between 1977 and 1982, swift and huge changes occurred in answers to the question, "Are we spending too much, too little, or about the right amount on the military, armaments, and defense?" *Source:* Richard G. Niemi, John Mueller, and Tom W. Smith, *Trends in Public Opinion: A Compendium of Survey Data* (New York: Greenwood Press, 1989), pp. 53 and 87, Tables 2.2 and 3.19.

Data are estimated for missing years and averaged for years with multiple polls.

on whether the United States was spending too little on defense shifted dramatically over the course of almost 20 years.

Of course, in addition to issues related to material politics, the political agenda contains enduring and highly symbolic issues involving religion, crime, and reproduction. For example, support for Supreme Court decisions prohibiting school prayer is not high, although it has increased slightly over the years. Moreover, a large majority favors the death penalty, and its numbers have gradually increased. Finally, a large and stable majority believes abortion should be legal if there is a strong chance of the baby's having a serious defect.

KNOWLEDGE, INTENSITY, AND STABILITY

Regardless of its content, American political opinion exhibits several important characteristics that affect how the general public enters the play of power. We will discuss three: the level of knowledge about politics possessed by Americans; patterns in the salience and intensity of opinions; and the degree of stability exhibited by these opinions.

Several characteristics of the context of American politics help explain the patterns found in our descriptions of these three. Americans take politics less seriously than do people in many other countries. The success of the U.S. economy, the absence of many deep social cleavages, the moderating effects of cross-cutting social attributes, and the nonideological nature of the civil religion all contribute to low-tension politics.

Lack of knowledge about and interest in politics Junior high school civics texts glorify the image of the "democratic citizen," the responsible person who always votes and pays careful attention to political issues and developments. However, opinion surveys have revealed for decades that many Americans know little about the structure of government or the identity of important officials. For example, nearly everyone knows that Washington, D.C., is our nation's capital and that presidents serve four-year terms. But only about half of those surveyed remember that there are two senators from each state or that the Bill of Rights is embodied in the first ten amendments to the Constitution. More people can name and give some facts about popular sports and entertainment figures than know the name of their senator or the Speaker of the House. Surveys routinely report that one in five Americans cannot name his or her state's governor and almost two-thirds do not know who their representative in Congress is.[25]

Americans recognize that they know little about politics. In a 1984 poll, an average of only 38 percent of respondents said they had "most" or "all" the information needed on six highly visible issues.[26] At the same time, 56 percent accepted the view promoted in secondary schools that keeping fully informed was an important obligation for citizens. Lack of knowledge about politics and public policies may prevent many people from understanding the play of power or making well-informed decisions about policy, but scholars disagree about the significance of these shortcomings. Certainly most people live their lives without noticing whether their low level of political knowledge makes any difference.

Low salience and intensity Low levels of political knowledge result at least in part from the fact that for most people most of the time, neither politics nor the issues on its agenda are especially important. On most issues in the new media—indeed, on most questions asked by pollsters about politicians and public affairs—the majority of people show little interest. Respondents may indicate a *direction* for their opinion, but the matter carries little *salience* and produces no *intensity*.

Evidence of a lack of interest can be found in the number of surveyed people who express no opinion even when given the opportunity to express one. Unless a "no opinion" option is available, many people obligingly answer a question even when they have no idea what is being asked. In one survey, nearly one-third of the respondents expressed an opinion about the obscure Agricultural Trade Act of 1978. When the phrase "or do you not have an opinion on that issue" was added, the number expressing a view dropped to 10 percent.[27]

Good polls identify people holding no opinion by using *filter questions*— that is, questions that ask whether the respondent has an opinion. For example, after the hotly debated and highly publicized Gramm-Rudman budget balancing act was passed in 1985, the *Washington Post*/ABC poll posed a filter question to determine whether people had "read or heard anything about Congress passing the Gramm-Rudman bill, which requires the government to balance the federal budget by 1991." Forty percent said they had not and thus could have no informed opinion. But even among those who had heard of the proposal, four in ten said they had "no opinion." Thus, only 36 percent of those polled had heard of the bill and had an opinion. Among this group, 61 percent approved and 39 percent disapproved.[28] These findings show the value of filter questions. A poll that merely asked, "Do you approve or disapprove of the Gramm-Rudman bill recently passed by Congress?" would have reported an answer from most of the 60 percent of its respondents who hadn't heard of the bill or had heard of it but had no opinion.

Most commercial polls do not ask filter questions. Academic polls usually do, and their "no opinion" rate for most issues hovers around 20 percent.[29] In 1964, over one-third of the public had no opinion about the Vietnam War; even by 1968, with the war at its height, 10 percent still had no opinion.[30] A year and a half after George Bush became president, about one person in six had no opinion about how good a job Bush was doing; after 21 months of Bill Clinton's promotion of health care reform, about one person in ten had no opinion about how well he was handling the issue.[31]

Indeed, there is good evidence that even when people hold opinions, these frequently carry little intensity. A classic study of the shift in opinions on major issues involving a group of people interviewed three times in four years on the same two issues found that over one-half said they had no opinion in one interview, but expressed one in another. Among those who changed their answer, the shifts seemed random.[32] Thus, for many people much of the time, the opinions offered on most issues tap no salient concerns or intense feelings. If you ask, "What does *the* public think about issue X?" the answer often is, "*The* public doesn't think much of anything. Most people don't really care very much."

However, there are extremely important exceptions to this generalization. On virtually every issue, some people and groups exist who care very much, have intense opinions, and are willing to act on them. Members of such *attentive publics* provide the most important exception to the generalization about lack of interest in issues. Also, the most active participants in politics, the political elites, display a high level of interest in most issues. Elected decisionmakers, their key aides, news reporters, interest group leaders, corporate executives, and many other individuals maintain a high level of interest in and concern about a broad range of issues.

Furthermore, at any given time, the general public cares about one or two of the most prominent issues on the political agenda. Polls that ask respondents to name the problems that most worry them provide fairly reliable clues about what these issues are. Sometimes a moderate degree of intensity accompanies this broad interest. However, such intensity displays little stability. Before long, other issues capture the limited interest and attention of the general public. Finally, as we noted earlier, both symbolic and material issues dealing with race and sometimes with other matters (e.g., prayer and crime) generate many intense and salient opinions.

Patterns of stability and change in opinion   As many of the figures in this chapter indicate, views on some issues can shift dramatically and quickly. However, on long-term domestic policy issues such as gun control, views change more slowly. Of course, even small yearly shifts can produce substantial changes.

Also, what appears to be a lack of change in the general public may mask large shifts in individual views, as demonstrated by one *Washington Post/*ABC News poll. At the start of an interview, the poll asked, "Do you approve or disapprove of the way Ronald Reagan is handling his job as president?" It then posed specific questions about Reagan's performance on a number of domestic and foreign issues before returning to the initial question. The breakdown in *direction* appeared to show little change—59 percent of respondents approved of Reagan's performance both times, and the number disapproving inched up from 37 percent to 39 percent. But an astonishing 16 percent of those interviewed evidently held unstable views of Reagan's performance, shifting between approval and disapproval during the interview itself![33]

This mixed pattern of stability and change in the direction of opinion does not say anything about shifts in salience and intensity. Figure 7.5 (on p. 231) confirms what you might expect: The salience of most issues changes as events unfold. There are exceptions, however. A few issues touching on basic beliefs (gun control, abortion, school prayer, communism) with a heavy symbolic component can remain salient for some people for years. We know less about the behavior of the intensity level of opinions over time. For instance, the *Washington Post/*ABC News poll described in the preceding paragraph also found large shifts in intensity (measured by whether people "somewhat" or "strongly" approved or disapproved of Reagan's performance) between the start and conclusion of the interview: 31 percent changed their evaluation from "somewhat" to "strong" or vice versa.

Thus, our discussion so far suggests that the general public's opinions on various issues change independently; there are few, if any, links among them. However, there is intriguing evidence that the general public has real preferences about the *general direction* of domestic policy.[34] Of course, most people lack stable and salient opinions backed by information on narrow issues. But this evidence suggests that people do hold global views concerning whether

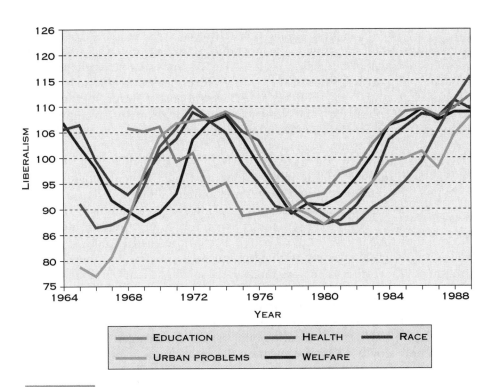

FIGURE 7.9

PUBLIC MOOD REGARDING GOVERNMENTAL INTERVENTION IN FIVE POLICY AREAS This figure shows changes in direction of opinion on whether government should do more or less regarding education, health care, welfare, racial issues, and urban problems. Opinions favoring more government activity result in higher scores on the "policy liberalism" scale on the left axis. Not only did opinion in all five policy areas vary considerably, but the five opinion lines generally moved together, creating a broad policy mood about how active government should be. For example, the urban riots of the mid-1960s reduced for several years the previously high level of public support for government activism. Support rose from about 1968 to 1973, dropped again to a low about the time Ronald Reagan became president, then began to rise again. *Source:* James A. Stimson, Michael B. MacKuen, and Robert S. Erikson, "Opinion and Policy: A Global View," *PS: Political Science & Politics* 27 (March 1994), p. 32. Reprinted by permission of the University of Michigan Press.

or not the government should be involved in a variety of domestic issues, and that these views change over time. Furthermore, attitudes toward involvement in the separate policy areas move in tandem, producing a changing *policy mood*. Figure 7.9 plots the movement of public attitudes toward governmental intervention in five policy areas over 25 years, showing that the public's general preference for governmental activism in domestic politics carries across a number of issues.

HOW INDIVIDUALS' OPINIONS DIFFER

Can people's political opinions be predicted? Does knowledge of a person's sex, race, political party identification, age, or income help in predicting the direction, intensity, salience, and stability of that person's political opinions?

The answer is yes. The opinions of blacks and whites, men and women, rich and poor, liberals and conservatives, old and young, and Republicans and Democrats are not identical. But social attributes, childhood socialization, and the other factors that shape opinions do not absolutely determine political opinions. You probably know people whose views on an issue just do not "fit," prompting you to wonder, "Where did they get that notion?" Further, some attributes more reliably predict opinions than others, and some, such as gender and race, may interact. Finally, people sharing characteristics often disagree on particular issues.

How well can opinions be predicted? We begin our answer by considering one of the most powerful predictors of opinion: race. We then examine the effects of gender, other socioeconomic attributes such as education and income, and, finally, ideologies such as conservatism and liberalism. You will see that these factors help explain some differences in opinions but leave much unexplained. Indeed, the connections between opinions and social attributes turn out to be very complex.

RACE AND POLITICAL OPINION

The central role of race in American society and politics explains the salience and intensity of opinions on racial issues. Consequently, it is not surprising that blacks and whites differ sharply on many matters, including the *extent* of racial discrimination, the *reasons* for it, and appropriate *remedies* to overcome it.[35] When asked to what extent discrimination had hindered blacks and other minorities in getting ahead, 53 percent of blacks chose "a lot" compared to just 26 percent of whites.[36] Fully 60 percent of whites but only 34 percent of blacks agreed that many blacks' inferior jobs, income, and housing resulted in part "because most blacks don't have the motivation or willpower to pull themselves out of poverty."[37] Should colleges and universities reserve admission slots for qualified blacks and other minorities? Sixty percent of whites agreed (5% strongly), compared to 84 percent of blacks (10% strongly).[38]

These differences, however, convey only part of the complex pattern of opinion on issues involving race. Earlier, we mentioned that widespread support for *equality of opportunity* did not extend to support for *equality of result* through governmental action. One might expect that blacks would strongly favor policies seeking equality of result, such as busing to integrate schools and affirmative action in hiring and college admissions. But, although blacks support such policies more than whites, the gap is much smaller than in the examples in the previous paragraph. In fact, in some cases blacks overwhelmingly oppose such policies. When asked whether women and minorities should be given preferential treatment in getting jobs and in college admissions (rather than being selected on the basis of test scores), about 10 percent of whites expressed support. Among blacks, only 13 percent more, a total of 23 percent, supported such a policy.[39]

Furthermore, both blacks and whites disagree to some extent within their own racial group on most issues. Recall that 34 percent of blacks cited blacks' lack of willpower as a cause of black poverty and that 26 percent of whites felt that blacks faced "a lot" of discrimination. Something else besides race must shape the attitudes of blacks and whites toward the level of discrimination, its causes, and its remedies. Although blacks give most of their votes to Democrats, some identify with the Republican Party and support its candidates. This, too, suggests that factors other than race shape opinion.

GENDER AND ETHNIC DIFFERENCES

Do women's opinions differ significantly from men's? The general answer is: yes, but usually not by very much. Few differences between men and women emerge where we might most expect them—on gender issues, such as whether women enjoy equal opportunity. When asked if a woman's chances of working full-time were as good as a man's, 38 percent of men and 40 percent of women chose "worse than average" or "much worse than average."[40] Men were only slightly more likely than women to think the chances for women to get ahead had "improved greatly" (62% versus 52%). Men were just as likely to feel that preferential treatment of women was fair (44% versus 41%), and similar percentages of women and men strongly opposed passage of the Equal Rights Amendment (24% versus 20%). Surprisingly, 34 percent of women disagreed or strongly disagreed with the statement that women should have the right to compete with men in every sphere of activity, as compared to 14 percent of men. Women also do not necessarily support abortion rights more than men do. In fact, a poll conducted in 1989 found that about as many women as men believe abortions should either (1) not be permitted at all (11% versus 8%), or (2) only be allowed in cases of rape or incest, or to save the mother's life (40% versus 38%).[41]

Before 1980, students of public opinion found few differences by sex. Since then, a pattern of small but fairly consistent differences (called the *gender gap*) has emerged. Often, women give somewhat more "liberal" opinions on a variety of issues. More than men, women oppose the death penalty, favor banning handguns, and support increased spending on health care, Social Security, and other governmental services. But they less often take a strong "civil liberties" position: More women than men disapprove of X-rated movies, the legalization of marijuana, and allowing Communists to teach in colleges.[42] Finally, women are a little more likely than men to consider themselves Democrats and, since 1980, to support Democratic presidential candidates.

Similar patterns emerge in other social groups. For instance, as a group Hispanics strongly support the Democratic Party and its candidates. But important differences exist among the groups encompassed by the broad "Hispanic" label. Cuban Americans are more likely to vote Republican than are Mexican Americans and Puerto Ricans. And, although the three groups differ little in their strong support for bilingual education, they disagree sharply on some social issues. For example, over half the Cubans polled in one survey said that college admission and jobs should be awarded on the basis of merit only, with no government quotas, whereas only about one-fourth of Mexican Americans and Puerto Ricans held this opinion.[43]

OPINIONS AND OTHER SOCIAL CHARACTERISTICS

Students of opinion examine the association between citizens' political views and social and economic characteristics such as religious affiliation, depth of religious belief, age, region, occupation, and "generational cohort."

(An **age cohort** is a group with common characteristics that is identified by age.) All show some relationship to opinion, although the precise pattern of the direction, intensity, and size of differences associated with each varies.

The effects of two characteristics—education and income (*learning* and *earning*) are especially noteworthy. Much to the relief of college students, the two seem to be related. The more education people obtain, the higher their income is likely to be. But learning and earning have independent effects on opinion. People with the same income but different levels of education, among and/or the same education but different incomes, hold somewhat different views. Table 7.1 illustrates this for one issue—support for abortion. It shows that the higher one's education, the higher one's support for legal abortions in all circumstances is likely to be at each level of income. For example, among people with a medium family income, 39 percent of those without a high school diploma express support, whereas among college graduates support rises to 62 percent. Likewise, comparing attitudes on this issue among people with the same level of education but different income shows that income shapes opinion independently of education. For example, 79 percent of low-income college graduates support abortion rights, whereas only 58 percent of high-income college graduates do.

To understand the dynamics of public opinion, one must account for the twin effects of *age* and *generational cohort*. Most younger college students recognize the effects of age and experience on older adults' attitudes. Older college students can recall how their own views have changed. It is hardly surprising that people with young families hold salient and intense opinions about schools, employment, and taxes, or that older people care deeply about

TABLE 7.1			
THE EFFECT OF EDUCATION AND INCOME ON WHITES' OPINIONS ON ABORTION			
	% BELIEVING WOMEN SHOULD BE ABLE TO GET AN ABORTION AS A MATTER OF PERSONAL CHOICE		
INCOME LEVEL	**LESS THAN HIGH SCHOOL EDUCATION**	**HIGH SCHOOL GRADUATE**	**COLLEGE GRADUATE**
Low family income	36%	41%	79%
Medium family income	39	44	62
High family income	—[1]	53	58

[1] Too few cases for analysis.

Source: Robert S. Erikson and Kent L. Tedin, *American Public Opinion: Its Origins, Content, and Impact,* 5th ed. (Boston: Allyn and Bacon, 1995), p. 182, Table 7.3. Erikson and Tedin used data from a National Election Study survey conducted in 1992.

Social Security and health care. Thus, people in the same age cohort are likely to share at least some attitudes.

The political and economic upheavals experienced by people throughout their lifetime, especially during their childhood socialization, also produce similarities in the attitudes of people of the same age. Each generation experiences defining events: the Great Depression and World War II; the Cold War and the prosperity of the 1950s; the assassinations and the Vietnam War in the 1960s; the Watergate scandal of the 1970s; and Ronald Reagan's dominance of politics in the 1980s. The replacement of earlier generational cohorts with those born later guarantees steady, gradual shifts in the overall distribution of public opinion even if individual views remain constant.

Finally, changes in the social characteristics of the U.S. population contribute to gradual but significant shifts in opinion over time. The population's average level of education has changed dramatically since World War II. In 1952 only one white American in six spent at least some time in college, whereas nearly four in ten did not go beyond eighth grade. These proportions changed only a few percentage points every four years, but the cumulative effect by 1980 resulted in a sharp reversal of the earlier pattern. In 1980, almost four in ten whites had at least some college education, and just one in ten had an eighth-grade education or less.[44] Since education strongly affects opinions, increased schooling produces shifts in opinion. In coming decades, racial minorities and Hispanics will constitute a steadily increasing portion of our nation's adult citizens. These changes will very likely alter the content of public opinion.

Thus, public opinion shifts over time as the result of *cohort replacement* and *change within cohorts*. Table 7.2 shows how the effects of both can be assessed by looking at answers in 1972 and 1983 to the question "If your party nominated a woman for president, would you vote for her if she was qualified for the job?" Older cohorts said "yes" less often in both years. But between 1972 and 1983, people in every age cohort except the oldest became more likely to say "yes." Of course, during this period the proportion of the population in each cohort changed. Fewer people remained in the older, less favorably disposed cohorts; this decrease contributed to the overall shift in opinion.

HOW WELL DOES IDEOLOGY ORGANIZE OPINIONS?

On the surface, at least, the distinction between "conservatives" and "liberals" divides the public in a meaningful way. The news media and other active participants in the play of power use these terms all the time. When asked whether they are liberals or conservatives, many citizens will make a choice, and their numbers remain fairly constant over time. But are the opinions of liberals and conservatives constrained by an *ideology?* The answer matters. If Americans *did* adhere to a true liberal or conservative ideology, they would express consistent opinions on a variety of issues.

Evidence indicates that most Americans adhere to no true ideology. If given a choice, almost half decline to classify themselves as liberals or conservatives. In a 1988 poll, for example, 22 percent of respondents called themselves moderates and 22 percent had no opinion about what to label themselves.[45] The same poll showed that although the political opinions of

TABLE 7.2

GENERATIONAL SUPPORT FOR FEMALE PRESIDENTIAL CANDIDATES, 1972 AND 1983[1]

COHORT (BY YEAR OF BIRTH)	1972 RESPONSES		1983 RESPONSES	
	% ANSWERING "YES"	% OF SAMPLE	% ANSWERING "YES"	% OF SAMPLE
1955–65	—	0%	89%	23%
1945–54	78	20	89	24
1935–44	73	19	87	15
1925–34	70	17	84	14
1915–24	69	18	78	13
1905–14	64	13	74	8
Before 1905	60	12	50	3
Total	70	99[2]	84	100

[1]This table shows responses to the question "If your party nominated a woman for president, would you vote for her if she was qualified for the job?"
[2]Does not total 100 because of rounding.

Source: Adapted from William G. Mayer, *The Changing American Mind: How and Why American Public Opinion Changed between 1960 and 1988* (Ann Arbor: University of Michigan Press, 1992), p. 148, Table 7.1.

self-identified conservatives and liberals differed, the gap was smaller than one would expect if they adhered to a developed ideology. Although 65 percent of the liberals favored a "government insurance plan that would cover all medical and hospital expenses for everyone," as compared to 39 percent of conservatives, 35 percent of liberals nonetheless *did not* support a health plan for everyone, and 39 percent of conservatives *did*.[46]

Are liberals and conservatives consistent in their opinions across issues, as one would expect if they possessed a true ideology? In a 1988 survey, people were divided into five categories based on their answers to ten questions about political issues. Approximately one-fourth of the sample turned out to be "liberal" on these issues. But 20 percent of those who gave "liberal" answers on the ten issues considered themselves conservatives![47]

The fact that self-identified liberals and conservatives hold a number of "mistaken" positions is less surprising when one looks at how such people define these terms. According to research in the 1960s, only about half the electorate could correctly define *liberal* and *conservative*. Although the proportion has since increased, the most frequently cited difference between liberals and conservatives is still the starkly simple idea that liberals like to spend money on governmental programs and conservatives do not.[48] The half of the population claiming to be conservative or liberal thus apparently has a poor understanding of these terms.

However, about one-fifth of the public does think ideologically. For these people, ideology is important. Furthermore, many active participants in politics have coherent "liberal" or "conservative" ideologies that produce a consistent set of political opinions.

PUBLIC OPINION IN THE PLAY OF POWER

One definition of public opinion emphasized that it is something that politicians find it "prudent to heed." But something more than prudence based on self-interest is at work. The American civil religion, to which politicians usually subscribe, teaches that the wishes of the people *ought* to guide politicians' decisions. At the same time, we have noted that many people do not believe the public's views guide policy in practice.

This raises questions about the role of public opinion in the play of power that go to the heart of the nature of America's democracy. How much does public opinion really shape policy? When is it more powerful, when less so? How responsive are politicians to it, and how does their responsiveness vary? What other mechanisms besides leaders' responsiveness translate opinion into policy?

These questions are as difficult to answer as they are significant. There is still much to learn about how public opinion and policy interact. The relationship continues to evolve, making the search for answers both challenging and unceasing. Although we cannot provide definitive answers, we can offer some partial explanations of how the opinions of both the general public and attentive publics affect policy.

THE IMPACT OF GENERAL PUBLIC OPINION

We begin by identifying three ways in which the opinions of the general public affect policy. Public opinion can be **determinative**—that is, so strong that it becomes an "irresistible force" (as did the public opinion that compelled defeat of the congressional pay raise). Alternately, it may only be **suggestive**, providing decisionmakers with knowledge of the public's broad preferences and some reasons for heeding these views, but leaving many of the important details to be decided by the play of power. Finally, when the general public is uninformed and uninterested, public opinion can be considered **permissive**, allowing wide latitude in the range of policies that can be adopted without fear of any significant public reaction.

When public opinion determines action   When public opinion emerges as an "irresistible force" there is not much subtlety about it, as seen in "Outcomes: Watergate, Public Opinion, and the Fall of Richard Nixon." The effects are direct and visible to all. But several conditions must be met before public opinion becomes *determinative.* A high proportion of the public must know about the issue and consider it salient. Further, opinions must be intense and the distribution of direction must result in a strong majority rather than an even split.

These conditions rarely exist. A dramatic example occurred during the first year of Ronald Reagan's presidency.[49] Seeking to slash governmental spending, the administration proposed reducing Social Security benefits for

WATERGATE, PUBLIC OPINION, AND THE FALL OF RICHARD NIXON

To deal with the growing controversy over newspaper stories linking the Nixon White House to a 1972 burglary of Democratic headquarters in Washington's Watergate Hotel and the subsequent coverup, Richard Nixon appointed a special prosecutor and pledged to respect his independence. When the prosecutor, Archibald Cox, sought a court order for copies of tape recordings made in the president's Oval Office, Nixon ordered him to drop the suit. Cox refused, whereupon Nixon ordered Attorney General Elliott Richardson to fire Cox and abolish the office of special prosecutor. The attorney general refused and resigned instead. Nixon then ordered Deputy Attorney General William Ruckelshaus to fire Cox. He too refused and resigned. That left the solicitor general, Robert Bork, in charge of the U.S. Department of Justice. Bork agreed to do the deed.

The two resignations and the firing took place on a Saturday night, causing the episode to be known as "the Saturday Night Massacre." The general public was outraged and directed an astonishing avalanche of phone calls, letters, and telegrams to the White House and Congress. The resignations of the top two Nixon appointees in the Department of Justice highlighted the political sin committed by Nixon: He had broken his promise to give the special prosecutor a free hand, violating a widely shared and strongly held belief that the president ought not to break a solemnly and earnestly made promise to the American people. The public uproar forced Nixon to relent and appoint another special prosecutor to continue the investigation.

A later series of Watergate-connected events generated a similar reaction and contributed to Nixon's ultimate downfall. Having denied knowing anything about a coverup of the burglary, Nixon spoke to the nation with a stack of supposed transcripts of the White House tapes

After President Richard Nixon ordered special Watergate prosecutor Archibald Cox to drop his lawsuit seeking access to tape recordings of conversations in the White House Oval Office, Cox called an emotionally charged press conference to announce he would not obey the order. His announcement triggered a series of events known as the "Saturday Night Massacre."

behind him, saying that they would vindicate him. Nixon assured his audience that he knew nothing of the burglary or a coverup, that he was not a crook, and that he was not lying. It turned out that the transcripts, which the new special prosecutor obtained after going to the U.S. Supreme Court, had been altered to remove incriminating portions. When several tapes produced "smoking guns" in the form of statements by Nixon showing he had indeed known about and approved the coverup, another massive outburst of public anger occurred. Nixon lost the backing of all but a tiny handful of his most ardent supporters. Both the general public and political elites, believing Nixon had lied while professing to tell the truth, were infuriated. Support for him collapsed, and impeachment, conviction, and removal from office seemed inevitable. In the face of mounting public anger, Nixon resigned the presidency.

people who retired before age 65, delaying the effective date for a cost-of-living adjustment in benefits, and tightening requirements for receiving disability payments. Congressional Democrats, sensing widespread public anger, immediately heaped criticism on the proposals. A poll taken for the Republican National Committee discovered that two-thirds of the respondents did not want any reductions in Social Security benefits and that 63 percent felt the Democrats would do a better job than the Republicans of "helping the elderly and retired." Congressional Republicans, fearing a public backlash in the next election, urged Reagan to eliminate a defense of his proposals in a major speech on taxes. Stung by Democratic criticism and public reaction, the administration backed off from many of its proposed changes, never to revive them.

The ramifications of such incidents extend far beyond their immediate effects. They teach important lessons to active participants in the play of power that shape their subsequent behavior. For years after the Republican poll, members of Congress and policymakers in the executive branch alike refused to propose any changes in Social Security benefits despite the expense of automatic cost-of-living adjustments that were driving up the federal deficit. But such incidents also reinforce a more general principle: that public opinion can awaken and *determine* policy, and prudent politicians had better learn to think ahead and *anticipate* when it just might do that.

The Watergate scandal and Nixon's subsequent resignation suggest another way in which public opinion is determinative in addition to blocking wildly unpopular proposals and reversing unpopular actions. It prevents such proposals and actions from occurring at all through the mechanism of *influence through anticipated reactions.* Active participants in the play of power monitor public opinion, seeking advance warning of what they must not do if they want to avoid angering the general public.

Instances in which participants refrain from acting because they anticipate an adverse public reaction are difficult to detect. However, the most active participants in politics follow certain "rules of thumb." In areas where there is strong policy consensus, such as Social Security, Medicare, and the need for a strong national defense, conventional wisdom instructs participants to steer clear of challenging these policies. For example, few successful politicians advocate cutting defense by two-thirds, slashing Social Security, or taking giant steps toward a socialist economy. Such positions are unthinkable because participants know that proposing them would produce instantaneous and fierce opposition. In such cases, general public opinion leaves decisionmakers little discretion to change policy.

Public policy and "suggestive" public opinion   The ability of strong and united opinion to prevent governmental action is no mean accomplishment. But what happens when public opinion is not so strong or so united? Important decisions about policy continuously emerge from the play of power, and they do not trigger massive public protest. A short list of major legislation passed during the first two years of the Clinton administration (the Family Leave Act, the Deficit Reduction Bill, the Crime Bill, the North American Free Trade Agreement) suggests that these decisions are hardly trivial. So the question remains: What role do the opinions of the general public have in the play of power? Our previous discussion of *policy moods* provides an answer. Active participants in the play of power respond to broad movements in opinion. Their actions rest in part on their belief that decisionmakers in a democracy ought to heed such opinion.

However, another mechanism, the electoral process, links opinion to policy. To the extent that voters' opinions about candidates' stances on issues determine winners and losers, elections can affect the policies adopted. Perhaps more crucial, officials change their behavior according to their anticipation of how their actions might affect their re-election chances. Sensing that voters might become aroused enough about issues to defeat them, elected officials monitor the public mood and modify their words and actions to conform to the public's preferences. Thus, two attributes of the play of power provide a way for the general public's opinions to affect outcomes: (1) politicians' belief that they ought to heed these opinions, and (2) the capability of elections to replace officials who disregard such opinions.

The process through which public opinion shapes decisions has several distinctive qualities. First, it operates indirectly and subtly over a long period of time. Modifications in the behavior of participants in the play of power are rarely announced or widely publicized. No newspaper headline or television news program announces that "Twenty-five key players modified their position today in anticipation of possible electoral defeat next year." Second, the role played by issues in deciding elections is limited and changes from one election to the next. Finally, the degree to which public opinion determines the *specific* details of governmental action is very limited. As three students of public opinion's impact on policy point out, "policy making is highly specific, detailed, and informed, while mass electorates choose not to attend closely enough to politics and policy to have preferences that are similarly specific, detailed, and informed."[50] That is why we refer to this type of influence on public opinion as *suggestive*.

How can one confirm that the broad suggestions about governmental actions provided by policy moods actually get acted on? One method is to devise a measure plotting how the behavior of the House, the Senate, and the presidency changed over time when dealing with social issues that constitute the "policy mood." The measure for all three ought to rise and fall in the same general way, and the swings in all three ought to follow the movement of the general public's mood. Figure 7.10 (on p. 260) presents the results of just such an analysis. As explained in the caption to this figure, independent measures of policy liberalism in the behavior of the House, the Senate, and the president indicate that they generally rise and fall in accordance with the public's mood on the issues as plotted in Figure 7.9 (on p. 250).

When the public remains uninvolved: "Permissive" opinion and policy
What role does public opinion play on the vast majority of occasions when a decision is made on an issue in which opinions lack salience, intensity, and stability? Clearly, active participants enjoy wide discretion. The outcomes emerge from a play of power in which the public does not participate either directly or indirectly (i.e., through the anticipated reactions of those who are politically active). Many items on the political agenda deal with matters most citizens know little about and on which they have no strong opinion. For example, should banks be allowed to open offices in any state they wish? What changes should be made in regulations governing the telecommunications industry? What rules should govern how commercial real estate is

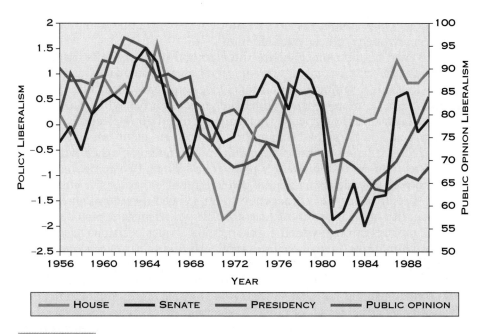

FIGURE 7.10

**PUBLIC MOOD, POLICY LIBERALISM, AND CRUCIAL DECISIONMAKERS'
BEHAVIOR** The red line summarizes the public mood regarding the five domestic
policy areas shown in Figure 7.9. The lines for the House and Senate indicate the
degree to which roll-call votes produced liberal or conservative outcomes, and the
line showing the president's policy liberalism is derived from announced positions
on key votes, the stance of presidential supporters in congressional roll-call votes,
and the positions taken in briefs the administration submitted to the Supreme
Court. The results confirm the conclusion that general policy moods suggest the
direction of policy to key decisionmakers, but only loosely. *Source:* James A.
Stimson, Michael B. MacKuen, and Robert S. Erikson, "Opinion and Policy: A
Global View," *PS: Political Science & Politics* 27 (March 1994), p. 34.

depreciated for tax purposes? In these situations public opinion is *permissive*,
that is, willing to accept a broad range of outcomes without changing its
level of inattention. Yet the cumulative significance of these decisions for
who gets what is immense.

SPECIAL PUBLICS' OPINIONS

Earlier, we termed those who know and care about an issue the *attentive*
or *special* public. The salience and intensity of the opinions of attentive
publics increase the likelihood that they will actively seek to influence the
outcome of the issue. It also causes other participants to take account of
these opinions. After all, the attentive public contacts public officials, contributes to their election campaigns, and keeps track of how they vote and act.
To understand the role of special or attentive publics in the play of power,
you need to learn about interest groups, political parties, and elections. In
this text we devote a separate chapter to each. Our discussions of interest
groups, parties, and the electoral process—indeed, most of the chapters that
follow—can only be understood in light of the role played by attentive

publics. In a sense, then, these chapters can be thought of as an elaboration of how the opinions of special publics shape political outcomes.

Here we can draw several preliminary conclusions about the conditions under which attentive publics have the greatest influence. First, the larger the role of the general public, the less influential attentive publics will be. Where public opinion is *determinative,* special publics' role will diminish. Special publics may and do try to shape the content of public opinion, as suggested in our "Strategies" feature earlier in this chapter. But they cannot stem the progress of the sporadic yet irresistible tides of general opinion that dictate some policies. Where public opinion is *suggestive,* attentive publics influence specific details within the broad outlines of acceptable policy defined by the public. Because attentive publics often split on these details, some of the fiercest battles in politics involve the clash of views among them. Finally, attentive publics exert the most influence on the wide range of issues where public opinion is *permissive.*

Second, the role of attentive publics varies by *arena.* At the national level, the fact that all 535 members of Congress face re-election makes them more attuned to the concerns of attentive publics. Only the president and vice-president in the executive branch are elected, and their constituency is national in scope. Attentive publics influence the behavior of the executive branch, but probably less effectively than they do Congress. Finally, the links between attentive publics and the judicial branch, whose members are appointed, are even looser. However, as our later discussions of the federal courts will show, even here special publics are not completely powerless.

Third, the influence of attentive or special publics changes by *level.* Because we focus on the national government in this text, we can do little more than suggest that the identity and effectiveness of attentive publics in the play of power at the state and local levels vary widely.

THE INTERACTION OF INSTITUTIONS, PUBLIC OPINION, AND POLICY

Public opinion interacts with the institutions and policies of government in complex ways. The general public determines some policies, suggests the general direction of others, and provides little input on still others. Furthermore, the content of its opinions depends in part on the actions and words of governmental officials and other active participants in the play of power, and in part on reactions to the adoption of policies. In addition, the relationship between opinions and a given policy changes over time.

The question of whether the government ought to provide old-age pensions illustrates these dynamics well. At the end of the nineteenth century, public opinion was so against this idea that few politicians dared to advocate it. Public opinion then was *determinative* and deterred politicians from raising the issue. Opinion shifted over time, however, especially after the Great Depression began in 1929. By the time President Franklin Roosevelt proposed establishing the Social Security system in 1935, enough support existed to put the issue on the agenda. Its passage did not eliminate strong opposition, though. As Social Security became established and seemed to work well, support for it became as overwhelming as opposition had been at the beginning of the century. Public opinion thus turned full circle.

The rate at which participants' opinions change also varies. The direction, salience, and intensity of the general public's views concerning issues on the agenda often change faster than do those of attentive publics and long-term active participants. Issues involving the regulation of business in particular exhibit this pattern. A dramatic event—an oil spill, sickness caused by contaminated food or medicine, or windfall profits for an industry—may produce short-term general support for the enactment of new regulations. But after they pass, the general public turns to other issues, paying little heed to how the regulations are implemented. However, the special publics directly affected by the regulations remain interested and active, and they seek to influence how bureaucracies implement the regulations. The result often is that the general public receives *symbolic rewards* in the form of reassurance that "regulations have been passed," and the interests supposedly regulated continue to enjoy *material rewards* thanks to their ability to affect the way in which the bureaucracy enforces the regulations.

SUMMARY

Political opinions evaluate public matters. Many opinions are consistent with underlying values and beliefs or with attitudes that predispose people to approve or disapprove of related issues. In addition to varying in direction, opinions differ in intensity, salience, and stability. Intensity is an important parameter since intense opinions impel people to act. People hold political opinions about the nature of the political regime, groups, political leaders, and major public policies. Opinions about the same general topic, known as opinion clusters, are often not very consistently related; however, some people adhere to an ideology that imposes considerable consistency both within and between opinion clusters.

Public opinion consists of the distributions of the political opinions of a public. General public opinion consists of the views of ordinary citizens. Many citizens know little about most issues, and the intensity, salience, and stability of their views are low. On some matters, however, the general public cares deeply and holds strong views. On virtually every aspect of politics and public policy, a smaller group of people cares very much and pays close attention. These special, or attentive, publics also contribute to the opinions that shape the play of power.

Citizens acquire many basic political values and beliefs, as well as political attitudes from a variety of agents of socialization, during childhood. However, political socialization continues into adulthood. Life experiences, significant public events, reactions to the success or failure of public policies, and the words and deeds of groups and leaders also mold adult opinion. Four significant characteristics of contemporary adult opinion are especially relevant to understanding the play of power: the decline of trust in government; the erosion of confidence in politicians and political institutions; the growing acceptance of women, African Americans, and other minorities (albeit accompanied by lingering prejudice); and broad support for major policies such as those dealing with Social Security and a strong defense. Finally, despite the fact that citizens know little and care little about many specific issues, it is possible to identify shifts in the public mood about the extent to which government should act to solve social and economic problems.

Opinions differ according to race, gender, ethnicity, and a host of other factors such as education and income. However, there are significant splits of opinion within these categories, and social characteristics predict opinions with only moderate success. Furthermore, differences between liberals and conservatives, at least in the general public, are muted. As the population ages and age cohorts grow older and shrink, the composition of the public will change, and with it public opinion.

Public opinion has been defined as something that politicians find it prudent to heed. The extent to which opinion shapes policy is critical to the very notion of democracy, but it is difficult to assess. Occasionally the general public expresses intense, salient opinions, determining what elected officials do. More often the public mood is merely suggestive, indicating the general direction policy should take but saying little about specifics. Politicians heed such opinion both because they believe they ought to and because they anticipate possible electoral defeat if they do not. Finally, on many issues, opinion is permissive because most people have no strong view one way or the other. For these issues, however, there is likely to be an attentive public, and the direction and intensity of its opinions shape policy. In short, the opinions of both the general public and special publics take an important part in the play of power.

APPLYING KNOWLEDGE

INTERPRETING PUBLIC OPINION POLLS

 The only way to know the content of general public opinion is through published polls. Learning to interpret polls correctly is therefore crucial. Consider the information in the table on the following page, taken from a typical published poll—this one regarding opinions on abortion. The poll results included a detailed breakdown of answers to the following question: "Should abortion be legal as it is now, or legal only in such cases as rape, incest, or to save the life of the mother, or should it not be permitted at all?"

How does one go about interpreting such a poll? Most readers would probably immediately look at the breakdowns of opinion to draw conclusions about who supports what position. However, a better approach is to first ask what the poll does *not* reveal about opinion on this issue. Using the concepts presented in this chapter, discuss what attributes of opinion the poll does not reveal. How might a more sophisticated poll obtain and present this missing information? How would the information not provided help in assessing the probable effect of opinion on this issue? Finally, how do the results presented here illustrate the chapter's information about the nature of public opinion?

HOW GROUPS DIFFER ON ABORTION

% SAYING ABORTION SHOULD BE:	LEGAL AS IT IS NOW	LEGAL ONLY IN CERTAIN CASES	NOT PERMITTED AT ALL
Total adults	49	39	9
Age			
18–29 years	56	35	8
30–44 years	49	49	9
45–65 years	45	39	12
65 years and over	39	45	9
Sex and marital status			
All men	51	38	8
Unmarried men	60	31	7
Married men	46	41	8
All women	47	40	11
Unmarried women	54	37	7
Married women	42	42	14
Education			
Less than high school	37	41	16
High school graduate	47	41	9
Some college	56	35	7
College graduate	58	35	5
Race			
White	49	39	9
Black	45	42	13
Religion			
All Protestants	44	44	9
Religion very important	34	49	13
Religion not so important	61	36	1
All Catholics	48	36	13
Religion very important	28	49	22
Religion not so important	72	22	4
Political philosophy			
Liberal	65	28	5
Moderate	54	38	6
Conservative	38	46	13
Exposure			
Women saying they had an abortion	79	12	9
People knowing someone who had an abortion	58	34	7
People not knowing anyone who had an abortion	39	44	12

Source: New York Times, April 26, 1989, p. A25. Copyright © 1989 by The New York Times Company. Reprinted by permission.

Key Terms

attitudes (*p. 226*)

opinions (*p. 226*)

political opinions (*p. 227*)

direction of opinion (*p. 228*)

salience of opinion (*p. 229*)

intensity of opinion (*p. 229*)

stability of opinion (*p. 230*)

opinion cluster (*p. 234*)

ideology (*p. 234*)

public opinion (*p. 236*)

general public (*p. 236*)

attentive public/special public (*p. 236*)

political socialization (*p. 236*)

agents of socialization (*p. 236*)

age cohort (*p. 253*)

determinative public opinion (*p. 256*)

suggestive public opinion (*p. 256*)

permissive public opinion (*p. 256*)

Recommended Readings

Abramson, Paul R. *Political Attitudes in America.* San Francisco: Freeman, 1983. A very good review of the academic research on public opinion.

Erikson, Robert S., Norman R. Luttbeg, and Kent L. Tedin. *American Public Opinion: Its Origins, Content, and Impact,* 4th ed. New York: Macmillan, 1991. A good general introduction to the topic.

de la Garza, Rodolfo O., Louis DeSipio, F. Chris Garcia, John Garcia, and Angelo Flacon. *Latino Voices: Mexican, Puerto Rican, and Cuban Perspectives on American Politics.* Boulder: Westview, 1992. A rare treatment of the political views of Hispanic groups.

Key, V. O. *Public Opinion and American Democracy.* New York: Knopf, 1961. A classic and influential treatment of public opinion and its place in the American political system.

Mayer, William G. *The Changing American Mind: How and Why American Public Opinion Changed between 1960 and 1988.* Ann Arbor: University of Michigan Press, 1992. A survey of the major trends in American public opinion in recent decades.

Sigelman, Lee, and Susan Welch. *Black Americans' Views of Racial Inequality: The Dream Deferred.* Cambridge, Eng.: Cambridge University Press, 1991. A comprehensive treatment of the differences in the attitudes of blacks and whites.

The Role of the Mass Media

IN 1991 THE PERSIAN GULF WAR provided some of the most gripping television images ever seen in this country. Indeed, the opening of the clash was covered live as television viewers were treated to nighttime shots of the Iraqi capital, Baghdad, under air attack by United Nations planes and missiles. In another broadcast, General Norman Schwarzkopf, commander of U.N. forces, appeared alongside a television monitor showing grainy black-and-white footage from a U.S. warplane's "smart missile." What began with a bird's-eye view of the missile heading toward a blurry vehicle scurrying across a desert road ended abruptly as the vehicle, its driver, and the camera were obliterated by the missile's exploding warhead. General Schwarzkopf returned to explain that we now had one fewer Iraqi mobile Scud missile launcher to worry about. The power of such images led to the belief that the war was overwhelmingly successful owing to the effectiveness and "surgical" precision of American "smart weapons." We knew this to be true, of course, because we had seen it with our own eyes.

However, later we learned that what we had seen on television was not as clear as we had thought. It turned out that what Schwarzkopf had said was a mobile Scud missile launcher was more likely a Jordanian truck trying to break through the U.N. blockade. An Air Force study concluded that there was no evidence that U.N. forces had destroyed even a single mobile missile launcher, and it found that Iraq had as many missile launchers at the end of the war as it had had at the beginning.[1] In fact, virtually all the images Americans saw of the war were carefully censored and controlled by the military. This image management created a vision of a war dominated by "smart" weapons that could "surgically" destroy military targets while sparing civilians. In actuality, "smart" weapons constituted only a tiny portion of all the armaments used, most of which were decidedly "low-tech" artillery shells and iron bombs dropped from B-52s. Further, the massive bombardment of Baghdad and other areas in Iraq did considerably more damage than was revealed by the images we saw.[2]

The lesson we can draw from this example is that video images of an event, even live broadcasts, cannot be accepted as if they were the event itself. Any information we receive through the mass media must be critically analyzed. If you hope to understand American politics, you must always consider the ways in which the mass media affect the play of political power. In fact, the media's impact on politics is itself best thought of as a struggle among various political participants to influence the political content of media coverage.

It is often hard to keep this struggle in mind, especially when dealing with the role of television, our most pervasive medium. Because television shows us images of political events, sometimes as they actually occur, there is a strong tendency to believe that we are

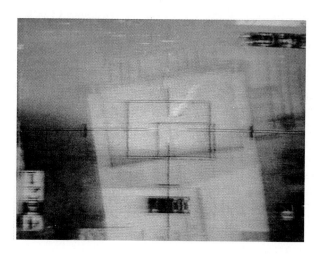

What are we seeing here? According to one source, this is a huge Iraqi ammunition and missile dump in Kuwait City as seen through the guiding scope of an American missile seconds before the site was destroyed. While the tremendous advances in communications technology enable images to be transmitted instantaneously, the power of these images sometimes allows their meaning to be manipulated. This was certainly true during the Persian Gulf War.

seeing the totality of those events and don't need to understand the strategies and players that produce them. This preconception must be resisted if we are to understand the impact of the mass media on politics. The tools we offer you in this chapter will help you with such an analysis and allow you to better understand the media's role in the play of power.

COMMON BELIEFS

THE MASS MEDIA AND THE PLAY OF POWER

To understand why the media are so central to an understanding of the play of political power in the United States, we can start by asking a simple question: Where do you get your information about American politics? Most citizens don't actually communicate directly with governmental officials. They don't often attend congressional committee meetings, read the text of Supreme Court decisions, or plow through administrative regulations. For most of you, knowledge about politics—especially on a day-to-day basis—comes almost entirely from the mass media: newspapers, radio, magazines, television, and so forth. In fact, more than half of all Americans say they get most of their political information from television.[1] It is clear that the media play a central role in democratic politics.

Of course, not all of you believe everything you see on television or read in newspapers or magazines. Many citizens do not share the belief that the media provide us with a neutral "window" on the world. Most do, however, adopt another common belief: They assume that the media are *supposed to* provide a neutral and objective picture of the political world. So, when journalists fail to live up to this standard, citizens conclude that the media are engaged in an attempt to mislead the American people, either for liberal or for conservative ends. For example, almost 70 percent of respondents in a recent *Los Angeles Times* survey agreed with the statement "The news media give more coverage to stories that support their own point of view than to those that don't."[2] We believe that this conclusion is unwarranted.

First, a conclusion that the media ought not to be biased is based on an impossible standard. Simply put, we believe that it is impossible to provide a neutral or objective picture of political events. The facts of political life never stand on their own, but must always be interpreted. This means that anyone—a journalist, a politician, even the author of an American government textbook—who recounts or analyzes political events must adopt a perspective. In this text, for example, we have decided to approach politics as a play of power in which participants use resources and strategies to achieve their ends. We feel that this is a useful way to teach you about the American political system, although it is certainly not the only way to think about politics.

Second, simple assumptions about the media's responsibility to be objective distort and overstate the ways in which the mass media work in American politics. In fact, the belief that the media manipulate the public for nefarious purposes is as misleading as the first conclusion—that media images can be trusted as if they allowed us to see the totality of the events they portray.

268

Consider the 1988 and 1992 presidential elections. In 1988, controversy over broadcast images swirled around the campaign: the Bush campaign ad featuring the picture of a glowering Willie Horton, a black man who had raped a white woman while on weekend parole from prison; Democratic candidate Michael Dukakis wearing an oversized helmet and riding around in a tank; Republican candidate George Bush angrily confronting CBS anchor Dan Rather over his questions about the Iran-Contra affair. The role played by images such as these is certainly an important part of the way in which elections are conducted today. Using the image of Willie Horton to raise deep-seated fears about black men among white voters is especially troubling. However, it is important not to overestimate the significance of such images in election outcomes, as we discuss in "Strategies: Willie Horton and the Grammar of Campaign Ads" (on p. 272). A focus on media coverage that assumes an all-powerful and manipulative media structure can obscure other significant aspects of elections. In 1988, for instance, George Bush was an incumbent vice-president running in times of economic expansion; such candidates in such times do not often lose. This fact was a more important determinant of the election's outcome than anything the media did or did not do in the campaign. Indeed, when George Bush ran in 1992 as an incumbent president seeking re-election during a recession, the outcomes of the same media strategies and of the election were quite different, and Bill Clinton won a solid victory.

In short, although it is important to analyze the role of the media in politics, it is also important to keep this role in perspective. Part of keeping this perspective requires an understanding that the press has always been the target of harsh criticism, because it has been so central to politics since our nation's beginnings. In fact, Presidents George Washington and Thomas Jefferson both attacked newspapers for printing lies. Likewise, the press has long been accused of oversimplification and of pandering to a lowest common denominator, as when nineteenth-century newspapers began to print pictures and were accused of being "frivolous."[3] Rather than holding the press up to impossible standards, seeing it as all-powerful, or considering it simply a passive mechanism through which to view the political world, the approach we develop in this chapter recognizes the media as a significant and complex, but hardly all-powerful, influence on American politics.

Our task is to help you understand how reporters, political elites (e.g., elected politicians such as the president and members of Congress), and other participants struggle to influence the perspectives used in the stories that the mass media tell about politics. Remember that political events are always open to alternative interpretations, and that choosing among these interpretations can have important implications for public policy.

---

[1] Michael X. Delli Carpini and Bruce A. Williams, "Fictional and Non-Fictional Television Celebrate Earth Day: or, Politics Is Comedy Plus Pretense," *Cultural Studies* 8: 1 (January 1994).

[2] Poll cited in Jon Katz, "AIDS and the Media: Shifting Out of Neutral," *Rolling Stone*, May 27, 1993, pp. 31–32.

[3] "Magazines and Newspapers," in Eric Foner and John A. Garraty, eds., *The Reader's Companion to American History* (New York: Houghton Mifflin, 1992). ∎

PARTICIPANTS IN THE STRUGGLE OVER MEDIA COVERAGE OF POLITICS

Who are the principal participants shaping the role of the media in American politics? We identify six broad categories of participants: *the owners of the mass media, journalists, elected politicians, unelected policymakers, organized interests,* and *the general public.* These players are involved in an ongoing struggle to influence the sort of political information that is provided and the interpretation of this information. As with our other discussions of participants in politics, it is always important to consider those who are excluded from or have great difficulty entering the play of political power. Here again, race, class, and gender play a significant role in shaping who can and who cannot effectively influence media coverage of politics. We shall see that the ways in which journalists and media organizations cover the political world make it quite difficult for the poor and people of color to speak for themselves and participate directly in the struggle to shape media coverage of politics in the United States. For example, "Strategies: Willie Horton and the Grammar of Campaign Ads" (on pp. 272–73) describes an instance where the silent image of an African-American male was used to inspire fear in voters. Similarly, the Senate campaign of Jesse Helms in North Carolina and the gubernatorial campaign of Kirk Fordyce in Mississippi used silent images of young African-American women to inspire voter anger over welfare policies.

THE OWNERS AND CONTROLLERS OF MEDIA ORGANIZATIONS

To understand how those who own and control media organizations affect the role of the media in the play of power, you must first understand the basic structure of the mass media in this country. Although some television and radio stations, such as National Public Radio (NPR) and Public Broadcasting Service (PBS), are operated through governmental funding and voluntary contributions, the vast majority of media outlets (virtually all print media and most electronic media) with the largest audiences are privately owned and operated for a profit. Even though most other democratic societies follow this pattern for the print media, their approach to the electronic media differs significantly from that of the United States in that they rely more heavily on publicly owned information outlets (e.g., the British Broadcasting Corporation in Great Britain and the government-supported broadcasting system NHK in Japan). In theory, private ownership of the mass media provides a marketplace of ideas where a great deal of political information from a variety of perspectives is offered to citizens, who can then sift through the competing ideas and form their own opinions about politics.[3] In this way, we avoid the dangers of governmental control of the media as well as the limited number of electronic outlets characteristic of publicly owned systems. Arguments of this sort are being made by the Republican majority in Congress as it criticizes what it sees as the elitist and liberal bias of public television and radio and questions the need for continued public support of the Corporation for Public Broadcasting.

Limited diversity of ownership   Viewed solely in terms of the number of outlets for information, the marketplace of ideas is thriving. Great diversity exists in this system of private ownership, with approximately 1,600 daily newspapers, 11,000 magazines, 9,000 radio stations, 1,000 television stations, 3,000 book publishers, and 7 major movie studios.[4] On the surface, there appear to be thousands of sources that citizens can use to get information.

However, when we examine who owns these outlets, the sense of diversity changes. Over the last 50 years, while the number of outlets has increased, their ownership and control have become increasingly concentrated in the hands of a relatively small number of corporations. This trend has affected both print and electronic media. At the end of World War II, 80 percent of daily newspapers were independently and locally owned; by 1989, 80 percent were owned by outside corporations. Today, more than 50 percent of all newspapers in the United States are owned by 14 corporations. In addition, 68 percent of daily newspapers, with 76 percent of all the circulation, are controlled by chains. Ninety-eight percent of all cities now have only one daily newspaper. To put it another way, in 1923 there were 503 cities with more than one separately owned daily newspaper; now there are only 49 (and in 20 of these cities competing papers have joint business and printing arrangements). The same pattern of ownership holds for magazines. In 1981, 20 corporations owned 50 percent of the nation's 11,000 magazines; by 1992, two companies owned more than half the magazines in the country (see Table 8.1).

In the electronic media, four out of five television stations in the 100 most densely populated markets, serving almost 90 percent of the nation's households, belong to multiple owner groups. Although cable television has given us a vast array of new channels, of the 11,000 cable companies in the country, only seven account for more than half of all cable subscribers. In

TABLE 8.1
THE INCREASE IN CORPORATE MEDIA OWNERSHIP, 1983–1992

TOTAL NUMBER OF MEDIA OUTLETS (APPROXIMATE NUMBER IN 1992)	NUMBER OF COMPANIES CONTROLLING HALF OF TOTAL OUTLETS	
	1983	1992
Newspapers 1,700	20	11
Magazines 11,000	20	2
Book publishers 3,000	11	5
Cable television services 11,000	—	7
Total for all media 25,000[1]	50	20

[1] Total includes additional types of media outlets (e.g., movie studios and television and radio stations).

Source: Ben H. Bagdikian, *The Media Monopoly,* 4th ed. (Boston: Beacon, 1992), pp. ix–xvii. © 1983, 1987, 1990, 1992 by Ben H. Bagdikian. Reprinted by permission of Beacon Press, Boston.

WILLIE HORTON AND THE GRAMMAR OF CAMPAIGN ADS

In the 1988 presidential campaign between Massachusetts governor Michael Dukakis and Vice-President George Bush, campaign ads by the latter demonstrated the power that video images have and the ways in which they can be used in political campaigns. Especially troubling were ads focusing on crime that catapulted escaped murderer William Horton to national prominence. The Horton ad, run by the National Security Political Action Committee (NSPAC), a conservative group supporting George Bush, linked Dukakis to a convicted murderer who had raped a woman and beaten her fiancé while on a weekend furlough from a Massachusetts prison. Horton was a black man; the victims were white.

In her book *Dirty Politics: Deception, Distraction, and Democracy*, Kathleen Hall Jamieson shows how the NSPAC ad and follow-up ads by the Bush campaign itself carefully juxtaposed images with voice-overs to invite viewers to make misleading inferences about Dukakis that played on fears of crime and racial stereotypes:

> The [NSPAC] ad opens with side-by-side pictures of Dukakis and Bush. Dukakis's hair is unkempt, the photo dark. Bush, by contrast, is smiling and bathed in light. As the pictures appear, an announcer says, "Bush and Dukakis on crime." A picture of Bush flashes on the screen. "Bush supports the death penalty for first-degree murderers." A picture of Dukakis. "Dukakis not only opposes the death penalty, he allowed first-degree murderers weekend passes from prison." A close-up mug shot of Horton flashes

onto the screen. "One was Willie Horton, who murdered a boy in a robbery, stabbing him nineteen times." A blurry black-and-white photo of Horton . . . appears. "Despite a life sentence, Horton received weekend passes from prison." The words "kidnapping," "stabbing," and "raping" appear on the screen with Horton's picture as the announcer adds, "Horton fled, kidnapping a young couple, stabbing the man and repeatedly raping his girlfriend." The final photo again shows Michael Dukakis. The announcer notes, "Weekend prison passes. Dukakis on crime."

The Bush campaign followed up with a series of ads on crime that, although not mentioning Willie Horton by name, invited viewers to recall the NSPAC ad.

> [One] stark black-and-white Bush ad opened with bleak prison scenes. It then cut to a procession of convicts circling through a revolving gate and marching toward the nation's living rooms. By carefully juxtaposing words and pictures, the ad invited the false inference that 268 first-degree murderers were furloughed by Dukakis to rape and kidnap. As the bleak visuals appeared, the announcer said that Dukakis had vetoed the death penalty and given furloughs to "first-degree murderers not eligible for parole. While out, many committed other crimes like kidnapping and rape."
> But many unparoleable first-degree

short, although there are more than 25,000 media outlets, in 1986 only 29 corporations owned more than half of them. In 1992 the figure was down to 20 corporations.

In addition to being increasingly concentrated, the owners of media outlets show little diversity in race and gender. Of all electronic media

BUSH & DUKAKIS
ON CRIME

Supports
Death Penalty

Opposes
Death Penalty

Allowed Murderers
to Have Weekend
Passes

Willie Horton

Horton Received
10 Weekend Passes
From Prison

During the presidential election campaign of 1988, an advertisement developed by
George Bush's media consultants linked Massachusetts governor Michael Dukakis
to Willie Horton, a convicted murderer who raped a woman after being released
under a prison furlough program. Note how the sequence of images visually links
Willie Horton with Michael Dukakis without ever explicitly stating whether
there is a connection, or what it might be.

murderers did not escape. Of the 268
furloughed convicts who jumped furlough during Dukakis' first two terms,
only four had ever been convicted first-
degree murderers not eligible for parole. Of those four not "many" but one
went on to kidnap and rape. That one
was William Horton. By flashing "268
escaped" on the screen as the announcer speaks of "many first-degree
murderers," the ad invites the false
inference that 268 murderers jumped

furlough to rape and kidnap. Again, the
single individual who fits that description is Horton.

The lesson here is that the development of
political ads and their ability to juxtapose video
and audio messages create new and potent strategies in campaigns.

Source: Kathleen Hall Jamieson, *Dirty Politics: Deception,
Distraction, and Democracy* (New York: Oxford University
Press, 1992), pp. 17, 19–20.

outlets in 1992 (AM and FM radio stations plus television stations), African
Americans owned 1.8 percent, Hispanics .9 percent, Asians .1 percent, and
Native Americans 0 percent. In short, of the 10,834 electronic media outlets,
minorities own only 308, or 2.8 percent.[5] Although there are no comparable
figures on the number of female owners, we do know that few women

occupy top decision-making positions. A 1987 survey found that of all vice-presidents, general managers, and presidents at television news stations, only 6 percent were women. This figure was only a slight improvement over 1978.[6] The picture is similar for the print media. A 1987 survey of publishers and general managers found that only 5.4 percent (79 out of 1,454) were female. Worse, one-quarter of the total were employed at a single company.[7]

The dangers of concentrated ownership What difference do the private ownership, increasing concentration, and lack of racial and gender diversity make? One danger is that the large corporations that own most media outlets will tend to slant the news in favor of their own interests. In some areas of the news there is evidence that this is happening. For example, media researcher Ben Bagdikian cites many examples of newspapers' and local television stations' slanting coverage of events to boost local business fortunes. They do this because newspapers and television stations depend on the economic vitality of the cities and states in which they operate. The more growth a local area experiences, the more circulation and advertising revenues the mass media gain. There is evidence that industries with many advertising dollars to spend, such as the automobile and real estate industries, are even able to prevent critical coverage from appearing in local newspapers.

For instance, in 1990 the *Washington Journalism Review* (now the *American Journalism Review*) reported that many local newspapers go to great lengths to curry favor with auto dealerships that pay for expensive advertising supplements. These papers refused to print reviews critical of new models or columns that give buyers advice on bargaining effectively and avoiding the deceptive practices of automobile salespeople. Moreover, the advertising supplements, complete with glowing reviews and favorable news releases from the new- and used-car industries, were formatted to make it difficult for readers to distinguish between paid advertisements and normal newspaper copy.[8]

In similar ways, the lack of women in ownership and supervisory roles at newspapers and electronic media outlets may subtly affect news coverage. Investigative journalist Susan Faludi argues that the American mass media have encouraged a conservative backlash against the gains made by feminists in the 1970s.[9] She argues that the limited number of women editors and owners, coupled with the pressure that men are feeling from the women's movement, made the media much more likely to publish stories highlighting the negative effects of the gains made by women in the workforce. Some of the stories that received prominent coverage, such as the limited marriage opportunities for older women and the dramatic declines in fertility rates for women who postpone having children for the sake of a career, turned out to be either false or dramatically overstated. The lack of minority-owned outlets may also lead to a slighting of coverage of events that are of specific concern to members of those groups.

Nevertheless, some commentators, although concerned about the growing concentration of patterns of media ownership, still argue that two factors prevent media outlets from simply providing information that serves the interest of their owners: *the demands of the audience* and *the norms and values of journalists.* As media scholar Doris Graber suggests, "American journalists in large organizations, like their colleagues in small, independently owned enterprises, are interested in appealing to their audiences, and therefore, their stories usually reflect the values of mainstream American society, regardless of the journalists' personal political orientation."[10] Before we reach any conclusions about the impact of private and concentrated ownership on politics, it is important to examine the other players.

JOURNALISTS

There is a very old joke that asks, "Where does a 500-pound gorilla sit?" The answer, of course, is, "Wherever he wants." As with the gorilla, when it comes to the content of the mass media, those who own media outlets get what they want. Studies of the mass media cite many examples of specific changes in news coverage that occurred as a result of direct intervention by owners. One of the most often cited examples occurred in 1961, when *New York Times* reporter Tad Szulc got information that an invasion of Cuba by CIA-trained Cuban exiles was imminent and that the Kennedy administration was sponsoring the invasion. Szulc filed his story and it was set to run on page one of the *Times*. However, the story was dramatically altered at the insistence of Orvil Dryfoos, publisher of the *Times,* who had close ties to the administration.[11] The final version revealed neither the plans to invade Cuba nor the U.S. government's involvement. This decision, made over the objections of the reporters and editors at the *Times,* became especially controversial after the resulting Bay of Pigs invasion proved to be a dismal failure and an embarrassing foreign policy defeat for the Kennedy administration.

Yet most careful studies conclude that such direct interference does not happen very often.[12] Those who own media corporations do not directly determine the specific content of newspapers or news broadcasts on any given day. This is the job of journalists who staff these organizations. To understand how news coverage is shaped, we need to understand the rules that journalists follow in gathering and reporting the news. "The news" and "political events" are not in and of themselves labeled as such. Rather, what becomes the news on any given day is shaped by the rules and practices that journalists use in doing their job. Further, these rules have specific implications for the play of power in American politics because they favor certain interests and disadvantage others.

Journalists: Neutral observers or watchdogs?   What rules do journalists follow? First and foremost, almost all American journalists are committed to providing fair, neutral reporting of political events. Except on the editorial page, newspapers' reports are not supposed to espouse any particular political viewpoint. This was not always the case, and American journalists did not always strive for neutrality. Until the early twentieth century, both political reporting in the United States and the newspapers themselves were fiercely partisan. Newspapers were associated with specific political parties and used their pages to further their own political position and to attack, often in the most scurrilous manner, the positions and even the private lives of their opponents. Only in the twentieth century did journalism, along with many other areas of American life, become professionalized and adopt the goal of objective reporting of political events. This shift to an attempt to report the news objectively does not always have happy results, as is illustrated in "Outcomes: Media Coverage of the Holocaust" (on pp. 276–77).

MEDIA COVERAGE OF THE HOLOCAUST

As we have noted, American journalists did not always strive to be objective. There are many reasons why the press shifted from its earlier partisan perspective to an attempt to be neutral, fair, and objective. One important influence on this change was the unprecedented and effective propaganda used by both sides during World War I. Stories of German soldiers in Belgium impaling babies on bayonets and raping civilian women were reported in American newspapers as if they were true. These stories were believed by readers and helped to shift U.S. public opinion against the Germans. After the war, it was discovered that most of these stories were part of a well-organized Allied propaganda effort to mobilize public support for the war. It is unclear how much has really changed; recall our discussion of similar governmental manipulation during the Persian Gulf War. Nevertheless, following World War I, journalists began to reevaluate their relationship to the information provided them by government and other political sources.

The ideal of **objectivity**, in this context defined as extreme skepticism toward information provided by governments and other political sources, emerged as the ideal for newspaper reporting.[1] This skepticism led to a commitment to tell both sides of every story. However, is this skepticism always warranted? Are there always two sides to every story? Is it, in fact, "objective" to tell two sides to all stories?

> ### POLES SPURN POSTS UNDER NAZI REGIME
> ---
> #### COOPERATION BID FALLS UPON DEAF EARS, DOCUMENT FROM UNDERGROUND LEADERS SAYS
> ---
> #### ARDENT PLEA TO ALLIES
> ---
> #### REPORT READ IN LONDON PUTS GERMANY'S TOLL OF VICTIMS AT 3,000,000 MARK

Three-quarters of the article accompanying this headline dealt with a Polish underground report on the fate of 3 million Poles and Jews who had been murdered or deported by the Nazis. The report noted that 3,000,000 Jews had been exterminated. Only a small portion of the story, at the very end, dealt with the refusal of Poles to serve in the Nazi regime. However, the headlines tell a very different story and never mention Jews at all.[4]

This journalistic commitment to skepticism had tragic consequences with respect to stories of the Nazi extermination of European Jews from 1933 to 1945. Although we like to believe today that news of the death camps in Eastern Europe didn't reach the United States until after the war, this is not true. Eyewitness reports of the Holocaust reached the Allies by 1942, a time when most European Jews were still alive. Further, these accounts were available to newspapers and regularly appeared in major newspapers such as the *New York Times.*

Objectivity is not the only goal that most journalists strive for. They are also committed to the idea that the press should serve as a public watchdog, exposing wrongdoing in government and other circles of power.[13] From the muckrakers of the early twentieth century, such as Ida Tarbell and Lincoln Steffens, who exposed corruption in various areas of American life, to Bob Woodward and Carl Bernstein's dogged investigation of the Watergate scandal and coverup, American journalists celebrate a history of investigative journalism. Similarly, Seymour Hersh gained prominence as an investigative journalist for his exposé of American atrocities at My Lai during the Vietnam War. However, as many journalists realize, the desire to expose wrongdoing can sometimes conflict with the desire to be fair and neutral.

Why didn't the public demand that the Allies do more to help save European Jews? One reason was anti-Semitism, both among high officials (especially in the U.S. State Department) and among the American public.[2] However, another reason was the *way* in which the press covered the story. Because the press was fearful of being manipulated, as it had been in World War I, the stories were rarely given much prominence. This is shown in a headline from the July 9, 1942, *New York Times*. Further, the stories were always qualified as being "allegations" or "assertions" by interested parties (e.g., international Jewish relief organizations). As a result, the stories were easy to dismiss.

Ironically, this commitment to objectivity led to a general unwillingness to believe and report the stories as true.[3] Reports of the Holocaust were received by American journalists throughout its horrible course, but they were largely dismissed as World War I–style propaganda efforts by politically motivated sources to stir up anti-German sentiment. So reports were often attributed to specifically identified Jewish, and presumably suspect, sources. (After all, as we will see later in this chapter, "sources make the news.") Articles about the Holocaust tended to be buried deep in newspapers where they attracted less attention than more prominently placed stories. Little mention was made of any systematic Nazi policy of extermination.

"Distrusting much of what they could see and, of course, even more of what they could not see, reporters and the public greeted news of the persecution of the Jews skeptically."[5]

As a result, little pressure was brought to bear on governmental officials to intervene more forcefully: President Franklin Roosevelt never issued a warning to the Germans that they would be held accountable, after the war, for the murder of Jews; the railroad lines serving the death camps were never bombed; Auschwitz itself was never bombed. Although it is clearly wrong to blame the American press for allowing the extermination of the European Jews to continue, coverage of the Holocaust is a cautionary tale. It shows us that skepticism and other ways of putting into practice a commitment to objectivity do not always have neutral political implications.

[1] Michael Schudson, *Discovering the News* (New York: Basic, 1978).
[2] On the degree of prewar anti-Semitism, both among the public and among high-level State Department officials, see David Wyman, *The Abandonment of the Jews* (New York: Pantheon, 1984); and Benjamin Ginsberg, *The Fatal Embrace: Jews and the State* (Chicago: University of Chicago Press, 1993).
[3] Deborah E. Lipstadt, *Beyond Belief: The American Press and the Coming of the Holocaust, 1933–1945* (New York: Free Press, 1989).
[4] Headline reproduced in Lipstadt, *Beyond Belief*, p. 9.

Too liberal or too conservative? The endless (pointless?) debate   Are journalists neutral or fair? Some individuals and organizations accuse them of having a liberal bias. Surveys of journalists have indicated that, by and large, they *do* tend to be more liberal and more likely to consider themselves Democrats than are members of the general population.[14] This leads to the belief that the media have a liberal slant and that journalists are more critical of conservative politicians than they are of liberals.

However, for every individual and organization accusing the press of having a liberal bias, there is an individual or an organization that accuses the press of having a conservative bias. These critics argue that, given the close connection between journalists and the political elites they cover, the press generally supports the status quo. The result is press coverage that provides only the most limited critique of the structures of economic and political power in the United States and thus encourages citizens to accept these institutions as legitimate and inevitable.[15]

Even as those accusing the press of a liberal bias point to the political attitudes of journalists, those who accuse the press of a conservative bias point to the economic and social circumstances of journalists. Over the last few decades, journalists have become increasingly well paid. Each of the anchors on the nightly television news earns well over $1 million a year. Print journalists for the leading newspapers (e.g., the *Washington Post* and the *New York Times*) are also extremely well paid. This (so the argument goes) makes them sympathetic to the perspectives of the rich and powerful.

In addition, political leaders recognize the significance of the mass media and try to develop ongoing friendly relationships with journalists, especially with reporters from leading newspapers and networks. Sometimes it becomes difficult to draw a line between journalists and the political elites they cover. For example, David Gergen served as a communications adviser to Republican presidents Richard Nixon and Ronald Reagan before becoming an editor at *U.S. News and World Report* and a highly visible figure on the television political talk show circuit. Then, to much surprise, in May 1993 he was appointed communications chief for President Clinton! It has been argued that American politics has moved past the two-party system and adopted a new system, called "insiderism." Gergen's movement between parties and ideologies, and across the divide between the press and politics, is a good example of this.[16] In short, mainstream political journalists are clearly a part of the Washington establishment.

How diverse is the press corps?   As with ownership patterns, the makeup of the journalism profession indicates that minorities and women are underrepresented. Many critics believe this explains why the press does a poor job of reporting issues that are of special concern to women or minorities. Issues such as sexual harassment, child care, and institutional racism are unlikely to be given prominent or sympathetic treatment by a profession made up primarily of white males.

This pattern of exclusion is most severe in the national television news, the most watched and most influential source of political news. In 1992, of the 50 television reporters with the most on-air time, 45 were white, 3 were African American, 1 was Asian American, and 1 was Hispanic.[17] However, there is evidence that the pattern is quite different for local news broadcasts. A study of 13 of the largest markets found that 11 local stations employed at least one African-American anchor.[18]

The pattern is changing slowly in the print media. The level of minority employment in newsrooms has increased dramatically, from 3.9 percent in 1978 to 10.5 percent in 1994. However, the rate of change has slowed significantly. Newsrooms are unlikely to reach the goal set by the American Society of Newspaper Editors to achieve, by the year 2000, a news staff that reflects the diversity of the rest of the workforce (which is 25 percent minority). Moreover, 45 percent of newspapers still employ no black, Hispanic, Asian-American, or Native American journalists.[19]

The pattern of exclusion is also changing, albeit slowly, for women. Journalism schools now enroll more women than men. At newspapers, 35

Over the past two decades, David Gergen has worked as President Ronald Reagan's communications adviser, as editor of *U.S. News and World Report*, as a talk show host, and most recently as an adviser in the Clinton administration. What does this reveal about political influence in the age of video? Is Gergen the voice of the mainstream or simply the ultimate political insider?

percent of the workforce is female and more than half of the new employees are women. (Of course, these figures include employees at all levels, not just reporters and managers.) As we move up the newspaper hierarchy, the number of women declines precipitously. In 1989, 90 percent of the editing jobs at newspapers were held by men and 76 percent of newspapers had no women in any editorial positions.[20] The pattern in broadcast journalism is even worse: In 1992, women constituted a mere 24.8 percent of the journalistic workforce in television newsrooms.[21]

Has the growing number of women in the media made a difference in the kinds of stories told? Two writers believe that the answer to this question "must be a cautious maybe." At the same time, they argue that although the number of articles written by women has increased, the mass media continue to slight or ignore stories that are of primary concern to women. Overall, "the evidence compiled over the past two decades indicates that

MINORITY WOMEN IN JOURNALISM

Given their status as double outsiders, minority women journalists are in a position to see more clearly than most the ways in which issues of race and gender are often ignored in the business-as-usual media coverage of politics. Ethel Payne, an African-American woman, worked as a journalist for four decades. Payne—who worked for the *Chicago Defender*, a minority-owned and -operated newspaper, and later for CBS radio and television—"distinguished herself by her prodigious coverage of civil rights and human rights and her commentary on public affairs." Four years before her death in 1991 at the age of 79, she was interviewed by Kathleen Currie for the Washington Press Club Foundation's oral history project. The interview, excerpted in Maurine Beasley and Sheila Gibbons's *Taking Their Place: A Documentary History of Women and Journalism*, provides eloquent testimony to the unique problems faced by women of color in journalism trying to cover politics in Washington:

Payne: Male chauvinism was alive and well. That was a gender thing, not a race thing. . . . I can remember that aggressive people like Sarah McClendon were almost ostracized and rebuked at times by some of [their] male peers. I think May Craig stood out because she was a senior member of the White House press corps, and she was highly respected in her own right. But I think Sarah took the heat a lot of times. . . . There was always that haughty air about males in the press corps. They had names and reputations. It was almost like they were holier than thou. . . .

Currie: I think you were describing, too, that when you went to apply for credentials at the House and Senate galleries, you got kind of a—

Payne: I really got a cold reception, as if, "What do you need this for? Why are you here?" But I had White House press credentials, so there was hardly any reason for them to turn me down, but I don't think they liked it at all . . . it was almost—well, it was an all-white preserve, in the first place, and it was almost an all-white male preserve. It was an old boys' network, and they didn't want any stray critters trespassing there. "What are you here for?" You know. You get that. It's unspoken, but you

the influence of women journalists on coverage decisions lags far behind their presence in the profession."[22]

In 1992, women of color constituted 7 percent of the workforce at newspapers. The higher up the ladder one goes, the smaller the percentages become: 3 percent of executives and managers; 6 percent of news-editorial employees (up from 5 percent in a 1990 survey). Broadcast journalism reveals a similar pattern. In a 1990 study of the national news networks (including CNN and PBS), only 2 percent of all stories were filed by minority women. One source estimates that, among news directors, minority women constitute 3.2 percent of television news directors and 2.8 percent of radio news directors.[23] "Participants: Minority Women in Journalism" describes how this pattern affected the experiences of Ethel Payne, an African-American woman and longtime journalist.

feel it. I know sometimes, I would go up there when there wasn't really any hard news to cover, but I'd just go so it would be a practice that they would see me. Then one time, one of them came up to me and said, "Well, I don't think there's anything here that you'd be interested in. It's not in your arena."

I said, "What is my arena?"

"Well, uh, ah, I just didn't think it was of any interest." You know—like that.

I said, "Well, who are you to tell me what my arena is? How do you get off with that?" And I just stared him down. I never had any more trouble out of him, but that was gratuitous, and it was nasty, and it was a put-down. So you had to encounter the fact that you were black, you were a woman, and you were from a minor press. So you were just like some little flea wandering in. They just didn't like to be annoyed that way.

Source: Maurine H. Beasley and Sheila J. Gibbons, *Taking Their Place: A Documentary History of Women and Journalism* (Washington, D.C.: American University Press, 1993), pp. 309, 311–13.

The first black woman to be accredited to the White House press corps, Ethel Payne worked for more than 40 years as a journalist covering public affairs for newspapers, radio, and television.

Although there is compelling evidence that the news media slight many issues of concern to women and minorities, it is difficult to attribute these gaps specifically to the demographics of newsrooms. We believe that much of the popular criticism of journalists—they are too liberal; they are too conservative; they do not fairly represent the diversity of the general population—is, in many ways, pointless. In particular, the argument falters on the lack of convincing information connecting the personal beliefs or characteristics of journalists with the sorts of stories they tell. Instead, it is more useful to look at the institutional and structural patterns of journalistic practice to explain possible biases or slants in coverage of politics. If we are to understand the media's role in American politics, we must have at least a basic knowledge of how journalists define "the news." Whatever the political beliefs or racial, ethnic, or gender characteristics of journalists, their approach to reporting the news is best understood as being shaped by the way they define what it means to be fair and neutral.

How journalists define fair and neutral reporting   What does it mean to be neutral and fair? Rather than argue this question in the abstract, it is more important to learn what journalists are trained to define as fair reporting and then see how this particular definition of neutrality affects the way in which politics is reported in the mass media. It is important to understand that there is no "correct" way to report on politics. The rules that journalists use are not the only way politics can be covered, nor are they the only way to determine what is worthy of coverage. As one newspaper editor puts it, "What the press covers isn't what the press must cover."[24]

Perhaps the most significant rule used by journalists in their coverage of news stories is that "sources make the news." Journalists cannot simply write what they know. They cannot start a story by saying, "I know for a fact that . . ." News is not what happens or what journalists may know to be the case, but rather what someone says has happened or will happen. This means that before any political event can be reported, reporters must find someone who is regarded as a credible source to say it is so. Read any news article in any newspaper and you will see that it is peppered with phrases such as "The president said yesterday. . ." or "Informed sources in the State Department reported that . . ."

Because credible sources make the news, reporting is dominated by political elites who are, by definition, credible sources. Some media critics argue that this fact alone gives the news a decidedly elitist slant. Media organizations concentrate their reporters in Washington so that they will be in constant contact with these sources. Since media organizations have limited resources, organizing reporters' beats around Washington limits the number of other stories that can be covered. Consequently, news stories in the national media focus heavily on what goes on within the Washington Beltway. As a result, in one study of the *New York Times* and the *Washington Post*, the most important sources for stories were found to be governmental officials; the president, presidential candidates, House and Senate members, and other federal officials constituted more than half the sources used in all stories reported. Further, the higher up in government a source is located, the more likely it is that he or she will make the news.[25]

Why should we care that political elites dominate the news? At the most general level, this means that the news itself focuses on issues that are mostly defined by, and of concern to, elites, who are overwhelmingly wealthy, white, and male (see Table 8.2). Groups that are not well represented among political elites have a difficult time getting their issues and concerns covered on a regular basis in newspapers and on television. In an election campaign, for example, if the major candidates—who are part of the political elite—do not want to talk about issues that concern inner-city blacks, it is difficult for journalists (even black journalists) to write about these concerns. Similarly, until quite recently, when more women were elected to Congress (even though their numbers remain small), many issues of central concern to women—child care, sexual harassment, and so forth—were not raised in the political news sections of newspapers and television broadcasts. Instead, such issues were consigned to the "women's pages" or feature stories.

TABLE 8.2
THE TOP TEN NOTABLE QUOTABLES

1. William Schneider	58	appearances
2. Ed Rollins	51	appearances
3. Kevin Phillips	43	appearances
4. Norman Ornstein	42	appearances
5. Harrison Hickman	33	appearances
6. Robert Beckel	27	appearances
7. David Gergen	24	appearances
8. Fouad Ajami	21	appearances
9. Stephen Hess	19	appearances
10. Robert Squier	18	appearances

These 10 individuals, all of them white men, were the most frequently used experts or analysts on the network news. The appearance tallies refer only to the ABC, CBS, and NBC evening news programs for the 104 weeks in 1987 and 1988.

Source: Adapted from a table that originally appeared in Laurence C. Soley, "The News Shapers," a manuscript published by the University of Minnesota School of Journalism (1989). All rights reserved. Copyright © 1992 by Lawrence C. Soley.

Journalism professor Mercedes Lynn de Uriarte shows how journalists' definition of credible sources steers news coverage away from the concerns of minorities:

> The journalism profession tends to be very slow to recognize that there is more than one way to see anything. . . . The editor defines as objective only his/her own perception in looking at a story. . . . If you look at sociological work done on the media, you find that 85% of sources used in U.S. mainstream journalism represent government or other official sources—white sources. When you turn to other sources, you begin to work with a pool of people who are unknown, suspect. You have to try to credential them to gain a legitimacy that is automatically accorded to white people who are in jobs that carry titles that are familiar—no matter who they are.
>
> Editors need to look at how often they look to people as sources, and for story ideas that they can depend on to be predictable, familiar, reliable—and how often that excludes minorities and the minority community.[26]

If sources determine what issues make the news, it is also important to understand how journalists' definitions of fair and neutral coverage lead them to use information from these sources. For journalists, being fair means providing two sides to every story. This means that sources must be balanced against one another. If, for example, President Clinton makes a claim about a legislative proposal, reporting of the president's viewpoint will usually be balanced by a quotation from a source from the Republican leadership criticizing the president's claims. But is this the only way to define fairness or neutrality? Might there be more than two sides to a story? Might there sometimes be only one side to a story?

The notion that sources make the news and that fairness means giving two sides of every story is not the only way to define objectivity or neutrality. Many social scientists, for instance, also believe their own work to be fair and neutral. However, the quest for fairness does not require them to provide two sides to every research question. Instead, they are usually committed to discovering the single answer they think is the correct one. They typically do not rely on interviews with individual sources; rather, they tend to report the findings of aggregate surveys, experiments, or other forms of quantitative analysis. Neither the journalistic nor the social scientific approach to being fair and objective can be demonstrated conclusively to be better than the other. They simply represent different ways of finding out about the social and political world around us, and they may lead to different conclusions about the political world. For instance, the scientific method used by social scientists aims to produce definitive answers about social processes. This may be a more appropriate technique than telling two sides to every story when, for instance, we try to answer questions about what factors most influence a citizen's voting decision. However, the journalistic approach may be more appropriate if we are trying to answer questions such as "Was the Gulf War a good thing?"

One implication of the way in which journalists define fairness is that it focuses attention on individuals—the sources that make the news—rather than on the institutions of government. As the prominent syndicated columnist David Broder puts it, "Most [journalists] are a lot more comfortable thinking about and writing about individuals than institutions. 'Names make news' is almost the first commandment of journalism, and the gaudier the individual, the better."[27] The result is that the press has a difficult time effectively covering complex institutions like Congress. In contrast, when an institution—like the presidency—is embodied by a single individual, the press has a much easier time. So, whereas journalists tend to focus on colorful, individual members of Congress, political scientists, because of their own definitions of neutrality and objectivity, tend to focus more on the institutions and structures of Congress (e.g., rules, committee meetings, hearings).

Another implication is that, in general, politics tends to be portrayed as a clash of individual wills. Political issues are reported as partisan struggles for advantage between two forces. One danger of this approach is that it tends to emphasize the struggle itself, as well as the strategies used by the two sides, over the substance of the political issues being fought over and their implications for those who are not themselves, or are not well represented by, credible sources.

Given that sources make the news, we must examine how another set of players who disproportionately serve as sources—political officials, both elected and unelected—pursue their interests in the struggle to influence the media agenda.

In many ways, the rise of the electronic media has transformed the world within which elected politicians operate. The first president to fully understand and use the potential of the emerging electronic media was Franklin Delano Roosevelt. His "fireside chats," broadcast over the radio, allowed him to speak in a personal and direct way about the troubling issues of the day and helped him to forge a link between his presidency and the American people. This connection proved to be a powerful resource in political struggles for Roosevelt and for subsequent presidents who were able to exploit it. Before the emergence of the electronic media, presidential power operated primarily through parties and Congress. By appealing directly to the people, Roosevelt was able to bypass these other centers of political power and establish a direct connection between the president and public opinion. In the 1980s Ronald Reagan, often called the "Great Communicator," was quite successful at bucking congressional leaders by giving tightly scripted televised speeches that made his case directly to the people, thus shaping public opinion about issues. More recently, Ross Perot in 1992 ran the most successful third-party presidential campaign since Theodore Roosevelt's by using his enormous wealth to purchase large blocks of network television time and make his appeal directly to the American people.

The creation of a direct link between presidents (or other elected leaders) and the general public makes the standing of leaders in public opinion polls important to their ability to achieve their policy goals and get re-elected. Nowadays public opinion polling that tracks the popularity of presidents is a staple of news reports, and it affects the ability of the president to influence Congress and other political leaders. Indeed, David Broder says, "In reality, the play of public opinion, as reported and magnified by the press, has grown so powerful in the United States that it has become the near preoccupation of government. . . . When a president loses popularity, he also loses the ability to govern."[28]

One result of the need to maintain public support is that elected politicians, especially presidents, must always keep one eye (and sometimes two) on the impact their actions will have on short-run changes in public opinion. This pressure leads to a tendency for chief executives to behave in office as if they were still running for office. The result is what is sometimes called "the never-ending campaign."

Elected officials other than presidents have seen the potential of the mass media, especially television, for bypassing existing structures of power and forging political careers. In the 1950s, Senators Joe McCarthy and Estes Kefauver both used the emergence of television—albeit in very different ways—as an important new medium for appealing directly to the American people and building prominent national bases of support. Kefauver used his chairmanship of prominently televised Senate hearings into organized crime as a springboard to national prominence and to the vice-presidential slot on the national Democratic presidential ticket in 1956.

McCarthy's career is especially interesting because it illustrates the ways in which television and its impact on public opinion, journalistic rules about fairness and neutrality, and the value journalists place on investigative reporting can all work to make and break the reputations of elected politicians. McCarthy, the Republican junior senator from Wisconsin, took advantage of widespread public anxiety about the threat of communism. Along with other Republicans, McCarthy condemned the Truman administration for the "loss" of China to the Chinese communists and blamed the Soviet development of an atomic bomb on the transfer of atomic secrets by spies. The result was a wave of anti-communist hysteria in the United States that led thousands of Americans to lose their jobs after being accused of harboring secret communist sympathies, or of having once belonged either to the Communist Party or to communist "front" organizations.[29]

McCarthy achieved prominence by making widely publicized claims that the State Department and the Pentagon were infiltrated by communists. He would wave official-looking documents in the air and proclaim that the documents proved that there were—and here he would name a specific number—so-and-so many known communists in the State Department, or in the Army. That these claims were publicized at all, when most responsible journalists knew them to be false, is explained by the way journalists define news and sources, as we have already discussed. Remember that for journalists, being fair means presenting two sides to every story—and that the two sides are defined by the statements of credible sources. Even though journalists had their own information about the veracity of McCarthy's claims and about his personal failings (he was widely known to be an alcoholic), they could not simply report what they knew. As a U.S. senator, McCarthy was, by definition, a credible source. Reporters were bound by their professional training to report these claims. Further, the climate of the times made it unlikely that other equally credible sources (i.e., other prominent politicians) would go on the record to challenge his claims. Even such challenges would merely have been reported as the other side to McCarthy's claims.[30]

Just as journalistic definitions of fairness and neutrality, coupled with the use of television, allowed Senator McCarthy to build a prominent, if infamous, public career, competing journalistic goals of investigative reporting and exposing wrongdoing combined with the emergence of television to lead to the senator's downfall. When McCarthy attacked the U.S. Army as being infiltrated by communists, the hearings were again televised. This time, however, McCarthy was effectively challenged and was exposed as a bullying demagogue. Further, in one of the most widely watched television broadcasts up to that time, the well-known journalist Edward R. Murrow delivered a withering and effective exposé of McCarthy and his methods. The combination of these events and the increased willingness of other politicians to challenge McCarthy led to his ultimate censure by the Senate.

More recently, the Anita Hill–Clarence Thomas hearings had a dramatic impact on politics. In 1991 the accusations of law professor Anita Hill that she had been sexually harassed by Supreme Court nominee Clarence Thomas when he was her superior at the Equal Employment Opportunity Commission produced a widely watched series of Senate hearings into the matter. These hearings, with their images of an all-white, all-male group of senators passing judgment on a sexual harassment dispute between two African Americans, raised questions in the minds of viewers about the ability of white male political elites to deal fairly with issues of concern to women and/or African Americans. Indeed, several women candidates, including Carol Moseley Braun, the first African-American woman to win election to the Senate, specifically cited those hearings as a factor in their decision to run for office in 1992.

Since the advent of television, highly publicized congressional hearings—such as the Army-McCarthy hearings in 1954 and the Clarence Thomas confirmation hearings in 1991—have transfixed the nation and focused attention on specific public issues and problems. How does the presence of television cameras alter Congress's investigative role? Do such public hearings help or hinder the democratic process?

The presence of television cameras in Congress and the broadcast of catchy sound bites from speeches on the floor on the nightly news have created new career paths in Congress. For example, Congressman Newt Gingrich (R-Ga.) used these new developments to build a national reputation as a fiery spokesman for conservative Republican causes. He was able to build this reputation through his frequent speeches, televised on C-SPAN, parts of which were often picked up in brief sound bites on the national evening news. These sound bites constituted the conservative side (remember, for journalists every story has two sides) of many political stories. Gingrich, sometimes sarcastically called "the congressman from C-SPAN," used this national standing to achieve a leadership position (first as House minority whip and then, following the 1994 elections, as Speaker of the House) much faster than he would have had he used more traditional, seniority-based ways of building power.

While effective when they occur, the opportunities for politicians to communicate directly with the public are rare. How often does the president actually make a speech on national television? Very few members of Congress can find the sorts of issues that McCarthy and Kefauver developed to justify televised hearings. Rather, elected politicians still depend on journalists to transmit what they say and do to the public. However, the nature of this dependence varies, as we shall see. This dependence creates a struggle, best thought of as a play of power, between politicians and journalists to control the content of the news we read or view in the mass media. But before we talk about the strategies, resources, and outcomes of that game, we must first describe two other kinds of players.

UNELECTED POLICYMAKERS

Unelected policymakers, such as career civil servants, appointed governmental officials, and others who staff large governmental bureaucracies, also wield great power. However, we are often not aware of the important relationship between the mass media and unelected governmental officials. In fact, the media's ability to influence public opinion provides one of the few counterpoints to the growing power of bureaucratic organizations. Journalists' desire to serve as public watchdogs and expose wrongdoing in government runs counter to the bureaucratic penchant for secrecy and isolation.[31]

In a careful study of the impact of the media on the policy-making process, Martin Linsky and his colleagues demonstrate the degree to which journalists can influence the normal operating procedures of bureaucratic agencies.[32] In most cases, even though elected politicians make public policy, implementation is generally overseen by bureaucratic agencies that operate outside the awareness of most members of the public. As a result, the policy process advances incrementally and public policies tend to change slowly over time.

However, Linsky found that when media attention focuses on a policy issue, the policy process changes quite dramatically. Two changes are especially important. First, the process speeds up as pressure is created for quick and dramatic solutions to problems that journalists bring to the public's attention. This pressure is even more intense when the issue makes it into television news, as opposed to print media. Second, press attention tends to take responsibility away from lower-level officials and move it to the tops of organizations, even to the White House. In short, when the *Sixty Minutes* cameras roll into a bureaucrat's office to expose some alleged governmental wrongdoing or incompetence, there is enormous pressure on officials higher up to respond quickly.[33]

Linsky and his colleagues found both these effects when they explored press coverage of the Love Canal story during the Carter administration. In 1978 it was discovered that many homes in the Love Canal neighborhood of Niagara Falls, New York, had been built on top of a leaking hazardous waste storage site left by the Hooker Chemical Corporation years earlier. Families living closest to the leaking facility were evacuated and their homes purchased by the government. However, many families living in the same neighborhood, but further from the site, were not evacuated. Instead, the Environmental Protection Agency (EPA) chose to proceed slowly and incrementally. Scientific studies were ordered to determine the danger posed by the leaking wastes, and this information was to be used to determine where to draw the boundaries for further evacuation at the government's expense. These carefully laid plans were upset when the Love Canal case became the focus of national media attention.

Linsky uses this case to illustrate the ways in which media attention imposes a "doomsday clock" on policymakers. Press coverage emphasized the horrors of the leaking wastes and their health effects. The prominence of these stories increased the pressure on the Carter administration to "do something" and do it fast. Responsibility for making decisions about Love Canal evacuations moved from the EPA to the White House. The result was a quick decision, without waiting for the scientific studies, to evacuate many more of the families and purchase their homes at the government's expense. Whether this was a wise decision or not, the impact of media attention forced the president himself to take responsibility for the decisions away from unelected decisionmakers at the EPA and to find a quick and dramatic solution.

Because unelected governmental officials have come to understand the importance of the press, they now spend a great deal of time devising press strategies for influencing journalists. Linsky found that the more recently the presidential appointees he interviewed had served in government, the more time they reported spending on dealing with the press. This means that if you are to understand the impact of the mass media on American politics, you must consider unelected policymakers as significant players in the game.

ORGANIZED INTERESTS AND THE PUBLIC

So far, the players we have considered are those who struggle over the content of the news media. It is important to remember a central reason for this struggle: that coverage of political events in the mass media has a significant impact on citizens and public opinion. Consequently, we need to consider citizens as important players in the struggle over control of media coverage of politics.

"Ordinary" citizens have a difficult time influencing media coverage of events. One reason has to do with the way in which journalists define credible sources and the pattern of journalistic coverage that flows from this definition. As we have seen, media organizations tend to assign their reporters to cover the activities of governmental officials with whom they develop ongoing relationships. This is no less true for established interest groups. So, for example, since journalists define fairness as providing two sides to every story, any story on the dangers of cigarette smoking must include a statement by the tobacco lobby. In anticipation, the tobacco industry supports a sophisticated publicity organization and a network of spokespeople who are routinely consulted by reporters. Similarly, when stories are written about environmental issues, virtually no story is complete without appropriate quotes from representatives of environmental groups such as the Sierra Club or Greenpeace.

Part of the play of power surrounding issues covered by the media revolves around the struggle of newly organized groups to become credible sources and, as a result, be given a voice in news stories. In debates over AIDS policies, the theatrical displays of a group like ACT UP are partly an attempt to have their views become part of the discourse conveyed by journalists.

In contrast to the struggle among organized groups and the routine and familiar relationships that result, those citizens and interests that are less well organized have a much more difficult time becoming a credible source for journalists. Newcomers to the political fray, or one-time participants, have a difficult time winning or even understanding the struggle to influence media coverage of politics. They often fail to gain the type of coverage they seek. During the 1960s, for instance, college students at several campuses created Students for a Democratic Society (SDS), an organization devoted to a wide range of left-wing causes—including civil rights for African Americans, anti-apartheid protests, and opposition to the Vietnam War—and a radical program for establishing grassroots democracy. Sociologist Todd Gitlin analyzed the history of SDS, showing how the group's unfamiliarity with the media and its inability to communicate its broader concerns effectively to journalists led journalists to identify SDS as a single-issue, anti-war group. The broader concerns of SDS members were ignored in media coverage.

Because it was through this media coverage that college students across the country learned about SDS, the organization began to attract students who only wanted to protest the Vietnam War. SDS itself changed as a result of its inability to effectively influence media coverage of its own activities and positions.

When the news does focus on nonelite sources, there is evidence that the resulting coverage works to the disadvantage of minorities. One study of local television news coverage found that half the stories involving blacks (49.8%) were about either crime or politics.[34] In both areas, blacks were portrayed significantly differently than whites. These differences were not neutral, but rather worked in ways that were likely to reinforce what social scientists call "modern racism"[35]—a general hostility toward black people, a rejection of black political aspirations, and a denial that discrimination is still a problem for blacks. For example, there was a tendency to treat black and white accused criminals differently in the news. Blacks were less likely to be named than whites; they were more likely to be shown being physically restrained by police officers; and they were less likely than whites to be allowed to speak in their own defense. All these differences are likely to reinforce false beliefs in the general dangerousness of people of color, especially African-American males. With respect to political participation, media scholar Robert Entman found that

> black activists often appeared pleading the interests of the black community, while white leaders were much more frequently depicted as representing the entire community. News about blacks who acted politically conveyed the notion that they spoke and behaved more than whites to advance "special interests" against the public interest.[36]

These differences are likely to undermine the legitimacy of black political leaders and their demands as being more parochial and self-interested than those of white political activists. Other studies have found similar differences between the ways in which stories are told when they involve blacks as opposed to whites.

Aside from crime stories and occasional accounts about local activists, when "ordinary" citizens enter media coverage of politics it is primarily through the constant reporting of the results of public opinion surveys. Yet, in a somewhat circular fashion, media coverage of politics influences public opinion. What do we know about the ways in which media coverage of politics affects public opinion? The effective use of the mass media for propaganda purposes by totalitarian regimes during World War II caused much concern about the increasing dependence of citizens in democratic societies on the mass media. However, early postwar social science research revealed that the media had only minimal effects on public opinion and were far from being an all-powerful influence. These studies found that other influences—especially education, friends, family, and co-workers—were more important influences on political opinion than the media.[37]

More current research indicates that the mass media have an important, but not unlimited, impact on citizens' political attitudes and beliefs.[38] Perhaps their most important influence is their power to set the political agenda for citizens. **Agenda setting** means that, although the media cannot tell

Joseph Goebbels, the Nazi regime's minister of propaganda, prepares one of his frequent radio broadcasts. The Nazis made effective use of radio to mobilize German public opinion in support of their policies. Fears of the potential impact of this sort of propaganda inspired much of the early research about the impact of the mass media on public opinion in America. While this research found little evidence that the mass media have a profound effect on public opinion in the United States, the media's role in politics is still a source of much concern.

people what to think, they do influence what people think about. Thus, issues prominently featured in the media become the issues that citizens think are most important. For example, when, in 1990, on the twentieth anniversary of the first Earth Day, the television news featured many stories on the environment, citizens focused their attention on that issue and public opinion polls revealed that the environment was a number-one priority. A similar dynamic may be at work with public concern over crime. Even though most statistics indicate that violent crime rates are declining in most areas, public concern remains high. One reason for this may be that the issue receives prominent coverage in both the national and local news.

Inevitably, however, media attention will shift to other issues—perhaps the economy, or the latest events in the Middle East—and there will be fewer and fewer stories on crime. At that point, citizens will shift their attention to these new issues and public opinion polls will reveal that they are of overriding importance to citizens, who now seem to be much less concerned about crime. These shifts in public attention result not from changing political conditions (the crime rate or the environment has not improved or worsened) but rather from the changing focus of media coverage.

There is a second way in which the media influence public opinion: News coverage helps to shape the criteria that citizens use to make political judgments about the issues they consider important. This influence is called **priming**. For example, if news stories emphasize the connection between presidential decisions and the economy, individuals are more likely to evaluate the president in terms of economic performance and attribute to him responsibility for the economy. On the other hand, if stories about the president emphasize issues of character and trust, then viewers tend to evaluate the president's performance in terms of these criteria, as opposed to policy performance or the state of the economy.

At a more general level, the way journalists tell stories also suggests to citizens where to place responsibility for various political conditions. For example, if stories about the economy focus on governmental policies and aggregate statistics about national unemployment rates, inflation rates, interest rates, and so on, then viewers tend to assign responsibility for economic performance and well-being to governmental officials. In contrast, when stories emphasize individual workers and specific companies, then viewers tend to attribute responsibility for the economy and economic well-being to individuals and not to political officials.[39]

What this means is that journalists, elected politicians, and unelected governmental officials all struggle over the content of media coverage of politics because they understand that this coverage has an important impact on public opinion. However, although priming and agenda setting are important influences, the media do not constitute an all-powerful institution: They can influence public opinion, but they cannot determine it. So, for example, agenda setting suggests that even though the media can place environmental protection on the political agenda, this coverage does not change those who favor less regulation and interference with business into rabid environmentalists.

Perhaps more important, researchers find that the media exert the most impact on citizens who are the least well informed. Priming and agenda setting, for instance, least affect members of the audience who have the background and information needed to rebut the media messages they receive.[40] Thus, well-informed citizens are able to use and critically evaluate the bits and pieces of information they receive to form their own opinions about issues.[41]

Let us now see how the strategies and resources used by the various players we have identified shape media coverage of the political world.

ECONOMIC AND POLITICAL ARENAS OF COMPETITION

The best way to understand the political information produced by the mass media is to think of two separate competitions being waged to influence the media agenda.[42] Political news is the outcome of both political and economic competitions. In these competitions, the players we have identified compete to attain desired outcomes in the struggle over the content of political coverage.

THE ECONOMIC MARKETPLACE AND THE COLLAPSE OF THE "OLD NEWS" MONOPOLY

As we noted, the overwhelming majority of American media outlets are privately owned and operated for a profit. Competition in the economic marketplace encourages news organizations to minimize costs and maximize profits. This pressure has increased as media outlets are more likely to be owned by large corporate conglomerates setting profit targets to be met by the news divisions of networks and newspapers that have been absorbed into national chains. How do the strategies for maximizing profits and minimizing costs help us understand the news we receive?

First, we must always remember that, because the competition occurs in a private market, the goal of a newspaper or a network or cable news program is to make money as well as to inform us about events. Profit depends on the size of the audience that a network or newspaper can attract, so the pressure to maintain an audience makes it very risky for any particular media outlet to break with the expectations of its audience. This pressure accounts for the great similarities among network news broadcasts and the sometimes superficial quality of coverage of political events in the print and electronic media outlets with the largest audiences. As one scholar points out, because there is assumed to be little demand for first-rate political news coverage, news organizations cannot produce it; to do so would risk losing part of their audience. Ironically, however, only sustained production of such coverage would lead to a sophisticated audience that might demand it.[43]

Economic competition also leads to strategies that minimize the costs of producing the news as well as maximize the size of the audience. Especially in television news, network budget cuts and the rising salaries of the best-known anchors and reporters have made it necessary to cut back the number of bureaus and reporters employed.[44] This process increases journalists' reliance on elite sources. Political elites, especially national elites in Washington, D.C., can be counted on to produce news stories every day. Reporters assigned to Washington beats—the White House or Congress, for example—can be counted on to provide stories on a daily and predictable basis. In contrast, assigning reporters to Denver, Colorado; Madison, Wisconsin; or New Delhi, India, may produce a story deemed nationally newsworthy now and then, but not on a regular basis. To the news producer who is pressured to produce the daily news at the lowest possible cost, it makes economic sense to assign reporters to locations that will produce a steady flow of news stories. This is one more reason why the national news has come to be dominated by political elites within the Beltway.

Yet, even though economic pressures lead to stable and predictable news production, other forces have introduced dramatic changes in how and where we receive our information about political events in the world. One advantage of seeing media coverage of politics as the outcome of a competition is that it alerts us to the changing process of news production. In any competition, players constantly seek to take advantage of new strategies and resources as these become available. This is nowhere more true than in the mass media, where changes in communications technologies have dramatically affected the way in which the players compete in their struggle to control media coverage of politics.

New media technologies and the emergence of the "new news"   In the last decade or so, almost unbelievable changes have occurred in communications technologies: the advent of cable television; fiber optics; hand-held minicams; personal computers; VCRs; and so forth. New technologies have altered the structure and dynamics of the competition for audiences in which the major news producers engage. In 1980 only 22 percent of all households in the United States received cable television. By 1991 the percentage had tripled and more than 60 percent of homes were wired for cable, which brings large numbers of stations into the home. VCRs make it possible for viewers to watch what they want, when they want, and most American households now have one.[45] These developments have shattered the monopoly of the three major networks as suppliers of political information in the electronic realm and, indirectly at least, have reduced the audience for newspaper coverage of politics as well (see Figure 8.1).

Changing media technology has fragmented the audience for national news. Before cable television, during the 7 to 8 P.M. time period the three network news broadcasts had as their audience the entire television-watching public. Whatever the three networks broadcast became the central images and definitions of the news shared by most of the nation. Most of the television-inspired images of the last 30 years that became the staples of our political consciousness were produced by this media system: the Kennedy assassination; scenes from the civil rights struggles in the South; urban riots; the Vietnam War and Vietnam War protests; the Watergate hearings. The audience for the nightly network news used to approach 90 percent of all homes with television sets; this figure is now down to less than 60 percent.[46]

An obvious change from the "golden age" of the network news is the emergence of 24-hour news stations—C-SPAN, CNN, and so forth—that compete with the major networks. The 30 minutes devoted each night to the network news, and even the extensive coverage of events in the *New York Times*, must now compete with the constant barrage of political talk shows (e.g., *Washington Week in Review, Capital Gang, The McLaughlin Group*), live news conferences, and direct broadcasts of congressional debates. Although these new sources attract only tiny audiences of news "junkies," this audience includes the political elite in Washington. The major networks' news broadcasts also face competition from other non-news "public affairs" programming—for example, daytime talk shows (such as those hosted by Oprah Winfrey and Phil Donahue) and sensationalized parodies of news programs broadcast during the traditional news hour, such as *Hard Copy* and *A Current Affair*. Even MTV regularly broadcasts a news program aimed at younger viewers. Moreover, during the presidential campaign in 1992, the Comedy Network covered the Democratic and Republican conventions from its own humorous perspective. These alternative sources of information can attract large audiences and are usually quite inexpensive to produce. Indeed, one of the reasons for the failure to expand the nightly news broadcasts to one hour is that local television stations make much more money from the alternative programming (game shows, reruns, or shows like *Hard Copy*) that currently runs in that time slot.

Jon Katz, media critic for *Rolling Stone*, argues that the proliferation of less traditional outlets produces a conflict between the "old news" and the "new news." The "old news" is provided by journalists in more traditional outlets such as newspapers and nightly news broadcasts. In contrast, the "new news" is found in less traditional outlets—talk shows, MTV, docudramas, music, and films—and is provided by figures who sometimes mimic, but do not adhere to, the practices of professional journalists.

Radio talk shows, especially those hosted by conservatives such as Rush Limbaugh, have become a particularly potent force among the "new news" outlets. With an audience of millions every day, these shows play an increasingly important role in both agenda setting and priming. It is no surprise that when public opinion turned against President Clinton's plan to provide loan guarantees to Mexico, Alan Greenspan, chairman of the Federal

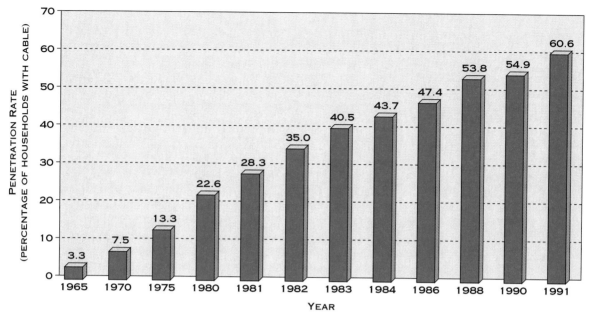

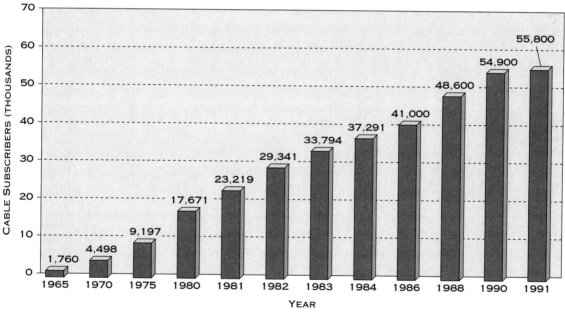

FIGURE 8.1

THE CABLE TELEVISION EXPLOSION As more and more homes are wired for cable television, the number of channels available to most viewers has skyrocketed. What are the implications of this increase for political life in America? Does it improve American democracy by providing a greater diversity of sources of information? Or does it fragment and divide the polity by reducing the audience for the network news? *Source:* Data from *Television and Cable Factbook* (Washington, D.C.: Warren Publishing, Inc., 2115 Ward Court, N.W., Washington, D.C., 20037, 1991), as quoted in Doris Graber, *Mass Media and American Politics,* 4th ed. (Washington, D.C.: Congressional Quarterly Press, 1993), p. 415.

Reserve Board, called Limbaugh to try to educate him about the administration's position.

Although the increasing prominence of the "new news" is often decried, it offers important opportunities for groups usually neglected by mainstream news outlets. The "new news" tends to be less constrained by "old news" notions of who constitutes a credible source or how one goes about telling a news story. For example, black rap artists claim that their music is the true news of the ghetto (see "Applying Knowledge: Gangsta Rap—Menace to Society or News from the Ghetto?" on p. 313). Indeed, the Los Angeles riots could not have come as a surprise to anyone familiar with the music of rappers such as Ice-T, Ice Cube, or Dr. Dre. "New news" outlets sometimes raise issues that are ignored by more traditional news outlets, as with Oliver Stone's provocative exploration of the Kennedy assassination in the movie *JFK*. Talk shows, although sometimes sensational and pandering, also raise issues—especially issues of concern to women—that have traditionally been defined as private and ignored by traditional political elites and mainstream journalists. It is no accident that the predominantly female audience for such shows leads them to address issues such as child abuse and sexual harassment. The increased diversity of information sources provides great opportunities for those who are willing to sift critically through these new and often conflicting sources of information.

The economic competition to maximize audiences has produced changes in how traditional news outlets gather and report the news. In fact, there has been a marked tendency for even the more conservative news outlets to become more sensational and less focused on traditional definitions of the news. Thus, for example, at least one-third of the *New York Times* is now devoted to "lifestyles" topics such as arts and entertainment, fashion, computers, and sports. Consequently, less space than before is devoted to coverage of political events.

Changes in the economic marketplace facing television journalists are even more dramatic than those facing their print counterparts. Indeed, changes in the broadcasters' economic marketplace may have obliterated the boundaries between what we traditionally think of as news broadcasts and the fictional or entertainment programming on television.[47] First, there is a growing tendency for entertainment television to reflect real-world issues and events (e.g., docudramas that portray actual events or series that deal with current political or social issues). Since entertainment shows are consistently watched by large audiences, it seems reasonable that such shows will influence the ways in which viewers understand the issues presented. Politicians certainly seem to think so. For example, in 1992 Vice-President Dan Quayle attacked the show *Murphy Brown* as undermining family values by portraying the main character's decision to have a child out of wedlock.

Second, there is a subtler tendency for nonfiction television to use the form and substance of fiction—staging events, using graphics and movie clips to dramatize issues, employing the narrative conventions of fictional storytelling, giving newscasters celebrity status, and so forth.[48] One reason nonfiction programming borrows conventions from entertainment programming is that the latter type of programming dominates. Thus, most people's expectations about what they will see on television and how it will be presented require that public issues be presented in an entertaining fashion. This has clear implications for the ways in which public issues can be raised even on nonfiction programs. For example, the news program *Dateline NBC* was the subject of a storm of controversy after its producers rigged tests of a GM pickup truck, suspected of having a faulty and dangerous gas tank design, to produce a fiery explosion.

Third, television blurs the line between fiction and nonfiction by presenting fictional accounts of issues or events while they are still topical, and by tying these entertainment broadcasts into the news itself, often using each to promote the viewership of the other.[49] For instance, following the Los Angeles riots in 1992—which were sparked by the acquittal of police officers accused of beating an African American, Rodney King—many entertainment shows had plots that revolved around the riots. A growing television genre, the docudrama, which portrays real-world events in fictional accounts, further blurs the lines between fact and fiction, entertainment and traditional news. The time between actual events and their treatment as fodder for docudramas is growing shorter and shorter. For example, the networks rushed to be the first to air docudramas about the government's assault on the Waco, Texas, compound of David Koresh's Branch Davidian religious sect. (In fact, this show was put into production before the fiery resolution of the standoff.) The O. J. Simpson trial (which is still going on as we write) has been extensively covered by the network news, shows such as *Entertainment Tonight* and *Hard Copy,* cable television's *Court TV,* and several prime-time docudramas. It is an open question whether the "old news" coverage of these events or the "new news" docudramas will have more of an impact on shaping public perception and opinion about these and other public issues.

CONTROLLING THE MEDIA AGENDA IN A CHANGING ENVIRONMENT

Just as changes in communications technologies have altered the dynamics of the economic competition for audience and profit, so too have they altered the strategies and resources for influencing the content of media coverage of the political world. This struggle, however, takes place not among media outlets but among journalists, political elites (both elected and unelected), and, to a lesser extent, the public. Figure 8.2 (on p. 298) traces one measure of this struggle—the length of sound bites in which presidential candidates speak for themselves. As the figure shows, the average length of such sound bites has been steadily decreasing since 1968. This means that candidates have become more dependent on journalists to interpret their speeches to the public.

Each "side" has something that the other side needs. Journalists' primary resource is their control over access to most types of media. In the end, it is reporters who write political stories. And most people learn about political activity through the news media. However, as we will see shortly, there are other ways to get political information to the polity. Campaign advertising and the emergence of new forms of media—infomercials, talk shows, and so forth—make it possible to bypass journalists. Nevertheless, in most cases the story must be told by journalists if the message is to reach the American public in a credible manner. Any story that is deemed unnewsworthy, too complicated, or too marginal for the newspapers and/or television news will not get told.

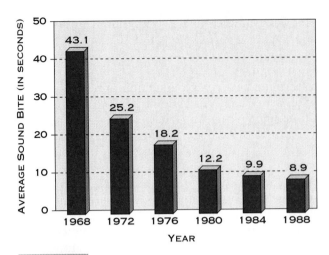

FIGURE 8.2

AVERAGE LENGTH OF CANDIDATE SOUND BITES IN TELEVISION COVERAGE OF ELECTIONS, 1968–1988 By broadcasting shorter candidate sound bites, television news gives journalists more power to interpret campaigns for viewers. As reporters become more influential in the play of power, increasing amounts of "horse-race" coverage may crowd out more substantive campaign coverage. *Source:* Data from Daniel C. Hallin, *We Keep America on Top of the World: Television and the Public Sphere* (New York: Routledge, 1994), p. 145.

On the other hand, political elites—and sometimes average citizens—have an important resource in this competition. They are the sources with the information that reporters need to write their stories. Remember: For journalists, "sources make the news." Journalists cannot simply write what they know to be true; they need someone to say that what they know is indeed true. Just as political elites depend on journalists to tell their stories, journalists need these elites to provide the source material for their stories. The result is a struggle in which sources and journalists look for the best ways to control the political information that is transmitted in newspapers and the electronic media.

The outcome of this struggle has all the characteristics of the sorts of games we have mentioned throughout the text. Sometimes journalists "win" and at other times political elites "win." It has been suggested that the competition among elites is certainly a game, but the game is more like professional wrestling than like football.[50] Since the struggle is limited and often more symbolic than real, the usual result is a sort of draw. Further, the strategies and resources change over time as the changing technology of communications alters both the structure of the game and the usefulness of particular strategies and resources. Especially important are the ways in which, for example, new communications technologies change the ability of journalists to control access to the mass media.

CONTROLLING THE AGENDA IN ELECTIONS

The way in which this game is played and the strategies used by both sides are best illustrated by the relationship between candidates and the press during an election campaign. During the course of a campaign, candidates and their staffs enter into a protracted struggle with reporters to positively shape the stories journalists tell about them and their campaigns. Campaigns are interesting because each one constitutes a separate competition, and in competitions each side tries to profit from what worked and what didn't work in the past. Just as the successful Monopoly player learns from success and failure in past games, so too do journalists and politicians try to learn from past experiences. This means that the strategies and resources shift from campaign to campaign. Further, as with Monopoly, this game has a scorecard and a final tally that candidates, journalists, and the public all pay attention to: the candidates' standings in the polls as the campaign progresses, and the final outcome of the election.

As we noted at the beginning of this chapter, it is possible to overestimate the impact of media coverage on the outcome of a specific election. Indeed, all participants in the game often do this, attributing victory and defeat to the outcome of the struggle over media coverage. However, victory and defeat can also be attributed to other causes, especially the economy, campaign spending, the advantage of incumbents, and so forth. Nonetheless, although the media may not determine the outcome of specific elections, modern mass media have surely altered the way in which campaigns are conducted.

For example, in 1984, when Walter Mondale decisively lost the presidential campaign to Ronald Reagan, the Democrats pointed a finger at the clever media campaign waged by the Reagan camp. They argued that what cost them the election was Reagan's ability to avoid the issues and to run a primarily symbolic campaign oriented around images of flags and sunrises.[51] Indeed, Reagan surrounded himself with an experienced cadre of media advisers who were able to emphasize their candidate's strengths and minimize his weaknesses. For instance, the Reagan campaign held few press conferences, tried to limit the number of face-to-face debates, and chose to rely on stump speeches and campaign commercials that were carefully staged to reinforce patriotic values. However, it is important not to see this media campaign as decisive in the election. Remember that Reagan was an incumbent president running for re-election during times of perceived economic expansion, and that incumbents in such times typically avoid getting bogged down in substantive debates with their opponents.

Although it is doubtful that media coverage cost Mondale the election in 1984 or Michael Dukakis the election in 1988, there is little doubt that the modern mass media have changed the ways in which political parties, candidates, and their advisers think about and conduct their campaigns. Consider the selection of candidates. In a television age, candidates must pass minimum standards of physical attractiveness. It is unlikely that an unusually unattractive person (perhaps even some of our nation's former presidents) could be nominated by either of the major parties today. Other impacts of the struggle over media coverage are less obvious.

Campaign advertising Voters receive two different kinds of information about campaigns through the mass media: **paid coverage** and **free coverage**. The first kind, coverage paid for by the candidate, consists of advertisements that are designed by the campaign to get the candidate's message out to the public. These advertisements present the candidate to voters in the most favorable light possible. Sometimes they inform voters about a candidate's stand on various issues. Often, however, they follow the logic of commercial advertisements and simply associate the candidate with positive images that have little substantive content. These campaign ads are sometimes called **feel-good ads**.

Think about most commercials you've seen. Although some commercials make specific and concrete claims about their products, many others have the following "logic": We see a pleasant image—say, attractive young people frolicking on the beach and having a good time, or happy families spending time together. Then we see the logo of the product being sold—a soft drink, laundry detergent, candy bar, or automobile. The ads make no logical connection between the product and the image. Happy, attractive people don't get that way because they drink one soft drink rather than another. Instead, drawing on psychological theories of operant conditioning, the ad is trying to get us to associate the positive images with the product logo. The advertiser hopes that later, when we are in a supermarket or mall and see the logo, a positive reaction will be triggered and we will be more likely to buy the product.

"Feel-good" campaign commercials follow the same logic. Ronald Reagan's "Morning in America" commercials are good examples. Shot by the same company that produced Pepsi-Cola commercials, they showed us scenes of American flags being run up flagpoles, neighbors gathered in small-town eateries to drink coffee, a fire station complete with a Dalmatian lying peacefully on the ground, and an array of other images designed to make us feel good. Then we were shown a picture of Reagan and were told to vote for him. Such campaign ads make no logical claims about the connection between the images shown and the policies that the candidate supports. Indeed, no political issues are raised at all. Some political analysts object to feel-good ads precisely because of this. Instead of trying to make logical arguments about specific positions and their consequences, the ads condition viewers to feel a positive association with the candidate shown at the end of the ad. Presumably, when we see his or her name in the voting booth, we will experience that positive feeling and be more likely to vote for the candidate.

A second type of campaign advertisement, the **attack ad**, has become both more prevalent and more controversial in recent elections. Unlike ads that try to increase support for a candidate by conveying positive information or feelings, as do feel-good ads, attack ads attempt to raise doubts and questions about the candidate's opponents. Perhaps the most famous attack ad, and one of the most well-known television campaign ads of all time, was run in 1964 by the Lyndon Johnson campaign. The ad was designed to raise voters' fears about the willingness of Johnson's opponent, Barry Goldwater, to involve the United States in a nuclear war. It did so without even mentioning Goldwater or making a logical argument about the connection between his positions (or Johnson's, for that matter) and the likelihood of nuclear holocaust.

The ad began with pictures of a small girl picking petals off a daisy and counting the petals. Suddenly, her count was picked up by a male voice counting down, "Ten, nine, eight, seven . . ." As the countdown progressed, the camera closed in on the girl's face; at zero, her eye, which by now took up the entire picture, dissolved into a nuclear explosion and then a mushroom cloud. We then heard Johnson saying, "These are the stakes. To make a world in which all of God's children can live. Or go into the darkness. We must either love each other, or we must die." The ad caused an instant sensation and was attacked as unfair and inflammatory. In fact, it was immediately withdrawn by the Johnson campaign. However, the ad and its claims became part of media coverage of the campaign. Since journalists covering the controversy over the ad discussed the charges it made about Goldwater, those charges about his willingness to push the nuclear button were constantly repeated. In short, the ad had done its job. As the media campaign waged by candidates becomes the topic for journalists' analyses, campaign ads become part of the news about campaigns. As a result, the line between paid advertising and journalists' coverage of campaigns becomes blurred.

Some analysts argue that attack ads pose a danger to the integrity of political campaigns.[52] Rather than using logical arguments subject to refutation by opponents if the ads make unfair or false claims, attack ads often appeal to voters' unexamined fears and emotions. Instead of making an argument about the foreign policy proposals advocated by Goldwater, the Johnson campaign's daisy ad simply played on the deep-seated anxieties most Americans had about nuclear war. More recently, many individuals pointed to the Willie Horton ads used by the Bush campaign in 1988 as appealing to white voters' deep-seated fears about crime and their racist attitudes about African-American males. Television ads make their arguments by subtly combining the visual images they show (e.g., a nuclear explosion, or a mug shot of an African-American male) with carefully scripted audio tracks (e.g., the comment by Lyndon Johnson or a reference to Michael Dukakis's record on crime as the governor of Massachusetts). These kinds of arguments are both difficult to refute and often difficult for viewers to examine critically (see "Strategies: Willie Horton and the Grammar of Campaign Ads" earlier in this chapter).

Which advertising campaigns are more successful—"feel-good ads" such as Ronald Reagan's "Morning in America" series in 1984 or "attack ads" such as the "daisy girl" ad used by the Lyndon Johnson campaign against Barry Goldwater in 1964? What happens when a controversial ad is aired only a few times, as the Johnson ad was, but is then discussed endlessly by the news media?

At first glance, then, campaign ads seem to be a way for candidates to avoid struggling with journalists for control of media coverage of their campaigns. As long as they are able to raise the money needed to produce and air these ads, candidates are almost entirely free to use campaign advertising as a way to get their message directly to voters without any filtering by journalists. Yet the impact of campaign ads is limited. They are successful at making voters familiar with a candidate's name and some of the broad themes of the campaign. However, as with commercial advertising, viewers seem to be skeptical of the claims made. After all, they know that the advertising is paid for by the candidate since ads must inform viewers of who is paying for them. A goal of campaign reform legislation is to make it more difficult for candidates to distance themselves from the attack ads that they run. For example, one reform proposal would require candidates themselves to speak the claims about opponents made in attack ads.

Even though paid ads are an important part of modern election campaigns, they (1) are expensive, (2) may be discounted by voters, (3) are undergoing more careful criticism by reporters, and (4) are becoming the target of reform proposals. Much more desirable than paid advertising are positive stories by the journalists who cover campaigns. This sort of coverage is called "free coverage," as contrasted with paid advertisements.

Free campaign coverage   The distinction between paid and free media coverage is not hard and fast. Often campaign advertisements become the topics of journalists' election stories. Whether and how these ads will be covered in "straight news" stories is part of the struggle over controlling the media agenda. Sometimes this can work to the advantage of a campaign, as with the Johnson campaign's daisy ad, where even critical coverage of an ad manages to remind voters constantly of the troubling claims the ad makes. The same phenomenon may have been at play with the media coverage of Bush's Willie Horton ads. Some argue that because the ad was constantly repeated on the nightly news, despite their criticism journalists actually reinforced rather than undermined the ad's effectiveness.[53]

This last example suggests just how complicated the struggle between campaigns and journalists becomes with respect to television coverage. Unlike print media, television transmits both visual and audio information, and the two may work at cross-purposes. David Broder relates a story about Lesley Stahl, CBS News White House correspondent during the Reagan presidency, to show how "pictures can override even the most biting commentary":[54]

> During the 1984 campaign, she put together a four-and-a-half-minute piece that showed how far the White House had gone in staging events for Reagan. "It was a very tough piece," Stahl said later, "showing how they were trying to create public amnesia about some of the issues that had turned against Reagan. I was very nervous about going back to the White House the next day. But the show was no more than off the air when a White House official called me and congratulated me on it and said he'd loved it. I said, 'How could you love it?' And he said, 'Haven't you figured it out yet? The public doesn't pay attention to what you say. They just look at the pictures.' Stahl said she looked at the piece

again, with the sound off, and realized that what she had shown was a magnificent montage of Reagan in a series of wonderful, up-beat scenes, with flags, balloons, children, and adoring support-ers—virtually an unpaid campaign commercial.[55]

The outcome of the struggle for control is not always a triumph for the candidate. In 1988, Michael Dukakis staged a campaign stop at a military installation and rode around in an Army tank. The event was staged to provide visual images that would counter claims by the Bush campaign that the Democratic candidate was weak on defense. But the event backfired. Images of Dukakis in the tank, wearing a large helmet and looking a little like the cartoon character Snoopy in his Sopwith Camel, were actually used to ridicule the Democrat in Bush attack ads. Further, images of the event became the subject of a number of damaging news stories about the Dukakis campaign. Coverage of this event shifted the focus from the actual policy positions of the two candidates to a focus on the strategy and symbolism of the campaign. That is, the stories were less about Dukakis's record on defense issues, which was actually much stronger than Bush claimed, than about the unintentionally humorous visual images in the Dukakis ad. In this way, reporters were able to gain the advantage in the struggle to control media coverage by reinforcing the importance of their role as the public's neutral and informed interpreter of strategies of the two campaigns:

> The tank ride invited reporters to play out the full range of strate-gic interpretation. Dukakis wants you to believe he is strong on

When newscasts repeatedly show the video images of controversial advertise-ments (such as this one, from the 1988 campaign, of presidential candidate Michael Dukakis riding in a tank), do viewers listen to what the stories say about candidates? Or do journalists, by using these video clips, simply reinforce visually whatever preconceptions viewers may already have?

defense, said the reports, but the polls show that the public perceives he is instead weak. Appearance versus reality. Here the reality is not the positions the Democrat has articulated but rather the perception of him in the polls. The story also gave journalists a chance to explain how politics works. Politicians craft visuals to create false impressions. Reporters, by contrast, reveal "what's actually going on."[56]

Of course, journalists do not win every time. To understand media coverage of campaigns, or of politics more generally, it is important to think about why a given story was written the way it was. To do this we need to think carefully about the resources that journalists and political elites have and how they are used in any particular story.

Since sources make the news and fairness means having two sides to a story, it is very difficult for journalists to write campaign stories without having comments from the candidates or their staffs. During campaigns, the candidates and their advisers become virtually monopoly suppliers of news and information. This gives campaigns a great deal of power. In his influential book *The Selling of the President 1968*, journalist Joe McGinnis describes how Richard Nixon's campaign managers were able to take advantage of reporters' dependence on candidate access to carefully reconstruct their candidate's image.[57] Rather than allowing the informal give-and-take between reporters and candidate that had characterized past presidential contests, Nixon's handlers restricted reporters to meticulously managed appearances by the candidate. This meant that the quotations and visuals used by reporters in their stories could be more carefully controlled by the campaign. Because reporters needed Nixon as a source, this careful control gave the candidate an advantage in shaping the sorts of stories that came out of his camp. Other candidates learned from the Nixon strategy and have become much more guarded about access to reporters.

Yet, as in any game, as one side becomes aware of what the other plans to do, a learning process occurs. Reporters, in the end, have a great deal of power in deciding the sorts of stories they will write and how they will handle the source material from the candidates. As candidates developed more careful strategies for influencing stories, journalists began to cover the strategies themselves rather than simply passing on the material provided by the candidate. Today this focus on strategy makes it difficult for the candidate to get media stories that focus on his or her issue positions.

In 1984, Colorado senator Gary Hart was a candidate for the Democratic presidential nomination. To contrast himself with front-running Walter Mondale, Hart tried to portray himself as a "new" Democrat with innovative and complex solutions to the nation's problems. However, it was difficult for Hart to convince reporters to write about his proposals. Instead, they tended to focus on the symbolic and strategic importance of Hart's casting himself as a "new" Democrat:

> [J]ust after his first successes, the newsmen began to peg their stories on this burning rhetorical question: "Will Hart be able to get his message across?" The answer was implicit in the question, and the answer was "no." So long as the TV newsmen are wondering aloud about any candidate's so-called "message," that message

isn't going to come across, since it's up to them to find it out and tell us about it. Otherwise the candidate will need a lot of dimes, because he'll have to use the telephone to fill the voters in.[58]

One result of this focus on campaign strategy is the dominance of what is often called **horse-race coverage**. Here the story is not about where the candidate stands on the issues, but rather about where he or she stands in the polls. As access to candidates is restricted to strategically planned speeches or visuals, reporters increasingly focus on whether these strategies are working with the voters. Indeed, most studies of campaign coverage reveal that two-thirds of the stories are of the horse-race variety, whereas the remaining one-third deal with other topics, including candidates' issue positions. For example, a study of the *CBS Evening News* during the 1980 election found that from January to October, 15.5 hours were devoted to campaign coverage. Of this, 10 hours were devoted to so-called horse-race coverage and about 5.5 to information about the candidates, including, but not limited to, their platforms.[59] Another study, using a slightly different methodology, found that more than half of the coverage of the 1988 election was devoted to horse-race stories.[60]

The focus on standing in the polls, characteristic of horse-race coverage, may even affect the way in which reporters tell their stories about other aspects of the campaign. One study found a connection between a candidate's standing in the polls and the ways in which reporters describe his or her issue stands and character. This researcher concludes that "reporters . . . fit candidates with images consistent with their positions in the races."[61] For example, after the 1988 Democratic Convention, candidate Michael Dukakis was well ahead of George Bush in the polls. He was favorably portrayed by *Newsweek* as a competent and "credible candidate" who was "relentless in his attack." Yet as the campaign wore on and Dukakis slipped in the polls, he was described by the very same magazine as "reluctant to attack" and only "trying to present himself as a credible candidate." Did Dukakis change, or was this simply a case of reporters' adjusting their stories to fit with the Democrat's standing in the polls? The same thing happened with coverage of George Bush. When he was behind in the polls, he was labeled a "wimp." But as his standing in the polls improved, the "wimp factor" was, according to *Newsweek*, "banished, as long as Bush is on the attack." The rise and fall of Bill Clinton's "character issue" exhibited a similar dynamic in the 1992 election.

A focus on the polls, rather than on the issues—either those raised by candidates or those of concern to voters—may be disturbing, but it is unclear how it affects the outcomes of specific elections. Once again, we must be careful to distinguish between the ways in which media coverage affects the conduct of campaigns—here, the increasing focus on strategy rather than issues—and the outcome of a specific election. As we have argued, it is difficult to conclude that media coverage won George Bush the 1988 election or cost him the 1992 election. In general elections where voters have many cues to guide their decisions, media coverage probably does not decisively affect most electoral outcomes.

Media and primaries Media coverage may not determine who wins in November, but it can have a powerful influence on who runs in the election. Recall that research suggests that media effects are most pronounced when citizens have little information or have not formed strong opinions. In primaries, voters do not have the cue of party affiliation (since all candidates in a primary are in the same party) and are often undecided until right before the election. Because they are less informed about the races, and because they may have less well formed opinions, the effect of media coverage is likely to be significant.

How does media coverage actually affect the primary process? One important way is through setting expectations for candidate performance. It is often less important whether a particular candidate actually wins than that he or she is able to live up to the performance expectations established before a particular primary. Journalists and campaign managers struggle over setting these expectation levels. Much as coaches of a favored team will try to downplay expectations in order to take the pressure off their players, so too will campaign managers try to downplay the percentages their candidates are expected to achieve on primary day. However, there is a significant difference between sports and the general election, on the one hand, and primaries on the other. In games and elections, whether expectations are exceeded or not, the team that wins on game day and the candidate who wins on election day are the victors. In contrast, candidates can actually lose during the primary season even if they win. This can happen if they do not satisfy expectations set by the media before the primary.

The history of primaries is littered with the failed campaigns of candidates who couldn't exceed press expectations even while winning. In 1972, Maine senator Edmund Muskie was heavily favored in the Democratic New Hampshire primary. However, he was reported as having broken down in tears of emotion while angrily attacking the editor of the *Manchester Union Leader*, who had published false accusations about Muskie's wife. (To make the story even better, there is some question about whether the droplets running down his cheeks were actually tears, or just melting snow.[62]) The tears were taken by reporters as a dangerous sign of instability and lack of toughness. When Muskie garnered "only" 46 percent of the vote rather than the 50 percent or more that he was expected to attract against his major opponent, George McGovern (who received 37 percent), his candidacy effectively ended. Similarly, during the 1992 New Hampshire primary campaign, Bill Clinton was hurt by allegations of marital infidelity and draft dodging. This led to the press's expectation that he would be knocked out of the race for the nomination in the opening primary. When Clinton, the former frontrunner, finished second to Massachusetts senator Paul Tsongas, he had exceeded expectations. Declaring himself "the comeback kid," Clinton was on his way to the Democratic nomination.

Why do expectations play such an important role? First and most simply, they follow from the implicit similarity between sporting contests and elections contained in the horse-race metaphor. Just as we handicap horse races, journalists—in negotiation with campaign managers—handicap primaries. In addition, by eliminating candidates for not meeting or exceeding expectations, press coverage can help to winnow the field down to a small number of candidates. Indeed, journalists themselves have an interest in narrowing the field. It is easier and cheaper (remember the struggle to minimize costs!) to cover a small number of candidates than an unruly field of seven or eight. This desire to narrow the field helps explain why in 1988, when there were still a large number of Democratic contenders for the nomination, the press dismissed them as "the seven dwarves" rather than covering the positions of all the candidates.

The struggle between reporters and politicians does not stop with primary or election victories or losses. As we have noted, presidents and other elected officials must constantly strategize about how to get their messages to the public through the mass media. Presidents are constantly aware of the need to maintain public support for their actions. Likewise, even unelected officials know that unwanted and hostile press attention can disrupt their carefully, or not so carefully, laid policy plans. On the other hand, journalists struggle to write stories that live up to their notions of fairness and neutrality. This means that they must maintain relationships with credible sources while avoiding becoming mouthpieces for these sources.

Politicians, their supporters, and their detractors, as well as many political analysts, often attribute positive and negative media coverage to the political biases of journalists. In this chapter we have rejected that explanation. Yet it certainly is true that some political figures seem to receive much more positive media coverage than others. What explains these differences in the outcome of the political struggle to control the media agenda? Thinking about media coverage as the outcome of a political struggle helps us answer the question. This perspective requires us to think about how stories are written, the sources that journalists use, the central actors' popularity in the polls and with other elites, and so forth. As you read newspapers or watch television, we hope that, rather than simply absorbing the reports as if they were the "truth" or rejecting them as the product of hopelessly biased media, you will begin to analyze them by using the tools we provide.

These contrasting *Time* magazine covers demonstrate the dramatic difference between coverage of the suicide bombing of the American military headquarters in Lebanon in 1983 and coverage of the attempted rescue of American hostages in Iran in 1980. Given that the general circumstances of the failed hostage rescue mission and the bombing in Beirut were similar, what accounts for the differences in coverage? Is it the ideology of the press, or is it the way in which journalists go about their jobs?

Robert Entman gives an informative example of how one can develop a deeper understanding of both politics and the mass media through carefully analyzing why stories have the slant they do. He analyzes the substantial differences in press coverage of two similar events in two different presidential administrations. In April 1980, President Jimmy Carter authorized an attempt to rescue Americans held hostage for five months in the U.S. Embassy in Teheran, Iran. The raid was a failure and nine U.S. servicemen died in the aborted attempt. The press coverage of this event was heavily critical of the president, blaming him personally for the failure. His approval ratings in public opinion polls dropped dramatically following the raid. In 1983, Ronald Reagan stationed U.S. Marines in Lebanon in an attempt to intervene in the civil war that was tearing that country apart. Two hundred forty-one of those Marines were killed in a suicide bombing of their Beirut barracks. Press coverage of this event was far less negative than had been the case for the failed raid in 1980, and Reagan's popularity ratings increased dramatically after the tragedy.

What accounts for the dramatic differences in the coverage of these two events? Although there were differences in the circumstances surrounding the events, Entman writes: "In both the Iran rescue and the Beirut bombing, Americans died. . . . Different presidential decisions might have averted the deaths. Both crises raised grave questions about the wisdom of U.S. policy and the competence of the U.S. military—and thus about the President."[63] What accounts for the differences in the coverage of the two stories is not any objective differences or similarities, nor the ideological bias of journalists, but rather the characteristics of the struggle for control of the media agenda and the strategies of the players in that struggle.

Recall that two of the key rules for journalists are that sources make the news and that fairness involves telling two sides to every story. In analyzing media coverage of political events, it is always important to think about who the sources are and how the two sides are defined. It is usual for Republicans to criticize the actions of a Democratic president. Journalists—and readers—are likely to discount such criticism as partisan bickering. However, journalists are likely to make much more of a story that involves critical comments from credible sources within a president's own party. In fact, it has been argued that the main reason for the emergence of critical media coverage of American policy during the Vietnam War was the increased willingness of officials within the Johnson administration to go on the record with their criticisms of the war[64] (see "Outcomes: Media Coverage of the Vietnam War" on pp. 310–11).

Jimmy Carter was a president who had vocal critics within his own party. This made it possible and likely that reporters would produce extremely critical stories of the failed rescue attempt in Iran. In contrast, there were fewer credible sources willing to criticize Ronald Reagan. First, he was more popular within his own party than was Carter, and Republicans were reluctant to criticize his actions. Second, and somewhat ironically, the greater loss of American life in the Beirut bombing made politicians of both parties less willing to criticize the president and thus risk being seen as being unpatriotic in a time of tragedy.

A third reason for the difference in the coverage was the disparate press strategies followed by the two presidents after the failures. In the struggle to influence media coverage of political events, some strategies are more effective than others. Carter went to great lengths to take personal responsibility for the decision to attempt the rescue. This may be an admirable characteristic in a commander-in-chief; indeed, the taking of personal responsibility may have been part of a strategy to help portray Carter as a strong leader willing to take responsibility for his actions. However, the strategy failed because it reinforced the linkage between the failure and the president himself. Media coverage therefore "primed" the public to judge the president responsible for the failed rescue. In contrast, Reagan distanced himself from the events in Beirut by attributing the bombing to an insane, and therefore unpredictable, act of madmen. This strategy proved much more successful because it "offered a clear villain other than Reagan, along with a simple, symbolic explanation matching familiar audience stereotypes: fanatical Middle Eastern terrorists."[65] In this case, media coverage "primed" the audience to hold responsible the instability and unpredictability of Middle Eastern politics and not Reagan.

Finally, chance can affect media coverage of an event. Both print and broadcast journalism have limited space to cover political events, and the amount of coverage devoted to any one event is determined by the number of other events in the world that are also deemed newsworthy. The Beirut bombing came only two days before the invasion of Grenada. The latter event quickly pushed the former event out of the headlines and limited the amount of critical scrutiny. In contrast, the Iranian rescue did not compete with any other dramatic U.S. foreign policy event and therefore remained in the headlines.

What accounted for the negative coverage of one president and the benign coverage of another was not simply "the facts" of the two incidents, nor was it the ideological biases of journalists. Both events were dramatic policy failures that could have been connected to the specific decisions made by the two presidents. Rather, the differences in coverage are explained by the dynamics of the struggle to control media coverage of political events: the rules of the competition (e.g., reporters' rules that sources make the news and that a fair story has two sides); the strategies used by players; and, as with most games, luck and timing.

Changes in the struggle   The primary dynamic of the political struggle to control media coverage of political events is the interdependence of reporters and political elites. However, changing communications technologies may be rapidly altering this interdependence. Political elites have been taking advantage of new media outlets to try to reduce their dependence on journalists. This process seemed to speed up in the 1992 presidential race. Bill Clinton, in both his primary and election campaigns, appeared in a variety of unconventional settings: He put on dark glasses and played the saxophone on *The Arsenio Hall Show*; he appeared on *Larry King Live*; he answered questions from young people on MTV. Even though George Bush at first criticized such unconventional appearances as being beneath the dignity of the office of president, he was soon regularly appearing on talk shows and the Nashville Network. Ross Perot avoided answering journalists' questions by simply buying large chunks of network time for his homespun "infomercials" (programs, paid for by advertisers selling a product, that imitate other types of shows—e.g., news broadcasts, talk shows—in order to seem less like commercials and more like sources of neutral information).

MEDIA COVERAGE OF THE VIETNAM WAR

The Vietnam War still haunts American politics. In 1988 and 1992, Dan Quayle's and Bill Clinton's military records, or lack thereof, during the Vietnam War became problems for them. Even in 1994 there was much speculation about how Clinton's former draft avoidance and opposition to the Vietnam War would affect reception of his speech as U.S. president at the fiftieth anniversary of the Allied invasion of Normandy. Virtually every military action by the United States since 1975 has been analyzed in terms of how it was affected by the Vietnam legacy. The American media have similarly been implicated in the American failure in Vietnam. In general, two diametrically opposed interpretations of the press's role in the war have come to dominate public discourse.

The first is the idea, held mainly by defenders of U.S. policy in Vietnam, that press coverage created a mistaken view of the military situation that undermined public support for the war. Because it was the first war to receive nightly television news coverage, so this argument goes, the American public was constantly bombarded with graphic combat footage showing American casualties and the war's brutality. Some argue that, had the public been fed similarly graphic pictures of combat in World War II, support for that war would have rapidly eroded as well. Beyond this, media critics argue that a liberally biased press created a purposefully distorted picture of the American military effort in Vietnam.[1] This is the lesson of Vietnam that Pentagon officials drew as they justified extreme censorship of press coverage of the invasions of Grenada and Panama and the Persian Gulf War.[2]

For many people, images such as this one—of American soldiers burning a Vietnamese village— are what they remember from television coverage of the war. Indeed, such scenes are often blamed for turning American public opinion against the war. But how representative are they of total media coverage of the war? What accounts for the ways in which this or any war is covered in the mass media? Does government need to censor or otherwise control the ways in which journalists cover such events?

The opposing idea, held by the press itself, is that crusading journalists challenged the lies and distortions of the U.S. government. This interpretation sees young journalists in Vietnam as heroically uncovering the truth about the failures of the American military, a truth that the government tried to cover up. It was through this dedicated, and often dangerous, news gathering that the American public began

All these communications channels allowed the candidates to create the illusion that they were talking directly to "the people." Further, in the future, these channels may provide us with new and diverse insights into politicians and the political process. Much remarked on was the televised presidential debate in which "average" citizens, not journalists, asked the questions. In this debate all three candidates were called on to avoid the personal assaults and pat answers that had characterized the campaign up to that point.

to get a clear picture of the quagmire of Vietnam. As a result, American public opinion turned against the war.

Proponents of each of these interpretations continue to defend these two views and apply them to more recent issues, such as the government's censorship of the press during the Persian Gulf War. However, in *The "Uncensored War": The Media and Vietnam*, Daniel Hallin offers an explanation of press coverage of the war that is both less ideological and more consistent with the perspective on the relationship between the media and politics we have been developing in this chapter. He carefully examined the actual coverage of the war in newspapers and on television between 1961 and 1973 and found that the overwhelming bulk of coverage prior to 1968 was positive and followed the government's and the military's perspective. Although a small number of stories were somewhat critical of the war (these are the ones journalists like to recall), most stories supported the government's position and employed story lines quite similar to those used in previous wars, such as World War II. Although many reporters in Vietnam came to understand that American strategy was deeply flawed, they could not report what they knew, from personal experience, to be the case. Rather, since "sources make the news," they were limited to what "credible" sources were willing to say for the record. Few military or governmental officials were willing to speak against U.S. policies, so journalists wound up simply reporting the official line. Further, even when reporters filed critical stories, these were often edited or altered by editors and producers back in the United States.

Why did press coverage turn against the war in 1968? Hallin's answer is that by then the Johnson administration was itself badly split over the war. Senior officials such as Clark Clifford, the secretary of defense, were turning against the war. For the first time, credible sources were beginning to speak to reporters about their doubts regarding the war. As a result, the boundaries of what Hallin calls the "sphere of legitimate controversy," defined by the disagreements among political elites who serve as credible sources, expanded to include debate over the war itself. Consequently, press coverage came to reflect the skepticism about the war that political elites themselves were feeling.

The important point is that press coverage followed, rather than led, changes in elite and mass opinion about the war. Even though student demonstrations and anti-war sentiment began to rise prior to 1968, press coverage did not become critical until that date. Moreover, Hallin's analysis weakens the case for censorship and suggests that governmental officials have less to fear from press coverage than they do from their own policy disagreements.[3]

[1] See Peter Braestrup, *Big Story: How the American Press and Television Reported and Interpreted the Crisis of Tet 1968 in Vietnam and Washington* (Boulder: Westview, 1977).
[2] John McArthur, *Second Front* (Berkeley: University of California Press, 1993).
[3] Daniel C. Hallin, *The "Uncensored War": The Media and Vietnam* (Berkeley: University of California Press, 1989).

The appeal of such new outlets is no doubt connected to widespread suspicion about the media. Recall that we argued that this suspicion comes from the belief that the media *should be* providing us with an objective, neutral picture of the political world, but that they fail to do this. Infomercials and the direct questioning of candidates by average folks may be attractive because they seem to provide us with a direct view of the candidate and the political world. However, the approach developed in this chapter concludes that this is impossible. Despite appearances, the seemingly direct connection of an infomercial or a town hall meeting is a carefully crafted illusion. Perot's infomercials were self-consciously designed to appear folksy and homespun. Bill Clinton's appearance with a saxophone was part of a well-developed campaign strategy. Many politicians prefer to answer the questions of average citizens or adoring talk show hosts because those doing the asking are likely to be unskilled at following up questions and forcing candidates to move beyond their prepared and pat answers. Rather than accepting these new ways of viewing political elites as bringing us closer to the objective truth of political events, we need to evaluate them critically in terms of the strategies and resources that govern their content, appearance, and impact.

SUMMARY

The modern mass media constitute a central actor in the play of power in American politics. Media sources provide most of us with most of the information we receive about the political world. Research suggests that while the media cannot tell you what to think about political issues, they can certainly set the agenda and tell you what to think about. Comprehending American politics and your role as an active citizen requires a sophisticated understanding of how the mass media go about telling stories about politics.

In order to develop this understanding, you may need to abandon some of your initial beliefs—for example, that the media are too liberal or too conservative, or that reporters systematically attempt to insert their own prejudices into stories. We argue that such beliefs are based on the impossible standard that journalists should provide a fair, neutral, and objective account of the political world.

Instead, a better understanding can be achieved by analyzing the significant players who struggle over control of the media agenda. Owners and controllers, journalists, elected politicians, unelected policymakers, organized interests, and the public all wield resources that can be used to shape media coverage of the political world. Especially important is understanding how journalists go about their jobs—the rules they use to decide what constitutes a story, and how they go about telling that story. In this chapter we saw that the ideas that "sources make the news" and that there are "two sides" to every story are central to what journalists mean by being fair and objective. However, the journalistic definition of who constitutes a credible source means that stories tend to focus on certain political players, especially political elites in Washington, and to ignore or distort the actions of other players, especially women and minorities. Further, given that journalists define their job as finding sources on two sides of a story, the role of institutions (e.g., the bureaucracy) and political structures (e.g., federalism) tend to be ignored in most media stories about politics. Understanding the media is made even more difficult by the fact that constant and dramatic changes in the structure and technology of communications make the media a "moving target."

Journalistic practices establish the boundaries of the ever-changing struggle between journalists and political elites to determine the contour of media coverage. This struggle is most intense during political campaigns as candidates and journalists try to shape public opinion through paid and free campaign coverage. Media coverage is especially important during primaries because voters have less other information to go by. The impact of media coverage may be much more limited during the general election campaign. However, the balance of power between journalists and politicians varies over time and across campaigns.

It has been argued that the way the media present the political news and most people's expectations about the role of journalists together constitute a "news prison":

> As a result of the political and journalistic factors that shape the news, the public is placed in a difficult bind. On the one hand, those who pay serious attention to the news run the risk of absorbing its subtle political messages, accepting its familiar stereotypes, and adopting its rigid modes of thinking. On the other hand, people who avoid the news may suffer the social stigma of ignorance, the guilt of being poor citizens, and the confusion of not knowing what is happening in the world.[66]

The tools we have developed in this chapter provide one way out of the "news prison." Fortunately, you will have many opportunities to develop your skills at using these tools as you go about your everyday life.

APPLYING KNOWLEDGE

GANGSTA RAP—MENACE TO SOCIETY OR NEWS FROM THE GHETTO?

 In the 1990s, especially in the wake of the Los Angeles riots of 1992, the rap music genre called "gangsta" rap became politically and culturally controversial. Gangsta rap is performed most often by young African-American males—NWA, Snoop Doggy Dogg, Ice Cube, Ice-T—who adopt the persona of inner-city gang members. Their raps purport to chronicle the life experience of inner-city African Americans. Reflecting this experience, the music is full of references to violence, drug use, drug dealing, and sex.

On the one hand, the genre has been vigorously criticized by politicians and journalists of all political persuasions and races. In the 1992 presidential campaign, Bill Clinton criticized the female rap artist Sister Souljah for advocating, he claimed, the killing of whites by blacks. Also, Jesse Jackson criticized gangsta rap for its negative images of African-American males and its misogyny. In general, critics argue that gangsta rap cashes in on the misery of inner-city African Americans, sanctions violence and drug use, glamorizes gang violence, and encourages the most sexist stereotypes about women and the most racist stereotypes about African Americans. Dan Quayle, among others, called for large record companies to refuse to distribute so-called "hard-core" gangsta rap; and, in fact, Warner Brothers dropped Ice-T from its stable of artists.

On the other hand, there are many defenders of gangsta rap. The artists themselves claim that they are simply describing conditions that exist in inner cities in America. These conditions are seldom chronicled in the mainstream press from the perspective of inner-city residents themselves. Instead, the artists say, the news portrays young inner-city African-American males as criminal monsters who cannot and should not be understood sympathetically, but should simply be "dealt with" by the police and the criminal justice system. The artists also argue that their critics are motivated by racism and a desire to avoid the unpleasant truths of inner-city life. As Ice-T has pointed out, few people confuse Arnold Schwarzenegger with the role he plays as the cop-killing Terminator.

In a sensitive analysis of gangsta rap, historian Robin Kelley argues that the music must be understood from the perspective of the long history of African-American music and the black experience in the United States, and especially today's inner city. Through the personas they adopt, rap artists try to tell the story of gang life from the inside, writes Kelley.[1] The music is troubling because it suggests that young African-American males who turn to crime and drug dealing are simply following the American dream of upward mobility by taking advantage of the few opportunities available to inner-city black males. Ice-T's "New Jack Hustler" expresses this perspective and its paradoxes well:

> I had nothin' and I wanted it, you had everything and you flaunted it, turned the needy into the greedy. With cocaine, my success came speedy. Got me twisted, jammed into a paradox, every dollar I make, another brother drops. Maybe that's the plan and I don't understand. You've got me sinking in quicksand. . . . imagine me working at Mickey D's. That's a joke, 'cause I'm never gonna be broke. When I die, it'll be bullets and gunsmoke. . . . cool out and watch my new Benz gleam. Is this a nightmare, or the American dream?[2]

Given these choices, it's no wonder that the Hustler comes down with "a capitalist migraine."

Taking another tack, media critic Jon Katz argues that the mainstream press has always been hostile to any youth culture that is critical of the status quo. He likens the largely negative reaction of journalists to gangsta rap to the hysterical moral condemnation of Elvis Presley and rock and roll in the 1950s.

What do you think?

Questions

1. Given what you know about the ways in which journalists go about their jobs, do (can) they do a good job covering the conditions in American inner cities?
2. Is there a connection between criticisms of rap music and racism?
3. Given the conflict between the "new news" and the "old news," is rap music a form of news?
4. Should this music be regulated by the government (e.g., with warning labels)?
5. Is gangsta rap a form of political expression?

6. Is it different from other forms of youth-oriented music and popular culture?

[1] Robin D. G. Kelley, "Kickin' Reality, Kickin' Ballistics: 'Gangsta Rap' and Postindustrial Los Angeles," in *Race Rebels: Culture, Politics, and the Black Working Class* (New York: Free Press, 1994).
[2] "NEW JACK HUSTLER." Written by Tracy Marrow and Alphonso Henderson. Copyright © 1991 PolyGram International Publishing, Inc. and Rhyme Syndicate Music (and as designated by co-publisher). Used by permission. All rights reserved. ◼

Key Terms

objectivity (*p. 276*)

agenda setting (*p. 290*)

priming (*p. 291*)

paid coverage (*p. 299*)

free coverage (*p. 299*)

feel-good ad (*p. 300*)

attack ad (*p. 300*)

horse-race coverage (*p. 305*)

Recommended Readings

Bagdikian, Ben H. *The Media Monopoly*, 4th ed. Boston: Beacon, 1992. A classic and very readable study of economic changes in the American news media and their impact on content.

Broder, David. *Behind the Front Page: A Candid Look at How the News Is Made.* New York: Simon and Schuster, 1987. A well-respected syndicated columnist reflects on his years as a reporter and offers a host of interesting suggestions for reforming media coverage of politics.

Diamond, Edwin, and Stephen Bates. *The Spot: The Rise of Political Advertising on Television*, 3rd ed. Cambridge, Mass.: MIT Press, 1992. The best study of the history and dynamics of political advertising on television.

Entman, Robert. *Democracy without Citizens: Media and the Decay of American Politics.* New York: Oxford University Press, 1989. A solid, insightful, and readable academic study of the causes and consequences of bias in media coverage of politics.

Hertsgaard, Mark. *On Bended Knee: The Press and the Reagan Presidency.* New York: Schocken, 1989. A critical study of the ways in which understanding how journalists go about their jobs helped Ronald Reagan's advisers to effectively control coverage of his administration.

Jamieson, Kathleen Hall. *Dirty Politics: Deception, Distraction, and Democracy.* New York: Oxford University Press, 1992. An informative analysis of the ways in which television advertising and campaign coverage can distort and degrade the electoral process, written by a respected media scholar.

Manoff, Robert Karl, and Michael Schudson, eds. *Reading the News.* New York: Pantheon, 1986. An extremely interesting series of essays on how the questions journalists ask shape what we consider to be the news.

Postman, Neil. *Amusing Ourselves to Death: Public Discourse in the Age of Show Business.* New York: Penguin, 1985. A polemical but delightful jeremiad on the implications for public discourse of the rise of television as our dominant medium.

Organized Interests
in the Political Arena

INTEREST GROUPS ARE the most important nongovernmental participants in the play of power. Without a thorough understanding of their role, no attempt to comprehend American politics can succeed. The story of how these groups shaped policy toward the savings and loan industry provides a revealing introduction to the critical role they play.

Beginning in the 1930s, a healthy savings and loan industry offered millions of depositors a safe haven for their money and provided millions of families the mortgages that allowed them to own their own homes. These institutions became so ingrained in the fabric of American society that the letters "S&L" served as a well-known shorthand for the industry.

Trouble began in the late 1970s, when interest rates soared.[1] Existing regulations set limits on the rates S&Ls could pay, so depositors took their money elsewhere. Although Congress removed the cap to encourage depositors to stay, this presented the S&Ls with a new problem:

How could they afford to pay depositors the new, higher rates when most of their income came from long-term home mortgages with low interest rates?

The industry's organized interest groups, especially the U.S. League of Savings Institutions, implored Congress to lift restrictions on the kinds of investments they could make. Partial success came in 1980 when Congress passed the Depository Institutions Deregulation and Monetary Control Act. This act granted S&Ls broad powers to attract deposits and increase their earnings through schemes such as offering consumer loans and investing in commercial real estate.

When these changes failed to ease the financial crunch on S&Ls, Congress in 1982 passed the Garn–St. Germain Act. This act allowed the institutions to make riskier loans for which they could charge higher rates. Problems continued, however, and regulations were further relaxed. Both traditional S&L officials unfamiliar

This headline announces one of the most serious political scandals in recent memory. The savings and loan scandal, which ultimately would cost taxpayers billions of dollars, was produced in part by government policies that sought to accommodate the demands of interest groups representing the savings and loan industry.

with the new loans and eager entrepreneurs who bought into the industry felt free to make risky loans with other people's money. Some also committed outright fraud and spent institutions' money with unbridled enthusiasm on their own salaries and on vacations, cars, yachts, and other expensive toys.

Edwin Gray, head of the Federal Home Loan Bank Board (the agency charged with regulating the S&L industry), warned of trouble before he resigned in 1987. A former savings institution executive who owed his job to the support of the U.S. League of Savings Institutions, Gray should have been credible. But the industry denied that problems existed, and Congress took no action. An article in the *Wall Street Journal* quoted Gray's summary of the role the U.S. League played at the time: "When it came to thrift matters in the U.S. Congress, the U.S. League and many of its affiliates were the de facto government. What the League wanted, it got. What it did not want from Congress, it got killed."[2]

One thing the League wanted was more governmental funding for the Federal Savings and Loan Insurance Corporation (FSLIC), which guaranteed that depositors would not lose their money if an S&L went broke. The League claimed in 1987 that a loan of just $5 billion from the Treasury, to be paid back later by industry earnings, would solve the problem.[3] Congress obliged, providing the FSLIC with what turned out to be a woefully inadequate $10.5 billion. Meanwhile, Congress's investigative arm, the General Accounting Office, reported that the FSLIC's deficit was $13.7 billion.

The S&L industry also sought to influence the choice of Edwin Gray's replacement as head of the Federal Home Loan Bank Board. Again it won, successfully lobbying the Reagan administration to select Danny Wall, the minority staff director of the Senate Banking Committee and a protégé of ranking minority member Senator Jake Garn.

Even as more and more S&Ls failed, and as FSLIC losses mounted, the new Federal Home Loan Bank Board head continued to maintain that all was well, and the U.S. League continued to echo these assurances. Congress not only bought their story but again, in 1988, passed legislation benefiting the industry and rejected proposals to raise the industry's insurance costs. In fact, the former president of the U.S. League modestly noted about 1988, "We worked with the Bank Board, and everything we tried to do we were successful at."[4] Members of Congress agreed. Representative Henry Gonzalez, who became chair of the House Banking Committee in 1989, complained in a public hearing several years earlier that "Everything the industry has wanted, the Congress has rolled over and given . . . to them."[5]

The immense dimensions of this problem emerged, not by accident, only after the 1988 election. In February 1989, President George Bush announced his S&L bailout plan, with estimates that it would cost a truly astonishing $148 billion in 11 years, and even more over the next 30. By the time the Senate passed its version of the bill in April, the estimated cost had grown to $157 billion in the first 10 years.

The U.S. League of Savings Institutions and its related industry groups were not exclusively responsible for the horrible and costly mess that emerged over time. However, the policies that led to this problem cannot be understood without acknowledging the influential role these interest groups played. This chapter provides the information, conceptual tools, and examples needed to understand how political interest groups like the U.S. League succeed and fail in the play of power.

AMERICAN AMBIVALENCE ABOUT ORGANIZED INTERESTS

 Like many Americans, you may come to the study of inter-est groups with concerns about the role they play. Indeed, from our nation's earliest days, Americans have felt great ambivalence about organized interests in politics. James Madison's analysis in *The Federalist*, No. 10, remains the foundation of American political thinking about interest groups.[1] To Madison, it was in-evitable that people would organize into "factions" to pursue common in-terests. These groups, he cautioned, constituted a potential threat to popu-lar government and to the public, or general, interest. Although Madison worried about factions, he believed that restricting people's freedom to pursue selfish interests was a remedy "worse than the disease," and he called instead for "controlling the mischiefs of faction" by designing a government with dispersed power and plenty of checks and balances.

The ambivalence expressed by Madison is widely held by today's Americans, as we can see from contemporary survey results and popular writings. Almost three-quarters of the students surveyed in a large introduc-tory American government class agreed with the statement "Interest groups exert tremendous influence over the content of public policies in the United States." And, although 28 percent thought that the role played by interest groups was beneficial, 33 percent disagreed and the rest were unsure. Fi-nally, almost 70 percent agreed with the statement that the government "is pretty much run by a few big interests looking out for themselves" rather than "run for the benefit of the people." Surveys of the general public have also found that, since the 1960s, large majorities have chosen the "run by a few big interests" answer when polled on this topic.[2] Newspaper editorials reinforce such views with frequent unflattering references to the role played by "special interests." For example, the *Philadelphia Inquirer*, commenting on Congress's ethics rules, observed that such rules should keep lawmakers "from getting easy money from special interests hoping to make friends and influence people on Capitol Hill."[3]

In common usage, the term *special interests* merges with *special in-terest groups* or just *interest groups*. Such expressions generate in the minds of many people two characterizations: (1) Interest groups are power-ful in shaping politics and policy, and (2) when that power is exercised, the results are harmful. However, such negative views compete with the belief, held by many others, that interest groups play a beneficial role in politics. Since neither set of beliefs dominates popular attitudes, ambiva-lence about organized interests persists.

In this chapter we explore the role of interest groups in the play of power. We define interest groups, examine their goals and place in the po-litical system, discuss how rules shape their activities, describe the diver-sity in their structure and organization, catalogue their resources, and summarize the conventional and unconventional strategies they pursue.

---

[1] This discussion is drawn from Jeffrey M. Berry, *The Interest Group Society*, 2nd ed. (Glen-view, Ill.: Scott, Foresman, 1989), pp. 1–4.
[2] Harold W. Stanley and Richard G. Niemi, *Vital Statistics on American Politics*, 4th ed. (Washington, D.C.: Congressional Quarterly Press, 1994), p. 169, Table 5.9.
[3] "Buying Congress," *Philadelphia Inquirer*, April 21, 1989, p. A24. ■

INTEREST GROUPS AND ORGANIZED INTERESTS DEFINED

The word *group* conveys a variety of meanings. In the course of a single day, you might use the term to describe the people riding on your bus, standing in line at the registrar's office, sitting at a dorm cafeteria table, or attending a meeting of the association of women students on campus. To fit our definition of an **interest group**, a group must exhibit the following characteristics:

1. Significant *shared attitudes*, interests, or attributes
2. *Awareness* on the part of the members of what is shared
3. A *minimum frequency of interaction* in an organization with a *formal structure* and stable patterns of interaction

Of the preceding examples, only the women students' group meets these criteria. It has a name, a formal constitution, officers, and a regular meeting time. The members share an interest in matters that touch on their status as women on campus. Indeed, it is this shared interest that accounts for the group's existence.

The notion of *shared* attitudes or interests is crucial. It explains the reason for an interest group's existence and determines the activities in which the group will engage. The women students' group would not focus its efforts on campus recycling or tuition levels any more than an environmental group would lead a rape prevention campaign or seek better lighting on campus. To predict what activities a group will undertake, we must identify what its members have in common. The attitudes or characteristics shared constitute the group's *interests*.

Not all interest groups participate in politics. Stamp-collecting clubs, prayer groups, the Girl Scouts, and countless other organizations that meet the criteria in our definition of an interest group play little or no role in politics. We will focus on **political interest groups**—organized bodies of individuals who share political interests and policy goals and act to influence public policy. When interest groups make claims on government, they engage in the play of power to seek benefits that affect the shared interests at the core of the groups' existence.

As we will soon show, however, a surprising number of groups whose primary activities have little to do with politics or government sometimes find that their shared interest impels them to enter the play of power—even if only for a brief time, for limited goals, and as a small portion of their total activity. Thus, many groups may act as political interest groups to varying degrees.

If we limit our attention to political interest groups, many significant participants will be ignored. *Social movements* arise from widespread discontent coupled with demands for political change. Particularly in their early stages, such movements lack structure and spawn few organized groups. However, organized interest groups soon form as social movements gain momentum. The role of the civil rights, gay and lesbian, Chicano, and women's rights "movements" in politics cannot be captured solely by looking at the full-blown political interest groups they have generated.[6]

Political interest groups are not necessarily highly organized bodies of individuals acting as members of a formal organization. Often individuals such as those shown here band together briefly to publicize a common concern.

Over the past 20 years or so, a growing number of organizations have begun to participate in politics in Washington. They include nonprofit foundations, public interest law firms, state and local governments, foreign governments, and individual business corporations. These participants differ from membership organizations because they do not necessarily speak for their employees or need to consult them before acting.[7] Indeed, such groups' growing importance leads some scholars to use the more inclusive term *organized interests* in place of *interest groups.* Therefore, this chapter examines a variety of organized interests, not just formally organized political interest groups.

HOW RULES AFFECT ORGANIZED INTERESTS

Few rules constrain the activities of organized interests. The three exceptions we discuss—tax laws, federal laws that regulate lobbying, and campaign contribution laws—do not invalidate the conclusion that such groups operate virtually unhindered.

Some organized interests—such as think tanks, hospitals, charitable organizations, and churches—enjoy tax-exempt status. This encourages contributors by allowing their donations to be deducted from their income taxes. To retain tax-exempt status, however, these organizations must not engage in overt political activities such as lobbying legislators on pending bills; thus, tax-exempt status does restrict political activity somewhat. In practice, many such organizations totter on or cross the line of prohibited activities, although elected officials shrink from removing the tax-exempt status of groups such as the Roman Catholic Church or the Red Cross.

Political interest groups rely heavily on professional lobbyists to achieve their goals, as we will discuss later in this chapter. In 1946, Congress enacted the Federal Regulation of Lobbying Act to limit their activities. In practice it is more loophole than law, requiring only that groups whose principal purpose is to lobby Congress register with the clerk of the U.S. House of Representatives and report quarterly on salaries paid, expenses, and Washington-based activities. Many groups argue that their principal purpose is not to lobby and, thus, they refuse to register. Because the 1946 law contains no enforcement provisions, failure to comply with it results in no penalty.

Federal law also regulates the interest groups' strategy of contributing to federal political campaigns. Direct contributions from labor unions, businesses, and other groups are prohibited. These contributions flow through political action committees (PACs), whose important activities we describe in more detail later in this chapter. A 1974 law limits PAC contributions to any candidate to $5,000 for the primary election and another $5,000 for the general election. In addition, federal laws prohibit outright bribery. The fact that such practices are outlawed does not, of course, guarantee that they never take place. In 1993, for example, Deborah Gore Dean, who had served as an official in the Department of Housing and Urban Development (HUD) during the Reagan administration, faced criminal charges of conspiracy, lying to Congress, and accepting illegal gifts in connection with the award of HUD housing grants.[8]

Constitutional guarantees of freedom of speech and assembly allow organized interests wide latitude barely touched by the exceptions we have noted. Virtually any group can organize if it wishes. It can advertise for members, solicit funds by mail or telephone, establish a headquarters and field offices, contribute to politicians' campaigns through PACs, and urge members to contribute, stage demonstrations, or engage in numerous other activities.

Rules can facilitate the activities of organized interests as well as restrict them. These rules, of course, help some more than others. For instance, until 1993 business corporations could claim most lobbying expenses, including the costs of hiring lobbyists and running a Washington-based office, as tax-deductible business expenses. Citizens who donate to civil rights, environmental, or women's advocacy groups enjoy no similar tax break; their members pay dues with after-tax money.[9] Rules also can facilitate the creation and maintenance of some groups. For example, not-for-profit groups, especially those depending on mass mailings to recruit members, enjoy reduced postal rates. The policies embodied in formal rules thus determine the number, types, and strength of participants in the organized interest universe.

THE STRUCTURE AND ATTRIBUTES OF ORGANIZED INTERESTS

Although the number of organized groups is so large that scholars are wary of offering an estimate, numerous shared interests in society are unrepresented by formal groups. Thus, for example, depositors in S&Ls remain unorganized, and most college campuses have no "Students United against Higher Tuition" or "Students Demanding Easier Exams" even though many people share an interest in these causes. Why do some shared interests in society form groups likely to make claims on government while others remain apolitical or completely unorganized? How can we classify the organized interests that do participate in the play of power? What attributes do they possess in terms of social diversity, goals, and relations with other groups?

A number of explanations have been offered for why groups do or do not form.[10] Until the 1960s, scholars called "group theorists" or "pluralists" believed that Americans naturally formed groups whenever they recognized a common problem or threat. In this view, disruptions in social or economic life, especially those caused by economic growth, created problems that in turn led to organization. The organization of some groups posed a new threat that spurred other groups to form. Thus, pluralists believed that organization bred counterorganization.

Constant changes in economic and social life guaranteed an expanding, dynamic, and diverse universe of groups. If some potential groups remained dormant, it meant their grievances were not strong enough to prod them into action. However, pluralists recognized that some shared interests produced organized groups more readily than others. In particular, potential groups whose members share *economic interests*, such as those drawing S&Ls to the U.S. League of Savings Institutions, form quickly. This was the pluralists' explanation for the fact that groups seeking economic benefits account for the lion's share of significant and active political interest groups.

A book by economist Mancur Olson, *The Logic of Collective Action* (1965), challenged the group theorists' explanations. Olson argued that three related factors determined whether a potential group with existing shared interests would form: (1) the nature of the benefits sought, (2) the size of the potential group, and (3) its geographical distribution.

House Speaker Newt Gingrich addresses a meeting of the American Medical Association. At such meetings, the AMA debates positions it will take on public policy issues affecting its membership. Although its lobbying on behalf of physicians may be important to some of the organization's members, most join for the selective benefits it provides, such as subscriptions to the AMA's journal.

For Olson, the nature of the benefits sought was crucial. He focused on the concept of **collective benefits**. These exist only if they are provided through governmental action. Moreover, they are indivisible; if a collective benefit such as clean air or a strong national defense is produced, everyone reaps the rewards. Olson argued that groups seeking collective benefits would find it difficult to organize, especially if the number of potential members was large and geographically dispersed. When groups seeking collective benefits succeed, they do so not because members join to obtain collective benefits but, rather, because members desire **selective benefits**— that is, things of direct and personal value. For instance, physicians join the American Medical Association not because it protects the interests of all physicians through its political actions but because membership brings benefits such as hospital privileges, a subscription to the *Journal of the American Medical Association*, and other personal advantages.

Large size, geographical dispersion, and collective benefits operate together to explain why S&L depositors, consumers, and even students find it difficult to organize groups that engage in political activity. Olson's view predicts that relatively few groups that lobby for noneconomic benefits will form, and those that do will attract only a small portion of those who would benefit from their success. Similarly, few groups working on behalf of the economic interests of large groups of people will form; when they do, their membership too will usually be small. In other words, when such groups do emerge, most potential members do not join, instead remaining **free riders**. Because these groups owe their existence to the selective benefits they provide rather than their potential for shaping politics, they must limit their political activity. Furthermore, when they do enter politics, they must confine their actions to matters involving the shared interests of their members. This explains why a campus women's organization does not delve into environmental issues.

Although noneconomic political interest groups provide some selective benefits (e.g., a magazine subscription), these organizations exist primarily to seek collective benefits through political action. Perhaps best known and most effective are environmental groups such as the Environmental Defense Fund, the Sierra Club, and Greenpeace. Some groups focus their efforts so narrowly that they have become known as **single-issue interest groups**— organized bodies of individuals devoted to one particular political issue. The National Right to Life Committee, for example, devotes itself to fighting abortion. A typical noneconomic group, however, works on a closely related set of issues—for example, the American Civil Liberties Union (ACLU) on civil liberties and civil rights issues, and Ralph Nader's Public Citizen on consumer and environmental issues. More broadly ideological groups—such as Americans for Democratic Action (ADA) for liberals and Americans for Conservative Action (ACA) for conservatives—take stands on a wide range of issues. Somewhere in between are groups like Common Cause and the League of Women Voters, whose main focus is on "process issues" involving the structure and operation of governmental institutions, but which sometimes take on other causes. "Participants: Profile of a Noneconomic Public Interest Lobbying Group" profiles one such group, Common Cause.

Although much of Olson's basic argument remains valid, it overstates the obstacles to formation of interest groups by focusing on people's desire for material benefits and undervaluing the power of symbolic rewards. Our

PROFILE OF A NONECONOMIC
PUBLIC INTEREST LOBBYING GROUP

In 1970 former secretary of health, education, and welfare John Gardner founded Common Cause, a nonpartisan organization, to lobby on behalf of citizens on issues of common concern. Overcoming the obstacles to organizing cited by economist Mansur Olson, the organization quickly attracted 100,000 members. By 1995 it had approximately 250,000 members and a budget of approximately $6 million. Common Cause relies on a unique source of funds: Roughly 90 percent come from dues and contributions of $100 or less from members themselves. The members elect a national governing board by mail ballot.

Common Cause uses the money it receives to employ a Washington-based staff of about 50 people. The staff manages the organization's finances and operations, undertakes research on campaign finance and other issues, conducts "inside" lobbying on Capitol Hill on the handful of issues on its current agenda, and coordinates the "outside" grassroots lobbying efforts of its volunteer members throughout the country.

Throughout the history of Common Cause, the core of its agenda has dealt with issues involving how government works. Its unofficial motto has been "to open up the system and to make government more open, accountable, and responsive to the people." Many issues relate to the impact of money, especially in Congress. Common Cause has led the fight to reform the financing of federal election campaigns and to reduce the role played by political action committees (PACs). It has also sought to end the payment of speaking fees and to strengthen conflict-of-interest and financial disclosure laws regarding public officials. The organization supports most major civil rights issues and worked actively to ratify the Equal Rights Amendment, usually by joining but not leading broad coalitions. (See Chapter 3 for a discussion of efforts to win ratification of the ERA.) Occasionally it works on issues beyond this core agenda. For example, it lobbied for legislation to end the Vietnam War and vigorously opposed development of the MX missile, the "Star Wars" defense system, and anti-satellite weapons.

Common Cause helped create some of the new "political technologies" widely used today. Roger Craver, a major mass mailing fundraiser for liberal causes, developed and refined his techniques through his work for the group. The organization's "outside lobbying" techniques usually mobilize members to write to their Washington representatives, much as the National Rifle Association does. Common Cause also organizes members to monitor their representatives' public appearances in their district and ask them questions about Common Cause issues. When combined with "internal lobbying" by professionals, these outside strategies have led to considerable success for the organization. So far, Common Cause has managed to combine idealism with pragmatism, limiting its efforts to a few issues and using sophisticated legislative strategies to build coalitions and make timely compromises. Consequently, it has acquired a reputation as one of the most influential and important public interest lobbies.

discussion of the function of opinions earlier in this book noted that some opinions are held because they express deeply held values ("externalization") or help people to fit into social groups ("social adjustment"). Joining groups to better the status of women, African Americans, or Hispanics or to protect the environment can be attractive for similar reasons. Thus, the nonmaterial selective benefits of membership in such groups turn out to be more powerful than Olson thought.

How do broadly based noneconomic interest groups manage to overcome the obstacles to organization? It takes more than the existence of a number of people willing to join such groups when asked. Specifically, it takes bold leadership from "entrepreneurs" who recognize conditions favoring the formation of a group, use effective symbols to mobilize support, and succeed in finding outside financial sponsors to provide start-up funds and continuing support.[11]

THE UNIVERSE OF ORGANIZED POLITICAL INTERESTS

The conglomeration of organized interests in the United States has several attributes worth examining in detail. For instance, it contains *high and growing numbers* of a *diverse* set of organizations, most of them pursuing *economic benefits* and with a *strong bias* toward the socially and economically better off in society. In addition, it displays pronounced disagreement on issues, sometimes even between groups that seem to have nearly identical interests. We will now discuss each of these attributes of organized interests in American politics.

The large and growing number of organized interests   The *Encyclopedia of Associations* lists 21,000 U.S.-based national and international nonprofit associations, including trade organizations such as the U.S. League of Savings Associations, professional groups such as the American Bankers Association, and social welfare, public affairs, religious, sports, and hobby groups. It lists more than 50,000 other groups organized on a regional, state, or local basis. Although many of these groups never engage in political activity, many others do, making the total number of political interest groups truly impressive.

Almost all academics who study interest groups remark on a striking increase since the 1960s in the numbers of organized interests active in politics. The increase in **citizens' groups**—organizations that seek collective benefits and membership in which is open to anyone—has received the most attention. As Figure 9.1 indicates, most of these citizens' groups, such as those active in environmental and women's issues, formed beginning in the mid-1960s. Consider that in 1960, just 16 groups sought to shape policy involving Native Americans, and that by 1980 there were 48 such groups.[12] Interest groups organized around church denominations, hospitals, and state and local government have also proliferated, as have think tanks and foundations. But a survey of 7,000 organizations with a presence in Washington found that despite the attention citizens' groups receive, organizations representing business, particularly individual corporations, grew the most in number.[13]

The diverse world of organized economic interests   Economic interests (i.e., "for-profit" interests) account for most of the significant and active political interest groups, as is indicated in Figure 9.2 (on p. 328). The size and complexity of the American economy generates a bewildering variety of economic enterprises and professions and a somewhat more limited but nonetheless extensive group of labor unions. Virtually all professions and many occupations have generated one, and sometimes many, organizations. Although

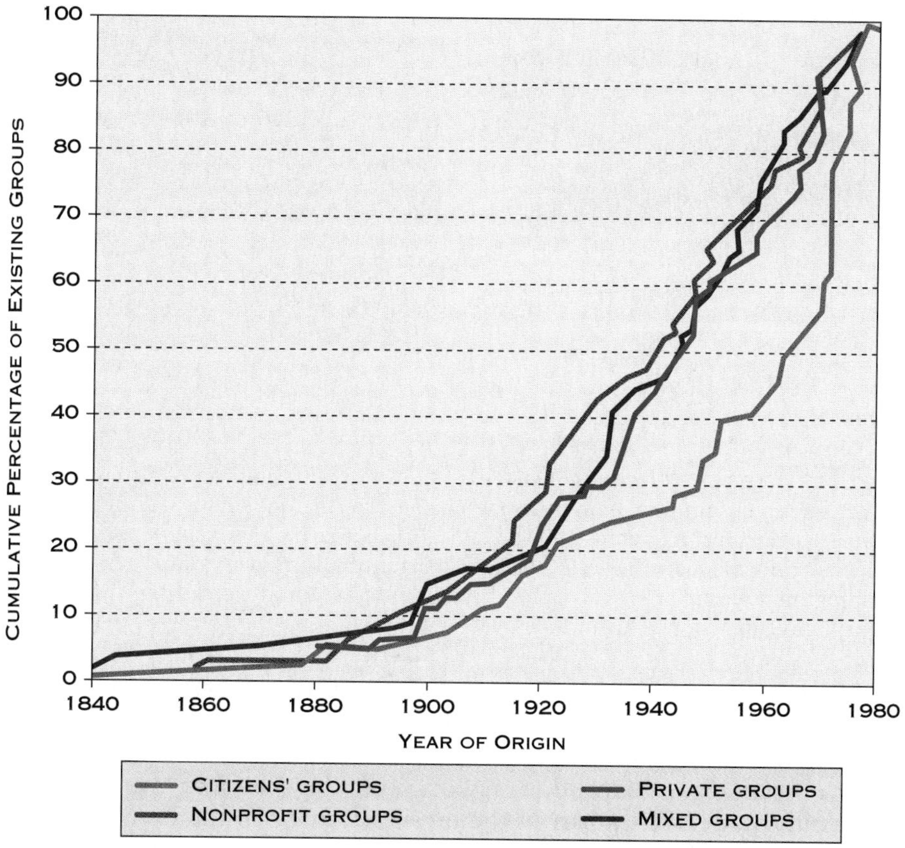

FIGURE 9.1

ORGANIZED INTEREST GROUPS This graph shows the enormous growth over time in the number of organized interests. Most of this growth has occurred since 1960 along with increased awareness of environmental, women's, and civil rights issues. Although the proliferation of citizens' groups is impressive, research suggests that the most dramatic increase was in the number of organizations representing business. *Source:* Survey conducted by Jack L. Walker, "The Origins and Maintenance of Interest Groups in America," *American Political Science Review* 77 (June 1983), p. 395. Reprinted with permission of *APSR*.

The "mixed" category is composed of groups that have members from both the public and private sectors.

profession-based interest groups do engage in political activity, most people who join these organizations probably do so to obtain selective benefits rather than to shape policy. The same is true of labor unions, where the potential benefits to be gained from the union's political activities play an even smaller part in drawing members.

In addition to occupation-based interest groups, individual economic enterprises, particularly the larger corporations, engage in direct political activity. A study of 7,000 organized interests in Washington, D.C., found that 45 percent were corporations whereas less than 5 percent represented citizens' groups and the disadvantaged.[14] For example, major defense contracting firms hire their own lobbyists and send their executives and employees to meet with

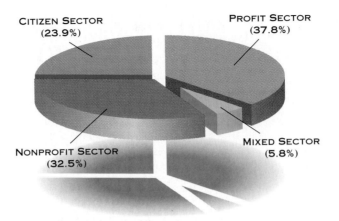

FIGURE 9.2

TYPES OF INTEREST GROUPS This chart shows the distribution of different types of interest groups. The profit sector, made up of groups representing business organizations, represents the largest percentage of active groups. Citizens' groups constitute less than one-quarter of all interest groups active in Washington. *Source:* Jack Walker, *Mobilizing Interest Groups in America* (Ann Arbor: University of Michigan Press, 1991), p. 59.

The "mixed" sector is composed of groups with members from both the public ("nonprofit") and private ("profit") sectors.

public officials. Like many other organizations, they also form a **political action committee (PAC)**, a part of the organization given the task of raising money and dispersing it to political candidates in the hope of gaining effective access to them if they are elected to office. The role of PACs in the play of power is the focus of "Strategies: The Use of PACs by Organized Interests."

Noneconomic interest groups A complete roster of political interest groups would include a number of noneconomic groups with shared interests revolving around ethnic or racial identity, religious beliefs, and issues such as abortion, the environment, and even how government works. Indeed, the social pluralism exhibited by the American population provides the basis for a rich mosaic of noneconomic organizations and associations open to general membership.

Virtually any social characteristic can serve as the basis for a shared interest spawning an organization. Ethnicity (e.g., the Ancient Order of Hibernians, for the Irish), race (e.g., the National Association for the Advancement of Colored People—NAACP; the League of United Latin American Citizens—LULAC), religion (e.g., the Knights of Columbus), and ancestry (e.g., Daughters of the American Revolution) all produce a variety of organizations. People sharing an interest in the same hobby account for the existence of groups such as the American Philatelic Society and the National Rifle Association (NRA). Common views on political and social issues produce groups such as the National Right to Life Committee, the National Organization for Women (NOW), Handgun Control, the League of Women Voters, and Common Cause.

The diversity of such associations can be illustrated by looking at the history of groups seeking to promote the interests of women. Stereotypes

THE USE OF PACs BY ORGANIZED INTERESTS

Most political action committees owe their existence to a "parent" political interest group such as a corporation, labor union, or professional association. They raise most of their money from their own members. For example, executives of S&Ls give to PACs associated with S&Ls; bank executives give to the American Bankers Association PAC; beer distributors give to SIX-PAC, and chiropractors to BACK-PAC. A few PACs depend on contributions from citizens who share their ideological perspective. For instance, some conservative people give to Senator Jesse Helms's National Congressional Club; liberals give to the National Committee for an Effective Congress; right-to-lifers contribute to LIFE-PAC, and their opponents to the National Abortion Rights Action League (NARAL).

The Federal Election Campaign Act's 1974 amendments extended the right to form PACs and permitted associations, corporations, and labor unions to pay the costs of organizing and running PACs. This, along with reforms in the presidential campaign financing system that diverted funds to PACs, produced an explosion in the number of PACs. In 1975 there were 608 PACs registered with the Federal Elections Commission; in 1992 there were 4,195.[1] Most dramatic was the increase in PACs associated with corporations—from 89 in 1974 to more than 1,700 by 1992.[2] More significantly, the amount of money collected and spent on elections soared. For example, PACs gave $12.7

million to congressional incumbents in 1974; by 1992, the figure had grown to $187.9 million.[3] During the 1992 election campaign, corporate PACs gave more than $68 million to candidates for Congress; labor PACs, $41 million; associations mostly composed of business-people and professionals, close to $54 million; and a mixed bag of "unconnected" PACs, $18 million.[4]

PACs employ a variety of mechanisms to decide to whom contributions should be given. Some rely on recommendations of local affiliates and contributors; others use a Washington-based committee. Most PACs, however, rely to a significant extent on the advice of their lobbyists. Thus, PAC contributions become an important element in overall strategy. Contributions overwhelmingly go not just to incumbents but to incumbents whose leadership position or committee assignments make good access to them especially important. The timing of contributions, some of which come during consideration of legislation important to the PAC's parent, provides further evidence of the integral part PAC money plays in interest groups' lobbying strategies.

[1] Harold W. Stanley and Richard G. Niemi, *Vital Statistics on American Politics*, 4th ed. (Washington, D.C.: Congressional Quarterly Press, 1994), p. 175, Table 6.1.
[2] Stanley and Niemi, *Vital Statistics*, p. 175, Table 6.1.
[3] Stanley and Niemi, *Vital Statistics*, pp. 182–83, Table 6.7.
[4] Stanley and Niemi, *Vital Statistics*, pp. 182–83, Table 6.7.

of women as traditionally apolitical contrast sharply with the history of women's groups. In fact, a national conference of feminists convened in 1848 at Seneca Falls, New York, and adopted a Declaration of Sentiments and Resolutions that asserted the equality of the sexes and demanded a variety of rights, including property rights and the right to vote.[15] Women's groups organized to advocate the abolition of slavery, the prohibition of alcohol, and—most important—the right to vote (woman suffrage).

A list of the activities such groups undertook conveys the scope and vigor of the battle for woman suffrage. Victory came after 56 referendum campaigns, 480 efforts to get state legislatures to allow suffrage referenda,

Prior to the ratification of the Nineteenth Amendment in 1920, women's groups engaged in grassroots political struggle for the right to vote. In this photo, women hang posters urging support for their cause.

47 campaigns at state constitutional conventions for suffrage, 277 attempts to include suffrage in state party platforms, and 19 campaigns to get the Nineteenth Amendment through Congress.[16]

Today women's groups include national feminist organizations such as the National Organization for Women (NOW), the National Women's Political Caucus, the Women's Rights Coalition, the National Abortion Rights Action League (NARAL), and grassroots feminist groups providing shelters for battered women, as well as women's health clinics, traditional community women's organizations such as the YWCA, Hadassah, and the League of Women Voters, and women's labor groups such as the National Association of Working Women. Feminist groups face opposition from "new right" women's groups such as STOP-ERA, which opposed the Equal Rights Amendment, and the Eagle Forum, the Pro-Family Forum, and Concerned Women of America, which advocates the "pro-life" position on abortion.[17]

Our roster of significant noneconomic organized interests includes a number of other types of organizations. Among them are major **foundations** such as the Ford Foundation and the Scaife Foundation, which do not lobby directly or advocate positions on major policy. Rather, they shape the play of power indirectly by funding academic policy research and demonstration projects whose results affect policy debates. For example, the Head Start program for prekindergarten children living in poverty began as a demonstration project under Ford Foundation funding.[18] The John M. Olin Foundation, known for its very conservative views, gave close to $55 million in 1988 to support conservative research.[19] Other foundations support research and projects with a more moderate policy slant. In 1993, more than 6,300 foundations, most of them giving some grants for research or projects dealing with policy, reported assets of more than $151 billion and gave almost $8.4 billion.[20] Foundation decisions about which research or projects to fund and which to reject exert a subtle but significant long-term impact on policy.

Foundations also affect current thinking about policy through their support of another type of organization, **think tanks** (private research groups that explore policy questions through social science research and advocate specific proposals). Some think tanks, such as the Brookings Institution, concentrate on academic research and avoid vigorous advocacy. Others, the majority of which have a conservative orientation, focus on devising and promoting policy alternatives. The Institute for Policy Studies, funded by labor unions, is one of the few visible think tanks with a liberal perspective. The American Enterprise Institute, the Hoover Institution, and the Heritage Foundation form the core of the conservative think tanks. With the exception of the Brookings Institution, which has a large endowment, most think tanks depend heavily on contributions from foundations and corporations; some raise considerable revenue from the sale of books and publications; and a few, such as the Urban Institute and the Rand Corporation, receive governmental grants to conduct policy research.

A final category of groups must be introduced to complete the universe of organized interests: lobbyists in Washington-based lobbying firms. This is the focus of "Resources: How Professional Lobbyists Do Their Job" (on pp. 332–33). Although the best-known organizations specializing in lobbying are law firms, a major study of lobbyists active in four issue areas found that lawyers and outside consultants constituted only 20 percent of the total.[21] The rest were nonlawyers working for public relations, consulting, and nonlawyer lobbying organizations.

The "imbalance" favoring the better off Our description of the roster of organized groups so far suggests that organizations representing the advantaged, especially those who are better off economically, are much more numerous than those representing the disadvantaged. Despite the recent growth in the number and types of noneconomic groups, including citizens' groups, the number of organizations actively promoting economic interests has increased even faster. And relatively few economic groups act on behalf of large numbers of people of modest means, such as consumers. Labor unions, whose membership usually does not include people commonly regarded as rich, suffer from declining membership and resources. At the same time, the increase in the number of corporations acting on their own behalf has been, in the words of one interest group scholar, "particularly striking"; the number of corporations with Washington-based offices increased tenfold between 1961 and 1982.[22]

A final aspect of the diversity among organized interests deserves special emphasis. The predominance of corporations among organized interests does not mean that perfect unanimity reigns among businesses. Far from it. Business interests are deeply split over many important issues. Sometimes industries that most people view as monolithic in fact disagree vehemently. The Clinton health care plan produced sharp splits within the pharmaceutical and liquor industries and between large and small health insurers. For example, when the Distilled Spirits Council of the United States (DISCUS) learned that taxes would be increased on hard liquor but not on beer or wine, its president wrote to President Bill Clinton arguing that beer and

HOW PROFESSIONAL LOBBYISTS DO THEIR JOB

Broadly defined, the term *political lobbying* refers to efforts to exert influence over political decisionmakers. Citizens writing a letter to the mayor or their U.S. senator engage in lobbying just as the head of a local savings and loan does when urging regulatory officials to change the interpretation of lending rules. The word **lobbyist**, however, commonly has a more restricted meaning. It refers to people who make a living by receiving money to exert political influence on behalf of the individuals, organizations, or associations that hire them.

Professional lobbyists work for organized interests such as corporations, labor unions, universities, and trade and professional associations, or for "political law firms." Although the term *lawyer-lobbyist* is often used, many lobbyists do not have law degrees. They usually claim some other expertise: a graduate degree, or experience working as a legislative staffer, bureaucratic official, or appointed governmental official. Former senators and House members often become professional lobbyists after leaving office.

Because both the news media and scholars of American government focus most attention on lobbying of the U.S. Congress, we too will focus our discussion on legislative lobbyists. Political scientist William Keefe, who studies Congress, describes lobbyists' activities in the legislature:

Representative Patricia Schroeder (D-Colo.) (*left*) laughs with lobbyists for a family leave bill, including Nancy Zirkin (*center*) of the American Association of University Women and Judith Litchman, president of the Women's Legal Defense Fund. Despite the common impression that lobbyists twist arms to get what they want, they more typically seek influence by gaining the trust of public officials who come to rely on their knowledge of issues and on the information they provide.

Lobbyists collect and organize data, fashion arguments, produce speeches, trade information, join members in devising strategies, arrange for meetings with other interested parties, and assist in the drafting of legislation.[1]

wine taxes should be raised if liquor taxes were. The president of the National Beer Wholesalers Association expressed his displeasure in an angry letter to DISCUS, vowing, "I don't expect that we will be doing anything on a cooperative basis with them in the near future."[23]

THE RESOURCES OF ORGANIZED INTERESTS

Without resources, no participant in the play of power can exercise influence. It is precisely because organized interests have formidable resources that they play such a central role. Our catalogue of principal group resources includes *money* and *size; members' social status, prestige,* and *public image;*

Lobbyists also influence and sometimes determine which members get campaign contributions from their organization's PAC. Most attend fundraising breakfasts, brunches, lunches, cocktail parties, and midnight snacks. In addition, they recommend who should be invited to speak to the parent organization or to attend a corporate retreat, and they help clients devise strategy, such as deciding which legislators to contact and what to say to them.

Whether they are lobbying face-to-face or guiding the efforts of others, professional lobbyists' primary resources are *information* and *knowledge* of how things work. As Hale Boggs Jr., a prominent lawyer-lobbyist, noted in a *New York Times* "op-ed" piece, lobbyists' first source of power is facts.[2] The second is money. "Lobbyists help," he wrote, "by raising money from clients, colleagues, and allies. And the help brings influence, connections, and returned phone calls."

The combination of knowledge, time, and willingness to assist legislators can translate into services that legislators and their staff members appreciate. A student of interest groups talked to a senator's staff member who expressed this view:

My boss demands a speech and a statement for the *Congressional Record*

for every bill we introduce or cosponsor—and we have a *lot* of bills. I just can't do it all by myself. The better lobbyists, when they have a proposal they are pushing, bring it to me along with a couple of speeches, a *Record* insert, and a fact sheet. They know their clout is tripled that way.[3]

Thus, lobbyists' activities go far beyond the proverbial "twisting of arms." Lobbyists with knowledge of congressional procedures can suggest to their allies ways to get their bill heard first, to get a vote when the opponents are away, or to prevent anything from getting done. Those with good contacts in Washington and an understanding of who's who can help forge the legislative coalitions with other groups that are essential for success.

[1] William J. Keefe, *Congress and the American People*, 3rd ed. (Englewood Cliffs, N.J.: Prentice Hall, 1988), p. 174.
[2] Thomas Hale Boggs Jr., "All Interests Are Special," *New York Times*, February 16, 1993, p. A17.
[3] Norman J. Ornstein and Shirley Elder, *Interest Groups, Lobbying, and Policymaking* (Washington, D.C.: Congressional Quarterly Press, 1978), p. 85, quoted in David J. Vogler, *The Politics of Congress*, 5th ed. (Boston: Allyn and Bacon, 1988), p. 289.

the group's *access* to other important players; the *intensity* of its members; the *cohesion* of the organization; and the *skill* of its leadership.

MONEY AND SIZE

Obviously, groups with more money and a larger membership enjoy an advantage over poorer and smaller groups. Citizens' groups with many members, particularly if they are skillfully mobilized, possess a potent resource. For example, the National Rifle Association, with more than 3 million members, has earned a well-deserved reputation as one of the most powerful groups in American politics. It developed techniques to get hundreds of thousands of its members to write or call members of Congress and object vociferously to any bill restricting the sale of guns or ammunition, and it

Members of the National Rifle Association oppose the Brady Bill, legislation regulating the possession of handguns. The NRA has been the most effective interest group in our nation's history. Despite public opinion polls that show a majority of Americans favoring some type of gun control, the NRA has prevented most legislation from passing. The Brady Bill was its first major defeat in some 25 years.

distributed large amounts of money every year to political candidates at every level of government. Until passage in 1993 of the Brady Bill, which requires a five-day waiting period for the purchase of a handgun, the NRA won virtually every battle in Congress over a 25-year period.

STATUS, PRESTIGE, AND IMAGE

The high *social status* of an organization and its members also serves as a resource, leading to other advantages including better *access* to decisionmakers and the mass media and higher *prestige*. The group's *public image*, a function of the social status and prestige of its members and the nature of its interests, can also serve as a resource. Organizations composed of corporate executives or Nobel Prize–winning scientists, for example, fare better than those made up of significantly less privileged individuals.

Another important component of a group's public image is its *reputation* among other active participants in politics. A group that, like the NRA or the U.S. League of Savings Institutions, has a powerful and effective image more convincingly shapes perceptions and behavior. Conversely, a group with a reputation for past failure has fewer chances of winning new victories.

ACCESS

Access refers to the ability of an organization's members or representatives to get the attention of decisionmakers. Although it is usually thought of in terms of face-to-face meetings, access applies to telephone calls and mail as well. Without it, direct influence through authority, exchange, or persuasion cannot occur because each of these methods of influence requires communication. When the telephone is busy, or no one returns your calls, or you can't get an appointment, your effort to influence will fail. In contrast, if the target of your influence attempt is receptive and the channels of communication are open, the likelihood of influence is enhanced.

INTENSITY

Interest groups rarely deploy all their resources in any given battle. The more important the issue—that is, the greater the *intensity* of opinion about the issue—the more likely it is that available resources will be deployed at a high rate. Groups with few resources to mobilize therefore require a high degree of motivation in order to be effective.

The S&L crisis showed how powerfully a threat to financial well-being can inspire interest groups to mobilize their resources. For example, a corporation named MacAndrews & Forbes Holding, Inc., invested $315 million at the end of 1988 as part of the Federal Home Loan Bank Board's effort to rescue failing S&Ls. In return, the corporation was to receive a truly impressive $900 million in tax breaks over 10 years. According to the *Wall Street Journal*, the company hired no fewer than six top lobbying firms. A spokesman for the company explained: "There was a deal, an investment of $315 million, and anyone making that kind of commitment is eager to see the integrity of the arrangement preserved."[24]

ORGANIZATIONAL COHESION

Cohesive groups can focus their efforts to maximize their influence. On the other hand, groups riven by internal dissension find it difficult to act effectively. Such conflict absorbs time and energy that could be better spent elsewhere, and it complicates the task of deciding what the group's short-term goals and strategies will be.

One reason why the U.S. League of Savings Institutions enjoyed success for many years was its high degree of cohesion. It spoke with one voice, pursued a clear and consistent set of goals, and concentrated its considerable resources to implement its strategies. The proposals made to confront the S&L crisis in 1989, however, split the industry. The interests of the healthy institutions diverged from those in financial trouble. For instance, healthy S&Ls were lukewarm about imposing higher charges to help pay the costs of rescuing their weaker counterparts.[25]

SKILLFUL LEADERSHIP

Skillful leadership can enhance the efficient use of resources, raise the group's level of intensity and its motivation to pursue its goals, and preserve cohesion. Bungling leadership produces the opposite results. When events undermine cohesion, skilled leaders can minimize the extent of its erosion and hasten its rebuilding. Identifying the characteristics of skillful leadership remains an elusive task, but these differences clearly affect outcomes in the play of power.

IS BUSINESS IN A "PRIVILEGED POSITION"?

Organized interests vary in the amount and quality of resources they possess. Although many scholars believe that the proliferation of organized groups creates a relative balance in the play of power,[26] political scientist Charles Lindblom presents a powerful argument that business groups, more than any other interests, occupy a "privileged position."[27] Lindblom argues that business interests enjoy such substantial advantages in the play of power that their disproportionate influence calls into question whether American politics is truly democratic.

For Lindblom, business's power rests not just in its overwhelming resources, substantial though they may be. Rather, the active use of these resources merely supplements a far more subtle and powerful form of influence. Businesses perform crucial economic functions—such as investing money, creating jobs, and driving economic growth—that determine the health of the economy. If businesses fail to invest and produce jobs, elected politicians incur the wrath of voters. Politicians know this. They also know they cannot force businesses to invest and create jobs. Thus, if elected officials want to win re-election, they must induce businesspeople to produce.

Holding the success of the economy in its hands, business exerts influence through anticipated reactions, inspiring solicitous concern among politicians. Politicians provide business with access and consult it in making decisions. They also enact policies that seek to induce business to prosper. These policies include providing and maintaining a banking and transportation system, lowering business taxes, providing support for research and a variety of direct and indirect subsidies, and protecting markets with tariffs and military force abroad.

Business exerts a subtler form of control, Lindblom maintains, through its ability to shape the content of the belief structure. The sacred status of such symbols as freedom, liberty, free enterprise, and free markets stems not from coincidence but from continuous, long-term reinforcement using the resources at the disposal of business. These symbols reinforce the privileged position of business by helping to keep issues that challenge business interests off the political agenda. Such symbols also help persuade citizens to oppose those policies harmful to business that do make it onto the agenda.

Lindblom's sophisticated argument does not claim that business gets everything it wants. But it does, according to Lindblom, get everything it needs. Few politicians risk confronting business, for doing so would result in a loss of confidence and in reluctance to produce. Instead, nearly all politicians, including most Democrats, pursue both rhetoric and policies designed to court business. And in cases where business does not get what it wants without asking for it, its political clout results in frequent victory when it enters the play of power. Business may be split on relatively minor issues, in Lindblom's view, but on the really important questions that relate to its continued dominance, business—especially giant corporations—virtually always prevails.

THE CONVENTIONAL STRATEGIES
OF ORGANIZED INTERESTS

337

CHAPTER 9

ORGANIZED INTERESTS

IN THE POLITICAL ARENA

Lacking the formal status of governmental officials, political interest groups can rarely influence the play of power through the exercise of authority. Instead, they rely heavily on *persuasion* and *exchange*. They use persuasion to mobilize their own members and allies, to sway public opinion, and to win general support. They use exchange to build coalitions. They combine exchange and persuasion in lobbying key decisionmakers.

MOBILIZING SUPPORTERS

The National Rifle Association's leadership did not like a provision in certain gun control legislation pending in 1988 in the U.S. House of Representatives that would establish a seven-day waiting period for gun purchases. The NRA faced formidable opposition. Sara Brady, the wife of President Ronald Reagan's press secretary James Brady, who had been permanently disabled when he was shot in the head during an attempted assassination of the president in 1981, received extensive press coverage of her efforts on behalf of the measure. Police organizations lobbied for it, and even Reagan, a consistent opponent of gun control legislation, expressed support. A purely political citizens' lobbying group, Handgun Control, spent $250,000 on newspaper advertisements supporting it. Moreover, public opinion polls indicated that consistent majorities favored requiring a police permit in order to purchase a gun, a measure more stringent than the provision the NRA disliked.

Yet opponents of gun control, many of whom belonged to the NRA, held intense, salient, stable opinions. Unlike the lukewarm majority, this intense minority possessed the attributes that make group members easy to mobilize, which is precisely what the NRA did. It spent almost $3 million in a massive lobbying effort, much of it on mailing up to three different letters to each of its 2.8 million members.[28] According to the *New York Times*, "Many members, heeding advice in these letters, wrote, telephoned and, in some cases, buttonholed their representatives, imploring them to vigorously oppose the legislation."[29] The NRA succeeded, by a 228–182 vote, in getting the House to substitute a much weaker alternative to the seven-day waiting period provision. The sponsor of the ill-fated measure reflected with some bitterness on what happened: "We were outgunned by the NRA in a campaign of deceit and distortion."[30]

The NRA, which had refined the strategy of mobilizing its members to the status of high art, exploited several resources at its command—money, internal cohesion on the issue, a large membership, and intensity of members' opinions. The U.S. League of Savings Institutions enjoyed similar advantages. What it lacked in sheer numbers it made up for in the high status of the S&L executives who belonged. Most groups, however, cannot mobilize their membership on most issues most of the time as well as these two did. Attempts to do so cost money and time. If they fail, as when student groups calling for mass demonstrations get only a handful of people to attend, the strategy of mobilization can actually weaken a group's efforts. Skilled group leaders, then, employ the strategy of trying to mobilize their members only after careful calculation.

Until the late 1980s, grassroots mobilization of ordinary citizens remained the specialty of groups such as the NRA, labor unions, citizens' groups in the environmental and consumer movements, the religious right, and Common Cause. According to one corporate lobbyist, only a handful of the largest corporations employed grassroots strategies.[31] But by the early 1990s, use of grassroots strategies had spread. Enterprising consulting firms in Washington, D.C., employed new technology to mobilize not only group members but the general public as well. Some associations use fax machines to instruct member groups to urge their officers and employees to write or call Congress; the American Trucking Association has linked its headquarters with state affiliates by satellite. The management of IBM used electronic mail to urge 110,000 employees to inform their senators and representatives that they opposed two Democratic health care bills pending in Congress.[32] Television advertisements promoting certain policies give an 800 telephone number. Callers are screened and then connected directly with their representative's Washington office.[33]

MOLDING PUBLIC OPINION

Proponents of gun control can testify that favorable public attitudes toward an issue hardly guarantee success. Nonetheless, positive public images of a group and support for its overall policy positions provide a favorable environment in which to seek both short-run and longer-run group goals. Consequently, some political interest groups with the money to do so engage in general public relations campaigns.

Interest groups seek to mold public opinion in ways that benefit their cause. Some wealthy groups buy expensive full-page advertisements in newspapers and magazines to take their case to the American public. This ad, an example of "advocacy advertising" by Mobil Oil, extols the virtues of free-market capitalism and portrays government regulation as mistaken and harmful.

Interest groups sometimes mount more focused public relations campaigns in the midst of intense battles. In the mid-1980s, for example, Congress was about to enact legislation requiring banks and S&Ls to withhold federal tax from interest payments to depositors. The banks and S&Ls mounted a furious public campaign in opposition. They placed postcards with a preprinted message of opposition addressed to a member of Congress on bank counters and enclosed them in statements mailed to depositors. The campaign produced an avalanche of mail to Congress and helped defeat the proposal. Few public relations campaigns attain this level of success. In fact, most probably have only a marginal effect. But the financial industry's achievement illustrates the potential value of this strategy and helps explain why interest groups continue to use it.

BUILDING COALITIONS WITH OTHER GROUPS

Coalitions are so important in American politics that we introduced the concept in Chapter 1 and will use it again in the chapters on political parties and Congress. Here we focus on how political interest groups follow the strategy of seeking allies in coalitions less as an option than as a necessity. Indeed, the pervasiveness of the strategy explains why one so often hears that "politics makes strange bedfellows."

Often, broad coalitions of interest groups consist of compatible bedfellows whose interests on a particular issue intersect. For instance, the Leadership Conference on Civil Rights brings together 180 organizations that share an interest in civil rights. By pooling its members' resources in a relatively permanent but loose coalition, the Leadership Conference can wield substantial influence. For this to happen, however, members must be united. The Leadership Conference, for example, effectively opposed Ronald Reagan's nomination of Robert Bork to the Supreme Court in 1987 in part because its members united behind the effort. But such broad coalitions encounter great difficulty in establishing a common position and maintaining cohesion. Thus, when the Leadership Conference opposed George Bush's nominee to head the Department of Justice's civil rights division in 1989, some members refused to participate.

Most interest group coalitions are less permanent than the Leadership Conference. Rather, many coalitions are ad hoc and temporary. For instance, police organizations joined with Handgun Control to lobby for a waiting period for would-be handgun purchasers. Other issues that police organizations support, such as reform of federal criminal law or aid to local law enforcement, find Handgun Control indifferent and inactive.

Some interest groups find that success demands forming coalitions. This is especially true of citizens' groups facing obstacles to organization and lack of money. Feminist groups often find themselves in this situation and consequently form coalitions on most issues involving women's rights, including ratification of the Equal Rights Amendment, educational equality, women's health, women and poverty, and labor issues. For example, 300 groups, including feminist and labor organizations, formed the Coalition to End Discrimination against Pregnant Workers, which succeeded in getting pregnancy covered as a disability.[34] Over time, activists in feminist groups develop contacts with leaders in a number of other groups. The experience gained from previous joint efforts facilitates forming a series of new coalitions on a variety of related women's issues.

Sometimes interest groups form coalitions even when their interests do not intersect. Their goal is to exchange support on unrelated matters of vital concern to each. For decades the American Medical Association (AMA) lobbied on behalf of the position of the American Farm Bureau Federation on agricultural policy in return for the Farm Bureau's support in the AMA's fight against national health insurance. Farm policy was not a central concern of the AMA, nor was medical policy central to the Farm Bureau. Their mutual back-scratching was a marriage of convenience.

GETTING ACCESS TO DECISIONMAKERS

Political interest groups, especially their leaders, understand the critical importance of access to key decisionmakers. It is little wonder, then, that they persistently pursue several strategies to obtain, preserve, and expand good access. They do so by trying to influence who makes decisions, by employing people who already have good access to decisionmakers, and by ensuring that their contact will be welcomed (i.e., they will enjoy "friendly access").

Shaping who makes decisions   Interest groups participate in the politics surrounding the recruitment of key decisionmakers in a variety of ways. Trying to influence who gets elected provides the most obvious example. Many, but by no means all, interest groups use a combination of the following techniques to shape election outcomes: helping to recruit candidates; urging their members to support particular candidates with contributions, campaign work, and votes; providing public endorsements; and contributing the organization's funds (see "Strategies: The Use of PACs by Organized Interests" on p. 329).

Interest groups also try to influence who gets appointed to critical administrative positions. Earlier in this chapter we described the U.S. League of Savings Institutions' success in getting Danny Wall appointed head of the Federal Home Loan Bank Board. Pursuing such a strategy requires considerable resources, a long-range perspective, and political skill. If a group can succeed in getting one of its own members or someone favorable to its perspective appointed, it virtually guarantees friendly access.

Hiring lobbyists who already have access   The strategy of hiring lobbyists has grown in popularity as the number of organized interests in Washington has increased. Between 1975 and 1985 the number of registered lobbyists doubled, and the number of attorneys tripled.[35]

No educational institution offers a degree in political lobbying. Lobbyists learn by experience. Many, however, get a head start by acquiring such experience in the work they do before they become lobbyists. What do lobbyists offer to prospective employers? Many have learned how the political process works and acquired substantive knowledge of policy while serving in Congress or working in the bureaucracy or on the staff of a member of Congress or a congressional committee. Others gain these skills after becoming lobbyists. As useful as an understanding of politics and policy is, acquaintance (and especially friendship) with key players such as administrators, legislators, newspaper reporters, and other lobbyists is equally valuable.

Examples of the crucial importance of friendly access abound. For instance, when the S&Ls decided that the U.S. League of Savings Institutions could no longer speak for their interests, they hired lobbyists with knowledge and contacts whose access contributed to their value and effectiveness. Former U.S. congressman Thomas Evans (R-Del.), who at one time regulated S&Ls, gained five S&L clients. The former staff director of the House Banking Committee received $150,000 a year from the National Council of Savings Institutions. One S&L hired six lobbying firms, including the law firm headed by the former chair of the Democratic National Committee, Robert Strauss.[36]

These examples illustrate very nicely the concept of the **revolving door**, the movement from governmental employment to the influence industry. If a group has the money to employ these people, it can buy or rent their knowledge and good access to current decisionmakers. Of course, access does not go just to those who have recently left government. The major lobbying firms possess expertise and access developed over many years of political activity. These firms, however, constantly replenish their workforce with newly "revolved" personnel and often nourish access with their ability to direct campaign contributions to officeholders.

The system persists despite efforts at reform. On assuming office, President Clinton announced new rules prohibiting key departing aides from directly lobbying their former agencies for five years. Yet, at the end of 1993, two top staff members found a large loophole. The deputy chief of staff and the chief liaison with Congress took jobs paying $500,000 a year that involved supervising others who would lobby their former colleagues and subordinates.[37]

Enhancing "friendly access"   Several practices widely employed by political interest groups help ensure friendly access. The patterns of PAC contributions to candidates for Congress show that many appear to be given to ensure good access to incumbents, not to affect who wins office. Congressional incumbents, who usually win re-election by comfortable margins, receive the lion's share of PAC money. Further, they get PAC money whether they need it or not, even when they face no opposition. In addition, PAC contributions disproportionately go to congressional leaders and members

Many analysts refer to the "revolving door" in Washington between government employment and lobbying jobs. Robert Strauss, shown here, was a member of several presidential administrations and has also worked as a lobbyist. During the savings and loan scandal, Strauss's law firm represented one of the key institutions.

who serve on committees that address issues of special concern to the PAC. Finally, PACs make contributions immediately after an election and at just about the time that legislation vital to their interests is being considered, as well as in the heat of a campaign.

The savings and loan industry used the strategy of enhancing access through campaign money with a vengeance. Former Home Loan Bank Board chief Edwin Gray told television interviewers that "Congress listened too well to the organized thrift industry and did nothing, year after year," despite his warnings of impending catastrophe. Asked why Congress ignored his warnings, he replied: "One of the big reasons was that the powerful thrift lobby, with all of those [campaign] contributions going to the members of Congress, were stopping any action by the Congress to come to grips with it."[38] The *Wall Street Journal* reported that as the S&L industry tottered toward the brink, the contributions of its PACs mushroomed. The 163 S&L-related PACs upped their contributions 42 percent after the 1984 election, giving $1,850,000 to candidates running in 1988. Nearly all of it (89%) went to incumbents. The contributions focused on members of the House and Senate banking committees, particularly the chair of the House Banking Committee, who received almost $150,000.[39] Such tactics continued during the battle over George Bush's bailout plan in early 1989.

The pattern of heavy and increased PAC contributions coinciding with major legislation affecting an industry can be found surrounding the play of power over virtually every major piece of economic legislation, whether it involves health care, energy, or a free-trade agreement. For example, PACs associated with the health and insurance industries became more active after President Clinton's election suggested that health care reform would move to the top of the political agenda. According to the public interest group Public Citizen, its PACs increased contributions to congressional candidates (mostly incumbents) in the first seven months of 1993. They gave $3.9 million, $700,000 more than in the first seven months of 1991. About half of the total went to incumbents sitting on the committees that would consider the president's health care package.[40] Ultimately, Clinton's health care reforms faced increasing opposition in the country and in Congress, and the legislation was abandoned. Of course, scholars disagree about the significance of financial contributions for legislative outcomes. At times, however, the relationship appears strong, as in the case described in "Outcomes: The NRA, Financial Contributions, and Legislative Outcomes."

LOBBYING DECISIONMAKERS: PERSUASION, REWARDS, THREATS

Members of Congress and their staffs find that the services provided by lobbyists are useful. But how do organized interests and their lobbyists use their resources to exert influence and shape outcomes?

Instances in which lobbies exert influence through authority in Congress exist but are probably rare. Members of Congress report that some of their colleagues "adopt a rule to obey" in return for money. A reporter quoted one member of Congress as observing, "There are a few guys up here who vote a certain way because they're bought."[41] Occasionally, outright bribes come to light through indictments and convictions. However, even the most

THE NRA, FINANCIAL CONTRIBUTIONS, AND LEGISLATIVE OUTCOMES

During the summer of 1994, President Clinton had his hands full seeking to pass health care legislation and a crime bill. The crime bill became controversial for many reasons, not the least of which was its ban on several types of assault weapons. Excerpts from an Associated Press story show how the National Rifle Association used its considerable financial resources to encourage a number of legislators to switch their previous votes:

> The National Rifle Association donated tens of thousands of dollars to lawmakers in the weeks just before they cast deciding votes against consideration of the crime bill. Among the biggest beneficiaries of the NRA's largesse between June and early August were a handful of Democrats who abandoned President Clinton last week after voting for an earlier version of his crime bill, according to a review of campaign reports. Among them: Bart Stupak of Michigan ($1,950 on June 27), Martin Lancaster of North Carolina ($2,500 on June 1), Charlie Wilson of Texas ($2,500 on

June 29), and Bill Orton of Utah ($4,450 on June 28).

> A few Republicans who reversed course last week also were big beneficiaries, including Gary Franks of Connecticut, who got a $4,950 general election donation from the NRA in June.

> One NRA contribution was made the same day as the vote: Alaska Republican Don Young got $3,500 on Aug[ust] 11. A few days earlier, Republicans Wayne Allard and Dan Schaefer of Colorado got similar donations. All three voted against both versions of the crime bill.

> An AP computer analysis of NRA contributions to the House since the start of the 1994 election cycle found the group gave nearly 88 percent of its $621,000 in donations to lawmakers who opposed the crime bill. Those figures include nearly $60,000 in donations AP identified as coming in the weeks immediately before the vote.

Source: "Crime Bill Set NRA in Motion," *Portland Press Herald,* August 19, 1994, p. A4.

severe critics of Congress do not believe that bribery is standard operating procedure. From a lobbyist's viewpoint, it is extremely risky as well as expensive.

Defenders of Congress in general and the lobbying system in particular argue that persuasion, and gentle persuasion at that, characterizes the overwhelming proportion of interactions. They identify three types of persuasion: (1) *reinforcement* of the support of those already favorably disposed, (2) *activation* of members already favorably inclined to make sure they vote and perhaps seek to convince colleagues to support the lobby's viewpoint, and (3) *conversion*, efforts to change the views of opponents or those who are neutral. Nearly every study of lobbying concludes that reinforcement and activation dominate. Lobbyists rarely try to convert opponents.

What about the use of threats? Explicit threats rarely surface. Vows to defeat incumbent legislators in the next election lack credibility and anger legislators in a way they don't soon forget, as do threats to contribute financially to an election foe.

Implicit threats, however, appear more often. Going "all out" on an issue conveys how strongly the group feels about it. Legislators understand that rejecting such pleas can jeopardize campaign contributions and electoral support. Simply alerting group members in the field to their representative's voting record through a published "scorecard" or a special letter can pose an effective threat. The interest group's people in Washington need not be nasty; they can leave that up to their grassroots members. The effectiveness of such tactics is illustrated by representatives' comments anticipating their NRA constituents' likely reactions if they were to support the proposed seven-day waiting period to purchase a handgun:

> "You just have a ton of grief in your district if you don't vote with them," said Representative Terry L. Bruce, a Democrat from southeastern Illinois who opposed the waiting period. "They really turn up the heat locally. You just don't want to spend every district appearance talking about that one vote."[42]

The exchange of benefits—legislators' support in exchange for the services lobbyists perform—is more typical of the legislator-lobbyist relationship. Trading information, writing speeches, offering testimony, talking substance and strategy on issues, and even discussing topics other than politics and policy characterize relations between legislators and lobbyists, intertwining exchange and persuasion. Free hunting trips, golf outings, and campaign contributions account for only some transactions. Further, lobbyists rarely offer such things openly in return for a vote or other legislative act. To do so would make the transaction sound like, look like, and smell like a bribe. Experienced participants in the play of power are too sophisticated to be so crude. But, while many legislators and some lobbyists avoid even sophisticated and subtle exchanges of support for money, undoubtedly others engage in them routinely. It is a fact of life in Washington.

A vehement opponent of gun control and a supporter of the anti-environmentalist and states' rights movements, Representative Helen Chenoweth (R-Idaho) was accused of accepting the backing of the extremist United States Militia Association in the wake of the Oklahoma City bombing in early 1995. Chenoweth's support of such movements raised questions about the influence she may have exerted in Congress on their behalf.

It was the mid-1970s, and many students at Penn State University in central Pennsylvania were displeased. Final exams were approaching, and the university had shortened the library's hours because of a severe shortage of funds. The graduate student association accepted the library administration's invitation to send its economics and accounting grad students to examine the financial records, and they came away convinced that the money to keep the library open past 10:00 P.M. simply wasn't there. The undergraduate academic assembly also failed to get extended hours when it used conventional strategies to complain and seek longer hours. But the head of the student government's academic assembly had a bright idea. She organized a "study-in," convincing hordes of students to come to the library with books, notebooks, and flashlights. When closing time came and the lights were turned off, she suggested people turn on their flashlights and keep working. The news media were informed. School administrators could, of course, have lobbed tear gas into the stacks; alternately, they could have sent in the police and dragged out students screaming, "I want to study! I want to study!" in front of television cameras. It would have been a scene guaranteed to make the network evening news. Somehow, the library found the money to extend its hours.

THE CRUCIAL ROLE OF UNCONVENTIONAL STRATEGIES

Unconventional tactics such as those used by Penn State's students rarely surface in American politics. After all, the United States has enjoyed a sustained period of *low-tension politics* since the Great Depression of the 1930s. Despite some turns in the business cycle, we have experienced no mass economic or social disruptions on the scale of those witnessed in the 1930s. Domestic political violence, although not unknown, has claimed relatively few lives. With few exceptions, organized interests use the conventional strategies expected of "insiders," obey the formal and informal rules of the game, and direct most of their efforts toward formal decisionmakers.

Exceptions do exist, however, and they are extraordinarily important. Looking only at the conventional strategies of established interests produces a narrow, static, and distorted understanding of the play of power. Less established groups frequently resort to *unconventional strategies* such as strikes, boycotts, sit-ins, demonstrations, protest marches, and violent or other illegal acts.

At times in our nation's history, unconventional strategies have been enormously successful in encouraging policy changes. For example, both the end of slavery and women's winning of the right to vote came about because disadvantaged and unpopular groups employed unconventional strategies.

WHY GROUPS USE UNCONVENTIONAL STRATEGIES

When and why would unconventional tactics be used? Several conditions usually precede them. First, a group of people needs to feel intensely about an issue that can be addressed by government. Living as a slave, being denied the right to vote, watching abortions continue, suffering discrimination on the basis of sexual preference—these are among the many conditions leading some to express an intense desire for change. Second, such groups find that conventional strategies are unavailable or, if available, unlikely to succeed alone. Sometimes adoption of unconventional strategies comes only after conventional actions fail. Third, members of such groups usually lack many of the resources needed to exert influence in conventional ways. In these circumstances, the group must give up or resort to unconventional tactics.

Consider the obstacles faced by women seeking the right to vote during the nineteenth century. They began with a huge disadvantage—the resource of the vote being denied to them. In addition, wives could not sign contracts and their earnings legally belonged to their husbands.[43] Few women worked or owned property; they depended on men to give them the small sums they did obtain. Women therefore lacked another crucial resource—money. Decisionmakers, overwhelmingly men who deemed it inappropriate for women to participate at all in politics, gave the suffragists little friendly access.

Nonetheless, many aspects of the suffrage movement's strategy would be considered conventional today. Its leaders employed political symbols in speeches, pamphlets, and articles. They held meetings to organize, and they sought to build coalitions with other groups working for the interests of the disadvantaged, particularly abolitionists of both races. Many of these activities were unconventional, especially because women conducted them, and even radical given nineteenth-century attitudes. Some qualify as unconventional even by today's standards. One of the best-known suffragists, Susan B. Anthony, engaged in civil disobedience by going to a polling place

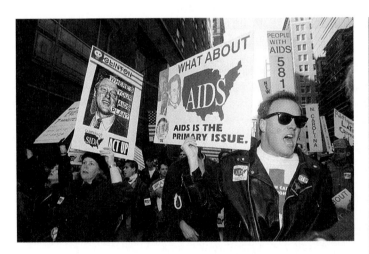

Some organized interests use unconventional political strategies to further their cause. These photos depict two such cases—a demonstration by the AIDS awareness group ACT UP, and a group of disabled Americans crawling up the steps of the Capitol building to support the Americans with Disabilities Act. Such actions bring to the public's attention issues of concern to groups lacking the resources needed to engage in traditional political strategies.

and trying to vote in 1872. She was prosecuted and convicted for her efforts. Finally, as conventional efforts repeatedly failed, the more militant suffragists staged mass demonstrations and protest marches and picketed the White House.

Although success ultimately came when the Nineteenth Amendment was ratified in 1920, the movement had suffered repeated prior defeats for at least 70 years. By 1913, after decades of struggle, only nine states had adopted full voting rights for women.

For such groups to organize and engage in any activity requires *intensity* of commitment, a willingness to use the few resources available (especially time), a committed core of activists, and skilled leadership. Even so, disadvantaged groups face enormous obstacles. The story of the National Welfare Rights Organization, whose membership came predominantly from the ranks of African-American women on welfare, provides a recent example. Combining traditional lobbying with protest strategies, the organization quickly developed internal rifts, lost support from major financial supporters, faced declining membership, and finally disbanded.[44]

BENEFITS AND RISKS OF UNCONVENTIONAL STRATEGIES

Many groups have no choice but to rely heavily on unconventional strategies. The fewer the resources available—such as the vote, money, or access—the more necessary unconventional action becomes. Yet the benefits of such strategies are few. Perhaps the biggest advantage for deprived groups with strongly held convictions is simply that the alternatives are worse: to do nothing, or to pursue conventional strategies doomed to fail.

To make matters worse, the costs of unconventional strategies are high. Because many resources are lacking, the few available ones must be used at a high rate. This takes a serious toll on time and physical energy. Such strategies sometimes invite physical risks from violence as well. Employing violence incurs the risk of physical harm, arrest, and imprisonment—hardships few people are willing to undergo.

Pursuing unconventional strategies can also lower the public's opinion of the group and its members. The very activities that must be undertaken given the lack of alternatives—sitting in, blocking traffic, even marching in a mass demonstration—arouse hostility from many other people and organizations. To motivate supporters to undergo such hardships, leaders often use symbols that generate extreme hostility from the larger community. For example, protesters who burn American flags may be engaged in self-defeating activity, since many people in American society find this act threatening and may increase their opposition to the protesters' demands.

Social movements composed mostly of groups pursuing conventional strategies often contain a few groups that do not. For instance, the environmental movement includes groups such as Earth First!, which expressed its opposition to logging old-growth forests by driving metal spikes into trees to break loggers' chain saws. Likewise, most groups opposing abortion pursue conventional strategies—political advertising, legislative lobbying, litigation, peaceful demonstrations, and mass marches. However, some, such as Operation Rescue, picket or blockade abortion clinics, heckle women entering them, and make threatening phone calls to employees.

Militant groups may damage the image and effectiveness of more moderate organizations. At the same time, their very existence can increase the leverage of the moderate groups, who point to the activities of militants as evidence of worse trouble to come if their more reasonable demands are not met.

THE ROLE OF ORGANIZED INTERESTS IN THE PLAY OF POWER

We conclude our analysis of organized interests by examining the arenas and levels of government in which they act, the timing of their activity, and the balance of material and symbolic components in the issues they address. Finally, we look at the overall influence of organized interests in the play of power by assessing their role in the formation of public policy.

ARENAS AND LEVELS OF ORGANIZED INTEREST ACTIVITY

The news media's coverage of the activities of organized interests focuses almost exclusively on their efforts to influence Congress. In reality, these groups pursue their interests everywhere in the political system. They seek to influence not just legislative bodies but political executives (and the bureaucracies under them) and courts as well. In other words, they act in all three arenas—legislative, executive, and judicial. Further, they do so at all three levels—local, state, and federal.

Studies of interest groups in various states find that the number and types of groups active there are growing, just as at the national level.[45] As more important decisions shifted to the states in the 1980s, the significance of state-level decision making rose, attracting the increased attention of organized interests with a stake in those decisions. Indeed, the opportunity to shift the arena and level of activity offers interest groups a rich variety of strategies from which to choose in seeking to exert influence.

TIMING OF INTEREST GROUP ACTIVITY

When are interest groups active? In one sense, the answer is *always.* As we just noted, these groups communicate continuously with their members, seeking to retain their membership, educate them about group activities and goals, and receive their assistance and financial contributions. But focused efforts to influence specific decisions vary in intensity over time. Just as the activity of some participants, such as political parties, rises and falls with the cycle of elections, so too does interest group activity run in cycles.

The factors that determine the rise and fall of interest group activity are numerous and complex. The legislative cycle is one such factor: At certain times of the year, decisions on budgets must be made. Another occasion for intense activity coincides with the expiration date of legislation. For instance, when the Clean Air Act is up for renewal, the groups interested in it will have to become more active. Vacancies in key decision-making posts also generate activity. Thus, when Ronald Reagan nominated Robert Bork in 1987 to fill the Supreme Court vacancy created by the resignation of

This oil slick on the coast of Alaska was caused by the spill in 1989 from the *Exxon Valdez* tanker. Environmental groups used this event and the resulting dramatic pictures of environmental devastation to publicize their cause, attract new members and financial contributions, and push for new legislation.

Justice Lewis Powell, his action triggered a frenzy of activity among a variety of groups.

Events, too, can stimulate interest group activity. For example, accidents involving nuclear power plants or oil tankers, or a series of S&L bankruptcies will force some groups to become active and entice others to try to seize the opportunities presented. After the massive oil spill from the tanker *Exxon Valdez* in Alaska's Prince William Sound in 1989, media coverage of and political attention to a variety of environmental issues rose. Environmental groups seized the opportunity to solicit increased contributions by referring to the spill in their fundraising.[46]

INTEREST GROUPS IN SYMBOLIC POLITICS

Political interest groups deal with virtually every kind of issue. It is not surprising that because the majority of political interest groups pursue economic interests, most of their issues touch on some aspect of material politics. Most issues in material politics, however, also contain elements of symbolic politics. The S&L bailout, for example, raised questions such as "Where did the money go?" and "Who were the crooks who made this mess?" Such questions, coupled with reactions roused by reports that billions of "taxpayer dollars" would be spent on the bailout, evoked emotional responses among members of Congress and some attentive publics.

The balance between material and symbolic components varies from issue to issue. For example, the struggle over the legality and availability of abortions certainly affects material politics by shaping who gets abortions, when, and at what cost. But, as the very names used by the contending groups to characterize the issue suggest ("pro-life" versus "freedom of choice"), the controversy is heavily symbolic.

The balance between the symbolic and material components of an issue affects the strategies of contending interest groups and their chances for success. You can better understand how and why the strategies of interest groups succeed and fail by asking a simple question: What is the balance between the symbolic and material aspects of the issue? Emotion-laden symbolic issues such as abortion, crime, and school prayer produce more *intense* opinions than other issues do. They more often attract intense public attention and media coverage, and they involve groups with large memberships mobilized to activity by the emotional pull of the symbols involved.

ORGANIZED INTERESTS IN THE POLICY-MAKING PROCESS

In terms of policy formation, a recurring pattern has been found in a number of issue areas. A small number of interest groups, members of relevant congressional committees and subcommittees, and government bureaucrats have determined policy together. The terms **iron triangle** and **subgovernment** have commonly been used to designate these three central actors' control over policy. For example, agricultural policy flowed from the joint interaction of the agriculture committees, career bureaucrats in the Department of Agriculture, and a few farm groups, with the Farm Bureau Federation dominating the process.

Although some iron triangles still remain in control of narrow issues (e.g., policy toward veterans' hospitals and rice growers), many have declined in significance since the 1970s. More groups participate, and they represent a greater diversity of interests, including citizens' groups. In addition, the process is more open to outside scrutiny, and the level of conflict is higher. The resulting fragmentation has spelled the end of the largely unchallenged domination of the Farm Bureau Federation and the American Medical Association. In fact, a study of the 1985 farm bill found more than 200 organized interests involved.[47] Where the AMA used to prevail, a welter of groups now contend, including hospital associations, medical insurers, drug companies, and a variety of citizens' groups such as those representing the elderly.[48]

To further complicate the picture, the patterns of interest group activity now vary by issue area, making generalizations impossible. Thus, in energy policy, where oil companies once dominated, a variety of "producer" groups in oil and gas, nuclear and electric, and major trade associations cooperate with one another and battle with environmental groups; in labor policy, associations of unions and business organizations engage in protracted, two-sided ideological conflict.[49] Indeed, as the power of once-dominant interest groups has declined, the relative influence of other participants in Congress, the bureaucracy, and the White House has increased.

The term *iron triangle* remains useful and somewhat relevant. But the broader notion of issue networks now encompasses many of the changes just described. An **issue network** is "a shared knowledge group having to do with some aspect . . . of public policy."[50] Issue networks contain a number of participants who, although acquainted with one another, bring varied interests, perspectives, and knowledge to the play of power. The mixture of groups active at any one time changes, as do the coalitions built among them. Thus, issue networks are somewhat amorphous and fluid.

What generalizations characterize the overall impact of interest groups in the play of power? Clearly, interest groups wield considerable influence in the play of power. But recognizing that organized interests are key participants leaves open the question of how important they are and which ones are most important. Experts hold different views on this question. Even if one focuses on narrower questions about the role of interest groups, the study of politics provides few clear-cut answers.

SUMMARY

Interest groups play a central role in the play of power, especially in the give-and-take over the formation of policy concerning issues on the national political agenda. A political interest group is an organization with a minimum frequency of member interaction and some formal structure that (1) emerges because its members share certain attitudes or attributes, and (2) makes claims on government to further the shared interests of its members. Although economic interests account for most of the thousands of political interest groups that organize and participate, noneconomic groups based on such characteristics as ethnicity, race, religion, leisure interests and hobbies, and political beliefs also form. Some groups are interested in a number of issues, others are single-issue interest groups. We use the term *organized interests* to emphasize the role of nongovernmental players such as foundations, think tanks, corporations, and even social movements.

Few rules significantly restrict the activities of political interest groups, although tax laws help shape the content of the universe of organized interests. The rules permit groups to form freely and to pursue a variety of strategies. The context of American politics—the belief structure, the social structure, and the economic structure—creates the potential for far more groups to organize and act in politics than actually do. The larger the size of the potential group, the more difficult the task of organization. Citizens' groups seeking collective benefits shared by members and nonmembers alike also have difficulty forming, as many potential members may act as free riders, enjoying the fruits of a group's efforts without bothering to join. Groups may gain members by offering selective benefits of direct and personal value to those who join.

Organized interests possess varying amounts of resources, including money, membership, status and prestige, access, organizational cohesion, and leadership skill. Money probably outranks the others in importance, but access, group cohesion, high intensity, and skilled leadership all contribute to success.

Leaders decide on the strategies pursued. Conventional strategies include mobilizing supporters and members, molding public opinion, building coalitions with other interests, and getting good access to decisionmakers. Access strategies are especially important. Groups seek to determine who makes decisions and to ensure friendly access by intervening in the appointment process and giving campaign contributions (through political action committees), endorsements, and other assistance to elected officials' campaigns.

They also hire lobbyists who already have access and may have gone through the revolving door from government to lobbying organizations.

Interest groups rely primarily on exchange and persuasion rather than authority to exert influence. They do this through direct efforts while working inside government; they also seek to influence indirectly, sometimes by going to those who in turn influence the actual governmental decisionmakers. Furthermore, many groups employ "outside" strategies by mobilizing public opinion, their own members, or allies, who in turn contact decisionmakers. Disadvantaged groups sometimes find it necessary to supplement conventional strategies with unconventional strategies such as strikes, mass demonstrations, boycotts, sit-ins, and even violence.

Popular beliefs that organized interests exert considerable influence in American politics are basically accurate. However, interest groups exert less influence than they did several decades ago, when iron triangles dominated policy in many areas. The policy-making process is increasingly characterized by issue networks containing a broader and more diverse group of participants. Yet it is difficult to determine just how powerful organized interests are. Whether their role is good or bad depends on the political values of the person making the judgment. Nevertheless, interest groups will continue as crucial participants in the play of power.

APPLYING KNOWLEDGE
SYMBOLIC APPEALS IN DIRECT MAIL

Organized interests, especially citizens' groups, make important use of mailings to solicit funds and members. Identify the symbols and their use in the following two solicitation letters, the first from Jerry Falwell when he headed the Moral Majority and the second from the American Civil Liberties Union after the election of George Bush as president.

JERRY FALWELL OF THE MORAL MAJORITY: Moments ago I was told my family and home would be blown up today. Sometimes I actually feel that the Falwell family has been living in a foxhole for the last several years. . . . Militant homosexuals, pornography peddlers, and abortionists have often threatened our lives. . . . But since my return from South Africa, communist terrorists are openly threatening to kill me and my family because of my campaign to stop the Soviet Union from taking over the vital minerals, strategic sea lanes, and naval bases of South Africa.

I am willing to confront the death threats, the twisted and biased media coverage—but only if I have the help, prayers, and support of my friends. . . . I must ask you to immediately send an emergency gift of $100, $50, $25.[1]

AMERICAN CIVIL LIBERTIES UNION: Not only does *Ronald Reagan's legacy of court appointments haunt us*—as it will for years to come—but George Bush has proven himself an even more determined opponent of liberty than his predecessor!

For the first time in our history, we have a President who explicitly campaigned for office against civil liberties . . . who

attacked the ACLU repeatedly . . . who tried—and nearly succeeded—in using a trumped-up "flag-burning" issue to change the
Constitution and *reduce the free speech guarantees of the First
Amendment* for the first time in history . . . and who has promised to *appoint even more judges to the Supreme Court* who share
his hostility to civil liberties.

[1] Quoted in Jeffrey M. Berry, *The Interest Group Society*, 2nd ed. (Glenview, Ill.: Scott,
Foresman, 1989), p. 60.

Key Terms

interest group (*p. 320*)

political interest group (*p. 320*)

collective benefits (*p. 324*)

selective benefits (*p. 324*)

free rider (*p. 324*)

single-issue interest group (*p. 324*)

citizens' group (*p. 326*)

political action committee
(PAC) (*p. 328*)

foundation (*p. 330*)

think tank (*p. 331*)

lobbyist (*p. 332*)

revolving door (*p. 341*)

iron triangles/subgovernments
(*p. 350*)

issue network (*p. 350*)

Recommended Readings

Berry, Jeffrey M. *The Interest Group Society*, 2nd ed. Glenview, Ill.: Scott, Foresman,
1989. An examination of the roles and activities of interest groups and the
tremendous growth of interest group politics.

Cigler, Allan J., and Burdett A. Loomis, eds. *Interest Group Politics*, 3rd ed. Washington, D.C.: Congressional Quarterly Press, 1991. A collection of essays focusing
on the activities of a variety of interest groups in the policy-making process.

Gelb, Joyce, and Ethel Klein. *Women's Movements: Organizing for Change.* Washington, D.C.: American Political Science Association, 1988. A study of the relationship between the women's movement and political interest groups that have
formed around women's issues.

Lindblom, Charles S. *Politics and Markets.* New York: Basic, 1977. In this broad-
based study of the relationship between democracy and capitalism, Lindblom
focuses attention on the dominant role of business and corporate interests in
Western democracies.

Olson, Mancur, Jr. *The Logic of Collective Action.* New York: Schocken, 1965. A
classic study of the reasons why people join organizations. Olson, an economist,
argues that material and selective benefits are crucial in the formation and
maintenance of groups, and that organizations seeking collective benefits must
overcome the problem of free riders.

Schattschneider, E. E. *The Semi-Sovereign People.* New York: Holt, 1975. A classic
study of how interest group politics benefits business interests and limits citizen
involvement in the political system.

Schlozman, Kay Lehman, and John T. Tierney. *Organized Interests and American
Democracy.* New York: Harper & Row, 1986. An important, comprehensive
text on interest groups' activities and impact, based in part on interviews with
175 representatives of organized interests in Washington, D.C.

Political Parties

in the Governing Process

PRESIDENT BILL CLINTON was elected in 1992 largely because of voter dissatisfaction with the U.S. economy. When he presented his economic reform proposals to Congress on February 17, 1993 (less than a month after his inauguration), Democratic legislators leaped repeatedly to their feet, roaring approval. Republicans remained seated and silent—some even heckled the president. These very different reactions foretold the sharp partisan conflict that would culminate in two dramatic votes later that year.

The most spirited confrontation following the 1992 elections came in an August 1993 vote on the federal budget. Although the budget bill no longer contained many of the president's initial proposals, it preserved the core of his plan by increasing taxes on the wealthy and reducing spending in order to cut the federal deficit. The House approved the measure by just two votes; while Democrats overwhelmingly supported it, every Republican voted against it. In the Senate, solid Republican opposition in combination with nays from a few Democrats led to a 50–50 tie that had to be broken by Vice-President Al Gore. Without his party's overwhelming support, Clinton and the centerpiece of his economic policy would have suffered a humiliating defeat.[1]

Barely two months later the new president faced another crucial leadership test when he undertook a last-minute high-profile campaign to win congressional approval of the North American Free Trade Agreement (NAFTA) with Mexico and Canada.[2] Clinton's commitment to NAFTA raised the stakes: Defeat would seriously undermine his ability to win passage of other legislation. Once again the president won, this time by unexpectedly large margins. However, two of the top three House Democratic leaders had worked actively to defeat NAFTA; and House Speaker Tom Foley, although he did not oppose it, offered only lukewarm support. Among the Democrats who had backed Clinton on the budget vote, 135 deserted him and voted against NAFTA. Republicans provided 132 of the 234 votes for passage, saving the day for the president.

These two contrasting episodes suggest that the role of political parties in American politics is both complicated and important. Without parties, in fact, the play of power in Congress and most state legislatures would change dramatically and elections would be drastically different. Indeed, the two major parties in the United States perform a crucial role in contesting elections, forming policy, and linking citizens to government.

COMMON BELIEFS

AMERICAN POLITICAL PARTIES

Americans' understanding of political parties resides on the fringes of political awareness. In a survey of our own students, we asked whether they thought the Republican and Democratic parties were cohesive, centralized national organizations. Almost 30 percent of our students had no opinion or simply did not know. Among the rest, opinions were about equally divided. But a strong majority were able to say that the two parties differed in some degree: More than 77 percent disagreed with the statement "There's not a dime's worth of difference between the Republican and Democratic parties." In addition, the students recognized to some extent the importance of American political parties. Only 22 percent agreed with the statement that parties are "largely irrelevant to the operation of the political system," whereas 64 percent disagreed.

Think about what you hear, see, and read about the Republican and Democratic parties. You may receive conflicting impressions of the way they operate. You learn that House Republicans vigorously opposed the president's economic package but that Democratic Party leaders failed to support their president on NAFTA. Moreover, when several politicians vie for their party's presidential nomination, the clash of negative campaign commercials, as well as encounters with passionate supporters of each candidate, presents yet another image of American political parties.

When the media refer to "the" Democrats or "the" Republicans, they appear to suggest the existence of well-organized, disciplined, cohesive national organizations. Is such a conception accurate? Just how different are Republican and Democratic officials and voters? Most important, how do parties shape the play of power in American politics? In this chapter we address these questions by focusing on the Republican and Democratic parties, but we include a brief look at the nature and role of third parties as well. ■

THE DIVERSE ROLES AND GOALS OF POLITICAL PARTIES

In this book the term *political parties* is used in three ways: party as symbol, party as contester of elections, and party as governing mechanism. Besides examining these alternative conceptions of American political parties, we will distinguish and discuss two different party goals: winning elections, and controlling the machinery of government.

THE PARTY AS SYMBOL

The many ways in which political parties and their members are spoken of make it difficult to understand the parties' nature and role. In any given week a newspaper might refer to "the national Democratic Party," "the Republican National Committee chairman," "congressional Democrats," "the Orange County Republican Women's Club," "Democratic voters," or "the Republican candidate for drain commissioner." These varied usages suggest that political parties are *not* single, united, highly structured organizations with strong central leadership and one clear statement of principles.

Something else holds these individuals and organizations together under one umbrella: a common symbol evoked by the party name. For most Americans, the words "Republican" and "Democrat" call up a complicated set of emotions, images, and associations. Indeed, *American political parties are powerful symbols used by politicians to attract support and win elections, and used by citizens to help them organize their thoughts about politics and decide how to vote.*

Candidates who can invoke the words "Democrat" and "Republican" enjoy tremendous advantages. Voters who are confused by politics rely on the party symbols to help them form opinions and make voting decisions. Moreover, because many voters identify with one party and support it year

Symbols representing our two major political parties, especially the donkey and the elephant, appear frequently during election campaigns. The symbols worn by these national party nominating convention delegates leave no doubt as to which participant is the Democrat and which one is the Republican.

after year, candidates can always draw on a stable core of support. Political leaders respond even more strongly to the party symbol.

Our definition of the **party as symbol** emphasizes the widely shared, complex, and intense feelings and images it evokes. The symbol derives its power from U.S. political history, which resonates with the personalities and activities of its major political parties. Of course, Americans associate past presidents as well as significant past events and policies with particular parties.

Presidents and party symbols   Both parties' symbols draw heavily on the memory of revered or reviled presidents. For example, Democrats hold an annual Jefferson-Jackson Day Dinner, recall the great deeds of Franklin Delano Roosevelt (FDR), and invoke the memory of John Fitzgerald Kennedy (JFK). Indeed, the initials FDR and JFK are themselves recognized symbols. The Republican Party's nickname, "the GOP," stands for the post–Civil War phrase "Grand Old Party," used by Northerners to express their fondness for Abraham Lincoln and the Union's deeds. Republicans call on the memory of Lincoln, Dwight D. Eisenhower ("Ike"), and Ronald Reagan. (On the other hand, the Democrats rarely mention Lyndon Johnson, and for years the Republicans ignored Herbert Hoover and Richard Nixon.)

The electorate, for its part, remembers both heroes and goats from both parties. Table 10.1 (on p. 358) presents the results from a 1986 opinion poll about public figures whom voters associated with each party. Young voters' views of political parties often relate disproportionately to their images of the individual who served as president when they matured. Thus, Reagan exerted a powerful effect on current college students' images of the Republican Party.

TABLE 10.1
THE TWO PARTIES' "POLITICAL PERSONALITIES"

DEMOCRATIC FIGURES	% OF RESPONDENTS MENTIONING	REPUBLICAN FIGURES	% OF RESPONDENTS MENTIONING
Question: "Which person—from either the past or the present—do you associate most with [each] party? Who else?"			
John Kennedy	46%	Ronald Reagan	54%
Franklin Roosevelt	20	Richard Nixon	21
Jimmy Carter	13	Dwight Eisenhower	18
Harry Truman	10	Abraham Lincoln	6
Walter Mondale	6	Gerald Ford	4
Edward Kennedy	4	George Bush	3
Lyndon Johnson	3	Herbert Hoover	2
All others	20	All others	13

Source: Data are from Larry Sabato, *The Party's Just Begun: Shaping Political Parties for America's Future* (Glenview, Ill.: Scott, Foresman, 1988), p. 140, Table 4.9. Copyright © 1988 by Larry J. Sabato. Reprinted by permission of HarperCollins College Publishers. The survey was conducted in 1986. Percentages total more than 100 percent because respondents were allowed to name more than one political figure.

The parties and politically significant events   The party holding the presidency inevitably becomes associated with significant political events that occur during its term in office. Traumatic events can color a party's image for decades. For example, Herbert Hoover was president when the Great Depression struck, earning the GOP the title "party of depression." The Democratic Party, which occupied the White House during World Wars I and II as well as the Korean and Vietnam Wars, bore the onus of being the "party of war." These images, although durable, are not unchangeable. The prosperity of Ronald Reagan's last six years diminished the GOP's reputation as the harbinger of hard times. However, the Democrats remain at a disadvantage in public judgments about which party can best handle foreign relations and keep the peace.

Less momentous events have a briefer effect. The Watergate scandal shaped views of Richard Nixon and the Republicans in the mid-1970s; Iran's seizure of American hostages in 1979 colored views of Jimmy Carter and the Democrats; and the victory over Iraq in the Persian Gulf War in 1991 and the signing of an agreement between the Palestine Liberation Organization (PLO) and Israel in 1993 affected views of George Bush and Bill Clinton and their respective parties for a time.

The parties and public policies   Major policy changes produce an association between those policies and the party in power. Franklin Roosevelt's "New Deal," especially its creation of Social Security, helped establish the image of Democrats as being attuned to the needs of the average person, whereas Republicans were seen as favoring big business and the wealthy. Reagan's increases in defense spending and his patriotic rhetoric led people to associate patriotism and a strong national defense with Republicans. A list of comments about the two parties from a 1986 public opinion survey, reproduced in Table 10.2, reflects the persistence of these images. Current and recent events and policies may temporarily affect the public's view of the two parties, but these short-term factors do not dislodge more deeply seated images that derive from great events and major policy shifts.

The varying content of party symbols   The preceding description oversimplifies the complex patterns of citizens' responses to the two parties' symbols. For instance, as we will see, in recent years women have viewed the Democratic Party a little more favorably and supported its national candidates slightly more often than have men. Furthermore, the impact of the links between presidents and parties depends on a person's party identification. For committed Republicans, the association between the GOP and former presidents Ronald Reagan and George Bush strengthens their party loyalty, while for strong Democrats, this same link provides another reason to dislike the Republican Party. Finally, although most older Americans disliked Richard Nixon, some older Republicans still regard him as one of the greatest presidents ever.

Minorities also view the parties differently. Until the 1960s, African Americans held the Republican Party, the party of Lincoln, in high esteem. Its party symbol evoked positive responses, even when the Depression led many minorities to support Roosevelt. Not until 1964, when the GOP nominated Barry Goldwater (who had voted against the 1964 Civil Rights Act) for president, did African Americans overwhelmingly identify with the Democratic Party. Since then, blacks have viewed the Democrats more favorably

TABLE 10.2
POPULAR POSITIVE IMAGES OF POLITICAL PARTIES

DEMOCRATIC PARTY	REPUBLICAN PARTY
"Looking out for the small guy."	"Hard-working, patriotic, and conservative."
"More idealistic and more concerned for the disadvantaged."	"They're doing a good job and turning things around."
"Fair and responsive to all the people."	"They promote individual enterprise and are against government giveaway programs."
"They consider the average American."	"Breath of spring; they brought back good living without spending money indiscriminately."
"Camelot."	"Pragmatic, like business."

Source: Based on Larry Sabato, *The Party's Just Begun: Shaping Political Parties for America's Future* (Glenview, Ill.: Scott, Foresman, 1988), p. 137, Table 4.8. Copyright © 1988 by Larry J. Sabato. Reprinted by permission of HarperCollins College Publishers.

and the Republicans more negatively than have whites.[3] Hispanics, except for Cubans, also overwhelmingly identify with the Democrats. A sophisticated understanding of the nature of the party as symbol recognizes differences in citizens' images and realizes that these images change over time.

PARTY ORGANIZATIONS AS CONTESTERS OF ELECTIONS

Our second definition of the **party as contester of elections** focuses on what we maintain to be the primary function of the major parties in the United States: to contest and win elections.

Why do the parties so fervently stress winning? Party leaders covet victory for a variety of reasons. Some find the symbolic rewards of playing and winning the elections game attractive. Others are attracted to the material benefits of victory—appointments to well-paying jobs, lucrative contracts, or access to decisionmakers. Still others are particularly interested in implementing certain policies and hope to be able to do so if their party wins. However, the desire to enact specific policies is not the central goal of American parties. The parties act as an interest group or an association of groups. The shared interest in each party is the desire to win elections, and this overriding interest often suffices to overcome internal differences that may threaten the party's cohesion.

PARTY ORGANIZATIONS AS GOVERNING MECHANISMS

One third definition views **parties as governing mechanisms**. The central goal of most parties in the world's democracies is to organize and control government and implement a set of policies. In the American political system, as we shall see, these aims take a back seat to the goal of winning elections. Party leaders "win" symbolically when their candidates are elected and hold a majority in legislative bodies. Nonetheless, with victory comes a chance for the party to serve as a governing mechanism and an implementer of specific policies. Control of a legislative body means occupying leadership roles that help achieve legislative victories. As indicated by the vote on Clinton's deficit reduction plan (described in the opening pages of this chapter), legislative parties sometimes do unite to promote and adopt new policies. Beyond the fact that parties may tend to favor certain policies, successful measures enacted by a party's officeholders can increase public support for its candidates in the next election. Furthermore, serving as a governing mechanism can bring material rewards that help win elections. For example, the distribution of governmental contracts, the introduction of certain tax measures, and other decisions by officeholders can produce contributions for subsequent campaigns.

A BRIEF HISTORY OF POLITICAL PARTIES

Our discussion of parties in the play of power concentrates on the present. However, a brief history of the parties will provide perspective and context, demonstrating how the play of power changes.

Nothing in the Constitution suggests a crucial role for political parties in the governmental system. The Framers' strongly negative attitudes toward political parties precluded their mention, although parties could not be prohibited without infringing on individual political liberty. George Washington's farewell address conveys the depth of such feelings at the time: "Let me now . . . warn you in the most solemn manner against the baneful effects of the spirit of party generally . . . in [popular governments] it is seen in its greatest rankness and is truly their worst enemy. . . ."[4]

Yet shortly thereafter, groups of like-minded politicians led by Alexander Hamilton and Thomas Jefferson began to organize to nominate and elect candidates. Since then, dramatic changes have occurred in the United States and in the workings of its political system. The right to vote expanded from a narrow base of propertied white males to eventually include virtually all adult citizens. The Civil War nearly destroyed the country. Massive immigration swelled the population. The United States evolved from a largely agrarian society to the world's leading industrial power. The size and scope of government exploded. Throughout this truly dramatic series of transformations, political parties remained a constant and significant force.

The party system did not remain static, of course. One classification divides it into five distinct eras: the initial period of Democratic-Republicans versus Federalists (up to 1815); Democrats versus Whigs (1828–1850s); the "first" Republican versus Democrat era (1860–1896); the "second" such era (1896–1932); and the "New Deal" party system (1932–present).[5] During each era a stable coalition of interests and social groups supported each party; except for the current era, each ended with a **political realignment**, or shift in the interests and groups supporting each party.

An alternative classification focuses on the characteristics and behavior of party organizations that predominated in four distinct eras.[6] According to this view, the *pre-party era* (1790s–1830s) saw the ad hoc formation of factions around issues but little permanent organization or public acceptance of the parties. By the mid-1830s, public distaste for parties waned and elite efforts to build party organizations accelerated. Although the Republicans replaced the Whigs and the nature of the Democratic Party changed during this *party era,* the central role of two parties persisted into the 1890s. One scholar observes, "Each party's popular support was rooted in the intense, deep, persistent loyalty of individual voters to their party home. The electoral pattern furthered such commitment. Party warfare split Americans decisively and evenly."[7] During this time, party organizations actively contested elections at all levels. Newspapers blatantly supported one party. Party conventions were held regularly at all levels of government. Even third parties held national conventions, adopted platforms, and built organizations.

Several developments led to the demise of the party era in the 1890s. The rise of the Progressive movement and its associated reforms attacked the party's base of control, instituting the secret (or "Australian") ballot, primary elections to choose nominees, and nonpartisan elections. Civil service–based hiring on merit replaced patronage jobs awarded on the basis of party loyalty. Republicans' inability to win elections in the South and Democrats' weakness in some New England states sapped the energy of party organizations. Finally, as government increasingly provided welfare benefits, the urban party mechanisms that previously had provided such services weakened. This *post-party era* lasted into the 1950s.

In this view we are now in the *nonparty era.* The importance of party nominating conventions continues to wane. Candidates rely largely on their own resources rather than on those of parties to raise money, hire consultants, and communicate with voters through television. The number of citizens strongly attached to the party symbol has declined, and party cues exert less influence on vote choice.

PARTICIPANTS IN THE PARTY SYSTEM

In accordance with the three definitions of political parties presented previously, we now examine three major groups of party participants: rank-and-file party members (who are attached to the party symbol), party leaders and activists (who are occupied in contesting elections), and elected officials (who control the machinery of government). We conclude this section with a brief examination of third parties.

RANK-AND-FILE PARTY MEMBERS (VOTERS)

Unlike European parties, American parties do not formally recruit members, issue membership cards, or engage in other activities to solidify members' commitment to a party. Instead, membership in the Republican or Democratic Party requires only a psychological attachment to the party symbol. The degree of this attachment, called **party identification**, is usually measured by a series of questions:

"Generally speaking, do you think of yourself as a Republican, a Democrat, an independent, or what?"

[If Republican or Democrat] "Would you call yourself a strong [Republican/Democrat] or not very strong [Republican/Democrat]?"

[If independent] "Do you think of yourself as closer to the Republican or the Democratic Party?"

The answers produce seven categories: strong Republicans, weak Republicans, independent Republicans, independents, independent Democrats, weak Democrats, and strong Democrats.

About 60 percent of Americans call themselves strong or weak Democrats or Republicans; another 25 percent consider themselves independent Republicans or Democrats. Many individuals acquire their party identification in childhood, especially if their parents express strong adherence to a party. Moreover, party identification is fairly stable and strengthens with age if an individual continues to vote for that party's candidates. Figure 10.1 plots the distribution of party identification from 1952 to 1992. As Chapters 11 and 12 will show, voting and elections in the United States cannot be understood without considering the powerful effects of party identification.

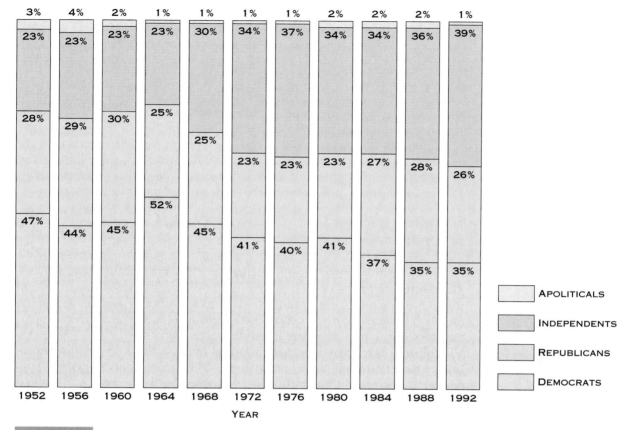

3%	4%	2%	1%	1%	1%	1%	2%	2%	2%	1%
23%	23%	23%	23%	30%	34%	37%	34%	34%	36%	39%
28%	29%	30%	25%	25%	23%	23%	23%	27%	28%	26%
47%	44%	45%	52%	45%	41%	40%	41%	37%	35%	35%
1952	1956	1960	1964	1968	1972	1976	1980	1984	1988	1992

YEAR

APOLITICALS

INDEPENDENTS

REPUBLICANS

DEMOCRATS

FIGURE 10.1

PARTY IDENTIFICATION IN PRESIDENTIAL ELECTION YEARS, 1952–1992 This graph shows party identification among the electorate in presidential election years from 1952 through 1992. The "independents" category combines "independent Republicans" and "independent Democrats" with "independents." Three key attributes of party identification can be derived from the patterns shown. First, the distribution of party identification among Republicans, Democrats, and independents does not change rapidly. Second, the number of independents increased over the 40-year period, rising steadily until 1976, dipping in 1980, then resuming its rise. Third, despite a decline in Democratic identifiers, the Democrats continued to enjoy an advantage over the Republicans. *Source: Harold W. Stanley and Richard G. Niemi, Vital Statistics on American Politics,* 4th ed. (Washington, D.C.: Congressional Quarterly Press, 1994), p. 158, Table 5.1.

Because of rounding, percentages may not total 100.

PARTY LEADERS AND ACTIVISTS

Although most citizens identify with a political party, only a tiny percentage do anything more than vote for its candidates. It is the party activists who fill leadership positions, from precinct captain to chair of the national committee. These activists concentrate on recruiting candidates, raising money, campaigning, and getting out the vote. Moreover, the activists in each party tend to have more well-developed political beliefs than the rank and file, as we shall see.

Like other political elites, party leaders hardly resemble a cross-section of the population. A 1980 study of county party leaders found that fewer than 3 percent were nonwhite and that women held only 19 percent of leadership positions in the Democratic Party and 24 percent in the Republican Party; also, approximately 33 percent of the leaders had some graduate education.[8] Delegates to the parties' presidential nominating conventions have come disproportionately from the ranks of the well educated and wealthy. However, minorities are beginning to be represented at rates close to their share of the population. In 1992, blacks accounted for 18 percent of Democratic convention delegates and 5 percent of Republican delegates.[9] As "Participants: The Changing Role of Women in Political Parties" suggests, women, too, are becoming increasingly influential in the leadership of both parties.

ELECTED OFFICIALS

American parties as governing mechanisms involve a third set of players. The elected members of each party in a city council, a state legislature, and the U.S. Congress form a caucus, elect party officers, and, increasingly, establish fundraising committees. Such **legislative parties** are important participants in the play of power in representative bodies.

Political executives also lead informal organizations of party members that support their candidacies and policies. The party leaders and groups supporting an incumbent president make up a **presidential party**; governors and mayors assemble similar coalitions at lower levels. The existence of party organizations in the legislative and executive branches is captured by the term **party in government**, as distinguished from the **party in the electorate** (ordinary voters) and **electoral party organizations** (which manage elections).

THIRD PARTIES

Every presidential election has candidates from parties other than the two major ones. In 1992, minor-party candidates received approximately 770,000 votes out of more than 104.2 million cast, about .6 percent.[10] Most minor parties, including the Communist Party USA, the Socialist Workers Party, and the Libertarian Party, possess a well-defined ideology. They usually run candidates for only a few offices, including U.S. president, and elect few members to any office. They do not form structures organized around precincts or other electoral boundaries because they run candidates to broadcast their ideas, not to win. Only a tiny number of voters identify with third parties. The committed supporters, activists, and officials of these parties, while engaging in a very high level of political activity, play mainly on the fringes of American politics.

Although such minor or third parties receive little attention, their advocacy of new policies provides insights into the complex and varied nature of the play of power. For example, free African Americans began sending delegates to conventions as early as 1830 to discuss their status and security.[11] In 1853, 114 black delegates came from nine states to discuss responses to the Fugitive Slave Act of 1850. In 1864, a meeting of 140 delegates formed the National Equal Rights League; headed by Frederick Douglass, it established

THE CHANGING ROLE OF WOMEN IN POLITICAL PARTIES

Women no longer occupy only the lower-echelon positions in American politics. Now spearheading political campaigns are participants such as Mary Matalin (*left*), who served as George Bush's campaign director in 1992, and Susan Estrich (*right*), the campaign manager for Michael Dukakis's presidential bid in 1988.

For years, women provided much of the labor required in contesting elections. They addressed and stuffed envelopes, distributed literature door to door, phoned voters, and kept records. But few held important party office or campaign posts. Furthermore, women activists were less politically ambitious than men, so fewer used their party work as a springboard to higher positions. A study of California's national convention delegates from both parties from 1964 to 1986, for example, consistently found women less interested than men in seeking elective or appointive office; however, by 1986 the level of political ambition among women had increased.[1]

The 1972 Democratic convention adopted reforms that required state delegations to include women, young people, and minorities in reasonable proportion to their presence in the state's population. Soon thereafter the Republicans also sought to raise the number of minority and female convention delegates. The results were dramatic. Between 1948 and 1968, women accounted for approximately 15 percent of delegates at both conventions.[2] By 1992, 50 percent of Democratic and 43 percent of Republican delegates were women.[3]

Women also are increasingly assuming top

positions as campaign professionals and have recently filled such posts as executive director of the National Republican Congressional Committee (Nancy Sinnott in 1982), national presidential campaign manager (Susan Estrich for the Democrats in 1988), and party chairs in 12 states (in 1991).[4] In the 1992 presidential campaign, Dee Dee Myers served as Clinton's press secretary and pollster Celinda Lake as a senior adviser; Mary Matalin, formerly chief of staff to Republican National Chairman Lee Atwater, became the Bush campaign's highly visible and outspoken campaign director. However, men still run for and win more elective offices than women.

[1] See Edmond Constantini, "Political Women and Political Ambition: Closing the Gender Gap," *American Journal of Political Science* 34 (August 1990), pp. 741–70, especially pp. 752–53, Table 3.
[2] Barbara C. Burrell, "Party Decline, Party Transformation, and Gender Politics: The USA," in Joni Lovenduski and Pippa Norris, eds., *Gender and Party Politics* (London: Sage, 1993), p. 297.
[3] Harold W. Stanley and Richard G. Niemi, *Vital Statistics on American Politics*, 4th ed. (Washington, D.C.: Congressional Quarterly Press, 1994), p. 149.
[4] Burrell, "Party Decline, Party Transformation, and Gender Politics: The USA," pp. 298–99.

organizations in many major cities to further blacks' right to vote. In 1866, the league merged with women's suffrage organizations to form the American Equal Rights Association, which advocated the right to vote for both women and African Americans. The 1960s saw the formation of the Mississippi Freedom Democratic Party and the National Democratic Party of Alabama. Although these movements were short-lived, they contributed to the gradual move toward political equality for African Americans.

From time to time, third parties and independent candidates mount serious electoral challenges. Sometimes these parties benefit by promoting an issue ignored by the major parties; occasionally, dynamic personalities enhance the appeal of such issues and attract voters in their own right. Their leaders and presidential candidates then play in the center, not on the fringes, of politics. In 1968, Alabama's Governor George Wallace, running on the American Independent Party ticket, won 14 percent of the popular vote; in 1980, former congressman John Anderson captured 7 percent as an independent; and in 1992, independent Ross Perot won nearly 19 percent, a proportion second only to former president Theodore Roosevelt's showing in 1912.

In close elections, strong third-party or independent candidacies can change the outcome of a contest between the two major-party candidates. Theodore Roosevelt, through his Bull Moose Party in 1912, split the Republican majority and caused President William Howard Taft's defeat by Woodrow Wilson. Similarly, Ross Perot's entry, withdrawal, and re-entry contributed to George Bush's defeat in 1992.

Especially when third-party candidates attract widespread voter support, one or both major parties may adopt portions of their platform. Although this usually spells electoral doom for the third party, the party's impact lives on through the policies it championed. For example, the Populist Party, spurred by economic disruption and hardship toward the end of the nineteenth century, won electoral votes and a few seats in Congress in the 1892 election. Its success contributed to the Democrats' nomination of William Jennings Bryan in 1896 and the subsequent adoption of Populist Party policies such as the graduated income tax, women's suffrage, and the direct election of U.S. senators.[12]

The changes advocated by third parties help compensate for the rigidity and unresponsiveness that sometimes characterize the major parties. Furthermore, a charismatic third-party or independent leader (such as Perot) can inspire nonvoters to vote and can mobilize the previously inactive to participate in politics. In addition, by eventually capturing positions and nominations for office in one of the established parties, insurgents who begin as members of a third party can inject new blood into the ranks of major-party cadres. Third parties may not be key players in the play of power, but they introduce change into the electoral system and bring new ideas to the political agenda.

THE IMPACT OF RULES ON POLITICAL PARTIES

Structural rules creating a federal political system with a separation of powers profoundly affect political parties. *Procedural rules,* especially those controlling the conduct of elections, also shape them. Together, these rules have produced a stable two-party system in which Republicans and Democrats monopolize the struggle for elected office.

SELECTION OF THE PRESIDENT BY THE ELECTORAL COLLEGE

Most democracies do not have two major political parties; they usually have some form of multiparty system.[13] The most distinctive feature of the U.S. Constitution's structural rules—an executive branch headed by a president chosen by the Electoral College—best explains why only two major parties compete for power in the United States.

If the office of president did not exist, neither would the two-party system. The presidency is the grand prize of American politics. Winning it provides party leaders, activists, and strong supporters with the symbolic satisfaction of "victory," hope for the adoption of favored policies, and the promise of material benefits through access to key decisionmakers or appointment to high federal posts.

Winning the presidency so entices state and local party organizations that they forge a broad coalition to do so. This fact is acknowledged in one definition of national political parties:

> [D]espite rule changes which have tended to nationalize the parties, they remain in many respects coalitions of state parties which meet every four years for the purpose of finding a candidate and forging a coalition of interests sufficiently broad to win a majority of electoral votes.[14]

Clearly, the incentive to capture the presidency sustains the two parties. If Congress picked the president, a number of small parties organized around race, region, or policy might bargain among themselves to make the choice. Instead, the framers of the Constitution created the **Electoral College** to choose the president. States have one electoral vote for each of their two senators and each representative. To win the presidency, a candidate must receive a majority of all the votes in the Electoral College. If no one does, the House of Representatives chooses from among the top three finishers, with each state delegation, regardless of size, receiving a single vote. Under this system each of two broad coalitions stands a better chance of assembling an Electoral College majority than could one of a number of narrower parties.

The procedural rules adopted by states for choosing electors (as voting members of the Electoral College are called) also encourage the development of only two major parties. State legislatures once chose electors. Now they are selected by popular vote in every state. When you cast a vote for president, you actually vote for a slate of electors pledged to vote for your candidate in the Electoral College. Every state (with the exception of Maine) awards all of its electoral votes to the one candidate who wins a plurality—that is, more votes than any other candidate. A candidate winning a sizable percentage of the popular vote, but not a plurality, in a state gets none of its electoral votes. Having more than two candidates on the ballot decreases each one's chances of winning a plurality. Thus, the rules establish a powerful incentive to form two and only two broad coalitions.

SINGLE-MEMBER DISTRICT RULES FOR OTHER ELECTIONS

Procedural rules governing other elections reinforce the two-party system. With few exceptions, states create **single-member** congressional **districts**; a state entitled to 20 House seats will create 20 districts with equal population. Theoretically, it could create five large **multimember districts**, each electing its top four vote recipients. The third (and sometimes fourth) strongest party would stand a chance of winning one seat in such a multimember district. But in single-member districts electing only the one candidate who receives a plurality, only the two strongest parties stand a good chance. Elections for statewide executive offices such as governor and attorney general (and, in many but not all states, for state legislature) use single-member districts with plurality voting, again reinforcing the advantage of the two strongest parties and discouraging additional ones.

OTHER RULES AFFECTING PARTIES

Other rules also hinder the ability of third parties to compete. All states guarantee a ballot position to the presidential candidate of each of the two major parties. However, a new party's candidate must comply with cumbersome requirements that are time-consuming and expensive to meet; furthermore, the content of these requirements varies from state to state, posing formidable organizational challenges to new parties. In addition, under federal presidential campaign financing laws, it is hard for third-party candidates to qualify for public money, and these funds are paid only after the election. State laws specify the structure of party governing bodies; establish a pyramidal structure of local, county, and state bodies; stipulate how leaders from precinct captains to state chairs are chosen; and sometimes even determine internal operating rules.[15] Despite their low visibility, these rules are significant because they help maintain the dominant position of the two established parties and discourage third parties.

PARTY ORGANIZATIONAL STRUCTURE

The organization of the two major parties reflects the federal structure of government and the separation of powers into legislative, executive, and judicial branches. Americans elect close to half a million officials at all three levels of government. Most of these contests are **partisan elections**, in which the parties select candidates who appear under the party label, rather than **nonpartisan elections**, in which no party labels appear next to candidates' names. The task of contesting thousands of elections requires an elaborate party structure. Although even casual observers of politics experience a barrage of campaign advertisements, mailings, and perhaps even rallies, the complex web of organizations behind them remains nearly invisible.

STATE AND LOCAL PARTY ORGANIZATION

A bewildering array of state laws determines the structure and procedures of the 50 state and thousands of local party organizations. As Figure 10.2 shows, local organizations are built around particular voting districts or

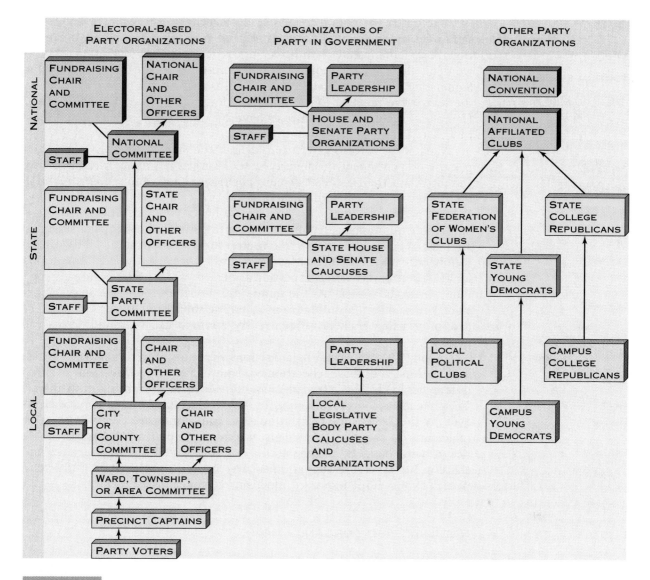

ELECTORAL-BASED PARTY ORGANIZATIONS | ORGANIZATIONS OF PARTY IN GOVERNMENT | OTHER PARTY ORGANIZATIONS

FIGURE 10.2

THE ORGANIZATIONAL STRUCTURE OF POLITICAL PARTIES Three types of formal party organizations can be found at all levels of government. Lower bodies select the members of higher ones, and all but the lowest select a chair and other officers. Above the local level, formal party organizations have staff resources and increasingly effective fundraising committees. Not shown are informal party coalitions and the financial ties between national committees and presidential campaigns on the one hand, and state and local electoral organizations on the other.

precincts or around larger political jurisdictions such as cities or counties. At the neighborhood level in many states, precinct captains distribute campaign literature, recruit poll watchers, and mobilize party supporters on election day. Precinct officials often serve on ward, city, or county party committees. In some states, county party committees select representatives to the state party committee; other states let party voters select state party committee members.

The state party committee itself elects a chair, other officers, and representatives to the national party committee. A number of ancillary organizations complement state committees' activities, including federations of local women's clubs and the Campus Young Democrats and College Republicans. Some state legislatures have party caucuses that exert a powerful effect on the legislative process. Recently, state House and Senate Republican and Democratic campaign committees have also raised large sums to fund their candidates' campaigns.[16]

Most local party organizations below the state level are weak. They possess no permanent headquarters, staff, or stable source of funds. Many cannot fill all their precinct captain positions. They cannot control who is nominated or the behavior of their elected officeholders. Before World War II, local party organizations—particularly big-city Democratic "machines" and their suburban Republican counterparts—exhibited considerable strength. A few held onto power long after the war. For example, the Chicago Democratic machine run by Mayor Richard J. Daley controlled nominations for local, state, and federal offices and dispersed jobs, contracts, and campaign money into the 1970s.[17] At the same time, a powerful Republican organization exercised similar influence in adjoining suburban DuPage County. Today, only a few pockets of local party strength can be found across the nation.[18]

In the 1970s state party organizations were no stronger than local parties. They consisted of loose federations of county and city parties, whose sharp differences in ideology, strength, and interests made them difficult to unite.[19] They had few resources (including money) to overcome their diversity. However, by the 1980s state party strength began to increase. Many state party organizations acquired permanent headquarters, professional staffs, stable and growing budgets, and the ability to provide sophisticated services and training for candidates.[20] National party organizations helped build state parties by providing increased funds and technical assistance.[21]

NATIONAL PARTY ORGANIZATION

National party organizations resemble their state affiliates in many ways. The hierarchy begins with the Democratic and Republican National Committees. The Democratic National Committee consists of 403 members, more than 80 percent of them drawn from the states on the basis of population and record of support for Democratic candidates in each state. The 165-member Republican National Committee includes the chair as well as a national committeeman and committeewoman from each state, from Washington, D.C., and from the territories.[22]

Each national committee elects a national chair, a fundraising chair, and other officers. The winning party's national committee always accepts the newly elected president's choice for national chair. After each presidential election the losing party's chair usually resigns. Its national committee elects a new chair to mediate among the party's factions, build its organizational and financial resources, and speak for the party in the media. Both national chairs help to plan the party's fundraising and organizational strategies, make arrangements for the next presidential nominating convention, and oversee the activities of the national committee staffs.

The national presidential nominating convention, held in July or August of presidential election years, is part of each party's organization. Delegates from each state vote to nominate a presidential candidate, choose (or, in effect, ratify the nominee's choice for) the vice-presidential nominee, adopt a party platform, and establish delegate selection rules for the next convention. The convention's significance has eroded substantially. Since at least 1972, the successful candidates have sewn up their nominations in advance by winning a majority of delegates in presidential primaries.

Affiliated party organizations—such as federations of Democratic and Republican women and the Young Democrats and College Republicans— also exist at the national level. Republicans and Democrats in both the House and Senate organize legislative party organizations, including increasingly successful fundraising committees.

Like their state counterparts, national party organizations have become more vigorous. Despite these developments, however, the parties still provide no enforceable central direction to the thousands of local party officials and organizations, exert little effect on policy, do not control who runs for office, and have little influence on the issue stands taken by candidates.

THE DISTINCTIVE CHARACTER OF AMERICAN POLITICAL PARTIES

The decentralization of government produced by federalism explains much about the organizational characteristics of the two major parties. So does social pluralism, along with cross-cutting memberships and the nonideological nature of the American civil religion. In the following pages we show that the two major parties are noncentralized, nonideological, pluralistic coalitions that seek to contest and win elections with the help of the party symbol.

THE "NONCENTRALIZED" NATURE OF AMERICAN PARTIES

The multiple levels of government in the American political system result in a hodgepodge schedule of elections. A centralized party structure attempting to prepare for elections for city councils in April in one state, for state supreme court justice in May in another, and for county drain commissioner in June in a third would face an impossible task. Instead, local party people organize to contest these elections.

The separation of powers also encourages noncentralized parties. Mayors, governors, and most other executives are elected by voters rather than by legislators. Executives and legislators consequently assemble separate electoral coalitions, represent different geographical constituencies, and rely on independent bases of power. This enables them to resist control by any single organization, including a political party. By giving elected officials a degree of autonomy, the separation of powers disperses power in parties as well as in government. As a result, political parties display few of the characteristics of centralized organizations.

Anyone employed by a national corporation knows firsthand about centralized control. Corporations dominate large, far-flung operations from the center, establishing and enforcing policies, controlling budgets, and so forth. They may choose to decentralize functions, allowing local managers to make more decisions, but they can re-centralize authority at any time.

If both parties were centralized, what would they look like? At a minimum, their national leaders would either control or have the power to recapture control over the following:

- Who could join the party, who would hold lower party offices, and who would be the party's nominees
- The campaign and political strategies used, the policy positions that members and officeholders would publicly support, and the way legislators would vote on key issues
- Imposition of sanctions (e.g., expulsion from the party or denial of renomination or campaign funds) against party officers and elected officials who violated party policy or campaign strategy guidelines or refused to implement the party's program

Of course, neither party has ever possessed authority or control over those identifying with it. You can consider yourself a Republican and register as one, but vote for Democrats. Any voter can seek a party's nomination for office. Candidates are free to pursue any campaign strategy they choose, to advocate any policies they desire, and to raise money on their own; party officials can only threaten to withdraw support and withhold whatever campaign funds they may have available.

David Duke, a former official of the Ku Klux Klan and a former Nazi sympathizer, provided well-publicized proof of the powerlessness of higher party officials. In 1989 Duke registered as a Republican and sought the GOP nomination for a Louisiana legislative seat. Horrified, national Republican Party officials rushed to the district to work against the nomination. Nevertheless, Duke won both the nomination and the election; the next year, he was the GOP nominee for a U.S. Senate seat and nearly won.

Finally, officeholders can behave as they wish without risking effective sanctions from their party. Efforts to deny renomination to mavericks usually fail, because such people rely on local support and their own sources of campaign funds. Threats to deny committee assignments to renegade legislators probably deter some behavior to which party leaders object. Sometimes, however, efforts at sanctions fail, demonstrating the lack of party discipline. For example, when the Democrats stripped Texas representative Phil Gramm of his committee assignments for sponsoring Ronald Reagan's economic proposals in 1981, Gramm resigned his seat, ran as a Republican, and won.

Clearly, then, the two major parties are not centralized. We discuss one reason for this in "Rules: Why the Parties Are Open to New Activists." But it would be inaccurate to describe the major parties as decentralized. This would imply that they once were centralized and then devolved power downward, and that the center could recall that power, or re-centralize. The truth is that American political parties have never exhibited great centralization. In fact, according to one book, "most close observers of American political parties [believe] that the national committees are stronger today than at any time since their creation."[23]

WHY THE PARTIES ARE OPEN TO NEW ACTIVISTS

Structural rules, including federalism and the separation of powers, produce noncentralized parties. Procedural rules govern matters such as how one runs for a party's nomination and who can vote in its primary. Both types of rules are permissive and have important consequences.

National party organizations wield relatively little influence over state and local parties. They not only lack the power to prevent people whom they reject (such as David Duke) from capturing party nominations, they also cannot prescribe what new types of people emerge as nominees or as local party officials. They cannot, for example, require local parties to nominate more women or minorities. The level of influence exerted by other party bodies over lower units varies from place to place. Sometimes a state, county, or city party organization achieves a degree of cohesion, especially when headed by a skillful and astute leader. Even then, its power may be short-lived, giving way when a leader dies or encounters scandal or electoral defeat.

For their part, most local party organizations depend on a handful of activists to hold party offices, attend meetings, and recruit candidates. It therefore takes only a few participants to win control of local party organizations. In addition, the permissiveness of procedural rules makes it easy for new, highly motivated, committed, and well-organized individuals to do so. A handful of people organized around a common interest (e.g., gender, race, ethnicity, or religious belief) can mobilize followers and win lower party office with little opposition.

The political mobilization of the religious right in Republican Party affairs in the early 1990s demonstrates this. The Christian Coalition, founded in 1989 by Pat Robertson (a television evangelist who ran for the Republican

High motivation, intense commitment, effective strategizing, and strong organizational skills have made the Christian Coalition a major force in the Republican Party. Founded by Pat Robertson (*right*) and directed by Ralph Reed (*left*), the coalition is now a significant power broker by virtue of its membership of more than 1.5 million and a budget exceeding $14.8 million.

nomination for president in 1988), had by 1994 organized more than 800 chapters nationwide. In 1994 the coalition's supporters and allies won effective control of Republican state party organizations in Texas, Virginia, Oregon, Iowa, Washington, and South Carolina, with a significant presence in Florida, New York, California, and Louisiana as well.[1]

[1] See, for example, Richard L. Berke, "Religious Right Gains Influence and Sows Division in the G.O.P.," *New York Times*, June 3, 1994, p. A1; Berke, "Wielding Fiery Words, Right Conquers Texas G.O.P.," *New York Times*, June 12, 1994, p. 20; and Berke, "In Oregon, Christian Right Raises Its Sights and Wins," *New York Times*, July 18, 1994, p. A10.

As we noted in Chapter 5, most people in the United States do not possess a well-developed political ideology. America's major political parties are almost as nonideological as its citizens. The parties' leaders and, especially, their rank-and-file members lack a well-defined, consistent set of beliefs that meets the requirements of an ideology. In fact, in each party the sharp internal splits on some issues coexist with broad agreement on other issues. The internal splits can even cause some Democrats, for example, to resemble Republicans more closely than they do other Democrats, and vice versa.

The answers that rank-and-file Republicans and Democrats give to questions about a broad range of issues display a muddled pattern of moderate agreement and disagreement, as Table 10.3 indicates. As we will see, activists—including nominating convention delegates—tend to differ sharply on

TABLE 10.3
REPUBLICAN AND DEMOCRATIC VOTERS' OPINIONS, 1988

	DEMOCRATIC VOTERS	REPUBLICAN VOTERS	DIFFERENCE
Political philosophy			
Describe own political views as conservative	22%	43%	21%
Describe own political views as liberal	25	12	13
Size of government			
Prefer smaller government providing fewer services	33	59	26
Prefer bigger government providing more services	56	30	26
Domestic policy			
Favor increased federal spending on education programs	76	67	9
Favor increased federal spending on day care and after-school care for children	56	44	12
Say abortion should be legal, as it is now	43	39	4
Say the government is paying too little attention to the needs of blacks	45	19	26
Foreign and military issues			
Favor keeping spending on military and defense programs at least at current level	59	73	14
Are more worried about communist takeover in Central America than about U.S. involvement in a war there	25	55	30
Support use of military to stop inflow of drugs	61	63	2

Based on *New York Times*/CBS News polls conducted in 1988.

Source: New York Times, August 14, 1988, p. A32. Copyright © 1988 by The New York Times Company. Reprinted by permission.

issues, depending on their party affiliation. But the views of leaders within each party are still far from monolithic. Furthermore, leaders and followers of both parties generally subscribe to the same underlying beliefs about government and policy.

THE PLURALISTIC COMPOSITION OF THE TWO PARTIES

The diversity of the American population not only inhibits the formation of coherent ideologies but also causes both parties to seek support from a number of groups. In most jurisdictions the parties must build electoral coalitions composed of people with diverse social attributes. The larger the population and geographical area represented by an office, the broader the coalition needed to attain an electoral majority. The Electoral College encompasses the entire nation, thereby guaranteeing that to capture the presidency the two parties must assemble very broad coalitions of people with different racial, ethnic, religious, regional, and economic characteristics.

In fact, both parties draw at least some support from virtually all social groups. Table 10.4 (on p. 376) reports how various groups split their votes between Bill Clinton and George Bush in 1992. Clear patterns emerge, with some groups heavily favoring one party. For example, individuals earning less than $15,000 voted almost three to one for Clinton. But no group totally supports either party. Bush received more than one in every four votes from low-income people. Similarly, the richest voters gave Bush a majority of their votes, but fully 42 percent supported Clinton. Likewise, among blacks (who strongly favored Clinton), 12 percent voted for Bush; and among white born-again Christians (who strongly favored Bush), 27 percent backed Clinton.

Diversity can also be found among party activists and leaders, and among officeholders of each party. In fact, elected officeholders and party officials coexist within the same party who would never think of having one another to dinner and who disagree passionately about issues ranging from abortion to arms control. Some Republican officeholders, for instance, enjoy strong support from labor unions and advocate greater regulation of business on behalf of consumers. Even more Democrats win the avid backing of conservative business interests; some vehemently oppose abortion and always vote for new weapons systems.

The pattern is clear. Although each party has core bases of support and its candidates may focus campaigns especially on certain issues and images, both parties must make broad appeals to win support from every group. Two political scientists explained this point in a classic study of American parties:

> First, the parties draw their leaders, workers, and candidates from all strata of American society. . . . Second, the parties direct their appeals for votes at . . . all the strata, and so give everyone the feeling that the rest of the society is concerned about the welfare and prestige of people in his [or her] stratum. . . . Third, the parties promise . . . to each group *some* but never all of what it wants. . . . No group, in other words, has reason to feel that the rest of society is a kind of giant conspiracy to keep it out of its legitimate "place in the sun." . . . [No group] lines up unanimously behind one party or the other.[24]

TABLE 10.4

SOCIAL CHARACTERISTICS OF THE 1992 PRESIDENTIAL CANDIDATES' SUPPORTERS

GROUP	% OF VOTERS IN 1992	% OF TWO-PARTY VOTE FOR BUSH	% OF TWO-PARTY VOTE FOR CLINTON
Sex and marital status			
Married men	30%	52%	48%
Married women	35	48	52
Unmarried men	15	42	58
Unmarried women	19	40	60
Income			
Under $15,000	14	28	72
$15,000–$29,999	24	44	56
$30,000–$49,999	30	48	52
$50,000–$74,999	20	51	49
Over $74,999	13	58	42
Education			
Not a high school graduate	6	34	66
High school graduate	25	46	54
Some college education	29	47	53
College graduate	24	51	49
Postgraduate education	16	42	58
Race and region			
Whites			
East	21	44	56
Midwest	25	50	50
South	24	59	41
West	16	49	51
Blacks	8	12	88
Hispanics	3	28	72
Religion			
White Protestant	49	58	42
Catholic	27	45	55
Jewish	4	13	87
White born-again Christian	17	73	27

Source: Adapted from Gerald M. Pomper, "The Presidential Election," in Gerald M. Pomper, ed., *The Election of 1992* (Chatham, N.J.: Chatham House, 1993), pp. 138–39. Copyright © 1993 by Chatham House Publishers, Inc. Reprinted by permission.

The pluralistic nature of the electoral support attracted by each party, and the strange bedfellows that politics makes, suggest yet another characteristic of the two major parties: their *coalitional nature.* Indeed, as "Strategies: Forming Party Coalitions" (on p. 378) indicates, forging coalitions may constitute the most important and effective strategy pursued by parties. The major parties have been described as

> a flexible and somewhat undisciplined pool of active groups and individuals . . . recruited to the party for different reasons . . . activated by different candidates, different issues, and different elections . . . a reservoir of organizations and activists from which are drawn the shifting organizational coalitions that speak and act in the name of the party.[25]

THE RESOURCES OF POLITICAL PARTIES

The role of political symbols in describing political parties has surfaced repeatedly. The party symbol constitutes a principal resource that plays an especially critical role in elections. Candidates acquiring the right to have the designation "Republican" or "Democrat" beside their name on the ballot immediately enjoy tremendous advantages. First, the party label usually guarantees the support of a strong majority of voters who identify with the party. Second, the media typically focus attention on the major-party candidates. Third, appearing as a major-party candidate brings party volunteers' time and money to the candidate's campaign.

At one time, the parties' control of patronage jobs and their classic ability to provide a bucket of coal or a basket of food during hard times gave them important material resources. Later, the replacement of patronage with civil service jobs and the growth of the welfare state depleted the value of these resources and weakened the parties. Now, as "Resources: The Growth of National Party Organization Strength" (on p. 379) illustrates, national party organizations draw on growing institutional and financial resources to provide campaign services. However, the parties remain noncentralized, non-ideological, pluralistic coalitions. Candidates often emerge on their own, organize their own campaigns, and do not depend on the party for campaign money. In 1986, for example, party committees accounted for only 1 percent of all money received by congressional candidates;[26] in 1994, they still accounted for only 6 percent.[27]

PARTIES IN THE POLITICAL PROCESS

The role of parties in the play of power depends not only on the size and nature of their resources but also on how these are used. Despite their substantial resources, American political parties are weak when compared with parties in most other democracies. Just what role do American political parties play?

We begin our discussion by considering the difficult question of whether political parties are continuing to decline or have begun a significant rebirth.

FORMING PARTY COALITIONS

The social pluralism and economic diversity of the United States predispose participants in the play of power to construct coalitions. Four conditions are generally required for the formation of coalitions. First, there must be a set of independent and autonomous or semiautonomous participants. Second, each participant must possess at least some resources. Third, the coalition members must share some common goals or interests. Finally, the participants can achieve shared goals only by pooling their resources and acting together.

Forming a coalition does not guarantee success and is not without costs. Members must sacrifice some individual goals to achieve shared ones. Moreover, it takes time and resources to hammer out a common position and strategy. Some coalition members may fail to deliver, or may even defect, as conditions change. However, forming a coalition is often the only way participants can achieve their goals.

American parties easily meet all four requirements. Local or state party organizations form the component groups that make up the coalition. Each contributes electoral resources such as volunteers, money, general support, and votes. Winning the election constitutes the coalition's common goal. Finally, no one party organization can win by itself. The larger the population and geographical area encompassed by an office, the broader the electoral coalition that must be assembled to capture it.

A study of the Daley machine in Chicago expresses the coalitional nature of an organization that was widely regarded as one of the most centralized and effective local political parties in the 1970s:

> [T]he machine in Chicago is not really one citywide organization but, rather, a composite of approximately 3148 local precinct organizations, each under the control of an individual responsible for his organization.[1]

Electoral party organizations are not the only ones that rely on coalitions. The parties in government also encompass a broad range of opinions and interests. For example, both parties in Congress contain some people with views that diverge widely from more prevalent views in the party. By joining together under a single umbrella, party members can organize the machinery of government and (if they are in the majority) allocate committee leadership positions to fellow party members Finally, the parties in the electorate also unite a diverse group of people around the party symbol.

[1] Milton Rakove, *Don't Make No Waves . . . Don't Back No Losers: An Insider's Analysis of the Daley Machine* (Bloomington: Indiana University Press, 1975), p. 116.

We then return to the question of differences between the Democratic and Republican Parties, considering especially the views of party activists and the degree to which parties actively promote different policies. Finally, and most important, we consider whether the two parties facilitate or inhibit the operation of representative democracy.

ARE PARTIES LOSING OR GAINING STRENGTH?

In seeking to understand changes in the dynamics of American politics, we will find it useful to consider whether political parties have reversed their long decline in the twentieth century. We can evaluate changes in party

THE GROWTH OF NATIONAL PARTY ORGANIZATION STRENGTH

National parties began acquiring new material resources several decades ago. The Republicans built a national headquarters in 1970, the Democrats in 1985. Both parties increased the size of their staffs and the scope of their activities. By 1988 the Republicans' staff numbered 425, the Democrats' 160.[1] National and congressional campaign committees' staffs also grew; the parties' congressional campaign committees increased from about 10 people each in 1972 to 130 for the Democrats and 168 for the Republicans by 1988.[2] Impressive gains in money raised (which are shown in the figure below) enabled both parties' national committees to provide increased electoral services and campaign support.

In 1993 the chair of the Democratic National Committee (DNC) proclaimed, "We must shift our fundraising focus to attract millions of small contributions." But even though the DNC collected $8.9 million in contributions of $200 or less in 1993, it had to spend $8 million to raise it.[3] The Republican National Committee (RNC) did better than the DNC in getting small contributions. More than three-quarters of the $34 million it raised in 1993 came from contributions of $100 or less.[4]

[1] Paul S. Herrnson, "Reemergent National Party Organizations," in L. Sandy Maisel, ed., *The Parties Respond: Changes in the American Party System* (Boulder: Westview, 1990), p. 51.
[2] Paul Allan Beck and Frank J. Sorauf, *Party Politics in America*, 7th ed. (New York: HarperCollins, 1992), p. 105.
[3] James A. Barnes, "Greener Acres," *National Journal*, April 23, 1994, p. 950.
[4] Barnes, "Greener Acres," p. 952.

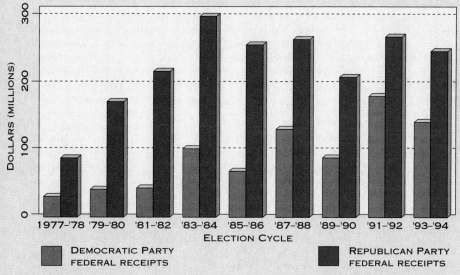

TOTAL PARTY RECEIPTS RAISED FOR USE IN FEDERAL CAMPAIGNS, 1977–78 TO 1993–94 (IN MILLIONS OF CURRENT DOLLARS) Contributions received by political parties for use in congressional campaigns have risen substantially. The Republicans consistently raise more funds than the Democrats, although the gap has narrowed. The GOP increased its total contributions dramatically between the 1978 and 1984 elections. Notice the dip in recent years in money raised for the midterm congressional elections as compared to receipts for the previous presidential election. *Source:* Federal Election Commission, "FEC Reports on Political Party Activity for 1993–94," press release, April 13, 1995, p. 1.

strength in three places: (1) in the electorate, (2) in government, and (3) in party organizations themselves.

The decline of the party in the electorate If most voters failed to identify with a party, the party symbol's value as a principal resource would evaporate. Figure 10.1 (on p. 363) showed that the number of people identifying with one of the two major parties remains high. Nevertheless, a number of factors, including rising levels of education and the increasing importance of the news media as providers of information about politics, have reduced the number of identifiers, especially committed ones, while the number of independents has risen. Further, the remaining strong party identifiers are less willing to vote a "straight party ticket" for all their party's candidates.[28] These measures of party strength suggest continued decline rather than resurgence.

The strength of the party in government For much of the twentieth century, one party or the other has dominated the politics of many states. But today there is a growing Democratic vitality in the formerly Republican New England states of Maine, Vermont, and New Hampshire, and a growing Republican strength throughout the hitherto solidly Democratic South. These trends are reflected in state government. In recent years only Mississippi has remained a one-party state; and in 20 states, both parties appear to compete fairly equally.[29] Moreover, the degree of organization, cohesion, and party voting in state legislatures varies tremendously. In some state legislatures, the power of party caucuses exceeds that found in the U.S. Congress; in others, party organization matters hardly at all.

In Chapter 13 we will examine in detail the nature of party organizations in Congress. Considering the amount of money raised by the House and Senate party campaign committees, party strength seems to be growing. Two other measures of party strength are frequency of party voting and degree of party unity. A **party vote** occurs when a majority of one party votes differently in a roll-call vote from a majority of the other party. **Party unity** can be defined as the percentage of legislators in each party who vote with their colleagues on party votes. Party voting reached fairly high levels during the 1920s but declined until the 1960s, when it began to rise, especially among Democrats.[30] Table 10.5 shows that, beginning in 1981, party voting generally rose in the House; in the Senate it fluctuated widely, jumping to its highest level in 1993 and then beginning to decline. Party unity scores in both chambers also peaked in 1993 before beginning to drop. These patterns suggest a strengthening of the role of party in government. One analysis of party leadership concludes that House Democratic leaders became stronger and more policy-oriented in the 1980s, whereas (at least as of the late 1980s) the Senate continued to lack strong party leadership.[31] More recently, House Speaker Newt Gingrich, a Republican, began his tenure by amassing considerably greater power than had been wielded by the Democratic Speaker he replaced.

The strength of party organizations Compared to the state of affairs in 1900, party organizations today dispense few patronage jobs, must compete with PACs and independent consultants in providing campaign assistance to candidates, and exert little control over who gets nominated or renominated. For these reasons, party influence has declined. On the other hand, as we have seen, both national parties have a permanent headquarters, a growing professional staff, a stable and increasing financial base, and an expanding set of financial and other relationships with local and state parties.

TABLE 10.5

**DIFFERENCES BETWEEN THE LEGISLATIVE PARTIES:
THE FREQUENCY OF "PARTY VOTING" AND
"PARTY UNITY" SCORES**

YEAR	"PARTY VOTE" FREQUENCY[1] (% OF TOTAL VOTES)		"PARTY UNITY" SCORE[2] (HOUSE-SENATE AVERAGE)	
	HOUSE	SENATE	DEMOCRATS	REPUBLICANS
1981	37	48	69	76
1982	36	43	72	71
1983	56	44	76	74
1984	47	40	74	72
1985	61	50	79	75
1986	57	52	78	71
1987	64	41	81	74
1988	47	42	79	73
1989	55	35	81	73
1990	49	54	81	74
1991	55	49	81	78
1992	64	53	79	79
1993	65	67	85	84
1994	62	52	83	83

[1]A "party vote" occurs when a majority of each party's members votes together to oppose a majority of the other party. For example, if a roll-call vote found 260 Democrats split 200 for and 60 against, and 175 Republicans split 75 for and 100 against, this would be a party vote. The last row shows that in 1994, 62 percent of all roll-call votes in the House were party votes.

[2]The "party unity" score is calculated just for party votes as defined above. It measures the percentage of party members who vote together. In the example in note 1, Democrats' score would be 77 percent (200/260); the Republicans' would be 57 percent (100/175). The table lists average scores on all party votes for both the House and the Senate.

Source: Party vote and party unity scores are from *Congressional Quarterly Weekly Report,* December 31, 1994, pp. 3658–59.

These developing trends present a mixed and uncertain picture of party strength today. Although party identification appears to be declining, there are signs that party organization has in some respects improved. Moreover, congressional parties show signs of growing stronger. As of the mid-1990s, there is no clear-cut answer to the question of whether American political parties are growing stronger or weaker.

HOW DIFFERENT ARE THE TWO MAJOR PARTIES?

Lewis Carroll, in *Alice in Wonderland,* concocted the nearly identical twins Tweedledum and Tweedledee to satirize the lack of meaningful differences between the major British political parties of the day. Today, the common belief that "not a dime's worth of difference" separates the Republican and Democratic parties echoes Carroll's view. But partisan passions at election time and the verbal barbs that legislative leaders exchange suggest sharp divergence. Just how different are the two major parties? The answer makes a difference. If today's voters are choosing between Tweedledum and Tweedledee, their ability to affect material politics plummets. If, on the other hand, important differences exist between the two major political parties, voters can shape outcomes by changing the party that controls Congress and the presidency.

In assessing the differences in the ways parties function, we examine five areas: (1) the composition of party supporters in the electorate; (2) the policy views and ideology of party officials and activists; (3) the coalitions that unite behind the two parties' presidential candidates; (4) the ideology and behavior of the congressional or legislative "parties in government"; and (5) the policies pursued by the two parties, especially with regard to women and race.

Differences in the party in the electorate   Table 10.4 (on p. 376) revealed the coalitional nature of parties by showing that they draw some support from all groups. However, it also revealed substantial differences in the degree to which various groups support each party. For example, in the 1992 presidential campaign the less well off—especially members of minority racial and religious groups—leaned heavily toward Bill Clinton, the Democratic candidate. White Protestants and males formed the core of George Bush's Republican support.

What about the policy views of rank-and-file party identifiers? As we noted earlier, Republicans and Democrats agree on most basic rules of the game as well as on the general direction of policy. In Table 10.3 (on p. 374) we compared the views of Republican and Democratic voters on a variety of issues. Although Democrats as a group disagree with Republicans, the differences are usually small. For example, Democrats favor increased federal spending on education, but only by 9 percent. On only four issues in Table 10.3 did the Democratic-Republican gap exceed 25 percent.

Differences between party officials and activists   The policy views and ideology of party activists differ sharply. A comparison between the views of delegates to the Republican and Democratic national conventions in 1956 revealed consistent, sharp differences on a variety of issues.[32] The differences persist to this day. Table 10.6 compares the views of Republican and Democratic convention delegates surveyed by the *New York Times* in 1988 on the same questions for which rank-and-file voters' responses are reported in Table 10.3. The gap between Republican and Democratic delegates' views on every issue (except the use of the military to stem drug traffic) is greater than the largest difference found between rank-and-file voters in the two parties, as Figure 10.3 (on p. 384) shows. Although equivalent data for Republican delegates to the 1992 convention are not available, the differences in the opinions of Republican and Democratic delegates in 1992 were probably even larger than in 1988.[33]

Differences between the "presidential parties"   Although both major parties appeal to all groups for support, each has particular groups on which it tends to rely, especially in presidential elections. The coalitions of voters and party leaders that lend crucial support to presidential candidates form what are called the *presidential parties.*

The rules for electing the president, especially the "plurality-winner-take-all" method of allocating electoral votes, make large states and specific

TABLE 10.6
REPUBLICAN AND DEMOCRATIC LEADERS' OPINIONS, 1988

	DEMOCRATIC DELEGATES	REPUBLICAN DELEGATES	DIFFERENCE
Political philosophy			
Describe own political views as conservative	5%	60%	55%
Describe own political views as liberal	39	1	38
Size of government			
Prefer smaller government providing fewer services	16	87	71
Prefer bigger government providing more services	58	3	55
Domestic policy			
Favor increased federal spending on education programs	90	41	49
Favor increased federal spending on day care and after-school care for children	87	36	51
Say abortion should be legal, as it is now	72	29	43
Say the government is paying too little attention to the needs of blacks	68	14	54
Foreign and military issues			
Favor keeping spending on military and defense programs at least at current level	32	84	52
Are more worried about communist takeover in Central America than about U.S. involvement in a war there	12	80	68
Support use of military to stop inflow of drugs	54	67	13

Data are based on telephone interviews with 739 Republican delegates and 1,059 Democratic delegates.

Source: New York Times, August 14, 1988, p. A32. Copyright © 1988 by The New York Times Company. Reprinted by permission.

blocs of voters within them attractive targets. Democratic presidential candidates rely heavily on votes from African Americans, Hispanics, Jews, and labor union members to help carry large states with concentrations of these groups. Thus, the Democrats' presidential party includes liberal intellectuals, labor leaders, and most minority group leaders. Republicans, by contrast, draw on a base of support in southern and mountain states. The

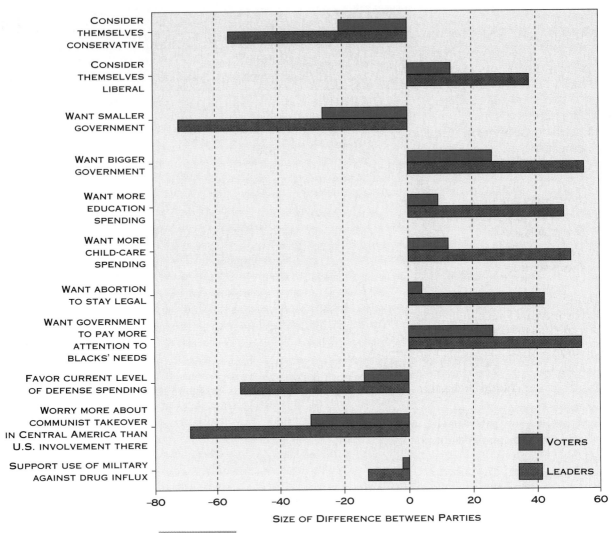

CONSIDER THEMSELVES CONSERVATIVE

CONSIDER THEMSELVES LIBERAL

WANT SMALLER GOVERNMENT

WANT BIGGER GOVERNMENT

WANT MORE EDUCATION SPENDING

WANT MORE CHILD-CARE SPENDING

WANT ABORTION TO STAY LEGAL

WANT GOVERNMENT TO PAY MORE ATTENTION TO BLACKS' NEEDS

FAVOR CURRENT LEVEL OF DEFENSE SPENDING

WORRY MORE ABOUT COMMUNIST TAKEOVER IN CENTRAL AMERICA THAN U.S. INVOLVEMENT THERE

SUPPORT USE OF MILITARY AGAINST DRUG INFLUX

VOTERS

LEADERS

SIZE OF DIFFERENCE BETWEEN PARTIES
−80 −60 −40 −20 0 20 40 60

FIGURE 10.3

SIZE OF DIFFERENCES IN OPINION AMONG DEMOCRATIC AND REPUBLICAN VOTERS AND LEADERS On every question dealing with policy in Tables 10.3 and 10.6, the level of disagreement between rank-and-file Democrats and Republicans is much smaller than that between party leaders. For example, 13 percent more Democratic voters than Republican voters called themselves liberals (25% − 12%); among leaders the difference was 38 percent (39% − 1%). *Source:* Totals are calculated from data in *New York Times,* August 14, 1988, p. A32.

GOP presidential party consists of conservative intellectuals, the religious right, executives from many large corporations and financial institutions, and noticeable numbers of white Protestant males. Finally, candidates in both parties must win in primaries, where Republican voters tend to be more conservative and Democratic voters more liberal than those who vote in the general election. This accentuates the differences between the presidential coalitions in both parties.

Party platforms clearly reflect the differences in composition of the presidential parties. An analysis of the 1984 platforms concluded that, "on a host of foreign and domestic issues, the two parties could not have taken more divergent substantive positions—and there is nothing mushy or tepid about the rhetoric employed, either."[34] Table 10.7 (on p. 386) summarizes the sharp differences in the two parties' 1992 platforms.

Differences between the parties in Congress   The congressional Republican and Democratic *legislative parties* also differ. Table 10.5 (on p. 381) illustrated a sharp contrast in the voting habits of Republican and Democratic members. The "report cards" that liberal and conservative interest groups compile on members' voting records also reveal substantial party differences. One study calculated the difference between the ranking that a liberal group, the AFL-CIO's Committee on Political Education (COPE), gave Republicans and Democrats and the ranking assigned by Americans for Conservative Action (ACE). It found that between 1968 and 1982, Republicans in both the House and the Senate were consistently much more conservative in their voting than Democrats were.[35]

Finally, a "presidential support score" can be calculated by looking at the rates at which senators and representatives vote for measures advocated by the president. These scores indicate that, as might be expected, members of the president's party consistently support the chief executive's program more often than the opposition does. Ronald Reagan fared especially well in 1981, when House Republicans supported him 72 percent of the time; Democrats did so only 46 percent of the time.[36] Toward the end of his term, support from Republicans dropped to 61 percent; from Democrats, to 28 percent. George Bush, during his last year in office, received support from Senate Republicans 73 percent of the time, whereas Senate Democrats supported him just 33 percent of the time. Bill Clinton's 1993 Democratic support scores of 77 percent in the House and 87 percent in the Senate helped him achieve an overall presidential success rate that was the highest in 40 years.[37] However, as the vote on NAFTA showed, presidents sometimes fail to win their party's support on some important issues, and the level of a party's support for its president can change both from issue to issue and over time.

Changes in policy stands   The extent of policy differences between parties changes substantially over time. One study, which examined party platforms from 1844 to 1968, found that the extent of divergence increased greatly during periods of political crisis and party realignment—that is, when the major sources of support and the policy positions of the parties underwent rapid major shifts.[38] The realignment in the 1860s saw the emergence of the Republicans, who drew their strength from the North, whereas the Democrats relied mostly on the South. In 1896, Republicans solidified their hold on the North while the Democrats became more entrenched in the South. The Depression elections of 1932 and 1936 broke the pattern, rearranging the composition of the parties by adding industrial workers, poor farmers, Catholics, Jews, and African Americans to the Democrats' base. During each of these bursts of realignment, sharp differences in the policies advocated by the parties prevailed. Following each realignment, differences declined until the next realignment. In 1964, when the GOP nominated a strong conservative, Barry Goldwater, and the Democrats outlined Lyndon Johnson's "Great Society,"[39] party platform differences rose to their highest level since 1932. With the nomination of Ronald Reagan in 1980, the party

TABLE 10.7
MAJOR POLICY DIFFERENCES BETWEEN THE 1992 PARTY PLATFORMS

ISSUE	REPUBLICAN PLATFORM	DEMOCRATIC PLATFORM
Abortion	We believe the unborn child has a fundamental individual right to life that cannot be infringed. We therefore reaffirm our support for a human life amendment to the Constitution, and we endorse legislation to make clear that the Fourteenth Amendment's protections apply to unborn children.	Democrats stand behind the right of every woman to choose, consistent with *Roe v. Wade*, regardless of ability to pay, and support a national law to protect that right.
Gun control	Republicans defend the constitutional right to keep and bear arms. We call for stiff mandatory sentences for those who use firearms in a crime.	We support a reasonable waiting period to permit background checks for purchases of handguns, as well as assault weapons controls. . . . We will work for swift and certain punishment of all people who violate the country's gun laws and for stronger sentences for criminals who use guns.
Education	The president's proposed GI Bill for Children will provide $1,000 scholarships to middle- and lower-income families, enabling their children to attend the school of their choice.	We oppose the Bush administration's efforts to bankrupt the public school system—the bedrock of democracy—through private school vouchers.
Energy	We will . . . [allow oil and gas drilling] access, under environmental safeguards, to the coastal plain of the Arctic National Wildlife Refuge and to selected areas of the Outer Continental Shelf.	[We will] oppose new offshore oil drilling and mineral exploration and production in our nation's many environmentally critical areas.
Voter registration	We support state efforts to increase voter participation but condemn Democrat attempts to perpetrate vote fraud through schemes that override the states' safeguards of orderly voter registration.	We need new voter registration laws that expand the electorate, such as universal same-day registration.
Health care	We endorse President Bush's comprehensive health care plan, which will make health care more affordable through tax credits and deductions that will . . . make health care more accessible, especially for small businesses, by reducing insurance costs and eliminating workers' worries of losing insurance if they change jobs.	We will enact a uniquely American reform of the health care system to control costs and make health care affordable; ensure quality and choice of health care providers; cover all Americans regardless of pre-existing conditions.

Source: Adapted from *Congressional Digest*, October 1992, pp. 234–55.

platforms again diverged sharply. As Table 10.7 shows, significant differences also existed in 1992.

The recent divergence in policy is evident in issues involving race.[40] In 1960, race was not a partisan issue. The Democratic Party's southern wing contained strict segregationists, whereas many Republicans took a decidedly liberal stance. The two parties helped transform American politics as their positions on race changed. Segregationist Democrats largely disappeared; some, including Senator Strom Thurmond of South Carolina, switched to the GOP, and Republican racial liberals and moderates declined in number. The Democrats became the party of racial liberalism, supporting federal intervention to guarantee civil rights (e.g., through school integration), and the Republicans became the party of racial conservatism, opposing such intervention. The policies pursued by the two parties' congressional wings and presidential administrations increasingly reflected the divergence.

The parties' policies on issues involving women, especially abortion, have diverged as well. The Republicans have strongly endorsed restricting abortions, whereas the Democrats have advocated maintaining abortion rights. Republicans have also emphasized differences with Democrats over other social issues, including crime, school prayer, and patriotism. Traditional differences have persisted regarding governmental regulation of economic activity and who should bear the heaviest burden of taxes. The Democrats have generally favored more regulation and heavier taxes on the wealthy than have the Republicans. The parties' positions on all these issues have found expression in both congressional and presidential decisions.

HOW DO PARTIES FACILITATE DEMOCRATIC REPRESENTATION?

The two major parties perform five useful functions for the political system: recruiting and selecting certain key decisionmakers; structuring conflict and defining alternatives; moderating and channeling conflict; organizing the decision-making apparatus of government; and, through a combination of these functions, providing a crucial link between citizens and government.

Recruiting some key decisionmakers   Not all key players in the play of power acquire their formal authority by election. Media executives and reporters, interest group leaders and lobbyists, corporate executives and bureaucrats come to power in other ways. But many key participants win office in partisan elections for which the parties help to recruit, finance, and run the candidates' campaigns; to provide election officials; and to mobilize voters. Parties and their symbols thus can be said to play a central role in selecting certain important decisionmakers.

Structuring conflict and defining alternatives   Regularly scheduled elections provide opportunities for controlled, structured political conflict. Elections that coincide with the major party realignments, like those of 1896 and 1932, offer the electorate an opportunity to change party control of Congress and the presidency as well as to alter the course of policy. Other elections offer voters a clear choice when the two parties nominate candidates and advocate policies that differ significantly. When party labels can be used to separate "ins" and "outs," voters can oust the "ins" by supporting candidates from the "out" party. In such ways, parties can help voters make meaningful decisions at the polls.

Moderating and channeling conflict   Formal rules determine when elections are held in the United States. When the parties compete in regularly scheduled elections, they operate within a mechanism for controlling the timing and nature of conflict. Groups that might otherwise stage protests, undertake public relations campaigns, or lobby legislators and executives instead focus their energy on the upcoming election.

The fact that only two parties compete in most elections creates a powerful incentive for organized interests, party organizations, and officials to compromise over differences and build broad coalitions *before* the election. As a result, the appeals of the two parties usually converge. Both parties appeal for votes from all groups and seek the middle ground where a majority can be won. In multiparty democracies, on the other hand, the process of compromising and assembling a coalition typically takes place *after* parliamentary elections. Thus, electoral competition between the two parties in the United States may be said to help mute the intensity of political conflict.

Organizing government's decision-making apparatus   Once elected, city council members, state legislators, senators, and other officials need to organize themselves to staff and operate the machinery used to make decisions. Who will preside over legislative sessions or chair committees? The party symbol provides the answers. House members always vote for a member of their own party to be Speaker of the House, and each party fills a variety of other leadership posts. Moreover, the party leadership in the legislature exerts considerable influence over how its members vote.

Throughout this book we describe how the structure of government is divided into the executive, judicial, and legislative arenas. Such fragmentation poses an enduring question: How can the fragmentation be overcome so that the government can get things done? Political party organizations in government can perform part of this crucial task, bringing some coherence to a fragmented decision-making structure.

Linking citizens to government   Parties link citizens to a vast and distant governmental decision-making mechanism in several ways. Voters can reward or punish those who carry the party symbol by giving or withholding their votes. Knowing this, party elites often promote policies that will win public approval. The "out" party can craft an alternative set of policies or merely present itself as "*not* the incumbent party." Finally, the parties' open and loose structures provide mechanisms that permit previously underrepresented and unorganized groups to put their concerns on the political agenda. For example, the Democratic Party responded to demands from the women's and gay rights movements in the 1970s and 1980s, and the Republican Party incorporated many of the demands of social and religious conservatives in the 1980s and 1990s.

HOW DO PARTIES INTERFERE WITH DEMOCRATIC REPRESENTATION?

Critics assert that both major parties actually impede the process of democratic representation by restricting access to public office, limiting policy options, and failing to link citizens effectively with government.

American political parties are relatively open to attempts by previously inactive or marginal groups to increase their influence in party affairs. "Pro-life" groups, for example, have succeeded in convincing the Republican Party to adopt their position on abortion. In 1992 Phyllis Schlafly, chair of the Republican National Coalition for Life, exhibited boxes said to contain nearly 100,000 signed pledges urging the Republican Platform Committee to re-adopt its strong anti-abortion plank of 1984 and 1988.

Restricting access to public office   Despite dramatic increases, minorities and women still fail to win office in proportion to their presence in the population. Women held only 20 percent of state legislative seats in 1993 but constituted 52 percent of the voting-age population; equivalent figures were 6.9 versus 11.4 percent for blacks and 2.1 versus 7.8 percent for Hispanics.[41] Similar underrepresentation of women, blacks, and Hispanics in the U.S. House and Senate has persisted despite dramatic increases in their representation after the 1992 election.

What role do the parties play in affecting the number of women elected to office? Because party organizations do not control nominations, their behavior cannot explain this situation fully. Furthermore, most incumbents continue to be white males who win renomination easily.

However, political party leaders do play some role in recruiting and funding challengers and open-seat candidates, and their actions provide a partial explanation. A survey of more than 1,200 women candidates in 1976 explored the role of party leaders in candidate recruitment.[42] Party leaders were found to play a greater role in recruiting women to run for state legislatures than in recruiting them for Congress. Furthermore, they often encouraged women to seek those offices where the chances of winning were slim to none. A more recent survey of party policies in recruiting women found that, as of 1990, their "informal practices in some locales may still hinder women's opportunities" to seek office.[43]

Party leaders dislike taking risks with candidates, especially when they think the party has a good chance of winning. Unwilling to take risks, they often opt for nominees who do not differ greatly from those who ran previously. "Different" people—the young, minorities, and women—generally stand a poorer chance than others of being recruited. Yet as more women and minorities enter occupations (such as law) that serve as stepping-stones to elective office, and as voters become more willing to vote for women and minority candidates, party leaders' reluctance to encourage such candidates has been diminishing. In fact, in both 1992—touted as "The Year of the Woman"—and 1994, record numbers of women received a major-party nomination.

Limiting policy options   The Republicans and Democrats are usually not enthusiastic about proposing new and starkly different policies, fearing that to do so would hurt their electoral chances. In general, the parties' watchword is "Caution!" It usually falls to third parties to propose bold policy initiatives; only after such ideas become better known and more widely accepted may one or both of the major parties embrace them. As a result, both the candidates nominated by parties and the policies they advocate frequently fail to provide a wide or meaningful range of choices. The Tweedledum-Tweedledee nature of the options offered to voters deprives them of a meaningful opportunity to shape policy. The party system also limits the options of interest groups representing minorities, as is illustrated by "Outcomes: How the Two-Party System Limits Minority Options."

Failing to link citizens effectively with government   Only about half of all eligible adults vote for president; in many state and local elections, fewer than four in ten go to the polls. Some critics of the system blame the parties for failing to link these nonvoters to the decision-making process. Nonvoters' views count for little in the parties' eyes, because they do not contribute to winning elections. Furthermore, in some areas one party enjoys such an overwhelming advantage that the minority party is unelectable, thus providing no realistic alternative. When this is the case, the potential of elections to provide a mechanism for accountability and for retaliating against the "ins" is not realized.

Critics claim that the major parties inhibit the process of change even in times of crisis, stifle the timely consideration of new alternatives, and ensure that the same types of people continue to hold office. They deride the parties' failure to present clear policy alternatives and their inability to ensure that elected officials carry out their party's platform. Because these failures stem from the parties' lack of ideology and centralization, some critics propose they should become more ideological and centralized. Doing so would produce *responsible parties* like those found in Great Britain. In a responsible party system, the two parties present contrasting ideologies with clear and very different policy alternatives, and party discipline ensures that members will carry out the party program if elected. However, America's social diversity, nonideological belief structure, and method of electing the president probably would make it impossible for responsible parties to thrive.

HOW THE TWO-PARTY SYSTEM LIMITS MINORITY OPTIONS

What happens when the Democrats or Republicans retreat from a position dear to the hearts of core supporters? The answer, quite simply, is that those core supporters face a dilemma.

The situation arises for a simple reason: Both parties gravitate to the political center to assemble a broad electoral coalition. In 1995, for example, the Clinton administration reevaluated federal affirmative action policies that were strongly supported by black civil rights and women's organizations. It did so to limit possible damage to Democrats in the 1996 election from a growing backlash against such policies. Many Republican leaders thought that the central role of anti-abortion activists at the 1992 GOP convention and the emphasis on abortion in the ensuing campaign had hurt the party at the polls. So they vowed to downplay both the role of these activists and the significance of the abortion issue in 1996.

When a major party takes such an approach, abandoned groups face a difficult choice. They can form a third party or join an existing third party. But third parties don't win elections. Another option is to threaten to withhold support for the party ticket. In 1995, black civil rights leader Jesse Jackson began to consider running against President Clinton in the primaries. And the political director of the Christian Coalition, Ralph Reed, announced early in 1995 that his group would support the 1996 GOP ticket only if both the presidential and vice-presidential nominees committed themselves to the pro-life position on abortion. However, such a strategy increases the chances of victory for the other party, whose position is farther removed from the group's stand. The alternative is for the aggrieved groups to take their lumps and continue to support their traditional party, even though the reasons for supporting it have weakened.

Labor unions faced this choice in the 1994 midterm elections. Organized labor had vehemently opposed ratification of the North American Free Trade Agreement (NAFTA) and had pledged to withdraw support from Democrats who voted for it. When NAFTA passed with substantial Democratic support, one union leader vowed: "We're going to make sure we get even at the polls."[1] But as time passed, many unions came to realize that a Democratic Congress would be better for them than a Republican one. In fact, AFL-CIO president Lane Kirkland concluded that "the general record of this Administration is a good one. The thrust of their objectives is the same as ours."[2] Barely three months after the NAFTA vote, many unions returned to the Democratic fold.

[1] Kirk Victor, "Friend or Enemy?" *National Journal*, November 5, 1994, p. 2576.
[2] Jack W. Germond and Jules Witcover, "Labor and the Democrats: Together Again," *National Journal*, February 26, 1994, p. 492.

SUMMARY

American political parties participate in and affect the play of power in three different ways. First, they are symbols. Politicians use the party symbol to win elections, and citizens use it to understand the political scene and make voting decisions. The emotions and beliefs captured by party symbols form during childhood and evolve as citizens learn about the presidents, policies, and events associated with each. Second, parties affect politics as organizations that seek to contest and win elections. Third, they are governing mechanisms that organize within the structure of government.

The history of American political parties reveals dramatic fluctuations in both their nature and importance, as is shown by the periodic political realignments that have occurred. From the 1830s to the 1890s, citizens formed strong partisan attachments, and parties played a central role in electoral politics. Since then, their strength and importance have declined substantially.

Their significance today can be assessed in part by looking at three types of participants in party affairs: (1) party identifiers (loyal members of the party in the electorate), the 60 percent of rank-and-file voters who identify with one of the two major parties; (2) major-party leaders and activists (the electoral party organization), consisting mostly of people of higher socioeconomic status, such as party officials and convention delegates; and (3) Democratic and Republican elected officials, who make up the party in government in the legislative and executive branches. A fourth group of participants, those involved in third parties, plays on the fringes of politics. Occasionally, third-party candidates or independents manage to place new issues on the political agenda or lure away enough voters to affect the election outcome.

Structural rules (especially those establishing the Electoral College) and procedural rules (providing for partisan rather than nonpartisan elections in single-member rather than multimember districts) affect the party system in several ways. Most crucial is their encouragement of a stable two-party system in which parties are basically noncentralized, nonideological, pluralistic coalitions seeking to contest and win elections with the help of party symbols.

It is not clear whether the parties' decline, which began in the 1890s, is continuing. A drop in the number of party identifiers suggests a weakening of the party's power over the electorate. However, party organizations have recently showed renewed vigor. And, although the strength of party in government varies from state to state, in Congress the party organizations appear to be gaining strength as their ability to raise funds increases and the level of party voting and party unity shows signs of rising.

How different are the parties? Rank-and-file Democrats hold views that are similar to, but not identical with, those of rank-and-file Republicans. However, activists in the two parties differ sharply. Elected officials in Congress display significant differences along party lines in their roll-call voting, and party platforms have become polarized in recent years.

American political parties exercise an important role in the play of power. They help to choose certain important decisionmakers, structure political conflict and define alternatives, channel and moderate conflict, organize governmental decision-making structures, and link citizens to government. However, critics argue that the parties restrict minorities' and women's access to office, present a narrow range of policy options, and fail to link enough citizens effectively to their government.

APPLYING KNOWLEDGE

LINKING CHARACTERISTICS OF PARTIES TO NEWS REPORTS

The Democratic Party in Pennsylvania had some good news and some bad news following the 1994 general elections. The good news was that it appeared to have retained control of the Pennsylvania House, by a single seat.[1] The House Democratic Campaign Committee had wisely invested $35,000 in the campaign of its three-term incumbent, state representative Tom Stish, and Governor Robert Casey (a Democrat) personally campaigned for Stish's re-election. Stish won his race by a narrow margin.

The bad news was that several days after the election, Stish announced he was switching to the Republican Party to give it control of the House. A year earlier Stish had flirted with changing parties but had then distributed a signed statement confirming his loyalty to the Democratic Party. In announcing the change, he said his voting record "more clearly reflect[ed] Republican principles rather than Democratic principles."[2]

Such incidents occur often. They reflect many of the characteristics of parties described in this chapter. See how well you can apply your knowledge by writing an essay about characteristics and activities of political parties as illustrated by this incident. Then locate an article from this semester's news about political parties and perform the same type of analysis.

[1] Beth Wagner, "Republicans Grab Majority in Pa. House," *Centre Daily Times,* November 15, 1994, p. A1.
[2] Robert Mora and Robert Zausner, "Switch Gives GOP Control of Legislature," *Philadelphia Inquirer,* November 15, 1994, p. A1.

Key Terms

party as symbol (*p. 357*)

party as contester of elections (*p. 360*)

party as governing mechanism (*p. 360*)

political realignment (*p. 361*)

party identification (*p. 362*)

legislative party (*p. 364*)

presidential party (*p. 364*)

party in government (*p. 364*)

party in the electorate (*p. 364*)

electoral party organization (*p. 364*)

Electoral College (*p. 367*)

single-member district (*p. 368*)

multimember district (*p. 368*)

partisan election (*p. 368*)

nonpartisan election (*p. 368*)

party vote (*p. 380*)

party unity (*p. 380*)

Recommended Readings

Beck, Paul Allan, and Frank J. Sorauf. *Party Politics in America,* 7th ed. New York: HarperCollins, 1992. A good standard textbook on American political parties.

Carmines, Edward G., and James A. Stimson. *Issue Evolution: Race and the Transformation of American Politics.* Princeton: Princeton University Press, 1989. An insightful analysis of the changing politics of race in America and the role of political parties in bringing about these changes.

Maisel, L. Sandy, ed. *The Parties Respond: Changes in the American Party System.* Boulder: Westview, 1990. A collection of articles by leading scholars on the most important aspects of the American party system.

Sabato, Larry. *The Party's Just Begun: Shaping Political Parties for America's Future.* Glenview, Ill.: Scott, Foresman, 1988. An upbeat and optimistic assessment of the current status of parties and their prospects for renewal.

Walton, Hanes, Jr. *Black Political Parties: An Historical and Political Analysis.* New York: Free Press, 1972. A description of the little-known history of early black efforts at political organization.

Political Participation and Voting

THE ACT OF VOTING normally attracts little attention. Voting is so routine in democratic political systems that only extraordinary events dramatize its crucial nature.

One extraordinary election did take place in the spring of 1994, and the immense coverage it received in the media conveyed the crucial importance of voting. After decades of rule by a white minority, the Republic of South Africa held its first "all-race" election to choose a government under an interim constitution. Although most of the system of white domination known as apartheid had already been dismantled, political authority still rested in the hands of officials chosen only by whites. Months of hard bargaining between Nelson Mandela (leader of the African National Congress, which led the struggle for political freedom), F. W. de Klerk (prime minister of the white-dominated regime), and Chief Mangosuthu Buthelezi (head of the largely Zulu-based In-

katha Freedom Party) finally produced a framework for a transition of political power. But it took an election in which all citizens could vote to complete the transition from a racist apartheid regime to the new South Africa.

The election not only selected new political authorities but symbolically confirmed the end of one era and the beginning of another. A remarkable aspect of the transition came in the recognition by all South Africans—but especially the country's black citizens—of the significance of the act of casting a ballot. The voting received front-page treatment in American newspapers and prominent television coverage. For example, the lead article in the *Philadelphia Inquirer* on April 28, 1994, began:

> In dusty townships and exclusive suburbs, South Africans of all races stood for hours in long lines yesterday to cast their votes. At some polling stations,

Newspapers throughout the United States prominently displayed photographs showing people lined up to vote in the first all-race free elections ever held in South Africa. The patience and determination shown by South Africans demonstrate that what would be just another election in most countries can, under some circumstances, become an extraordinarily significant event in the lives of citizens.

the queues stretched nearly a mile, and voters were lined [up] two and three abreast.[1]

This article, which displaced coverage of former president Richard Nixon's funeral as the day's lead story, described how a rich white woman drove to the polls with her black maid, as well as how Nobel Peace Prize winner Bishop Desmond Tutu voted for the first time in his life.

Most elections do not reveal so clearly the central role voting plays in democratic political systems. In fact, in the United States about half of all eligible voters fail to go to the polls even in presidential elections—a sharp contrast to the massive turnout in South Africa in 1994. Yet even routine elections with low turnout

profoundly shape the play of power. No other mechanism for making decisions occupies a more central place in the ideology or practice of democracy. One can hardly imagine calling a political system democratic that prohibits citizens from voting and fails to conduct elections. It is not surprising, then, that elections occupy a central place in the American civil religion.

In this chapter we focus on voting and other forms of participation by ordinary citizens. We begin with a brief examination of existing beliefs about voting. We then look at voting and other forms of political behavior on the part of citizen participants in the play of power. We assess the rules that govern voting and other activity and conclude by inquiring into why people vote as they do.

COMMON BELIEFS

VOTING AND ELECTIONS

 Students' views about elections, as expressed on the first day of an introductory American politics class, confirm how important elections are to people. By an overwhelming 70 percent majority, the students agreed that presidential elections "constitute pivotal political events vital to the future of the nation." Only 12 percent disagreed. Furthermore, 30 percent agreed with the notion that voting was the *only* way they could have a say in what government does. Surveys of the general population produce similar results. For example, in a 1994 survey 66 percent of respondents agreed that "voting gives people like me some say about how government runs things."[1]

Many of your classmates, as well as many other Americans, believe it is important, perhaps pivotal, to vote. Yet a significant minority disagree, and many, especially among the young, do not bother to vote or engage in any political activity whatsoever. In this chapter we examine how and why ordinary citizens vote and undertake other types of political action, and we discuss the difference it makes to the play of power.

[1] Andrew Kohut, *The New Political Landscape* (Washington, D.C.: Times Mirror Center for The People & The Press, 1994), p. 132. ■

CITIZENS AS PARTICIPANTS IN THE POLITICAL PROCESS

A great variety of people participate in the electoral process, including candidates, campaign managers, party officials, political action committees (PACs), the news media, and pollsters. All direct their efforts to influencing the most crucial participants of all—citizens who possess the right to vote. It is appropriate, then, to devote an entire chapter to the nature of their participation.

For most people, the act of voting is the most common form of political behavior beyond talking about politics. But citizens can also engage in other activity that shapes the play of power. Before looking at who votes, we will consider the other ways in which ordinary citizens participate in politics.

LEVELS OF CITIZEN PARTICIPATION

Most people find at least some aspects of politics important and hold some intense, salient opinions about them. But typically they fail to act on such beliefs by engaging in activities such as donating to a campaign, writing a letter to a public official, or becoming active in a community organization.

Table 11.1 (on p. 398) provides information about the level of citizens' political participation in the early 1970s. It identifies three types of political participants by degree of involvement: *apathetics, activists,* and *passive supporters* (or *spectators*).[2]

Apathetics Most people rarely if ever engage in political activity beyond voting and talking about politics. Indeed, between 20 and 25 percent of the population engage in no political activity whatsoever, never voting, hardly ever paying attention to politics. These **apathetics** account for much of the "no opinion" category in public opinion polls. When they do express an opinion on current issues, it often carries little intensity, salience, or stability.

More recent research confirms that the number of apathetics has remained stable. In a 1993 survey, citizens were asked whether they undertook any of 15 activities beyond voting, including writing a letter, attending a meeting, giving money, or joining an organization in support of a political cause. The survey found that 24 percent had not participated in any of these activities in the recent past.[3]

Activists Individuals who undertake more demanding actions are called **activists** and engage in one or more of five modes of participation: *protest, community politics, party or campaign work, political communication,* and *contacting officials.* As Table 11.1 indicates, the more demanding the activity in each area, the fewer the people who engage in it. Thus, this study found that 28 percent of those surveyed had tried to persuade others to vote a certain way, but only 19 percent had attended a meeting or rally, 13 percent had given money to a candidate, and 3 percent had been willing to become a candidate.

Another study found that most activists specialized in a narrow range of political behavior.[4] For example, a campaign worker who gives money to

TABLE 11.1
FREQUENCY OF POLITICAL PARTICIPATION BEYOND VOTING

		% OF RESPONDENTS WHO ENGAGE IN SPECIFIC POLITICAL ACTIVITIES
Activists		
Protesters	Refuse to obey unjust laws	16%
	Attend protest meetings	6
	Join in public demonstrations	3
	Riot if necessary	2
Community activists	Work with others on local problems	30
	Contact officials on social issues	14
	Form groups to work on local problems	14
	Are active in community organizations	8
Party and campaign workers	Persuade others how to vote	28
	Actively work for party or candidate	26
	Attend meetings, rallies	19
	Give money to party or candidate	13
	Run as candidates for office	3
Communicators	Engage in political discussions	42
	Send support or protest messages to political leaders	15
	Write letters to newspaper editors	9
Contact specialists	Contact local, state, and national officials about personal, not general, problems	4
Spectators	Keep informed about politics	67
	Vote regularly in elections	63
	Show patriotism by flying the flag, attending parades, etc.	70
Apathetics	Do not engage in any political activity	22

Source: Based on Lester Milbrath and M. L. Goel, *Political Participation,* 2nd ed. (Chicago: Rand-McNally, 1977), pp. 18–19, Figure 1.1.

a candidate is more likely to engage in other campaign activities such as trying to persuade others how to vote than to undertake the tasks of community activists. Only about 10 percent of the population qualify as complete activists by engaging in a variety of more demanding activities such as donating to a campaign, writing a letter to a public official, or becoming active in a community organization.

The rate at which citizens engage in the more demanding activities listed in Table 11.1 has not changed much since the 1970s. A 1993 survey found that only 8 percent had given money to a candidate in the previous year, with 17 percent having done so in the recent past; 4 percent had written a letter to a newspaper editor in the past year, with a total of 12 percent having done the same within the recent past.[5] Data from a study of the 1988 presidential election show similar rates of participation. For instance, just 7 percent of those surveyed had attended a political meeting and just 9 percent had given money to a party or candidate.[6]

Activists willing to undertake "unconventional" protest activities are quite rare. As Table 11.1 indicates, very few people will attend a protest meeting, demonstrate, sit in, or go on a hunger strike. The very fact that such activities are called "unconventional" suggests what research reveals—that most people disapprove of them. At the height of the demonstrations against the Vietnam War in the late 1960s and early 1970s, only 18 percent of a general sample approved of peaceful demonstrations, and fewer than 10 percent approved of disruptive protests.[7] The general disapproval that such unconventional tactics arouse is demonstrated by the fact that while secondary school civics classes teach the value and mechanics of voting, very few describe how to organize a sit-in or a protest march.

Spectators   Activists and apathetics together account for about one-third of the population. The remaining two-thirds are **spectators**, keeping somewhat informed about politics, voting regularly, and flying the flag or attending parades. Occasionally spectators undertake some of the less strenuous actions listed in the top part of Table 11.1. About one person in three tries to persuade others how to vote, and 40 percent engage in political discussions. For the most part, however, spectators remain passive observers, and only half-interested ones at that. For example, in every presidential election since 1952, opinion polls have asked whether people are "very much," "somewhat," or "not much" interested in the campaign. In each election at least one-third replied they were only "somewhat" interested, and their numbers rose to more than 40 percent beginning in 1972.[8] Spectators thus follow the game of politics with half-hearted attention from afar, occasionally cheering when the action is exciting. But they rarely enter the field of play.

This picture of political activity contrasts sharply with idealized notions of the active citizen in a democracy. Clearly, *Americans engage in a low level of political activity.* Apathetics do nothing. Most people, the spectators, use only a tiny fraction of their resources to engage in politics. Only a small minority can be considered political activists. However, the United States is not alone among the world's democracies in this regard. With the exception of voting rates, citizens in many other democracies participate at about the same or even a lower frequency.[9]

SOCIAL CONDITIONS AND LEVEL OF POLITICAL ACTIVITY

Two characteristics of citizens' political activity, including voting, shape the play of power in critical ways. The first is the *general level* of political activities and the direction and speed of its change. The second is *the types of people* who choose to engage in these activities.

In part, the general level of political participation reflects the context of American politics as described in Chapter 2. Usually the United States

Desperate circumstances increase the likelihood of unconventional political activity. The severity of the Great Depression in the 1930s led many workers to intensify their efforts to organize labor unions. In Flint, Michigan, workers at an auto plant engaged in a "sit-down" strike, refusing to leave the factory. Sit-down strikes provided the fledgling labor union, the Congress of Industrial Organizations (CIO), with its most effective means of organizing.

enjoys low-tension politics, with relatively little intense political activity. However, changes in general economic and social conditions can produce sharp upturns in the general level of political activity. During severe economic depressions and times of war, the intensity and salience of opinions needed to sustain high levels of political activity increase. The Great Depression of the 1930s witnessed numerous spontaneous and organized protests, sit-down strikes, seizures of welfare offices, and mass demonstrations.[10]

Unconventional protest activities are not confined to massive upheavals caused by war or economic crisis. The evolution of social movements can raise issues that tap deep emotions. In the decades leading up to the Supreme Court's landmark decision in *Brown v. Board of Education,* which declared segregated schools unconstitutional, the National Association for the Advancement of Colored People (NAACP) led the struggle for equal rights for African Americans by using conventional tactics, primarily lawsuits. Although some African-American leaders such as Marcus Garvey (who led a Harlem-based black nationalist movement that had several million adherents by 1920) attracted a considerable following in the first half of this century, it was not until the mid-1950s under the leadership of Martin Luther King Jr. and others that the struggle for civil rights began to mobilize large numbers of people to engage in protest activities. A variety of factors, including the NAACP's successes in court, the movement of many African Americans to the North, and the experiences of black veterans in World War II, combined to convert long-standing grievances into a mass political movement.

Other issues stirring passions can arise more quickly. The Supreme Court's 1973 *Roe v. Wade* decision, which ruled state laws banning abortion unconstitutional, spurred the formation of groups strongly opposed to abortion. By the late 1980s the conflict over this issue produced intense opinions on both sides, but especially among those opposed to abortion. Increasingly, protesters demonstrated in front of abortion clinics, shouted at women trying to enter them, and tried to block their entry. Some not only willingly faced arrest but remained in jail rather than give their names to the police after their arrest. The most militant protesters used violence, planting bombs at a number of abortion clinics and even killing several physicians and staff employees working there.

Of course, many other issues also inspire those who hold strong views to undertake political action. Sometimes these protests are small and have little impact on policy. For years, small groups of dedicated opponents of the death penalty have staged vigils at prisons when executions are scheduled, but their number has not increased, and their protests have failed to bring about change. Occasionally, however, an activist emerges whose skill and dedication result in significant accomplishments. "Participants: Cesar Chavez Organizes the 'Unorganizable' " (on pp. 402–3) describes such an individual.

HOW SOCIAL CHARACTERISTICS AFFECT POLITICAL PARTICIPATION

Apathetics, spectators, and activists are not distributed randomly in the population. A variety of social characteristics affect whether people vote and engage in more demanding types of political activity. As we have noted in earlier chapters, the scope of politics, *who* participates *how much*, affects outcomes in the play of power.

Two social characteristics, *education* and *income* ("learning and earning"), substantially affect political opinions (see Chapter 7). They affect participation rates just as much. Across a range of activities, the higher one's level of education and income, the greater the level of participation. For example, surveys conducted in 1991 and 1992 found that almost three out of every four college graduates read a newspaper every day, as compared to four out of ten people who did not graduate from high school; identically large differences in newspaper reading habits separated people whose household income was more than $75,000 and those earning less than $10,000.[11] Of course, as many college students happily anticipate, learning and earning are closely linked. But each affects political participation independently. In other words, people with the same incomes but different levels of education differ in their participation rates, as do those with the same education but different incomes.

Level of participation also varies with other social characteristics, such as *place of residence.* In part because of their isolation, farmers and others who live in rural areas are less politically active than urban dwellers. In addition, people living in the South, whose traditional political culture does

CESAR CHAVEZ ORGANIZES THE "UNORGANIZABLE"

From at least the 1880s, the working conditions and pay of the farm workers who picked America's fruits and vegetables were deplorable.[1] Of course, industrial workers such as coal miners and garment makers toiled under equally grim circumstances. But by the 1940s they had managed to unionize and improve their pay and working conditions, whereas migrant agricultural workers' efforts to organize had failed repeatedly. Organizers had to work with people who were poor, exhausted by the back-breaking work, constantly on the move, and often barely familiar with English. Furthermore, federal laws protecting the right of workers to unionize did not cover agriculture.

A remarkable man, Cesar Chavez, led the movement that overcame these formidable obstacles. Born in 1927, he was a farm worker himself; his grandparents had come to the United States from Mexico. When the family lost its small farm in 1938, its members took to the road as migrant laborers. Cesar finished eighth grade but left school for good at the age of 14. Inspired by Father Donald McDonnell, a Roman Catholic priest ministering to the migrant community in San Jose, California, Cha-

Cesar Chavez, himself a migrant worker, provided inspired leadership to other migrant farm workers during the long grape pickers' strike and boycott. As this photo suggests, when Chavez spoke to migrants about forming a union, his audience could easily identify with him.

vez was hired in 1952 at $35 a week as an organizer for the Community Service Organization (CSO), a group seeking to organize farm workers. By 1960 his organizing skills led to his selection as the general director of the CSO.

not encourage political activity, engage less actively in politics. *Age* is another factor, with participation increasing with age until the infirmities of old age force a decline. *Religious affiliation* can also make a difference: Jews are somewhat more politically active than Catholics, and Catholics are more active than Protestants.

According to numerous studies, African Americans engage in less political activity than whites do. One study of 15 separate activities, including contacting officials, campaigning, and participating in groups, found blacks to be less active than whites in all but one.[12] However, the full story is more complex. First, on average African Americans are less educated and earn less, which accounts for much of the difference. When one examines the rate at which whites and blacks with the same education or income register to vote, blacks do so more frequently. For example, among those with a fourth-grade education or less, more than 51 percent of blacks compared to 33 percent of whites are registered to vote. In fact, blacks with less than a high school education or with a yearly income of less than $20,000 consistently

Chavez resigned in 1962 after the CSO failed to adopt his proposal to focus its efforts on union organizing in the fields. He took his life savings, $900, and moved his family to Delano, California. With the help of several associates from the CSO, he began to organize California's farm workers into the National Farm Workers of America (NFWA). Sometimes working in the fields by day, and meeting with families one by one at night, Chavez and his associates assembled a small organization by 1965. Although Chavez thought the NFWA was too weak to undertake a major strike, when Filipino grape pickers went on strike, their action forced the NFWA's hand. Under Chavez's leadership the grape strike grew in strength. In addition to picketing and withholding labor, the strike employed such unconventional tactics as calling for a nationwide boycott of grapes, organizing a 230-mile protest march to the state capitol, and provoking highly publicized arrests merely for picketing and for displaying the word *huelga* (strike) after the local sheriff forbade it. Chavez ultimately succeeded in winning recognition and contracts from several major grape growers; his union won additional contracts in the next few years, aided by its ability to win nationwide support for consumer boycotts of grapes.

Although Chavez's union met with setbacks as well as victories, the fact that he was able to organize farm workers at all is remarkable. A study of the social movement to organize California's agricultural workers noted that

> Chavez is a Chicano (Mexican American) who worked in the fields as he organized; who looks and talks and acts as farm workers do; who is trusted by farm workers as perhaps no other leader of the movement has ever been.[2]

Chavez continued his struggle on behalf of farm workers until his death in 1993.

[1] This discussion is based on Ronald B. Taylor, *Chavez and the Farm Workers* (Boston: Beacon, 1975), especially Chapters 4–7; and Joan London and Henry Anderson, *So Shall Ye Reap* (New York: Thomas Crowell, 1970), Chapter 7.

[2] London and Anderson, *So Shall Ye Reap*, p. 141.

register more often than whites in the same categories.[13] As Figure 11.1 (on p. 404) shows, the overall gap between black and white voting rates narrowed considerably from 1948 to 1988, reflecting the removal of barriers that prevented blacks from voting until the mid-1960s. Furthermore, African Americans participate at significantly higher rates in cities where they have attained political representation by electing an African American as mayor.[14]

Ethnicity is also associated with participation rates. Hispanics register and vote less often than do whites or blacks; their participation rate, like that of blacks, increases where their political power increases.[15] Hispanics, like African Americans, also generally have less education and income than whites, which lowers participation. The proportion of Hispanics who vote is lower in part because many, although residing legally in the United States, are not yet citizens. Illegal immigrants, regardless of ethnicity, tend to shun any activity that might draw attention to them and result in deportation. Finally, many Hispanic immigrants come from nondemocratic countries where voting was not an important or meaningful traditional activity.

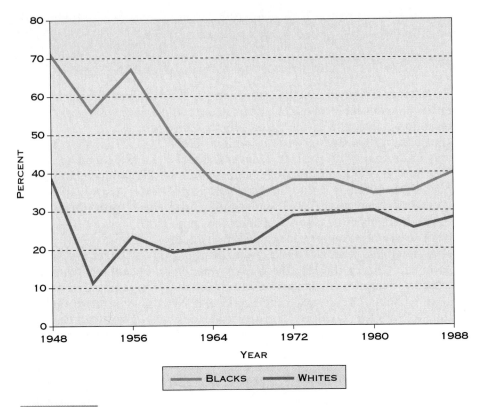

FIGURE 11.1

PERCENTAGE OF BLACKS AND WHITES WHO REPORTED NOT VOTING IN PRESIDENTIAL ELECTIONS, 1948-1988 As this graph indicates, the gap between black and white voting rates in presidential elections persisted from 1948 until around 1960, when the proportion of nonvoting blacks began to drop rapidly. By 1968, when the full effects of the Voting Rights Act of 1965 began to be felt, the gap narrowed to a few percentage points. However, more blacks than whites continued to report not voting for president. *Source:* Edward G. Carmines and Robert Huckfeldt, "Party Politics in the Wake of the Voting Rights Act," in Bernard Grofman and Chandler Davidson, eds., *Controversies in Minority Voting: The Voting Rights Act in Perspective* (Washington, D.C.: Brookings Institution, 1992), p. 127, Figure 5. Reprinted by permission.

At one time *gender* differences existed, with men being more politically active than women, especially among people of lower socioeconomic status. However, these differences have all but disappeared today. Where differences exist in participation rates between men and women, they are usually quite small. Women are a little less likely to try to persuade others how to vote, and they engage in fewer types of campaign activities.[16] But they have voted at higher rates than men since 1980.[17]

Table 11.2 summarizes many of the general patterns of participation by presenting the characteristics associated with voting for president. One generalization encompasses these patterns of how participation and social characteristics relate: *The higher the socioeconomic status, the greater the rate of political participation.* Although there are exceptions, this generalization summarizes research findings on who participates.[18]

TABLE 11.2

VOTER TURNOUT IN THE 1992 PRESIDENTIAL ELECTION, BY SOCIAL CHARACTERISTICS

SOCIAL CHARACTERISTICS	% VOTING	SOCIAL CHARACTERISTICS	% VOTING
Gender		*Education*	
Male	60%	8 years or less	35%
Female	62	1–3 years high school	41
		4 years high school	58
Race/Ethnicity		1–3 years college	69
White	64	4 or more years college	81
Black	54		
Hispanic (U.S. citizen)	48	*Age*	
		18–20	38
		21–24	46
Employment		25–34	53
Employed	64	35–44	64
Unemployed	46	45–64	70
Not in labor force	59	65 and older	70

Source: Harold W. Stanley and Richard G. Niemi, *Vital Statistics on American Politics,* 4th ed. (Washington, D.C.: Congressional Quarterly Press, 1994), pp. 87–88, Table 3.1. The data on voting percentages are drawn from the U.S. Bureau of the Census, *Current Population Reports.* These reports tend to overestimate voter turnout somewhat. However, the differences in turnout by social group shown here accurately reflect actual differences.

PERSONAL CHARACTERISTICS AND POLITICAL ACTIVITY

In addition to social characteristics, several individual attitudes help explain differences in the rate at which people vote and engage in other political activities. One, the **sense of political efficacy**, stands out in study after study as especially important. This attitude refers to how effective people feel they can be in politics. Most studies measure efficacy[19] according to how people react to the following four statements:

1. "Sometimes politics and government seem so complicated that a person like me can't really understand what is going on."

2. "Voting is the only way that people like me can have any say about how the government runs things."

3. "People like me don't have any say about what the government does."

4. "I don't think public officials care much about what people like me think."

The first two statements measure **sense of internal efficacy** (feelings based on people's evaluations of themselves). The last two measure **sense of external efficacy** (assessments of the political system's responsiveness).

The relationship between both internal and external efficacy and political activity is straightforward. The greater a person's sense of efficacy, the more likely that person will be to vote and engage in the other political activities listed in Table 11.1 (on p. 398). Thus, in 1988, among people who agreed that politics was too complicated to understand, 15 percent fewer voted than among those who disagreed; among those who believed public officials didn't care what they thought, 18 percent fewer voted than among those who felt officials did care.[20]

Feelings of political efficacy differ among social groups. For example, blacks score somewhat lower than whites on both internal and external efficacy. People with more education score higher on both. Of course, the reported attitudes reflect reality in the sense that blacks and poorly educated people generally find it more difficult to influence government. Furthermore, self-reported levels of efficacy reflect the impact of political events. Thus, as more people participated in civil rights and anti-war demonstrations during the 1960s, the number of people agreeing that voting was "the only way" to exert influence declined.

Changes in levels of perceived external and internal political efficacy reflect not only the impact of events but the gradual shift in social characteristics within the population over time. The combined effects of these forces on people's sense of efficacy can be substantial. Figure 11.2 shows, for whites, how (1) external efficacy declined substantially between 1952 and 1980, and (2) the more educated consistently felt more efficacious than the less educated. If educational levels had not been rising, the drop in efficacy would have been even larger.

Other attitudes, many linked to political efficacy, also help predict the level of participation. For example, people with a *strong party identification* or *psychological involvement* in politics engage in more political activity. People with a high *sense of citizen duty*—that is, a belief that they ought to engage in political activity regardless of whether it makes any difference or not—participate more, especially by voting. Further, the lower the level of *trust in government*, the less often people engage in conventional political activities.[21]

Is *political ideology* related to participation? Are conservatives or liberals more active? Research has found that those who think governmental spending should be lower participate about as much as those who favor higher spending, and that liberals engage in about the same level of activity as conservatives.[22] However, a survey by pollster Andrew Kohut found conservatives more likely to listen to and call radio talk shows than liberals. For example, 11 percent of liberals listened regularly to such shows, as compared to 24 percent of conservatives; 16 percent of conservatives tried to call, as compared to 12 percent of liberals.[23]

DECLINING TO PARTICIPATE: HOW MANY PEOPLE VOTE?

The role played by voting and elections in politics is aptly described in a classic study of who votes:

> Elections are at the core of the American political system. They are the way we choose government leaders, a source of the government's legitimacy, and a means by which citizens try to influence public policy. And for most Americans, voting is the only form of political participation.[24]

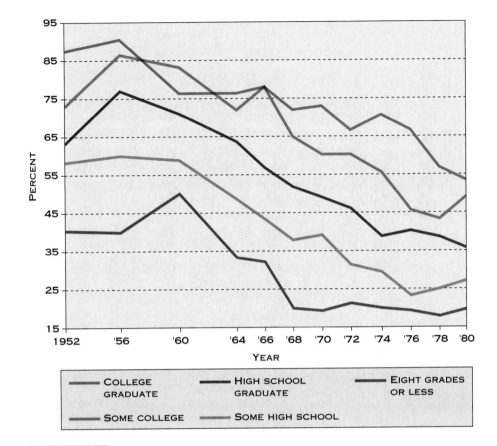

FIGURE 11.2

PERCENTAGE OF WHITES REPORTING HIGH EXTERNAL POLITICAL EFFICACY, BY LEVEL OF EDUCATION, 1952–1980 The graph shown here demonstrates two important attributes of external political efficacy for whites between 1952 and 1980. First, regardless of the survey year, the more education one had, the higher one's sense of external efficacy. In 1956 only 40 percent of those with no more than an eighth-grade education scored high, compared to 90 percent of college graduates—a gap of 50 percent. In 1980 the gap was still large, about 35 percent. Second, regardless of education, levels of self-reported external efficacy generally declined steadily and by large amounts during this period. *Source:* Paul Abramson, *Political Attitudes in America* (San Francisco: W. H. Freeman, 1983), p. 181, Figure 10.4. Copyright © 1983 by W. H. Freeman and Company. Reprinted by permission.

It is striking, therefore, that millions of citizens who are eligible to vote do not engage in this fundamental and relatively simple political activity. Another study of voting rates begins by observing that in the United States "the sheer magnitude of nonvoting is staggering."[25] It found that the average voter turnout among 20 other democracies during the 1980s was 78 percent, fully 25 percent above that in the United States during the same period.[26] Figure 11.3 (on p. 408) shows that the number of nonvoters has almost equaled the number of voters in recent presidential elections. Even the voter turnout in the 1960 presidential election, 62.8 percent, fell far below what most other democracies achieve. For the next 20 years, the voting rate

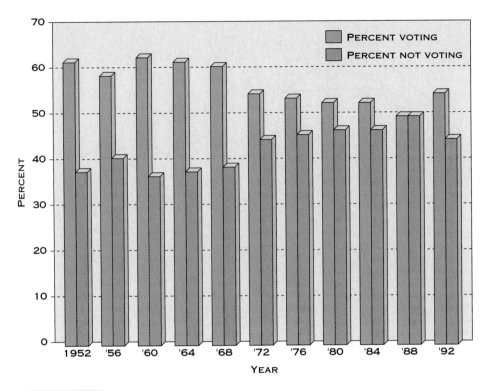

FIGURE 11.3

TURNOUT IN PRESIDENTIAL ELECTIONS, 1952–1992 As this graph shows, only between 50 and 63 percent of those eligible to vote actually did so in presidential elections over a 40-year period. After hovering between 59 and 63 percent for two decades, turnout began to decline in 1972, reaching a low of just over 50 percent in 1988 before jumping five points in 1992.

declined steadily. It rose slightly in 1984, dipped by almost 3 percent in 1988, and then jumped significantly for the first time in decades to 55.2 percent in 1992.

These figures actually overestimate voting rates. In recent "off-year" elections for Congress, voter turnout fell approximately 15 percent below the presidential election turnout rate of two years earlier. Even fewer people vote in primary elections, where candidates seek the two parties' nominations for president and Congress. In 1992, for example, fewer than 30 percent voted in the first 20 presidential primaries.[27] When elections for local offices do not coincide with presidential or off-year congressional elections, turnout is usually lower still. In statewide judicial elections in Pennsylvania, held in odd-numbered years, the spring primaries attract only about 20 percent of the electorate.

The average turnout figures conceal considerable variation from place to place and over time. For example, in the South, turnout in presidential elections consistently falls 7 to 10 percent below that in the rest of the country. Furthermore, non-southern states differ widely in the number of citizens who do vote. In fact, as Figure 11.4 shows, there are differences between counties in the same state. The fact that some elections in some places draw most citizens to the polls does not invalidate the broad generalization that turnout in the United States is low.

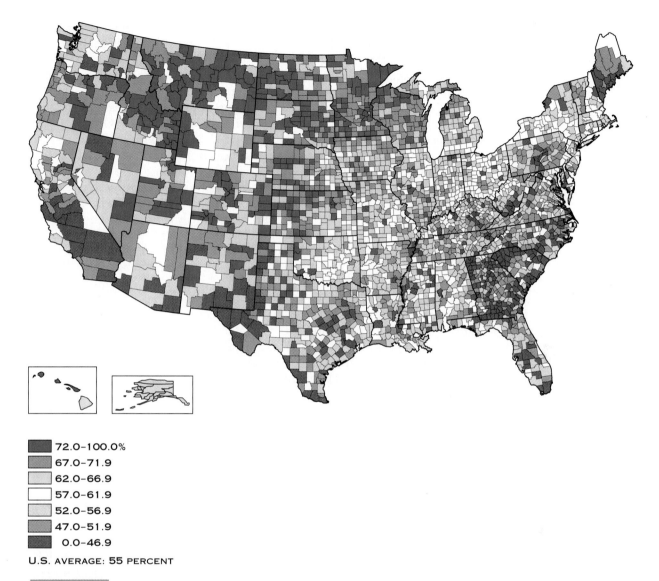

■	72.0–100.0%
▨	67.0–71.9
▨	62.0–66.9
□	57.0–61.9
▨	52.0–56.9
▨	47.0–51.9
■	0.0–46.9

U.S. AVERAGE: 55 PERCENT

FIGURE 11.4

VOTER TURNOUT IN THE 1992 PRESIDENTIAL ELECTION (BY COUNTY) The lowest voting rates, shown here in red, are found throughout the South, in the Appalachian regions of the East, and in counties in the West and Southwest with high minority or rural populations. Counties with the highest voting rates, shown in blue, are largely in the upper Midwest, Midwest, and Mountain states. *Source:* Peirce Lewis, Casey McCracken, and Roger Hunt, "Politics: Who Cares?" *American Demographics,* October 1994, p. 25. *American Demographics* magazine © 1994. Reprinted with permission.

WHY VOTING RATES IN THE UNITED STATES ARE LOW

Several characteristics of the American political system explain its low voting rates.[28] First, a number of *rules* (discussed later in this chapter), especially those requiring citizens to register before they can vote, increase the costs of voting and decrease both its symbolic and its material benefits.

Furthermore, rules about how geographic districts are drawn and how votes are counted in the United States make it likely that votes for probable losers will be wasted; this discourages their supporters from bothering to vote.

Second, the weakness of American political parties (in comparison with parties in other democracies) makes them less able to get citizens to the polls. Furthermore, the weaker attachment to parties that Americans feel reduces the power of the symbolic benefits of voting. In the United States fewer people care passionately whether "their" party wins or loses, in part owing to the low level of tension in American politics and many citizens' lack of interest in politics.

Finally, a number of other factors contribute to reduced voter turnout in the United States. In many places the lack of competitive elections where each candidate has a chance of winning discourages voting. The frequency of elections, long campaigns, and the negative character of television campaign advertisements also reduce interest in voting. Moreover, whereas in some countries elections are held on Sunday and voting is mandatory, the United States has no such provisions.

The cumulative effects of these factors produce the comparatively low voting rates in the United States. Their effects are strong because voting rates remain low even though the personal characteristics of Americans, especially their high levels of education and income, tend to increase turnout. Voting levels would be even lower if these personal characteristics did not operate to counteract the effects of the system characteristics that depress voting rates.

A final question about voter turnout is: Why did turnout decline for so many years even as educational levels rose, and why did it jump in 1992? One explanation for the decline is the increase in the numbers of young voters, single people, those not participating in organized religion, and those generally less connected socially. Decreasing trust in government (external political efficacy) provides a second reason.[29]

It may be too early for scholars to explain the jump in turnout in 1992. Disaffection with government and politicians persisted. No surging trust in government or feelings of political efficacy surfaced. The combination of concern about the economy, the closeness of the election, and the interest generated by Ross Perot's candidacy may account for most of the increase. If so, the increase may have been only temporary, which suggests that the prevailing pattern of low voter turnout will continue. The 1994 midterm elections, characterized by high levels of voter dissatisfaction with government, saw a small rise in voting rates as compared to the 1990 midterm elections. Regardless of small changes in turnout, any effort to understand the role taken by voters and elections in the play of power must begin with the recognition that almost half of those who are eligible to participate either merely watch or pay no attention whatsoever.

PARTICIPATION AND VOTING AS RESOURCES

In the United States no other political resource is distributed as widely or equally as the right to vote. However, possession of a resource does not lead to influence unless the resource is mobilized. Differences in the *rate of use* of the vote carry the potential to affect the play of power. What would happen if millions of nonvoters suddenly began to participate? We will return to this question at the end of the chapter.

Two resources most ordinary citizens can bring to political participation in addition to voting—their physical presence and time—are also distributed evenly. However, people who are struggling just to "get by" may lack both the time and energy to do anything else. Furthermore, the low level of political tension in the United States and the low priority of politics means that relatively few people hold sufficiently intense opinions to cause them to participate much beyond voting. Therefore most people devote little time and effort to politics. To the extent they do, it comes in the form of membership in organized interest groups where they do little more than pay dues.

For members of disadvantaged groups, however, both the vote and their time and physical presence assume greater significance. Their disadvantaged status confers on them fewer resources, including money and education. Whether and how they use the resources they do have becomes especially telling.

Such groups rarely achieve sustained success from conventional participation in the play of power. For them, unconventional strategies may offer one of the few available alternatives to shape political outcomes. Because of their very lack of resources, their deprived status generates the intensity needed to make the commitment of time and energy required by unconventional strategies. If a group's membership is large, it has the potential to mobilize large demonstrations and to become a mass movement. This is precisely the situation in which African Americans find themselves, as is explained in "Strategies: Using Limited Resources to Achieve Success in the Play of Power" (on p. 412).

HOW RULES AFFECT PARTICIPATION

In South Africa the dramatic effect of granting equal voting rights to all, a practice termed **universal adult suffrage**, provided a rare instance in which the usually unnoticed effects of rules on citizen participation became obvious. Yet regulations governing citizen participation shape the play of power in every political system. In the following discussion we identify the rules that structure how Americans participate in politics. First we examine rules that affect voting; then we consider how rules affect other forms of political activity.

THE REGULATION OF VOTING

The story of how the right to vote expanded to include more and more people is familiar to most readers. At the time the Constitution was ratified, most states allowed only white male property owners to vote. Gradually property ownership requirements disappeared, and by the 1830s all white male citizens could vote. This meant that black men, whether free or enslaved, and all women regardless of race could not vote. These severe restrictions strike most people today as profoundly undemocratic. But when we compare the United States to other political systems at the same time, we

USING LIMITED RESOURCES TO ACHIEVE SUCCESS IN THE PLAY OF POWER

The strategies used by participants to exert influence must be chosen with an eye to available resources. Actions that require unavailable or insufficient resources will fail. But the resources that are available change over time. Furthermore, the way they change depends in part on the success or failure of earlier strategies.

Having been denied the right to vote in most of the South; lacking wealth and education in the North; and facing discrimination in employment, housing, and other social spheres, African Americans faced a severely limited choice of strategies at the end of World War II. The leading organized interest working on their behalf, the NAACP, therefore used the talents of its small group of outstanding attorneys to challenge the segregation of public education in the federal courts. A crowning victory came in 1954 when the Supreme Court, in *Brown v. Board of Education*, ruled the system of state-mandated "separate but equal" schools to be in violation of the U.S. Constitution.

This victory and subsequent cases affirming and implementing it did little to end school segregation or to change the conditions of life for blacks. But the Supreme Court's symbolic affirmation of the equality of African Americans' constitutional rights further energized a growing mass movement among blacks. The first highly visible victory won by this movement resulted from the boycott of the bus system in Montgomery, Alabama, an action begun in 1955, the year following the *Brown* decision. Led by Martin Luther King Jr., the successful boycott inspired other unconventional strategies that made use of committed followers, including sit-ins at segregated restaurants and bus stations, boycotts of segregated businesses, and mass demonstrations and protest marches. As one scholar has observed, "Movement-protest politics provides an attractive option to those who lack the resources needed to make effective use of electoral politics, litigation, or other strategies to achieve their objectives."

Groups with limited political resources can exert influence if they are willing to use whatever resources they do have, including their time and their bodies, in a committed way. But the price in terms of fear and physical danger can be high. These civil rights demonstrators in Birmingham, Alabama, in 1963 were savagely attacked by police.

The civil rights movement played a pivotal role in bringing about passage of the Civil Rights Act of 1964 and the Voting Rights Act of 1965. The changes in rules governing voting led to massive increases in the number of blacks both registering and voting in the South. As a result of the newly acquired resource of the vote, the number of African-American elected officials in the South skyrocketed. In the North, the changing social composition of the urban core cities and the mobilization of African Americans in general brought increasing success in races for mayor, city council, and the state legislature.

Source: Lucius J. Barker, "Limits of Political Strategy: A Systemic View of the African American Experience," *American Political Science Review* 88 (1994), pp. 1–13. Quotation from p. 3.

find that few if any other countries in 1800 permitted a comparably large segment of their population to participate in the selection of governmental officials.

African Americans struggle for the vote   The struggle over the right of African Americans to vote reveals that decisions about rules are central to the play of power. The Fifteenth Amendment, ratified in 1870, declared that "The right of citizens of the United States to vote shall not be denied or abridged by the United States or by any State on account of race, color, or previous condition of servitude." Former slaves began voting throughout the South in large numbers immediately afterward; they elected a number of officials, including U.S. senators and members of the House of Representatives. Soon, however, a systematic and successful campaign was initiated to deny the vote to black men. The intimidating and violent activities of the newly formed Ku Klux Klan caused the numbers of voting African-American males to plummet to almost zero throughout the former Confederacy.

For nearly 100 years, the dominant political interests governing southern politics fought tenaciously to keep blacks from voting—and succeeded. Only the passage and subsequent enforcement of the 1965 Voting Rights Act brought significant increases in the number of African Americans voting in the South.[30] It is easy to overlook the significance of these changes. Yet it is remarkable to realize that within the lifetimes of some of our readers, and virtually all of their parents, millions of African Americans could not engage in the most basic form of citizen participation.

Hispanics and women struggle for the vote   Blacks were not the only minority denied the right to vote in the early post–World War II era. Hispanics also faced discrimination and obstacles to voting. In recognition of this fact, the 1975 amendments to the Voting Rights Act extended voter registration provisions to Spanish-speaking jurisdictions.[31] In 1982 Congress passed additional amendments requiring bilingual ballots and assistance to voters, and applying "preclearance" provisions to rules changes in districts in Texas and elsewhere that systematically blocked Hispanics from voting.[32]

In previous chapters we have referred briefly to the role played by women's organizations in winning for themselves the right to vote. Although a number of states had extended the right to vote to women before passage of the Nineteenth Amendment in 1920, only its passage guaranteed that, in the words of the amendment, "The right of citizens of the United States to vote shall not be denied or abridged by the United States or any State on account of sex." However, not until the Great Depression of the 1930s did white women begin voting at rates approaching those of men. Black women in the South remained excluded from the electoral process until the 1960s.

Two other constitutional amendments address the right to vote. At one time, most southern states imposed a small yearly "poll tax" of a dollar or two, due months before the election. In order to vote, one had to present a receipt showing the tax had been paid. The poll tax suppressed voting by poor whites as well as blacks, but it was directed primarily at blacks. By 1964 only five states still required payment of a poll tax.[33] The Twenty-Fourth Amendment, ratified in 1964, prohibits any state from denying citizens the right to vote for failure to pay a poll tax or any other tax. The Twenty-Sixth Amendment, ratified during the Vietnam War, lowered the minimum voting age to 18.

The effects of limits on voting by women, blacks, and Hispanics extended beyond excluding them from choosing local officials. It also prevented them from seeking and winning elected office themselves.

How voter registration rules affect turnout We noted earlier that the United States imposes some of the strictest requirements on voting. State **voter registration** rules require citizens to fill out a form that allows election officials at their assigned polling place to ascertain their identity and record that they have voted. Such procedures deter people from voting more than once and reduce other types of vote fraud.

However, these procedures impose extra burdens on would-be voters that reduce turnout. People who move must reregister and learn where their new polling place is. They must take the time and effort to register before their interest in an election is likely to peak. (Only a few states, such as Minnesota, allow people to register on election day.) Until 1993, when Congress passed the National Voter Registration Act (commonly called the Motor Voter bill), in many states a person's voter registration expired if he or she failed to vote at least once every two years; this meant that people who only voted for president had to reregister every time (see "Rules: The Motor Voter Bill—Battles over Rules in the Play of Power"). Finally, registration rules can be manipulated and unequally enforced to suppress voting. Before passage of the Voting Rights Act, blacks found it almost impossible (as well as physically dangerous) to register. Registration took place in distant courthouses on only a few days each month for only a few hours each day. Moreover, officials administered "literacy requirements" that were extraordinarily difficult and unfair.

States enact other rules affecting who can vote in primary elections that select each party's candidates for the general election. Most states require voters to indicate a party preference when they register. In a **closed primary** only people registered with a party can vote in its primary. Other states conduct an **open primary** where voters can choose to vote in either party's primary, regardless of which party they selected when they registered. A few states do not allow citizens to indicate a party preference when they register; in these states, voters can vote in either party's primary.

REGULATING OTHER TYPES OF POLITICAL ACTIVITY

A variety of rules regulate other types of political participation (e.g., contributions to political campaigns). In general, however, relatively few restrictions apply to conventional forms of political participation.

Unconventional political activity faces more regulation. In 1994 the militant and unconventional tactics of people protesting at abortion clinics prompted Congress to pass and President Bill Clinton to sign legislation making it a federal offense to use threats, physical obstruction, or force to interfere with people performing or seeking abortions at clinics.[34] Moreover, shopping mall operators have generally succeeded in convincing courts to let them ban political demonstrations, speeches, and leafleting on their property. The civil rights demonstrations of the 1950s and 1960s often led to arrests based on assertions that rules restricting the conduct of such activities had been broken.

THE MOTOR VOTER BILL—
BATTLES OVER RULES IN THE PLAY OF POWER

President Bill Clinton praised the National Voter Registration Act of 1993 in a bill-signing ceremony on May 20, 1993. Calling it "our newest civil rights law," the president claimed that "registration for federal elections will become as accessible as possible, while the integrity of the electoral process is clearly preserved."[1] But less than a year earlier, President George Bush had explained his veto of similar legislation by asserting that it would "impose unnecessary, burdensome, expensive, and constitutionally questionable Federal regulation . . . [and] also expose the election process to an unacceptable risk of fraud and corruption."[2]

Like many decisions about basic procedural rules, the struggle to reform voter registration rules attracted little public or media attention. These rules' importance to the play of power remained unclear to most people. But key participants, including party officials, members of Congress, and leaders of organized interests, understood the voter registration bill's significance and fought fiercely over its passage.

The bill requires that as of January 1, 1995, states must allow citizens to register to vote when they apply for or renew a driver's license. This provision accounts for the act's informal name, the "Motor Voter bill." It also requires states (1) to permit citizens to register by mail, and (2) to provide registration forms at certain state agencies (e.g., welfare offices) that provide public assistance. Many states had already enacted some of these provisions. For example, 27 states had some form of "motor voter" provisions in effect in 1992,[3] and about half of all states permitted registration by mail.

The arguments between Republicans and Democrats over the registration issue echoed the positions stated by Presidents Clinton and Bush. Republicans also accused the Democrats of trying to change the rules in a way that would increase the number of people voting Democratic; and Democrats accused the Republicans of seeking to preserve obstacles to voting that would benefit the Republican Party. The intensely partisan conflict, which included a Republican filibuster in the Senate, resulted in a number of votes that found nearly all Republicans voting together in opposition to nearly all Democrats.

The full impact of this change in rules will not be clear for some years. Proponents hail it as possibly the biggest advance for voting rights since the early 1960s, claim that it has brought even higher voting rates in states that had already adopted its provisions, and hope that it will substantially increase overall voting rates. Opponents hope that there will be minimal effects on turnout and on election outcomes. In part, the act's impact will depend on how states implement the legislation. Some states, including Pennsylvania and California, had failed to comply with some of its provisions by the January 1, 1995, deadline and were sued by the Department of Justice to force their compliance. Like any change in rules affecting the play of power, this one will surely continue to generate conflict.

[1] "Remarks by the President at Signing of the National Voter Registration Act of 1993," *Weekly Compilation of Presidential Documents* (Vol. 29), May 24, 1993, pp. 914–16.
[2] "Message to the Senate Returning without Approval the National Voter Registration Act of 1992," *Weekly Compilation of Presidential Documents* (Vol. 28), July 6, 1992, p. 1201.
[3] " 'Motor Voter' Bill Enacted after Five Years," *Congressional Quarterly Almanac 1993* (Washington, D.C.: Congressional Quarterly Press, 1994), p. 199.

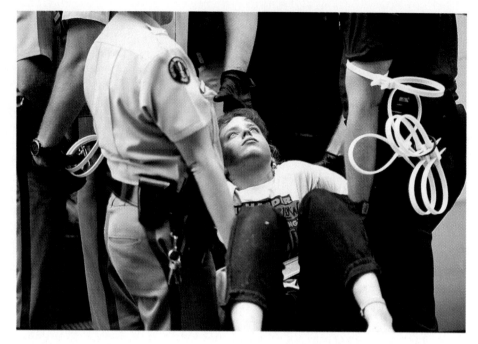

This abortion foe demonstrates the lengths to which committed activists will go in using unconventional political tactics. Placed under arrest, she refused to walk into custody and had to be carried. Such incidents bring to a head the conflicts that unconventional tactics often raise between free speech, on the one hand, and the upholding of order and the rights of others, on the other hand.

The task of balancing competing values involved in unconventional behavior poses some knotty problems. Should candidates' freedom of speech allow them to campaign door-to-door in college dormitories? At what point does a protest march disrupt traffic and public order to the point where it can be banned? It is difficult to determine whether rules about unconventional political behavior expand or restrict the activities of those who want to engage in it, partly because the rules evolve as courts continue to make decisions on such matters.

HOW PEOPLE MAKE VOTING DECISIONS

Efforts to understand elections ought to begin by determining the number of people who *do not* vote, who they are, and what the effects of their inactivity are. Any interpretation of the outcome of the next presidential election that does not do this will be missing a crucial part of the story. At the same time, many people do cast ballots, and their decisions clearly make a difference, for they determine who wins office. The central role taken by elections in the play of power guarantees that the question "Why do people vote the way they do?" will continue to be pivotal.

How do voters decide whom to support? Arriving at a good answer presents several challenges. One problem is that elections for all sorts of offices, each conducted under differing rules, take place at all levels of government almost every year. Another problem is that many states and small jurisdictions elect judges, town council members, and a host of other officials in *nonpartisan elections*, where candidates do not run under the Democratic or Republican party label. Deciding how to vote in these elections involves very different decision-making techniques than those used in voting for Congress or president. Because most research and media attention focuses on presidential elections, our discussion will do so as well, although we also take a quick look at voter choice in congressional elections.

DECIDING WHOM TO VOTE FOR IN PRESIDENTIAL ELECTIONS

Political scientists vigorously argue the merits of several competing explanations for how people decide whom to vote for in presidential elections. Like economists' various theories for explaining the course of the economy, these explanations are neither fully compatible nor contradictory, and their predictive ability remains imperfect.

Our discussion unavoidably reflects this somewhat uncertain state of knowledge. Rather than presenting a single explanation for voting in presidential elections, we present several useful approaches. One draws heavily on the concept of *party identification*, a central component of political parties as symbols. The second identifies a *hierarchy of societal conditions*, headed by the state of the economy in the months preceding the election, that strongly predispose voters to either elect or defeat the candidate of the party occupying the White House.

The role of party identification   Party identification involves a psychological attachment to a political party. Although citizens' attachments to the Republican and Democratic Parties have weakened somewhat in the last several decades, a majority of Americans still identify with one party or the other. Many people acquire their party identification in childhood; others form it later in life. Most people stay with the same party and even strengthen their ties as they grow older. Although both parties draw some support from virtually every social group, distinct differences in their social composition remain. In the following discussion we examine how party identification shapes the frequency and content of citizens' entry into the play of power.

By far the most critical attribute of party identification is its ability to predict how people behave politically. Strong party identifiers, Democrats and Republicans alike, differ from everyone else in a number of ways. For one thing, in election after election, strong party identifiers express more interest in the outcome, vote at higher rates, and engage in more activities beyond voting. In 1988, for example, 84 percent of strong identifiers reported voting; only 68 percent of weak identifiers, 64 percent of "leaners," and 50 percent of independents did.[35]

More telling, however, is the fact that party identification strongly influences vote choice. *Year after year, the stronger the party identification, the more likely is a vote for that party's presidential candidate.* In 1992 virtually all strong Democrats (97%) supported Bill Clinton, whereas 84 percent of weak Democrats did so; similarly, strong Republicans gave 98 percent of their votes to George Bush, whereas weak Republicans gave 80 percent.[36] A 1980 survey reported that 74 percent of strong Democrats claimed they *always* voted Democratic, whereas only 47 percent of weak Democrats did so; 58 percent of strong Republicans, but only 27 percent of weak Republicans, said they always voted for Republican candidates.[37]

How does party identification affect people's votes? In what are called *low-stimulus elections,* where interest in the contest and knowledge of the candidates is very low, the relationship is *direct.* Suppose you found yourself in a voting booth looking at the names of two candidates for the post of county drain commissioner. (The state of Michigan actually elects county drain commissioners in partisan elections, and nearly every jurisdiction in the United States elects some equally obscure officials.) Most voters know nothing about candidates for such a post. If you are like most voters, your vote for drain commissioner will go to the candidate running under the party label with which you identify.

However, in higher-stimulus elections such as presidential contests, the effects of party identification are *indirect.* Party identification provides a perceptual screen that shapes how voters view the candidates and the issues in a particular election. Because the candidates and issues change, this approach aptly calls the nature of each election's candidates and issues **short-term partisan forces**. These forces are partisan because voters view them as favoring one party or its candidate.

The Center for Political Studies at the University of Michigan, which has conducted careful studies of every national election since 1952, identifies six *partisan attitudes* shaped by these short-term forces.[38] These are the voters'

- Evaluation of the personal attributes of the Republican candidate;
- Evaluation of the personal attributes of the Democratic candidate;
- Attitudes toward domestic issues;
- Attitudes toward foreign policy issues;
- Judgment on which party is best for groups with which the voter identifies; and
- Judgment on which party can best manage the running of the government.

Party identification strongly influences, but does not absolutely determine, voters' attitudes on each of these short-term forces. In other words, it predisposes people to view issues and candidates identified with "their" party more favorably. Thus, party identification acts as a pair of "political glasses" through which one watches the play of power. Anyone who has observed a fanatical sports fan evaluating referees' decisions knows how the process works. Team loyalty causes fans to view calls against their team as wrongheaded and calls against the opposition as absolutely right. When fans with different loyalties watch the same game, they often reach diametrically opposed conclusions about the same sequence of events.

Likewise, citizens' party loyalties color, simplify, and help make sense of politics. The "party glasses" used to view the world of politics are useful. They reduce the time people spend arriving at a position on issues and candidates, and they simplify difficult voting decisions.

The power of party identification in shaping political judgments provides one of the most consistent and useful devices for understanding politics.

On any question for which a "Republican" and "Democratic" position can be identified, differences will exist between adherents of the two parties. For example, party identification clearly affected students' assessment of President Clinton's performance, as Figure 11.5 indicates.

During presidential campaigns, both Democrats and Republicans are likely to view their party's candidate more favorably, glossing over weaknesses, emphasizing strengths, even misperceiving the candidates' issue stands when these clash with their own. Of course, as we noted earlier, party identification does not absolutely determine what people think about issues and candidates. Individuals with strong views about abortion will not accept a candidate from their party whose views on abortion differ from their own. Nor will they automatically like all candidates from their party. Party identification colors perceptions only to a degree. For example, many Republicans did not like Barry Goldwater, their party's candidate for

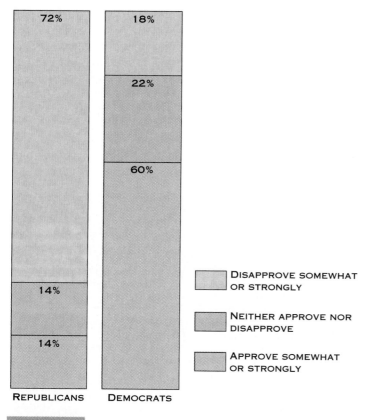

FIGURE 11.5

REPUBLICAN AND DEMOCRATIC STUDENTS' RATING OF BILL CLINTON'S JOB PERFORMANCE, 1995 Students in a large introductory American government class were asked, "Do you approve or disapprove of the job Bill Clinton is doing as president?" The survey also asked about students' party identification. The diagram confirms the powerful effects of party identification on political opinions. Among those classified as Republicans ("strong" and "not very strong" Republicans combined with independents "leaning" to the Republicans), 72 percent disapproved of Clinton's job performance. Democrats were almost as likely to approve of it.

president in 1964, and many Democrats rejected their party's 1972 candidate, George McGovern. Likewise, strong partisans who lose their jobs while their party controls the presidency may find it difficult to ignore their predicament in evaluating how well the party is managing the economy.

How well do citizens' opinions on the six short-term partisan forces predict their vote? Surprisingly well. This was first demonstrated by the Survey Research Center at the University of Michigan in 1956. The Center's researchers interviewed voters both before and after the election, predicting which way respondents would vote based on measures of each of the six partisan attitudes examined in the pre-election survey. When these predictions were checked against the respondents' answers to a post-election survey about whom they actually voted for, 86 percent were found to be correct. Interestingly, the prediction was slightly better than that made by respondents themselves when they were asked before the election whom they would vote for.[39] Nor was the 1956 election a fluke in this regard. An analysis of voters' attitudes toward the candidates, parties, and issues in the nine elections from 1952 to 1984 produced correct predictions for the votes of 85 to 90 percent of respondents.[40]

Furthermore, voters' short-term evaluations of candidates and issues continue to relate strongly to vote choice. Among voters surveyed as they left the polls in 1992, 19 percent mentioned health care as an important issue. Bill Clinton, who campaigned hard on this issue, got 67 percent of their votes; 87 percent of voters who mentioned foreign policy as important voted for George Bush.[41] Finally, as Table 11.3 shows, voters who felt that the economy was poor and that their personal finances had deteriorated over the past four years voted for Democratic challenger Bill Clinton far more

TABLE 11.3
HOW VOTERS' VIEWS OF THE ECONOMY AFFECTED THEIR PRESIDENTIAL VOTE IN 1992

VIEW OF NATIONAL ECONOMY[2]	CHANGE IN PERSONAL FINANCES IN PAST FOUR YEARS[1]		
	BETTER OFF (24%)	ABOUT THE SAME (34%)	WORSE OFF (41%)
Excellent/good (20%)	7 (7)	11 (10)	40 (31)
Not so good (47%)	34 (29)	52 (42)	75 (56)
Poor (33%)	71 (55)	80 (63)	90 (67)

[1]The first number in each column is Bill Clinton's percentage of the two-candidate vote; the number in parentheses is Clinton's percentage of the three-candidate vote.

[2]Voters' evaluation of the state of the economy emerged as the most important domestic issue in the 1992 election. When asked whether they were better off, about the same, or worse off economically than they had been four years ago, those who said "worse off" voted heavily for Clinton. Reading *across* all three rows shows that the percentages supporting Clinton go up as evaluations of personal finances become gloomier. Looking *down* the columns shows that the worse the economy seemed, the more votes Clinton received. Combining answers to both questions shows the power of this short-term factor. Only 7 percent of those who considered themselves better off and thought the economy was good to excellent voted for Clinton; 90 percent of those who believed they were worse off and that the economy was poor voted for him.

Source: Adapted from Gerald M. Pomper, "The Presidential Election," Chapter 5, p. 147, Table 7.4, in Pomper, ed., *The Election of 1992* (Chatham, N.J.: Chatham House, 1993).

often than for incumbent president George Bush. Thus, the dynamics of politics and the unfolding of events shape perceptions—but only after they have been interpreted by and filtered through the perceptual screen of party identification.

The hierarchy of issues   The second approach to understanding presidential election outcomes focuses on the nature of the political environment, or the "condition of the nation" at the time of the election. Foremost is the state of the domestic economy. Political scientist Edward Tufte summarizes the **theory of the impact of economic conditions** on election outcomes as follows:

1. Economic movements in the months immediately preceding an election can tip the balance and decide the outcome of an election.

2. The electorate rewards incumbents for prosperity and punishes them for recession.

3. Short-run spurts in economic growth in the months immediately preceding an election benefit incumbents.[42]

The short version of this theory states simply, "As goes economic performance, so goes the election."[43] As another scholar who sees the economy as crucial in elections puts it,

> In terms of an issue variable that moves the electorate, there's nothing that's more powerful or reliable than the economy, except war. In peacetime, how the economy goes determines how well the President and his party are going to fare in the next election.[44]

This theory is partly confirmed by the strong association between how well voters believe the economy is doing and whether they vote for the incumbent party (the "ins") or the challenger (the "outs"). Table 11.3 shows this for 1992. The same general pattern existed in 1988. Most of those surveyed in a *New York Times*/CBS News poll (in fact, 73%) who saw the economy as improving voted for George Bush, and 59 percent of those who said it was good and would remain that way did so; however, only 25 percent of those who saw a worsening economy supported Bush.[45] Furthermore, those who said they were better off financially than a year ago supported the "ins" much more strongly than the "outs" in 1968, 1972, and 1976.[46]

The power of a sour economy to influence citizens' evaluations of politics extends to other areas. Assessments of a president's performance plunge during hard times. One of the most popular presidents since World War II, Ronald Reagan, found that his standing dropped shortly after unemployment rose between 1981 and the middle of 1982, as is shown in Figure 11.6 (on p. 422). The unprecedented free fall of Bush's popularity from its 90 percent level during the Persian Gulf War was likewise driven by a deteriorating economy.

Of course, even these relationships bend to the persistent tug of party identification. For example, people in similar economic circumstances are more likely to think that the economy is doing well if they identify with

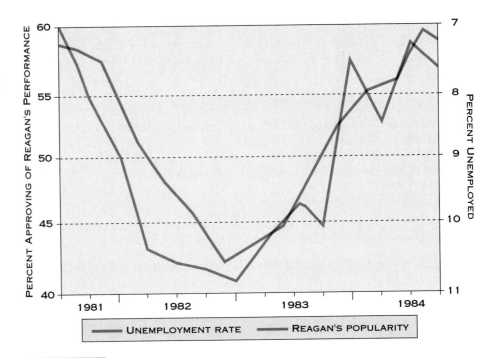

**UNEMPLOYMENT AND APPROVAL OF RONALD REAGAN'S JOB PERFORMANCE,
1981–1984** As this graph indicates, the severe recession that began in 1981 had a
devastating impact on President Ronald Reagan's popularity. His approval levels
are measured on the left vertical axis. The unemployment rate is measured on the
right vertical axis, with higher levels (worse unemployment) reflected at the
bottom. By mid-1982, when unemployment was approximately 10.5 percent,
Reagan's popularity was at a low of just over 40 percent. When the economic
recovery began in 1983, Reagan's popularity climbed rapidly as unemployment
fell. *Source:* Scott Keeter, "Public Opinion in 1984," in Gerald M. Pomper, ed.,
The Election of 1984: Reports and Interpretations (Chatham, N.J.: Chatham
House, 1985), p. 94, Figure 4.2.

the "in" party, and vice versa. Thus, as a group, Republicans will view the
economy somewhat more favorably than Democrats when a Republican sits
in the White House and, therefore, they will be more likely to vote Republican. But even though people's evaluations bend under the influence of party
identification, they cannot completely ignore reality. The worse the economy, the less party identification blinds supporters of the "in" party to the
fact that times are bad. Even staunch Republicans and Democrats, when
facing severe economic hardship, will resist the pull of party identification
and acknowledge a poor economy.

Very good times provide almost as great a boost to the prospects of
the "ins." When the economy's performance is neither terrible nor terrific,
however, other issues become more important. The more compelling these
other issues are, the greater their ability to counteract the influence of
economic conditions.

Most potent among these other issues are ones that relate to foreign
affairs in general and questions of war and peace in particular. Both the

Korean War in 1952 and the Vietnam War in 1968 aroused considerable concern in the electorate and contributed to the Democrats' loss of the presidency. Perceptions that the world is at peace and that the foreign affairs of the nation are going well help the "ins," but probably not as much as an unpopular war hurts them.

Other issues—including abortion, crime, health care, welfare, taxes, and foreign policy—play a greater role for most voters when there is no recession or war to displace their perceived importance. However, such issues affect some voters even when the economy dominates a campaign. The state of the economy drove the 1992 election campaign more than any other issue, but the candidates spoke about a number of the aforementioned other issues as well.

Of course, a relatively small number of **single-issue voters** feel so intensely about an issue that they will support any candidate who clearly favors their position. Although any issue can attract a small number of people who feel passionately enough about it to base their vote on it, the two most potent issues of this type are gun control and abortion. Usually, single-issue voters belong to an organized interest group (e.g., the National Rifle Association, or a group opposing or supporting abortions) that urges its members to vote on the basis of its issue position. Sometimes votes determined by the candidates' positions on a single issue can affect the outcome in a local election. In elections for Congress or the presidency, though, the net gain or loss from such votes is usually not large enough to determine the outcome.

Although each election campaign presents voters with a somewhat different combination of issues, their response to the state of the economy exerts the strongest and most persistent influence on how they vote. Whether the balance of short-term partisan forces favors the incumbent or challenging party depends more than anything else, over a series of elections, on the state of the economy. Edward Tufte explains:

> No doubt the state of the economy affects different elections somewhat differently—depending upon the particular mix of economic conditions, the salience of economic issues in relation to other issues in the campaign, and the dozens of idiosyncratic influences to each campaign. Despite all the short-run "noise" in presidential contests, the extent of prosperity prevailing in the election year remains a regular and significant determinant of the vote won by the nominee of the in party.[47]

Do voters look back or ahead?   Theories of how the state of the economy and the president's popularity affect voters' decisions assume that voters engage in **retrospective voting**—that is, that they look backward and make judgments.[48] Positive evaluations cause voters to reward the incumbent party with a vote, and negative ones produce punishment in the form of a vote for the "out" party. Of course, retrospective judgments cannot ordinarily be made about challengers. This suggests that **prospective voting**, based on evaluations of what candidates promise to do and what voters think they might really do, also takes place. Certainly candidates act as if prospective voting might occur, because they make assertions about the good things they will do if they are elected. Furthermore, some research has revealed a relationship between citizens' expectations about the future of the economy and the level of presidential popularity, suggesting a forward-looking perspective.[49] A study of the 1984 election, for example, found that "prospective economic evaluations [had] an effect at least as strong as that of retrospective evaluations."[50] Apparently both kinds of judgments play a role.

VOTING IN PRESIDENTIAL PRIMARIES

The rules governing the selection of the Republican and Democratic candidates for president place primaries at the heart of the process. To receive the nomination, candidates must win in the primaries—especially the early ones. Thus, primary voters in these elections perform a key role.

Voting in presidential primaries differs from voting in the general election in several important ways. First, party identification cannot help voters decide, because they can only choose among contenders from the same party. Second, because the primary comes before interest in the presidential election is very high, many voters know relatively little about the candidates. Third, primaries attract few voters—typically between 15 and 30 percent of those eligible to vote.

The characteristics of those who do vote differ from each party's rank and file. (For example, about 20 percent of Democrats nationally are African American, but only 2 percent of Democrats in New Hampshire, a crucial primary state, are; Democrats in this state are also wealthier, better educated, and more likely to be Catholic than in the country as a whole.[51]) In general, primary voters tend to be more interested in the election, better informed, older, more active, better educated, and somewhat wealthier than the average voter.[52] Consequently, voters in Democratic primaries are more liberal and

Voters in presidential primaries cannot rely on partisan cues in choosing a candidate, so they often seek guidance from television, including televised debates. Here Democratic candidates debate each other in the 1992 New Hampshire primary. The brief exchanges between the candidates and the media's interpretation of each candidate's performance provide voters with limited but direct information.

Republican primary voters are more conservative than voters in the general election.[53]

Perhaps the most notable feature of voter choice in primaries is voters' heavy reliance on information from the news media. Voters know relatively little about the candidates, and they cannot draw on their party identification to help interpret what they do know. Consequently, the media serve as the principal intermediary between candidates and primary voters. Reporters and editors decide which candidates to cover, and they interpret the status of the contest in terminology commonly used to describe a horse race. Primary coverage frequently discusses who the "frontrunner" is and who is "gaining" or "losing" ground. The media also focus on the dramatic and entertaining aspects of the contest. Voters may well learn more about purported scandals in the candidates' personal lives or accusations the candidates are making against one another than they do about the contenders' positions on important issues. Unfortunately, despite the primaries' importance, comparatively little research has been done on why people vote as they do in them.

VOTING FOR CONGRESS

How do people decide which candidate for the House or Senate will get their vote? For many years, students of voting behavior largely ignored this question, making congressional elections the "orphans" of U.S. election research.[54] Although congressional elections have attracted more attention recently, scholars have not yet untangled the complex factors that interact to determine votes for Congress.[55] But some general characteristics of these elections can be presented.

How congressional elections differ from presidential elections   Half of the time, congressional candidates run in non–presidential election years. These **midterm elections**, so named because they fall in the middle of a president's term, exhibit two patterns. First, the president's party loses congressional seats. Second, voter turnout falls by about 15 percent from the level of the preceding presidential election. Thus, voting in congressional elections follows a different pattern in midterm elections than it does in presidential election years.

How voters decide in congressional elections   Even in presidential election years, contests for the House and Senate attract less voter interest than does the presidential race. Approximately 10 percent of people voting for president, for example, do not bother to vote for a House candidate on the same ballot. Given the lower visibility of and interest generated by these elections, it is not surprising that many voters know little about the candidates. In fact, fewer than half can name their incumbent representative, and only slightly more can name their senators.[56] Challengers are much less likely to be known. Only about 20 percent of the electorate know the name of House challengers or can recall anything else about them.[57]

In light of these facts, one possibility is that party identification primarily determines votes for Congress. Given the short-run stability in party identification, a 50–50 split among independents would produce virtually identical results from one election to the next. But the proportion of the vote garnered by each party's congressional candidates differs from year to year, as does the number of seats won by each. Clearly, party identification alone does not determine votes for Congress.

Alternatively, voters might base their vote for Congress on their presidential vote. In that case, no districts or states would favor one party's presidential candidate and elect a congressional candidate from the other. In reality, every presidential election since 1956 has seen at least 100 House candidates elected from districts that supported the other party's presidential candidate.[58]

If voters *always* supported incumbents, challengers would *never* win. Although most incumbents do indeed win, between 5 percent and 10 percent do not; in 1992, 43 House and 5 Senate incumbents (12% and 17%, respectively, of those seeking re-election) lost.[59] In 1994, just 2 Senate and 35 House incumbents lost,[60] or 9 percent and 10 percent of incumbents running, respectively.

The factors just mentioned—*party identification, attitudes toward the president,* and *incumbency*—combine to determine votes for Congress. Incumbency plays a strong role. Between 70 and 76 percent of the votes cast for House candidates from 1982 through 1992 followed the voter's party identification, with only about 20 percent defecting.[61] But two other categories of factors also influence outcomes: **national factors** that operate irrespective of local candidates or issues; and **local factors** unique to the district and the candidates.

National factors shaping votes for Congress   Two national factors *directly* affect whom voters choose for Congress. In presidential election years, votes for Congress are affected by the strength of each party's presidential candidate and the state of the economy. A weak economy causes some people to vote for congressional candidates from the "out" party. A strong showing by a presidential candidate produces the **coattail effect**, in which some voters decide to cast their congressional vote for candidates from the same party as their choice for president.

In midterm elections the popularity of the incumbent president joins the state of the economy to affect votes for Congress.[62] (Of course, presidential popularity depends to a considerable extent on the economy.) When voters disapprove of the job the chief executive is doing, their unhappiness rubs off on congressional candidates from the president's party. Although both presidential popularity and the economy influence votes for Congress, some scholars believe that the degree to which they determine votes declined during the 1980s.[63]

In addition to these direct effects, presidential popularity and the economy *indirectly* affect outcomes through their impact on the quality of challengers. Savvy, strong potential challengers calculate their chances by assessing the economy and the president's popularity.[64] When prospects look good for candidates from their party, more of these strong candidates—called **strategic politicians**—decide to run. When national factors suggest that their party will not fare well, strategic politicians refrain from running, allowing weaker candidates from their party to do so.

Local factors and congressional voting   Some political scientists emphasize the importance of purely *local* factors in determining congressional election outcomes. One is the quality of challengers in each district. Research documents the proposition that better candidates get more votes.[65] The behavior of incumbents also plays a role in getting the vote. For example, some researchers have found that incumbents who paid more attention to meeting requests for help from constituents got more votes. Such "constituency service" is one of many ways by which incumbents become better known. Although many voters still cannot name their representative, even fewer can name the challengers. When voters only know one candidate, nine times out of ten it is the incumbent.[66] And people tend to vote for candidates with whom they are familiar. For example, one study found that 40 percent of those familiar with House challengers voted for them, but only 16 percent of those unfamiliar with them did. The comparable figures for Senate challengers were 55 percent and 27 percent, respectively.[67]

Other local factors, termed *special conditions*, powerfully affect outcomes.[68] Most crucial is whether the incumbent seeks re-election. As we show in Chapter 12, incumbents rarely lose. When the incumbent does not run—in what is called an **open-seat election**—a mad scramble for the nomination among strategic politicians in both parties usually occurs. Moreover, when candidates (especially incumbents) become embroiled in a scandal, face arrest or prosecution, develop serious health problems, or reach an advanced age, voters take note and vote accordingly. Voters also tend to support candidates who share their race, ethnic identity, or religion. Recall the example given in Chapter 2, in which Cuban-American and Jewish voters alike overwhelmingly supported the congressional candidate who shared their background. The fact that African-American candidates find it difficult to attract many white votes when running against a white opponent means that to win, African Americans must run in districts where a majority of voters are black.

Controversial referenda on the ballot can likewise affect the outcome of congressional races. In 1994, for example, the presence on the California ballot of Proposition 187—a measure that called for denying welfare, health, and education benefits to illegal immigrants—increased voter turnout and probably helped candidates (mostly Republicans) who supported the measure.

CITIZEN PARTICIPATION AND THE PLAY OF POWER

The American civil religion places citizen participation at the very heart of democratic government. Without effective citizen participation, democracy does not exist. It is that simple. Assessing the nature and effect of citizen participation, however, is far from simple. To make such an assessment, we must consider forms of participation besides voting. We also need to examine how voting shapes outcomes in politics.

POLITICAL BEHAVIOR BEYOND VOTING

As we mentioned earlier in this chapter, about one-fourth of citizens engage in virtually no political activity whatsoever. When the remaining three-fourths engage in political activity other than voting, it usually takes a conventional form. Many fewer undertake what we have called *unconventional political behavior*.

Conventional political behavior The image of the lone citizen moving the political system through sheer determination and effort captures the imagination and attention of both the media and the public. Examples of those who succeed provide evidence for the proposition that "One person can make a difference." Such people exist, of course, and their stories inspire the public and politicians alike. But they are rare, and their efforts usually focus on *local* rather than state or national government.

Most citizens who undertake conventional political activity beyond voting do so *collectively*, as part of an organized group. Examples of such action at the local level abound. A group of parents organize to oppose the closing of a local elementary school or to demand a crossing guard at a dangerous intersection. A handful of neighbors bring a petition to their local government opposing the building of an all-night convenience store in their neighborhood. A few students organize a recycling program on a college campus. When it comes to national politics, however, most such activity takes the form of membership in a large organized interest group. Citizens participate by donating money, responding to calls to write letters to governmental officials, or volunteering their time to make phone calls on election day or to distribute campaign literature. Without such participation, many broadly based organized interests would be much weaker or would not exist at all, and party organizations would be less effective in contesting elections.

The role of unconventional political behavior Far fewer people engage in unconventional political behavior. However, when they do, they typically focus on local government. Although protest activities undertaken by a single individual can attract attention and have the desired effect, collective action stands a better chance of making a difference. A protest march by a lone person carries less weight than a march by a thousand people. Usually, unconventional political behavior constitutes only a small part of the play of power. In times of political crisis, however, the numbers who engage in it and its impact on the political process can rise significantly, as during the Great Depression and the Vietnam War era.

VOTING AND THE PLAY OF POWER

How do voting and nonvoting affect politics? Our answer reveals a good deal about how and by how much ordinary citizens shape policy in a nation of more than 260 million people. We examine both the *symbolic* and the *material* impact of voting.

The symbolic politics of voting and nonvoting Voting, which is so central to notions of democracy and citizenship in the American civil religion, carries immense symbolic significance. Indeed, in many respects the act of going to the polls and casting a ballot resembles the most meaningful of religious rituals. Millions of people simultaneously engage in a prescribed sequence of ritual behavior on presidential election day. On this special day, "the most powerful leader in the world" is chosen. Citizens must first have established their eligibility to participate in the ritual by registering. On election day they journey to a special place, affirm their eligibility by signing a voting card, and have an official certify their right to vote. Voters then walk to a specially enclosed booth and, shielded from the eyes of others, pull levers, punch holes in cards, or make marks in little boxes.

Many depart satisfied that they have fulfilled an important obligation and participated in a momentous event. The ritual continues into the evening, when citizens gather around two-dimensional screens and watch the distant images of the high priests of the media announce the results of the voting, anoint the leader, and divine the meaning of it all.

Three psychological responses to engaging in the ritual behavior of elections shape the play of power. First, voters (and even some nonvoters) gain a *sense of participation* in the governing of their lives. Second, voting *legitimizes government* and its actions. Participation in elections produces a "democratic coronation effect," evoking positive feelings toward the government and increasing the number of people who feel they have a say in government.[69] Third, voting *regularizes and channels political activity* in predictable, acceptable, and peaceful directions. Electoral activity probably displaces other forms of participation, including protest and violence. The common response to an unpopular policy is to "wait until the next election." After voting, many people consider their political activity to have been completed, freeing them to return to normal pursuits.

However, about half of all eligible citizens fail to vote in presidential elections. This means that close to 100 million people do not undergo the experience of voting or the symbolic legitimizing effects usually produced by it. This fact has implications for the stability of the political system. With so many people psychologically detached from the existing play of power, the legitimacy of the political regime rests on a narrower base. Given the nonvoters' lack of attachment, might they be available for mobilization by new and potentially undemocratic movements?

The answer depends in part on the reasons why so many citizens fail to vote. Political scientists debate these reasons without arriving at a consensus. One explanation assumes that nonvoters are satisfied with how things are going and see no need to vote. According to this view, lack of voting suggests that political stability is not in jeopardy. Rather, the system performs so well that citizens feel they can afford to leave politics out of their lives. Another view takes the opposite tack, arguing that nonvoters are revealing their deep cynicism and mistrust of politics, consciously refusing to vote because they see little meaningful difference among the candidates. A third view takes a middle position, arguing that although the failure to vote reflects no conscious rejection of politics, it does mean that the agenda of politics in general, and the issues fought over in elections, simply fail to address the issues that nonvoters find compelling.

Voting and material politics   Obviously, voters determine who wins and loses elections. They drive the process that selects elected officials. They provide the vehicle through which the major issues of the day (including the state of the economy), the personalities of the candidates, and the performance of the political parties determine who holds office. The changes in outcomes that occur when groups previously prevented from voting begin to do so attests to one important effect of voting (see "Outcomes: The Impact of the 1965 Voting Rights Act on Southern Elections" on pp. 430–31).

THE IMPACT OF THE 1965 VOTING RIGHTS ACT ON SOUTHERN ELECTIONS

During the early 1960s the civil rights movement increasingly turned its attention to the denial of the right to vote to African Americans in the South. For decades, racist local officials had used existing state rules to prevent African Americans from registering. Even though federal rules embodied in the Civil Rights Acts of 1957, 1960, and 1964 empowered the Department of Justice to bring suit against election officials who discriminated in registration, these rules required filing suit on a case-by-case basis and proved to be slow, expensive, and largely ineffective. Courageous civil rights workers who tried to promote black voter registration faced jail, beatings, and outright murder. The killing of Viola Liuzzo, a white middle-class woman from Michigan who was working on voter registration, and the violent suppression of a march for voting rights in Selma, Alabama, galvanized support for passage of the Voting Rights Act of 1965.

The act addressed many of the obstacles to voting faced by blacks living in the South.[1] It declared that violations of voting rights had occurred in any state or county where turnout in the 1964 election fell below 50 percent and where a literacy test had been used. It prohibited any further use of literacy tests in such "covered jurisdictions" and called for dispatching federal registrars to enroll voters. Anticipating that recalcitrant southern politicians would change other rules to deny African Americans the chance to win office, it required "preclearance" approval by the U.S. Department of Justice of *any* change in voting rules in covered jurisdictions. Finally, to ensure that

southern judges did not block implementation, the act provided that the federal courts in Washington, D.C., would hear litigation arising from the new law. When Congress renewed the Voting Rights Act in 1970, it added a ban on literacy tests as a precondition for voting that applied everywhere in the United States.

The act had a dramatic material impact. In the seven southern states falling under the 1965 act's provisions, the proportion of registered African Americans jumped from 29.3 percent in 1965 to 56.6 percent by 1972.[2] In 1968 the gap between the proportion of whites and blacks registered in five states in the deep South was 47 percent; by 1988 black registration rates there actually exceeded those of whites by more than 4 percent.[3] The number of black elected officials also rose dramatically. The seven states originally targeted by the 1965 act elected only 100 blacks in 1965; by 1989 the figure had increased to more than 3,200.[4] One student of the act summarized these changes:

> With black enfranchisement, of course, came a radical shift in the rules of the political game. Racist politicians either changed their tactics or bowed out, and in many places the new voters swept new faces (both black and white) into office.[5]

The act also wrought a huge change in the symbolism of politics in the South. Before the act's passage, white politicians in many jurisdictions benefited by making racist appeals in

The effect of voting on material politics, then, depends on the extent to which elections determine material outcomes in the play of power. If elections are mostly symbolic, barely affecting how the interaction of other participants affects who gets what, then voting is an ineffective means of citizen participation. If, on the other hand, elections shape outcomes in material politics, then voting offers citizens a powerful means of participation. We discuss the mate-

Martin Luther King Jr. helped organize and lead a march from Selma, Alabama, to Montgomery, the state capital, to dramatize African Americans' demands for voting rights. The march, widely covered in the news media, served as a potent symbol that mobilized support for passage of the 1965 Voting Rights Act.

their campaigns. But now the situation changed dramatically. "Almost nowhere, in fact, is explicit racism still politically advantageous."[6] Finally, the racial composition of the two parties in the South changed. Blacks overwhelmingly registered Democratic, and whites increasingly voted for Republican candidates. A number of white officials elected as Democrats switched to the Republican party. These developments help explain the growth of Republican strength in the South (as described in Chapter 12).

[1] This discussion is based largely on Abigail M. Thernstrom, *Whose Votes Count? Affirmative Action and Minority Voting Rights* (Cambridge, Mass.: Harvard University Press, 1987), Chapter 1.
[2] Chandler Davidson, "The Voting Rights Act: A Brief History," in Bernard Grofman and Chandler Davidson, eds., *Controversies in Minority Voting: The Voting Rights Act in Perspective* (Washington, D.C.: Brookings Institution, 1992), p. 21.
[3] Davidson, "The Voting Rights Act," p. 43.
[4] Davidson, "The Voting Rights Act," p. 43.
[5] Thernstrom, *Whose Votes Count?*, p. 2.
[6] Thernstrom, *Whose Votes Count?*, p. 3.

rial and symbolic effects of elections on the play of power in more detail in Chapter 12. For now, you should note that elections give citizens less influence over material politics than is implied by the central role the American civil religion attributes to them. Nonetheless, elections do provide citizens with a means to make a difference. Finally, the extent of the effect on material politics varies considerably from one election to another.

What would happen if the *rate* of voting increased and millions of nonvoters suddenly went to the polls? Would outcomes change? One way in which students of elections approach this question is to ask whether the drop in voter turnout beginning in 1964 and continuing to 1988 may have affected outcomes because more people of lower than of upper socioeconomic status withdrew. If so, the change in the "class bias" of the electorate would hurt Democrats and help Republicans. It turns out that, when measured by income, the class bias in the electorate did not change; although people of higher socioeconomic status consistently vote at higher rates, voter turnout in all income groups dropped at about the same rate.[70]

But what if voter turnout jumped suddenly? Would the millions of new voters, disproportionately poorer and less educated than the existing electorate, provide the Democrats with huge victories? The answer is by no means clear. Nonvoters have less interest and knowledge of politics than voters, and their party identification is weaker. They might well cast votes for whichever party's candidate benefited from the current short-term partisan forces. Furthermore, some dramatic and perhaps catastrophic events would have to occur to induce participation by huge numbers of nonvoters. The changes that an event such as the collapse of the economy would bring to politics would be so great that they would alter the play of power in ways that are difficult to anticipate.

SUMMARY

Political participation provides a crucial mechanism through which ordinary citizens enter the play of power. People believe that elections are pivotal political events and that voting provides them with a way to affect government's activities. But participation goes beyond voting. Political activists, who usually specialize in one form of participation such as local politics or campaign activities, are the most involved. Only about one person in ten, however, can be considered a complete activist. About two-thirds of American citizens participate primarily as spectators, following politics, voting, and only occasionally undertaking more demanding tasks. Significantly, a large minority of the citizenry—between 20 and 25 percent—are apathetic, caring little about politics and engaging in no political activity whatsoever.

Societal conditions such as the general level of social tension and the emergence of crises (e.g., economic disruption) shape the degree of political activity engaged in by citizens as well as how it changes over time. Social attributes affect which people engage in what levels of activity. Higher income and education levels in particular lead to higher levels of participation. Psychological traits such as citizens' sense of both internal and external political efficacy also shape their level of political participation.

Although the United States has universal adult suffrage, only about half of all Americans eligible to vote actually do so, even in presidential elections. In other elections, voter turnout typically falls another 15 percent. The rate at which citizens use other widely distributed resources in political participation (e.g., time, physical presence) is also low.

Rules affect political participation, especially voting, in a variety of ways. Voter registration rules reduce turnout somewhat. Until passage and implementation of the 1965 Voting Rights Act and subsequent amendments,

African Americans and Hispanics in many areas of the country were denied the opportunity to vote. Other state rules determine whether candidates seek party nominations in open or closed primaries.

Political scientists use several approaches to explain why people vote as they do. One approach finds a direct link between how people vote and their party identification in low-stimulus elections, and an indirect link in high-stimulus elections such as presidential ones. Party identification shapes, but does not absolutely determine, how voters perceive short-term partisan forces such as attitudes toward the candidates, issues, and parties. These partisan attitudes in turn affect the vote. Another approach, the theory of the impact of economic conditions, focuses primarily on the condition of the economy, predicting victory for the incumbent president's party in good times and defeat in bad times. In wartime the importance of economic issues is reduced. When the economy is neither booming nor slumping, other issues may play a role. Although voters mostly make retrospective judgments in deciding how to vote (assessing how well candidates and parties have performed), they also make prospective judgments about how candidates will behave if they are elected. Single-issue voters choose candidates on the basis of their stand on a particular issue.

Both national and local factors shape how people vote in congressional elections. National factors include the state of the economy, the strength of the winning presidential candidate's coattails, and, in midterm elections, the popularity of the president. These factors also shape strategic politicians' decisions regarding whether to run for Congress. Local factors include the quality of challengers, the characteristics of incumbents, and special conditions (e.g., an open-seat election or a scandal involving the incumbent).

Voting, the most common form of citizen participation, affects the play of power in both symbolic and material ways. Voting gives citizens a sense of participating in government, legitimizes government and its decisions, and regularizes and channels participation. Its effects on material politics are somewhat smaller than the civil religion claims, and they differ in magnitude from one election to another.

WHO VOTED HOW IN THE 1994 CONGRESSIONAL ELECTIONS?

Republicans captured control of both the Senate and (for the first time in 40 years) the House in the 1994 elections. Overall, about 53 percent of votes cast for the House went to Republicans. Remember that 1994 was a midterm election and that the Democrats entered the campaign controlling the presidency and both houses of Congress.

Based on the information in this chapter about why people vote the way they do, predict how each of the groups listed below voted in 1994 by choosing one of the following categories:

A. *Overwhelmingly Republican*—70 to 100 percent voting for Republican candidates

B. *Strongly Republican*—60 to 70 percent voting for Republican candidates

C. *Somewhat Republican*—50 to 60 percent voting for Republican candidates

D. *Somewhat Democratic*—50 to 60 percent voting for Democratic candidates

E. *Strongly Democratic*—60 to 70 percent voting for Democratic candidates

F. *Overwhelmingly Democratic*—70 to 100 percent voting for Democratic candidates

For example, we said that conservatives listen to radio talk shows more often than liberals do, and the Chapter 10 stated that conservatives are more likely to be Republican. A good guess about "Frequent listeners to political talk shows" would be B, "Strongly Republican." (They were, casting 64 percent of their votes for Republicans.)

1. Women	A	B	C	D	E	F
Blacks	A	B	C	D	E	F
Hispanics	A	B	C	D	E	F
2. Democrats	A	B	C	D	E	F
Independents	A	B	C	D	E	F
Republicans	A	B	C	D	E	F
3. Liberals	A	B	C	D	E	F
Moderates	A	B	C	D	E	F
Conservatives	A	B	C	D	E	F
4. People earning						
Less than $15,000	A	B	C	D	E	F
$15,000–$29,999	A	B	C	D	E	F
$30,000–$49,999	A	B	C	D	E	F
$50,000–$74,999	A	B	C	D	E	F
$75,000 or more	A	B	C	D	E	F
5. People whose financial situation in 1994 as compared to 1992 was						
Better	A	B	C	D	E	F
About the same	A	B	C	D	E	F
Worse	A	B	C	D	E	F
6. People who believe the condition of the economy is						
Excellent or good	A	B	C	D	E	F
Not so good	A	B	C	D	E	F
Poor	A	B	C	D	E	F
7. People who approve of President Clinton's job performance	A	B	C	D	E	F
People who disapprove of President Clinton's job performance	A	B	C	D	E	F

Key Terms

apathetics (*p. 397*)

activists (*p. 397*)

spectators (*p. 399*)

sense of political efficacy (*p. 405*)

sense of internal efficacy (*p. 405*)

sense of external efficacy (*p. 405*)

universal adult suffrage (*p. 411*)

voter registration (*p. 414*)

closed primary (*p. 414*)

open primary (*p. 414*)

short-term partisan forces (*p. 418*)

theory of the impact of economic conditions (*p. 421*)

single-issue voter (*p. 423*)

retrospective voting (*p. 423*)

prospective voting (*p. 423*)

midterm election (*p. 425*)

national factors (in congressional elections) (*p. 426*)

local factors (in congressional elections) (*p. 426*)

coattail effect (*p. 426*)

strategic politician (*p. 426*)

open-seat election (*p. 427*)

Recommended Readings

Abramson, Paul. *Political Attitudes in America.* San Francisco: Freeman, 1983. A good overview of research on political attitudes, especially political efficacy as of the early 1980s.

Campbell, Angus, Philip E. Converse, Warren E. Miller, and Donald E. Stokes. *The American Voter.* New York: Wiley, 1960. A classic and highly influential study of voting in the 1952 and 1956 elections. It fully develops and applies theories about how party identification affects short-term partisan attitudes and, consequently, votes for president.

Milbrath, Lester W., and M. L. Goel. *Political Participation: How and Why Do People Get Involved in Politics?* 2nd ed. Chicago: Rand-McNally, 1977. An early, classic, comprehensive overview of political participation.

Niemi, Richard G., and Herbert F. Weisberg. *Controversies in Voting Behavior,* 3rd ed. Washington, D.C.: Congressional Quarterly Press, 1993. A collection of excellent treatments of current academic research on crucial questions about voting behavior.

Piven, Frances Fox, and Richard A. Cloward. *Poor People's Movements: Why They Succeed, How They Fail.* New York: Pantheon, 1977. An influential and provocative analysis of the development of mass social protest movements, their successes, and the erosion of their effectiveness as they begin to wring concessions from the political system.

Pomper, Gerald M. et al. *The Election of 1992: Reports and Interpretations.* Chatham, N.J.: Chatham House, 1993. Seven chapters by different authors provide a broad summary of the 1992 elections for president and Congress, examining the candidates' strategies, the role of public opinion, and the meaning of the election.

Teixeira, Ruy A. *The Disappearing American Voter.* Washington, D.C.: Brookings Institution, 1992. A recent analysis of the extent of and reasons for the decline in voter turnout in the United States.

Wolfinger, Raymond E., and Steven J. Rosenstone. *Who Votes.* New Haven: Yale University Press, 1980. An influential and still timely analysis of voter turnout in the United States.

Elections in the Political Process

MOST AMERICANS WITNESS some 15 presidential and 30 midterm congressional elections during their adulthood. In the autumn of every leap year, the level of interest in politics rises to a four-year peak. Each presidential election's distinctive drama creates lasting memories.

The 1992 election was no exception. Incumbent George Bush's re-election seemed certain after his highly successful handling of the Iraqi invasion of Kuwait and the U.S. military victory over Iraq in early 1991. With a 90 percent approval rating, Bush watched as the strongest potential Democratic challengers decided not to run. Senators Bill Bradley, Jay Rockefeller, and Al Gore, Representative Richard Gephardt, New York governor Mario Cuomo, and the African-American leader Jesse Jackson all stayed on the sidelines.

As 1992 began, six lesser-known candidates (dubbed the "six dwarves") filled the vacuum. Among them was Arkansas governor Bill Clinton, a man virtually unknown to the general public. Clinton fared poorly in the Iowa caucuses, which were won by native son Senator Tom Harkin. Clinton survived the New Hampshire primary with a second-place finish despite a media story about an alleged extramarital affair, gained momentum by winning in Georgia, and swept the southern primaries on "Super Tuesday." Victories in Illinois and Michigan virtually ensured Clinton the nomination, although he had to weather a late surge by former California governor Jerry Brown.

Two other presidential candidates caused media attention to be directed elsewhere. Television commentator and former Reagan speechwriter Pat Buchanan vied for the Republican nomination with vicious attacks on George Bush for having violated his 1988 nomination pledge, "Read my lips. No new taxes." Buchanan won a surprising 37 percent in the New Hampshire Republican primary but fared less well in subsequent contests. By March 3, Bush had won three primaries and 148 delegates, Buchanan no primaries and just 20 delegates, virtually guaranteeing Bush's renomination.

H. Ross Perot, a multibillionaire from Texas with a gift for pithy sound bites and a strong appeal to citizens alienated from "politics as usual," was the media's other focus. Perot announced his willingness to run in a February appearance on the *Larry King Live* show and began organizing volunteers. Although he attacked the current system and its politicians generally, he directed his strongest criticisms at Bush. Perot's standing in public opinion polls rose steadily, and by June he led both Bush and Clinton in some polls. While Perot dominated the news media's coverage, Bush lost ground and Clinton quietly edged closer.

The turning point in 1992 came during the Democratic convention. With his campaign faltering, Perot abruptly announced his withdrawal and virtually endorsed Clinton by stating: "Now that the Democratic party has revitalized itself, I have concluded that we cannot win in November and that the election will be decided in the House of Representatives."[1] Clinton built a solid lead over Bush, and Perot's withdrawal caused his own support to plunge.

In a final dramatic reversal, Perot renewed his independent campaign in September. His support began to recover and he ultimately received 19 percent of the vote, more than any other independent or third-party candidate in modern political history. As Figure 12.1 (on p. 439) shows, support for him was very uneven geographically. Perot's presence was perhaps the most distinctive feature of the 1992 election, clearly changing its dynamics. As Perot and Bush traded charges in the spring, Clinton's support grew. During the fall debates, Perot's presence prevented Bush from focusing on Clinton. Indeed, on several occasions Perot blunted Bush's attacks on Clinton by coming to Clinton's defense or attacking the president.

Clinton, focusing on the weakness of the economy and the need for jobs, captured 43.3 percent of the vote as compared to Bush's 37.7 percent. His 53.5 percent of the votes cast for the two major-party candidates matched Bush's 53.9 percent margin over Michael Dukakis in 1988. Thus, what had looked like certain re-election for Bush turned into a three-candidate race whose surprise winner was the virtually unknown Bill Clinton.

437

COMMON BELIEFS

THE CRUCIAL ROLE OF ELECTIONS

It is inconceivable that a democratic nation would not conduct elections. After all, elections provide citizens with an important mechanism for exerting influence. No other component of the play of power attracts as much attention, stimulates as much activity, or produces as much drama as a presidential election.

Our own survey of student beliefs has confirmed the central role of elections in Americans' understanding of politics. A solid majority of student respondents (52%) agreed that presidential elections indicate what the American people want policy to be; 36 percent disagreed; and only 6 percent had no opinion. Furthermore, an overwhelming 90 percent felt that elections make government pay "a great deal" or "some" attention to what its citizens think.

Although many students saw elections as pivotal, a significant minority did not. Chances are that people in your class also disagree on the importance of elections. We will examine just how important American elections are after we discuss the participants, rules, resources, and strategies that shape them. ∎

THE PARTICIPANTS IN ELECTIONS

Earlier chapters focused on voters as well as the role of the mass media in elections. In this section we consider the most active participants in elections: candidates, contributors, and consultants and campaign aides.

THE CANDIDATES

Americans elect people to fill close to a half-million offices, from precinct election officials to president. This chapter discusses presidential candidates; in the next chapter we will consider candidates for Congress.

Who becomes president? All 41 presidents of the United States have been white, male, and (usually) Protestant, belonging to a high-status denomination such as Episcopalian or Presbyterian.[2] They typically have come from affluent families with British ancestry and have been educated at prestigious private institutions. Even Bill Clinton, whose background is more modest than that of most presidents, graduated from Georgetown University and Yale Law School.

Presidents' prior careers also tend to share certain characteristics. Twenty-five presidents, including Bill Clinton, were lawyers. Most held office in state or local government and then won higher office. About two-thirds served in Congress before becoming president. Fourteen presidents first served as vice-president, with nine assuming the presidency on the incumbent's death or resignation. Just three, all popular generals, held no prior public office: Zachary Taylor, Ulysses Grant, and Dwight Eisenhower.

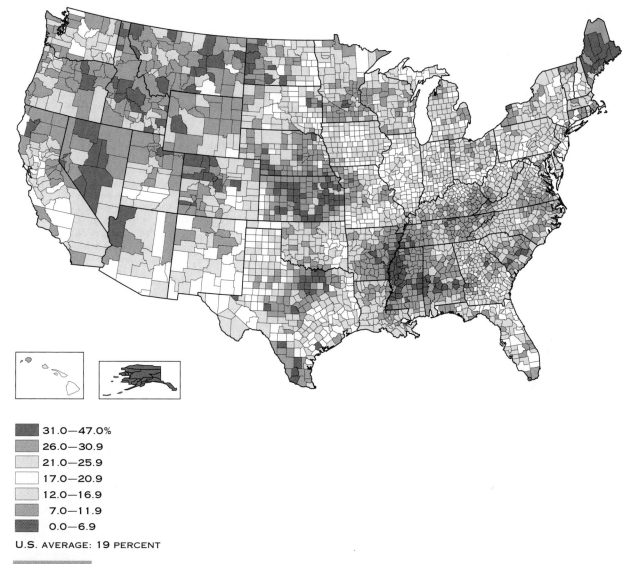

31.0—47.0%
26.0—30.9
21.0—25.9
17.0—20.9
12.0—16.9
7.0—11.9
0.0—6.9

U.S. AVERAGE: 19 PERCENT

FIGURE 12.1

THE GEOGRAPHICAL DISTRIBUTION OF THE PEROT VOTE, 1992 The authors of this map of Ross Perot's support describe it as depicting "the geography of anger." Counties in blue provided Perot with more than his national average of 19 percent of the vote, those in red with less. *Source:* Peirce Lewis, Casey McCracken, and Roger Hunt, "Politics: Who Cares?" *American Demographics*, October 1994, p. 22. *American Demographics* magazine © 1994. Reprinted with permission.

The grand prize of American politics has never been won by a woman, an African American, a Hispanic, an Asian American, or a Jew. Only one Roman Catholic, John F. Kennedy, has done so. Jesse Jackson, an African American, is the only nonwhite male ever to mount a significant challenge for a major party's presidential nomination.[3]

Why are women and minority candidates rarely nominated? The list of women who have sought the presidency is short.[4] The People's Party nominated the first, Victoria Claflin Woodhull, in 1872, and the National Equal

Rights Party nominated Belva Lockwood in both 1884 and 1888. However, it was 1964 before a major political party considered a woman as its nominee. At that time Margaret Chase Smith, the Republican senator from Maine, entered the presidential primaries and won 30 delegates. Shirley Chisholm, an African-American congresswoman from New York, ran in the 1972 Democratic primaries. Although Frances Farenthold was placed in nomination for vice-president at the 1972 Democratic convention, not until 1984 did a woman, Democratic congresswoman Geraldine Ferraro, actually receive a major party's nomination for that office.

Women have not been nominated or elected president for a number of reasons.[5] Until passage of the Nineteenth Amendment in 1920, the time, energy, and talent that women expended in politics were devoted to gaining the right to vote. Women today still occupy relatively few of the lower political offices that serve as stepping-stones to higher office. In 1971 they held only 5 percent of state legislative seats; by 1994 this number had risen to only 21 percent.[6] Moreover, women officeholders often cannot raise the money needed for a presidential campaign because major contributors are reluctant to support someone they see as an underdog.[7] Representative Patricia Schroeder estimated that she would need to raise $2 million to run in 1988; when she decided in September 1987 not to enter the presidential race, she had raised less than $1 million.[8]

Are voters willing to support a woman for president? As Table 12.1 indicates, over time Americans have become increasingly willing to do so "if she were qualified for the job." But as late as 1983, 20 percent said they

Victoria Claflin Woodhull (*left*) ran for president on a third-party ticket in 1872, almost 50 years before ratification of the Nineteenth Amendment gave women the right to vote. A hundred years later, Congresswoman Shirley Chisholm (*right*) became the first black woman to seek a major party's nomination.

TABLE 12.1
PUBLIC WILLINGNESS TO VOTE FOR A WOMAN FOR PRESIDENT

YEAR	% ANSWERING "YES"
Question: "If your party nominated a woman for president, would you vote for her if she were qualified for the job?"	
1937	31%
1958	52
1978	76
1983–87	85
1988–91	89
1993	91

Percentages are for those answering "yes." Figures for 1983–87 and 1988–91 are averages for questions asked during those periods.

Sources: For 1937, 1958, 1978: George Gallup Jr., *The Gallup Poll: Public Opinion 1984* (Wilmington, Del.: Scholarly Resources, 1985), pp. 120–21; for 1983–87, 1988–91, 1993: *The General Social Survey* (Chicago: University of Chicago, National Opinion Research Center, 1994), p. 251, Question 200.

would not vote for a woman. In 1984, when a Gallup poll asked whether a woman would do a better job than a similarly qualified man in handling six aspects of the presidency, women received higher marks than men in only one area—improving the quality of life in the country.

African Americans face obstacles as great as or greater than those encountered by women. They hold few of the lower offices needed to make a bid for the higher posts that are launching pads for presidential bids, and they find it difficult to attract campaign contributions. Moreover, a 1983 Gallup poll found that 16 percent of those surveyed would not vote for an African-American man who was "generally well qualified," while 7 percent expressed no opinion.[9] Like women, African Americans have most often sought the office as independent or minor-party candidates. In 1988 more than a dozen persons of color—including a Native American, Russell Means—ran under various party banners. In addition to Shirley Chisholm and Jesse Jackson, only two other blacks—Democratic Virginia governor Douglas Wilder in 1992, and conservative Republican talk show host Alan Keyes in 1996—have sought a major party's nomination. Only Jackson's 1988 effort, described in "Participants: Jesse Jackson's 1988 Primary Campaign" (on p. 442), attracted strong support.

CAMPAIGN CONTRIBUTORS

The cost of running for federal office has skyrocketed. Representatives elected in 1976 received $42.5 million in contributions. By 1994 they collected $245.8 million. Senate winners' total receipts increased just as much,

JESSE JACKSON'S 1988 PRIMARY CAMPAIGN

Never has anyone other than a white male done as well in seeking a major party's presidential nomination as the Reverend Jesse Jackson, a veteran of the civil rights movement and a gifted orator. In 1984, under the banner of his "Rainbow Coalition," Jackson captured 19 percent of the votes and 10 percent of the delegates in Democratic primaries. Despite this impressive showing, the press relegated him to the sidelines as someone whose support was confined to a core constituency of blacks.

In the 1988 election Jackson fared even better. He won 92 percent of black Democratic Party voters' support; his message, directed at the economically less well off, also attracted 30 percent of Hispanic and 12 percent of white voters.[1] As other candidates dropped out, Jackson's campaign gained momentum and attention. He won five states on "Super Tuesday," took second place in Illinois, and won a stunning victory over the front-running Michael Dukakis in the Michigan caucuses. At this point only three candidates (Jackson, Dukakis, and Tennessee senator Al Gore) remained in the race.

Party officials and the press speculated that a victory in New York would demonstrate Jackson's potential to capture the presidency. Although Jackson won 37 percent of the vote there, Gore's poor showing allowed Dukakis to win 51 percent and all but sew up the nomination. Ultimately Jackson received 6.6 million votes, 29 percent of Democratic primary votes cast. He carried many of America's leading cities, including New York, Chicago, Atlanta, Baltimore, Houston, Dallas, Detroit, New Orleans, and Milwaukee. He won 30 percent of the delegates and raised close to $27 million, second only to Dukakis. Accorded a central role during the first two days of the Democratic convention, Jackson gave a stirring speech that at-

Jesse Jackson's ability to attract votes and win delegates in his 1988 bid for the Democratic nomination strengthened his role as the leader of an important element in the Democratic Party's coalition. Other party leaders recognized his influence by scheduling him to speak before the 1992 convention during prime time.

tracted more television viewers than any other event at the convention. His impressive showing established him as a major player in American national politics.

[1] Stephen J. Wayne, *The Road to the White House 1992: The Politics of Presidential Elections* (New York: St. Martin's Press, 1992), p. 106, Table 4.4.

from $21 million in 1976 to $151 million in 1994.[10] Without money, candidates cannot afford the television time, postage, polling costs, and consultants' fees necessary to win. Senate candidates in California, for example, must reach 32 million diverse people. The cost of a 30-second prime-time television spot there in 1994 was $40,000.[11] The individuals, political action committees (PACs), and party organizations that provide campaign money play a critical role.

Individual contributors to national candidates Only about one in every seven people makes political contributions. The roughly 15 million people who contribute resemble other activists: They are highly educated, financially well off, likely to vote, and very interested in politics.[12] Contributions from individuals (about half at $500 or more per contribution) now account for almost half of the money congressional candidates receive. With the exception of union members giving to labor PACs, contributions of $50 or less now provide only a small portion of all campaign donations.

Federal law limits the amount a person can give to federal candidates in a two-year election cycle. However, a loophole in this law permits large contributions to state parties to fund registration and "get-out-the-vote" drives that directly benefit presidential campaigns. Consequently, presidential campaigns channel such contributions, called **soft money**, to their state parties. In 1988 at least 397 individuals gave $100,000 or more in soft money to state parties.[13] In 1992 the Clinton campaign benefited from approximately $40 million in such contributions; the Bush campaign, $25 million.[14]

Most individuals contribute voluntarily, primarily for political reasons. Some give out of a sense of duty or loyalty to a political party or candidate. Others contribute to gain access to and win influence over elected officials. Asked by a Senate committee whether he thought his contributions helped in getting senators to assist him in his troubled dealings with federal regulators, savings and loan executive Charles Keating replied, "I want to say in the most forceful way I can: I certainly hope so."[15]

PACs, party organizations, and other sources of campaign money Chapter 9 described the tremendous growth in the number and total contributions of PACs since 1974. Although PACs can and do give to presidential candidates' primary campaigns, most of their money goes to congressional candidates, especially incumbents. Furthermore, despite the two parties' fundraising prowess, individuals and PACs provide all but a small fraction of total campaign funds. However, state party organizations raise some money as well as serving as conduits for soft money.

Each party in the House and Senate also has a committee to raise money for its candidates. These committees raised only modest amounts in 1976. But in the 1994 election cycle, the Democrats' House and Senate campaign committees raised $45.8 million, and the Republicans' committees raised $92 million.[16] In addition to direct contributions, these committees provide many of the campaign services that private consultants offer, but at a lower cost. Their influence was demonstrated in 1994 when the Republicans won control of the House of Representatives. The chairman of the National Republican Congressional Committee contributed to this victory by rebuilding the committee's fundraising apparatus, recruiting strong candidates, and providing them with training and advice.

Several other sources of campaign funds exist as well. For example, some members of Congress with excess funds and little competition give to party candidates when the race is tight. Further, members seeking leadership posts make contributions from so-called leadership PACs to improve their chances of winning the recipients' support. Since 1978 about 50 such leadership PACs have been active.[17]

CONSULTANTS AND CAMPAIGN STAFFERS

Consultants play a major role in elections by using the new campaign technologies described later in this chapter. They conduct opinion polls, organize phone banks, direct media relations, write speeches, create mailings, organize fundraising and get-out-the-vote drives, devise and implement campaign strategy, and sometimes even help choose candidates.[18] Their efforts rarely win or lose races and can tip the balance only in close elections. But the consultants charge handsomely for their services, and the polling, mailing, and media advertising they recommend are expensive. In this way they contribute to the growing pressure on candidates to raise more and more money. In addition, their reliance on negative or "attack" ads most likely fosters citizens' negative views of politics and politicians. Finally, consultants often continue their relationship with winners after the election is over, conducting polls, advising on strategy, and even lobbying elected clients on behalf of private business clients.[19]

When political consultant Ed Rollins boasted that he had helped suppress votes for the Democratic candidate for governor in New Jersey in 1993, the resulting extensive media coverage indicated that reporters and editors recognized the important role of consultants in modern campaigns. This incident increased the visibility of such consultants in the eyes of the general public as well.

Campaign officials constitute another group of participants in elections. Presidential campaign staffs usually include some people who are not professional campaign consultants or media specialists. Perhaps the best known and most highly regarded in recent years has been James Baker, who played a central role in every Republican presidential campaign from 1980 through 1992.

Although men once dominated campaign staffs, women increasingly play key roles. For instance, Susan Estrich managed Michael Dukakis's 1988 presidential campaign, and Mary Matalin and Dee Dee Myers served as prominent spokeswomen for the Bush and Clinton campaigns, respectively, in 1992.

THE CRITICAL ROLE OF ELECTION RULES

Elections form an integral part of the political landscape. Yet the rules under which they operate attract very little attention. In South Africa's 1994 election (described in Chapter 11), a little-noticed rule provided for proportional representation of the parties in the new parliament. This ensured that the largely white National Party and the black Inkatha Freedom Party won seats. Different rules providing for single-member districts where the candidate with the most votes wins would have virtually excluded these two parties. Because they do make a difference, we examine rules governing the selection of the president and Congress, including those that determine the timing and financing of elections.

RULES OF PRESIDENTIAL SELECTION

Two sets of rules shape how presidents are chosen. One set specifies how the Electoral College votes and how its members are chosen. The other set guides how the party conventions nominate candidates and how the delegates to these conventions are selected.

The Electoral College and winner-take-all rules   The Constitution requires that a candidate win an absolute majority of the Electoral College's votes to be elected president. Chapter 10 showed how the "winner-take-all" rule for casting states' electoral votes encourages formation of two broad political parties. The **winner-take-all rule**, or **unit rule**, also places a premium on winning the largest states. For example, winning just one more vote than anyone else in California brings a candidate all of the state's 54 electoral votes. Consequently, candidates focus their efforts on large states where they have a chance, and winners pay special attention to large states' needs once they are in office. Bill Clinton lavished more attention on California, which he won in 1992, than on any other state. His 1993 economic stimulus package included $252 million for summer jobs for California alone; three of his cabinet members hailed from California; and his administration took a number of steps to stimulate the state's ailing economy.[20]

The winner-take-all rule also magnifies the influence of minority groups concentrated in large states. For example, attracting a high percentage of Jewish votes in New York or Hispanic votes in Texas improves a candidate's

Presidential campaign coverage often features photos of candidates campaigning in the states with the most electoral votes. Shown here is Bill Clinton during one of his many campaign trips to California, which has 54 electoral votes, 20 percent of the number needed to win the election.

chances of winning those big states' electoral votes. It also magnifies the winners' margin of victory in the Electoral College as compared to the popular vote. Bill Clinton's 43 percent of the popular vote translated into 69 percent of the Electoral College. Of course, losers suffer heavily under the rule. Ross Perot won 19 percent of the popular vote in 1992 but did not receive a single electoral vote.

The nominees of the two major parties automatically appear on the November presidential ballot, but third-party and independent candidates face a daunting and bewildering array of state requirements to get their names on the ballot. For example, to place Ross Perot's name on the New York ballot, his supporters had to gather 20,000 valid signatures, with a minimum of 1,100 from each of at least half of the state's 34 congressional districts.[21]

Nominating conventions The Republican and Democratic parties each choose a candidate whose name automatically appears on every state's presidential ballot at the party's national nominating convention, held in the summer before the November election. A complex and changing set of rules, summarized in Table 12.2, govern the selection of these delegates.

TABLE 12.2
ELECTION RULES AND THE SELECTION OF CONVENTION DELEGATES

MATTERS DETERMINED BY RULES	PROVISIONS IN EFFECT IN 1992 ELECTION
The total number of delegates and the number to which each state is entitled	Democrats had 4,286 delegates. Republicans had 2,203 delegates.
Whether delegates are chosen in primary elections or in party caucuses and conventions	More than 80% were chosen in primaries and just 20% in state conventions (e.g., Michigan's) or precinct caucuses (e.g., Iowa's).
Whether elected delegates are selected from districts or the state at large	Democrats required 75% of a state's delegates to be elected by district, 25% at large; Republicans permitted winner-take-all.
Whether there are quotas for the number of women delegates and affirmative action plans to increase the number of minority delegates	Democrats required that 50% of delegates be women, and they encouraged minority delegate selection; Republicans encouraged selection of both but set no quotas.
Whether delegates are allocated by proportion of the vote received, and, if so, the minimum threshold for receiving delegates	Democrats used proportional rule with a 15% threshold; Republicans allowed states to decide how to allocate delegates, with no national threshold.
Whether delegates must vote for their candidate, and whether a "preference poll" with candidates' names is on the ballot	Delegates could vote their conscience in both parties; each state decided whether there was a preference poll and how names got on the ballot.
When a primary election or caucus will be held	Democrats limited the period from the first Tuesday in March to the second Tuesday in June; Republicans had no rule.

Source: The information in this table is largely drawn from Stephen J. Wayne, *The Road to the White House 1992: The Politics of Presidential Elections* (New York: St. Martin's Press, 1992), p. 99, Table 4.2.

Because usually one candidate has "sewn up" the nomination by May, states choosing delegates later than April exert little influence on the nomination. This creates a powerful incentive to change the primary date. After the 1992 election a mad scramble occurred among states for influence, accelerating what is known as **front-loading**, that is, compressing the primary season into an early, short period. Ohio changed its primary from early May to the third Tuesday in March, and California moved its delegate-rich contest from June to late March. In response, Governor Mario Cuomo advocated changing New York's primary to early March. When Californians heard this, they began talking about moving their primary yet again. Consequently, by the end of March 1996, almost 60 percent of the Republicans' and two-thirds of the Democrats' delegates will have been chosen. This has prompted one journalist to comment that the process bears "more resemblance to a high-speed blender than to a deliberative process."[22]

RULES AND THE SELECTION OF SENATORS AND REPRESENTATIVES

Article I of the Constitution sets House members' terms at two years and senators' at six. Consequently, the entire House is selected every two years, as compared to one-third of the Senate. Article I also calls for direct election of representatives; but until the Seventeenth Amendment (ratified in 1913) required senators' direct election, state legislators could choose them.

The Framers allotted one House member for each 30,000 "persons" (each slave was counted as only three-fifths of a person). Applied today, this formula would result in a body with more than 8,400 members. In 1929, House membership was capped at 435. Each member now represents approximately 584,000 people.

By custom, every state elects its representatives from single-member districts with *plurality voting.* The plurality rule declares that the candidate who receives more votes than anyone else wins. A *single-member district* is a district that elects only one person. Other jurisdictions, including many large cities, elect some council members *at large,* producing *multimember districts* with two or more representatives from the same geographic area.[23]

The constitutional rule apportioning House seats on the basis of state population requires the recalculation of each state's number of representatives following each census (that is, every 10 years). This process, called **reapportionment**, affects states' number of seats in Congress and votes in the Electoral College. For example, following recent population shifts, states in the Northeast and Midwest lost seats to states in the South, Southwest, and West in the last apportionment. California gained seven seats and Florida and Texas three each, while New York lost three and Ohio, Michigan, Illinois, and Pennsylvania lost two each.

Even if the number of seats for a state is unchanged, internal population shifts require adjusting district boundaries to comply with Supreme Court decisions mandating "one person, one vote."[24] The redrawing of district lines, called *redistricting,* thus occurs after every census everywhere save in the smallest states, which are entitled to just one representative (who is elected statewide).

Because the way in which new districts are drawn affects election outcomes, incumbents and political party officials engage in a furious play of power to design boundaries that they think favor them and disadvantage opponents. Drawing districts to give one party or group an advantage is called **gerrymandering** after an early governor of Massachusetts, Eldridge Gerry, who reluctantly signed a reapportionment measure that created a district with tortured outlines resembling a salamander.[25] *Partisan gerrymandering* seeks to draw the lines to maximize the number of seats won by one party. *Incumbent gerrymandering* crafts districts to help incumbents retain their seats. *Racial gerrymandering* seeks to promote or hinder the chances of members of racial or ethnic minorities of winning office.

Politicians work hard at partisan gerrymandering because they think it will help their party win more seats, but their best efforts often fail. Drawing district lines in ways designed to protect incumbents, however, usually succeeds. Finally, efforts to facilitate or prevent the election of African-American, Hispanic, and other minority-group candidates through redistricting have been very effective, as "Rules: Drawing Congressional Districts to Increase Minority Representation" (on pp. 450–51) shows.

The rules establishing single-member districts with plurality voting strongly encourage the building of two broad coalitions in each district and thus reinforce the two-party system. Further, a losing party capturing close to 50 percent of the vote in a district wins nothing. Conceivably, a party whose candidates won 51 percent in every district would take all 435 seats. Usually, the party winning the House gets a higher percentage of seats than of total votes cast.

What do rules say about how the two parties select general election candidates? For most of this century, a primary election has been conducted in the spring. In presidential election years the primaries usually coincide with presidential primaries. Each voter may only participate in one party's primary. Furthermore, most states hold *closed* (rather than *open*) primaries, which means that anyone voting in a party's primary must be registered with that party. Usually, winning a plurality secures the nomination. In some southern states, when three or more candidates run and none receives an absolute majority, a *run-off primary* is held between the two top vote getters.

Recently the remarkable re-election success enjoyed by incumbents, coupled with growing anti-incumbent sentiment, has provided fertile ground for a movement to limit the number of terms served by legislators. Beginning with Colorado in 1990, 17 states rapidly enacted such restrictions, or **term limits**; in 1994 another 8 states approved proposals for limited terms, bringing the total to 25. Most sought to limit House members to no more than three or four terms and senators to two terms. In early 1995 the new Republican House majority tried and failed to pass a constitutional amendment limiting terms. Several months later, on May 23, 1995, the Supreme Court ruled that state laws limiting congressional terms were unconstitutional, dealing a huge setback to the term limits movement.

REGULATING FEDERAL CAMPAIGNS

Rules shape three other aspects of American political campaigns: when elections are held, how campaigns are conducted, and how candidates finance them.

Establishing electoral calendars   Each state chooses its own date for holding primary elections. But the Constitution sets the first Tuesday after the first Monday in November in even-numbered years for the general election. This means that elections take place whether momentous issues exist or not. Although many voters may wish for an immediate election when a scandal arises or a crisis cripples or bolsters a president's popularity, they must wait for the next scheduled election.

The calendar's control of elections in the United States prevents them from being used to choose among candidates and policies during times of economic or foreign crisis. By the same token, this system avoids snap judgments by the electorate. However, it also introduces into American politics regular *cycles* that are as inevitable as the effect of the moon on the ocean's tides. Every four years, in November, interest in politics and electoral activity reaches a peak during the presidential election; two years later, a lower peak coincides with congressional elections. This cycle affects many aspects of the play of power, including the willingness of Congress to pass necessary but unpopular legislation in an election year.

DRAWING CONGRESSIONAL DISTRICTS TO INCREASE MINORITY REPRESENTATION

The 1965 Voting Rights Act increased the number of African Americans registered to vote and the number of those elected to local office. It also generated intense conflict over the drawing of new congressional district boundaries following the 1990 census.

The act encouraged efforts to increase the number of African-American and Hispanic majority districts. The Department of Justice reviewed new districts drawn in states falling under the act's provisions and in some cases ordered states, including North Carolina, to produce black-majority districts. Democrats wanted to protect their minority incumbents, and some in the party wanted to increase their numbers. Many Republicans also sought more minority districts in order to concentrate Democratic voters in a few districts, reducing Democratic strength in other districts.

Because minorities did not live in compact configurations, some contorted district lines had to be drawn to encompass a majority of minorities in a single district. Nationwide, the redistricting process produced 54 "majority-minority" districts: 32 black, 20 Hispanic, and 2 Asian.[1] According to one analysis, reapportionment resulted in the election of 13 additional black House members and 6 Hispanics in 1992.[2]

The battle over these rules soon shifted to the judicial arena. Five voters in North Carolina's redrawn 12th District filed suit. In the case of *Shaw v. Reno*, the Supreme Court ruled that citizens can challenge the constitutionality of "bizarre" districts, but it did not invalidate the 12th District's boundaries. A lower court subsequently found the district's lines a reasonable way to increase minority representation. Suits challenging districts elsewhere followed. In June 1995 the Supreme Court ruled in *Miller v. Johnson* that Georgia's 11th District was unconstitutional because race had been used as a "predominant factor" in drawing its lines. Court challenges to other minority districts based on this case will force lower courts to decide when race has been "predominant," guaranteeing that the struggle over rules will continue.

[1] "Redrawn Minority Districts Face Challenges," *Congressional Quarterly Almanac: 1993* (Washington, D.C.: Congressional Quarterly Press, 1994), p. 22-A.
[2] Ronald Smothers, "Fair Play or Racial Gerrymandering? Justices Study a 'Serpentine' District," *New York Times*, April 16, 1993, p. B7. Smothers cited Frank R. Parker, voting rights project director for the Lawyers Committee for Civil Rights under Law.

Regulating the conduct of campaigns   Rules require purchasers of campaign advertisements to identify themselves and organizers of campaign rallies and parades to obtain a permit. They also prohibit the bribing of election officials and the buying of votes. Overall, however, rules impose few restrictions on what candidates can do or say. A rich American tradition of vicious campaigning, character assassination, and dirty tricks continues to the present day. The negative campaigning and "attack ads" of recent elections are hardly new. Indeed, campaigns during the nineteenth century were generally more vicious than today's.

The permissiveness of campaign regulations flows from democratic rules guaranteeing citizens the right to assemble peacefully, to enjoy freedom of speech, to engage in organized activities free of state control, and to run for political office. When nondemocratic systems decide to conduct free elections (as Hungary and Poland did in 1989 and 1990), they must loosen restrictions on speech, the press, and the right to assemble. Free elections

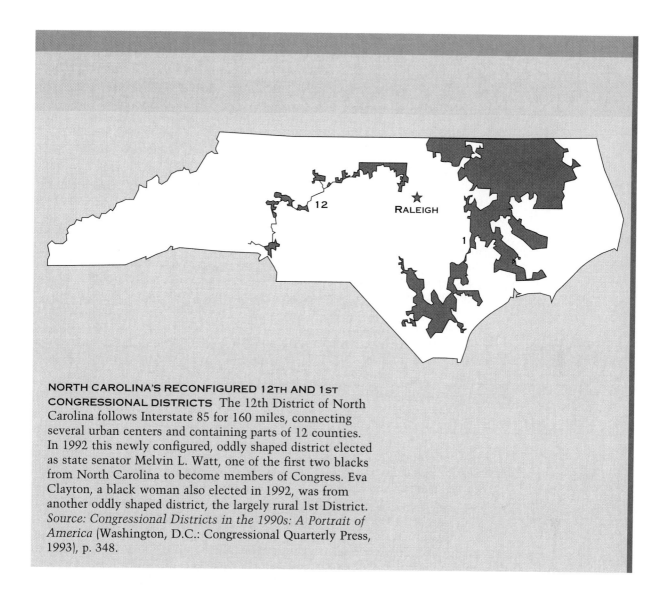

NORTH CAROLINA'S RECONFIGURED 12TH AND 1ST CONGRESSIONAL DISTRICTS The 12th District of North Carolina follows Interstate 85 for 160 miles, connecting several urban centers and containing parts of 12 counties. In 1992 this newly configured, oddly shaped district elected as state senator Melvin L. Watt, one of the first two blacks from North Carolina to become members of Congress. Eva Clayton, a black woman also elected in 1992, was from another oddly shaped district, the largely rural 1st District. *Source: Congressional Districts in the 1990s: A Portrait of America* (Washington, D.C.: Congressional Quarterly Press, 1993), p. 348.

thus carry many significant implications for the content of other rules that guide the play of power.

Regulating the financing of federal campaigns   Prior to passage of the Federal Election Campaign Act (FECA) of 1971, there were few restrictions on how federal candidates raised money, and even fewer were enforced. The FECA continues to regulate federal campaign finance. As amended in 1974, it provides for public financing of presidential elections, limits the size of contributions, and requires public disclosure of most sources of candidates' money. Its provisions are administered, interpreted, and enforced (often timidly) by the Federal Election Commission (FEC), which is composed of three Democratic and three Republican commissioners.

The FEC disperses public money to the two parties' presidential nominees for their general election campaigns. To receive this money, a nominee must agree to refuse contributions from any source, including individuals and PACs, and to limit expenditures to the amount of the public money granted ($56.5 million to each nominee in 1992). However, both parties' presidential campaigns rely on huge "soft money" contributions that go to state parties but help presidential campaigns. The FECA also allows the national political parties to spend on behalf of their presidential candidate. In 1992 each party spent the limit: $10.3 million.[26] Third parties that captured at least 5 percent of the vote in the previous election obtain smaller grants, and independent candidates or new parties meeting the 5 percent threshold receive grants after the election. Readers who have looked closely at the first page of the federal income tax form will know the source of this public money. The form contains a box allowing taxpayers to designate $3 ($6 for a joint return) of their taxes to go to the Presidential Election Campaign Fund.

The FECA also provides for partial public financing of presidential primary campaigns. Candidates who qualify by raising at least $5,000 in 20 different states (in contributions of $250 or less from individuals) have additional individual contributions up to $250 matched with public money. Recipients must agree to an overall spending limit as well as the limit in their state. In 1992 the candidates received more than $42 million in prenomination matching federal funds.[27]

Despite their exploitation of the "soft money" loophole, presidential candidates rely far less on large contributors and PACs than they did before public financing took effect in 1976. Both the overall and state spending limits in the primaries affect candidates' strategies. Candidates partly avoid these limits by creating "precandidacy" PACs.[28] Ostensibly organized to help their party's candidates and organizations, these PACs serve as shadow campaign committees for the future candidate, who serves as his or her PAC's honorary chair.

The 1974 amendments to the FECA limited the total amount candidates could spend in seeking election, the amount of personal money they could devote to their own cause, and the sums groups and individuals could spend independently to support or oppose a candidate. However, in 1976 the Supreme Court struck down these limits. The principal surviving restrictions on congressional campaigns mostly focus on limiting contributions. Attempts to enact public financing for congressional elections and limit the amount and size of PAC contributions since 1974 have met with repeated failure.

RESOURCES IN ELECTION CAMPAIGNS

Each participant brings resources to the electoral process. As we discuss later in this chapter, incumbents know how to campaign, raise money, and exploit the advantages of incumbency. Challengers usually lack these skills. Many open-seat candidates are strategic politicians familiar with voters, campaign techniques, and fundraising.

Campaign consultants and managers draw on their reputation and knowledge of how to run campaigns. Newspaper and television journalists and talk show hosts use their ability to shape the news and provide an audience. Their access to sources of information, including party officials and consultants, coupled with knowledge of the electoral process, allows journalists to exert influence.

Campaign contributors' principal resource, obviously, is money. People who possess wealth, and organizations whose members can afford to

HOW EMILY'S LIST GETS MONEY TO WOMEN CANDIDATES

Challengers to incumbents find it difficult to raise money. PACs overwhelmingly give to incumbents, most of whom are male. Add to this the fact that women candidates—even for open seats—do not attract as much campaign money as men, and the problems faced by women loom even larger.

Enter Ellen Malcolm, a political activist who adapted and refined the technique of "bundling" campaign contributions. By collecting individuals' checks—at or below the $1,000 contribution limit—made out directly to a candidate's campaign committee, a fundraiser can deliver a bundle of money far exceeding the $5,000 limit on PAC contributions. Malcolm figured that many women would willingly support Democratic candidates who supported abortion rights, so she started a PAC called EMILY's List—the acronym standing for Early Money Is Like Yeast (i.e., it makes the dough rise)—that listed endorsed women candidates and asked contributors to send it personal

checks made out directly to one or more of the recommended candidates. EMILY's List then bundled and delivered the checks.

By the 1990 election, EMILY's List had disbursed more than $1 million. Spurred by the anger many women felt at Anita Hill's treatment during the Senate confirmation hearings for Supreme Court nominee Clarence Thomas in 1991, contributions jumped fourfold in 1992. The PAC distributed more than $6 million to 55 House and Senate candidates, and it became the single largest donor to congressional campaigns. Twenty-five candidates supported by EMILY's List won, an amazing record given that virtually none were incumbents.

In 1994, a strong Republican year, only 9 of the 40 candidates supported by EMILY's List won. But the group raised even more money, approximately $8.2 million. Since its inception, it has enabled some 24,000 people (mostly women) to provide crucial support to women candidates for Congress.

contribute, dominate the financing of campaigns. Political parties, ideologically oriented PACs, and labor unions find it hard to pool small contributions; when they succeed, they face limits on how much they can contribute to any one candidate. Occasionally a participant devises a strategy that successfully overcomes such limits, as "Resources: How EMILY's List Gets Money to Women Candidates" demonstrates.

The role of money in elections can easily be exaggerated. Many other factors—the quality of candidates, incumbency, and economic conditions—make it impossible for any candidate to "buy" any election. Nevertheless, challengers cannot beat incumbents unless they raise and spend significant sums. Money provides access to a range of other resources—television time, direct mail, polling, sophisticated and knowledgeable campaign consultants—that are essential in campaigns for president and Congress.

THE DYNAMICS OF ELECTION CAMPAIGNS

Campaign managers and consultants guide candidates for Congress and the presidency in using their resources under existing rules. The media, party officials, and ultimately voters respond to these strategies; the resulting play of power produces election victories and defeats that have important consequences for politics.

In every recent election, the nomination process has placed a premium on a strong showing in the early primaries and the ability to raise huge sums of money. As a result, both the Democratic and Republican nominations have been virtually wrapped up early in the process.

Several elements in the strategic environment of the nominating process account for this pattern. These elements include the necessity of winning delegates in primaries; the front-loading of many primaries in a frantic six-week period from mid-February to the end of March; the need to begin fundraising and campaigning up to two years before the election; the front-loading of spending in the early primaries; and the crucial role of the news media as intermediaries between candidates and primary voters.

Deciding to run for president   Every election presents a unique "opportunity structure" that shapes presidential aspirants' calculations about whether to run. Most crucial is whether a first-term incumbent president sits in the White House. If that is the case, few if any of the strongest potential candidates from the president's party will run. No incumbent president in this century has been denied renomination, although some (Gerald Ford in 1976 and Jimmy Carter in 1980) have had their re-election prospects dimmed by strong challenges to their renomination.

Most candidates who seek the "out" party's nomination to challenge a sitting president are well-known politicians. Frequently drawn from the ranks of governors and senators, they assess the incumbent president's re-election prospects, the strength of their rivals, and their ability to raise enough money to wage a competitive race.

The opportunity structure attracts more candidates when the incumbent does not seek re-election. In 1988, five Democratic and four Republican "insiders" sought the nomination. In addition, two "outsiders" ran: the Reverend Jesse L. Jackson (a Democrat), and the Reverend Marian G. "Pat" Robertson (a Republican). Both drew heavily on religious support and built on their name recognition and ability to campaign full-time.[29] Although outsiders like Jackson and Robertson are unlikely to win, by running they can influence their party's platform and ideology.

Raising campaign money   Campaigns for the presidential nomination cost a lot of money. In 1992 Bill Clinton raised more than $25 million and George Bush more than $27 million.[30] Both used these contributions to qualify for and receive matching public funds, more than $12.5 million for Clinton and $10.5 million for Bush.[31] Almost all their money came from these two sources. Unlike congressional contenders, aspirants to the presidency receive almost no money from PACs.[32]

Together, the eight Democrats and three Republicans in 1992 received more than $85 million from individuals and another $40 million in matching funds. Thus, matching public funds accounted for just over one-third of all receipts, individual contributions for all but about 1 percent of the rest.

Presidential nomination campaign strategy   In June 1994, almost two and a half years before the 1996 presidential election, the front page of the *New York Times* carried an article headlined "Dole Takes First Real Steps toward

'96 Presidential Race."[33] The article quoted Senator Robert Dole's reasons for asking his advisers to solicit pledges of support so early:

> Whether or not [you] do it, you've got to be prepared. If you're going to get into this thing, you ought not to wait until after next year. If you wait until you decide to do it, you may be behind the curve.

The article reported that Dole's precandidacy PAC, Campaign America, had already sent field workers to Iowa and New Hampshire and had paid for his trips to speak on behalf of Republican Party candidates. His advisers projected a need for $15 to $20 million. The difficulty that several other potential Republican candidates (including former vice-president Dan Quayle) anticipated in raising this sum contributed to their decision to stay out of the 1996 race.

Dole's early move dramatically illustrates how the "front-loaded" nature of the nomination process requires serious insider candidates to make early decisions about running. An early start prevents other candidates from gaining an advantage through better organization and, if the candidate seems strong, deters possible rivals from entering. Candidates who wait too long don't have time to build the support and raise the money needed to compete in the early primaries. If strong candidates stay out, aspirants who are less well known see their chances improve and enter the race. It is doubtful that most of the "six dwarves" (the media's term) among the 1992 Democratic contenders would have run if Al Gore, Bill Bradley, Richard Gephardt, Jay Rockefeller, or Mario Cuomo had entered early.

Senate Majority Leader Bob Dole formally announced his candidacy for the presidency on April 10, 1995. However, his efforts to capture the Republican nomination had begun in earnest more than a year earlier, reflecting the front-loading of the nomination process.

Candidates make strategic decisions about how much to campaign in what places, and where to spend their money. Iowa and New Hampshire are small enough for candidates to campaign personally there. But they must buy television ads for Super Tuesday, when many southern states schedule their primaries. Spending large amounts does not guarantee victory in any primary, but few candidates other than the president or vice-president are well-known enough to win without spending at least close to what others spend.

A crucial role in primaries is played by the news media, especially in their focus on the "horse-race" aspects of the contest and their part in winnowing candidates after the early primaries. Candidates also seek to exploit "unpaid coverage," structure expectations to demonstrate "momentum" (or "Big Mo," as George Bush called it in 1980), and define the issues in a way that favors their candidacy. Frontrunners seek big victories to deliver a "knockout" blow, while others seek an early upset.

THE CAMPAIGN FOR THE PRESIDENCY

For people who love politics, nothing tops the drama of a presidential campaign or the joy of Monday-morning quarterbacking after the election is over. The juicy details of each election are different, of course, so remembering them does not help much in understanding future elections. There are, however, two enduring questions you can ask to get a better understanding of presidential elections: How much difference does the campaign make? What are the key strategic choices the candidates make?

Do campaigns make a difference? When short-term forces such as an economic recession or an unpopular war strongly favor the "out" party, is its victory assured? Although there are no guarantees in politics, the answer is probably yes, provided the party does not nominate an extraordinarily weak candidate. Thus, the economic crash of 1929 and the ensuing depression virtually guaranteed Democratic candidate Franklin Roosevelt's victory in 1932. Likewise, if the economy is booming and the nation is at peace, no matter whom the "out" party picks, what issue stands it takes, or how skillfully it wages the campaign, its only chance of victory rests in massive (and unlikely) blunders by the other side. Dwight Eisenhower's re-election campaign in 1956, Lyndon Johnson's race in 1964, and Ronald Reagan's re-election bid in 1984 each presented the "out" party with a nearly impossible task. Reagan's former acting chair of the Council of Economic Advisers observed wryly, "I think Jesus Christ would have had real trouble [defeating Reagan] in '84."[34]

The closer the balance of short-term forces, the more room there is for the candidates' character and the nature of the campaign to make a difference. However, there is a crucial difference between an election with an uncertain outcome and the degree of uncertainty in many voters' minds. Economic conditions and international relations, viewed through the lens of party identification, cause the vast majority of voters to decide how they will vote long before the election. In every election since 1952, at least half of the electorate made up its mind by August and only one person in six remained undecided in the last weeks of the campaign.[35] Thus, in every election, many voters have made up their minds before the campaigns swing into full gear.

Nevertheless, elections are won or lost after the votes of those who decide late are added to the votes of those who knew all along how they would vote. The real battle in close elections is over the undecideds, or **swing voters**. As the number of party identifiers has eroded and disaffection with politics has grown, so too has the number of swing voters. Close elections are decided not just on *how* those who go to the polls vote, but on *who* goes to the polls at all. The success of the parties' campaigns to mobilize their supporters, to convince the undecided swing voters, and to convert disaffected members of the other party can change the outcome of close elections.

What strategic choices are made by presidential candidates? Candidates must choose between two general strategies.[36] One strategy calls for consolidating core supporters, then building on this base. George Bush's campaign adopted this approach in 1992, swinging to the right (to meet Pat Buchanan's primary challenge) during the primary campaign, in the party platform, and in the party's nominating convention. The plan was to then capture undecided voters in the center. The other strategy targets swing voters from the start. Bill Clinton adopted this tack by distancing himself from core supporters among African Americans, liberals, and labor union leaders.

The choice of a vice-presidential candidate provides an early clue to strategy. In 1992 George Bush rejected suggestions that he dump his vice-president, Dan Quayle, who had solid backing among Republican conservatives but whose appeal to moderates was minuscule. Bill Clinton chose Al

When Pat Buchanan, a former Reagan speechwriter, conservative columnist, and TV commentator, opposed incumbent George Bush for the Republican presidential nomination in 1992, the president's campaign moved to the right in hopes of blunting Buchanan's challenge. Although Buchanan attracted considerable media attention and ran well in New Hampshire, he faded in later primaries. However, his savage attacks on Bush for raising taxes did little to help Bush's re-election prospects.

Gore, a moderate Southerner and Vietnam veteran whose stands on defense and "family values" would appeal to moderate white middle-class voters who had supported Reagan and Bush in the 1980s.

Election rules cause decisions on *where* to campaign and spend money to assume central importance. Campaigns avoid pouring resources into states where victory is assured or highly unlikely. The local television and newspaper coverage generated by a candidate's personal appearance boosts his or her standing in that region.[37] Clinton's campaign efforts focused on 32 states (he won 31 of them), although he did go to Florida and Texas as well to force Bush to defend his lead there.[38] When the Bush campaign determined in September 1992 that it would lose New York and California, it abandoned campaigning in those states, which allowed Clinton to cut back there as well.[39]

Campaigns also search for a unifying theme that will reinforce the advantages their candidates gain from the short-term partisan forces at work. For example, in 1992 Ross Perot tapped disillusionment with politics and politicians by emphasizing his "outsider" status, and he articulated voters' concerns about economic decline and the budget deficit. The Bush campaign focused on the president's experience in foreign affairs and attacked his rivals' "character." Sensing that they were not cutting Clinton's lead, Bush's campaign officials concluded, "We're not going to start moving until we rip the skin off the guy."[40] A series of negative attack ads followed, unmatched in frequency by the Clinton and Perot campaigns. Clinton's theme was "change" in areas such as health care, welfare, and the economy. His campaign consistently emphasized these themes, symbolized by a sign in the Little Rock headquarters that read: "Change vs. more of the same. The economy, stupid. Don't forget health care."[41]

When communicating directly with the party faithful, both parties invoke the party symbol to reinforce party ties. However, the two parties usually pursue different strategies in their general appeals. The Democrats, with more party identifiers, usually stress party ties. Until the advent of Ronald Reagan, Republican presidential candidates downplayed party. In 1992 Bill Clinton sought to preempt attacks labeling him a "tax-and-spend liberal" by claiming that he was a "new Democrat."

Finally, campaigns devise strategies to affect turnout. Both parties usually conduct voter registration drives, although with varying resources and results. In 1984, for instance, the Democrats launched an aggressive registration campaign, only to see the Republicans mount one of their own that—thanks to better organization and more money—registered even more voters.[42] Both parties also conduct "get-out-the-vote" drives, deploying volunteer and paid election-day workers, organizing phone banks to remind likely supporters to vote, and providing child care and rides to the polls.

Running campaigns Presidential candidates create their own organization to run their campaigns rather than relying on existing party structures. Although the structure and operation of campaigns vary, all employ a campaign director to oversee the entire operation, a campaign manager to direct day-to-day activities, and division chiefs to manage activities such as speech writing, polling, advertising, fundraising, scheduling, and making arrangements for the candidate's appearances and travel.[43]

Presidential campaigns utilize nearly every existing communications technology, both old (public opinion polling, mail, and telephones) and new

THE CLINTON CAMPAIGN'S USE OF NEW TECHNOLOGIES IN 1992

No campaign made use of new technologies as wholeheartedly or as skillfully as Bill Clinton's during his 1992 presidential bid. When the Bush campaign attacked Clinton, Clinton's campaign staff searched its computerized data banks for information to mount a counterattack and then faxed the media a rebuttal in time to make the news on the very day of the attack. The Clinton team also provided local news anchors with live interviews with their candidate through satellite feeds. The team's own cameras at campaign rallies sent live pictures to campaign headquarters in Little Rock; if they showed a small crowd, a sign blocking the audience's view of Clinton, or anything else detracting from the image, the campaign team called its staff on the scene by cellular phone and instructed them to correct the problem.

Satellites, fax machines, cellular phones, television, and computerized data transfer thus allowed Clinton's campaign headquarters staff in Little Rock to remain in constant contact with its field staff. Furthermore, its decisions about where to campaign, purchase television time, and try to increase voter turnout were all based on high-tech computer mapping and analysis programs. The Clinton campaign's success suggests that these techniques will be used even more intensively in future campaigns.

(cellular phones and satellites).[44] Computers are used to store, combine, sort, and apply information from voter registration lists, from names generated by the campaign itself or lists sold commercially, and from public opinion polling.[45] The information obtained from daily "tracking polls" guides the content of mail appeals, speeches, and television advertising. These new techniques have transformed political campaigns from labor-intensive operations run by party workers and volunteers to capital-intensive, high-tech enterprises relying on professional "hired guns." As "Strategies: The Clinton Campaign's Use of New Technologies in 1992" shows, these techniques reached new levels of refinement in 1992.

How the new political technology shapes campaigns   Voters rely primarily on television news and short campaign commercials for information about candidates. Although campaigns can control the content of their paid commercials, they have much less control over unpaid media coverage, especially the nightly news. During the 1980s the campaigns learned how to shape the nightly news; in 1992 they learned how to circumvent it.

In the 1980s the Republicans influenced nightly news broadcasts by changing Ronald Reagan's campaigning style. They restricted his activity to one important event each day, narrowed the message delivered, and carefully staged the event's setting and timing to make it visually attractive and symbolically powerful. Access to these events was restricted to cheering party loyalists. To ensure that the networks picked up the message of the day, campaign consultants concocted short, snappy, quotable phrases for their candidate to deliver. Known as *sound bites*, these phrases fit well into the 9 to 15 seconds the networks typically spent broadcasting the candidate's words directly.

Ross Perot shunned traditional campaign techniques when he reentered the 1992 presidential race. His half-hour campaign commercials, complete with simple charts that illustrated his points, attracted large television audiences.

Similar techniques, with new wrinkles, emerged in 1992. But this campaign also saw important innovations. For one thing, television news, stung by criticism of its superficial and candidate-manipulated coverage in 1988, did a somewhat better job of covering the issues, assessing the candidates' claims, and exercising its own judgment on what to air. More significantly, however, the candidates used new techniques to circumvent the nightly news. Ross Perot appeared frequently on television talk shows such as CNN's *Larry King Live.* Later in the campaign he purchased half-hour blocks of time to make "low-tech" presentations using simple charts, and he avoided giving live campaign speeches. Bill Clinton appeared in dark glasses playing his saxophone on *The Arsenio Hall Show* and fielded questions from citizens in "town meetings," including one aired by MTV. Finally, the candidates—ultimately including a reluctant George Bush—made themselves available to the network morning television shows. The extensive use of the free media provided by these outlets led one journalist to coin the phrase "the talk-show campaign of 1992."[46]

THE DYNAMICS OF CONGRESSIONAL CAMPAIGNS

For every president elected, more than 930 representatives and senators are selected. "Off-year" races differ from those that coincide with the election of a president. House races involve smaller districts and attract less money, media attention, and voter interest than do Senate races. These factors make it difficult to generalize about congressional elections. However, we can discuss several enduring and crucial general aspects of these elections.

Patterns of congressional campaign financing Significant changes have transformed the way in which candidates for Congress conduct and fund their campaigns.[47] During the first half of this century, congressional candidates required much less money than they do today, and it was raised within the district by local party organizations. As party organizations weakened, so did their role in running and funding congressional campaigns.

Beginning in the 1950s, campaigns for Congress became increasingly *candidate-centered*, run by ad hoc organizations put together by office-seekers themselves. Unable to reach voters through party precinct captains and workers, candidates turned to new and costly forms of mass communication, adopting the campaign technologies used by presidential campaigns. The rising cost of campaigning fueled the increase in the amounts House and Senate candidates had to collect.

Table 12.3 shows how much money was raised by incumbents, their challengers, and those contesting open seats in the 1994 general election. Clearly, incumbents collect more than challengers do. On average, for every dollar a House challenger received, the incumbent received $2.66. Senate incumbents raised $1.11 for every dollar their challengers obtained. This pattern appears to be firmly established and unlikely to change without campaign finance reform legislation.

Raising such enormous amounts imposes a heavy, unwelcome, and never-ending burden on incumbents. The 16 Senate incumbents who ran in 1994 raised an average of $5 million each. To do so, they needed to collect more than $16,000 a week for six years. The 232 House incumbents running in 1994 had to raise more than $5,800 a week to reach their average of $606,000.

TABLE 12.3
GENERAL ELECTION CONGRESSIONAL CANDIDATES' 1994 CAMPAIGN RECEIPTS

	HOUSE			SENATE		
	NO. OF CANDIDATES	**TOTAL RECEIVED (MILLIONS)**	**AVERAGE**	**NO. OF CANDIDATES**	**TOTAL RECEIVED (MILLIONS)**	**AVERAGE**
Incumbents	383	$220.42	$575,509	26	$113.33	$4,358,846
Challengers	347	75.08	216,369	26	102.21	3,931,154
Open seat	94	59.25	630,319	18	54.65	3,036,111
Total	824	$354.75	$430,052	70	$270.19	$3,859,857

Figures do not include $1.52 million and $2.74 million raised by third-party and independent candidates in the House and Senate respectively.

Source: Federal Election Commission, "1994 Congressional Fundraising Climbs to New High," press release, April 28, 1995, pp. 18–19.

Even though a senator's or representative's geographic base, or **voting constituency** (i.e., the individuals and groups who provide votes), remains crucial, Table 12.3 confirms that harvesting votes increasingly requires harvesting dollars. As a result, winning candidates must respond to two constituencies: their voting constituency and their **funding constituency** (i.e., the contributors who provide the money essential to conduct the campaigns).

The funding constituencies of incumbents vary widely in size. Senators, elected statewide, need more campaign money than representatives do. A few, those who face weak opponents or enjoy widespread support, do not have to raise much money. For example, in the years leading up to his retirement in 1988, Senator William Proxmire (D-Wis.) raised virtually no money and spent only the postage necessary to return contributions and thank supporters. At the other extreme, Senator Alfonse D'Amato spent more than $11.5 million to win re-election in New York in 1992; and Republican challenger Michael Huffington spent a record $29 million, much of it his own money, in a losing effort in California to defeat incumbent Democratic senator Diane Feinstein in 1994.

Individuals account for about half of incumbents' money and are the largest component of their funding constituency. Although we do not know what proportion of individual contributors reside in the candidates' districts, it is likely that increasing proportions come from out of state, especially in Senate campaigns.[48]

Political parties, on the other hand, account for only a small proportion of incumbents' funds. In 1986 less than 1 percent of incumbents' funds came from parties, whereas more than 55 percent came from individuals; the proportions have changed little since.[49] These figures underestimate the role of parties, because registration drives and other activities funded with soft money to assist the presidential campaign also benefit candidates for Congress. Nevertheless, party funds provide only a tiny fraction of incumbents' financial support. The rest comes from PACs. Total PAC contributions to congressional candidates have grown rapidly both in current dollars and after adjustment for inflation (see Figure 12.2). PACs' share of total funds contributed to congressional campaigns has also risen. In 1978, 17 percent of this total came from PACs; in 1994, 24 percent did.[50] Most PAC money goes to incumbents, sometimes when they face only token or no opposition. In 1994, for example, almost three-fourths of all PAC contributions to congressional candidates went to incumbents. For every dollar going to challengers, incumbents received $7.31.[51] Since incumbents usually win, the result is that PACs form a large part of the funding constituency of every Congress. For example, House members elected in 1994 received almost 40 percent of their money from PACs; for Senate winners, the proportion was more than 21 percent.[52] PACs also give disproportionately to congressional leaders in both parties as well as to members who serve on committees that address matters of concern to the PAC.

Stability and change in House election outcomes   The 1994 congressional elections produced a stunning victory for Republicans. For the first time since 1954, the Republican Party won a majority of House seats. Why did Democrats enjoy a virtual lock on the House for four decades? What finally caused it to end?

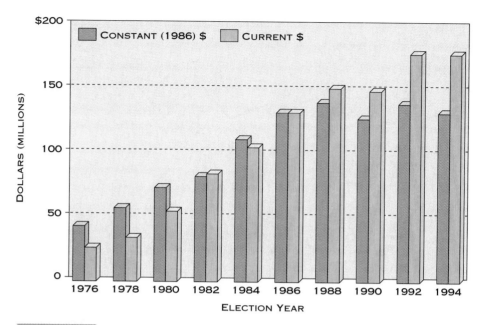

FIGURE 12.2

PAC CONTRIBUTIONS TO CONGRESSIONAL CANDIDATES, 1976–1994 (IN CURRENT DOLLARS AND CONSTANT 1986 DOLLARS) PAC contributions to House and Senate candidates grew at a steady and rapid rate between 1976 and 1988, both in current dollars and in constant dollars (i.e., dollars adjusted for inflation). In 1990 the increase halted; it resumed in 1992, then *declined* slightly in 1994 in terms of constant dollars. *Sources:* For 1978–94, Federal Election Commission, "PAC Activity in 1994 Elections Remains at 1992 Levels," press release, March 31, 1995, p. 2; for 1976, Harold W. Stanley and Richard G. Niemi, *Vital Statistics on American Politics,* 4th ed. (Washington, D.C.: Congressional Quarterly Press, 1994), p. 178.

During most of this period, a high proportion of incumbents sought re-election and won (see Figure 12.3 on p. 464). Until 1992, only once had more than 10 incumbents lost in a primary. Further, most primaries and many general elections were not competitive. Using the traditional measure of a "safe" seat (i.e., winning by at least a 55 percent margin), 91 percent of all House races were safe in 1986 and 71 percent were "supersafe" (i.e., having a winning margin of at least 65 percent). Only 3 percent of House races in 1986 were won by less than 52 percent. Congressional races have become tighter since. The percentage of Democrats elected in "marginal districts" (those that garner less than 60 percent of the vote) increased steadily: In 1988, it was 13 percent; in 1990, 23 percent; in 1992, 32 percent; and in 1994, 42 percent.[53] The proportion of Republicans elected by less than 60 percent similarly rose.

Because Democrats went into every election between 1956 and 1994 with a House majority, their chances of maintaining control were helped by the advantages of incumbency discussed in "Outcomes: Why House Incumbents Usually Win" (on p. 465). Thus, the factors that caused incumbents to seek re-election and win perpetuated the Democrats' advantage. Because turnover was low, hovering around 10 percent, it was hard for the

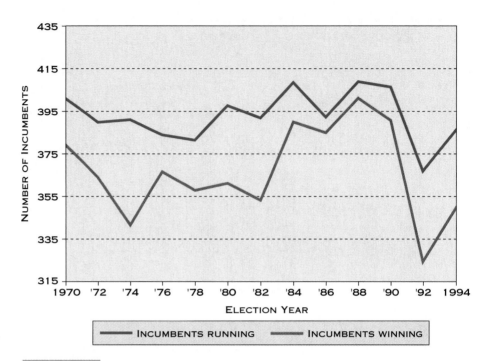

FIGURE 12.3

THE NUMBER OF HOUSE INCUMBENTS RUNNING AND WINNING, 1970–1994
Most incumbents seek re-election, and most win. Nevertheless, there was a
significant drop in 1992 in comparison with the four previous elections. The
rate at which incumbents ran and won in 1994 was also comparatively low. When
the partisan balance in the House is close, even small changes in these rates can
bring about momentous events, as the Republicans' capture of the House in 1994
demonstrated. *Source:* For 1970–1992, Harold W. Stanley and Richard G. Niemi,
Vital Statistics on American Politics, 4th ed. (Washington, D.C.: Congressional
Quarterly Press, 1994), pp. 206–7, Table 7.5; for 1994, author's calculations.

Republican Party to win many seats. The Democrats' advantage in party
identification helped them win more governorships and state legislative
seats, producing a larger pool of potentially strong candidates for Congress.

The 1992 election results foreshadowed trouble for the Democrats. A
widely publicized "anti-incumbent" fever, encouraged by the "outsider"
style of the campaigns of Pat Buchanan, Jerry Brown, and especially Ross
Perot, induced a number of House incumbents to retire. Redistricting and
a scandal involving members' misuse of the House bank helped lead to the
defeat of 19 incumbents in the primaries; another 24 lost in the general
election. In addition, the proportion of the two-party vote received by winners continued to decline: It was 68.4 percent in 1988, 64.5 percent in 1990
(lower than in any election since 1974), and 63.6 percent in 1992.[54] Only 325
incumbents returned after the 1992 elections, as compared to the previous
modern record low of 402 set in 1986.

End of an era: The Republicans capture the House   Chapter 11 identified
factors affecting midterm congressional election outcomes, including low
voter turnout, the loss of seats by the president's party, the state of the

WHY HOUSE INCUMBENTS USUALLY WIN

Even in 1992, when 43 incumbents lost their bid for re-election, their overall success rate exceeded 88 percent. In 1994, a bad year for Democrats, 191 of their 226 incumbents (84.5%) returned, and all 159 of the Republicans' incumbents won.

One reason why incumbents usually win is that very few fail to be renominated. Only once since 1970 have more than 12 lost a primary; few even face a primary challenge. In 1994 fully 68 percent of incumbents faced *no* primary opposition.[1] When incumbents do have an opponent, they usually win handily. In 1994 only 15 incumbents won by less than 60 percent, and just 4 (1% of those running) lost. Primary challengers cannot raise enough money to become known and can rarely count on the support of the party faithful, who are most likely to vote in primaries.

Incumbents also do not often face stiff opposition in the general election. In most years, potential strong challengers ("strategic politicians") forgo confronting an incumbent. Many incumbents represent districts with more adherents of their own party than of the opposition, making the challenger an immediate underdog. Thus, many challengers are weak and cannot raise enough money even to make themselves known.

Incumbents enjoy other advantages as well. They capitalize on numerous opportunities to make themselves known. They receive free media coverage when they announce federal grants and projects in their districts, tout their sponsorship of bills, explain their votes, or react to newsworthy political events. In addition, they have the **franking privilege**—that is, free use of the mails for "official business"—to send information about issues to people on specialized mailing lists. The general newsletter that goes to everyone in the district features the incumbent prominently and promotes his or her competence, integrity, and legislative prowess. Furthermore, incumbents appear regularly at meetings and dinners throughout the district, thereby expanding their network of acquaintances.

Incumbents also generate goodwill and a positive image through the *constituency service* performed by their paid staff in Washington and in offices scattered throughout the district. People remember when a member's office tracks down a lost Social Security check, expedites the release of a student loan, or mails a copy of a congressional report needed for a term paper.

Finally, incumbents generally run well-financed and effective campaigns. Serving in Congress familiarizes them with most public policy issues in a way that few challengers can match. Constant practice teaches them how to handle smoothly the tough questions that can twist a challenger's tongue into knots. And if, perchance, a strong challenger does appear, incumbents can raise even larger sums to wage an even more expensive campaign.

[1] *Congressional Quarterly Weekly Report,* October 8, 1994, p. 2908.

economy and presidential popularity, and the quality of candidates challenging incumbents and running for open seats. How can these factors be applied to explain the dramatic Republican victory in 1994?

In midterm congressional elections, voter turnout generally falls by approximately 15 percent from the previous presidential election. In comparison to the previous midterm election in 1990, turnout in 1994 increased slightly from 36.5 to 38.7 percent.[55] But it actually fell by 16.5 percent from the 1992 presidential election's relatively high turnout of 55.2 percent.

Another characteristic of midterm elections is the loss of seats by the incumbent president's party. In every midterm election from 1950 to 1990, the president's party lost from 4 to 48 seats, with an average loss of 23. This pattern held in 1994, but with a significant difference: The Democrats lost 52 seats, more than in any midterm election since 1946.

Political scientists have struggled to explain the existence and varying magnitude of the midterm loss of seats by the incumbent president's party.[56] The theory of **surge and decline** argues that in presidential election years, candidates from the president's party benefit from a surge of support from two sources: (1) independents who vote for both the winning presidential candidate and the congressional candidate from his party; and (2) partisans who identify with the losing presidential candidate's party but defect to vote for the winner *and* the congressional candidate from his party. Being less involved in politics, the independents do not bother to vote in the following midterm election. The partisans who defected return to their party-based vote.

A revision of the surge and decline hypothesis argues that partisans from the losing presidential candidate's party, sensing defeat or disliking their candidate, stay home, only to return to the polls at midterm. A related theory holds that these elections are referenda on the president's performance, and that low ratings translate into greater losses.

Whatever the reasons for the "normal" loss of seats, what explains the abnormally large drop in 1994 that gave the Republican Party its stunning victory? Consider the two national factors that *directly* affect voting for Congress at midterm: the state of the economy, and the president's popularity. In 1994 by most measures the economy was doing extremely well. Nonetheless, many people did not see it that way. Almost half, 44 percent, rated the economy as fairly bad or very bad.[57] More telling, Bill Clinton's popularity was especially low at election time: 48 percent disapproved of his job performance overall and 56 percent disapproved of his handling of the economy.[58] Among voters who disapproved, fully 81 percent voted for the Republican congressional candidate. The result was that, nationwide, Republican Party candidates captured 52.4 percent of votes for the House, as compared to 45.5 percent just two years earlier.

National factors *indirectly* affect outcomes by influencing decisions by strong candidates (strategic politicians) to run or not to run. Clinton's low standing in the polls encouraged an unusually strong group of Republican candidates to run in 1994. A Republican Party PAC called GOPAC, headed by Georgia congressman Newt Gingrich, also played a significant role in recruiting, funding, and training strong candidates.[59] At the same time, 20 Democratic incumbents retired, as compared to only 7 Republicans, thereby depriving the Democrats of the advantages of incumbency in more districts. There were 48 open-seat contests, more than in the previous three midterm elections. Republicans there benefited from favorable national factors without having to overcome the advantages enjoyed by incumbents. As a result, the Democrats lost 21 open seats, the Republicans just 4.

How much did the highly publicized Republican "Contract with America" affect votes for Congress? Probably not very much. According to a *New York Times*/CBS News poll, even a month after the election 73 percent of the public said they had neither heard nor read anything about the "Contract."

However, another factor, the Republicans' success in raising money, may have had more impact. Republican challengers spent 40 percent more on average in 1994 than in 1992, whereas Democratic challengers' spending hardly rose at all.[60]

One final factor with long-term significance affected the 1994 outcomes: Republicans continued to make significant gains in the South. Eighteen of the 52 seats gained by the Republican Party were in the South, giving it a majority of seats there for the first time. As Figure 12.4 suggests, there has been a regional *partisan realignment* in the South—that is, a steady change in the underlying distribution of party identification—that will continue to provide Republicans with a solid and lasting base of support. The new alignment breaks sharply along racial lines. In 1994 fully 65 percent of southern whites voted for Republican House candidates, but only 6 percent of blacks did so.[61]

Competition and stability in the Senate In all but one election between 1982 and 1994, incumbent senators seeking re-election won nearly as often as House incumbents. In 1994, for example, 24 of the 26 on the ballot (92%) won, a re-election rate exactly matching that of House incumbents.

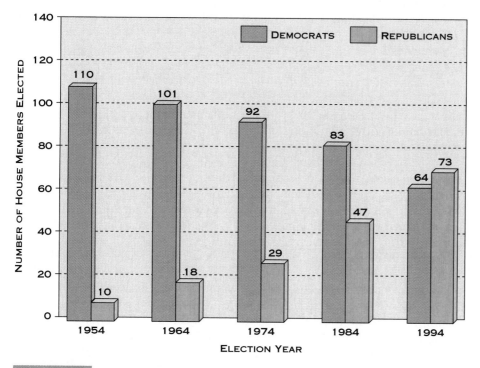

FIGURE 12.4

DEMOCRATIC AND REPUBLICAN HOUSE MEMBERS ELECTED FROM SOUTHERN STATES, 1955–1995 In 1994, for the first time, Republicans running in 13 southern states won a majority of House seats. In 1955 the South had just 10 Republican U.S. representatives. But their numbers nearly tripled in the next 20 years and continued to grow steadily thereafter, ultimately producing a 73–64 majority in 1995. *Source: Congressional Quarterly Weekly Report,* November 12, 1994, p. 3231.

However, more Senate than House elections are close. In 1994, 14 of the 24 incumbent winners (58%) received less than 60 percent of the vote. Thus, although Senate incumbents usually win, they often face stiff competition.

The Democrats' success in winning a majority of Senate seats has not quite matched their 40-year winning streak in the House. After resuming control in 1954, they held a majority in every year except the first six years of the Reagan administration. Usually, however, the size of their majority was smaller than in the House. In the 101st Congress, elected in 1988, Democrats held 55 seats as compared to the Republicans' 45; in the 102nd they held 56; and they began the 103rd with 57 seats. The Republican tide in 1994 allowed the Republicans to finally take control of the Senate. They gained 8 seats by defeating 2 incumbents and winning all 6 open seats. The switch of Senators Richard Shelby and Ben Nighthorse Campbell from Democrat to Republican increased the Republican Party's Senate membership to 54.

THE ROLE OF ELECTIONS IN AMERICAN POLITICS

Americans believe that elections have a central role in the play of power. How accurate is this view? Our answer relies on the distinction between *symbolic* and *material* outcomes.

THE SYMBOLIC EFFECT OF ELECTIONS

Presidential elections arouse strong emotions. The symbolic effects of all elections on both the mass public and elites have a significant impact on the play of power.

The impact of voting and elections on the mass public   As we observed in Chapter 11, the act of voting in a presidential election is rich in symbolism and ritual. Three psychological responses to this ritual shape the play of power. First, voters (and some nonvoters) gain a sense of participation in the governing of their lives. Second, voting legitimizes government and its actions. Participation in elections produces a "democratic coronation effect," evokes positive feelings toward government, and gives people a feeling of having a "say" in government. In 1968, for example, almost half of those who said they had little role in shaping governmental policies prior to the election became more positive about their impact after the election.[62] Third, voting regularizes and channels political activity in predictable, acceptable, and peaceful directions, displacing protest, violence, and other unconventional forms of participation. A common response to an unpopular policy is "Wait 'til the next election." Having voted, many people consider their political activity completed, freeing them to return to normal pursuits.

Symbolic effects on political elites   Political leaders respond to symbolism in politics just as strongly as ordinary citizens do. Elections allow leaders to reaffirm their commitment to the rules of the game, especially rituals for the transfer of office. The loser sends a congratulatory telegram or phones the winner, and gives a concession speech. In the speech, the loser pledges to continue to work for the noble principles he or she espoused but also acknowledges that "the people have spoken" and that the winner is now the legitimate leader. As with all effective rituals, when this one is violated by a loser's refusal to concede or by a mean-spirited concession speech, the news media and other politicians swiftly disapprove.

Victory affects behavior once the winner is in office. Winners who believe that the election gives them a mandate to pursue certain policies are emboldened to take vigorous action to implement those policies. Bill Clinton felt his election had made health care reform a major administration priority; likewise, the new Republican House majority elected in 1994 immediately set about enacting its "Contract with America," believing that victory gave it a mandate to do so.

ELECTIONS AND MATERIAL POLITICS

In Chapter 11, we noted that people vote both *retrospectively* (by making judgments about candidates' and their parties' past performance) and *prospectively* (by projecting what candidates will do in the future). Both behaviors affect material politics.

Elections as prospective judgments In theory, prospective voting provides the citizenry with a means to control policy. One well-known definition of democracy elevates this notion to a central position:

> Democracy is a competitive political system in which competing
> leaders and organizations define the alternatives of public policy
> in such a way that the public can participate in the decision-
> making process.[63]

Several conditions must be met for material policies to be controlled through prospective voting. First, candidates must differ consistently on important policies. Second, these differences must be communicated clearly. Third, voters' decisions must be based on their assessment of the two competing policy programs. Fourth, the election outcome must be decisive. Fifth, the winner must be capable of enacting the policies that led to victory.

Often, none of these conditions is met. Presidential candidates avoid giving specifics about many important issues. (For example, both Clinton and Bush avoided talking about the budget deficit in 1992.) Furthermore, the candidates' broad stands on many issues are similar. Even when clear differences emerge, they often display no consistent pattern across issues, thereby causing problems for issue-oriented voters. Whom do you support when candidate A adopts your position on some important issues and candidate B supports your position on others? Even if the candidates differed clearly and consistently, many people would vote on the basis of party affiliation and the candidates' personal qualities, paying little attention to issue stands.

Of course, many elections are close, hinging on the shift of a few votes in a few states; and millions of people do not vote at all. The citizenry, therefore, can hardly be said to have made a clear statement about the direction it wants material politics to take. Finally, the lack of centralization and cohesion in American political parties and the difficulty of uniting the congressional and presidential parties to enact a party program may thwart implementation of a policy direction indicated by a decisive victory.

There are other obstacles to control of policy through prospective voting. What happens when new issues arise after the election, or when conditions change so that what made sense at election time now looks like a road map to disaster? The conclusion is clear. For a number of reasons, elections fall far short of translating citizens' preferences into policy. Most elections simply do not provide a genuine "mandate" in the sense that "the public" chose one candidate over the other because it wanted a specific set of policies implemented.

This does not mean, however, that elections never provide policy direction through prospective judgments by voters. Two post–World War II elections involved candidates with very clear policy differences. The 1964 Republican candidate, Barry Goldwater, advocated a sharply conservative turn in policy, including making participation in the Social Security program voluntary. Many voters understood that his election would signify their desire for such a shift, so they voted for his opponent, Lyndon Johnson, in a landslide. The 1972 election saw a reversal in roles, with Democratic nominee George McGovern offering the prospect of much more liberal policies than those pursued by incumbent Richard Nixon. McGovern's message was understood, and he too suffered a massive defeat.

When such challenges fail, the electorate has made a prospective judgment to reject the losers' policies. However, the outcome says little about the policies that voters do favor, especially if the incumbent's positions are not clear. The electorate only indicates what it does *not* want.[64]

Elections as retrospective judgments Most politicians believe that, absent a state of war, "as goes economic performance, so goes the election." That is, voters retrospectively evaluate the performance of the economy, re-electing incumbents or their party's nominee when times are good, defeating them when they are not.

At first glance such retrospective judgments provide only the crudest control over future policy. The electorate in effect says, "Do something different," if times are bad (as they were in 1932) or "Keep on doing what you were doing," if times are good (as they were in 1984 and 1988). But such judgments shape future behavior because politicians know their performance will be evaluated at election time, in the same way that students are induced to study for an exam because they know their instructor will retrospectively evaluate their performance.

Anticipating that a poor economy will bring defeat, incumbent presidents usually pursue policies that improve the economy in the months before an election. Two principal tools are available to affect the economy: *fiscal policy,* which determines the level of spending and taxation; and *monetary policy,* which sets interest rates and the size of the money supply. One researcher argues that with the exception of the Eisenhower years, administrations have used these tools. Unemployment on election day was found to average one point below the rate 12 to 18 months before the election, and two points below its level at the equivalent time after the election.[65] Likewise, between 1946 and 1976 unemployment and inflation declined in half the presidential election years, but in only 2 of the 23 nonelection years.[66]

In 1964 the Republican presidential nominee, Barry Goldwater (*left*), held strongly conservative positions. In contrast, the 1972 Democratic candidate, George McGovern (*right*), was strongly liberal. Many voters understood each one's message and voted against them both. These two elections offer the best examples of prospective voting since World War II.

Politicians' beliefs and policies point to the same conclusion: An incumbent adminstration will pursue economic policies to try to guarantee favorable conditions at election time. Thus, the relationship between the economy and the electoral cycle, as summarized by one scholar, will continue to hold true: "Political life, then, is far more than an occasional random shock to a self-contained, isolated economic system; rather, economic life vibrates with the rhythms of politics."[67]

Some inescapable material effects of elections Elections determine who some of the key participants in the play of power will be, and this makes a difference. George Bush did not make the same decisions Michael Dukakis would have. Bill Clinton signed "motor voter" registration and family leave legislation vetoed by Bush. Although voters may not know how a challenger will behave if elected, changing who holds office introduces an inevitable sequence of transformations in the play of power. Changes in the presidency usually alter the agenda of politics. Interest groups, policy analysts, bureaucrats, congressional committee staffers, members of Congress, and other people with ideas for new programs regard a change in administration as an opportunity to move their issue up on the agenda. This opening of "policy windows" occurs regardless of whether or not the issue was mentioned in the campaign.[68] New presidents, even from the same political party, inevitably fill major policy-making posts with new people who bring their own preferences to office.

Elections also affect the importance of the voting and funding constituencies from which politicians draw support. When incumbents win, the identity of these constituencies remains stable. Newly elected officials rely on different coalitions of voters and contributors for support, thereby transfering access and influence to new participants.

These effects occur regardless of whether the election hinges on prospective or retrospective judgments. Elections introduce change into the dynamics of the play of power on a regular basis. Part of the complexity of politics, much of its fascination, and perhaps some of its adaptability to changing conditions derive from the unavoidable effects of elections.

SUMMARY

Elections lie at the heart of the play of power in American democracy. The major participants include not just voters but candidates, campaign contributors (especially wealthy individuals and PACs), political consultants and campaign staffers, and a variety of others such as newspaper and television reporters and radio talk show hosts.

Rules affect virtually every aspect of elections. The Electoral College, coupled with winner-take-all rules that allocate states' electoral votes, promotes the two-party system and enhances the importance of states with large populations. Both parties follow a complex set of rules for selecting the national convention delegates who choose presidential nominees. Increasingly, these rules "front-load" the process, forcing candidates to raise huge sums that must be spent in February and March of the election year. In addition, House members are elected by winning a plurality of votes in single-member (rather than multimember) districts, a procedure that encourages the two-party system. Population shifts necessitate the reapportionment of House seats among states every 10 years and, usually, the redrawing (redistricting) of district boundaries within states. Redistricting involves a play of power as efforts are made to gerrymander district lines to improve the electoral chances of party members, minority groups, or incumbents. Attempts to enact rules establishing term limits were blocked by the Supreme Court in 1995. Finally, the Federal Election Campaign Act provides public funds to presidential candidates and limits the size of individual, PAC, and party contributions to congressional candidates.

Incumbents enjoy clear advantages in acquiring the resources needed to win elections. In addition to the advantages of holding office (such as the franking privilege), which build support in their voting constituency, incumbents raise more campaign money from their funding constituency than challengers do, and they are more experienced campaigners.

The dynamics of presidential nominations in each party depend on (1) whether an incumbent seeks renomination, and (2) the perceived strength of the incumbent party going into the election. Ability to raise money early also affects who runs and how well they do. The front-loading of the nomination process results in an increasingly hectic nomination season, one in which media coverage plays a crucial role. General election candidates for president finance their race with public money from the Presidential Election Campaign Fund, although increasingly large amounts of "soft-money" contributions from private sources to state parties are used to further these campaigns.

Many presidential elections are over before they begin. The state of the economy largely determines whether the incumbent party will win or lose, although other short-term partisan forces play a role as well. Despite the media attention given to presidential campaigns, only in close elections do strategic choices about which issues to emphasize and how to campaign influence enough swing voters to make the difference between winning and losing.

Congressional campaigns with an incumbent running differ from open-seat contests. House incumbents raise more than their challengers by large margins, and they rarely lose. Senate incumbents are almost as successful. In midterm elections, voter turnout drops by approximately 15 percent and the president's party always loses seats. A variety of theories—for example, surge and decline—seek to explain this fact. Although both national and local factors affect outcomes, they play a larger role in open-seat races. A combination of such factors, including the unpopularity of President Clinton, produced the Democrats' stunning loss of control of the House in 1994 for the first time in 40 years.

Elections have both symbolic and material effects on the play of power. The act of voting gives citizens a sense of participation and helps legitimize government and its actions. Political elites accept electoral defeat and peacefully relinquish authority. Their beliefs about whether the election provided the winners with a "mandate" shapes their subsequent actions. Elections affect material politics primarily through the retrospective judgments made by voters about incumbents' performance. The extent to which such judgments indicate voters' preferences in policy is very limited. However, politicians anticipate that the electorate will judge their performance. Consequently, they pursue policies that will please voters. Although some voting anticipates what candidates might do if elected, the conditions for meaningful prospective choices between competing policy programs are rarely met. All elections, however, inject change into the play of power by replacing certain officeholders and stimulating a change in the agenda of politics.

APPLYING KNOWLEDGE
ASSESSING CLAIMS TO AN ELECTORAL "MANDATE"

Victorious politicians and other participants in politics recognize the symbolic potency of claims that there is a "mandate from the people" to implement certain policies. Here are four examples.

(1) In 1989, Congress was locked in a battle over whether to accept President George Bush's proposal to lower taxes on capital gains. A supporter of the proposal offered the following argument:

The matter should have been decided last year, when George Bush campaigned for the 15 percent rate, saying it would increase U.S. competitiveness, create jobs through new entrepreneurial activity and increase Government revenues. Michael Dukakis opposed the cut on fairness grounds. Since Mr. Bush won, he presumably had a mandate.[1]

(2) Several years later, President Bill Clinton argued that his election victory signaled a public mandate to reform health care. (3) Following the 1994 congressional election, Republican lawmakers repeatedly argued that their electoral victory constituted a mandate to enact their "Contract with America," a set of proposals that numerous Republican candidates pledged to work for if they were elected. (4) After the 1994 election, the Clinton administration's secretary of the interior indicated that he would henceforth be less vigorous in protecting endangered species on private lands, saying, "I understand the realities. I understand the mandate that the chairman and members of this committee have from the public."[2]

Choose one of these examples (or provide your own from a more recent election). Indicate the information you would want to have in order to assess the validity of a claim of a "mandate from the people." Draw on your knowledge of public opinion and voting behavior as well as material from this chapter in crafting your answer.

[1] Jude Wanniski, "To Aid the Poor, Cut Capital Gains Taxes," *New York Times*, July 25, 1989, p. A23.
[2] Katharine Q. Seelye, "In Attack on Gingrich, Democrats Use His Tactics and Reap Chaos," *New York Times*, January 19, 1995, p. A1.

Key Terms

soft money (*p. 443*)

winner-take-all rule/unit rule (*p. 445*)

front-loading (*p. 447*)

reapportionment (*p. 448*)

gerrymandering (*p. 448*)

term limits (*p. 449*)

swing voters (*p. 457*)

voting constituency (*p. 462*)

funding constituency (*p. 462*)

franking privilege (*p. 465*)

surge and decline (*p. 466*)

Recommended Readings

Carroll, Susan J. *Women as Candidates in American Politics.* Bloomington: Indiana University Press, 1985. A general treatment of the obstacles women face in winning political office.

Ginsberg, Benjamin. *The Consequences of Consent: Elections, Citizen Control and Popular Acquiescence.* Reading, Mass.: Addison-Wesley, 1982. A provocative, critical analysis of the weaknesses of the electoral process in enhancing citizen control.

Jacobson, Gary C. *The Politics of Congressional Elections*, 2nd ed. Boston: Little, Brown, 1987. A classic analysis of the dynamics of congressional elections.

Pomper, Gerald M., ed. *The Election of 1992: Reports and Interpretations.* Chatham, N.J.: Chatham House, 1993. A collection of essays on the many facets of the 1992 presidential election.

Sorauf, Frank J. *Money in American Elections.* Glenview, Ill.: Scott, Foresman/Little, Brown, 1988. A comprehensive survey of money in American elections through the mid-1980s.

Tufte, Edward. *Political Control of the Economy.* Princeton, N.J.: Princeton University Press, 1978. A classic study of how presidential policy seeks to shape the performance of the economy to enhance the president's re-election prospects.

Wayne, Stephen J. *The Road to the White House 1992: The Politics of Presidential Elections.* New York: St. Martin's Press, 1992. A comprehensive review of the politics of presidential nomination and election campaigns.

Congress

IN JANUARY 1995, after 40 years, Democrats gave up the gavel that symbolizes control of the U.S. House of Representatives. The image of House Democratic leader Richard Gephardt handing the gavel over to incoming Republican Speaker Newt Gingrich symbolizes the historic transfer of power that followed the Republicans' stunning victory in the 1994 midterm elections. But no single image can fully capture the high drama or significance of this transition as Republicans assumed control of both the House and the Senate, leaving a Democratic president to face a Republican-controlled Congress for the first time since 1946.

The very day Newt Gingrich grasped the Speaker's gavel, the new Republican majority adopted the most sweeping changes in House rules and structure since 1910. Changes were made in the House's committee structure, the distribution of power among its members, the procedural rules governing the flow of legislation and voting, and the content of many public policies that had been in place for decades. The effects of some of the internal reforms (e.g., the elimination of three standing House committees) were felt immediately; the impact of others (e.g., a new six-year limit on committee chairships) will not become clear for years.

In this chapter we will examine these and other significant changes in Congress. We will consider the full sweep of activities and dynamics in both houses of Congress, surveying our legislative body's structure and rules, resources, personnel, and legislative procedures, as well as the way in which it shapes the play of power in other areas of American political life.

As a result of the 1994 congressional elections, House Democratic leader Richard Gephardt handed the Speaker's gavel to Republican Newt Gingrich. This act symbolized the dramatic changes in Congress as the Republicans took control of both the Senate and the House of Representatives for the first time in more than 40 years.

In a survey, a majority of students in an introductory American politics class indicated that they accepted a central tenet of the American civil religion: the separation of powers. Two-thirds agreed that "by far, the overwhelming task Congress engages in is writing and passing legislation." Nearly as many (64 percent) believed that members of Congress faithfully represent their constituents' views because they will be defeated in the next election if they do not. And, by a margin of almost two to one (47 to 25 percent), they agreed that votes by members of Congress reflect rational consideration of issues and of the consequences of proposals. Nearly half thought that the real work of shaping legislation takes place on the House and Senate floor. Finally, more than 60 percent agreed that many members do the bidding of their campaign contributors.

These and other common beliefs about Congress are, in many cases, emotionally satisfying to those who hold them. But they are not entirely accurate, and they provide only a rudimentary framework for answering important but difficult questions. What exactly is the role of Congress in the play of power? How does Congress really work? How is Congress changing? In the following pages, we seek to provide a more elaborate and more complete understanding of this American political institution. ■

THE PEOPLE WHO WORK IN CONGRESS

Most of us think of Congress in terms of the 435 representatives and 100 senators who are elected to office. But Congress is a complex social institution with a budget of more than $2.3 billion and 23,000 employees.[1] To understand how Congress works, we need to look at these participants, too.

THE MEMBERS: WHO GETS ELECTED TO CONGRESS?

Until recently, the best answer to this question was "Incumbents!" Most incumbents (typically, white males) sought and won re-election. This fact contributes to the small number of women and minority-group representatives sitting in Congress. Denied the right to vote, unable to raise sufficient funds to undertake a competitive race, facing voters' reluctance to support them, and holding relatively few of the offices that normally serve as stepping-stones to Congress (e.g., district attorney, mayor, or state representative), these disadvantaged groups had few members in Congress for most of the twentieth century. Just 12 women, 12 African Americans, and 5 Hispanics sat in the 92nd Congress, elected in 1970.

The white males in that and succeeding Congresses have come disproportionately from the legal profession, business, and banking. Few come from the ranks of manual laborers, clerical personnel, and skilled craftspeople. Nearly all have college degrees. Virtually all are wealthier than average; in one recent year, approximately one in ten people elected to Congress was a millionaire.[2] Among Protestants, higher-status denominations are overrepresented. Both Roman Catholics and Jews are slightly overrepresented as well, as compared to their proportion in the general population.

However, in 1992 the number of incumbents seeking re-election dropped dramatically. This gave women and members of disadvantaged groups a better shot at winning. In what came to be known as "The Year of the Woman," women ran for congressional office in record numbers. As a result, after the election, the new House contained 48 women, up from just 29 the year before. The new Senate included six women, who were joined by a seventh after a special Senate election in Texas. African Americans and Hispanics, many running in newly configured districts with a majority of voters from their groups, also increased their numbers in Congress. The House now included 38 African Americans and 17 Hispanics. The new crop of senators included an African-American woman and a Native American man. Although there were no further such increases in the House in 1994, the increased presence of women and minorities in Congress has clearly benefited informal caucuses such as the Congressional Caucus for Women's Issues and the Congressional Hispanic Caucus, each organized to provide support services to its members. However, these caucuses have had trouble exercising influence, as is shown in "Participants: The Congressional Black Caucus" (on p. 480).

The congressional elections of 1994 were a sharp contrast to 1992, "The Year of the Woman." Although increased numbers of women ran overall, the increase did not match the surge of 1992.[3] The Senate gained one woman member, raising the number of women senators to eight. Five first-term House Democratic women lost, but overall the number of women in the House remained unchanged from 1992. Black and Hispanic representatives first elected from "majority-minority" districts (i.e., districts where minority-group members constituted a majority) created in 1992 survived the Republican tide. Although the House and especially the Senate still contain proportionately fewer minority-group members than does the population, their numbers have risen significantly since 1970, as Figure 13.1 (on p. 481) shows.

FIRST THINGS FIRST: GETTING RE-ELECTED

Almost everything members of Congress do directly or indirectly affects their re-election prospects. Thus, they constantly assess and adjust their behavior with an eye to its impact on their ability to win another term. The *influence through anticipated reactions* that voting and funding constituencies have over the men and women in Congress flows predominantly from their concern with re-election.

However, certain activities bear especially directly on election prospects. For example, constituency service, also known as case work, helps earn incumbents the support of voters and usually enables them to win re-election. Most members also travel frequently to their states or districts to appear at public meetings, speak before organized interest groups, or meet privately with key supporters, elected officials, and party activists. They send out newsletters and press releases, grant interviews, appear on local radio and television, and strive to be mentioned in newspaper articles. In Washington, D.C., they meet with individual constituents and groups who

THE CONGRESSIONAL BLACK CAUCUS

The mere presence of women and minorities in Congress does not guarantee their influence. Increasing numbers and votes help, but group cohesion, leadership skill, and the ability to form coalitions also affect power. The Congressional Black Caucus offers a case in point.

When it was formed in 1970, the Congressional Black Caucus had just seven members. By the 1980s the number had grown to 20, enough to begin to influence U.S. policy on civil rights and on imposing sanctions on the white apartheid regime in South Africa. After the 1992 congressional elections, the caucus's membership soared from 26 to 40 members, 37 of whom were House Democrats.

A solid bloc of 37 Democratic votes had the potential to shape legislative outcomes. But these African-American Democrats were a diverse group. Although a majority represented poor urban districts, 13 came from largely rural ones. They included women as well as men; some were strong liberals, others were more moderate. Some were powerful veteran lawmakers, including the chair of an important Ways and Means subcommittee (Charles Rangel, D-N.Y.); the chairs of the Government Operations Committee (John Conyers Jr., D-Mich.) and the Armed Services Committee (Ron Dellums, D-Calif.), and a chief deputy majority whip, John R. Lewis (D-Ga.). The aggressive leader of the Congressional Black Caucus, Kweise Mfume (D-Md.), displayed the energy and skill needed to achieve the cohesion necessary to exert influence. He also sought to form coalitions with other groups, most notably the Congressional Hispanic Caucus (which grew to 19 members in 1993).

The Black Caucus made its presence felt on a variety of issues in 1993 and 1994. Its successes included reducing the size of cuts in the Medicare program and winning approval of funding for two programs: immunizations for poor children, and summer jobs for urban youth. It also suffered some defeats, including (1) withdrawal of the Clinton administration's initial nominee to head the Department of Justice's Civil Rights Division, and (2) approval of the North American Free Trade Agreement (which the Black Caucus opposed). Moreover, the Black Caucus fought vigorously but unsuccessfully for a provision in the 1994 crime bill that would have permitted the use of statistical data on racial discrimination in appeals to death sentences.

The Black Caucus's ability to exert influence in the 103rd Congress was limited by the fact that it often had to support Democratic bills it did not like because the alternative was worse. In 1993, for example, members of the Black Caucus voted with their party 97 percent of the time on partisan votes.

The Democrats' recent loss of the House has considerably diminished the caucus's influence in the new Congress. The Republican majority has cut funding for all 28 special caucuses, including the Black Caucus. The group continues to exist, but without funds for salaries and operations.

come as tourists or to lobby for legislation. Finally, incumbents devote a large and increasing amount of time and effort to fundraising, not only in the midst of a campaign but throughout the year as well.

Although relations between representatives and their funding constituencies have grown increasingly important, voting constituencies remain critical. Richard Fenno's classic book *Home Style* asserts that representatives view their constituencies as a nest of concentric circles, from the *geographic* constituency to the *re-election* constituency (those who vote for them), the *primary* constituency (long-term, solid supporters), and the *personal* one

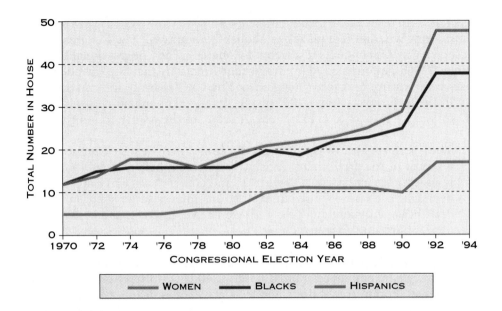

FIGURE 13.1

WOMEN, BLACKS, AND HISPANICS IN THE HOUSE OF REPRESENTATIVES, 1970–1994 In 1970 the number of women, blacks, and Hispanics elected to the House was very low. Although their numbers increased slowly during the next 20 years, only in 1992 did all three groups make significant gains. The 1994 election produced no changes, and the prospects for further increases for blacks and Hispanics were dimmed by Supreme Court decisions in 1995 that made it harder to draw districts with black or Hispanic majorities. *Sources:* For 1970–92: Harold W. Stanley and Richard G. Niemi, *Vital Statistics on American Politics,* 4th ed. (Washington, D.C.: Congressional Quarterly Press, 1994), p. 203; for 1994: "Minorities in Congress," *Congressional Quarterly Weekly Report,* November 12, 1994, p. 10.

(family and supportive friends).[4] Dealing with these constituencies requires many decisions: how to allocate time and other resources, how to present oneself to constituents, and how to portray one's activities in Washington.

Fenno's study found that most members seek above all to build a sense of trust, and that they often stress how they are different from the rest of Congress. This helps to explain why citizens think more highly of their own member of Congress than of the body as a whole, and why incumbents almost always "run *for* Congress by running *against* Congress."[5]

THE CONGRESSIONAL STAFF

You may not know that Congress has its own 1,200-member police force and its Architect's Office with more than 2,000 employees. However, most of the 23,000 people employed in Congress work in one of four support agencies or serve on a committee or a member's staff.

Much of the work in the four support agencies consists of gathering and analyzing information used by Congress. The *General Accounting Office* conducts field investigations and audits of executive agencies, and issues reports on its findings to Congress. The *Office of Technology Assessment* specializes in research on technical issues. (In 1995 House Republicans began considering whether to cut back or abolish both offices.) The *Congressional Research Service* conducts research for Congress. The **Congressional Budget Office (CBO)** performs so many important functions that we have devoted a separate feature to it (see "Resources: The CBO's Use of Information").

In 1992, approximately 17,000 people served on members' personal staffs or worked for leaders or committees. A study of their work found that

> They perform almost exclusively the constituent-service function; do most of the preliminary legislative research; help generate policy ideas; set up hearings, meetings, and conferences; carry out oversight activities—program evaluations, investigations, etc.; draft bills; and meet and talk with executive, interest, and constituent groups on substantive matters.[6]

In 1995, House Republicans cut committee staff by one-third, or 1,800 positions; the Senate employed approximately 1,100 committee staffers. Unlike their bosses, committee staffers and legislative aides devote their full time to narrow policy areas. They know a great deal about the substantive issues, the personalities of key congressional players, the principal lobbyists, and the way in which details of legislation emerge.

Staffers use their information and expertise to influence legislation. Of course, elected members still make major policy decisions and help work out details. But their legislative assistants, along with committee staff members, provide vital information and analysis and do the painstaking work required to transform members' decisions into legislative language and combine provisions of overlapping bills into a single package.

Congressional staffers can also translate their knowledge into financially rewarding career changes. They frequently move through the "revolving door" between government and private business into lucrative jobs as lobbyists or independent consultants. For example, the most far-reaching revision of the federal tax code in decades was passed in 1986; three years after its passage, half of the top personal staffers of members of the tax-writing committees and many committee staffers had left to lobby for major corporations seeking changes in the law.[7]

In recent years, women have come to occupy a number of key staff positions in Congress. A study of the four highest-ranking positions in House members' offices found women in 42 percent of them; in the Senate, they occupied 34 percent.[8] In 1993, 7 of the 21 people heading House standing committee staffs were women; however, when the Republican Party took control, women occupied just 2 of the remaining 18 top staff positions on standing committees.[9] Minorities have fared less well, holding only approximately 15 percent of all staff positions in Congress; few are high-ranking ones.[10]

THE STRUCTURE AND RULES OF CONGRESS

The process whereby legislation is produced in Congress reflects the combined effects of its division into a House and Senate, the role played by committees and subcommittees, the actions of major-party leaders, and

THE CBO's USE OF INFORMATION

Information fuels the play of power in Congress. Organized interests provide information in order to influence both members and their staffs. Members of the federal bureaucracy also use information to lobby Congress to enact legislation they draft and support. Finally, the support agencies and staff personnel in Congress find information crucial to doing their job.

A key information resource is the Congressional Budget Office (CBO), which was created in 1974. With its relatively small staff of 220, it provides independent estimates of economic growth, federal revenues and expenditures, and projected deficits. It also projects the costs of proposed federal programs, including the possible savings or additional expense of proposed amendments. Without this information, Congress would have to look to the executive branch—especially the Office of Management and Budget (OMB) in the Office of the President—or to the relatively small staffs of Republicans or Democrats. The CBO's influence depends on both the quality of its work (its economic projections have generally proven closer to the mark than the OMB's) and its ability to maintain the trust of both Republicans and Democrats.

The CBO and the role of its regular staff members go unnoticed and unmentioned in the news media's routine coverage of Congress. However, during the debate over health care reform in 1994, some glimpses of their contributions emerged. After the Clinton administration drafted its initial plan for health care

reform, the CBO conducted an analysis of the proposed plan's economic impact. It found (1) that the plan would increase the federal deficit by more than the president had projected, and (2) that it would cut the country's health spending by the year 2000.[1] Noted political columnist David Broder described the CBO's role in helping to draft health reform bills in committee as "a case study in the power and influence of one of the most anonymous parts of Washington."[2] He described how the Ways and Means Committee worked to have every new benefit matched by equivalent cuts in other parts of the program:

> As each change was proposed, committee staff member David Abernethy would get on the phone to someone in the Congressional Budget Office and—within minutes—put a price tag on the provision.

Along with other support agencies, such as the Government Accounting Office and the Office of Technology Assessment, the CBO provides Congress with a source of information that is not tied to other branches of government or to specific political interests. This type of information is crucial in the play of power.

[1] Viveca Novak, "By the Numbers," *National Journal*, February 12, 1994, p. 348.
[2] David S. Broder, "Drafting New Health-Care Legislation: A Study in the Power of Congressional Staffs," *Philadelphia Inquirer*, July 13, 1994, p. A10.

the elaborate formal and informal rules that guide each chamber's work. Understanding Congress requires knowing how each factor contributes to the play of power.

THE FRAMERS' LEGACY: TWO HOUSES

The Framers who sought to complicate the lawmaking process succeeded admirably by establishing two chambers in Congress. The Great Compromise (described in Chapter 3) was intended to provide both popular representation (the House of Representatives) and a body that reflected the broader interests of the states (the Senate). In keeping with their understanding of these roles, the Framers gave the House the responsibility of initiating all revenue bills; the Senate received the sole power to confirm appointments and ratify treaties. At the same time, in order to become a law, a measure must survive a dual obstacle course, and both House and Senate must ultimately agree to approve it in identical language.

This dual obstacle course has notable consequences. For example, it takes a long time to pass laws, few laws actually receive final approval, and the content of those that do often reflects compromises between divergent House and Senate versions.

STANDING COMMITTEES: THE WORKHORSES OF CONGRESS

Ask any college instructor how much of his or her time is devoured by committee meetings; or observe the operation of many major corporations, student governments, and dorm floors. Committees are everywhere. In Congress the permanent committees in each chamber, called **standing committees**, dominate the legislative process. Each addresses a specific set of policy issues and has a permanent budget, office and meeting space, and staff. In fact, the real work of crafting legislation takes place in standing committees rather than on the floor of each chamber.

Standing committees, which are not mentioned in the Constitution, played a small role until after the Civil War. Then they began to grow considerably in number and influence. Today Congress relies heavily on committees and subcommittees. After all, when faced with complex tasks, human organizations benefit from specialization and a division of labor. Dividing into small groups makes it possible to accomplish complex tasks that larger groups cannot. Furthermore, members of Congress like to serve on committees because doing so increases their influence over policies of special interest to them.

The present standing committee structure, depicted in Table 13.1, parallels the organization of the executive branch; nearly every major federal department and agency has at least one corresponding committee. However, committee responsibilities overlap, so that several committees share jurisdiction over the same issue. In 1994, for example, three House committees worked simultaneously on conflicting versions of President Bill Clinton's health care reform legislation.

Research in the 1960s and early 1970s classified House committees according to differences in the goals sought by their members.[11] The **power committees**—Ways and Means, Appropriations, and Rules—were said to deal with matters of vital concern to virtually all members, enhancing the personal power wielded by those who sat on them. Members interested in policy were found likely to seek seats on **policy committees** such as Banking, International Relations, and Education and Labor (now Economic Opportunity). Concern about re-election directed some members toward **constituency committees** such as Agriculture, Natural Resources (now Public Lands and Resources), or Armed Services (now National Security).

After the huge budget deficits of the early 1980s drastically reduced the number of new programs that could be established by committees, the

TABLE 13.1

STANDING COMMITTEES IN THE UNITED STATES CONGRESS

HOUSE	SENATE
Agriculture	Agriculture, Nutrition, and Forestry
Appropriations	Appropriations
National Security (name changed from Armed Services)	Armed Services
Banking and Financial Services (name changed from Banking, Finance, and Urban Affairs)	Banking, Housing, and Urban Affairs
Budget	Budget
Economic Opportunity (name changed from Education and Labor)	Labor and Human Resources
Commerce (name changed from Energy and Commerce)	Commerce, Science, and Transportation
International Relations (name changed from Foreign Affairs)	Foreign Relations
Government Reform and Oversight (name changed from Government Operations)	Governmental Affairs
House Oversight (name changed from House Administration)	Rules and Administration
Judiciary	Judiciary
Public Lands and Resources (name changed from Natural Resources)	Energy and Natural Resources
Transportation and Infrastructure (name changed from Public Works and Transportation)	Environment and Public Works
Rules	
Technology and Competitiveness (name changed from Science, Space, and Technology)	
Small Business	Small Business
Standards of Official Conduct	
Veterans' Affairs	Veterans' Affairs
Ways and Means	Finance

The following House committees were abolished in 1995: District of Columbia, Merchant Marine and Fisheries, Post Office and Civil Service.

importance of those dealing directly with taxation and the budget increased. Accordingly, the "new committee oligarchy" in Congress includes the House and Senate Budget, Commerce, and Appropriations Committees, as well as the Senate Finance and House Ways and Means Committees.[12] Also prized are assignments to any committee or subcommittee handling matters of concern to organized interests that make campaign contributions. For example, in 1989 the House subcommittee revising regulations on the savings and loan industry (known for its propensity to make campaign contributions) swelled to 47 members.[13]

The power of the chair's position   Committee heads—especially in the House—possess many resources, including prestige, access to information, control over the agenda and timing of meetings and votes, and authority over staff and budget. Their visibility, prestige, and influence reinforce one another. Other participants target them for influence, campaign contributions, and accolades.

House reforms enacted during the 1970s diminished the power of committee chairs, ending their ability to establish and abolish subcommittees. In addition, it became possible to remove autocratic, unpopular, or ineffective chairs. New rules adopted in 1995 under the Republicans include a six-year limit on a chair's term. Nevertheless, the heads of standing committees still exert considerable influence.

The seniority system   Prior to 1910, the House Speaker chose standing committee chairs. Abuse of the Speaker's many powers sparked a revolt that stripped the post of this and other duties, however; and it was agreed in both chambers that the majority-party member with the longest uninterrupted service (seniority) on each committee would automatically become its chair. The procedure became known as the **seniority system**.

For decades the House and Senate followed seniority as the sole criterion in appointing committee chairs. But in the 1970s the House Democratic Caucus (the entire body of Democratic representatives) decided to elect all committee chairs and chairs of appropriations subcommittees by secret ballot.[14] In 1975, House Democrats actually turned out three long-term incumbent chairs; in 1985, they deposed the Armed Services chair. For the most part, however, House Democrats have adhered closely to the criterion of seniority in electing chairs.

In 1995, Republican Speaker Newt Gingrich used his considerable influence (described in more detail later in this chapter) to play a major role in breaking the strict application of seniority. The ranking (i.e., most senior) Republican on Commerce did not become committee chair. And Gingrich

In 1995 Jan Meyers (R-Kans.) became the first female committee chair in 20 years when she was appointed head of the Small Business Committee by House Speaker Newt Gingrich. Gingrich's strategy was both laudable and politically shrewd, as polls have indicated that swing voters are more likely to vote for Republican women candidates than for Republican men.

passed over four more senior Republicans to select the chair of the crucial Appropriations Committee.[15]

The Democrats encountered trouble in 1995 when they tried to use seniority to unseat members on coveted committees. In response, the Republican majority cut the number of committee positions by 35, decreasing the Democrats' share of the remaining seats. Five of the seven female, Hispanic, and black Democrats on Appropriations were slated to lose their positions because of low seniority as newly elected members.[16]

In the final analysis, the seniority system's hold has been weakened in both the Senate and House, but seniority continues to help determine committee assignments and the chairship of committees and subcommittees.

The changing role of House subcommittees During the 1950s the number of House subcommittees began to mushroom, but the number of standing committees barely rose. The number of subcommittees peaked at 151 in 1975–76, then fell back somewhat to between 125 and 140.[17] A "Subcommittee Bill of Rights" adopted by the Democratic Caucus in 1973 also reduced the power of full committee chairs over subcommittees, mandated the establishment of some subcommittees, and required written rules for all committees. Later reforms accelerated the trend toward greater subcommittee autonomy and power.[18] The proliferation of subcommittees in part reflected a desire to match the increasing size and complexity of the executive branch. Congressional leaders also benefited from having more assignments and headships to distribute, and rank-and-file representatives saw increased chances to exert influence.

In 1995 the new Republican majority restricted most committees to no more than five subcommittees,[19] reducing their total by about 25. The House also cut the size of subcommittee staffs and returned control of hiring to the full committee chair. Thus, in the constantly changing balance of influence in the House, subcommittees appear to be losing ground. Nevertheless, they remain crucial to the legislative process.

CONFERENCE AND OTHER COMMITTEES

In addition to standing committees and their subcommittees, Congress uses several other types of committees. Joint committees draw members from both chambers and address topics of shared concern, such as the Library of Congress or the organization of Congress itself. Select committees conduct special studies or perform limited tasks, disbanding when their job is done.[20]

Conference committees are utilized when differences in House- and Senate-passed measures cannot be resolved informally either by acceptance of one chamber's language by the other or by an exchange of amendments.[21] Each conference committee is composed of members from the standing committees in each chamber that dealt with the bill.

The final content of major legislation often depends on negotiations in the conference committee; occasionally even provisions that did not appear in either chamber's bill will be accepted. Under this process each chamber's delegation determines its position on a measure by majority vote. When both delegations agree on final language, the conference report is sent to the floor of each chamber, where it is voted up (approved) or voted down (rejected), without the possibility of amendment.

Indeed, conference committees play a crucial role in crafting important legislation. A long and seemingly successful legislative effort can turn to ashes in the conference report; provisions deleted on the floor can be reinserted; a lobbyist's apparent early defeat can be turned into a late victory in conference; and failure to reach agreement, or defeat of the conference report in either chamber, can doom the legislation altogether.[22]

PARTY LEADERS AND ORGANIZATIONS

Most members of Congress feel a strong attachment to their political party, making them susceptible to appeals that evoke the party symbol. In an earlier chapter we suggested that a principal function of political parties is to serve as a governing mechanism that seeks to control and organize the machinery of government. The party organizations that do this constitute a crucial element in the structure of Congress.

Besides determining the chairs of all committees and subcommittees, the majority party in each chamber selects the leaders who manage the flow of legislation to the floor. In the House, the majority party nominates as its leader a **Speaker**, who is then formally selected by the entire House. Among other duties, the Speaker presides over the House, decides which bills to refer to which committees, nominates the chair and members of the influential Rules Committee, heads the committee that assigns members to standing committees, plays a key role in determining party issue positions, influences which bills are brought to a vote and when, and appoints the members of conference committees.[23]

In each chamber, routine tasks such as scheduling legislation and debate are overseen by a **majority leader** elected by the majority-party members of that chamber. A **minority leader**, also elected, performs similar functions for the other party. The Senate majority leader does not preside over the Senate (the formal but tedious and rarely performed duty of the vice-president) and in general possesses fewer resources than the House Speaker does. As a result, Senate majority leaders traditionally rely on their powers of persuasion and compromise in forging party positions on legislation. They also negotiate with the leader of the other party in the Senate on the scheduling of work.

In the House an elaborate structure of lesser party posts serves to assist party leaders. Among other posts, each party chooses *whips* and *assistant whips* who count noses on upcoming votes and distribute information about the content of legislation coming to the floor. Despite the name, however, they cannot coerce their party's members. Because the Senate is much smaller than the House, the structure of its party organizations is generally simpler.

The task of party leaders in both chambers is to enact the legislative programs of their parties. The leaders of the president's party generally assist in enacting the administration's legislative program. Leaders of the other party provide the focus for opposition to the president.

After losing some influence throughout the 1970s and 1980s, party leaders in Congress have begun to show increased ability to unite their membership. Later in this chapter we will examine how Newt Gingrich accelerated this trend on becoming Speaker in 1995.

FORMAL AND INFORMAL RULES

Crafting legislation and conducting the other work of Congress take place under an elaborate set of regulations. Formal rules, which differ between House and Senate in some important details, have five basic functions:

1. Regulating the introduction of bills and their referral to committees

2. Specifying how committees and subcommittees will consider proposed bills

3. Stipulating how legislation approved by a committee will be scheduled for floor action

4. Determining procedures for floor consideration, including what amendments can be offered, length of debate, and who will speak in what order

5. Outlining the steps used to reconcile differences between House and Senate versions in conference committees

By providing stable and authoritative procedures, formal rules facilitate and legitimize decision making. Rules can also reduce conflict and influence legislative outcomes—which is why debate over the rules and their interpretation is often so intense. After all, the rules sometimes produce bewilderingly complex situations. Consider the following description of a critical vote in 1980 from the official record of the Senate:

> So the motion to lay on the table the motion to reconsider the vote by which the motion to lay on the table the motion to proceed to the consideration of the fair housing bill (H.R. 5200) was rejected was agreed to.[24]

Not all situations are so complex, but the sheer number of rules is astounding; House rules fill more than 1,000 pages. No one could memorize them all. But members who acquire a good knowledge of rules and the skill to use them possess a powerful resource.

Formal rules profoundly shape the play of power in Congress. The emphasis on procedures provides opportunities to disguise substantive decisions on content as mere procedural votes. A vote to table a bill (remove it from formal consideration), to refer it back to committee, or even to adjourn may, in the debate's context, determine the outcome of the legislation. The use of rules, especially by the House leadership, to affect *when* votes occur can likewise be crucial. Finally, the rules provide ample opportunity for opponents of legislation to delay action, especially in the Senate.

The Senate, more than the House, tends to rely on informal agreement and consultation in the scheduling of business, reflecting both its smaller

A master of the filibuster, conservative Republican minority leader Everett Dirksen was responsible for the longest delay in the passage of a bill in Senate history—83 days—when it took the imposition of cloture to pass the Civil Rights Act of 1964. Here Dirksen, whose goal was to wear down southern opposition to the act, returns to the debate after a brief nap during the grueling proceedings.

size and its self-perception as "the greatest deliberative body in the world." Although the length of debate in the House is strictly controlled by the content of the "rule" adopted by the House Rules Committee, debate in the Senate continues until no one wishes to say any more, unanimous consent to stop is obtained, or an effort to postpone or prevent a vote by prolonging debate (known as a **filibuster**) is halted by a successful **cloture** motion—a motion to close debate. In order to pass, a cloture motion requires at least 60 votes.

The rough equality of status among members helps make both chambers collegial and nourishes the development of an elaborate set of informal rules, practices (sometimes called **folkways**)[25] and values that supplement the formal rules. We have already described seniority. Other highly valued practices and attitudes include the following:

- *Courtesy* The routine use of polite references such as "My good friend and distinguished colleague" even for bitter foes.

- *Reciprocity* Recognition of the expertise and jurisdiction of members and committees, and of the importance of keeping promises made to fellow members.

- *Legislative work* Appreciation of the importance of the time-consuming and painstaking attention to detail needed to formulate bills.

- *Specialization* Emphasis (especially in the House) on focusing one's efforts on a narrow area of policy linked to committee assignments.

- *Seniority* and *apprenticeship* A belief that new members should follow the lead of more experienced members.

- *Institutional loyalty*   A sense of pride in the traditions and values of one's chamber, coupled with a certain polite disdain for "the other chamber."

The specific content of these informal rules, and the extent to which they are adhered to, have differed between the House and the Senate. Senate folkways were especially strong in the 1950s.[26] In recent decades, folkways have begun eroding in both the Senate and the House. In fact, in 1995 several especially sharp exchanges on the House floor produced near chaos, led to official rebukes of members for their remarks, and forced apologies for intemperate remarks.[27] In the other chamber, when Senator Robert Byrd of West Virginia thought Majority Leader Robert Dole had breached the Senate's informal rules by moving to adjourn rather than vote on the Balanced Budget Amendment as agreed, Byrd rose in anger and intoned, "I deplore, I deplore this—this tawdry effort here to go over 'til tomorrow." Dole retorted, "We have every right to use the rules to determine if we have the votes or not or if we can pick up votes, and I intend to do that."[28]

A book aptly titled *The Decline of Comity in Congress* summarized the current status of informal norms:

> Most people are still polite to each other. Members share stories in the House and Senate gyms; they make deals on the pork barrel across party lines. Yet life is neither as uniformly civil nor as pleasant as it used to be. Tempers flare more often. Personal relations spill over onto the House and Senate floors. Members impugn the motives of each other and occasionally go even further. That trust that is essential for cooperation is in short supply.[29]

Our discussion of the structure and rules of Congress has presented important differences between the two chambers. A number of them are summarized in Table 13.2 (on p. 492).

RULES AND LEGISLATION: HOW BILLS BECOME LAW

Most congressional rules guide the process that transforms proposed legislation into law. Some, like those dealing with committee assignments, do so indirectly; others directly structure the formal process. Table 13.3 (on pp. 494–95) provides a simplified, step-by-step description of how a bill introduced in the House becomes law.

Many nuances add to the complexity of the formal process—far more than we can cover in this book.[30] To cite one example, a determined majority of House members can dislodge a bill stalled in a substantive committee or in the Rules Committee by signing a discharge petition that brings it to the floor for action. The rules governing the formal process provide the framework for the play of power in the struggle over legislation. A careful look at Table 13.3 will help you understand the dynamics of major legislative battles in Congress.

TABLE 13.2

THE HOUSE AND SENATE COMPARED

	HOUSE	SENATE
MEMBERS		
Number	435	100
Term	Two years	Six years
District size	About 580,000	Varies by state, from about 400,000 to more than 30,000,000
Elections	Higher incumbent success rate (90–95% or more) and fewer close elections; Democratic "lock" from 1954 to 1994	Lower incumbent success rate (75–85%), more close elections; Democratic majority slimmer and less secure
Goals and roles	Norm of specialization related to committee work is strong	More policy generalists and idea generators; some presidential hopefuls
STRUCTURE		
Leadership	Relatively stronger, more hierarchical	Weaker, more consensual
Committees	Center of legislative activity and expertise, strongly led, autonomous, concerned with success on chamber floor	Less central and important, less strongly led, having less expertise, less autonomous, less concerned with success on chamber floor
Influence	More hierarchical; leaders and committee and subcommittee chairs are influential	Less hierarchical; majority and leader roles are important but not dominant
RULES		
Control of floor	Proceedings strictly controlled by Rules Committee	Proceedings controlled loosely, largely through agreement among leaders and all members (unanimous consent)
Nature of debate	Formal, controlled, limited	Informal, unlimited

THE RESOURCES AVAILABLE TO CONGRESS

The principal resources that allow Congress to exert influence in the play of power stem from grants of authority to it in the Constitution. Article I grants Congress the power to tax and spend for a variety of purposes, to override a presidential veto of legislation, to impeach and try officers of the executive and judicial branches, to judge the qualifications of its members, and to make its own rules. The Senate alone is given the power to ratify treaties and presidential appointments of certain high officials.

Foremost among Congress's resources is the **power of the purse**—that is, the power to tax and spend. Money can be released from the U.S. Treasury only if Congress has passed legislation *authorizing* its use and has actually *appropriated* money to spend on a specific program. Substantive committees (e.g., Agriculture, Commerce) authorize programs, but the appropriations committees must actually approve the specific sums to be spent. Indeed, the power of the purse gives Congress tremendous leverage over policy. But this power has been weakened by "uncontrolled expenditures" resulting from, for example, legislation that calls for automatic cost-of-living increases or the provision of new benefits to anyone who qualifies under certain programs (e.g., Medicare or welfare). Congress also uses certain devices to circumvent the appropriations process (e.g., granting tax breaks that reduce revenues). Proposals to grant the president a **line-item veto** (i.e., the power to cut out specific items from a spending bill) would further weaken Congress's power of the purse.

Of course, the power to authorize new programs also provides Congress with tremendous influence. And Congress has developed an internal culture that celebrates its tradition of autonomy and institutional pride, which in turn helps maintain its effectiveness. Indeed, the word *Congress* and its pictorial representation are themselves powerful political symbols.

Individual members of Congress also possess resources, including (1) the *prestige* of serving in Congress; (2) an *independent base of power,* possessed by those members who are elected by their district with little or no party assistance; (3) *long tenure,* which allows many members to acquire considerable expertise in policy areas and in the rules and procedures that guide legislation; and (4) an *equal vote* in committees and on the floor.

THE DYNAMICS OF LEGISLATIVE BEHAVIOR

The work lives of representatives and senators resemble those of many Americans, including college students. Members of Congress have a heavy, varied schedule of daily activities; they face a bewildering array of demands and pressures but enjoy some freedom to choose what they will concentrate on and what they will let slide. In our description of legislators' behavior, we will consider two interconnected activities required of all members of Congress: performing committee work, and voting on legislation that reaches the floor. A third activity, seeking re-election, was discussed earlier in this chapter.

COMMITTEE WORK: WHERE THE ACTION IS

Although the most highly visible and controversial bills each year generate fierce floor debate and crucial votes, most deliberations that forge the content of legislation do not take place on the floor; they occur in committees and subcommittees, between committee members and their staffs, or between

TABLE 13.3
HOW A BILL INTRODUCED IN THE HOUSE BECOMES LAW

1. *Drafting and introduction*

   The draft of a bill is written at the urging of the president, a bureaucratic agency, a lobbying group, or a prominent constituent, and is dropped into the hopper of the House or on the Senate parliamentarian's desk.

2. *Assignment to a committee*

   The bill is given a number and is referred by the Speaker to a committee with jurisdiction over its subject matter.[1]

3. *Referral to a subcommittee*

   The committee chair refers the bill to one or more subcommittees, except in the case of major bills, when the full committee may conduct hearings and consider the bill without referring it to a subcommittee.

4. *Hearings and mark-up*

   If the subcommittee chair decides to consider the bill, hearings may be held and a *mark-up* session conducted. Here the bill is discussed line by line, provisions from bills stalled in other committees are inserted, new features are added, and existing provisions are changed. The mark-up may proceed by consensus or through a series of votes.

5. *Full committee action*

   After subcommittee approval the bill is reported to the full committee, where additional amendments may be made. Uncontroversial bills with bipartisan support may go directly to the *Committee of the Whole,* composed of all members of the House and needing only 100 members for a quorum. Major legislation, however, must first go to the Rules Committee, known as the House "traffic cop" because it controls the flow of traffic in bills.

6. *Rules committee action*

   The Rules Committee writes a *rule* regulating how the bill will be considered on the floor, including the time allowed for debate and which amendments, if any, can be proposed. A *closed rule* prohibits amendments; an *open rule* allows amendments; a *complex rule* allows amendments only to certain portions of the bill or places limits on the number of amendments that can be proposed by certain groups or legislators.

7. *Scheduling for full House action*

   The Rules Committee places the bill on either the "Union" or the "House" *calendar* (names given to lists of bills to be considered), and it is scheduled for debate. The chair of the committee or subcommittee reporting the bill serves as *floor manager,* allocating time among speakers supporting it, deciding whether to accept or fight amendments, and determining strategy.

8. *House action*

   The House debates and votes on the bill. A controversial bill with an open rule may require scores of separate votes on different amendments or procedural questions before the measure goes to a final vote. If considered first in the Committee of the Whole, the bill and any amendments permitted and offered must be considered again.

9. *Senate consideration*

   A bill passed by the House goes to the Senate, which may already have been working on similar legislation (which, if passed, is sent to the House).

(continued on facing page)

TABLE 13.3 (CONTINUED)
HOW A BILL INTRODUCED IN THE HOUSE BECOMES LAW

10. *Referral to a Senate committee*

The Senate majority leader, usually in consultation with the minority leader, determines which committees or subcommittees have jurisdiction.

11. *Subcommittee action*

The subcommittees consider the measure, hold hearings if necessary, mark up the bill, and send it back to the full committee for a final vote, much as House subcommittees do.

12. *Full committee action*

The full committee, after considering amendments, approves the bill.

13. *Scheduling for full Senate action*

The majority and minority leaders agree informally on the scheduling, length, and nature of debate on the bill. Their agreement performs the same function as the House Rules Committee's rule.

14. *Senate action*

The bill is debated and unlimited amendments are considered on the Senate floor. A senator or group of senators may hold the floor for as long as they desire or are physically able. *Filibusters* can be ended only by 60 votes for a *cloture motion*, which limits further debate to an additional hour for each senator.

15. *Reconciling House-Senate differences*

For a bill to become law, the House and Senate language of the bill must be identical. Often one chamber accepts the other's version or informal agreement results in both chambers' adopting identical language. On important bills with significant differences, a *conference committee* is appointed to reconcile differences.

16. *Approval of the conference committee report*

If the conference committee can agree on identical language, the conference report is scheduled for an up or down vote (i.e., a direct vote with no amendments permitted) in both chambers.

17. *Presidential action*

After a bill containing identical language passes both chambers, the measure goes to the president, who may sign it into law, veto it, or let it become law after 10 days without signing it. If a bill is passed at the end of the life of a Congress, the president can *pocket veto* it by simply refusing to sign it within 10 days of adjournment.

18. *Consideration of vetoed legislation*

If the bill is vetoed when Congress is still in session, each chamber must vote to override the veto by a two-thirds vote. If they manage to do so, the bill becomes law without the president's signature.

---

[1]In 1995 the new Republican House ended the practice of simultaneous referral of bills to multiple committees; under the new system, the Speaker decides which committee acts first. For a description of how the process had worked before, see Garry Young and Joseph Cooper, "Multiple Referral and the Transformation of House Decision Making," in Lawrence C. Dodd and Bruce I. Oppenheimer, eds., *Congress Reconsidered*, 5th ed. (Washington, D.C.: Congressional Quarterly Press, 1993), Chapter 9.

Although committee business is an essential part of congressional activity, committee and subcommittee hearings and meetings are often sparsely attended. There were many empty chairs when Secretary of Agriculture Dan Glickman testified before a committee meeting on commodity policy in June 1995. In such cases, decisions on important issues may ultimately be made by individuals with little knowledge of the particulars of a situation.

lobbyists on the one hand and members and staffers on the other. This is especially so in the House. Although any senator can participate in floor debate on any bill and offer any amendment he or she wishes, rules in the House restrict the role played by non–committee members on the floor. According to one description, the committee is the place where "the original legislative vehicle normally gets formulated, the central issues are debated, and most of the substantive and technical amendments are reviewed and adopted."[31]

In 1992 House members served on an average of 6.8 committees and subcommittees; senators, on 11.[32] This situation produces a scheduling nightmare and contributes to a workload too heavy for any senator or representative to handle. Thus, members of Congress inevitably pick and choose the issues they and their staff will address. This means that an *active core*, not all the members of a main committee or a subcommittee, determines the content of any given legislative proposal. Often, fewer than half of a committee's or subcommittee's members participate in crucial decisions on legislation, and even fewer attend committee hearings and other committee meetings. Final outcomes depend on which members are active on which issues; this, in turn, depends on the members' interests, the amount of time they can devote to them, and their skills.

Committees differ in how they go about their work. Partisan conflict is intense on policy committees that address labor and welfare issues, and relatively muted on others such as appropriations. Some committees are dominated by strong leaders, whereas rank-and-file members play a greater role in others. Some have members who specialize in a narrow set of issues; others exhibit less specialization. Committees whose members share com-

ASSESSING ROLL-CALL VOTES

The most common technique for assessing the behavior of members of Congress is to examine their floor votes on roll-calls. This narrow focus ignores other significant aspects of legislative behavior, particularly work on committees.

In addition, the actual roll-call vote can be misleading. One representative, faced in the early 1980s with a strong coalition of citizens' groups advocating a nuclear freeze (i.e., a halt to the production of nuclear weapons), pledged in a meeting with them to vote *for* the freeze resolution scheduled for floor action. But a telephone call from President Ronald Reagan a few days before the vote persuaded him to vote *against* the freeze resolution if it looked as though it might pass. When it became apparent that the vote was evenly split and the representative's vote would make the difference, he risked the wrath of some of his constituents and voted against the resolution.

But if the vote had not been so close, he could have cast a vote *for* the freeze. Roll-call analysis would have revealed his support of the freeze; however, the administration and his party leaders would have known and appreciated his pledge to vote against it "if needed." On many roll-call votes, a handful of these pledges may be made. Thus, because of the hidden dynamics behind such votes, roll-call analysis must be interpreted with care.

mon goals and agree on how decisions should be made (e.g., Appropriations, Ways and Means, and Interior) are generally more successful in persuading the entire House to approve their bills.

HOW MEMBERS MAKE THEIR CHOICES IN ROLL-CALL VOTES

The heavy and conflicting demands on members of Congress cause relatively low participation rates in committee mark-up sessions (in which proposed legislation is modified) and votes. However, approximately 95 percent of senators and representatives participate in **roll-call votes**, those for which the vote of each member is recorded and published.[33] Not all roll-call votes generate controversy or drama. Some are relatively unimportant; some bills pass by comfortable margins. However, on controversial issues—such as the assault weapons ban, President Clinton's first budget, and the North American Free Trade Agreement (NAFTA) in 1993—very close House votes produce high drama. In general, roll-call votes carry great symbolic significance for members and the organized interests affected, and they have become a major factor in evaluating members' performance (although their true significance may be exaggerated, as "Footnote: Assessing Roll-Call Votes" indicates).

Influence from within Congress Suppose your school required you to answer any question on any exam in any course, whether you were taking that course or not. You might find yourself pulled away from reading this text to take a test in differential calculus or the structures of intermediate molecules in glycolysis. What would you do? If you had friends in each class who understood the material well and you were allowed to consult them, you might well adopt their answers.

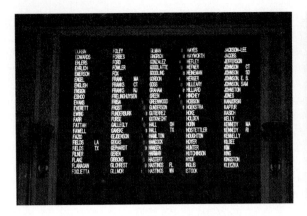

In the House, roll-call votes are indicated on an electronic tote board (*left*) that lists each member and his or her vote. These votes are then tallied and displayed on another board (*right*) indicating both the total number of voters present and how many votes were cast for and against a bill.

Members of Congress face a situation similar to this. In 1993, for example, House members faced 600 roll-call votes; senators, almost 400.[34] They had to vote "yes" or "no" on a wide range of issues, some complicated and difficult to understand. Years ago, a member lamented as he went to the floor, "I don't know a goddamn thing about the Amateur Sports Act of 1978. So I'll just have to ask someone on the floor. And, frankly, I couldn't care less."[35] In situations like this, **cue networks** develop among members who come to depend on a few colleagues known for their integrity, good judgment, or expert knowledge in a particular area.

Voting cues come from a variety of sources. In the House, the stance of others from members' home states, especially those from the same party, is often persuasive. Representatives from about half of the states meet regularly and take positions that help to determine votes. Sometimes the positions of members from other states with similar districts or ideology carry weight. On some issues, the stance of one of the House's many special caucuses (e.g., the Military Reform Caucus, the Steel Caucus, the Black Caucus, the Footwear Caucus) plays a role. Trusted members and friends who serve on the committee that drafted a bill also provide cues. Electronic voting in the House makes it possible to look at the vote board to see how cue givers have voted.

Other sources of influence exist within Congress. The fact that bills reaching the floor have passed muster in the committee process is usually sufficient to ensure passage. (Amendments opposed by the committee face an uphill battle.) Explicit bargaining also shapes some votes. For example, a member who doesn't care about the Amateur Sports Act might trade a vote with someone who does, in return for support on an unrelated bill. Trading votes on unrelated issues is so common that it has its own name: **logrolling**. Public works bills (which call for the building of post offices, dams, and other construction projects in specific districts) grow in length and balloon in cost as members vote for each other's pet projects. Tax bills suffer a similar fate. Spoilsports in Congress and the media refer to these measures as "Christmas tree bills" because of the many give-aways attached like ornaments to benefit constituents of various members.

The most important influence on roll-call voting is *party*; indeed, a majority of one party's members vote against a majority of the other party (in what is called a *party vote*) in 70 to 80 percent of all roll-call votes. One study of roll-call voting conducted during the 1970s found that voting cues associated with party were mentioned more frequently (in 41% of responses) than any others.[36]

Much of this party influence is indirect; the shared ideological predispositions of members of the same party incline them to vote similarly. But party influence is sometimes more direct. Party leaders in Congress cannot rely on mere *authority* to exert influence, but they can sometimes influence votes through *exchange*—by promising, for example, a good committee assignment, cooperation in scheduling a vote, appearances at fundraisers, or campaign funds.

More important, they can make use of their principal tool—*persuasion.* Appeals invoking the party symbol work best. For example, when Robert Dole was Senate minority leader early in the Bush administration, he pleaded for support on a crucial vote by admonishing his Republican colleagues as follows: "We need to recognize that this is a Republican initiative that we are about to dismantle here. This is a Republican administration, and the president has a right to expect some support from the Republican side of the aisle."[37]

Nevertheless, party leaders influence members' votes only to a limited degree. They are best able to do so when (1) the leadership is active, united, and committed, (2) the vote is on procedure rather than substance, (3) the issue's visibility is low, (4) members receive little pressure from constituencies, and (5) state delegations are inactive.[38] They are also better able to influence members when the voting is by voice, rather than by roll-call.

Despite the limited influence of party leaders themselves, if you were asked to predict how a member of Congress would vote on a number of issues and you could use just one piece of information, party affiliation would be the best by far. Furthermore, the role of party (especially in the House) seems to be growing. Thanks to the skill of recently elected Republican Speaker Newt Gingrich (described in "Strategies: Gingrich Strengthens the Power of the Speaker," on page 500), the GOP leadership in the House began to centralize more power in its hands after taking control in 1995.

Influence from the executive branch   The executive branch also influences the behavior of members of Congress. Members of the executive bureaucracy, including agency heads and their deputies, conduct most of the executive branch's lobbying of Congress. On crucial issues (e.g., the 1994 Crime Bill and NAFTA), department heads themselves go to Congress.

Sometimes presidents themselves play an important role. They can invoke the party symbol when appealing for fellow party members' support. When the issue involves foreign policy, presidents draw on their role as commander-in-chief of the armed forces and director of foreign relations, and they may cite national security requirements in support of their position. Presidents can also promise to support projects and contracts going to members' districts, endorse legislation that helps members' re-election prospects, nominate members' choices for federal appointments, and come to members' districts to campaign and raise money.

Of course, willingness and ability to lobby for votes vary from one president to the next and from one time to another. Such efforts can be

GINGRICH STRENGTHENS THE POWER OF THE SPEAKER

The policy changes advocated by the new Republican majority when it assumed control of the House in 1995 were controversial, but nearly everyone admired House Speaker Newt Gingrich's skill in devising and carrying out his strategy for bringing them to a vote. Gingrich quickly emerged as the most influential Speaker in decades. How did he do it?

One key element in his success was the support he mobilized among the large cohort of mostly very conservative and ideological first- and second-term Republican members. Most credited Gingrich's "Contract with America" (a detailed list of proposed actions that Republican candidates committed themselves to carrying out) with helping the Republican Party win the House. And many representatives owed him their cooperation because the political action committee Gingrich controlled (GOPAC) had recruited and guided them in seeking office and had contributed money to their campaigns.

The new Speaker made use of his support among rank-and-file members to weaken the role of committee chairs through rules changes. The new procedures required committees to open their meetings to the public, record all votes, and end the practice whereby absent members could give their "proxy votes" to someone else (often the committee chair). Another new rule ended the simultaneous referral of bills to several committees. Now the Speaker would decide which committee dealt first with any bill. Also, committee heads would henceforth be limited to six years as chair, a change that weakened the seniority system.

Speaker Gingrich struck another blow at seniority by breaking its hold on who would chair committees. He took a strong role in choosing chairs, managing to pass over more senior Republicans to tap his own supporters to head crucial committees such as Commerce and Appropriations. He further strengthened his position by seeing that an unusually large number of newly elected members received coveted spots on key committees.

Gingrich also recognized the importance of *when* things happen—timing—in devising his strategy. He moved as quickly as possible early in the session, while the flush of victory helped unify the Republicans and the Democrats were still somewhat disorganized.

undertaken on only a few issues, and not all members can be lobbied personally. Also, the probable vote outcome must be close enough for intervention to make a difference. When all these conditions are met, however, presidents often succeed. President Clinton's vigorous and effective efforts to persuade House members to support NAFTA in 1994, for example, turned a likely loss into a significant victory.

Influence from the geographic constituency   Many younger college students find that their parents or guardians resemble a sort of constituency. Parents may provide crucial resources (e.g., moral support, transportation, money) needed to stay in school, and they can hold their children accountable. But, as both students and parents can attest, parental control is incomplete. Parents remain largely ignorant of the details of their son's or daughter's daily activities and may be happy to remain so.

In many respects, the relationship between representatives and their constituents is similar. Some roll-call votes clearly reflect the constituency's influence, whereas others do not. Citizens learn about only a few of their representatives' activities and votes and often find it difficult to assess their significance. On many issues, most constituents do not care about outcomes, and no opinion poll exists to help representatives assess their district's views.

Of course, just as most parents care very much whether their children pass or flunk out of school, there are some issues about which constituents may feel strongly (e.g., the issue of congressional pay increases). Even when only a small attentive public feels strongly, if it is united and no pressure to vote the other way exists, most members will vote to support its position.

Why would a senator or representative heed the wishes of an attentive public under these conditions? After all, a handful of very angry people are unlikely to cause an incumbent's defeat. One member of Congress explained the logic of voting in accord with a united and attentive public's wishes:

> I suppose this one issue wouldn't make much difference. Any one issue wouldn't swing it. But you get one group mad with this one. Then another group—much more potent, by the way—gets mad about gun control. Then unions about compulsory arbitration. Pretty soon you're hurting. It doesn't take too many votes like this before you've got several groups against you, all for different reasons, and they all care only about that one issue that you were wrong on. A congressman can only afford two or three votes like that in a session. You get a string of them, then watch out.[39]

However, when a majority of the public takes one view but the member's *re-election* or *primary* constituency takes a different view, the majority view may not prevail. For example, although gun control legislation may be popular in one's district, National Rifle Association (NRA) members opposed to it are attentive and mobilized, and they often act as single-issue voters on gun control. They may represent a smaller proportion of the constituency but be influential enough to sway their representative's vote.

Another complication is the difficulty of assessing what "the district" desires. Members of Congress construct their own understanding of the district through their networks of acquaintances and sources of information.[40] But supporters more often write, return mailed opinion surveys, and meet with the incumbent than do bitter opponents.

What conclusions can be reached about the influence of constituencies on votes? On rare issues that arouse a united and broad public, members vote their districts' views. They also respond to the opinions of attentive publics or funding constituencies that feel strongly, especially when all agree and there is little counterpressure. (As we describe in "Outcomes: How Constituency Pressure Killed Catastrophic Health Coverage," on pages 502–3, such pressure from the elderly largely explains the repeal of the Medicare Catastrophic Coverage Act just one year after its enactment.) But on many issues, members do not face strong constituency pressures and thus are not greatly influenced by constituents.

The impact of organized interests and lobbyists Our earlier treatment of organized groups concluded that evidence on the extent of their influence on Congress was inconclusive. The matter is complicated partly because influence is difficult to assess, varying from group to group and from issue to issue. Groups that have roots in the district and represent attentive publics

HOW CONSTITUENCY PRESSURE KILLED
CATASTROPHIC HEALTH COVERAGE

In 1988 Congress passed the Medicare Catastrophic Coverage Act to cover the medical costs of catastrophic illness. But it reversed itself the very next year, when members came under heavy negative pressure from their constituencies. "When we passed this a year ago," observed Senator Robert Packwood (R-Ore.), "we all thought we had done a nice thing. [But] you almost have a sense today when you go home on this issue of being unwanted and unloved and unappreciated." Representative William Gradison (R-Ohio) commented, "It's hard to go home and have office visits without having someone come in to protest. It's reached the point in my town where I don't refer to this as the Gradison-Stark bill anymore. I refer to it as the Stark bill."

In describing what members encountered, media accounts used phrases such as "continuing flood of letters, phone calls and complaints at town hall meetings," "furious outcry," "continuing onslaught," and "verbal pummel[ing]." Senator John McCain (R-Ariz.) reported that his office had received 20,000 letters on the issue, with no more than 10 supporting the bill. Representative Mike Synar's (D-Okla.) staff referred to a tour of his district that encountered repeated protests from the elderly as "the catastrophic tour." After one meeting, the chair of the Ways and Means Committee, Dan Rostenkowski (D-Ill.), found his car blocked by angry seniors shouting "Liar!" and "Rottenkowski!" A picture of the scene ran in the *New York Times* and *Newsweek* and on two television networks.

The outcry was organized and inspired by a number of organizations that focused on the supplemental premium that the bill assessed against wealthier Medicare participants, calling it the "seniors-only surtax." The influential American Association of Retired Persons fought for preservation of the bill but found it could not counter the powerful negative appeal of the call to "repeal the seniors-only surtax." One congressional supporter, Senate Finance Committee chair Lloyd Bentsen (D-Tex.), said, "I happen to think we have done a good job, and the catastrophic-illness piece of legislation is a good piece of good legislation." But he also noted, "All of us have received thousands of letters from our respective states, and they indicate changes will be necessary."

with intense opinions and many political resources, such as the NRA, are generally more influential than those that do not. But interest groups without a strong presence in the district can hire Washington lobbyists, whose activities (e.g., providing information, producing speeches, drafting legislation) do not require a powerful district base. Some interest groups with few members in the district have political action committees (PACs) that become part of the funding constituency of incumbents.

As we have noted elsewhere in this book, interest groups' use of PACs has grown tremendously. The PACs generally seek to facilitate friendly access to incumbents, especially committee and subcommittee chairs and ranking minority members (i.e., people who influence the content of legislation and, through the cue network, the votes of others).

How much more than access is bought by PAC contributions? Reliable answers to this important and difficult question are hard to come by. Opponents of the current system of financing congressional elections, especially public interest groups such as Common Cause and Public Citizen, assert

Although this rarely happens, a mobilized and united public constituency can have a powerful impact on legislation. Such pressure from the elderly largely explains the repeal of the Medicare Catastrophic Coverage Act just one year after its enactment, as well as this attack by angry senior citizens on the car of Dan Rostenkowski (D-Ill.), at that time the powerful chair of the Ways and Means Committee and the bill's chief proponent.

The barrage of criticism and protest was too powerful to resist. House minority leader Robert Michel (R-Ill.) explained, "We're just getting too much flak. The only way to wash our hands, cleanse ourselves of the whole thing, is outright repeal." Representative Pete Stark (D-Calif.) reported that many members just said, "The hell with it. I don't want to deal with it anymore."

Source: This description and the quotes cited are based on an excellent series of articles by Julie Rovner in *Congressional Quarterly Weekly Report* in 1989 and Martin Tolchin's "Politics" column in the *New York Times*, August 30, 1989.

that contributions strongly affect members' behavior. These critics have released widely reported studies asserting a direct link between campaign contributions and roll-call votes. Most newspaper editorials support their view.

This conclusion is disputed by participants in the PAC contribution system (PAC officials, lobbyists, and some members of Congress) and by many scholars who study Congress and its elections. Denials from those participating in the PAC system obviously face a credibility problem. (No PAC official or lobbyist would admit publicly that campaign money buys influence.) Scholars who agree with them cite studies that find little or no relationship between PAC contributions and roll-call votes.[41] However, studies of PAC influence generally do not examine other aspects of the legislative process besides roll-call votes. One study that did so found that campaign contributions increased the likelihood of recipients' taking an active role on issues of concern to the donor in committee work rather than in floor votes.[42]

Regardless of the precise answer, few students of Congress deny that organized interest groups and their lobbyists play a significant role in shaping congressional behavior, including roll-call votes. Clearly, both lobbying and PAC contributions are elements in successful strategies for influencing Congress. The fragmented structure of Congress and the importance of committees in shaping legislation offer those who seek influence there (the president and the bureaucracy as well as interest groups) multiple points of access. Few social institutions are as open to as many influences from their strategic environment as is the U.S. Congress.

PATTERNS OF CONGRESSIONAL DECISION MAKING

Three features of congressional decision making merit particular notice: the difficulty of passing legislation, which we call the *defensive advantage;* the critical role *coalitions* play in the legislative process; and the *incremental nature* of most policy changes enacted.

The defensive advantage: How bills fail to become laws   Steering a piece of legislation through Congress is like playing a challenging video game whose object is to move a bill from introduction to passage. Legislation can be stymied at virtually every stage of the process. Failure to pass a stage either kills the bill or moves it back to an earlier stage.

Some bills die quickly because a crucial participant, the person introducing it, does not really care about its passage. Members perfunctorily introduce many bills merely to satisfy a constituency group or to project an image of competence, activity, and responsiveness. On the other hand, a dramatic event such as the terrorist bombing of a federal office building in Oklahoma City in 1995 can produce an avalanche of bills (in this case, to combat terrorism); the sheer volume of bills proposed guarantees that many will not be passed. At other times, issues at the top of the agenda, such as crime and health care, inspire many to introduce their own bills with no real intention of pushing them vigorously. Without the active support of its sponsor, a bill will die in committee.

Beyond this, many principal features of the structure and rules of Congress discussed earlier are intended to make passage of legislation difficult, and they succeed in doing so. Bills must pass through both chambers—and sometimes a conference committee—before the expiration of the two-year term of each Congress. Moreover, if a new program is to be established, it must run the legislative obstacle course twice: first to authorize the program, then to pass legislation appropriating money to fund it. Bills compete for attention and may fail because they are shunted aside by others with a higher priority.

Passage of a bill becomes harder when some members actively oppose it (the equivalent of switching to the "expert" version of a video game); indeed, opponents have many opportunities to kill a measure. They can get the Speaker to refer it to a hostile committee or convince the subcommittee chair to refuse to hold hearings, allowing it to die quietly. The full committee can kill the bill or gut its provisions. Opponents can use the rules to delay progress, especially in the Senate. A determined minority in the Senate can filibuster—a tactic used with increasing frequency and effectiveness. Major legislation can also be killed or mutilated in the conference committee. Even if these obstacles are cleared, the president can veto the legislation.

Thus, the dispersion of power that results from the joint effects of bicameralism, a strong committee system, a weak party structure, and a set of rules that confer advantages on opponents of action produces a formidable obstacle course. Legislation can be defeated at any one of the 18 steps in the process depicted in Table 13.3; each can be thought of as a *choke point* or bottleneck. Passage requires successful negotiation of every step. Finally, the sheer volume of bills introduced guarantees that even though Congress might want to pass many more bills, it lacks the time and energy to do so.

Together, these obstacles give opponents of proposed legislation a distinct **defensive advantage** in the congressional legislative process. Opponents need not agree on the provisions of a bill. All that is required is the assembling of a majority to oppose the bill regardless of the reason.

Certain underlying factors reinforce the effectiveness of the defensive advantage. Organized interests that benefit from the status quo possess more political resources and exert more influence than those that propose change; people also fight harder to preserve benefits they currently enjoy than they do to obtain new ones. In addition, the recruitment process promotes to the halls of Congress individuals who have benefited from the status quo, and most of them are skeptical of significant social change. And incumbents are naturally reluctant to jeopardize their seats by advocating new and controversial policies that might offend their voting and funding constituencies.

The result of the defensive advantage? Each year, thousands of bills are introduced, and only a tiny proportion become law, as Figure 13.2 (on p. 506) shows. In the 101st Congress (1989–90), for instance, almost 6,700 bills were introduced in the House and 3,700 in the Senate. Only 650 bills made it past both chambers.[43]

The central role of coalitions in passing legislation   The fragmented structure of Congress, coupled with the need to assemble a majority to pass a bill, guarantees that coalition building is the principal technique in forging the content of legislation. In Congress, coalitions consist of informal groups of members having both common interests and differences. They can achieve their goals if their resources are pooled and their differences minimized through compromise.

Once you understand that Congress acts through coalitions, a number of its operating characteristics make sense. For instance, the way Congress works is cumbersome because it is difficult and time-consuming to reach the compromises needed to forge a coalition. The wording used to bring about such compromises is sometimes deliberately unclear, because ambiguity is the cement of congressional coalitions. (Thus, sharp disagreement about a bill's *real* meaning sometimes arises among people who joined in voting for its passage.)

Another characteristic of legislative coalitions is their shifting composition from issue to issue and over time. Sometimes very unlikely allies work together to defeat legislation that they oppose, often for different reasons. The repeal of the Medicare Catastrophic Health Act in 1989 provided a dramatic example. For example, Henry Waxman (D-Calif.) voted with his most persistent opponent on health policy, William Dannemeyer (R-Calif.); Fortney Stark (D-Calif.) and William Gradison (R-Ohio),

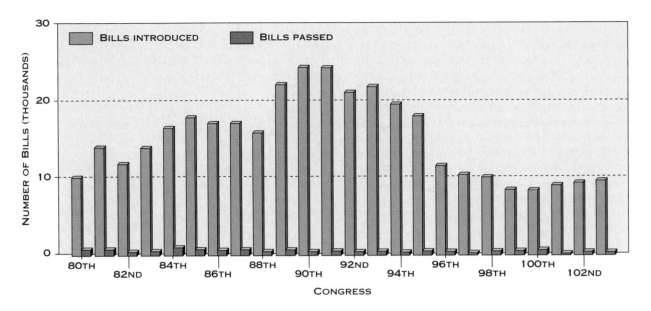

BILLS INTRODUCED AND PASSED, 80TH TO 103RD CONGRESSES The number of bills introduced varies considerably over time. Only a tiny proportion of bills introduced actually pass. *Sources:* For 80th to 102nd Congresses: Harold W. Stanley and Richard G. Niemi, *Vital Statistics on American Politics,* 4th ed. (Washington, D.C.: Congressional Quarterly Press, 1994), p. 218; for 103rd Congress: *Congressional Record* 140: 150, Daily Digest.

who had cosponsored the initial draft of the law, voted on opposite sides on the issue of repeal.

Finally, the notion of coalitions can enrich our understanding of the defensive advantage. Congress fails to legislate on certain important issues because a majority coalition cannot form around any proposal. For example, when the House tried to take up a conference committee version of the crime bill in August 1994, a negative coalition formed between some Black Caucus members and gun control opponents who were unhappy with different portions of the bill; they opposed the rule that would allow the measure to come to the floor. Also, for many years a negative coalition composed of Democrats and Republicans from the South blocked legislation on social programs and civil rights; although its strength waned as Republicans replaced Democrats in the South, the coalition still formed for 9 percent of recorded votes in both chambers in 1993.[44]

The incremental nature of policy changes The combined effects of the defensive advantage and coalition building help to account for another characteristic of congressional decisions, the incremental (i.e., gradual) nature of the changes wrought by most legislation. The best prediction of what the next policies formed by Congress will be like, most of the time, is that they will differ only marginally from policies of the past.

Even important electoral changes, such as the shift in party control of the Senate after the 1986 elections, need not cause dramatic changes in coalitions or policy. However, when very large electoral shifts take place—such as the one in 1994 that gave the Republicans control of the House as well as the Senate—the door is indeed open to nonincremental policy changes. Many of the new Republican initiatives introduced in the House in early 1995 departed from the established pattern of incremental change. However, some of the House-passed initiatives faced resistance in the Senate, which was less changed ideologically than the House. The two-chamber structure of Congress thus reinforces incrementalism.

The incremental nature of legislation has important implications. It reinforces the continuous nature of the play of power. The compromises hammered out by winning coalitions can be modified in the future if losing coalitions gain strength; on most issues, most of the time, there are no permanent winners and losers. Recognition of this fact colors decision making. Majorities recognize that they may become minorities in the future; consequently, they may modify the way they cement their gains.

Congress thus has a tendency to make only small changes and lacks a stable core of centralized power. With only a few short-lived exceptions—such as the political vacuum following the Watergate scandal in the 1970s and the activism of the 100 days following the Republicans' electoral landslide in 1994—Congress plays a secondary role to the president in determining major changes in policy. Even when it tries to initiate major change, Congress must confront possible opposition from the president.

CONGRESS IN THE PLAY OF POWER

Even though Congress's primary task is to write legislation, it also performs various other roles, both as a part of its legislative functions and in other areas.

REPRESENTING CONSTITUENTS' INTERESTS

One function performed by Congress is **representation**. One political scientist has defined the term as "acting in the interests of the represented, in a manner responsive to them."[45] But what does "acting in the interests of" mean? Elected officials answer this question in different ways. Some members of Congress consider themselves *delegates*, accepting the obligation to be guided by constituents' wishes regardless of their own views. According to one representative,

> I'm here to represent my district. This is part of my actual belief as to the function of a congressman. What is good for the majority of districts is good for the country. What snarls up the system is these so-called statesman-congressmen who vote for what they think is [in] the country's best interest.[46]

Others see themselves as *trustees*, chosen by their constituents to exercise their own best judgment after studying the issues, gathering information, and listening to experts and other legislators. For example, in explaining his vote to ratify the Panama Canal treaty during the Carter administration, then–Senate majority leader Robert Byrd expressed this viewpoint:

> There's no political mileage in voting for the treaties. I know what my constituents are saying. But I have a responsibility not only to follow them, but to inform and lead them. I'm not going to betray my responsibility to my constituents. I owe them not only my industry but my judgment. That's why they send me here.[47]

Still others, called *politicos,* fall in between, sometimes following their constituents' wishes, sometimes exercising independent judgment.[48] Finally, it may be added that some members of Congress—including many women and minority-group officeholders—regard themselves as providing representation in some sense for all members of their population group, whether part of their geographical constituency or not.

Regardless of the view adopted by members of Congress, they seek to convince constituents that they act on their behalf while in Washington. Earlier, we referred to the techniques used to establish this emotional response as "home style." Political scientists use the term *symbolic representation* to refer to the emotional or psychological response among constituents that occurs when legislators successfully build trust and convince people they are acting on their behalf.

Of course, not all members of Congress succeed in establishing widespread trust among their constituents.[49] Active partisans of the opposition party often feel strongly that the representative from their district does *not* represent their views. And citizens who care little about politics possess few of the feelings necessary for effective symbolic representation. Nevertheless, many people believe that their legislator represents their interests. Believing this helps bind them to the system, increases the legitimacy of government in their eyes, and contributes to political stability. It also increases the likelihood that these citizens will back the incumbent's re-election.

SHAPING THE AGENDA OF POLITICS

Much of the congressional agenda is dictated by external events—initiatives from the president or the federal bureaucracy, a major disaster, or the hostile actions of a foreign government. However, members themselves do initiate some proposals, and some of these rise to the top of the nation's political agenda.

The evolving composition of Congress has produced many changes in the agenda of politics. For instance, the Congressional Black Caucus raised the visibility of the issue of racial discrimination and the death penalty to new levels through its efforts to shape the 1994 crime bill. When the highest-ranking admiral in the Navy wanted to retire with four-star rank, women in the Senate, backed by women in the House, waged a fierce, highly publicized battle to reduce his rank to two stars because of his role in a sexual harassment scandal. They lost by a narrow margin but helped raise the visibility of sexual harassment as an issue.[50] Finally, the large group of first-term Republican representatives elected in 1994 helped push a number of conservative policy proposals to the top of the congressional agenda in 1995.

In an angry response to the Navy's decision to retire its highest-ranking admiral without a reduction in rank despite his alleged involvement in the Tailhook sexual harassment scandal, female senators and representatives staged a vehement protest. Increased diversity among members of Congress has influenced the types of issues raised on Capitol Hill. While these women ultimately lost the vote on this issue, because of their position as legislators they were able to focus public attention on the problem of sexual harassment in the military.

AFFECTING PUBLIC OPINION

The opinions of attentive publics sometimes exert a powerful influence on Congress. By the same token, Congress itself can help to influence public opinion and educate the citizenry. Its members speak as well as listen, and their views carry considerable weight with some voters and political elites, especially when they speak with authority on issues about which their audience knows little. Committees in Congress also affect opinions by issuing reports and conducting public hearings. Hearings affect public opinion through the publicity they generate and, in particular, through the testimony of interest groups.

OVERSEEING THE FEDERAL BUREAUCRACY

Anyone who has been asked, "Where are you going?" or "What did you do last night?" understands the notion of oversight. **Oversight** refers to Congress's monitoring of executive branch activities to ensure their conformity to policies established by law. This involves gathering information and taking action to ensure the laws it passes are implemented as intended.

Congressional hearings and investigations are an important form of oversight. In addition to generating reports by congressional committees and support agencies, hearings provide an essential ingredient of oversight: information about what is happening. Once obtained, such information can be used to revise existing legislation or enact new laws.

Constituent service also contributes to oversight. If something is wrong with the way the Postal Service is delivering the mail or the Veterans' Administration is handling checks, the flood of complaints from citizens alerts Congress. As any congressional college summer intern can testify, when a senator's or representative's office calls a federal agency on behalf of a constituent, the person answering the phone listens carefully. The cumulative impact of thousands of such interventions is not trivial.

Finally, the power of the purse provides an effective mechanism for inducing federal agencies to respond to requests for information, resolve a constituent's problem, or change policy. The threat of reduced appropriations is often only implicit and its exercise subtle. Sometimes, however, it is less subtle, as when a number of years ago a member of Congress concerned about the failure of the U.S. Forest Service to conduct a study of grazing fees on public lands mused to a Forest Service official, "I was wondering whether there was some place in this appropriation where we would make a substantial cut for the purpose of impressing upon you the desirability of making this study."[51]

LEGITIMIZING PUBLIC POLICY

Congress performs a significant symbolic function in the American political system when it places a "seal of approval" on many components of national policy. For example, George Bush's "War on Drugs," a major policy initiative of his first year in office, received rapid congressional endorsement. Bill Clinton's initiative to establish a public service corps likewise was legitimized when Congress approved his proposal.

Congressional resolutions, although lacking the status of law, often legitimize important policy. Bush's resistance to the Iraqi invasion of Kuwait in 1990 received immediate and strong congressional approval. The subsequent Senate debate on whether to give Bush the authority, as well as its approval, to use force to expel Iraq from Kuwait helped solidify public support for the Persian Gulf War.

Conversely, congressional failure to support a president's foreign policy initiatives, or its extending of lukewarm support, can reduce public support. Bill Clinton withdrew the U.S. troops committed by George Bush to a United Nations peacekeeping mission in Somalia because public support for the mission had decreased, partly as a result of congressional criticism and resolutions calling for troop withdrawal.

SENATE APPROVAL OF APPOINTMENTS AND TREATIES

The Senate's constitutional task of ratifying treaties negotiated by the president provides another vehicle for overseeing actions of the executive branch. The Senate's refusal to ratify a treaty can affect a president's foreign policy standing and general political standing. For example, the inability of President Jimmy Carter to win Senate ratification in 1979 of the second strategic arms limitations treaty (known as SALT II) damaged his prestige and complicated U.S. relations with the Soviets.

The Constitution also requires the Senate to confirm the appointment of key federal officials such as ambassadors, cabinet members, and judges; this task provides opportunities to engage in "anticipatory oversight." During confirmation hearings, senators sometimes seek to influence the nominee's sub-sequent behavior. Close questioning about nominees' views on abortion oc-curred routinely during the confirmation hearings of secretaries of health and human services and Supreme Court nominees in the 1980s and early 1990s.

The Senate's exclusive power to ratify treaties and confirm key ap-pointees may induce an administration to consult in advance with key Senate leaders and foreign affairs specialists during the negotiation of treaties or the process of selecting nominees. Also, the Senate sometimes rejects a president's nominee, especially when the president is from one party and the Senate majority is from the other. In a bruising battle, the Senate rejected George Bush's nominee for secretary of defense, John Tower. Coming as it did just a few months into the life of the new administration, the rejection demonstrated how much trouble the new president was having getting what he wanted from the Democratic Congress.

SUMMARY

The Constitution establishes the basic structure of Congress—with its two chambers, the House and Senate—and gives it the authority to pass legisla-tion, including the "power of the purse" to authorize and appropriate money. The Constitution says nothing, however, about the standing committees and subcommittees where most of the legislative work of Congress takes place. Committees in the House, which play a more important role than those in the Senate, can be classified as power committees, policy commit-tees, or constituency committees. The Constitution also does not mention conference committees, where differences between House and Senate lan-guage in important bills are reconciled. Although the influence of commit-tees and their chairs has declined somewhat in the past few years, power in Congress remains dispersed. Party leaders in both chambers can only partly overcome this dispersion.

No other institution in American politics relies as heavily on a complex set of formal rules as does Congress. Rules guiding the formal steps that bills must follow to be enacted are especially important. Both chambers also rely on informal norms or rules to reduce conflict and to get work done. In the Senate these informal norms, or folkways, are especially important. Senate rules permit a minority to block legislation through a filibuster that can be ended only if 60 senators vote for cloture.

Long dominated by middle-aged and older white men, Congress now in-cludes record numbers of women, Hispanics, and African Americans. How-ever, all three groups are still underrepresented. Incumbents continue to seek and win re-election at high rates, although the 1992 and 1994 elections re-sulted in many retirements and defeats. The 535 voting members of Congress rely heavily on the more than 20,000 people who staff members' offices as well as committees and support agencies such as the Congressional Budget Office.

Much of what members do—the committee assignments they seek, the issues they focus on in committee, the attention they pay to raising money and building support in their districts—is undertaken with an eye to winning re-election. Roll-call votes constitute only part of the process that determines the fate of legislation, but they receive the most attention from the media and scholars. A complex mixture of factors shape these votes. Sometimes, legislators trade votes (or logroll) on different issues. Cue networks also help legislators decide how to vote. However, party affiliation stands out as the most important factor in roll-call votes.

Decision making in Congress displays three important patterns: the defensive advantage, which makes it easier to block than to pass legislation; the need to form separate coalitions to pass (or defeat) legislation; and the incremental or gradual nature of the changes in policy that most legislation brings about.

The central task of Congress is legislating; through this and other activities, members of Congress provide representation for their constituents, shape the agenda of politics, affect public opinion, exercise oversight over the bureaucracy, and help to legitimize public policy.

APPLYING KNOWLEDGE

EXPLAINING HOUSE PASSAGE
OF THE ASSAULT WEAPONS BAN

 The prospects for House passage of legislation banning 19 semiautomatic assault weapons looked dim on May 5, 1994. Although a substantial majority of the public favored it with growing intensity, the NRA fiercely opposed it. It had organized its 3 million members into a cohesive lobbying force that bombarded Congress with letters, made substantial campaign contributions through its PAC, and deployed experienced lobbyists in Washington.

The NRA had defeated every proposed gun control measure but one in the previous 25 years. Through experience, members of Congress knew that failure to vote with the NRA usually led to "a ton of grief" in their district. Furthermore, the House leadership was not united behind the ban; and a vote on it was scheduled promptly, while leaders favoring the ban were still trying to line up support.

But when the final vote came, to the surprise of nearly everyone the ban passed by a two-vote margin. A majority of Democrats (177–77) joined a minority of Republicans (38–137) and the sole independent in passing it. Last-minute shifts turned a 214–213 defeat into a 216–214 victory—including "yes" votes from a retiring representative who had long supported the NRA but resented its failure to support his chief aide as his replacement, and from a first-term African American from a rural Georgia district who was strongly lobbied by members of the Congressional Black Caucus.[1] President Clinton contributed to the victory by making urgent phone calls, as did Treasury Secretary Lloyd Bentsen, a gun owner himself. Law enforcement officers mounted their own campaign on behalf of the measure, countering the NRA's pressure.

Does this incident illustrate how Congress works, as described in this chapter? Is the incident typical of how legislation is passed? Was this a "party vote"? If so, why? Which of the factors affecting roll-call voting might have been at work? What other concepts presented in this chapter help to explain the vote?

[1] Katharine Q. Seelye, "House Approves Bill to Prohibit 19 Assault Arms," *New York Times*, May 16, 1994, p. A1.

Key Terms

Congressional Budget Office (CBO) (p. 482)

standing committees (p. 484)

power committees (p. 484)

policy committees (p. 484)

constituency committees (p. 484)

seniority system (p. 486)

conference committees (p. 487)

Speaker (p. 488)

majority leader (p. 488)

minority leader (p. 488)

filibuster (p. 490)

cloture (p. 490)

folkways (p. 490)

power of the purse (p. 493)

line-item veto (p. 493)

roll-call vote (p. 497)

cue networks (p. 498)

logrolling (p. 498)

defensive advantage (p. 505)

representation (p. 507)

oversight (p. 509)

Recommended Readings

Berkman, Michael. *The State Roots of National Politics: Congress and the Tax Agenda.* Pittsburgh: University of Pittsburgh Press, 1994. An intriguing analysis of how political and policy changes in the states move to the national level as state legislators win seats in Congress.

Bianco, William. *Trust.* Ann Arbor: University of Michigan Press, 1994. An examination of the strategic calculations that members of Congress make regarding how to earn and keep the trust of their constituents.

Dodd, Lawrence C., and Bruce I. Oppenheimer, eds. *Congress Reconsidered,* 5th ed. Washington, D.C.: Congressional Quarterly Press, 1993. A number of top scholars provide chapter-length summaries of the latest research on nearly every aspect of Congress.

Fenno, Richard F., Jr. *Home Style.* Boston: Little, Brown, 1978. A classic study of how House members relate to their districts. For a good summary of Fenno's work, see Morris P. Fiorina and David W. Rohde, "Richard Fenno's Research Agenda and the Study of Congress," in Fiorina and Rohde, eds., *Home Style and Washington Work: Studies of Congressional Politics* (Ann Arbor: University of Michigan Press, 1989), pp. 1–15.

Keefe, William J. *Congress and the American People,* 3rd ed. Englewood Cliffs, N.J.: Prentice Hall, 1988. A lively general introduction to the politics of Congress.

Kingdon, John W. *Congressmen's Voting Decisions.* New York: Harper & Row, 1981. A classic study based on interviews with members of the House about how they decide how they will vote on legislation.

Ornstein, Norman, Thomas E. Mann, and Michael J. Malbin. *Vital Statistics on Congress: 1993–1994.* Washington, D.C.: Congressional Quarterly Press, 1994. A compilation of statistics on virtually every aspect of the membership and operation of Congress.

Sinclair, Barbara. *The Transformation of the U.S. Senate.* Baltimore: Johns Hopkins University Press, 1989. A description of the changing nature and politics of the U.S. Senate.

Vogler, David J. *The Politics of Congress,* 5th ed. Boston: Allyn and Bacon, 1988. A comprehensive text summarizing the major research findings on Congress.

The Presidency

LIKE A ROLLER-COASTER RIDE at an amusement park, President Bill Clinton's first year in office was characterized by striking highs as well as devastating lows. It included a little of everything: legislative failures, management gaffes, appointment fiascoes, questions about credibility, and a series of impressive legislative victories.

After a hard-fought campaign that had to overcome an incumbent president riding high after the Persian Gulf War, a number of personal charges, and a strong showing by independent Ross Perot, Clinton took the oath of office in January 1993 with enthusiasm, an agenda, and the strong public support that new presidents typically enjoy. Some 70 percent of the public viewed him as a strong leader, and nearly 60 percent approved of his performance during the presidential transition period and in the administration's early weeks.

Things quickly began to go wrong, however, as some of the new administration's actions alienated interest groups that had supported Clinton's election. Women's groups, for example, were outraged when Clinton facilitated the withdrawal from candidacy of two women nominated for attorney general—Zoë Baird and Judge Kimba Wood—when it became known that both had employed illegal aliens. This outrage intensified when the administration refused to withdraw the names of male nominees—such as Ron Brown for secretary of commerce—who had likewise employed illegal aliens. Clinton also offended many civil rights groups when he withdrew his nominee for head of the Justice Department's Civil Rights Division, Lani Guinier, an African-American woman highly regarded in the civil rights community but opposed by conservatives. Civil rights leaders and members of the Congressional Black Caucus protested Clinton's decision with angry words and with threats to withhold support on other matters.[1]

Early on, the new administration's reputation in Washington suffered from a perception

National newsmagazines reflect the intense monitoring of presidents' performance by the media. After a series of strategic mistakes, unnecessary compromises, and abrupt changes in policy position early in his administration, Bill Clinton's reputation in Washington began to suffer, as is shown by the unflattering images appearing on these covers.

that it was making strategic gaffes and unnecessary compromises. By announcing shortly after the inauguration its intention to end the ban on allowing gay men and lesbians in the military, the Clinton administration alienated major participants in Washington, such as the military establishment and key members of Congress. Clinton ultimately agreed to a compromise on this issue, offering his support for what became known as the "don't ask, don't tell" policy that permits gays to serve in the military if they remain "in the closet," hiding their sexual orientation. Clinton enraged gay and lesbian groups as well by arguing that the compromise was important "so that our country does not appear to be endorsing a gay lifestyle."[2]

Other issues continued to stand in the way of the president's traditional "honeymoon" period. Clinton's budget was subject to nearly daily changes in substance as he made concessions to a variety of interest groups and to members of Congress.[3] For instance, facing criticism from Republicans and drug company executives, the president agreed to drop crucial elements of a proposal for the government to buy and distribute all childhood vaccines in the United States.[4] Bowing to pressure from the business community, Clinton agreed to delay an increase in the minimum wage.[5] Finally, Clinton's reputation in Washington suffered a severe blow when his economic stimulus program, portrayed by opponents as filled with projects that would add unnecessarily to the federal deficit, was defeated in the Senate by a Republican filibuster.

Clinton's reputation with the general public also deteriorated, fueled by an impression that he had abandoned many of his campaign commitments. On the domestic side, Clinton did a dramatic about-face on his plan to provide a tax cut for the middle class. Instead of the promised tax relief, Clinton asked the middle class for sacrifices in a program of deficit reduction that required taxes to be raised.[6] In foreign affairs, Clinton backed away from campaign pledges to intervene in Bosnia and welcome Haitian refugees to U.S. shores. By the end of his first four months in office, only 38 percent of the public viewed him as a strong leader, down from the

nearly 70 percent who had perceived him in this way when he took office. Moreover, only 36 percent of the public approved of Clinton's job performance, a record low for a post–World War II president this early in his term.[7]

By offering a great variety of proposals almost immediately after taking office, the Clinton administration failed to focus public attention on any one proposal and appeared to lack a clear set of legislative priorities. The media began to raise questions about Clinton's presidency and his ability to govern. A *Newsweek* cover in early June showed a picture of a puzzled Clinton with the headline "What's Wrong?" *Time* magazine's cover of the same week showed a tiny, inch-tall Bill Clinton gazing up at the towering headline "The Incredible Shrinking President." Clinton himself told reporters that he and his staff needed to provide and communicate more of a focus to his agenda.[8] To help bring clarity to his administration, and to repair the damage to his reputation in Washington and among the general public, Clinton hired David Gergen—a former communications aide to Presidents Richard Nixon, Gerald Ford, and Ronald Reagan—to serve as a counselor to the president.[9]

Soon after Gergen's appointment, Clinton's fortunes began to change. By the end of the first year the new administration had put together an outstanding legislative record. The administration had successfully pushed through Congress the North American Free Trade Agreement, family leave legislation, a program of national community service, and, perhaps most important, a budget and economic program that would reduce the federal deficit. Although Clinton's legislative victories rested on many compromises, by January 1994 the president's public approval rating had climbed dramatically to more than 60 percent, higher than even Ronald Reagan's public approval rating at the end of his first year in office.

What happened when Bill Clinton became president helps to illustrate several important aspects of the role of the central participant in the play of power. This chapter elaborates on a number of the features touched on in our case study of Clinton's presidency, including the importance of the president's reputation in Washington and among the general public, the role of the media in helping to shape this reputation, the fragile nature of the resources available to the president, and the multiple constituencies faced by presidential administrations—including Congress, the bureaucracy, domestic interest groups, supporters and opponents, and foreign governments. We return to the case study throughout the chapter to highlight the many roles of the president, the importance of participants surrounding the president, and the strategies used by presidential administrations to advance their agenda.

COMMON BELIEFS
THE PRESIDENT AS SYMBOL

 The media's tendency to focus their stories on the president, as well as the images that such stories evoke, help to create and reinforce several common beliefs about the president and the presidency. For instance, the media's almost exclusive focus on the president—reporting on everything from policy proposals to tastes in junk food—affirms an association in peoples' minds between the president and the government. For many Americans the president personifies the government, acting as the primary symbol in the play of power. Newspaper headlines and broadcast news stories announce when a president makes a proposal; they are less likely to highlight congressional amendments or other changes to the proposal. Commenting on the bias in television coverage toward stories on the president, one presidential scholar observes, "If there is a balance of powers within the government, it rarely shows on television."[1]

The president also serves as a symbol of the nation. Having studied presidential rhetoric used in inaugural and major foreign and domestic policy addresses by presidents from Harry Truman to Ronald Reagan, Barbara Hinckley suggests that presidents themselves reinforce the association between the president and the nation.[2] Presidents rely heavily in their speeches on the collective pronoun "we," implying that the president, the public, and the nation are synonymous. The deliberate blurring of boundaries between the president and the public becomes a resource for presidents seeking to mobilize support for specific policy proposals.

As the primary symbol of government, the president is often viewed as acting alone. Phrases such as "It's lonely at the top" and "The buck stops here" describe the president's situation well. Presidents are thought to work alone much of the time, struggling with policy options to make the tough decisions. And they are quick to claim sole responsibility for actions and decisions that meet with public approval. Presidents cultivate the image of acting alone through the language they use in their speeches. In the speeches she studied, Hinckley found few references to participants in the play of power other than the president or the collective "we." Anywhere from 60 to 90 percent of references to specific participants in politics were to the president specifically or to the collective "we."[3] She concludes that because most presidential speeches focus on the president alone, "one gains little sense of a government—complete with advisers, congressional committees and leaders, cabinet secretaries and ambassadors—at work in the domestic or foreign policy sphere."[4]

517

Finally, presidents are viewed as very powerful. When they speak, others are expected to listen and obey. After President George Bush announced that he would veto a bill permitting federally funded abortions for poor women who were victims of rape or incest, Senator Brock Adams (D-Wash.) juxtaposed the position of poor women with that of the president. Illustrating conventional views of the president, Senator Adams argued: "I'm sorry the President of the United States, *the most powerful man in the world* [italics added], has chosen to veto this bill and thereby cause enormous additional suffering for some of the world's most unfortunate and powerless victims."[5] This sentiment—that presidents are among the most powerful individuals in the world—is widely shared among Americans.[6]

In this chapter we challenge the accuracy of these beliefs. We view the presidency as only one of many institutions seeking to influence outcomes in the play of power. Although presidents may on rare occasion act alone, more typically they are surrounded by many advisers who play key roles in decision making. Equally important, the president must interact with members of other branches of government to negotiate policy outcomes. Presidents are not always successful in influencing members of other branches, in part because of their relative lack of formal political resources.

Even if people's beliefs about the presidency are inaccurate, they have important political consequences. Images of the president as acting alone, as personifying the government and the nation, and as all-powerful, lead Americans to have high expectations of their presidents. For example, polls showed that when Bill Clinton took office in January 1993, a majority of Americans expected him to improve education, health care, the environment, and the economy; help racial minorities and the poor; keep the country out of war; increase respect for the United States abroad; and reduce unemployment.[7] Of course, a huge gap exists between these expectations and the actual power of presidents, a gap that often results in public perceptions of a failed and ineffective presidency.[8] The extraordinarily high public expectations of President Clinton, combined with early mistakes that reflected his inexperience in the national play of power, help explain the rapid and dramatic loss of public support in the early months of his presidency.

[1] Thomas Cronin, *The State of the Presidency*, 2nd ed. (Boston: Little, Brown, 1980), p. 96.
[2] Barbara Hinckley, *The Symbolic Presidency: How Presidents Portray Themselves* (New York: Routledge, 1990).
[3] Hinckley, *The Symbolic Presidency*, Chapter 2.
[4] Hinckley, *The Symbolic Presidency*, p. 52.
[5] "Senate Sends Bush a Bill to Broaden Abortion Aid," *New York Times*, October 20, 1991, p. A14.
[6] See Hinckley, *The Symbolic Presidency*, pp. 10–11.
[7] See Gallup poll data summarized in George C. Edwards III and Stephen J. Wayne, *Presidential Leadership*, 3rd ed. (New York: St. Martin's Press, 1994), pp. 96–99.
[8] Theodore J. Lowi, *The Personal President: Power Invested, Promise Unfulfilled* (Ithaca, N.Y.: Cornell University Press, 1985). ∎

Because media and popular attention focus on the office of the president, the individual occupying that position plays a crucial role in American politics. However, presidents can accomplish little on their own. In a highly complex and intricate play of power, they must surround themselves with numerous participants who help them to achieve their goals. In this section we examine attributes of the president and of these other important participants.

THE PRESIDENT

Of the 41 presidents from George Washington to Bill Clinton, a relatively homogeneous group of white, wealthy, almost exclusively Protestant males with substantial political experience have occupied this key position. Several groups of Americans, including African Americans, Hispanics, Asian Americans, Native Americans, and Jews, have had no members come even close to winning this important office. And women, constituting half of the population, have had no serious candidate for president and only one—Geraldine Ferraro of New York—nominated for vice-president on a major party's ticket. Part of the problem for women, according to some scholars, is that popular conceptions of leadership emphasize qualities commonly associated with males; this makes it difficult for women to be judged as qualified for the presidency (see "Participants: Must One 'Act Like a Man' to Be President?" on pp. 520–21).

THE PRESIDENTIAL OFFICE

Pick up a newspaper on any day and look at the vast array of reported events requiring some presidential response or action. The stories may describe national disasters such as floods, hurricanes, or earthquakes; civil wars in foreign lands; an increase in interest rates; the development of nuclear weapons by other nations; scandals within the administration; criticism by congressional leaders; and/or demonstrations by organized interests. The public's expectation that the president will improve their lives in a number of areas does not change to accommodate these breaking events. Presidents are expected to deal with all these situations while simultaneously pushing a legislative program that will produce tangible benefits for the nation.

But the president is only one person with limited time, knowledge, and resources. Indeed, a presidential commission studying the executive branch in 1937 concluded that "the president needs help."[10] Today that help is provided by many important participants—presidential staff who offer advice on political and policy matters. A presidential office has evolved since the late 1930s. Referred to as the **Executive Office of the President (EOP)**, it is composed of a variety of formal agencies that assist presidents in making administrative and political decisions. Also incorporated into the EOP is the **White House Office (WHO)**, staffed by advisers closest to the president (see Figure 14.1 on p. 522).

The presidential office has grown tremendously over the years.[11] Thomas Jefferson's supporting personnel consisted of one secretary and a messenger.

MUST ONE "ACT LIKE A MAN" TO BE PRESIDENT?

Although attitudes are slowly changing, many elements of American culture continue to reinforce stereotypical views regarding gender roles and behavior. These views apply to criteria that are often used to evaluate leadership, criteria that linguist Robin Lakoff suggests are gendered, favoring qualities commonly associated with males. Think about the adjectives we use to describe "good" leaders: Good leaders are *strong, tough, forceful,* and *aggressive.* This conception causes difficulties for women. To be judged as effective leaders, they must emulate men. Using the examples of Indira Gandhi in India, Golda Meir in Israel, and Margaret Thatcher in Great Britain, Lakoff suggests that women "must become honorary men" in order to succeed as leaders. According to Lakoff, "In politics it is currently axiomatic that a woman must act, if not speak, more aggressively than a man to convince the electorate that she can be trusted with their country." But women are confronted by a paradox. Those who successfully demonstrate that they are as strong, aggressive, and tough as men risk being seen as abrasive, strident, and overly aggressive.

Men who show traits typically associated with women may have difficulties in presidential politics as well. Perhaps the most notable example occurred in 1972 when Senator Edmund Muskie was reported to have wept on the steps of the offices of the arch-conservative New Hampshire newspaper, the *Manchester Union Leader.* Muskie, the frontrunner for the Democratic nomination, appeared to weep as he defended his wife from what he considered a vicious attack by the newspaper. Although Muskie denies that he actually shed tears, the news reports doomed his prospects for the nomination. Since 1972, public tears appear to have become more acceptable in male leaders—at least under certain circumstances. Ronald Reagan choked up on several occasions during his presidency. Massachusetts governor Michael Dukakis, the Democratic nominee for the presidency in 1988, wept publicly as his wife announced that she had been addicted to prescription drugs. Tears ran down President Clinton's cheeks as his newly announced nominee for the United States Supreme Court, Ruth Bader Ginsburg, spoke of her deceased mother. None of these events raised questions about the leadership capabilities of these men.

However, when Patricia Schroeder wept while announcing that she would not run for president in 1988, the media, along with many women commentators and activists, focused considerable attention on the tears. For example, a news analysis in the *New York Times* asked, "Did her tearful withdrawal damage the cause by reinforcing the fears of those who think that women are too emotional to govern?" A female pollster echoed this concern:

> I certainly sympathize with the fact that it was an incredibly emotional moment. But it seems to me her inability to command her emotions when she was making an announcement about the presidency only served to reinforce some basic stereotypes about women running for office—those . . . being lack of composure, inability to make tough decisions.

Although men may now publicly shed tears without necessarily suffering serious electoral consequences, the continuing significance of "manliness" is evident in George Bush's transformation during the 1988 campaign. One of his major tasks was to shake off the impression that he would not be a strong leader—that, as

Nowadays men may shed tears even if they are presidents or candidates, but it is much riskier for prominent women to weep publicly. When Bill Clinton wiped away tears while listening to Supreme Court nominee Ruth Bader Ginsburg speak of her deceased mother, his crying caused little comment. In contrast, when Patricia Schroeder wept as she announced that she would not run for president in 1988, she was criticized for demonstrating that women are too emotional to hold high office.

his opponents claimed, he was a "wimp." Bush had the misfortune of serving as vice-president under a president who had an unusually macho image. Ronald Reagan was routinely photographed and filmed chopping wood and riding horses on his ranch. He talked tough, referring to the Soviet Union as "the evil empire," for example. He enjoyed using expressions from macho movies, daring Congress at one point to pass a tax increase that he would veto with the words "Go ahead, make my day."

According to Robin Lakoff, Bush's advisers believed that his propensity for using "feminine" language patterns and gestures reinforced his image as a "wimp" and underscored the contrast between his image and Reagan's. Prior to the Republican National Convention, Bush's advisers felt that he spoke like a woman, with a deferential style and expressions that indicated a noncompetitive nature. Bush's sentences fre-

quently contained hedges, such as "I guess," "sorta," and "kinda." His speeches were delivered softly, with his voice frequently trailing off at the end of sentences. He smiled and giggled too much, it was thought. But by the time of the general election Bush talked tougher, raised his voice more frequently, and used more confrontational language. His gestures changed dramatically, becoming more aggressive and more "masculine"—pointing, chopping, and stabbing the air. His voice became tense and shrill. Bush now looked more "presidential"— tough, strong, forceful, aggressive.

Sources: Robin Tolmach Lakoff, *Talking Power: The Politics of Language* (New York: Basic, 1990); Bella Abzug, *Gender Gap* (Boston: Houghton Mifflin, 1984); Bernard Weinraub, "Are Female Tears Saltier than Male Tears?" *New York Times,* September 30, 1987, p. A26; Warren Weaver Jr., "Assessing the Lessons of Schroeder's Brief Run," *New York Times,* October 4, 1987, p. D5.

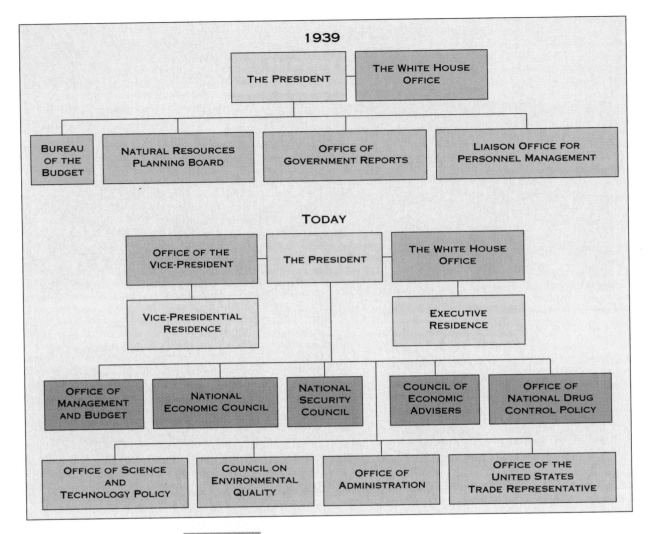

1939

THE PRESIDENT

THE WHITE HOUSE OFFICE

BUREAU OF THE BUDGET

NATURAL RESOURCES PLANNING BOARD

OFFICE OF GOVERNMENT REPORTS

LIAISON OFFICE FOR PERSONNEL MANAGEMENT

TODAY

OFFICE OF THE VICE-PRESIDENT

THE PRESIDENT

THE WHITE HOUSE OFFICE

VICE-PRESIDENTIAL RESIDENCE

EXECUTIVE RESIDENCE

OFFICE OF MANAGEMENT AND BUDGET

NATIONAL ECONOMIC COUNCIL

NATIONAL SECURITY COUNCIL

COUNCIL OF ECONOMIC ADVISERS

OFFICE OF NATIONAL DRUG CONTROL POLICY

OFFICE OF SCIENCE AND TECHNOLOGY POLICY

COUNCIL ON ENVIRONMENTAL QUALITY

OFFICE OF ADMINISTRATION

OFFICE OF THE UNITED STATES TRADE REPRESENTATIVE

FIGURE 14.1

THE EXECUTIVE OFFICE OF THE PRESIDENT AT ITS INCEPTION AND TODAY This chart depicts the tremendous growth in the Executive Office of the President. In 1939, when it was created, the EOP comprised five offices. Today there are 13 offices, many of which were developed to offer the president help with problems (e.g., drug control and the environment) identified as significant since 1939. *Source:* Adapted from George C. Edwards III and Stephen J. Wayne, *Presidential Leadership,* 3rd ed. (New York: St. Martin's Press, 1994), p. 176. Copyright © 1994. Reprinted with permission of St. Martin's Press, Inc.

Ulysses Grant employed three staff members, and Woodrow Wilson had seven aides. But as government and demands on the presidency grew during the 1930s, more help was needed. In 1939 Franklin Roosevelt issued an executive order creating the Executive Office of the President and the White House Office. Roosevelt staffed the WHO with some 45 people and the EOP with nearly 600 advisers. Today the White House Office staff has grown to approximately 500 members, and the EOP employs a staff of approximately 2,000 with a budget of more than $200 million.[12]

The president's closest and most trusted advisers constitute the WHO staff. Inevitably, some of the White House staff have worked for the president in his previously held positions and/or in the campaign that brought him to office. White House staff members serve at the pleasure of the president. Their positions are not subject to Senate confirmation and carry no tenure and few statutory constraints. The precise organization of the various units within the WHO varies among administrations. Presidents define the staff's tasks as they wish, may shift responsibilities among staff units, and may create or abolish units to suit their needs.

The White House staff The White House Office staff has diverse roles and responsibilities. For example, a national security adviser (an expert in international and military affairs) provides the president with daily briefings and advice. Domestic policy advisers assist in formulating a legislative agenda. Some staffers serve as political liaisons to organized interest groups, Congress, party officials, and various ethnic and religious groups. A press secretary acts as liaison to the print and electronic media; a communications director advises the president on how to deliver messages effectively to the public. Crises and emergencies may be met by the establishment of a new position. For example, to show his commitment to fighting the spread of AIDS, President Clinton created the new position of "White House AIDS coordinator," a position that quickly became known as "AIDS czar."[13] To demonstrate his commitment to dealing with the economy, Clinton created a National Economic Council composed of cabinet members and key White House staffers.

As the White House staff has grown, power and authority have been organized hierarchically and centralized in one or a few chief aides. The *chief of staff* performs a variety of roles; these include providing direction to the staff and helping to organize the president's day, determining who gains access to him. The chief of staff often handles the president's most difficult tasks, such as evaluating the performance of other aides and hiring and firing personnel. When mistakes are made by the administration, the chief of staff is expected to act as a "lightning rod," accepting blame and helping to divert controversy and criticism away from the president. Dick Cheney, chief of staff under Gerald Ford, described his role as follows: "If there's a dirty deed to be done, it's the chief of staff who's got to do it. The president gets credit for what works, and you get the blame for what doesn't work. That's the nature of the beast."[14]

When Bill Clinton's presidency appeared to be in serious trouble after the first four months, his chief of staff, Thomas McLarty, along with other members of his staff, received and accepted much of the blame. One response of the Clinton administration was to shake up the staff, firing a few aides, moving others to new or modified positions, and hiring David Gergen as the new communications director.[15] During a later troublesome period, Clinton replaced McLarty with Leon Panetta as chief of staff.

Finally, the chief of staff may protect the president by refusing to carry out potentially damaging presidential orders. In his memoirs Richard Nixon's chief of staff, H. R. Haldeman, remarked that "this president had to be protected from himself." Haldeman recounted the following stories to illustrate the point:

> Time and time again I would receive petty, vindictive orders. "Hugh Sidey is to be kept off Air Force One." (Sidey was *Time*'s man.) Or even, once or twice, "All the press is banned from Air

Force One." (Pool representatives of the press accompanied the President on every trip.) Or, after a Senator made an anti–Vietnam War speech: "Put a 24-hour surveillance on that bastard." And on and on. If I took no action, I would pay for it. He'd be on the intercom buzzing me ten minutes after such an order. "What have you done about Sidey?" I'd say, "I'm working on it," and delay and delay until Nixon would one day comment, with a sort of half-smile on his face, "I guess you never took action on that, did you?" "No." "Well, I guess it was the best thing."[16]

The centralized, hierarchical organization of power and authority within the White House staff may be necessary for decisions to be made expeditiously by an organization that has grown so large. However, these bureaucratic elements also present certain dangers. Presidents may easily become isolated, relying only on a small group of like-minded advisers. Moreover, the information and policy options received by presidents may be selectively filtered and interpreted by their closest advisers. Debates over policy and politics, thus, may occur within a very narrow range of opinion.

Agencies within the EOP   One step removed from the White House staff is a diverse set of formal agencies that form the rest of the EOP. Chief among these is the Office of Management and Budget (OMB), an influential body that is responsible for constructing the president's budget. Acting on agency requests, it advises the president on how much the administration should propose to spend for each governmental program, how it might pay for them, and where programs might be cut. The OMB also exercises "legislative clearance" by examining the budgetary implications of legislative proposals and the extent to which such proposals conform to the president's policy goals. Because of the office's critical functions, OMB directors can play significant roles in presidential administrations. Leon Panetta, President Clinton's first OMB director, was an influential voice in the administration, arguing that deficit reduction should be given a high priority. He so distinguished himself in this role that Clinton appointed him chief of staff. Panetta was replaced by Alice Rivlin, the first woman to direct the OMB.

The EOP also includes the Council of Economic Advisers, a small group of economists who advise the president on economic policy, and the National Security Council (NSC), a group of leading military and diplomatic experts. The president's national security adviser heads the NSC, which is charged with analytical and coordinating tasks. Occasionally, the NSC has gone beyond analysis and advising: During the Reagan administration, for example, some of its members secretly sold weapons to Iran in hopes of freeing American hostages and diverted the proceeds from the sale to the Nicaraguan Contras. Other units included in the EOP in recent administrations have been the Council on Environmental Quality and the Office of Science and Technology Policy.

THE CABINET

The president's **cabinet** consists of the heads of 14 executive departments, the OMB director, and the ambassador to the United Nations. Cabinet officials are appointed by the president with the Senate's consent.

Although the full cabinet could in theory have an influential advisory role, it rarely does so in practice. Most presidents take office intending to make good use of their cabinet, but after a short time they hold few, if any, meetings. This retreat from initial intentions occurs for several reasons. First, the cabinet is too large and unwieldy for focused and fruitful discussions. Second, unlike the White House Office staff, cabinet members with few exceptions have not enjoyed a previous personal relationship with the president. (A few exceptions include John F. Kennedy's attorney general, his brother Robert; and George Bush's secretary of state, longtime friend James Baker. These men did have significant advisory roles.) Most are selected for reasons other than loyalty to the president, such as their stature in the public view or their ability to please interest groups.

Furthermore, the advice of some cabinet members may be suspect if they are perceived to have conflicting interests. Indeed, their interests in carrying out the president's program may run counter to their own department's interest or those of important clientele groups that provide support to their department. Consequently, presidents and members of the inner circle (which may include cabinet members perceived as loyal) often suspect some cabinet members of "going native"—that is, developing the attitudes, predispositions, and interests of their executive branch agency.

In recent years presidents have sought to choose cabinet secretaries to reflect racial, ethnic, and gender diversity in the American population. President Clinton announced almost immediately after winning the White House that one of his goals in choosing cabinet department heads would be to increase the representation of women and minorities. Of Clinton's 14 initial cabinet nominations, four were women (Donna Shalala at Health and Human Services, Janet Reno at Justice, Hazel O'Leary at Energy, and Madeleine Albright as U.N. representative), three were African-American

The first Clinton cabinet reflected some of the racial, ethnic, and gender diversity in the American population. More than half of the cabinet secretaries were women or members of racial minorities. The composition of a president's cabinet has symbolic value; in this case the diversity suggests an inclusive administration.

men (Ron Brown at Commerce, Mike Espy at Agriculture, and Jesse Brown at Veterans' Affairs), and two were Hispanics (Henry Cisneros at Housing and Urban Development, and Federico Peña at Transportation).

Cabinet secretaries appointed by newly elected presidents are often drawn from the private sector or from positions in state and local government. Surprisingly, most have little, if any, federal executive branch experience. By and large, they come to Washington without close personal relations with other cabinet secretaries, members of the White House staff, or lower-level officials in their own agencies. Consequently, they tend to be "strangers with only a fleeting chance to learn how to work together."[17]

Cabinet secretaries occupy a precarious position in the play of power. When they and their departments achieve some success, they must share it with the president. But when failure occurs, they are held accountable and may be asked to resign. These characteristics do not make cabinet posts highly attractive; they lead to a pattern of appointees "getting in, gaining whatever experience and prestige are allowed, and getting out."[18] The high turnover rate among cabinet officials is shown in Table 14.1, which reports the average months of service for individuals occupying the various posts. Notice that few, on average, serve for a full presidential term (48 months).

Although presidents may come to distrust or ignore the full cabinet, individual cabinet members often provide important advice, assist in policy formation and implementation, and engage in other politically useful activities. For example, George Bush's Labor secretary, Lynn Martin, and his

TABLE 14.1
AVERAGE LENGTH OF SERVICE OF CABINET SECRETARIES, 1945–1993

DEPARTMENT	AVERAGE MONTHS OF SERVICE
Agriculture	44
Commerce	34
Defense	33
Education	31
Energy	32
Health, Education, and Welfare/ Health and Human Services	28
Housing and Urban Development	37
Interior	38
Justice	29
Labor	32
State	38
Transportation	30
Treasury	34
Veterans' Affairs	24

Source: Adapted from Charles O. Jones, *The Presidency in a Separated System* (Washington, D.C.: Brookings Institution, 1994), p. 65.

Health and Human Services secretary, Louis Sullivan, criticized congressional Democrats' health care proposals (which Bush opposed), suggesting that they would have "a devastating effect on workers."[19] Bill Clinton used his cabinet members as a sales team early in his term, sending them throughout the country to speak in favor of his economic programs.[20]

THE VICE-PRESIDENT

In addition to the White House staff, cabinet secretaries, and other advisers in the EOP, the president can rely for counsel on the vice-president. Historically, however, vice-presidents have rarely had significant advisory or policy-making roles. John Adams, the nation's first vice-president, remarked to his wife that "my country in its wisdom contrived for me the most insignificant office that ever the invention of man contrived or his imagination conceived." John Nance Garner, Franklin Roosevelt's first vice-president, went even further, suggesting that "the vice presidency isn't worth a pitcher of warm spit."[21]

Vice-presidents most often do political chores for the president, such as campaigning and fundraising. They also sit on the National Security Council, attend cabinet meetings, travel abroad as the president's representative, and make statements that presidents may agree with but wish to avoid making themselves.

Vice-presidents are rarely chosen because they excel at these activities. Typically, they are selected to help in the general election campaign by balancing the ticket. For example, John F. Kennedy and 1988 Democratic presidential candidate Michael Dukakis, both of Massachusetts, selected Texas senators Lyndon Johnson and Lloyd Bentsen, respectively, as running mates to provide geographical balance. Ronald Reagan, perceived as a member of the Republican Party's conservative wing, chose the more moderate George Bush to provide ideological balance. To the surprise of many political observers, Arkansas governor Bill Clinton selected another southern moderate, Al Gore, as his vice-presidential running mate. Although Gore provided neither geographical nor ideological balance, he did add significantly to the ticket in areas where Clinton was perceived to be weak. For example, Gore had served in the Vietnam War whereas Clinton had received a student deferment and was being accused of "draft dodging" by Republicans. Gore also had foreign policy expertise from his days in the Senate, something Clinton lacked. In addition, Gore's reputation as an environmentalist balanced concerns that environmental groups expressed regarding Clinton; and Gore's squeaky-clean image as a family man helped to balance any negative impressions of Clinton that resulted from stories that he had committed adultery.

During the past several administrations, the trend has been for vice-presidents to have a more expanded and important role in the play of power. Indeed, vice-presidents now have increased staff resources as well as greater physical proximity and more regular access to the president than ever before.[22] Under Ronald Reagan, George Bush chaired the crisis management team and was in charge of the Task Force on Regulatory Relief, an influential group that helped to abolish certain economic regulations. During Bush's presidency, Dan Quayle chaired the Council on Competitiveness, a group

Vice-President Al Gore has played an important role in the Clinton administration. Here he debates the passage of NAFTA with Ross Perot on the *Larry King Live* show. Tasks such as this cannot be performed by presidents lest they appear partisan and unpresidential.

that assessed the impact of federal legislation on the business community. In this position Quayle helped delay implementation of, or eliminated altogether, federal regulations that threatened to have an adverse effect on business. One of Bill Clinton's first acts as president was the abolition of this council.

Vice-President Al Gore has been an extremely important participant in the Clinton administration, wielding considerable policy influence in his areas of expertise, especially the environment and advanced technology. Gore has participated in selecting cabinet and some lower-level appointees; reviewed drafts of presidential speeches; headed a committee that drafted recommendations to "reinvent" or streamline the federal government; played a crucial role in lobbying Congress to pass some of the administration's legislative programs; and represented the administration admirably in a televised debate with Ross Perot on the North American Free Trade Agreement.[23]

FIRST LADIES AND OTHER INFORMAL ADVISERS

Not all presidential advisers serve in official positions. Presidents sometimes rely on advice from friends, close associates, and others whom they respect from outside the government. Andrew Jackson began a practice, utilized by many later presidents, of consulting with a group of unofficial advisers who became known as the "kitchen cabinet." Members included Jackson's private secretary, a longtime friend, and a newspaper editor. Jackson's critics tried to discredit him by depicting these advisers slipping into the president's study by way of the back stairs that ran through the kitchen.[24] Similarly, Theodore Roosevelt consulted with his "tennis cabinet"; Warren Harding with a "poker cabinet"; and Herbert Hoover, a physical fitness enthusiast, with a "medicine-ball cabinet." More recently Bill Clinton has consulted with informal advisers whom the media calls FOBs (Friends of Bill).

The president's spouse, the first lady, has the potential to play an important role in the play of power. Several have done so. Eleanor Roosevelt, Rosalynn Carter, and Nancy Reagan, for example, each served as her husband's closest adviser. Rosalynn Carter was the first presidential spouse to attend cabinet meetings and schedule weekly business lunches with the president. Many first ladies have devoted their time to particular policy and social welfare interests while in the White House. Rosalynn Carter worked on mental health issues, Nancy Reagan on drug programs, and Barbara Bush on literacy. Rosalynn Carter was the first presidential spouse to testify before a congressional committee, in her capacity as honorary chair of the President's Council on Mental Health.

Hillary Rodham Clinton has broken new ground in her role as first lady. Her participation in policy and personnel decisions of the Clinton administration has been more extensive than that of any of her predecessors. Designated by President Clinton to direct a task force on health care policy—a central position in the new administration—Hillary Clinton helped to write health care legislation and lobbied members of Congress, industry representatives, and policy experts. In addition, she spent considerable time meeting with the public at town meetings and held numerous interviews with the media.

COMPETITION AND CONFLICT AMONG PRESIDENTIAL ADVISERS

Staff members in the presidential office, cabinet members, and the vice-president participate in a play of power in which the victors win the president's ear and help shape presidential decisions. Competition and conflict are often quite fierce. Staff members compete for assignments and authority that are a measure of their standing and prestige as advisers. Advisers seek to make themselves indispensable to the president or those closest to the president, ensuring continued access, input, and influence. Within the White House, the competition has a physical dimension as staff members compete for office space, assessing relative influence in terms of proximity to the Oval Office in the West Wing.[25] When President Clinton took office in January 1993, office space in the White House was organized to maximize access to him by his communications team (see Figure 14.2 on p. 530). Communications director George Stephanopoulos, deputy communications director Ricki Seidman, and press secretary Dee Dee Myers maintained offices only steps away from the Oval Office.

To promote pet causes or personal careers, presidential advisers cultivate allies outside the White House among members of Congress, congressional staffers, lobbyists, and the press. Allies outside the White House may be called on to help or hurt other advisers within the White House. For example, when the press reported that John Sununu, George Bush's chief of staff, used the government's planes for personal travel, many speculated that the stories had been leaked by his opponents on the White House staff in order to

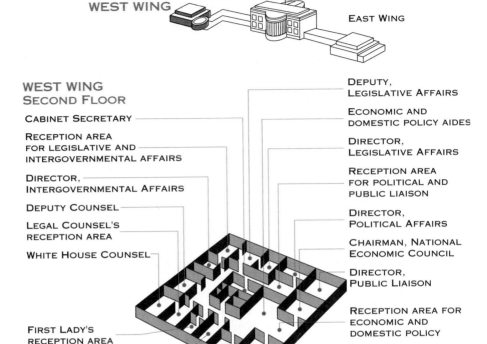

WEST WING

RESIDENCE AND STATE ROOMS

EAST WING

WEST WING
SECOND FLOOR

CABINET SECRETARY

RECEPTION AREA FOR LEGISLATIVE AND INTERGOVERNMENTAL AFFAIRS

DIRECTOR, INTERGOVERNMENTAL AFFAIRS

DEPUTY COUNSEL

LEGAL COUNSEL'S RECEPTION AREA

WHITE HOUSE COUNSEL

FIRST LADY'S RECEPTION AREA

FIRST LADY'S OFFICE

FIRST LADY'S CHIEF OF STAFF

DEPUTY, LEGISLATIVE AFFAIRS

ECONOMIC AND DOMESTIC POLICY AIDES

DIRECTOR, LEGISLATIVE AFFAIRS

RECEPTION AREA FOR POLITICAL AND PUBLIC LIAISON

DIRECTOR, POLITICAL AFFAIRS

CHAIRMAN, NATIONAL ECONOMIC COUNCIL

DIRECTOR, PUBLIC LIAISON

RECEPTION AREA FOR ECONOMIC AND DOMESTIC POLICY

ASSISTANT TO THE PRESIDENT, DOMESTIC POLICY

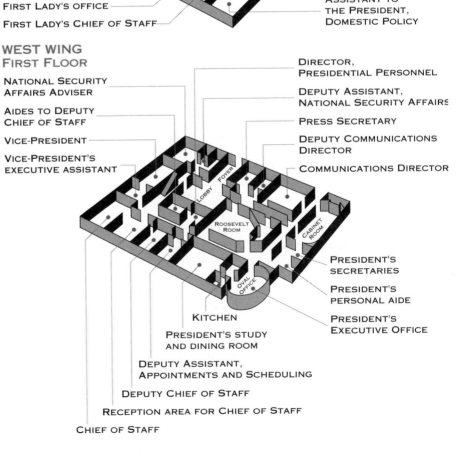

WEST WING
FIRST FLOOR

NATIONAL SECURITY AFFAIRS ADVISER

AIDES TO DEPUTY CHIEF OF STAFF

VICE-PRESIDENT

VICE-PRESIDENT'S EXECUTIVE ASSISTANT

LOBBY   FOYER

ROOSEVELT ROOM

CABINET ROOM

OVAL OFFICE

KITCHEN

PRESIDENT'S STUDY AND DINING ROOM

DEPUTY ASSISTANT, APPOINTMENTS AND SCHEDULING

DEPUTY CHIEF OF STAFF

RECEPTION AREA FOR CHIEF OF STAFF

CHIEF OF STAFF

DIRECTOR, PRESIDENTIAL PERSONNEL

DEPUTY ASSISTANT, NATIONAL SECURITY AFFAIRS

PRESS SECRETARY

DEPUTY COMMUNICATIONS DIRECTOR

COMMUNICATIONS DIRECTOR

PRESIDENT'S SECRETARIES

PRESIDENT'S PERSONAL AIDE

PRESIDENT'S EXECUTIVE OFFICE

FIGURE 14.2

THE WHITE HOUSE: CORRIDORS OF POWER The White House office arrangements have great political significance. Staff members compete for influence with the president, which is thought to increase with closer proximity to the Oval Office. The large number of communications staffers with offices on the West Wing's first floor when Bill Clinton took office suggests the growing importance of public relations to presidents. *Source:* Adapted from diagram by Dave Cook, *Washington Post,* national edition, February 5, 1993, p. A23. Copyright © *The Washington Post.* Reprinted with permission.

embarrass him. When damaging news stories began to appear emphasizing tax increases in Clinton's health care program, many suspected that the leaks came from his economic advisers, who were pushing for deficit reduction.[26]

Frequently, overlapping jurisdictions promote conflict between White House staff advisers and those in the cabinet. For example, overlapping foreign policy jurisdictions of the secretary of state and the president's national security adviser have resulted in open warfare in many administrations. Richard Nixon's secretary of state, William Rogers, and his national security adviser, Henry Kissinger, repeatedly battled over authority in making foreign policy. As one senator remarked, Kissinger won the game hands down: "They let Rogers handle Norway and Malagasy, and Kissinger would handle Russia and China and everything else he was interested in."[27]

HOW RULES AFFECT THE PRESIDENCY

Constitutional rules affect the presidency in a number of ways. First, rules setting up the Electoral College shape the strategies that presidential candidates use in running for office. Constitutional rules also specify the legal qualifications for office: The president must be a United States citizen, must be at least 35 years old, and must have been a United States resident for at least 14 years. Article II calls for a four-year term of office; and in 1951 the Twenty-Second Amendment limited the president's service to two full terms.

In a very general way, constitutional rules outline the president's roles, powers, and duties. Although Congress is given the power to declare war, the president serves as the military's commander-in-chief. The president also acts as the federal government's administrative head, being responsible to "take care that laws are faithfully executed." The Constitution gives presidents the authority to grant pardons and reprieves for federal crimes, except in cases of impeachment; make treaties with the Senate's consent; and nominate ambassadors, judges, and executive branch officials, also with the Senate's consent.

Rules in the Constitution offer the president a few resources that may be used in the play of power. Article I, Section 7 gives the president the authority to veto legislation passed by Congress—a veto that Congress may override with a two-thirds vote in each house. By requiring that the president "from time to time" inform Congress of the "state of the union," the rules offer the president a forum for presenting an agenda. This requirement has evolved into an annual State of the Union address in which the president outlines the administration's legislative priorities to a joint session of Congress and, perhaps more important, the nation. The Constitution also offers the president the opportunity to take further action on an agenda by proposing legislation to Congress. Finally, the rules permit presidents to convene both houses of Congress on "extraordinary occasions," although this has rarely happened.

Given the significance of the contemporary presidency, the constitutional rules describing the office's duties, functions, and powers are surprisingly vague and very brief. The presidency's resources, especially when compared to those of other institutions such as Congress, are relatively few and unimpressive. But the presidency's power and resources have evolved significantly over time, making this institution a crucial participant in the play of power. We now turn to a discussion of these resources.

RESOURCES OF THE PRESIDENCY: THE POWER TO PERSUADE

The colonies' experiences with the autocratic rule of King George III made the Constitution's framers wary of executive power. Consequently, the rules governing the play of power, as outlined in the Constitution, create a presidency with few formal powers and resources.

The Constitution's limited conception of the president's role has evolved over time through informal practices into a considerably more expansive view, one that offers presidents additional valuable resources. During wartime and other periods of national crisis, presidents have increased their authority dramatically by taking actions exceeding constitutional rules and commonly held notions of their proper powers. In such instances presidents have argued that the action in question fell within the office's **inherent powers**. From this much broader perspective, presidential power derives from constitutional provisions as well as from inferences that may be drawn from such provisions. In other words, presidents have argued that constitutional provisions imply powers and resources not explicitly stated in the Constitution. The trend has been for Congress and the federal courts to approve of most presidential action during crises. Once these expanded grants of authority have been claimed successfully by presidents, succeeding presidents may use the new grant of authority as a resource without further debate.

The primary means through which presidents exercise inherent powers is through the issuance of **executive orders** that have the force of law, actions whose legitimacy derives from liberal interpretations of the Constitution by presidents and the Supreme Court. Executive orders have played a major role in presidential policy making in areas such as national security, economic policy, and civil rights. For example, Franklin Roosevelt issued an executive order to mandate the internment of Japanese Americans during World War II. During the Great Depression of the 1930s, Roosevelt used executive orders to establish agencies with extensive authority over the prices, wages, and profits of private enterprises (e.g., the Office of Price Administration and the Office of Economic Stabilization). In 1948 Harry Truman ended racial segregation in the military by executive order. One of Bill Clinton's first acts as president was to issue an executive order rescinding the Bush administration's policy that family planning clinics receiving federal funds could not discuss or answer questions about abortion. Not all executive orders go unchallenged, however. When Truman tried to take control of the steel industry during the Korean War through an executive order, the Supreme Court struck down the action as unconstitutional.[28]

Presidential powers and resources also have grown as Congress has willingly delegated authority to the executive branch. Especially during times of national crisis, when the public demands swift and effective actions to resolve problems, Congress, through a process called **delegation of powers**, confers greater responsibilities on the president and the executive branch to create agencies and administer programs that address the crisis. For example, in the 1930s Congress delegated broad grants of authority to Franklin Roosevelt so he could deal with the Great Depression.

Because constitutional rules confer few formal powers and resources on the chief executive, presidents must develop informal resources to exert influence. In his classic study *Presidential Power*, Richard Neustadt argues that the relative lack of formal powers forces presidents to rely on one primary informal resource, their power to persuade.[29] The strength of this resource depends on the president's ability to convince others that what the White House wants them to do (e.g., vote for the president's legislative program) is what they ought to do for their own sake, using their own authority.

One effective way for the president to persuade is, in Lyndon Johnson's words, to "strike fear in people." Even though the probability of persuading others to act is enhanced by such things as reasoned argument, appeals to the "public good," and personal charm and charisma, effective persuasion is most likely to occur if people fear they may be hurt personally and politically by the president if they do not act in the desired fashion.[30] Members of Congress and other participants in governmental arenas are aware that at some point, the doing of their jobs and furthering of their ambitions may depend on the president. Effective presidents use the needs and fears of other players to their advantage. In particular, presidents play on such needs and fears as they persuade other participants to accept the substance of their policy proposals.

Here President Clinton confers with Democratic leaders in Congress. The president's party can be an important resource in winning legislative victories. Research shows that the party of the president offers the chief executive much more support than do members of the opposition. However, even members of the president's party cannot be taken for granted, and their votes are sometimes gained only after the president expends some of the office's resources.

Presidents are aided in their quest to persuade by a variety of important resources. Chief among them is the president's status as the most recognizable and prestigious participant in the play of power, the participant occupying the most respected office in American government. After all, few people are immune from the impulse to say "yes" to a presidential request. The impulse grows stronger when requests are made in the Oval Office or in the president's private study, "where almost tangibly he partakes of the aura of his physical surroundings."[31] The president may ultimately veto legislation that runs counter to the administration's agenda or may merely threaten to veto legislation so as to enhance the chances of gaining a suitable compromise. In some cases, appointments may be used as presidential resources. Presidents may reward allies by appointing those they support, or they may punish foes by opposing their preferred nominees. Popular presidents may persuade members of Congress by promising to campaign for them in the next election; those failing to show support may find the president campaigning for their opponents. Finally, presidents and other members of the administration may do personal favors, such as inviting other participants to state dinners, obtaining concert tickets, and offering photo opportunities that may help in the home district.

Even though political parties seem to be in decline, the president's party in Congress continues to be a significant resource. Table 14.2 indicates the support given to presidents by members of each party in roll-call votes in which the president registered a clear position. The data demonstrate substantial differences between the level of support presidents receive from members of the two major parties, both in the House and Senate. Although presidents from each party and within each party varied in their policy agendas, personalities, skills, and styles, members of their own party provided them with more support than did members of the opposition party, with the gap generally exceeding 25 percent.

TABLE 14.2
PARTY SUPPORT FOR PRESIDENTS, 1953–1990

	PRESIDENTS		DIFFERENCE IN SUPPORT
	DEMOCRATIC	REPUBLICAN	
House			
Democrats	69%	36%	33%
Republicans	38	66	28
Senate			
Democrats	61	34	27
Republicans	35	69	34

Figures are based on roll-call votes where the winning side was supported by fewer than 80 percent of those voting.

Source: Data from George C. Edwards III, *At the Margins: Presidential Leadership of Congress* (New Haven: Yale University Press, 1989). Data updated in George C. Edwards III and Stephen J. Wayne, *Presidential Leadership*, 3rd ed. (New York: St. Martin's Press, 1994), p. 300.

Where separate institutions share power, all participants have resources. This means that presidents do not always get what they want. To be effective, presidents must skillfully use the resources they have at their disposal.

PROFESSIONAL REPUTATION

The degree to which presidents are able to persuade depends on two characteristics. First, presidential success depends on the president's *professional reputation.* Those in Washington whom a president seeks to persuade must be convinced that the president has the skill and the will to use available resources to reward allies and punish opponents.

Because the president is the central participant in the play of power, the chief executive's behavior is closely observed and evaluated by other participants. The president's power to persuade depends in part on other participants' appraisals as they observe presidential behavior over time. By and large, these participants look for behavior patterns. Will the president hurt opponents? Will the president help those who provide support? Is the president reliable—do deeds live up to promises? Is the president able or inept? Is the president effective?

These appraisals are crucial because other participants calculate the costs and benefits of supporting and opposing the president, anticipating what the president will do if they behave in a certain way. Thus, if the president threatens to veto ten bills but makes good on the threat in only two cases, future threats will be taken less seriously. If the president announces support for legislation, then the administration must in fact provide vigorous support during congressional debate. If legislation supported by the president passes in Congress, the president benefits by appearing to be strong. If such legislation fails, the president may be damaged politically, appearing to be weak.

Bill Clinton's first year in office illustrates the administration's use of resources to persuade other participants. In pushing his legislative programs, Clinton used his status as the central participant in the play of power to his advantage. He made numerous courtesy phone calls to members of Congress, reaching one—Republican senator James Jeffords of Vermont—as far away as a hotel room in Damascus, Syria. Clinton thanked Jeffords for supporting his family leave legislation. According to the senator, "No matter how long we've been around the Hill, there's nothing more exciting than a call from the president." When asked if the call would make him more inclined to vote for Clinton's economic programs, he responded, "It doesn't hurt."[32]

During debates over the economic program, the Clinton administration also demonstrated that it knew how to offer inducements for support. While his popularity remained high, Clinton traveled to selected congressional districts to praise members whose support he needed. And he showed a willingness to trade resources at his disposal in return for votes.[33] For instance, he promised several members of the Congressional Black Caucus that in return for their support he would tighten sanctions on Haiti to increase pressure on the military regime that had taken power there. To garner support for the North American Free Trade Agreement (NAFTA), the administration made a number of similar deals and promises. For example, the 23-member Florida delegation announced its opposition to NAFTA, fearing its effects on Florida's citrus industry. Almost the entire delegation

Presidents must often negotiate with members of Congress to secure their support. In order to gain the support of the Congressional Black Caucus for his economic program, Bill Clinton promised its members that he would tighten sanctions on Haiti in hopes of restoring Jean-Bertrand Aristide's presidency.

changed its vote when the Clinton administration promised to impose stiff tariffs on Mexican shipments of frozen juices if they caused the price of juices produced in the United States to fall.[34]

On one occasion the Clinton administration sent a signal that it would punish its opponents. During a meeting with Vice-President Al Gore, Alabama's Democratic senator, Richard Shelby, embarrassed the administration by complaining to television reporters that Clinton's economic program was "high on taxes and low on cuts." In response the White House announced that it would move the management team for a space shuttle contract from Alabama to Texas, a loss of some 90 jobs for Shelby's state. Further, Shelby was given only one ticket to a White House ceremony honoring the University of Alabama football team, whereas Alabama's other senator, Howell Heflin, received eleven. According to one administration official, "We tolerate dissension here, but he embarrassed the Veep on national television instead of speaking to us privately."[35]

Because the Clinton administration was perceived to "tolerate dissension" (the Shelby case being an apparent exception), President Clinton's professional reputation suffered somewhat during his first year. Presidential administrations must be willing and able to strike deals and make trades in order to win votes for their legislative programs. Clinton, however, was perceived as being *too* willing to accommodate his critics and opponents. Other participants—members of Congress and journalists, in particular—wondered if Clinton would ever fight for something. The image of Clinton as too willing to compromise was reinforced by his abandonment of some campaign promises, such as a middle-class tax cut, his willingness to compromise on allowing gays in the military and increasing the minimum wage, and his withdrawal of nominees to a number of posts at the first sign of controversy. Other participants began to wonder whether they had anything to fear if they opposed this president.

At the same time, two events much discussed by members of Congress led some to perceive Clinton as an unreliable ally. First, California representative George Miller, chair of the House Interior Committee, mobilized support among other Westerners for a Clinton proposal to increase fees for mining and grazing on federal lands. Then, after being pressured by a few senators from western states, Clinton dropped the idea for a time, pulling the rug out from under Miller's feet.[36] Second, after Clinton promised House Democrats that he would not compromise on his controversial energy tax proposal, he appeared to abandon the tax and those who supported it when his economic program began making its way through the Senate.[37] House Democrats voting for the tax were enraged that the White House had backed away from the proposal, leaving them to feel the political heat back in their home districts.

Nevertheless, Clinton was able to win many legislative victories by the end of his first year, even if several were by very narrow margins. One way the administration convinced some Democrats to support its programs, especially the economic package, was to argue that the future success of this presidency depended on early victories. More than one congressional Democrat heard the president say of their vote on his economic plan, "You must save my presidency."[38] Although these victories were neither clean nor pretty, the administration could use the president's image as a winner and his increased public popularity to help in its quest to mend fences damaged during the battles over the first year's legislative agenda.

PUBLIC PRESTIGE

In addition to professional reputation, success in persuasion depends on the president's *public prestige.*[39] Whereas presidents' professional reputations depend on how they are viewed by participants in Washington, public prestige refers to how they are perceived outside of Washington. Other Washington participants anticipate the public's reactions (especially those of their constituents) to their support of or opposition to the president's program. Public prestige is assessed in various ways—by polls that are regularly taken and reported on in the media, by opinions expressed by visiting constituents, and by the words of political columnists and pundits. Popular presidents have an easier time persuading other participants than do unpopular presidents. According to one Senate staffer, "It's an absolute rule up here: Popular presidents get what they want; unpopular ones don't."[40] Bill Clinton's chance for success in passing his economic program was not helped, then, by his steadily declining popularity and the increasing number of critical editorials and news analyses appearing at the end of his fourth month in office. But the case study opening this chapter shows how quickly presidential popularity can change. Throughout 1993, Clinton received help from increasing popularity as measured in polls. The dynamic, ever-changing nature of presidential popularity is illustrated in Figure 14.3 (on p. 538).

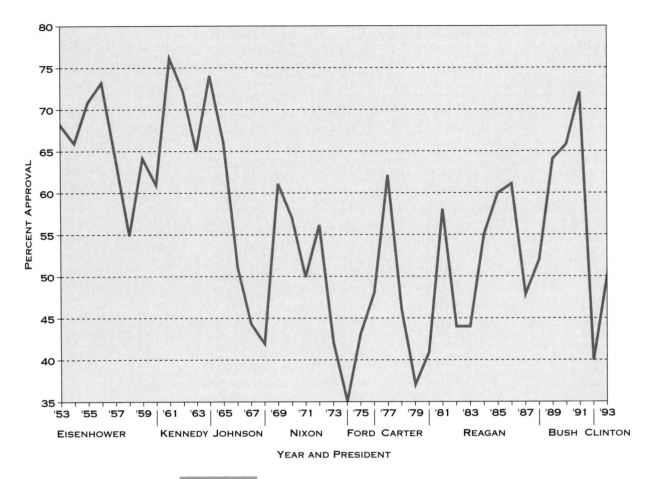

80

75

70

65

PERCENT APPROVAL

60

55

50

45

40

35

'53 '55 '57 '59 | '61 '63 '65 '67 | '69 '71 '73 | '75 '77 '79 | '81 '83 '85 '87 | '89 '91 | '93

EISENHOWER   KENNEDY JOHNSON   NIXON   FORD CARTER   REAGAN   BUSH CLINTON

YEAR AND PRESIDENT

FIGURE 14.3

PUBLIC APPROVAL OF PRESIDENTIAL JOB PERFORMANCE, 1953–1993 (YEARLY AVERAGES) Polls routinely ask respondents whether they approve of the job the president is doing. Presidents follow public approval figures carefully, as high ratings can be used as a resource in negotiations with Congress. Over the course of a four-year term, public approval can fluctuate wildly. It is sensitive to economic conditions, successes and failures in Congress, and international events, among other things. *Source:* George C. Edwards III and Stephen J. Wayne, *Presidential Leadership*, 3rd ed. (New York: St. Martin's Press, 1994), p. 106. Copyright © 1994. Reprinted with permission of St. Martin's Press, Inc.

Changes in presidential popularity can generally be explained by several factors. Economic conditions—inflation and unemployment levels, a recession—affect public evaluations of the president's performance.[41] Public evaluations are also likely to change as crises occur during a presidential term. The public "rallies 'round the flag" to support the president during unexpected events, particularly military crises. George Bush enjoyed enormous popularity in March 1991—with 90 percent of the public approving of his performance—in the wake of American involvement in the Persian Gulf War. Likewise, when Bill Clinton ordered a bombing raid on Iraq in retaliation for the Iraqi government's alleged involvement in an attempt to assassinate Bush, Clinton's public approval rating rose from 39 percent to 46 percent. An analysis of "rally events" during the past 10 presidencies showed that they produced, on average, an 8 percent increase in public approval and that they lasted an average of 10 weeks.[42]

The most significant factor affecting more lasting changes in public evaluations of presidential performance is the impression of presidential success or failure that the public gains from the news media.[43] News reports "prime" the public to focus on specific policy areas, and they provide the basis for public assessment of performance.[44] If the media report primarily on presidential policy successes (e.g., getting presidential initiatives passed in Congress), public approval will likely rise. Conversely, reports of presidential policy failures result in decreasing public approval.

The media, then, play a significant role in an administration's ability to achieve its goals. In response, modern presidents have increased their use of media and public relations advisers in the hope of controlling media output. Such advisers, sometimes called "spin doctors," seek to create conditions under which the media will report on the president's accomplishments and portray any failures in the best possible light. These tasks are often difficult. For example, George Stephanopoulos, a press secretary for Bill Clinton during the first few months, failed to convince the media that a particular week's major news story was Clinton's signing of family leave legislation rather than his dumping of Judge Kimba Wood as a nominee for attorney general.

Both the Nixon and Reagan administrations spent considerable time and energy trying to control media coverage. Advisers carefully planned each president's day so that all activities, when covered by the media, emphasized only positive features. David Gergen described the Nixon administration's approach:

> We had a rule in the Nixon operation that before any public event was put on the schedule you had to know what the headline out of the event was going to be, what the picture was going to be, and what the lead paragraph would be. You had to think of it in those terms, and if you couldn't justify it, it didn't go on the schedule.[45]

In a similar way, the Reagan administration organized the public portions of the president's day to dramatize a "story line of the day" that had been developed in advance.[46]

A president's public prestige and professional reputation interact in an important way. Presidents now spend considerable time seeking to improve or maintain their public prestige, because it may be used to win legislative victories. These victories, in turn, help to enhance the president's professional reputation in Washington.

STRATEGIES USED BY THE PRESIDENT

Presidents and their administrations use resources in designing strategies that will accomplish their goals. But the president operates in a difficult, often treacherous environment where success is elusive. Remember that because presidents personify government, they face extraordinarily high—even impossible—expectations. In addition, they face conflicting demands from multiple constituencies: the mass public, the bureaucracy, the party, allies and opponents on domestic policy, and friends and foes abroad. "Strategies: Bill Clinton Walks the Tightrope of Race" (on pp. 542–43) illustrates some of these demands and how a president responds to them.

Where separate institutions share power, presidents must deal with other institutions and participants to accomplish their goals. Although Americans tend to think of the president as personifying government and as acting alone, in reality presidents must rely on Congress to pass their legislative programs and on the bureaucracy to implement them. A great deal of evidence suggests that presidents have trouble influencing either institution.

STRATEGIES FOR DEALING WITH CONGRESS

In recent years the term *gridlock* has been used to illustrate the inability of the president and Congress to agree on a course of action. But presidents always have had difficulties in directing congressional action. Scholars who have examined presidents' relations with Congress suggest that, at best, presidents seeking policy change may influence a few critical participants and take advantage of opportunities for such change already in the environment. That is, presidents influence Congress at the margins.[47] This influence, however, may be significant on occasion, affecting enough votes to determine the outcome of heated legislative battles.

Many aspects of the presidency—available resources, prospects for legislative success, strategies likely to be successful, and the nature of the opposition—shift over time. An important strategy of successful presidents is to hit the ground running, moving swiftly during the first year to accomplish their legislative priorities. Presidents' prospects for success in dealing with Congress are best during the early months of the new administration, when public support is usually at its highest, the cabinet is fresh and enthusiastic, Congress is somewhat accommodating, the press is more supportive, and the administration has pushed for so little that it has yet to alienate any organized interests. During the second year, legislative proposals are complicated by typical losses in public popularity, growing opposition from organized interests hurt by administration action during the first year, and the upcoming midterm congressional elections. The final two years are devoted primarily to re-election, with members of the opposition party in Congress making it difficult for the president to pass any programs.

Successful presidents use the first year to set the national and congressional agenda by focusing attention on adminstration priorities. The legitimacy associated with winning a national election, combined with the media attention accompanying the office, permits skillful presidents to use their position in the play of power to set broad national priorities. If the president fails to focus congressional attention on administration priorities, the administration's legislative program is likely to get lost in the complex congressional process. In addition, presidents and their staffs can lobby Congress effectively on only one or a few proposals at a time. And presidents' political capital is so limited that they must decide where it is best spent.

Although presidents receive advice from a variety of sources and delegate important decision-making authority, images such as this photo of Jimmy Carter alone in the Oval Office reinforce the erroneous belief that presidents reach decisions on their own and are the sole participants in the play of power.

On taking office, the Reagan administration set the congressional agenda brilliantly by focusing attention on and then winning passage of three major priorities: a large tax cut, cuts in the rate of spending increases for domestic programs, and substantial increases in defense spending. In contrast, the administration of Jimmy Carter tried to accomplish too much too early in its term, proposing a variety of programs to Congress. As one observer commented, Carter "had so many priorities that he seemed to have none."[48] Consequently, Carter had great difficulty mobilizing public and congressional support for any particular program.

Negotiation and bargaining Many of the president's resources—such as modifying legislative proposals, offering policy inducements, making appointments, doing personal favors, promising to provide campaign support, and threatening to veto—are used in a strategy of bargaining or negotiating with members of Congress to pass legislation. When presidents engage in this strategy, it is particularly clear that they lack real authority and instead must use persuasion and exchange to get things accomplished.

Negotiations and bargaining are not conducted solely, or even primarily, by the president. The president has too little time and political capital to engage personally in negotiations over every piece of legislation before Congress. But presidents do expend much time and effort on major congressional votes regarding high-priority legislation. In most cases, negotiations take place between members of Congress and representatives of the president, such as the congressional liaison, chief of staff, other members of the White House Office, and cabinet secretaries. Journalist Bob Woodward tells the following story of how Secretary of the Treasury Lloyd Bentsen and

BILL CLINTON WALKS THE TIGHTROPE OF RACE

African Americans, along with other racial minorities, have come to occupy a prominent position in the Democratic Party's presidential coalition—the constellation of groups that have routinely voted for the Democratic nominee, especially since the 1960s.[1] In recent years African Americans, as well as Mexican Americans and Puerto Ricans, have been among the most reliable voting blocs for Democratic candidates.

How have racial minorities fared, given their loyalty to the Democratic Party and Democratic presidents? On the one hand, according to one historian, civil rights laws enacted during the 1960s and early 1970s, together with supporting court decisions and administrative enforcement, "broke the back of the system of racial segregation and destroyed the legal basis for denying minorities and women full access to education, employment, the professions, and the opportunities of the private marketplace and the public arena."[2] On the other hand, although these accomplishments are significant, they have done little to remedy stark and overwhelming inequalities in income and employment[3] and "savage inequalities" in resources for inner-city schools that disproportionately serve people of color, as compared to resources for predominantly white suburban schools.[4]

Ronald Walters suggests that policies to deal with racial inequalities have not been proposed or forcefully pushed, in part because African Americans have not developed effective strategies in the play of power. Walters argues that the unqualified support African Americans offer Democratic presidents is now taken for granted, with few material benefits being offered or delivered in return to ensure future loyalty.[5] Although African Americans have been rewarded with increasing numbers of positions in the national party apparatus, these jobs do not translate into influence with presidential candidates. Walters advises that African Americans and other racial minorities should think more strategically about how they participate in the play of power, seeking autonomy from

Although Bill Clinton knew Lani Guinier personally before her nomination for director of the Justice Department's Civil Rights Division, he withdrew his nomination as criticism of her position on racial quotas mounted. Clinton used this action to distance himself from the civil rights groups supporting Guinier and to demonstrate that he was a "new Democrat" committed to helping the middle class.

both political parties in an effort to transform their vote into a resource that can be used in bargaining over policy.

The problem for Democrats is that close ties to racial minorities make it difficult to win national elections. Several studies suggest that the Democrats' perceived alliance with African Americans has cost their party the support of certain segments of its traditional national electoral coalition. Following the New Deal in the 1930s, the Democrats stitched together a national coalition with an underlying class foundation. Democrats won national office by gaining the support of "have-not" segments of the populace: workers, Catholics, unionists, Jews, small farmers, urban dwellers, lower-income white Southerners, and people of color. By the mid-1970s many whites began to see the Democratic Party as allied too closely with African Americans, providing them with "special benefits" not shared, but paid for, by white members of the coalition.[6] Taxes came to be seen as imposing costs on white Americans while exclusively benefiting black Americans. Tax dollars were perceived to flow into a welfare system that disproportionately benefited African Americans even as the Democratic Party ignored the crime problem, perceived by many as an African-American problem. These feelings were exploited by the national Republican Party and its presidential candidates—in discussions of issues such as affirmative action, busing, crime, and welfare, and in advertisements that featured threatening symbols such as Willie Horton—to win over some white Southerners and the white working class in the North. By transforming the cleavage between parties from one of rich versus poor into one of taxpayers versus tax money recipients, of those believing in "merit" versus those believing in "special preferences," and of bearers of governmental costs versus beneficiaries of governmental programs, the Republican Party caused enough of a shift in electoral behavior to secure a virtual lock on the White House from 1980 to 1992.

The defection of traditional Democratic voters to Reagan and Bush in the 1980s helps to explain how the Clinton administration has navigated issues that touch on race. Being well aware of how racial and tax issues had hurt previous Democratic candidates, Bill Clinton distanced himself in the campaign from the traditional civil rights agenda advocated by leaders such as Jesse Jackson. Clinton's public criticism of Jesse Jackson for inviting Sister Souljah—a controversial rap artist who allegedly advocated violence against whites in the aftermath of the Rodney King beating case—to address Jackson's "Rainbow Coalition" convention, combined with his effort to portray himself as a "new" Democrat beholden to no one but "the forgotten middle class," paid electoral dividends by appealing to many of the so-called Reagan Democrats. Despite these moves, Clinton received some 83 percent of the African-American vote.

After the election Clinton got into trouble with moderate elements of his electoral coalition when he appeared to set quotas for appointments and failed to initiate actions in the areas of crime and welfare reform. Clinton abandoned a middle-class tax cut and began discussing the need for tax increases. Republicans in Congress labeled the proposals as "tax and spend"—a phrase that has become a code name for using taxpayers' money to fund programs benefiting African Americans and other disadvantaged groups—and complained bitterly, along with more moderate Democrats, about some of his nominees for administrative posts (e.g., Lani Guinier, the African-American woman nominated to direct the Civil Rights Division of the Justice Department). Senate Republicans accused Clinton's economic stimulus program of seeking to aid "big-city mayors," many of whom are African American and almost all of whom govern cities disproportionately inhabited by people of color. As Clinton's popularity slipped, the administration responded by withdrawing the nomination of Lani Guinier—dubbed a "quota queen" by the *Wall Street Journal*—and by emphasizing issues and positions of greater interest to more moderate segments of the electoral coalition.

[1] Katherine Tate, *From Protest to Politics: The New Black Voters in American Elections* (Cambridge, Mass.: Harvard University Press, 1993), Chapter 3.

[2] Hugh Davis Graham, *Civil Rights and the Presidency: Race and Gender in American Politics 1960–1972* (New York: Oxford University Press, 1992), p. 3.

[3] Andrew Hacker, *Two Nations: Black and White, Separate, Hostile, Unequal* (New York: Scribner's, 1992).

[4] Jonathan Kozol, *Savage Inequalities: Children in America's Schools* (New York: HarperCollins, 1991).

[5] Ronald W. Walters, *Black Presidential Politics in America: A Strategic Approach* (Albany: State University of New York Press, 1988).

[6] John R. Petrocik, *Party Coalitions: Realignments and the Decline of the New Deal Party System* (Chicago: University of Chicago Press, 1981); Edward G. Carmines and James A. Stimson, *Issue Evolution: Race and the Transformation of American Politics* (Princeton: Princeton University Press, 1989); Thomas Byrne Edsall and Mary D. Edsall, *Chain Reaction: The Impact of Race, Rights, and Taxes on American Politics* (New York: Norton, 1991).

Vice-President Al Gore worked to convince Texas Democrat Bill Sarpalius to vote for President Clinton's economic program:

> Right at Clinton's desk, Gore and Bentsen reached . . . Bill Sarpalius. . . . Gore made the first pitch. "We need you," he said. "This is it." Then Bentsen took the receiver. "I campaigned for you," Bentsen said sternly, holding the Oval Office rapt. "I've been down there for you. I have raised funds for you when you were behind and it looked like you were going to lose." Bentsen reminded Sarpalius that they had adjusted the budget to keep open a government helium plant in the congressman's district. . . . "I expect you to remember. . . . I'm telling you you're going to vote for this." Sarpalius finally agreed to vote yes.[49]

Because several members of the administration participate in negotiations, they must be careful to coordinate their strategy. Everyone must know when to trade, threaten, or hold firm. According to Woodward, the Clinton administration experienced many coordination problems in negotiating its economic program during the first year. For example, both Lloyd Bentsen and Howard Paster, Clinton's congressional liaison, were heavily involved in trying to persuade Congress to vote with the president. Bentsen's plan was to hold firm on the president's major commitments, such as the controversial energy tax. Paster was generally more accommodating and more willing to compromise. Consequently, Bentsen at one point complained to Clinton that "when he [Bentsen] held firm with members, they were simply going around him by dealing with Paster."[50]

Successful presidents minimize the amount of negotiation they do, especially when it means offering policy inducements and modifying legislation. Deals struck over policy to win votes may dramatically change the proposed legislation, sometimes in ways that run counter to an administration's goals. "Outcomes: How Bargaining Can Undermine Legislative Objectives" shows how negotiations between the Clinton administration and members of Congress over the passage of the North American Free Trade Agreement undermined some of the treaty's central objectives.

Too much bargaining can damage presidents' professional reputation in Washington and, thus, their ability to persuade. After President Clinton responded to concerns from western senators by removing from his budget the user fees for grazing and mineral rights on federal lands, a vast array of interest groups and members of Congress lined up at his door to demand other changes. Although Clinton's budget ultimately passed, in part because of the many deals the administration negotiated, he had much work to do to restore his reputation in Washington.

Going public To increase or maintain their personal popularity and public approval, presidents spend more of their time than ever before communicating with the American public. Sometimes presidential communications are a strategy used to avoid bargaining at all with other players. Referred to by Samuel Kernell as "going public," this strategy requires presidents to use speeches, television and personal appearances, visual communications, and

HOW BARGAINING CAN UNDERMINE LEGISLATIVE OBJECTIVES

To pass the North American Free Trade Agreement (NAFTA), the Clinton administration struck numerous deals with members of Congress, offering a variety of policy inducements. Following are excerpts from an article that appeared in the *New York Times* on November 17, 1993. This article shows how negotiations and bargaining on NAFTA undermined the treaty's central objective—to guarantee free trade by eliminating taxes on imports from Mexico and Canada.

WASHINGTON, Nov. 16—To pass the trade deal that eliminates taxes on imports from Mexico and Canada, the Clinton Administration has been forced to go this far: it has agreed to impose taxes on imports from Mexico and Canada.

The dealmaking on the eve of the decisive House vote Wednesday sounds like a grocery shopping list: beef, peanut butter, bread, sugar, orange juice, cucumbers, lettuce and celery.

But it's not that simple. To win the votes of legislators from farm states, the Clinton Administration has promised that if Mexican imports seriously cut into the American market, it will raise import taxes. . . . The White House arranged this week to limit Canadian peanut butter and wheat shipments to the United States, to prevent Australian and New Zealand beef from entering the United States through Mexico and to raise tariffs temporarily on Mexican tomatoes and other winter vegetables. The deals appeared to be effective, since up to a dozen House members from Oklahoma and Florida have agreed to support the President by voting for the trade pact. . . .

Wheat growers have been among the agreement's strongest critics for many months. But the Administration won them over today by agreeing to open talks with Ottawa on Canadian subsidies, and to begin in 60 days a legal process leading quickly to import limits if the talks are unsuccessful.

Source: Keith Bradsher, "Clinton's Shopping List for Votes Has Ring of Grocery Buyer's List," *New York Times*, November 17, 1993, p. A21.

radio addresses to encourage the public to pressure members of Congress to support their program.[51] Even if members of Congress are not persuaded by public pressure to vote for the program as presented by the administration, the public support helps the president negotiate a deal that is favorable to the administration's preferences.

Public appeals have not always been a part of the presidency. The Framers conceived of the president as an elite leader, someone relatively distant from the people. Presidents could interact frequently with other elite participants, but the Framers feared that presidents might incite popular passions if they routinely addressed the public. By and large, nineteenth-century presidents and presidential candidates accepted the wisdom of this limited conception. Presidents seldom made speeches to the public, averaging 10 or fewer public addresses per year.[52] Fewer than 1 percent of presidential statements during this time were spoken, and only 7 percent were addressed to the general public.[53] In contrast, 42 percent of presidential communications in the twentieth century have been spoken, with 41 percent directed to the public.

The frequency of presidential addresses has increased dramatically in recent presidencies. Although the number of major national addresses given by presidents has held relatively constant at about 10 per year, the number of minor addresses given to special audiences has soared. Presidents use speeches to special audiences to stitch together coalitions of varied interests on particular issues. Some presidents—Ronald Reagan may be the best example—request that individuals and groups listening to their address contact their representatives to express support for the president's program. On average, Nixon, Carter, and Reagan gave nearly five times as many minor addresses as did Hoover, Roosevelt, and Truman. George Bush doubled this already high level of targeted speeches. During his first three years in office he averaged a minor address nearly every other day.[54] Moreover, as Figure 14.4 shows, Bush did an unprecedented amount of traveling to deliver these speeches. He also used teleconferencing technologies to communicate electronically with targeted groups from behind his desk in the Oval Office.

Figure 14.5 (on p. 548) depicts how public appearances by presidents have proliferated as modern technology provides increased opportunities for such events. Presidential appearances, which are covered intensively by the national media, often convey positive symbolic visual messages that are likely to be captured on network news broadcasts and in newspaper photographs. George Bush hiked through the Grand Canyon to demonstrate his commitment to the environment (even though his record as president was sharply criticized by environmental groups). Ronald Reagan hoisted a mug of beer with workers in a Boston bar to show his commitment to America's working class (despite criticism from organized labor over his anti-union policies). And Bill Clinton walked through a low-income, predominantly African-American neighborhood in the District of Columbia to demonstrate his concern for urban social problems (amid criticisms by African-American leaders that he had abandoned cities in his quest to appeal to Reagan Democrats).

STRATEGIES FOR DEALING WITH THE BUREAUCRACY

Successful implementation of a president's program requires cooperative relations with the federal bureaucracy. But most presidents quickly learn that it is difficult to persuade the bureaucracy, particularly career civil servants not appointed by the president, to implement orders with which they disagree. Harry Truman is reported to have complained, "I thought I was president, but when it comes to these bureaucrats, I can't do a damn thing." Jimmy Carter, during the last year of his presidency, remarked during a news conference, "Before I became president, I realized and was warned that dealing with the federal bureaucracy would be one of the worst problems I would have to face. It has been worse than I anticipated."[55]

Career civil servants in the bureaucracy may impede presidential control over policy as a result of ideological differences. When Richard Nixon tried

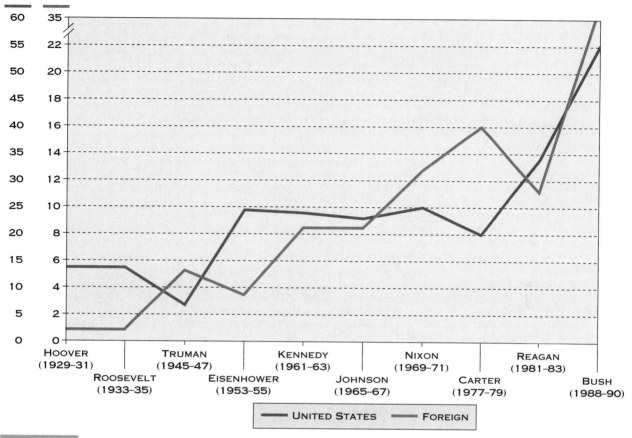

NUMBER OF
DAYS

U.S. FOREIGN

FIGURE 14.4

DAYS OF POLITICAL TRAVEL BY PRESIDENTS, 1929–1990 (YEARLY AVERAGES FOR FIRST THREE YEARS OF FIRST TERM) The increase in "going public" can be seen in trends in political travel by presidents. As this graph shows, Ronald Reagan and George Bush spent considerably more time traveling than any of their predecessors had. *Source:* Samuel Kernell, *Going Public,* 2nd ed. (Washington, D.C.: Congressional Quarterly Press, 1994), p. 105.

To eliminate public activities inspired by concerns of re-election rather than governing, only the first three years have been tabulated. For this reason, Gerald Ford's record of public activities during his two and a half years in office has been ignored.

to scale back social welfare programs created under Lyndon Johnson's Great Society program, he confronted large numbers of people in senior civil service positions who were hostile to his plan.[56] Indeed, so many were loyal to Democratic programs that even Nixon's tampering with the promotion process could not overcome the resistance.[57]

What can presidents do, then, to gain some measure of control over the bureaucracy? The following strategies have been suggested: (1) The president should choose cabinet secretaries and other executive officials who share the administration's goals and promise to pursue its policies enthusiastically.

547

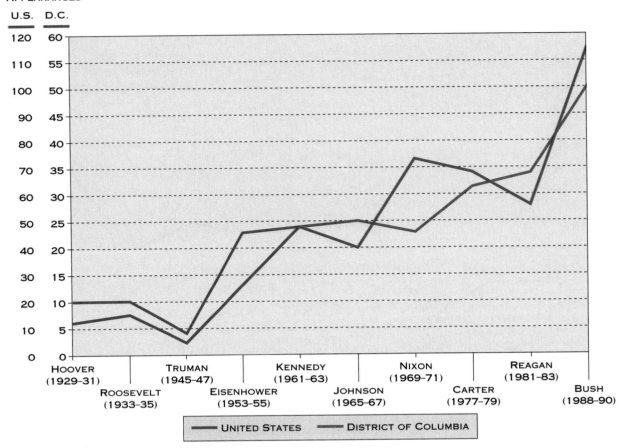

FIGURE 14.5

PUBLIC APPEARANCES BY PRESIDENTS, 1929–1990 (YEARLY AVERAGES FOR FIRST THREE YEARS OF FIRST TERM) As presidents come to rely increasingly on political strategies that involve gaining support for their policies, they are making more public appearances. These generally involve speeches to selected audiences, as well as posing for "photo opportunities" in the national media. This graph shows the dramatic increase over time in the number of public appearances by presidents. *Source:* Samuel Kernell, *Going Public,* 2nd ed. (Washington, D.C.: Congressional Quarterly Press, 1994), p. 102.

To eliminate public activities inspired by concerns of re-election rather than governing, only the first three years have been tabulated. For this reason, Gerald Ford's record of public activities during his two and a half years in office has been ignored.

Loyalty should be the first and foremost characteristic of those nominated for such positions by the president. (2) The White House may then motivate its appointees in the bureaucracy by rewarding loyalty, through resources such as invitations to state dinners, favorable mentions at news conferences, supportive phone calls from the president, and favorable budgetary decisions.[58] Most presidential administrations, however, continue to spend their time on such things as legislative agendas, electoral politics, crisis management, and foreign policy.

THE ROLE OF THE PRESIDENT
IN THE PLAY OF POWER

Presidents face conflicting demands as they play a variety of crucial roles in the play of power. Indeed, the president is the only individual in national politics who is asked to do so much. Some of these roles may conflict: For example, the president is both a symbol of the entire nation and the head of a particular political party. It is not surprising that few presidents meet the extraordinarily high expectations for their terms in office.

HEAD OF THE PARTY

As the only participants in the play of power who are elected nationally, presidents are expected to represent and advocate positions and take actions that reflect the national or general interest. But presidents are also expected to play the role of party leader, installing their own choices to chair the national party committee that reports to the president. Presidents typically have little difficulty gaining their party's nomination for a second term and controlling activities at the national convention. Moreover, in their role as head of the party, popular presidents make time to campaign for congressional candidates in off-year elections. In this way, presidents play a role in national electoral politics.[59]

NATIONAL LEADER

Being the sole elected official with a national constituency and being the participant in the national play of power who most commands the media's attention, a president has many opportunities to play an important role as national leader. Constitutional rules help focus attention on presidents in this role by requiring them to report periodically on the "state of the union" and encouraging them to propose legislation to Congress to deal with national ills.

Although we have seen that the office's formal powers are limited, other participants feel obligated to at least listen to the president's thoughts on national priorities. In the 1960s, for example, Lyndon Johnson redirected national resources to programs established to combat domestic social problems, such as poverty. In the 1980s, Ronald Reagan set new priorities, cutting back on many social programs established by Johnson and redirecting resources to national defense. Even though presidents may have difficulty influencing Congress and the bureaucracy, they have a significant role in the play of power by defining national priorities and the problems and issues that form the nation's political agenda.

Finally, presidents play a significant role as national leaders in their dealings with Congress. Presidential initiatives always receive a great deal of attention, if not respect. As we have seen, presidents do not always get everything they propose. But they are important players in the political negotiations that lead to public policy. In some instances they reject congressional action, exercising their power to veto legislation. This power, referred

Even with limited formal powers, skillful presidents can exert tremendous influence by using their national visibility to define governmental priorities and set the nation's agenda. When Lyndon Johnson announced his War on Poverty, the program signaled a major redirection of national resources to combat domestic social problems.

to by Senator George Mitchell as "the most underestimated power in the American political system," may be used by presidents to ensure that their priorities continue to guide national policy making.[60] At one point in his term, George Bush used the veto 22 times to contest legislation passed by a Democratic Congress, without once being overridden.[61]

INTERNATIONAL LEADER

Expectations for presidential leadership extend beyond the national to the international arena. Presidents play the roles of commander-in-chief and the nation's chief diplomat. In these roles they promote what they believe to be the national interest in world affairs by developing and implementing foreign policies. The policies define U.S. relations with allies and adversaries, as well as with neutral countries.

In most cases, presidents have greater independence in developing international policies than they do in the domestic arena. This is especially true during international crises and times of war, when Congress and the public generally support the president's actions and decisions. After the crisis passes, however, Congress may examine presidential action critically. For example, there were few criticisms of George Bush's decisions during the Persian Gulf War. However, almost immediately after the war, members of Congress began to call for a thorough investigation of the administration's actions.

Presidents can use intelligence information as a resource to further their autonomy in international affairs. Administrations can hinder or stifle debate by simply suggesting that their information, which comes from intelligence agencies and is highly confidential, leads them to act as they do. Moreover, presidential administrations use the media as a resource to shape the terms of public debate on issues of foreign policy. Despite the inability of the Bush administration to agree on what the overriding rationale was for American involvement in the Persian Gulf War—to protect American jobs, to protect oil supplies, or to resist aggression—the media did little to question the administration's policy or to explore alternative courses of action. Because of the intense attention paid to presidents, they are usually effective in mobilizing public support. With only a few exceptions, Congress, the media, and the general public support presidents who take strong stands and actions, even if American troops are involved.

Presidents may use foreign-policy and military successes as resources in the domestic play of power. Such successes inevitably lead to increased popularity and public approval, thereby providing presidents with a very favorable climate for persuading Congress to enact their programs. Thus, George Bush was sharply criticized after the Persian Gulf War for not using the leverage he had gained from the perceived American victory to push for an economic stimulus program to combat the recession. Bush had squandered precious political capital produced by a 90 percent approval rating.

SYMBOL OF THE NATION

Not only are presidents national and international leaders, they also play a crucial ceremonial role, serving as a symbol of the nation. In the various ceremonies presided over by presidents—Fourth of July celebrations, Veterans' Day commemorations, the lighting of the White House Christmas tree—they are perceived to act for the nation. Presidents are also a symbol of national concern and grief during crises and tragedies. For example, Ronald Reagan embodied the national sense of loss after the 1986 space shuttle explosion as he publicly praised the courage of those who had perished and comforted their grieving relatives. Likewise, Bill Clinton personified national concern for flood victims in the Midwest during the summer of 1993 as he toured the affected areas and encouraged those left homeless.

It has been suggested that the contemporary presidency now deals at least as much in symbolic politics as it does in tangible, substantive policy. According to this view, a significant and growing part of the presidency centers on "leadership as a spectacle." Actions that are spectacles are meaningful not for what they achieve but for what they signify. As the influence of the electronic media has increased, so too has this aspect of the presidential role.[62]

Because presidents are such symbolic figures, when they die in office people often react as if they had suffered a personal loss. Thousands wept openly when Franklin Roosevelt died. Some 100 million people watched the funeral of John F. Kennedy on television. In fact, research has shown that the Kennedy assassination produced the types of psychological and physical reactions that one expects from the loss of a parent or other close relative.[63] Many people can remember exactly where they were and what they were doing on November 22, 1963, when they heard the news that the president had been shot and killed.

SUMMARY

In this chapter we have emphasized the limitations on presidential power and the crucial role that the presidency plays in American politics. Although the president has tremendous symbolic and actual power, the presidency is only one of several participants seeking to influence outcomes in the grand game of politics. Other important participants in the executive branch include members of the White House Office, the Executive Office of the President, and the cabinet, as well as the vice-president, the president's spouse, and informal advisers. The growing complexity of national government and of the problems confronting presidents leads them to rely heavily for advice on an enormous staff of specialized experts. These advisers participate in their own play of power within the administration, trying to gain access to the president and, ultimately, to exert policy influence.

Constitutional rules affect the presidency, specifying legal qualifications for the office and the various functions to be performed, and offering a few formal resources. These resources include the veto, the State of the Union address as a forum for outlining an agenda, and the ability to propose legislation to Congress.

Although presidents lack a vast array of formal resources, the office has evolved into a potentially powerful one. The notion of inherent powers offers the president powers and resources not specified in the Constitution, such as the ability to issue executive orders. Other participants have increased presidential resources through the delegation of their powers to the president. In addition, the president has the power to persuade, a power that varies among presidents depending on their professional reputation in Washington and their public prestige.

Presidents use various strategies to influence Congress and the bureaucracy. They negotiate and bargain with members of Congress, using a variety of resources to win congressional votes. They increasingly engage in a strategy of "going public," communicating directly with the general public in the hope of bringing public pressure on Congress in support of their programs. In general, modern presidents spend much of their time communicating with the public, directly through speeches and other public appearances and indirectly through media coverage of their activities.

Although presidents do not act alone or independently, they are significant participants in the play of power. Presidents head their parties, set broad governmental and budgetary priorities, and work with and against Congress to shape public policy. Presidents serve as the nation's chief diplomat and as commander-in-chief. Finally, presidents serve as a symbol for the nation.

APPLYING KNOWLEDGE

GENDER STEREOTYPES AND
PRESIDENTIAL LEADERSHIP

 The following article appeared in the *New York Times* on July 17, 1994. After reading the excerpt, discuss how the article illustrates the ways in which gender stereotypes infiltrate discussions of presidential leadership.

WASHINGTON—President Clinton is a big man with a famously big brain, who runs an enormous country. Sufficiently riled, he could tell his army to rout any tinpot dictator who dared to smart off, and he obviously knows enough about practically everything to argue any enemy into intellectual submission. He has enough heft, and probably enough reach, to deck most Republicans with one punch, not to mention upstart foreign heads of state. . . . The point is, this is one mighty powerful leader. It raises a question: Why doesn't that intimidate anyone anymore?

Go figure. Presidential power is a queer thing; it attaches to the office, but sometimes, it seems, not necessarily to the person. When George Bush was Vice President, satirists laughed at his cloth watchbands and nasally whine and deference to Ronald Reagan, and said he had put his manhood in a blind trust. Then he became president and invaded Panama, crushed Iraq and, before the economy went south, reigned as the most popular leader in opinion-polling history. This from the fellow who reminded many a woman of her first husband.

Now comes the robust Mr. Clinton, and suddenly the image is more that of a certain type of ex-boyfriend, the one who seemed dashing until girls discovered they could walk all over him. Smart, persuasive, eminently likeable and eager to be liked, he believes the nation and the world should follow him to greatness. It's the leading part he hasn't mastered.

"We don't give him enough credit on one level for seeming to be a genuinely nice person," said Stephen Hess, a student of presidents at the Brookings Institution. "That doesn't fit our list of presidential qualities. We really do elect a President to be our very own sonofabitch. Clinton hasn't gotten it yet."

There have been many indications of this. Mr. Clinton could not wrangle his own Democratic Congress last year into approving a measly $20 billion in what he called economic stimulus, and some others called pork, and he had to plead—not bully—lawmakers to approve the first Democratic budget in a dozen years. He swept into office pledging to bring homosexuals into the military and to end handouts of Federal grazing rights to Western ranchers, and gave up both battles.

Source: Michael Wines, "Talk Often and Be a Soft Touch," *New York Times,* July 17, 1994, p. D1. Copyright © The New York Times Company. Reprinted with permission. ■

Key Terms

Executive Office of the President (EOP) (p. 519)

White House Office (WHO) (p. 519)

cabinet (p. 524)

inherent powers (of the president) (p. 532)

executive order (p. 532)

delegation of powers (p. 532)

Recommended Readings

Edwards, George C., III. *At the Margins: Presidential Leadership of Congress.* New Haven: Yale University Press, 1989. A study suggesting that presidents have minimal influence on congressional roll-call votes.

Hinckley, Barbara. *The Symbolic Presidency: How Presidents Portray Themselves.* New York: Routledge, 1990. A fascinating study of the rhetoric used by presidents to help create favorable public impressions of themselves and their administrations.

Jones, Charles O. *The Presidency in a Separated System.* Washington, D.C.: Brookings Institution, 1994. Shows how presidents are constrained by other political institutions and branches. Suggests that presidential success is variable, depending on available resources, advantages, and strategic position in relation to other institutions.

Kernell, Samuel. *Going Public: New Strategies of Presidential Leadership,* 2nd ed. Washington, D.C.: Congressional Quarterly Press, 1993. Demonstrates how modern presidents increasingly adopt strategies designed to mobilize the public in support of their policy agenda.

Lowi, Theodore J. *The Personal President: Power Invested, Promise Unfulfilled.* Ithaca, N.Y.: Cornell University Press, 1985. Argues that contradictory tasks and unreasonable public expectations make it very difficult for modern presidents to achieve success.

Nathan, Richard. *The Administrative Presidency.* New York: Wiley, 1983. Discusses the many problems presidents experience in controlling the activities of federal executive agencies and suggests strategies for more effective control.

Neustadt, Richard E. *Presidential Power: The Politics of Leadership, with Reflections on Johnson and Nixon.* New York: Wiley, 1976. Originally published in 1960, this is the classic study of how presidents seek to persuade others to follow their leadership.

Tulis, Jeffrey K. *The Rhetorical Presidency.* Princeton, N.J.: Princeton University Press, 1987. Shows how the relationship between presidents and the public has evolved from the Founders' conception of great distance to the modern closeness.

The Bureaucracy

IN OCTOBER 1962, during what came to be known as the Cuban Missile Crisis, the world moved to the brink of an all-out nuclear war.[1] After the United States discovered Soviet nuclear missile sites being constructed in Cuba, President John F. Kennedy issued an ultimatum to the Cubans that construction cease and the missiles be removed. In addition, he ordered a naval blockade of the island to ensure that no more missiles were shipped from the Soviet Union. Soviet premier Nikita Khrushchev refused to remove the missiles, claiming that American Jupiter missiles in Turkey posed the same threat to the Soviet Union as did the missiles in Cuba to the United States. Forced removal of the Soviet missiles, through invasion or air strikes, would likely be followed by forced removal of the missiles in Turkey; and the conflict might easily escalate into a nuclear exchange between the two superpowers. Kennedy and his closest advisers huddled in the basement of the White House considering their options while Soviet ships with more missiles steamed toward Cuba. In the end the Soviets "blinked," agreeing to remove their missiles, and the world backed away from the precipice of nuclear war.

The Cuban Missile Crisis raises many questions that can only be answered by understanding the inevitable role of bureaucratic organizations in modern government. Here we briefly consider two of these puzzles: Why was it so easy for the United States to discover the presence of Soviet missiles in Cuba even before they became operational? Why were there still U.S. Jupiter missiles in Turkey, despite the conclusion of a governmental study that they were worthless and Kennedy's subsequent order to have them removed in early 1962?

The Cuban Missile Crisis began when an American U-2 spy plane flew over Cuba and pho-

As the world teetered on the brink of nuclear war during the Cuban Missile Crisis in October 1962, President John F. Kennedy met with his cabinet and advisers while, in the Soviet Union, Premier Nikita Khrushchev and his advisers did likewise, each group struggling to maintain control of events. Although the focus of this dramatic incident was on the American and Soviet leaders and their staffs, the role of bureaucratic organizations and their standard operating procedures was significant in both the development and dynamics of the crisis.

tographed telltale signs of Soviet construction of nuclear missile sites. In particular, the photos identified the patterns of Soviet surface-to-air missiles (SAMs) deployed to protect nuclear missile sites. The Soviets were clearly aware of the U-2 flights, having shot down a U-2 over Russia in 1960. If they had wanted to keep the Cuban missile sites secret until the missiles became operational, it made little sense to undertake construction in a manner that was easily spotted by spy flights. This behavior may be explained by looking at the ways in which bureaucratic organizations perform complex tasks by developing routines, or standard operating procedures (SOPs), that they repeat again and again whenever the situation calls for that particular activity. Although Soviet intelligence agencies were aware of the U-2 flights and were indeed obsessed with secrecy, those agencies were not the specific ones assigned the task of actually constructing the missile sites. Responsibility for constructing the sites, and more specifically the SAMs that would protect those sites, was assigned to the Air Defense Command:

> For the manager of the SAM site, location of the sites and construction of the missiles were technical problems. Protection of a strategic missile installation required a particular configuration of SAM battalions. Thus Cuban SAM sites were constructed in the typical trapezoidal pattern for no better—and no worse—reason than that this is the way Soviet SAM construction teams position SAMs. Nothing in the organization's repertoire reflected any awareness of the possible clues this pattern might present to foreign intelligence.[2]

Although the SOPs of Soviet bureaucratic organizations help explain the discovery of missiles in Cuba, we must look to the behavior of American bureaucracies—specifically, the State Department—to understand why our own strategically useless nuclear missiles were still being maintained in Turkey. Kennedy directed his secretary of state, Dean Rusk, to order the missiles removed in early 1962. However, remember that a chief executive almost always delegates to others the authority to carry out orders. In the case we are examining, since the president cannot travel to Turkey to see with his own eyes that the missiles have been removed, he gives an order to subordinates (Rusk, in this instance), who in turn delegate their authority to the governmental bureaucrats assigned the responsibility for the specific tasks (here, informing the Turkish government that the missiles are going to be removed). Having given the order and delegated authority, the president goes on to his next task. But once authority is delegated, the actual carrying out of orders is affected by the interests of those who are supposed to oversee the tasks. In this case, State Department officials were concerned about maintaining good relations with the Turkish government and, consequently, let those concerns delay removal of the missiles. In the end, when Kennedy found out that his orders had not been obeyed, he exploded in anger and ordered immediate removal of the Jupiter missiles. Our point here is that it took the crisis of a nuclear showdown with the Soviet Union to focus the president's attention on whether an order he had issued had actually been carried out by the appropriate governmental bureaucracy.

THE ROLE OF BUREAUCRACY IN AMERICAN POLITICS

 Our opening case study of the role of bureaucratic organizations in the Cuban Missile Crisis was chosen for its drama and significance, but virtually no one in American society can spend even an ordinary day untouched by bureaucracy. Most students first encounter bureaucracy in junior high school. Hall passes, lunchroom monitors, and assistant principals generate fond memories for most of us. These experiences, and similar ones in high school, help prepare students for the college bureaucracy that registers them for classes, sends out tuition bills, records grades, and assigns dorm rooms. In fact, it affects students' lives so much that some claim that all American college students have dual majors—the one appearing on the transcript, and the one called "Dealing with Bureaucracy."

Most students bring to the study of bureaucracy their own stories that fall into standard categories such as "wrong line," "runaround," "It can't be done," and "catch-22." Joseph Heller's classic novel exposing the foibles of military bureaucracy, encapsulated in the book's title, *Catch-22*, has introduced the term into everyday language. Heller's protagonist, Yossarian, tries to avoid flying more bombing missions by invoking the rule that requires the doctor to ground anyone who is crazy. Doc Daneeka tells Yossarian:

> "Sure, I can ground Orr. But first he has to ask me to."
> "That's all he has to do to be grounded?"
> "That's all. Let him ask me."
> "And then you can ground him?" Yossarian asked.
> "No. Then I can't ground him."
> "You mean there's a catch?"
> "Sure there's a catch," Doc Daneeka replied. "Catch-22. Anyone who wants to get out of combat really isn't crazy."[1]

Unhappy encounters with bureaucrats, and the jokes and war stories often told about them, lead to images of bureaucracies as "incompetent, indifferent, bloated, and malevolent administrative departments of government."[2] In fact, when we asked our own students on the first day of class whether they agreed with the statement "When I see the words 'bureaucrat' and 'bureaucracy,' largely positive images and feelings come to me," only 10 percent agreed, whereas more than a quarter disagreed strongly.

Another widespread common belief is that bureaucracies do not work very well. When asked whether they agreed or disagreed with the statement "By and large, bureaucracies don't succeed very often in performing the tasks they are supposed to perform," 37 percent of our students agreed and only 23 percent disagreed; the rest were unwilling to agree or disagree. Yet another feature of common beliefs about bureaucracies is a lack of recognition that bureaucracies are important political players. Forty percent of students in our survey agreed with the statement "I do not know much about the role the federal bureaucracy plays in American politics," and fewer than 30 percent disagreed.

Still another common belief is that bureaucracies exist only in government. In fact, as we shall see, bureaucratic organizations are common in both the public and private sectors. Large private corporations such as General Motors are every bit as bureaucratized as large public organizations such as the Department of Defense. The idea that only government has become bureaucratized blinds us to the ways in which attempts to reduce the size and influence of public bureaucracies almost always increase the size and influence of private bureaucratic organizations. For example, although proposals to reform the health care system in 1994 were criticized and ultimately defeated on the grounds that they would expand the power of governmental bureaucracies, their defeat did not mean there would be less bureaucratic control over health care, but only that large private bureaucracies—insurance companies, health care providers, and so on—would expand their control in this policy arena.

We contend that these common beliefs interfere with understanding American politics. First, opinions about bureaucracy are a good deal more varied than negative images and jokes suggest. For instance, a Harris survey found that 73 percent of respondents who contacted the federal bureaucracy said the employee they dealt with was "helpful," and 46 percent said they were "highly satisfied" with their treatment.[3] Second, despite highly publicized stories highlighting slipups and mistakes, the fact is that most bureaucracies work fairly well most of the time. Most bills are accurate, most mail is delivered, and most students are registered for classes; indeed, many complex and difficult tasks get performed thanks to the activities of bureaucracies. Finally, the failure to recognize that bureaucracies are among the most important players in politics grossly distorts reality. When it comes to material politics, at least, the bureaucracy probably has more influence over outcomes than any other player does. Its decisions touch virtually every aspect of our lives, including those most crucial to our well-being.

[1] Joseph Heller, *Catch-22* (New York: Dell, 1961), pp. 46–47.
[2] Charles T. Goodsell, *The Case for Bureaucracy*, 2nd ed. (Chatham, N.J.: Chatham House, 1985), p. 1.
[3] Goodsell, *The Case for Bureaucracy*, p. 22.

WHAT IS BUREAUCRACY?

Given the central role of governmental bureaucracy in the play of power, it is all the more important to understand what bureaucracies are and how they (in theory) make decisions. **Bureaucracies** are the organizations that administer programs and policies, in both government and the private sector, and that exhibit characteristics such as large size, reliance on formal rules, and a hierarchy among permanent staff. This chapter focuses on just one type among the many bureaucracies that operate in American society: the *public* agencies in the *executive arena* at the *federal level* of government that administer governmental programs and policies. However, much of what we say here applies to state and local bureaucracies, and to the private bureaucracies found in corporations, law firms, labor unions, and nearly all other large organizations in our society.

Our discussion draws on the theories of the noted German sociologist Max Weber.[3] Writing early in the twentieth century, Weber argued that both the increasing geographic area controlled by the modern state and the growth in the populations under control of government led to dramatic increases in the tasks performed by government. The only way to meet such demands was to create larger and more complex administrative structures organized bureaucratically. Weber defined the characteristics, or "ideal types," that distinguish modern bureaucracy from all other forms of human organization.

Bureaucracies are generally characterized as colorless, humorless agencies run by faceless, compliant workers. However, the reality is that only through such large organizations, public and private, is our society able to provide on a mass basis the goods and services that we often take for granted.

Although no real-world bureaucracy exhibits all the "ideal type" characteristics, Weber's formulation captures the essential nature of this form of social organization.

THE FORMAL NATURE OF BUREAUCRACY

Bureaucratic organizations, according to Weber, have five defining characteristics: clear lines of authority, division of labor, specialization of tasks, hierarchy, and impersonality. These characteristics prevail not as a result of bureaucratic whim or a desire on the part of bureaucrats to be annoying and rigid, but rather because they work.

Bureaucracies establish written rules that define who has the authority to do what. Reflecting the definition of authority as existing when someone adopts a rule to be obeyed, bureaucracies establish internal rules about who must obey whom with respect to what. Thus, not everyone at a higher level in the bureaucracy can issue orders to everyone at a lower level. Only a person's designated superior can give orders to that person. Furthermore, there are established procedures for handing down instructions and sending up information. These *clear lines of authority* establish the "channels" that one must follow to get things done in a bureaucracy.

Bureaucratic organizations break down tasks into constituent parts and parcel out responsibility for accomplishing these tasks to separate units. One unit may handle the payroll and employee records; another, the hiring and promoting of employees; a third, public relations; a fourth, maintenance of equipment. Examine the structure of your own student government, college registration office, or employer and you will likely find a similar *division of labor.*

The division of labor permits employees to *specialize*—that is, to spend most or all of their time doing only one or a few things. Sometimes specialization is carried to extremes and produces less than fascinating jobs. Imagine spending a summer hand-filing the paid natural gas bills of people whose last name begins with the letter M. Furthermore, specialization explains why you stand in a long line only to learn when your turn finally comes that it is the wrong line. The person who tells you, "I'm sorry, that's not my department," is being truthful. However, specialization brings benefits as well. Specialists develop expertise. In fact, bureaucracies—such as college health services—effectively employ professionals who apply their specialized training.

The existence of clear lines of authority, a division of labor, and specialization leads to the fourth characteristic of a bureaucracy: distinct levels, or *hierarchy,* in the organization. Employees know what level they occupy, who is below them, and who is above.

Finally, bureaucracies operate *impersonally.* When you stand in line to register for classes, or when you go to the Department of Motor Vehicles to renew your driver's license, you are not treated as a unique individual but as a member of a predefined category: first-year student, or a person who wears eyeglasses and must pass a vision test before receiving a renewed

What do you see when you look at this photo of students waiting to register for classes? While most of us will immediately recognize our own frustration at waiting in long lines for class registration, a driver's license, a passport, or the like, can you think of a better medium than bureaucracy for serving large numbers of people who want to obtain similar goods and services?

driver's license. On the one hand, this impersonality allows bureaucracies to process large numbers of individuals. Imagine how long it would take everyone to register for classes if you had to first describe to a registrar exactly who you were, what your interests were, what times of day were the best for you, and so forth. On the other hand, as individuals we don't like to be treated impersonally. We are taught early that each of us is a unique individual to be valued for our own distinctive characteristics. Yet if bureaucratic organizations are to process large numbers of clients in a reasonable amount of time with limited resources, they must treat us impersonally. In fact, it is this clash between the imperatives of bureaucratic efficiency and the rhetoric of individualism that accounts for the intense frustration most of us feel when we deal with bureaucracies.

THE ROLE OF RULES IN BUREAUCRATIC DECISION MAKING

Bureaucrats make decisions according to predetermined, unambiguous, comprehensive rules. The rules specify conditions and the decision to be made when the conditions are met: If a letter to be sent first class weighs one ounce or less, charge 32 cents postage. If a student does not present an official copy of a paid tuition bill, do not let that student register. Because of these rules, the person who awaits you at the front of the line (once you find the right line) can say, "I'm sorry. Those are the rules. I don't make them. There is nothing I can do. Next!" Such rules also deter bureaucrats from serving whites before blacks, friends before strangers, and adults before teenagers who have been waiting longer.

THE CHARACTERISTICS OF BUREAUCRATIC DECISION MAKING

The ideal-type attributes of bureaucracies produce the characteristic style of bureaucratic decision making. The decisions are *impersonal*, being based on the application of technical rules regardless of who is involved or what explanations they may offer. The people who make such decisions often appear impersonal themselves, especially when they make the same decision over and over again every day. Decisions are rule based. They are often made in secret. Bureaucracies do not recognize the right of outsiders to observe decision making. Finally, bureaucratic decisions involve the gathering, use, and manipulation of information, much of it technical and specialized.

THE FEDERAL BUREAUCRACY

Our examination of the federal executive bureaucracy draws on the metaphor of politics as a play of power. In this section we introduce the role bureaucracies take on as the "fourth branch" of government. In subsequent sections we consider bureaucratic players and resources before examining the dynamics of bureaucratic decision making. Finally, we close the chapter by returning to the role of federal bureaucracies in the play of power.

A number of human characteristics combine in intriguing ways within our political system to guarantee that the bureaucratic organizations involved in accomplishing the goals of the state play a central role in determining who gets what. In fact, we assert that the federal bureaucracy plays a central role, perhaps the key role, in shaping the content of policy as it affects people. Its role in American politics and policy making looms so large that it deserves the title many give it, "The Fourth Branch of Government."

Although this discussion focuses on the federal executive bureaucracy, it explains much about the other bureaucracies that dominate contemporary society. Our approach utilizes the concept of **discretion**, or the ability to choose between significantly different courses of action without facing severe penalties for choosing one over the other, as the key to explaining why bureaucracies make policy. We examine two major types of discretion exercised by bureaucracies: **authorized** or **delegated discretion**, which results from explicit and legitimate grants of authority to make decisions; and **unauthorized** or **undelegated discretion**, which lacks official sanction and occurs without the knowledge or approval of other political players.

THE INEVITABILITY OF DELEGATION OF DISCRETION TO THE BUREAUCRACY

Assume that a bureaucracy is composed of conscientious, competent individuals who make every effort to follow to the letter the legitimate instructions of others. Could these bureaucrats avoid exercising discretion? To do so, they would need a complete set of instructions specifying what decision they should make in every circumstance they encounter. However, the nature of the tasks bureaucrats are asked to carry out makes such all-encompassing instructions impossible to create. The problems are simply too complex, and the variety of circumstances too diverse, to permit complete specification. Conditions never contemplated by those who devise policies arise constantly. Furthermore, policymakers would always need to know what decisions ought to be made in each new circumstance. As we saw in our chapter introduction, even though John Kennedy ordered Jupiter missiles removed from Turkey, he did not indicate to the State Department how quickly he wished this to be done or whether quick action was preferable to damaging relations with Turkey.

The problems policymakers wrestle with in implementing policy stubbornly refuse to be completely understood. For example, most people recognize serious shortcomings in the health care system and would dearly like to fix them. But, as the public debates over this issue in 1994 illustrated, the size and complexity of the problem make finding solutions daunting. The same can be said for social problems such as drugs, crime, poverty, racism, sexism, and falling educational standards. Indeed, the question of how to implement public policy preferences is one of the most important and most misunderstood dilemmas of public policy making.

Even if problems could be completely understood and appropriate responses in every set of circumstances could be spelled out, discretion would have to be exercised if the resources available (time, money, personnel) fell short of what was necessary. They nearly always do fall short. Not enough police exist to make an arrest for every violation of the law that they know about. The Environmental Protection Agency (EPA) has too few people to assess the safety of all agricultural pesticides on the market. Choices must be made regarding which arrests to make, or which pesticides to test first. The choices unavoidably affect outcomes. They therefore determine who gets what.

For these reasons, those who set policy for bureaucracies delegate discretion. Policymakers recognize that not all conditions can be anticipated, that the best response cannot always be determined beforehand, that choices about what to do and when to do it must usually be made in the field. Thus, they grant bureaucracies the freedom to exercise choice within the boundaries of the policies they establish.

these boundaries? In practice, not as narrow as they could be. Those who establish decision-making rules for bureaucracies frequently fail to give as much guidance as they could or should. In this they default in carrying out their responsibilities, leaving to the bureaucracy decisions they themselves could, in theory, render. Congress in particular fails to legislate policy that is as complete and specific as it could be.

Three reasons for the default of Congress (and other policy-making bodies) can be identified. The first is *buck passing*. Just as individuals tend to avoid making hard decisions, so does Congress. The second results from the *breakdown of decision-making processes*. Congress simply runs out of time, passing bills that fail to address important questions. The gaps in policy must then be filled by the bureaucracy. Third, legislation often contains *ambiguous instructions* because no agreement can be reached on clear language. After all, coalitions must be forged to pass legislation, and ambiguity is the glue that holds many coalitions together. Individual members of Congress know what they want; even groups of legislators have a clear idea. But various individuals and groups disagree about what should be done, and so produce vague language that anyone can interpret as expressing his or her own preference. The resulting ambiguous language forces bureaucrats to make the real choices during implementation. Those applying such statutes cannot faithfully execute the "intent of Congress" because Congress as a body has expressed no intent. In addition, warring legislative factions sometimes fail to arrive at even ambiguous wording on which they agree. Passage of legislation may hinge on the willingness of everyone to say absolutely nothing about the most controversial matters, again forcing bureaucrats to exercise discretion when such matters inevitably must be confronted.

Much of the legislation proposed by House Republicans in the 104th Congress was designed to reduce bureaucratic discretion. Yet the legislation that emerged still allowed considerable agency discretion. The "Contract with America," for example, contained provisions to reduce the discretion of agencies such as the EPA to impose regulatory burdens on private business. Yet the resulting requirement to impose strict cost-benefit requirements on regulatory agencies required much agency discretion, because the legislation was necessarily silent on crucial questions such as which costs and benefits would be calculated, what assumptions would be made when the data are not clear, and how quickly the analysis would have to be done.

Congressional vagueness in wrestling with political dilemmas can lead to bureaucratic discretion. The decided lack of an "intent of Congress" in early 1995 resulted in heated debates regarding term limits. Here Congressman Henry Hyde (R-Ill.) makes his appeal for the institution of term limits for members of the House and Senate before a nationwide audience.

THE INEVITABILITY OF UNAUTHORIZED DISCRETION

Detectives in murder mysteries identify suspects by applying the formula of "means, motive, and opportunity." Similarly, the *delegated authority* to make decisions, coupled with the resources to carry them out, provides bureaucracies with ample means to exercise unauthorized discretion. For example, health and food safety inspectors examine restaurants for sanitary violations. Provided a set of rules and the power to issue violations or even close a restaurant, they wield significant decision-making power. Most other bureaucrats possess similarly effective means to make decisions by virtue of their position.

Bureaucrats have many opportunities to exercise unauthorized discretion, especially because of the *low visibility* of many of their decisions. The size of the bureaucracy, the sheer number of decisions made, and the physical location of these decisions (behind closed doors, on the street, in the forest) shield them from the prying eyes of supervisors and the news media. Low visibility is essential for engaging in unauthorized behavior. Students planning to cheat on an exam seek seats far away from proctors precisely because they understand the importance of low visibility. Workers may loaf on the job when the boss is gone and become industrious when he or she reappears. When supervisors or outsiders see what is going on, the chances increase that sanctions will result from improper and unauthorized actions.

Another ingredient in the recipe for unauthorized discretion, *motivation*, comes to bureaucrats from several sources. Greed at times seduces lower-level employees in food and inspection services, police departments, and other bureaucracies to accept outright bribes. Such bribes may prove almost irresistible. Although federal bureaucracies enjoy a well-deserved reputation for honesty, sometimes their employees have accepted money or gifts in return for favorable decisions. Employees of the Food and Drug Administration, for example, were caught accepting bribes from generic drug manufacturers in the late 1980s.

When federal bureaucrats exercise unauthorized discretion, it more commonly consists of exploiting opportunities to further policy goals. After agency personnel have conceived a program, helped win enactment of legislation establishing and funding it, and participated in getting it started, their reluctance to change the main thrust of policy—even their willingness to defy legitimate orders to change—should come as no surprise.

BUREAUCRATIC GOALS

Federal bureaucracies' role in the play of power cannot be examined without understanding the nature of their goals. Like other participants, bureaucracies pursue their own interests and goals. Understanding their behavior requires examining the nature of these goals. The task is complicated by the fact that conflict often exists (1) over what an agency's mission and goal really is, and (2) between politically appointed agency heads and the career bureaucrats who serve under them.

Former policymakers offer useful insights into their goals in "kiss-and-tell" memoirs. Paul Craig Roberts, who was a "supply-side" economist in the Treasury Department under President Ronald Reagan, concluded from his experience:

> Once in government, aides and appointed officials always become involved in competing for access to policy levers, status, exposure in the media, and controlling power. The competition for power becomes much more important to the "players" than the substance of the policy, and policy concerns become subsidiary as players try to prevail over one another. Indeed, the widely used term "player" itself indicates that the policy process is viewed as a game. No doubt this sounds cynical, but it is merely the report of many who have observed and experienced political life in Washington.[4]

This observation probably understates the importance of substantive policy goals. Top players must satisfy both their own personnel and outside constituencies who care about policy. But it does suggest the compelling nature of a principal "organizational imperative"—the need to expand the agency's authority, budget, personnel, and influence.[5] As Roberts suggests, one explanation for the impulse to expand or "empire build" comes from the common belief that this is how the game is played. Policymakers and others in Washington evaluate performance by determining who is influential. Governmental officials' performance, unlike that of corporate executives, cannot be judged on the basis of profits earned. However, increases and decreases in the size of one's staff and budget can be measured precisely. Assessments of program scope and influence wielded (what is known as *turf*), although less precise, are regarded as equally important indicators of influence. Thus, administrators who want to be considered successful, influential, and important usually try to defend and expand their turf.

Organizational survival is an equally potent but less obvious goal. Employees and nongovernmental constituents of an agency naturally react vehemently to the prospect of its termination, both because jobs will disappear and because programs passionately believed in may end. Of course, real threats to an agency's survival are rare (although, as many agency heads who served during the Reagan administration can testify, they do occur). In the 1994 congressional election campaign, a variety of cabinet departments were mentioned as possible targets for elimination: Housing and Urban Development, Energy, Commerce, and Education, for instance. Even though pledges to dramatically reduce the size of the federal bureaucracy are staples of campaign rhetoric, they nevertheless are usually taken seriously by politically savvy bureaucrats.

Even the threatened termination of an obsolete component of an agency's workforce inspires fierce efforts at preservation. How else could one explain, for example, the Department of Agriculture's continued subsidies for the production of pitch pine, which was once needed for the nation's fleet of sailing ships?[6]

Proposals to transfer programs to other agencies or to grant authority over new programs to rival agencies are made all the time. They arouse anxiety about more serious challenges but are important in and of themselves. Because other agency heads also feel the need to expand their agencies' size and authority, conflict among bureaucracies with overlapping jurisdictions never ceases. Indeed, efforts to expand at the expense of a competing agency may be undertaken as a survival strategy. The idea is to strike first and capture control of policy before someone else attacks your turf. Likewise, newly created agencies, such as the Resolution Trust Corporation created by the savings and loan bailout bill, must fight for turf.

Campaigns to initiate new programs, expand existing programs and staff, or seize part of another agency's turf require substantial planning and calculation. However, proposals to reduce budgets and authority almost always elicit immediate and intense opposition. Typical is the response of Ron Brown, secretary of commerce in 1995, to a congressional proposal to eliminate his department and transfer many of its programs to other agencies. Brown immediately mounted a high-profile media campaign to contest the proposal and point to the new opportunities that the Department of Commerce had generated for American businesses. As head of his department and defender of its turf, the secretary had little choice. Career bureaucrats within his department and private interests such as the multinational corporations that benefited from Commerce Department programs (and that likewise denounced the proposal) expected him to protest. Failure to do so would have risked losing the confidence and support of his agency's personnel and of powerful outside groups, damaging his reputation and influence.

The goals of careerists within the bureaucracy may conflict with those of a political superior. Careerists' commitment to the continuation of their programs and policies is understandably intense. Strong and lasting relationships developed over many years with supporters in Congress and with interest groups reinforce career bureaucrats' willingness to protect ongoing programs and policies. For example, career employees in the Environmental Protection Agency and the Civil Rights Division of the Department of Justice protested privately and publicly when newly appointed political superiors tried to change long-standing policies. Such political appointees, on the other hand, must balance demands from their staffs to maintain policy with pressure for change stemming from their personal values and the broad policy goals of the president.

PLAYERS IN BUREAUCRATIC POLITICS

Public jobs constitute one of the most sought-after prizes dispensed by government. With the existence of almost three million federal jobs, political conflict over how they will be filled inevitably arises. The direct control by politicians of governmental jobs is known as **patronage**. Under patronage systems, elected politicians give governmental jobs to their supporters. Patronage is a potent resource that can be used to build political organizations and raise money. Such use of patronage reached its pinnacle in the urban political machines of the nineteenth and early twentieth centuries. Under the infamous Boss Tweed ring in New York City during the 1870s, people who wanted a city job—police officer, city clerk, or dog catcher—had to be solid Democrats (since the controlling machine in New York happened to

"'I stick my fist in as far as it will go, and pull it out as full as it will hold. I stick to my friends. That's me!' - There you have Tweed self-painted to the life."

This political cartoon from 1874 depicts New York's infamous Democratic Party leader William Marcy "Boss" Tweed, head of the scandalous Tweed Ring, with his hand in the public till. The development of the modern civil service, where jobs go to the well-qualified rather than to the politically well-connected, was a response to such use of patronage by political machines.

be Democratic), *and* they had to pay for the appointment as well as "kick back" a percentage of their salary to the Tweed ring.

Disputes over the selection of federal employees broke out among federal politicians in the early years of the Republic. The landmark Supreme Court decision *Marbury v. Madison* (which clarified the balance of power between state governments and the federal government) was handed down because the outgoing administration of President John Adams appointed scores of the party faithful to governmental positions. President Andrew Jackson, who deserves credit for elevating patronage to common practice, argued, "To the victor belong the spoils," and replaced as many holdovers from the previous administration with his own supporters as he could. Although critics pointed to the disruption caused by wholesale firings when administrations changed hands, as well as to the woeful incompetence of some loyal supporters of the winner, it took the assassination of President James Garfield by a disappointed office-seeker to win passage of a reform of the process for filling federal jobs.

The Civil Service Act of 1883 sought to replace the partisan considerations of patronage appointments with neutral criteria applied equally to all, hiring and promoting on the basis of merit. To administer the act, Congress established the Civil Service Commission. The commission developed tests to establish competence, and it ranked applicants. To guarantee a stable workforce of motivated employees, the system provided those hired with job security by carefully specifying the grounds for dismissal and establishing appeal procedures for those threatened with dismissal. Today nearly all federal employees enjoy the protection of the principles embodied in the notion of a competitive civil service. About 60 percent fall under the two successors to the Civil Service Commission: the Office of Personnel Management (OPM) and the Merit Systems Protection Board (MSPB). Another 30 percent or so work for agencies that administer their own merit system independent of the OPM and the MSPB (e.g., the Federal Bureau of Investigation and the Department of State).

Top career federal bureaucrats fall under the jurisdiction of three systems. The Senior Executive Service consists of some 7,000 higher officials, many drawn from the ranks of civil servants, who enjoy somewhat less job security than civil servants do, but receive higher pay. The Excepted Service, which does not follow competitive hiring practices or provide job security, covers approximately 100,000 employees, including attorneys, bank examiners, and some purely patronage appointees scattered throughout the bureaucracy. The Executive Schedule consists of some 700 high-level policy-makers appointed by the president, including cabinet heads, undersecretaries, and other agency officials.[7]

Thus, there is a fundamental distinction between (1) the *political executives* appointed by the president and his chief subordinates who head the major components of the federal bureaucracy, and (2) the *career executives* who serve just below them. The career executives take a long view, intend to remain in the bureaucracy, and care about the continued survival of their agency and its programs. Political executives plan only a temporary stay and worry much more about immediate policy choices and programs. One study found that the average political executive remains in office for only 18 months while more than two-thirds of all career executives spend their entire careers in a single agency.[8] Clearly, political and career executives depend on one another and often work for common goals. But they sometimes reflect competing perspectives, values, and interests that can produce conflict.

THE SIZE AND STRUCTURE
OF THE FEDERAL BUREAUCRACY

In this section we first examine the size of the federal bureaucracy and then describe its organization. Our discussion focuses on the 14 cabinet departments that employ most federal bureaucrats, but we also look at independent regulatory agencies, governmental corporations, advisory committees, and assorted commissions and boards.

THE SIZE OF THE FEDERAL BUREAUCRACY

A common preconception holds that the federal bureaucracy in Washington is huge and growing all the time. This view is misleading in several respects. First, seven of every eight federal bureaucrats work outside of Washington.[9] Second, as Figure 15.1 shows, the number of federal employees has remained just about level for the past 20 years. Since President Richard Nixon took office in 1969, state and local governments have accounted for the growth in public employment. When considered in the context of overall growth

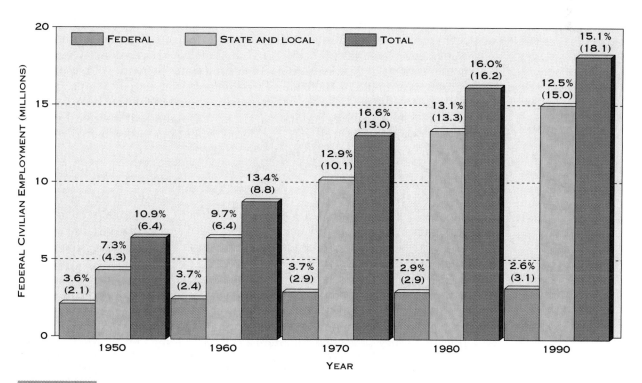

FIGURE 15.1

THE GROWTH OF FEDERAL CIVILIAN EMPLOYMENT, 1950–1990 While the number of public-sector workers has increased since 1950, most of this growth has occurred at the state and local levels. Since 1970, federal civilian employment has actually declined as a percentage of the nation's total workforce. *Source:* Data from B. Guy Peters, *American Public Policy: Promise and Performance,* 3rd ed. (Chatham, N.J.: Chatham House, 1993), p. 32.

Numbers in parentheses indicate percentage of total employment.

in the size of the American workforce, the percentage of federal civilian workers has declined, from a high of 3.7 percent of total employment in 1970 to 2.6 percent in 1990.

Of course, a longer time perspective confirms the growth of the federal bureaucracy. Approximately 3,000 people worked for the federal government in 1800. By 1881 the number reached approximately 95,000; by 1925, approximately 500,000.[10] By 1939 a million employees worked for the federal government; and by 1969, when growth slowed down, nearly three million did.

Whether the 3,143,000 federal employees in 1990 constituted a "huge" aggregation or not depends on how you think about it. To put this figure in perspective: They could fill up a large football stadium (accommodating 80,000 people) 39 times. Indeed, there are twice as many federal workers per capita in the United States as there are public employees in England, and four times as many as there were in the former West Germany. State and local governments employ almost 15 million more.

WHY HAVE PUBLIC BUREAUCRACIES GROWN?

Bureaucratic organizations may not perform as well as people expect or would like. Whether bureaucracies are running a major university, manufacturing automobiles, producing military weapons, or delivering the mail, their occasional failures attract attention and anger. Nonetheless most bureaucracies, including those in government, get things done. Of course, nonbureaucratic organizations can and sometimes do work well, especially if they are small and their tasks are modest. Many small businesses and service agencies, for example, use nonhierarchical methods to arrive at decisions and rotate tasks rather than specializing. But these alternative organizational forms do less well in carrying out ambitious tasks and in running large enterprises. They also take more time to reach decisions.

The strength of classic bureaucratic organizations in completing challenging tasks in a reasonably efficient fashion explains why we find them everywhere. To design a weapons system, distribute welfare checks, make and sell a billion hamburgers, or provide education to 35,000 students, bureaucracies' techniques of devising rules, recruiting specialists, dividing the tasks to be done into component parts, and mobilizing resources and people seem to work the best.

Max Weber notes that public bureaucracies have grown because the tasks government seeks to undertake have grown. Public expectations concerning what problems deserve "public" attention have changed. A hundred years ago, people without housing, employment, or enough food depended on private charity or fended for themselves. Thousands of children without parents lived on the streets of American cities.[11] And although the United States had an army and navy, their cost and size seem minuscule by today's standards. The development of technology went hand in hand with increasing expectations about the proper scope of government. The belief that government should involve itself in these and other problems led to the development of the *positive state.* A strong consensus exists in the United States that the government ought to engage in a wide variety of complex tasks to provide collective benefits such as national defense, a highway system, the licensing of those who drive on it, regulation of the production and sale of medicine and food, and the control of toxic waste. Public bureaucracies exist to provide these benefits. It is hard to imagine how it can be otherwise.

THE FOURTEEN CABINET DEPARTMENTS

Fourteen cabinet-level departments conduct most of the executive branch's bureaucratic activities. A department secretary heads each and serves in the president's cabinet by virtue of his or her position. The establishment of three such departments coincided with the formation of the new government in 1789. The *Department of State* conducts foreign relations; the *Treasury Department* engages in financial activities such as collecting taxes, printing money, and disbursing money to pay for the activities of government; and the War Department, renamed the *Department of Defense* after World War II, sees to the defense of the nation. The Navy Department, created in 1798, became part of the new Department of Defense in the 1940s.

Several other departments emerged as the growth of the nation and the federal government's activities made their functions increasingly vital. In 1849, programs dealing with Native Americans and public lands were consolidated into the *Department of the Interior*. The Judiciary Act of 1789 established both the office of attorney general and United States attorneys' offices in each judicial district. Not until 1870, however, did increasing litigation involving the United States lead to the creation of the *Department of Justice*, headed by the attorney general, who has authority to supervise all U.S. government attorneys.

The demands of a variety of interest groups who wanted the governmental programs touching their activities to enjoy cabinet status led to the creation of a number of "clientele" departments. These include the *Department of Agriculture* (1862), the *Department of Commerce* (1903), the *Department of Labor* (split from the Commerce Department in 1913), the *Department of Education* (1979), and the *Department of Veterans' Affairs* (1988).

Four other cabinet-level departments emerged in part through pressure from groups convinced that existing programs administered by non-cabinet-level bureaus would gain in importance and size through promotion to cabinet status. In 1965 President Lyndon Johnson created the *Department of Housing and Urban Development* (HUD) to signify his commitment to grappling with the problems of the nation's major urban centers. The *Department of Health and Human Services* (HHS), originally created in 1953, administers the non-education programs retained by the former Department of Health, Education, and Welfare after the creation of the Department of Education in 1979. The *Departments of Transportation* (1967) and *Energy* (1977) round out the cabinet. Table 15.1 (on p. 574) lists the 14 cabinet departments along with their estimated number of employees and their budget outlays (i.e., the amount they planned to spend) for fiscal year 1994. The simple listing of departments there obscures the truly magnificent complexity of the internal organization of each one.

The following example may convey the intricacy of the federal bureaucratic structure. An article appearing in the *New York Times* on September 13, 1988, under the headline "U.S. Would Ease Bone Labeling Rule for Meat" described existing federal regulations regarding "mechanically processed

TABLE 15.1
THE FOURTEEN CABINET DEPARTMENTS:
EMPLOYEES AND BUDGET

DEPARTMENT	CIVILIAN EMPLOYEES, 1993	BUDGET, FISCAL YEAR 1995 (IN MILLIONS)
Agriculture	108,500	$60.3
Commerce	35,800	3.6
Defense	854,900	290.1
Education	5,200	29.6
Energy	20,600	15.7
Health and Human Services	127,200	341.6
Housing and Urban Development	13,300	27.7
Interior	74,600	7.2
Justice	101,900	11.3
Labor	19,500	34.0
State	25,000	5.4
Transportation	67,500	37.3
Treasury	157,600	327.7
Veterans' Affairs	229,700	38.1

Sources: Employment data are estimates for 1995 from *Budget of the United States Government: Fiscal Year 1995: Analytical Perspective* (Washington, D.C.: Government Printing Office, 1994), p. 178, Table 12.1. Defense Department figures represent civilian employees. Budget figures are estimated outlays for fiscal year 1995 from *Budget of the United States Government: Fiscal Year 1995: Historical Tables* (Washington, D.C.: Government Printing Office, 1994), p. 62.

meat." It reported a proposal to revise these regulations to permit processed meat such as hot dogs and bologna to contain up to 10 percent bone without this fact's having to be indicated on the label. The proposal actually came from a bureau within the Agriculture Department, the Food Safety and Inspection Service, which itself fell under the jurisdiction of the assistant secretary of agriculture for marketing and inspection services and was buried deep inside the bureaucratic maze.

INDEPENDENT AGENCIES AND BOARDS

Virtually everyone knows the federal agency that has the most direct contact with the American people—the U.S. Postal Service. Regular newspaper readers will recognize other important agencies such as the Environmental Protection Agency (EPA) and the National Aeronautics and Space Administration (NASA), both of which have larger staffs and budgets than several smaller cabinet departments. The size and budget of another organization, the Central Intelligence Agency (CIA), remain secret. These organizations, along with about 60 others, are known as **independent federal agencies** because they do not fall under the organizational jurisdiction of either a federal cabinet department or the Executive Office of the President.

Many of these agencies, such as the EPA, the Federal Home Loan Bank Board, the Equal Employment Opportunity Commission, the Nuclear Regulatory Commission, and the Federal Trade Commission, engage primarily in regulation. The phrase *governmental regulation* has become a fairly potent political symbol, one that usually evokes threatening images. By the late 1970s both Democratic and Republican politicians made cutting "unnecessary" and "wasteful" regulations a priority. During the Reagan and Bush administrations, effort was expended to severely cut back on regulations by subjecting them to the most stringent cost-benefit review possible. These attempts were renewed in the "Contract with America." In 1995 the House of Representatives, with the support of both Democrats and Republicans, passed legislation that would make the passage of new regulations much more difficult. "Rules: Cost-Benefit Analysis and the Politics of Regulation" (on pp. 576-77) presents a discussion of one aspect of these attempts to limit the bureaucratic power to regulate. Regardless of citizens' emotional response to governmental regulation, this activity accounts for much of the significant work of the federal bureaucracy and will continue to do so.

The regular cabinet departments also engage extensively in regulation. Likewise, independent federal agencies whose primary task is regulation administer programs as well. A number of other independent agencies engage in little regulation. Some focus primarily on distributing money, such as the Small Business Administration, the National Science Foundation, and the National Endowments for the Arts and Humanities. Others, such as NASA, run their own programs. Several provide support services to the rest of the federal bureaucracy. The General Services Administration, for example, is responsible for sweeping the floors and emptying the garbage in federal buildings. The Smithsonian Institution's museums in Washington collect memorabilia for the nation.

The names and functions of many independent agencies are quite forgettable. What should you know about the Merit Systems Protection Board, the Board for International Broadcasting, or the Railroad Retirement Board? Agencies such as these engage in a wide variety of enterprises and activities, some important despite their obscurity, and some minor. Knowing they exist, however, conveys some sense of the tremendous scope of the federal government's activities.

The cumulative impact of independent agencies, although difficult to recognize, is substantial. Furthermore, their independent status often results from a complex play of power among competing interests. Established departments may lack enthusiasm for new programs and depend for support on interest groups opposed to such programs. Astute proponents of such programs therefore strive to grant independent status to the agency given the task of implementation. If the Environmental Protection Agency became part of the Agriculture or Commerce Department, or the National Science Foundation went to the Defense Department, the manner in which they went about their activities would change.

COST-BENEFIT ANALYSIS AND THE POLITICS OF REGULATION

As we noted in Chapter 2, government regulation is called for in a variety of circumstances. Nevertheless, the regulatory authority of bureaucratic agencies has been a source of controversy. During the Reagan administration and again during the 104th Congress, actions were taken to ostensibly limit the discretion of regulatory agencies by requiring that their actions be subject to strict cost-benefit analysis. Supporters argue that by requiring the benefits of regulation to exceed its costs, regulators will be forced to make maximal use of the resources that our society devotes to regulation. Further, cost-benefit analysis is touted as a "neutral" and "scientific" way of guiding regulatory policy. Couching actions in such terms is a powerful rhetorical device in political disputes. However, is cost-benefit analysis politically neutral? Can such rules be used to reduce bureaucratic discretion and also make government smaller and more efficient? While this may occur in some areas of regulation, in many other areas there are significant reasons to doubt the effectiveness of cost-benefit analysis.

One of the assumptions of cost-benefit analysis is that it is possible to accurately quantify (usually in dollars) both the costs and the benefits of proposed actions. In some regulatory arenas, this may be so. For example, price captures more or less the full value of an airline ticket, a train ride, or a kilowatt-hour of electricity; the challenge for regulators becomes to attach accurate price tags to such commodities when markets fail to do so. In such cases, cost-benefit analysis may be quite useful. On the other hand, price is less appropriate for capturing the full value of a healthy environment or that of avoiding cancer. While their value certainly has a component that is captured by money, they are not traded on the market, and estimating their price becomes a political rather than an economic determination.[1]

Furthermore, many areas of regulatory activity are characterized by scientific uncertainty, which renders any estimate of costs and benefits quite difficult. Efforts at estimating risks are further complicated by the fact that causes may be separated from effects by decades. Consider, for example, attempts to regulate carcinogens in the workplace. First, scientists are divided over the proper model for explaining carcinogenesis: Are one-shot or threshold models more appropriate? Disagreement also exists over the proportion of cancers that can be explained by environmental, genetic, or other causes. Consequently, there is a great deal of uncertainty as to what procedures are appropriate for estimating the relationship between given levels of exposure to a particular substance and actual cases of cancer. Resulting probability estimates can differ by several orders of magnitude, depending on assumptions about the relationship between tests on laboratory animals and human disease, or about the appropriate conclusions to be deduced from limited epidemiological studies.

The result of these problems is that in many areas of regulation the application of cost-benefit analysis is never a neutral exercise; it is always political and value-laden. The answers one gets depend on the assumptions one makes. These assumptions are fraught with political implications.

It is also unclear that imposing cost-benefit requirements will limit the scope of bureaucratic discretion. Instead, it simply shifts the authority for making decisions from one government agency to another. For example, the Reagan administration's efforts to impose cost-benefit requirements on regulations resulted in a dramatic expansion of the authority, size, and scope of the agencies assigned the task of conducting the analyses (usually the Office of Management and Budget):

How can we compare the value of clean air and water and human health on the one hand, and jobs on the other? When elected politicians and the public fail to resolve such issues, the task is left to bureaucratic agencies headed by people like Carol Browner, chief administrator of the Environmental Protection Agency. Whether one sees their regulatory efforts as being effective, ineffective, or an unwarranted intrusion on the private sector depends primarily on one's assumptions about market failures and government failures.

[T]he Reagan pursuit of a "new social contract," in the end, undermine[d] the prospects of enduring change. It . . . most obviously . . . put the advocates of "real economics" [i.e., cost-benefit analysis] in the uncomfortable position of carrying out a program to reduce the burden of government regulation through what amounts to unprecedented administrative aggrandizement. Regulatory relief, therefore, has led not to institutional reform that imposes restraint on government action, but to an important confirmation of a substantial government presence in overseeing social and economic activity.[2]

Whether regulation is "good" or "bad" depends on the interests and values of the observer. These interests and values can never be removed from the political process, even by seemingly neutral scientific techniques. As in many aspects of politics, a useful question to ask in explaining what positions interests take on regulation is "Whose ox is being gored?"

[1] The following discussion draws heavily on Bruce A. Williams and Albert R. Matheny, *Democracy, Dialogue, and Environmental Disputes: The Contested Languages of Social Regulation* (New Haven: Yale University Press, 1995), Chapters 1 and 2.

[2] Richard Harris and Sydney Milkis, *The Politics of Regulatory Change* (New York: Oxford University Press, 1989), p. 290.

OTHER FEDERAL BUREAUCRATIC ORGANIZATIONS

The federal government also contains bureaucratic organizations that do not fall within a cabinet department's control or qualify as independent agencies. Some engage in a wide range of economic activities—for example, selling stock and lending money; providing services such as insurance of bank and savings and loan deposits; and producing, distributing, and selling electric power. Run by bipartisan boards to insulate them from the normal play of power in politics, they resemble private corporations in many respects and in fact are commonly called **government corporations**. Examples include the Federal Reserve Board, the Federal Deposit Insurance Corporation, and the Tennessee Valley Authority. Although these groups may generate and spend substantial sums, their finances appear nowhere in the federal budget.

Advisory boards, commissions, and committees created and funded by the federal government offer expert advice and recommendations to the president or to federal agencies and departments. Some are permanent. (A search of the *U.S. Government Manual,* which provides a summary of the structure of the federal bureaucracy, will turn up, for example, the American Battle Monuments Commission, the National Mediation Board, and the Merit Systems Production Board.) In addition to performing designated functions, such agencies provide the president with a source of patronage appointments.

The activities of these boards and corporations sometimes become embarrassing to an administration. For instance, early in Ronald Reagan's administration, the Commission on Civil Rights issued reports that were critical of the president's civil rights policies. Reagan was able to end the commission's practice of issuing embarrassing reports by filling vacancies with people who were more sympathetic to his policies. Then civil rights supporters in Congress became so enraged that they slashed the commission's budget. More recently concern over the possibility of inflation led the Federal Reserve Board to raise interest rates several times during 1994 and 1995, actions that may have hurt President Bill Clinton by working to slow down the economy.

Presidents appoint some commissions in response to specific pressing political problems. Commissions dealing with crime and delinquency, racial riots, pornography, and AIDS, for example, were appointed and have issued special reports. Establishing such commissions serves several purposes. One is to delay action. Another is to generate publicity and support for a predetermined policy position, achieved by stacking the panel with committed supporters of the policy. Ronald Reagan's attorney general, Edwin Meese, promoted and endorsed the findings of a commission on pornography as part of one such concerted strategy to fight pornography.

When an agency cannot hire professional experts to help make technical decisions beyond the scope of its employees' expertise, it can establish permanent **advisory committees**. For example, the Food and Drug Administration relies on a number of advisory committees of health professionals, along with industry and consumer representatives, to help it make decisions about the effectiveness and safety of new drugs.[12]

Finally, the federal government contains a number of minor boards and committees that administer small programs, such as the Harry S Truman scholarships, or oversee one-time celebrations, such as the Commission on the Bicentennial of the United States Constitution.

THE FORMIDABLE RESOURCES
OF THE FEDERAL BUREAUCRACY

What resources do bureaucracies possess that allow them to play such a crucial role in governmental decision making? In the following section we discuss three principal resources that provide the answer.

AUTHORITY

The Social Security Administration (SSA) oversees specific federal programs, including Social Security, Medicare, and federal welfare programs. This gives the SSA, and units within it such as the Office of Hearings and Appeals, the power to issue regulations and make other decisions necessary to carry out its programs. Similarly, the Food Safety and Inspection Service of the Department of Agriculture, the EPA, and other agencies all have their own spheres of authorized activity. Without recognized authority, bureaucratic agencies would be powerless. Their authority exists both because Congress passed legislation granting it and because other players recognize it by obeying its decisions. Grants of authority provide bureaucracies with legitimacy. Try drawing up your own rules for Social Security appeals hearings and calling a press conference. Nothing will happen, because you lack the recognized authority to make such decisions.

This suggests that the nature and extent of an agency's authority shape its potential to determine outcomes. It also explains why struggles over authority (as when the jurisdiction of an agency is expanded or reduced)

Using the Food and Drug Administration's authority as a protector of public safety to increase his own public standing and prominence, FDA Commissioner David Kessler was able to retain his position under the Clinton administration even though he had originally been appointed by George Bush.

initiate some of the fiercest struggles in government. The conflicts that erupt when jurisdictions overlap provoke especially interesting confrontations. Classic battles for authority between the Bureau of Reclamation and the Army Corps of Engineers over water projects, and between the Army, Navy, and Air Force over weapons systems and missions, to name just two of many, punctuate American political history.

EXPERTISE AND INFORMATION

Bureaucracies possess expertise and information because their employees become recognized "experts" in the field of their authority. The sources of this expertise and information are not mysterious.[13] Bureaucratic employees devote all their time to issues that most other players consider only occasionally. Furthermore, bureaucracies specialize, and divisions within them focus on even smaller areas. Workers consequently develop a vast store of knowledge and experience. Even a well-informed member of Congress serving on a subcommittee dealing with Social Security likely knows little about procedures for handling citizen appeals stemming from denial of benefits, but staff members in the Office of Hearings and Appeals who have worked full-time for years on the topic know a lot. They not only acquire intimate knowledge of the process but also learn how and where to get additional information. They understand what important decisions must be made, and when; what happened in the past; and which interests, individuals, businesses, and congressional staffers hold what views. When bureaucrats have specialized professional training in law, accounting, or science, their education and professional contacts further bolster their expertise. Thus, a tremendous amount of information and expertise can be drawn on by bureaucracies.

Information and expertise, coupled with authority, allow bureaucracies to propose detailed regulations on obscure matters (e.g., procedures for hearings that review denial of Social Security benefits). They also serve as a potent resource in trying to exert influence through persuasion. Bureaucratic experts play a key role in developing many proposals, as well as lobbying for them. In fact, the initial drafts of many bills emerge from the word processors of career specialists buried in the depths of the bureaucracy. Information and expertise also contribute to the skill that bureaucracies use in pursuing their goals. Skill ranks as a crucial resource because it determines how effectively other resources will be used.

The career structure of bureaucracies obviously contributes to expertise. Tenure ensures that employees can, if they choose, stay with an agency as long as they wish. In practice, most spend many years in a single division. However, tenure provides the bureaucracy with resources that go far beyond its enhancement of expertise. One astute student of bureaucratic politics has summarized some of the advantages of permanence:

> [I]f the permanent employees can delay implementation of their nominal bosses' wishes, the bosses may be replaced by more congenial leaders or at least by newcomers initially unfamiliar with the programs they are in charge of, so the permanent staff can have their way for a while longer. Many civil servants think of

their temporary political superiors as birds of passage (and of themselves as the guardians of stability and continuity). They are, therefore, not motivated to leap immediately at the first instructions they receive if they regard the instructions as unwise.[14]

Political executives testify to the advantages that job security and long time frames bestow on bureaucrats. As we noted earlier, most political appointees remain in their jobs for 18 months or so, whereas permanent bureaucrats are likely to spend their entire career in the same agency. Thanks to the White House taping system, Richard Nixon's complaint on this point in the privacy of the Oval Office can be shared:

> We have no discipline in this bureaucracy. We never fire anybody. We never reprimand anybody. We never demote anybody. We always promote the sons-of-bitches [who] kick us in the ass.[15]

MOBILIZATION OF SUPPORT

Most bureaucracies deal with a variety of other organizations in their strategic environments. The groups and organizations affected by an agency's actions form its *constituency*. Some groups, which can be thought of as clients, benefit from these actions. Others, hurt by an agency's regulations, decisions, and penalties, become opponents. Later in this chapter we describe the close ties and ongoing exchanges of support and information that often develop between agencies and their clients. Here we note that the ability to call on these groups provides bureaucracies with a third potent resource.

The political symbols that bureaucracies use in mobilizing support from the general public provide another resource. Based on a common misperception that administration can be neutral and scientific, and drawing on the powerful images of "science," bureaucracies cloak their decisions in the symbolic rhetoric of scientific analysis and efficiency. Efficiency enjoys almost universal and uncritical acclaim. Efficiency is good, inefficiency bad. Claims that an "efficient" proposal results from "scientific analysis" carry great credibility. Bureaucracies recognize the power of these symbols and use them all the time.[16]

HOW BUREAUCRACIES WORK

What determines how bureaucracies actually make decisions? How do other participants in the game of politics shape their behavior? What strategies do bureaucratic agencies follow in using their resources to influence outcomes? The answers to these questions clarify the bureaucracy's crucial role in the play of power.

THE IMPACT OF BUREAUCRATIC CONSTITUENCIES

The term **constituency** usually refers to the residents of an elected official's geographically defined district. But a broader view of the nature of constituencies sharpens one's ability to understand politics. Applying the notion of constituency to bureaucracies demonstrates this well. As political scientist Eugene Lewis puts it, clients are people whom bureaucracies *work on,* whereas constituents are organized interests that bureaucracies *work for.*[17]

Consider, for example, what happened to the Reagan administration's proposals for cutting the size of federal programs shortly after it took office in 1981. It sought to eliminate all funding for the Legal Services Corporation (LSC), the agency that provides legal assistance to poor people in civil cases. It also proposed to eliminate the Small Business Administration (SBA), an independent federal agency, and to transfer a handful of its programs to the Commerce Department. In fact, the new administration sought to eliminate some 40 federal agencies.

Many of these agencies, including the LSC and the SBA, survived none-theless. How did they manage to do so when the president of the United States, the head of the executive branch, made their abolition an important part of his program throughout his term? An astute observer of bureaucracy

The reduction of the federal budget deficit, partly through downsizing of the federal bureaucracy, has been a goal of American presidents during the last two decades. In 1975 President Gerald Ford indicated to the public that he was drawing the line on further expansion of the deficit. Since then, despite the promises of elected officials, the deficit has continued to soar, and it continues to generate controversy in American politics.

provided the answer to a *New York Times* reporter in an article headlined "How 43 Agencies Keep Surviving the Reagan Ax":

> All of these programs, whether efficient or inefficient, have a constituency and it's hard to abolish them. To most other people they are invisible. And it is only the White House and its Office of Management and Budget that is opposing them. The people in the other departments and agencies and Congress represent the constituencies that want to keep them.[18]

Like an elected official's constituency, a bureaucratic agency's constituency has resources that can be used to support the agency's struggles in the play of power. This support takes many forms, including exercising influence on behalf of the agency or providing crucial information. Agencies come to depend on such support, just as elected officials depend on their voting and funding constituencies. With dependence comes responsiveness to the needs of constituents.

Bureaucratic agencies serve a complex mix of constituents, including congressional committees and subcommittees, other parts of the federal bureaucracy and state agencies, and the interest groups, businesses, labor organizations, and individuals affected by agency decisions. But the picture is even more complex than this. The composition of agencies' constituencies varies, as does the amount of influence wielded by their constituents. These differences help explain why some bureaucratic agencies achieve more success than others. The differences also alter the strategies that agencies use in mobilizing support. To understand the play of power in which bureaucratic agencies find themselves, you need to know something about the three categories of their constituents: Congress, executive agencies, and nongovernmental interests and organizations.

Congress as a bureaucratic constituency Congress has the potential to wield awesome influence over federal bureaucracies. Ultimately, their very existence depends on Congress's willingness to permit their continued operation. As we noted, their authority to administer programs comes from Congress, which delegates the discretion essential to carrying out the programs. Congressional subcommittees oversee agencies' performance and can acutely embarrass them by conducting hearings and issuing critical reports. Finally, Congress authorizes expenditures and appropriates money to pay for them. No one understands better than bureaucrats the basic political truth that without money, one can do very little to shape material politics.

The most crucial components of an agency's legislative constituency are (1) the substantive committee and subcommittee in the House and Senate that establish its programs and authorize the expenditure of money, (2) the appropriations committee and subcommittee that determine how much money will actually be spent each year, and (3) investigative arms of Congress such as the General Accounting Office (GAO) and the Office of Technology Assessment (OTA).

Executive-branch constituents of bureaucracies Anyone who has worked for a forceful supervisor knows why hierarchical superiors must be heeded. Less obvious, perhaps, is the dependence of superiors on their subordinates.

But it is no less important. Without the support of subordinates, the work cannot be done satisfactorily.

For any given level of the bureaucracy there are counterparts in outside agencies whose cooperation or resistance affects day-to-day administration. Top officials such as the secretaries of state and defense find that much of what they do is affected by the other's actions. Even subordinate agencies within departments find that counterparts elsewhere in the bureaucracy, called **lateral agencies**, administer overlapping programs. Lateral agencies can be useful allies or bitter enemies. If policy goals point in the same direction, mutual support and cooperation among agencies can develop. If policy goals clash, trouble is almost inevitable. All-out conflict can result in one agency's absorbing the personnel and programs of the other, with the loser going out of existence. During the 1970s the Bureau of Reclamation and the Army Corps of Engineers engaged in acrimonious and protracted conflict stemming from their overlapping responsibility to manage water resources. So did the Department of Justice's Drug Enforcement Administration and the Treasury Department's U.S. Customs Service during the 1980s.

The frequent and persistent conflicts between lateral agencies create problems that higher officials must confront. To cope with such problems, they may resort to the technique of establishing an interagency committee that draws members from the contending parties. The play of power in these committees, usually called **interagency groups (IGs)**, although rarely visible shapes many important outcomes. The first secretary of state in the Reagan administration, Alexander Haig, thought them especially significant:

> In organizational terms, the key to the system is the substructure of SIGs [Senior Interagency Groups] and IGs in which the fundamentals of policy (domestic as well as foreign) are decided. . . . IGs . . . can summon up all the human and informational resources of the federal government, study specific issues, and develop policy options and recommendations. . . . Built into it . . . is a high degree of bureaucratic competition and tension. . . . [H]e who controls the key IGs controls the flow of options to the President and, therefore, to a degree, controls policy.[19]

Of course, the president, if he takes the time to express interest in a bureau's activities, can become the most significant constituent of all. But his time and attention can almost never be devoted to a single agency's activities. For help in resolving such disputes, most presidents rely on one of the most important and powerful organizations attached to the Executive Office of the President, the Office of Management and Budget. Indeed, American presidents delegate a variety of essential tasks in coordinating the activities of the executive branch to OMB. Its location within the Executive Office of the President exempts it from close scrutiny and oversight from Congress. The news media also have difficulty following its activities. The low visibility of its role explains why few Americans know much about it. Nonetheless, as "Strategies: OMB—The Nearly Anonymous Power Center" explains, this agency plays an absolutely crucial role in shaping policy.

The public and interest groups as bureaucratic constituents   A few bureaucracies have enjoyed relatively high visibility and strong support from the

OMB—THE NEARLY ANONYMOUS POWER CENTER

No other agency, with the possible exception of the Federal Reserve Board, exerts as much influence without most people's realizing it as does the Office of Management and Budget. Most people don't have the faintest clue as to what the OMB does or how powerful it is. Nonetheless, understanding the play of power requires knowing this critical player.

The OMB has four principal tasks. First, it prepares the president's budget before this is submitted to Congress. Second, it reviews all agencies' proposals for legislation. Third, it reviews regulations proposed by all bureaucratic agencies. Fourth, it is charged with "reducing paperwork." Each of these tasks provides the OMB with awesome ability to shape policy.

In formulating the president's budget, the OMB scrutinizes every agency's and department's budget request. If it denies funds for an agency's new program or cuts an existing program, and if the agency cannot get the president to overrule it, the agency cannot formally ask Congress for restitution. In practice, the OMB rarely loses appeals of its decisions. Furthermore, Congress usually accepts the president's proposed budget for the overwhelming proportion of agencies. Control over agency budgets allows the OMB to influence the content of agency programs.

The OMB also calculates total revenues and expenditures, and therefore the size of deficits. To do this, it must estimate the rate of economic growth and the unemployment, inflation, and interest rates. As we show later, the content of these calculations can affect the entire course of policy.

The OMB reviews legislative proposals as well. Any agency of the federal government that wants to propose a change in legislation, even if it entails no additional expenditures, must submit the proposal to the OMB. This procedure, called **legislative clearance**, seeks to ensure coherence in the administration's overall policy. Feuding agencies cannot push contradictory policies in Congress because the OMB has the power to reconcile differences.

In addition, the OMB reviews proposed federal regulations. In 1985 the Reagan administration authorized the OMB to review new governmental regulations to produce "substantial improvements in Federal regulatory policy" and to "help ensure that each major step in the process of rule development is consistent with Administration policy."[1] Proposed rules could be delayed or modified by OMB action. The OMB used its power to modify hundreds of regulations, including those governing warning labels on aspirin, limits on the amount of grain dust (which is highly flammable) in grain elevators, and permissible levels of exposure of workers to ethylene oxide fumes. More recently, legislation passed by the House of Representatives in 1995 as part of the "Contract with America" required that the OMB complete substantial new cost-benefit analyses before any new regulation could take effect.

Finally, the OMB collects information about regulations. Building on a law passed at the end of the Carter administration to expand the OMB's authority to regulate the collection of information by federal agencies, the Reagan administration used the OMB to extend its control. For example, agencies must receive approval from the OMB's Office of Information and Regulatory Affairs on the content of questionnaires they use to assess the impact of regulations. Thus, the OMB removed questions from a federal agency's study of the health effects of video display terminals on the grounds that they were unnecessary and burdensome; critics complained that the resulting survey was too vague to assess the health effects.[2]

These duties, coupled with the increasing importance of budgeting in the face of huge deficits, have produced an agency whose importance cannot be overestimated.

[1] David Burnham, "Reagan Authorizes a Wider Role for Budget Office on New Rules," *New York Times*, January 5, 1985, p. A1.
[2] "Watching the Watchdogs: How the OMB Helps Shape Federal Regulations," *Newsweek*, February 20, 1989, p. 34.

general public. NASA and the Federal Bureau of Investigation (FBI), for example, long basked in media attention and general support. In fact, their leaders carefully cultivated their agencies' public image and support as valuable resources for dealing with Congress and the rest of the executive branch.

The majority of executive bureaucracies, however, mostly ignore the general public and the mass media, mirroring the media's and public's lack of attention to them. The bureaucracy is too large for the media to cover, and the public is too uninterested in the details of policy. Only when scandal, disaster, or controversy engulfs an agency does the average citizen read or hear anything about it. When the National Weather Service fails to predict destructive tornados or hurricanes, when the National Park Service finds itself fighting immense forest fires in Yellowstone National Park, or when an agency's head resigns amid allegations of corruption, the media and the public pay close attention for a brief time.

If the general public and the news media remain happily oblivious to most agencies' activities, the same cannot be said for special or attentive publics. Virtually every agency makes decisions of consuming interest to special publics. Groups and individuals directly affected by an agency's activities routinely scrutinize its decisions. For example, coal mining companies and the United Mine Workers know what the Bureau of Mines is doing. Lumber companies closely follow the decisions made by the Agriculture Department's Forest Service regarding the cutting of timber from national forests. And groups such as the American Association of Retired Persons and the National Senior Citizens Law Center know what the Office of Hearings and Appeals in the Social Security Administration is and what it does.

Executive agencies pay special attention to the nongovernmental groups that form their attentive publics. Such groups possess the motivation, skill, and resources to exert influence. Their support or opposition helps determine the outcome of struggles involving every aspect of an agency's activities: the enactment of legislation expanding or cutting back on its programs; its budget; and sometimes its very existence. In fact, agencies find such groups so useful that they not only cultivate their favor and encourage their growth but occasionally take the initiative in creating them. For example, the Farm Bureau Federation, which provides strong support to the Extension Service in the Department of Agriculture, was initially organized with the active participation of the Agriculture Department's county extension agents at the beginning of this century.[20] The strength of the nongovernmental groups in the Army Corps of Engineers' constituency helps explain why it so often won its battles with the Bureau of Reclamation, as in the case discussed in "Participants: The Army Corps of Engineers—A Study in Bureaucratic Constituency Power."

Sometimes the influence attempts of interest group constituents are so successful that the group in effect co-opts an agency or parts of its program. When this happens, the co-opted agency acts as the interest group's representative in the play of power.[21] One political scientist, discussing the Tennessee Valley Authority's responsiveness to agricultural groups' interests, noted, "Any public agency may thus find it necessary to yield control over a segment of its program to a significant interest group in order to buy the support of that group for more important goals."[22] However, research has found that agencies primarily engaged in economic regulation often become *captured* by

THE ARMY CORPS OF ENGINEERS—
A STUDY IN BUREAUCRATIC CONSTITUENCY POWER

The president of the United States, commander-in-chief of the armed services, and chief administrator had made a decision. Franklin Delano Roosevelt, one of the shrewdest and most effective presidents in our nation's history, faced with a dispute between the Army Corps of Engineers and the Bureau of Reclamation over who would build a dam, left no doubt about what was to be done when he wrote to the secretary of war. "I want the Kings and Kern River projects to be built by the Bureau of Reclamation and not by the Army Engineers."

Ultimately, however, the Army Corps of Engineers built the dam. Much of the explanation rests in the powerful support the Corps enjoyed from a diverse constituency. Its long history of building expensive public works projects in virtually every congressional district brought powerful support for it within Congress. Local chambers of commerce and groups formed to promote the building of large projects (e.g., the Atlantic Deeper Waterways Association and the Ohio Valley Conservation and Flood Control Congress) provided another source of support. Local interests, especially construction firms and barge companies, also had long-standing traditions of strong support for the Corps. The heads of each of the many district offices of the Corps of Engineers cultivated, cemented, and mobilized the support of these local interests.

Roosevelt wasn't the only president to learn of the political strength of the Corps. Presidents Warren Harding and Herbert Hoover wanted its civilian projects transferred to the Interior Department; so did several governmental reorganization commissions. But, thanks to the strengths of its constituencies, the Corps continued to build these projects.

The success of the Corps in preserving its autonomy illustrates how a combination of a

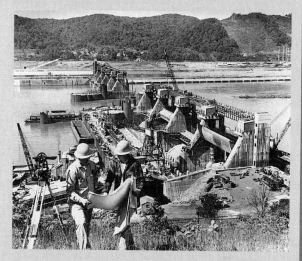

Although Franklin Roosevelt wanted the Kings and Kern River dam built by the Bureau of Reclamation, even the president of the United States was no match for the ability of the Army Corps of Engineers to mobilize powerful constituency groups to support its claim on the project.

strong constituency and political skill in mobilizing it can provide bureaucratic agencies with formidable political clout.[1] It demonstrates why the ability to mobilize constituents deserves to be ranked as one of the key resources administrative agencies can possess.

[1] This description draws on the analysis of the Corps of Engineers' success provided by Grant McConnell, *Private Power and American Democracy* (New York: Vintage, 1966), pp. 211–30. The classic study of the Corps is Arthur Maass, *Muddy Waters* (Cambridge, Mass.: Harvard University Press, 1951).

the very interests they are supposed to regulate. State and national agencies regulating insurance, utility rates, railroads, and other activities provide many examples of such capture.

HOW BUREAUCRACIES MAKE DECISIONS: "MUDDLING THROUGH"

Like most people, bureaucrats try to maximize their freedom of action and discretion. One effective way to do so can be found in official descriptions of how bureaucracies make decisions. Drawing on the usually negative feelings aroused by the symbol of "politics," such descriptions claim that bureaucratic decisions stem from following nonpolitical, neutral "administrative procedures." In addition to supposedly avoiding "politics," such decisions are said to be "scientific" and "rational." Thus, the **rational-comprehensive model of bureaucratic decision making** specifies the following sequence of steps in its formula for how bureaucracies reach decisions:

1. Specify clearly the goal to be achieved and the set of values, ranked in order of importance, that underlie the selection of that goal.
2. Identify every alternative method of achieving the goal.
3. Evaluate the outcomes that are likely to result from adopting each of these methods of achieving the goal by using the set of values.
4. Choose the method that best achieves the goal (i.e., maximize attainment of the goal).
5. Throughout, rely on extensive information, analysis, and explicit theory.

The rational-comprehensive model fits nicely into the ideal bureaucracy of Max Weber discussed earlier in this chapter. Specialization develops expertise in gathering information. Adherence to formal rules ensures that there are clearly stated goals and values that guide the evaluation of alternatives. The hierarchical structure allows rational judgments to be made at the top, utilizing the information and analysis gathered from below and sent up through the proper channels.

Organizational charts, proudly displayed by bureaucracies to explain how things work, express the ideology embedded in the rational-comprehensive model. They imply that it is possible to learn who will make what decision by examining the division of labor and authority depicted in the organizational chart. Responsibilities appear to be clearly defined; lines of division are sharp.

"Kiss-and-tell" renditions of how high policy actually comes to be made paint a very different picture of bureaucratic decision making. For example, an official in the Treasury Department during the Reagan administration observed that "extraneous, fortuitous events . . . impinge heavily on policy outcomes and often determine them":

> The professor in the classroom explains to his students the importance of careful statistical measurement, properly specified models, and sound economic analysis. But these things . . . count for little toward the making of policy. . . . If [decision makers] are

presented with a choice, they will choose on the basis of personalities, or according to who tells them what they want to hear, how it will play in the press, who[m] they want done in and who[m] elevated, who is smooth and who is strident, or who is the best manipulator of their own personalities.[23]

This description expresses the sharp contrast between the theory of rational decision making and the reality of the play of power. It also accurately conveys the prevailing view among scholars who study decision making in bureaucracies. Charles Lindblom, in a classic article entitled "The Science of Muddling Through,"[24] argued that bureaucracies in the real world depart from each of the five steps in the rational-comprehensive formula by adhering to the **incremental model of bureaucratic decision making**. Incrementalism exhibits the following characteristics:

1. *Goals normally remain unclear, and the values needed to assess alternatives are unclear and unranked.* Three reasons explain why this is so. First, policymakers, including those in Congress, don't know what to do about most problems and consequently often fail to spell out clear goals. Second, conflicts over values in society make it impossible to rank goals. There is simply too much disagreement over what ends government should promote. Third, troublesome conflicts between two (or more) attractive goals make choosing a policy difficult. We want low inflation but also high employment, freedom of the press but fair trials for defendants, the ability to move quickly in foreign crises but also a chance to engage in close consultation with allies before acting.

2. *Several factors limit the number of alternative policies receiving serious consideration to a few.* People tend to continue doing what they have done in the past. Radically different approaches rarely come to mind. The more an alternative policy departs from what has been done, the more intense political opposition to it will be. Policymakers usually reject any proposal that is not politically feasible. Some solutions encounter technical obstacles, at least in the view of experts, and are rejected out of hand. Usually, time pressures demand a decision before extensive analysis of very many possible solutions can be conducted. Often, the resources needed to conduct such analyses simply are not available. Consequently, most alternatives depart only incrementally from what is already being done.

3. *Only very limited analysis of the impact of the few alternatives considered is undertaken.* Unclear goals and values, consideration of only a few alternatives, and time pressures severely limit analysis of the probable results of choosing one policy over another. Partial analysis examines only a few of the most obvious outcomes.

4. *Policymakers end the decision-making process when they find an alternative that is "good enough."* Limited assessment of a few alternatives precludes selecting the "best" alternative the rational-comprehensive method calls for. It is impossible to "maximize." Instead, policymakers "satisfice"—that is, they choose a satisfactory rather than an ideal option. A common rule of thumb in identifying an option that fits the bill is to choose the first option that all participants agree on as being good enough.

5. *Following the procedures just listed, bureaucratic decisionmakers "muddle through."* Rather than making a rational, comprehensive, scientific calculation of the extent to which each of a wide assortment of alternative policies will realize clearly defined goals, bureaucratic decisionmakers select a "good enough" choice in the face of unclear goals. Furthermore, they look at only a few alternatives, and they examine only a few consequences of the limited set of choices. Little explicit theory is used. Calculation is ad hoc. Rather than drawing exclusively on extensive information produced through "channels" by bureaucratic specialists, decisionmakers rely on limited information from a variety of sources. Elaborate informal communication networks evolve over a long period of time. These networks ignore the rigid boundaries of organization charts and formal position. Personal acquaintances formed in college, in former jobs, and in past battles link bureaucrats, newspaper reporters, congressional staffers, members of Congress, lobbyists and interest group leaders, tennis partners, and other assorted types in a series of intertwined grapevines.

THE STRATEGIES BUREAUCRACIES FOLLOW

Like interest groups and other central players in politics, bureaucrats use a variety of strategies in deploying resources to achieve their goals. Watching the complex interplay among competing agencies' strategies provides some of the best entertainment in American politics. We will discuss and illustrate the following strategies here: controlling information; affecting a decision's visibility; affecting the scope of participation by shaping how many and what kinds of players enter the fray; affecting the direction of the conflict (what the decision or dispute is about); and affecting the timing of decisions and of the execution of the other four strategies.

Controlling information   Expertise, specialization, and authority combine to make information a critical resource for bureaucracies. The effectiveness of efforts to exert influence through persuasion depends on the use of information. Strategies that affect the availability and content of information, therefore, loom large in the arsenal of bureaucratic maneuvers. Students, parents, professors, friends, and lovers employ similar strategies all the time, including withholding information, inundating with too much information, and distorting the content of information.

Most agencies try to keep their internal deliberations from public view in keeping with bureaucracies' penchant for secrecy. Like almost everyone, they also try to conceal foul-ups, blunders, and waste. Agencies dealing with sensitive matters—national defense and intelligence, commercial transactions, diplomacy—can justify restricting information by classifying it as "secret" on the grounds of national security. The temptation to draw the national security blanket over information that is merely embarrassing can rarely be resisted.

Many decisions are presented as if they had been made according to the rational-comprehensive ideal. Complete withholding of the scientific, empirical data on which they purport to be made is impossible. Imagine the OMB announcing that it had calculated estimates of the rate of economic growth, of interest rates, of unemployment, and of governmental revenues,

THE OMB'S "ROSY SCENARIO" INFORMATION STRATEGY

In the early years of the Reagan administration, David Stockman, the president's first director of the OMB, refined the manipulation of projections based on economic data to such an art that his economic forecasts acquired their own name—"Rosy Scenario." The argument that the budget could be balanced while military spending rose and taxes were cut depended on a rosy economic forecast. Stockman describes how the Rosy Scenario forecast came to be constructed:

> So we got out our economic shoehorn and tried to jimmy the forecast numbers until all the doctrines fit. It was almost inherently an impossible task. . . .
>
> The forecasting sessions, which formerly had been crucibles of intellectual and ideological formulations, degenerated into sheer numbers manipulation. The supply-siders yielded a tenth of a percentage point toward lower real growth; the monetarists yielded a tenth of a percentage point toward higher money GNP.[1]

This still wasn't good enough, because it predicted inflation lower than the head of the Council of Economic Advisers, Murray Weidenbaum, could stand. So, in return for building a higher estimate of inflation into the model, Weidenbaum accepted a higher economic growth rate. According to Stockman, "The massive deficit inherent in the true supply-side fiscal equation was substantially covered up."[2] But the "information strategy" pursued went further. Even when it became clear that Rosy Scenario was dead wrong, the administration

David Stockman served as director of the Office of Management and Budget during the Reagan administration. His candid discussions of the strategies employed during his tenure have revealed much about the inner workings and behind-the-scenes struggles that characterize bureaucratic politics in Washington.

continued to rely on it. Stockman minces no words, asserting that a decision was made to "cover up what we knew to be the true budget numbers."[3] Stockman concludes that "the truth had been buried, and this time the White House had done it deliberately, in order to win the tax battle."[4]

[1] David A. Stockman, *The Triumph of Politics* (Boston: G. K. Hall, 1987), pp. 93, 96.
[2] Stockman, *The Triumph of Politics*, p. 97. The forecast bore no relation to what really happened. It predicted 5.2 percent real growth, but gross national product (GNP) actually fell 1.5 percent. Instead of 7.7 percent inflation, it was 4.4 percent. The huge budget deficits of the Reagan years began here.
[3] Stockman, *The Triumph of Politics*, p. 329.
[4] Stockman, *The Triumph of Politics*, p. 332.

and that the budget deficit would therefore be only $15 billion in the upcoming year—but that its estimates for each could not be revealed. Of course, release of some information to justify decisions does not preclude manipulation of data, as "Strategies: The OMB's 'Rosy Scenario' Information Strategy" shows.

Affecting visibility   Decisions differ depending on who participates, and who participates depends in part on visibility. Whether players want high or low visibility depends on whether the resultant changes in who participates will help or hurt their chances. If new participants come to the fray as visibility increases, will they be enemies or allies?

Restricting information by invoking secrecy provides the best way to guarantee low visibility. When information must be released, it can be done when other events, such as an election or an international crisis, increase the chance that it will be overlooked. But how does one go about raising visibility? Frequently, it is accomplished by "leaking" information to someone who can broadcast it widely. Other techniques for attracting media attention, such as holding a press conference or issuing a report, also come into play. Struggles over the definition of an issue determine the symbols used, thereby affecting visibility. For example, arguments that a fetal medical research program involves the abortion conflict have succeeded in raising the visibility of what might otherwise be a low-visibility issue.

Affecting the scope of participation   Restricting or expanding visibility usually alters the cast of participants, and hence the balance of contending forces and the outcome of the play of power. Changing the composition of the group of players is such an effective strategy that several techniques are used for bringing it about.

In 1971 Daniel Ellsberg (*left*), an intelligence analyst, leaked documents revealing that the government had systematically lied to and withheld information from the American people regarding the Vietnam War. The publication of these documents, known as the Pentagon Papers, in newspapers such as the *New York Times* hastened the end of the war.

Bureaucracies may resort to mobilizing supporters among their interest group, bureaucratic, and congressional constituencies. Indeed, the conventional view that interest groups always initiate contact with the bureaucracy is wrong: The reverse often occurs. Bureaucrats use their network of contacts among interest group leaders and lobbyists to spur them to action. Sometimes their message contains the plea "Please pressure us." At other times it is a call to arms: "Look, bad things are going to happen," or "Why don't you contact Congress?"

Making sure key supporters show up at crucial meetings, arranging with them in advance what will be said, and affecting who gets invited constitute another way to manipulate the alignment of players. A frank way of describing this strategy often emerges from the lips of losers: "This meeting was rigged!" For example, David Stockman has described how Drew Lewis, secretary of transportation in the Reagan administration, used a short cabinet meeting to advance a campaign to impose quotas on imports of Japanese cars:

> Drew Lewis was the first to speak, and what he had to say made my jaw drop. The time had come, he said, ... to restrict Japanese auto imports. ... After Lewis had said his piece, a drumbeat started. [Malcolm] Baldridge talked it up for import controls on Japanese autos, too. After Baldridge, Bill Brock spoke. He didn't speak, he echoed. Then Labor Secretary Raymond Donovan was recognized, and he started echoing. It was a Swiss chorus. ... The Cabinet Council meeting, in fact, had been rigged by Lewis.[25]

Affecting the direction of conflict   In discussing Japanese auto imports, Drew Lewis argued that the problem involved "American jobs" and "fair competition"; David Stockman argued that the integrity of "free markets" and the principles of the "free enterprise system" were at stake. The struggle they waged over the definition of the issue affected which symbols emerged in the discussion.

Because symbols can be used to persuade, astute players try to shape how issues are framed. This not only affects how participants can be influenced once they enter the play of power but also determines who will be motivated to join the fray. Being full-time players in the game of politics, skillful bureaucrats know this well.

Another way in which bureaucracies affect the direction of conflict is by influencing the content of policy proposals at the drafting and issuing stages. The bureaucracy's role in determining the content of proposed legislation remains much less evident. However, remember that bureaucracies possess the crucial resources of *information* and *expertise.* Who can better offer new proposals and draft provisions to guide their implementation than the people who have worked with a program for years? Thus, lower-level bureaucrats write the initial drafts of much legislation submitted both by the executive branch and by individual members of Congress. Further, the revisions that Congress adopts reflect the formal input of the bureaucracy (in the form of public documents and testimony) as well as informal, face-to-face interaction.

The bureaucracy's role in determining the content of legislation reinforces the basic premise of this chapter. More than any other element in the American political system, the bureaucracy determines policy. Earlier in this chapter we argued that it exercises both authorized and unauthorized discretion in *implementing* policy. Now we argue that the very *content* of the policies it implements reflects its interests. Thus, the bureaucracy takes a "double shot" at policy making. It influences the formal content of policy as embodied in legislation, and it exercises wide discretion in carrying out this legislation.

Affecting the timing of decisions   Good politicians pay close attention to when things happen. Because most bureaucrats become good politicians, they try to control the timing of events whenever they can. Timing can be used to affect how favorably proposals are received, how visible or invisible decisions are, and who else will be involved—that is, the *scope* of the conflict.

Bureaucrats know that events happening at the time a proposal is made affect how it will be received. For example, the middle of a big budget crunch is no time to propose new spending. Likewise, the end of an administration is not the time to announce a major policy initiative. But the onset of a new administration opens, at least for a time, a number of "policy windows." Proponents of ideas for new programs or major changes who have been biding their time make fresh attempts to put them on the agenda at the beginning of an administration.[26] Plans to seize program authority from a rival agency see the light of day when the rival becomes embroiled in scandal. Major events—a massive accident, an accomplishment by a foreign rival—also provide occasions for announcing new initiatives.

Timing affects visibility. A controversial proposal or an unpopular decision is less likely to be widely noticed if it can be held for release to coincide with an event that dominates the news (e.g., an election, the outbreak of a military conflict, an inauguration, or the culmination of a major legislative battle in Congress). The days between Christmas and New Year's Day, when the attention of the public and news reporters is elsewhere, are also a good time for low visibility. A proposal to change policies regarding the strip mining of coal in national parks and wildlife refuges is a good example: It appeared in the *Federal Register* on December 27, 1988, two days after Christmas and at the very end of the Reagan administration. Moreover, in an election year the incumbent administration's bureaucracy usually postpones announcing controversial decisions until after the election.

By the same token, information likely to generate support is issued when publicity is likely to be higher and have greater impact. Bureaucrats wait for "slow news days" to announce new programs, and studies that are likely to help win legislative battles are cited just as crucial votes come up. For years, readers of the *New York Times* knew when the House Appropriations subcommittee, which reviews the Pentagon's budget, was about to consider spending requests. As regularly as clockwork, the *Times* would carry a story quoting "Pentagon sources" revealing the results of a Defense Department study showing significant advances in Soviet military power that posed a direct and serious threat to the United States. Committee members and interest groups who are lobbying for the defense budget pay closer attention to such stories when they have to make decisions on military spending.

The lesson for understanding bureaucratic politics is simple but powerful: Timing makes a difference.

BUREAUCRACY IN THE PLAY OF POWER

The rational-comprehensive model of how bureaucracies work leaves no room for conflict. In this view, cooperation reigns; rational discussion and accommodation based on scientific analysis prevail. Formal organizational charts describe the structure of influence. Especially during a change of administrations, public relations specialists use a rhetoric of cooperation extolling how well the "wonderful new team" will work in harmony to achieve common goals. But the reality of bureaucratic decision making departs sharply from this picture.

CONFLICT EVERYWHERE

Although close cooperation among top officials and agencies sometimes occurs in the federal bureaucracy, conflict predominates. It occurs every-where—within different divisions of an agency, between superiors and subor-dinates, and among agencies with overlapping jurisdictions in different de-partments. Disagreements between entire cabinet departments and their heads either simmer quietly or boil over in widely publicized battles.

Conflict in the bureaucracy is normal and unsurprising for a variety of reasons. As noted, individual agencies and departments pursue a complex set of goals such as organizational survival; expansion of personnel, budget, and program authority; and advancement of their heads' careers. Add to this the demands of myriad constituencies who themselves reflect the full range of disagreement and conflict in American politics, and turmoil is virtually guaranteed. Usually these struggles take place behind the scenes and are revealed only years later in the participants' memoirs. But careful readers of the national press can glimpse them in progress through stories leaked by participants to damage their rivals.

In 1977 and 1978, for example, there was controversy within Jimmy Carter's administration over whether to proceed with development of the neutron bomb, a weapon that produced high levels of radiation but less blast than ordinary nuclear weapons. Development of the bomb became controversial when it was portrayed in the press as a weapon designed to kill people while preserving buildings. Hidden from initial public scrutiny, supporters of the bomb (e.g., Secretary of State Cyrus Vance, National Security Adviser Zbigniew Brzezinski, and the Pentagon) struggled with Carter's political advisers over whether to proceed with the weapon's development. A critical moment in the behind-the-scenes battle came when a story was leaked to the *New York Times* that Carter had given in to public pressure and was canceling development of the bomb. An astute reader would have glimpsed the hand of bureaucratic opponents trying to expand the scope of conflict and alter the decision. Indeed, Carter was attacked by Republicans and many Democrats for being weak on defense and, in the end, decided to delay rather than cancel the neutron bomb's development.[27]

Legitimate policy-based clashes of interest generate sharp conflict; the clash of personalities, egos, and ambition intensifies it. The president's advis-ers all seek increased influence over the president and his administration's

Although usually hidden from the public, battles among key members of the government administration are sometimes revealed through strategic leaks to the press as rivals seek to alter the scope of the conflict. During the Nixon administration, Secretary of State William Rogers (*top left*) waged a bureaucratic turf battle with National Security Adviser Henry Kissinger, who is shown here (*top right*) talking to Alexander Haig. When Haig served as secretary of state in the Reagan administration, he struggled with Attorney General Edwin Meese (*bottom left*) over control of foreign policy.

policies, but the desire for influence exceeds the amount that can be exerted. Not everyone can play the leading role. The memoir of Alexander Haig, the first secretary of state for the Reagan administration, illustrates this sort of conflict well. In his first press conference, Haig unwisely said that he saw himself as the "vicar" of foreign policy. The self-asserted role of "vicar," he notes, "played its part in creating first the impression, and finally the uncomfortable reality, of a struggle for primacy between the President's close aides and myself."[28] Soon an unflattering news story appeared about Haig's "grab for power" in a meeting in the White House on Inauguration Day. Haig complains:

> I could not conceive that any of the seven of us who had firsthand knowledge of the circumstances—the President's three senior aides, two members of his Cabinet, his advisor for national security, and myself—would be so mischievous, so numb to the requirements of the Presidency, as to plant such a story.[29]

Later, Haig describes how Edwin Meese, counselor to the president, began to erode his authority in foreign policy decision making by assuming control over the flow of documents from Haig to the president.[30] Indeed, it should come as no surprise that the members of the president's team do not always like one another.

WHY AGENCY INFLUENCE VARIES

Bureaucratic agencies differ in their resources, goals, constituencies, and strategies. Consequently, their success in the play of power differs as well. To further complicate the picture, the relative influence of agencies changes over time.

Some agencies exercise little independence, being dominated by powerful elements of their constituency. As we previously noted, regulatory agencies at both the federal and state levels sometimes become "captives" of the very interests they are supposed to regulate. Others, such as the Office of Management and Budget, serve such a limited constituency (in this case, the president) that they must respond to its wishes to maintain their effectiveness.

At the opposite extreme are agencies that for a variety of reasons enjoy great independence from Congress, clients, and the executive branch. The Federal Bureau of Investigation under J. Edgar Hoover (from 1924 to 1972) provides the most striking example. The general public, a vast and unorganized client interested in effective law enforcement, held Hoover and his agency in high regard partly because of Hoover's great skill at publicizing his agency's exploits. He was able to use the newly emerging mass media— first radio and later television—to publicize the heroics of FBI G-men (i.e., government men) as they waged battles against well-known criminals. The Department of Justice found that the agency's activities and budget could hardly be controlled—in part because of support provided by Congress, which reflected the public's esteem for the FBI. Indeed, Hoover's public standing made it virtually impossible for presidents to dismiss him, even though several wanted to very much.

Most agencies fall between the two extremes. Their strategic environments display greater complexity, finding both supporters and enemies in their constituency; and they walk a tightrope between the contending forces seeking to control them. Sometimes, however, the most powerful components of their constituency themselves agree on what the agency should do. When a cohesive set of interest groups forges strong ties with the congressional committees (and especially subcommittees) that address its issues, they join with the corresponding bureaucratic agency to exercise substantial control over policy. Together, they decide what program authority the agency should have, determine its budget, and influence how it implements policy. These *subgovernments*, or *iron triangles*, form their own formal communications network and informal grapevine and share a common perspective on what the content of policy in the issue area ought to be. Each of the three interacts with the other two. None of the three points in the triangle (agency, interest group, subcommittee) dominates the others. Rather, agreements emerge from continuous bargaining over the content of incremental adjustments to existing policy and over the routine administration of programs. When they unite, they usually exercise a near monopoly of influence over policy; hence the term *subgovernment*. The iron triangle that seized control over the savings and loan industry, described in "Applying Knowledge: The Savings and Loan 'Iron Triangle,' " illustrates how such arrangements come to dominate aspects of policy in the play of power.

SUMMARY

Americans' aversion to bureaucracy, born of unhappy personal encounters and reinforced by the news media's "bureaucratic horror stories," builds resistance to even thinking about bureaucracy, much less learning about its role in politics. The sheer size, number, and complexity of public bureaucracies, and their intrusive presence in nearly every aspect of daily life, compound the reluctance and difficulty we experience in comprehending its role.

Yet American politics cannot be understood without solid knowledge of the role of bureaucracy. After all, much of what government does is performed by bureaucracies. No other social institution seems able to accomplish what they do. In addition, bureaucracies make policy as they implement it. Politics and bureaucratic administration are thus one and the same.

The notion of bureaucracy incorporates the idea of a set of workers who remain in their jobs and move up a career ladder. The organizations in which they work are characterized by a division of labor, specialization of tasks, formal hierarchy, clear lines of authority, and established rules for making decisions. The decisions are usually made in secret and depend on the impersonal manipulation of information in accordance with rules.

The federal bureaucracy experienced tremendous growth from the nation's founding through the mid-1960s. Since then, federal civilian employment has held steady at approximately 3 million. In addition to the 14 cabinet departments, the federal government contains a number of independent agencies and boards, public corporations, commissions, interagency groups, and advisory panels. Most employees are career civil servants; at the top are senior executives and political executives appointed by the president.

Federal bureaucracies make policy through authorized (delegated) and unauthorized (undelegated) discretion. Authorized discretion flows from the complex nature of problems, the shortage of resources needed to address these problems, and the vagueness of instructions and legislation. The combination of low visibility, decision-making power, and commitment to program goals accounts for the exercise of unauthorized discretion.

Bureaucrats possess potent resources, including authority, expertise and information, and the ability to mobilize support from constituencies such as Congress, other parts of the executive branch, and interest groups. They pursue goals including (1) their own survival, and (2) protecting and expanding their authority, personnel, and budgets. They pursue the same strategies interest groups do, including controlling information; affecting the visibility of issues and decisions; changing the scope of conflict; changing the direction of conflict, that is, the definition of what the issue involves; and affecting the timing of events and influence attempts.

Those who believe politics and administration are distinct adhere to the rational-comprehensive model of bureaucratic decision making. This model requires clear goals and values, a survey of alternative ways to achieve each goal, evaluation of each alternative using a clear set of values, and choice of the best alternative. Throughout, rational calculation and theory guide the process. However, bureaucrats rarely make decisions in this way. Instead they muddle through, which results in incrementalism in policy making. Goals are unclear; only a few alternatives are examined, and most closely resemble what is already being done. The search stops when a solution that is "good enough" is found—often meaning one on which agreement can be reached. Rational calculation and theory play little role in the decision-making process.

Several characteristics of the play of power within and among federal bureaucracies result. First, conflict is ceaseless and found everywhere, although its visibility to the general public may be low. Second, the internal play of power displays more hierarchy, specialization, and use of authority than in other parts of the political process, particularly legislatures. Third, the influence wielded by agencies varies considerably depending on their resources, mission, constituencies, and political skills. Furthermore, the fortunes of individual agencies rise and fall. Fourth, one pattern of interaction among (1) a bureaucratic agency, (2) the interest groups that form its constituency, and (3) the congressional committees that oversee its work is so prevalent and important that it has its own name: the iron triangle, or subgovernment. Within its sphere of activity, the three components of an iron triangle constitute the government.

These characteristics of bureaucracy, coupled with the obvious fact that bureaucrats never face the voters in an election, cause concern that bureaucracies are too autonomous and unaccountable—and hence undemocratic. There is some basis for these fears, especially when agencies become "captured" or join a subgovernment with narrowly based participants. However, most executive bureaucracies operate in a complex strategic environment in which a variety of interests infringe on decision making. Rather than working in isolation from the play of power, executive bureaucracies operate at its very center.

APPLYING KNOWLEDGE

THE SAVINGS AND LOAN "IRON TRIANGLE"

 The full consequences of the savings and loan (S&L) crisis became apparent at the outset of the Bush administration. President George Bush's proposed plan to bail out failed savings and loan institutions would cost every family at least $1,000 in additional taxes, according to some observers at the time.

Why did so many savings institutions go bankrupt? The iron triangle consisting of (1) key members of Congress, including those serving on the banking committees, (2) the agency charged with regulating the savings industry, and (3) the savings industry and its lobbying groups explains a great deal. Chafing under regulations that restricted the size of insured deposits and the kinds of loans they could make, S&L institutions won the support of key members of Congress for legislation removing these restrictions. Once freed from past restraints, a number of S&Ls made commercial real estate loans and other loans in areas they knew little about; in some cases new owners bought out established S&Ls, raised interest rates to attract new money, and expanded their operations rapidly. Because deposits were insured, S&Ls could gamble without fear that bad debts would result in depositors' losing their money.

What about the regulators? The former chair of the Federal Home Loan Bank Board, Edwin Gray, claimed that he had warned for years that trouble was brewing. "Congress listened too well to the organized thrift industry and did nothing, year after year," claimed Gray. Why? "One of the big reasons was that the powerful thrift lobby, with all of those [campaign] contributions going to the members of Congress, were stopping any action by the Congress to come to grips with it."[1]

The *Wall Street Journal* supplied striking evidence of the contributions. It found that as the S&L industry fell into disaster, the contributions of its political action committees (PACs) mushroomed. The 163 S&L-related PACs increased their contributions by 42 percent after the 1984 election, giving $1,850,000 to candidates in 1988. Nearly all of it (89%) went to incumbents. The contributions focused on members of the House and Senate banking committees—particularly the chair of the House Banking Committee, who received almost $150,000.[2]

Some critics went further, blaming the iron triangle of "political fund raisers, corrupt bank and thrift owners, and compliant senior regulators" and asserting that the Federal Home Loan Bank Board had been captured by the "worst elements" in the savings industry.[3] What do you think?

Questions

1. Why are bureaucratic agencies, such as the Federal Home Loan Bank Board, so easily captured by the industries they are supposed to oversee?
2. What other areas of public policy are characterized by an "iron triangle" relationship between federal agency, congressional subcommittees, and interest groups?
3. What sorts of reforms might improve bureaucratic decision making and focus attention on the broader public interest? How can the broader public interest be defined? Would your reforms require rational-comprehensive decisions, or could they be incorporated within existing patterns of incrementalism?

---

[1] Quoted in Tom Webb, "Former Regulator Blames Lobbyists and Congress for S&L Problems," *Philadelphia Inquirer*, February 6, 1989, p. 3.
[2] Brooks Jackson, "As Thrift Industry's Troubles and Losses Mounted, Its PACs' Donations to Key Congressmen Surged," *Wall Street Journal*, February 7, 1989, p. A22.
[3] James Ring Adams, "The Bank Dicks' Dirty Linen," *Wall Street Journal*, February 15, 1989, p. A1.

Key Terms

bureaucracy (*p. 560*)

discretion (*p. 564*)

authorized discretion/
delegated discretion (*p. 564*)

unauthorized discretion/
undelegated discretion (*p. 564*)

patronage (*p. 569*)

independent federal agency
(*p. 575*)

government corporation (*p. 578*)

advisory committee (*p. 578*)

constituency (*p. 581*)

lateral agency (*p. 584*)

interagency group (IG) (*p. 584*)

legislative clearance (*p. 585*)

rational-comprehensive model of
bureaucratic decision making
(*p. 588*)

incremental model of bureaucratic
decision making (*p. 589*)

Recommended Readings

Allison, Graham. *Essence of Decision: Explaining the Cuban Missile Crisis.* New York: Little, Brown, 1971. A well-written and fascinating study of what we learn when we take the perspective of bureaucratic politics.

Heclo, Hugh. *A Government of Strangers: Executive Politics in Washington.* Washington, D.C.: Brookings Institution, 1977. An analysis of the problems that all presidents and their political appointees face in trying to harness the power of the federal bureaucracy.

Jackall, Robert. *Moral Mazes: The World of Corporate Managers.* New York: Oxford University Press, 1988. Although his focus is primarily on private organizations, Jackall's study of the moral dilemmas faced by high-level executives operating within bureaucracy has many lessons for public organizations as well. His case studies are fascinating.

Lewis, Eugene. *Public Entrepreneurship: Toward a Theory of Bureaucratic Political Power.* Bloomington: Indiana University Press, 1984. A study of the impact of some of the most powerful public bureaucrats. The stories of Robert Moses (perhaps the most powerful public official in the history of New York City), J. Edgar Hoover (longtime director of the FBI), and Hyman Rickover (the "father" of the nuclear navy) are interesting in and of themselves. However, Lewis does much more than tell their stories; he also makes a powerful argument about where the real power to change public policy lies.

Lipsky, Michael. *Street-Level Bureaucracy: Dilemmas of the Individual in Public Service.* New York: Russell Sage, 1980. By studying the jobs of bureaucrats who deal directly with citizens (e.g., police officers, welfare workers, teachers, and judges), Lipsky provides insight into the impact that public agencies have on our lives. He shows the ways in which, under the guise of helping the disadvantaged, public bureaucracies may actually reinforce their powerlessness.

Wilson, James Q. *Bureaucracy: What Government Agencies Do and Why They Do It.* New York: Basic, 1989. An excellent overview of the role of bureaucracy in American politics. Wilson provides a comprehensive analysis of the ways in which public organizations work and how they might better be controlled by elected politicians and the public.

The Federal Courts

As American flags waved and balloons floated down from the ceiling at the 1984 Republican National Convention in Dallas, Gregory Johnson and several dozen demonstrators marched outside to protest the policies of the Reagan administration and certain corporations located in Dallas.[1] As Ronald Reagan secured the enthusiastic nomination of his party for a second presidential term, demonstrators moved through the city chanting a variety of slogans, including "Reagan, Mondale, which will it be? Either one means World War III" and "Ronald Reagan, killer of the hour, perfect example of U.S. power." At several stops the demonstrators engaged in political theater, conducting "die-ins," simulations of people perishing in the aftermath of nuclear war.

When the march reached the Mercantile National Bank, one demonstrator pulled down an American flag and handed it to Johnson, the group's leader. The march ended in front of the Dallas City Hall when Johnson covered the flag with kerosene and ignited it. As the flag burned, the others chanted, "America, the red, white, and blue, we spit on you."

Subsequently Johnson was arrested, charged, and convicted in a Texas trial court for violating a state law that renders it illegal to "knowingly desecrate" a "public monument . . . a place of worship or burial; or . . . a state or national flag . . . in a way that the actor knows will seriously offend one or more persons likely to observe or discover his action." The trial court sentenced Johnson to one year in jail and a fine of $2,000, the maximum penalty allowed under this law.

Johnson appealed his conviction to the Texas Court of Appeals, which affirmed the

Gregory Johnson poses with the flag he burned as part of a protest of U.S. policy. His arrest raised important questions about whether such unconventional expression was protected by the First Amendment. Ultimately his conviction was reversed by the Supreme Court, which viewed his action as "symbolic speech."

conviction. He then appealed to a second state court, the Texas Court of Criminal Appeals, which reversed the lower court's decision, arguing that Johnson's conduct constituted "symbolic speech" protected by the First Amendment.

In the fall of 1988, as George Bush campaigned for the presidency in flag factories and criticized his opponent, Massachusetts governor Michael Dukakis, for vetoing a measure compelling public school teachers to lead their students in the Pledge of Allegiance, the Texas attorney general appealed Gregory Johnson's acquittal to the United States Supreme Court. As Bush sought to "outflag" Dukakis, the Supreme Court accepted the appeal. The case, *Texas v. Johnson*,[2] presented the Court with important questions of constitutional interpretation: What is covered by the First Amendment? What does the word *speech* in the First Amendment actually mean? Does burning a flag in the context of a political demonstration constitute "speech" protected by the First Amendment?

Because the case addressed such significant issues, it attracted a great deal of attention. Groups with an intense interest in the case's outcome submitted amicus curiae, or "friend of the court," briefs.[3] The American Civil Liberties Union and a group of 23 organizations (including the Christic Institute, the Community for Creative Non-Violence, the Lambda Legal Defense and Education Fund, the National Conference of Black Lawyers, the National Lawyers Guild, People for the American Way, the American Indian Movement, the War Resisters League, and the Writers Guild of America) filed legal briefs supporting Johnson. The conservative Washington Legal Foundation, joined by the National Flag Association, Veterans of Foreign Wars, Amvets, the Air Force Association, and the Allied Educational Foundation, submitted a brief on behalf of Texas.

After lawyers for the parties made oral arguments, the justices met at a private conference. As is customary, the chief justice had the floor first.[4] Chief Justice William Rehnquist spoke on behalf of upholding the Texas law; he was followed by Justice William Brennan, the associate justice with the greatest seniority, who advocated overturning the law. After each justice had spoken, the decision was reported as 5–4 in favor of Johnson. Brennan, Thurgood Marshall,

and Harry Blackmun—the three most consistently liberal justices on the Court—joined with Antonin Scalia and Anthony Kennedy to overturn the law.

In the aftermath of a presidential campaign that focused so much attention on patriotic symbols, the Court's majority opinion provoked a response of outrage from many political leaders. Shortly after the Court announced its decision, George Bush, now president, traveled to a nearby statue of Marines hoisting the flag during the 1944 invasion of Iwo Jima to demand that Congress reverse the Court's decision; Bush sought a constitutional amendment giving Congress and the states the power to prohibit flag desecration. Opinion polls showed that 70 percent of the American public supported a constitutional amendment.[5]

Within a week of the decision, 39 separate resolutions for an amendment had been proposed. After intense debate and a series of votes on a variety of amendments, resolutions, and statutes, Congress—unable to mobilize a two-thirds majority on behalf of a constitutional amendment—passed the Flag Protection Act of 1989. This statute, requiring a simple majority for passage, made it illegal to "knowingly mutilate, deface, physically defile, burn, maintain on the floor or ground, or trample upon any flag of the United States."

Challenges to the new law were immediate. Only minutes after the law took effect, several demonstrators in Seattle, Washington, burned a flag; they wrote in leaflets publicizing the event that "blind patriotism must not be the law of the land. Unlike the flag-kissers, we will Rock and Roll in a Festival of Defiance."[6] Shortly thereafter a group of demonstrators that included Gregory Johnson burned several flags on the east steps of the U.S. Capitol. Following their arrest, federal district court judges June Green in Washington, D.C., and Barbara Rothstein in Seattle overturned the Flag Protection Act, relying on the *Texas v. Johnson* precedent. The Supreme Court, in *United States v. Eichman*, affirmed the lower courts' decisions despite Solicitor General Kenneth Starr's argument that the federal law differed from the Texas statute.

The ink had barely dried on the Court's opinion when George Bush announced that he would continue to press for a constitutional amendment, calling for Congress to pass it by the Fourth of July. But with public support dwindling, opponents dug in and shifted the terms of debate. The issue was not one of banning flag burning, they argued, but one of tampering with the Bill of Rights. Using the American public's veneration of the Bill of Rights as ammunition, Representative William Ford (D-Mich.) exclaimed that he "would rather amend the Ten Commandments than the Bill of Rights."[7] Senate Majority Leader George Mitchell (D-Maine) echoed this theme, arguing that "the question before us is whether or not, after 200 years, the most effective statement of individual liberty in all human history is to be changed for the first time. I do not believe we

should tamper with this sacred document."[8] The amendment ultimately fell 34 votes short of a two-thirds majority in the House and 9 votes short in the Senate.

Throughout this chapter we return to the story of *Texas v. Johnson* and the controversy over flag burning to illustrate some of the characteristics of judicial politics in the United States. We pay special attention to the important role that federal courts play in the making of public policy, assessing their resources, operating procedures, structure, and decision-making rules. By examining the attributes of key participants and their interactions with one another, we situate the institutions in which they operate—courts—in the broader play of power.

 Courtrooms are full of powerful symbols. The raised bench where the judge sits, the black judicial robe, references to the judge as "Your Honor," the flags and Bibles that decorate the room, and the use of a specialized language understood only by those with legal training—all help to shape and reinforce a set of widely shared beliefs about law, courts, and judges.

The distinctive aura of the courtroom reinforces popular notions about the special characteristics of law. **Law**, the set of written and unwritten materials used by lawyers and judges in formulating arguments and rendering decisions, is perceived to have a number of distinguishing attributes. Law is thought to be *certain*—that is, it exists and may be discovered and applied. In this view, law is also *uniform*, the same everywhere; it is *pure*, untainted by extraneous and irrelevant forces; it is *continuous*, not something recently devised; it is *equal*, ignoring the distinctions people make on the basis of wealth, religion, gender, race, ethnicity, and so on; and it is *general*, being applicable to all.[1]

These important attributes produce a second common belief—namely, that law and the courts are separate from politics. After all, politics is often perceived as anything but noble and exalted. Rather, politics is seen as messy, inconsistent, impure, and ever-changing. Many people believe that political decisions are based on such things as personality, status, wealth, values, greed, and ambition. Courts, then, are perceived to employ an alternative and superior method for making decisions, a method that in many important respects is "above" politics.[2]

Finally, the symbolism surrounding the American legal system is reflected in expectations about judges. Known variously as the "cult of the robe" or the "cult of the judge," these views portray the judge as a "high

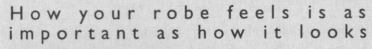

The judicial robe is an important symbol surrounding the operations of courts. This advertisement for judicial robes appeared in *Judicature,* a journal read by many judges. Notice how the ad draws a connection between wearing a robe and the image of dignity and authority the robe is meant to project.

priest of justice with special talents for elucidation of 'the law.'"[3] Legal decisions are announced by judges, whose job it is to find the law, not make it. Judges are thought to be specially trained to discover and apply the law objectively and neutrally. The wise, nonpartisan, unbiased judge renders decisions based solely on the law or on **precedent**—that is, on previous decisions announced by judges. These decisions are unaffected by politics, economics, or the nature of the times.

This view of judges is closely related to the notion that they are independent, totally divorced from the world of politics, and uninfluenced by the legislative and executive branches of government. The potency of this view is nourished by a belief that a separation of powers characterizes the government established by the Constitution. Although Congress makes the laws and the president and executive branch enforce them, it is the separate and distinctive task of courts and judges to interpret and apply them without influence from politics or political pressure.

The view of judges as neutral and nonpartisan arbiters of justice is reinforced by participants within the judicial system. Out-of-court speeches by judges and lawyers often underscore these themes. And Senate confirmation hearings for federal judges (with some notable exceptions) often include statements by nominees that they will bring no special policy biases to the bench. For example, Clarence Thomas, who had expressed strongly held political views in speeches and writings during his tenure as a top civil rights official in the Reagan administration, asked members of the Senate Judiciary Committee to discount his previously announced views. As a Supreme Court justice, Thomas argued, he would strive to be impartial. He described the transformation that occurs when one becomes a judge as being similar to the experience of a sprinter or long-distance runner at a track meet removing a warm-up suit. According to Thomas, "when one becomes a judge, it's an amazing process. You want to be stripped down like a runner."[4] To Thomas, this "stripping down" is akin to peeling away layers of personal values, attitudes, and long-held predilections, so that judicial decisions are based solely on law and precedent.

All these common beliefs about law, courts, and judges confer an enormous legitimacy on law and the operations of judicial institutions. Americans tend to accept the authority law invokes and to believe that laws should be obeyed. In this chapter we challenge many of these common beliefs. Our view of law suggests that it is often *uncertain* and *varying*; is regarded by some as *impure* and *changing* rather than continuous; is often *unequal* in its application and effects; and, far from being applied uniformly everywhere, is *specific* and *contingent*. Further, we suggest that law and politics are inextricably intertwined, that the judicial system is an important component of the larger play of power, and that judges are influenced in their decision making by their own values, attitudes, and political views and by the political climate in which they work. Moreover, we seek to show how important links between the judiciary and other governmental arenas constrain and otherwise influence the work and decisions of federal court judges.

[1] Jerome Frank, *Law and the Modern Mind* (Garden City, N.Y.: Anchor, 1963), p. 53.
[2] Stuart Scheingold, *The Politics of Rights* (New Haven: Yale University Press, 1974), pp. 63–71.
[3] Walter Murphy, *Elements of Judicial Strategy* (Chicago: University of Chicago Press, 1964), p. 13.
[4] Linda Greenhouse, "In Trying to Clarify What He Is Not, Thomas Opens Questions of What He Is," *New York Times*, September 13, 1991, p. A19.

THE STRUCTURE AND ORGANIZATION OF THE FEDERAL COURTS

The cases about flag burning that open this chapter illustrate some of the complexity of the American judicial system's structure and organization. The cases of *Texas v. Johnson* and *United States v. Eichman* were heard in no fewer than five different courts: *Johnson* in a Texas trial court, the Texas Court of Appeals, the Texas Court of Criminal Appeals, and the United States Supreme Court; and *Eichman* in a federal district court and the Supreme Court.

Largely because of the influence of federalism on the structure of American government, the United States has multiple court systems. Each state has its own court system, no two of which are exactly the same. Article III of the Constitution creates a national, or federal, judicial system. Federal courts have jurisdiction over subject matter limited to the resolution of "federal questions," questions whose resolution depends on an act of Congress, a United States treaty, or a provision of the Constitution. In addition, federal courts may hear cases that arise under **diversity of citizenship**—that is, cases involving parties who reside in different states.

State court jurisdiction covers a much broader range of subjects and includes **concurrent jurisdiction**—that is, jurisdiction over many matters that are also appropriate for resolution in the federal courts. Citizens fall within the jurisdiction of both state and federal judicial systems. They may sue or be sued in either system, depending on what issues are raised in the case. In some cases parties can choose which level of court to use. This situation creates certain strategic possibilities for participants, such as **forum shopping**, or carefully selecting the level of court most sympathetic to the claim. For example, African-American litigants challenging segregation in the South in the 1960s often sought to place their cases in federal, rather than state, courts because they believed that federal court judges would be further removed from the local culture supporting segregation. Most legal disputes are resolved in state courts. Whereas federal judges adjudicate more than 300,000 cases a year, state courts handle well over 90 million a year.[9]

The fact that state courts have jurisdiction over some matters that federal courts may also handle produces conflict between the two systems. This has expressed itself most typically in disputes between state courts and the United States Supreme Court. Anticipating such disputes, the Constitution's framers included a *supremacy clause* (Article VI, Section 2) that established federal law as superior to the laws of individual states. However, the Constitution is silent on the question of who resolves conflicts between state and nation. The First Congress sought to clarify the situation when, in Section 25 of the Judiciary Act of 1789, it authorized the Supreme Court to review

This African American protests the justice system in Los Angeles in the aftermath of the Rodney King police brutality case. The sign proclaiming, "Criminal justice, justice that's criminal," the phrase "Guilty until proven innocent" under the faces of African Americans, and the shirt with the slogan "Arrest the president" show that beliefs in the equality of law and the impartiality of court personnel are not shared by everyone.

state court decisions involving federal questions. And the Supreme Court in the case of *Martin v. Hunter's Lessee* established the principle that for the sake of promoting national uniformity in legal interpretation, state courts were bound by the decisions of the United States Supreme Court.[10]

The only federal court explicitly established by the Constitution is the United States Supreme Court. However, the Constitution also gives Congress the authority to establish other national courts—authority Congress has used to create a three-tiered judicial system. Along with the Supreme Court, Congress has created a set of trial courts, known as federal **district courts**, and intermediate level appellate courts, the **courts of appeal**.

The most numerous of the federal courts are the trial courts. Currently, 94 district courts are distributed around the country. Each state has at least one district court, and no district covers territory in more than one state. In 1993, some 649 full-time district court judges heard nearly 267,000 criminal and civil cases.[11] Although districts vary in the total number of judges hearing cases, only one judge is assigned to hear an individual case. In general, district courts have **original jurisdiction**—that is, jurisdiction to hear a case for the first time, rather than on appeal—for civil disputes between citizens of different states, federal questions arising under the Constitution (e.g., the constitutionality of the Flag Protection Act of 1989), and criminal cases involving federal law (e.g., tax fraud or illegal immigration).

Decisions of the district courts, along with decisions rendered by federal regulatory agencies, may be reviewed by courts at the second tier—the courts of appeal. Each of 12 courts of appeal serves one of the 12 judicial circuits into which the country has been divided. Judges in some 179 courts of appeal hear approximately 37,000 cases a year.[12] Courts of appeal typically convene in three-judge panels and seek to correct errors in lower-court proceedings, rather than deciding factual questions.

At the top of the federal judicial hierarchy is the **United States Supreme Court**. Composed of nine justices, the Supreme Court has **appellate jurisdiction**—that is, jurisdiction to review decisions rendered elsewhere—over cases decided by federal district courts, courts of appeal, and the highest state appellate courts. Although it is requested by petitioning parties to hear thousands of cases annually, the Supreme Court in a typical year writes full opinions in only roughly 150 cases.[13] Occasionally the Supreme Court also receives cases through its original jurisdiction, which requires it to give the first hearing to cases that involve two or more states, the United States and an individual state, a state and a citizen of a different state, and foreign ambassadors and other diplomats.

Native American tribes have their own court systems, which function parallel to the federal and state court systems. Although the specific courts and their jurisdiction vary among tribes, most have the equivalent of a trial court and an appellate court.[14] Some tribes have elaborate court systems. The Blackfoot tribe of Montana, for example, has a small claims court, a traffic court, a juvenile court, courts of general civil and criminal jurisdiction, and an appellate court. The Navajo tribe, too, has an elaborate and sophisticated judiciary, processing more than 45,000 cases per year and publishing its decisions in an official reporter.[15] Most cases taken to tribal courts may also be taken to state and federal courts. Moreover, many decisions in tribal courts may be appealed to federal district courts. The various courts that

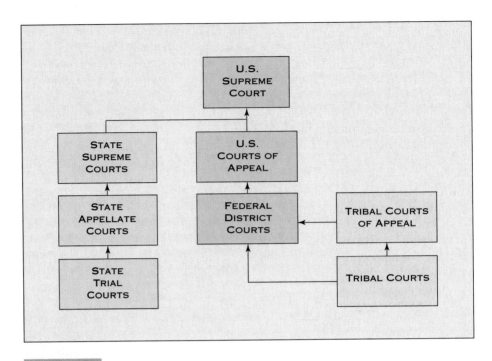

THE AMERICAN JUDICIAL SYSTEM This organizational chart depicts the flow of cases among the different levels of tribal, state, and federal courts. Cases make their way into the federal court system from tribal courts, tribal courts of appeal, and state supreme courts. The Supreme Court hears cases that originate in tribal and state courts, as well as cases that begin in federal district courts.

hear and decide cases in the United States, and their relationship to one another, are depicted in Figure 16.1.

Finally, in addition to the three levels of federal courts we have described, Congress has created a number of other federal courts to handle special types of cases. These courts include the United States Court of Claims, the Customs Court, the Court of Customs and Patent Appeals, the Tax Court, and the Court of Military Appeals. The decisions of these specialized courts may be appealed to the Supreme Court.

THE PARTICIPANTS

An enormous number of participants engage, directly and indirectly, in American judicial politics. In this section we focus on three major participants who are directly involved in and directly affect what takes place in federal courts. We look first at federal judges, examining the politics of federal judicial appointments and the characteristics of those sitting on the federal bench. We then examine salient attributes of lawyers and the legal profession. Finally, we examine the role and selected characteristics of litigants and other interested parties to litigation, such as interest groups.

Among the central participants in the federal judicial system are the judges and justices who hear disputes and render decisions. Their decisions often have important policy implications because they may nullify laws made elsewhere (e.g., the decision in *Texas v. Johnson*) or interpret the meaning of ambiguous laws.

The significance of the judicial appointment process For a variety of reasons, many lawyers aspire to sit on the federal bench. For one thing, the yearly pay is not bad: in 1994, district judges earned $133,600, appeals judges $141,700, associate justices of the Supreme Court $164,100, and the Supreme Court's chief justice $171,500. Life tenure is an added perquisite. Most desirable of all are the prestige and power these positions carry. Because federal judicial positions are among the most important and most prized in government, vacancies in federal judgeships typically unleash a torrent of political activity.

Judicial appointments attract the interest not only of lawyers but of all participants in the play of power who understand that judges make significant decisions affecting who gets what. For example, because the Republican Party understands the connection between judicial appointments and policy-making content, language included in several presidential platforms has called for the appointment of judges who respect "traditional family values and the sanctity of innocent human life." An understanding of the policy implications of appointments was also evident in presidential candidate Bill Clinton's remark that if he became president, he would appoint only judges who respected a woman's right to choose abortion.

Interest groups from all points on the political spectrum recognize the stakes involved in who becomes a federal judge. The necessary ingredients for a contest are all present. There is a prize that many seek: nomination by the president and approval by the Senate. A number of participants care which contestant prevails, both for material reasons (the judicial decisions that will be rendered) and for symbolic reasons. Some groups seek symbolic representation through the appointment of a group member to the bench, similar to the symbolic representation members of Congress provide. Further, struggles over who will get appointed sometimes test players' ability to influence outcomes, raising the stakes and affecting subsequent appointment battles. The result is that resources are used to persuade, to bargain, and to invoke authority in strategies designed to control who gets appointed.

Formal rules and procedures of the judicial appointment process The Constitution establishes the basic rules of the judicial appointment process: The president makes a nomination by submitting to the Senate the name of a candidate to fill a federal judicial vacancy, and the nominee must be confirmed by a majority of those senators present and voting. A number of additional rules and procedures fill in the details.

The Department of Justice performs the task of gathering names of potential nominees, screening the individuals, and making recommendations to the president. Recommendations flow to the Justice Department from a variety of sources: state political figures, party officials, bar associa-

Federal court vacancies, especially on the Supreme Court, attract a great deal of attention. Groups representing women and racial minorities usually lobby presidential administrations to consider members of their groups as potential nominees. Administrations may float the names of candidates from particular groups as a way of showing that members of those groups are under consideration. Jose Cabranes, pictured here, was mentioned by the Clinton administration as a Hispanic candidate for a Supreme Court nomination in 1994.

tions, law schools, the candidates themselves, even sitting and retired judges and justices. However, by far the most significant source of recommendations, especially for district judges, are senators belonging to the president's party from the state in which a vacancy exists.

The Justice Department submits those names that remain viable after initial screening to the American Bar Association's (ABA) Standing Committee on the Federal Judiciary. Based on an examination of the candidate's record as well as phone calls to lawyers around the country, the ABA committee rates each potential nominee as "exceptionally well qualified," "well qualified," "qualified," "not qualified," or (in recent years) "qualified/not qualified." (The last rating is used when a minority of the committee finds someone not qualified.)[16] The FBI conducts background investigations of the leading candidates, looking for personal or professional characteristics that might disqualify the candidate or prove embarrassing during the Senate confirmation process. A recommendation is made to the attorney general, who in turn submits the name to the president for a final decision.

The Senate's procedures significantly strengthen the hand of senators of the president's party from the state in which a district court judgeship vacancy occurs. Because the districts are wholly within one state, senators from the president's party have a proprietary interest in influencing district court appointments made in their state. Should they not be consulted, a declaration that the candidate being nominated is "personally obnoxious" to them spells doom for the appointment. Other senators, fearing that they too may not be consulted, vote "no" in support of their colleague. A strong tradition of respecting one another's wishes in these matters has evolved, a tradition referred to as **senatorial courtesy**.

To facilitate the operation of senatorial courtesy, a procedure has been developed to solicit senators' input. When a candidate is nominated, a blue slip is sent to senators from the state where the vacancy exists asking for any information concerning the nominee to be sent to the Senate Judiciary Committee (the Senate committee that reviews the appointment first). If the senator fails to return the slip, hearings are either delayed or never held at all. In practice, senatorial courtesy gives senators from the president's party veto power over nominees to district court vacancies in their state. Consequently, determined senators can insist that their choice receive the president's nomination. One former Justice Department official suggests that the constitutional provision for the appointment of district court judges should read: "The Senators shall nominate, and by and with the consent of the President, shall appoint. . . ."[17] When no senator from the president's party represents a state where a district court vacancy arises, the role of other participants (including state and local party officials) increases.

Senatorial courtesy plays a smaller part in appointments to courts of appeal because each of these courts extends over several states. Thus, it is more difficult for a senator to invoke courtesy. However, a senator's strenuous objection to a nominee from his or her state still carries some weight. In these cases the problem for senators is that success in blocking the appointment might lead to the nomination of someone from another state in the region covered by the court. In these nominations, the role of individual senators lessens as the role of the Senate Judiciary Committee increases.

Appointments to the United States Supreme Court differ in a few important respects from those to lower federal courts. First, senatorial courtesy plays no role in the nomination process. Second, the president typically plays a much more active and personal role in selecting Supreme Court justices. The nomination of Clarence Thomas by George Bush is a good illustration. After Thurgood Marshall retired, Bush worked with a few administration officials to make a choice. After compiling a short list of candidates, Bush tentatively selected Thomas, invited him to visit the Bush compound at Kennebunkport, Maine, and offered him the position during a private conversation.[18] Likewise, President Bill Clinton played a key role in the nominations of Ruth Bader Ginsburg and Stephen Breyer.[19]

Finally, the intense publicity and senatorial scrutiny that Supreme Court nominees undergo distinguish this selection process from that for lower federal courts. Although the presumption is always that the Senate will confirm a presidential nominee, the occasional defeats serve as reminders that the Senate acts as more than a mere rubber stamp. Indeed, the Senate has failed to confirm 28 Supreme Court nominees, approximately 20 percent of the total it has considered. One observer points out that "this proportion of defeats is higher than for any other position to which the president makes appointments. For instance, only nine cabinet nominees have been defeated."[20]

Supreme Court nominees are rejected by the Senate for a number of reasons, including poor qualifications and insufficient experience. But the chief reason is partisan politics. Several nominees lost their appointment bids when the party in control of the Senate anticipated victory for its presidential candidate in the next election and denied the choice of the incumbent, "lame-duck" president of the opposing party. Most recently the Senate rejected on partisan and ideological grounds Ronald Reagan's

This full-page advertisement ran in the *New York Times* while Clarence Thomas stood for nomination to the Supreme Court. Such lobbying has become more prevalent in recent years, beginning with the campaign to defeat Robert Bork in 1987. Some scholars argue that campaigns surrounding the nomination process have politicized it and have damaged the integrity of the federal courts. But others maintain that advocacy ads help to inform the public and legislators about a nominee's position on critical issues (e.g., reproductive rights) that may come before the Court.

nominee, Robert Bork, who had taken clearly conservative positions on a number of controversial issues in his writings and speeches.

Since the Bork experience, presidents have been careful to select nominees with either undetectable political views or reputations as political moderates. For example, George Bush nominated David Souter, whom the media named the "Stealth candidate" after the bomber designed to elude radar. Unlike Bork, Souter had left no "paper trail" of speeches and writings spelling out his political or judicial views. Bush's nomination of Clarence Thomas had a greater potential for controversy, even before Anita Hill's charges of sexual harassment, because of politically conservative statements he had made and written on topics such as affirmative action while he was an official in Republican administrations. But the Bush administration calculated that Thomas's African-American heritage would split Democrats—some would support him because of his race, whereas others would oppose him because of his views. Bush administration advisers instructed Thomas to emphasize to the Senate Judiciary Committee his humble origins in Pine Point, Georgia, and to argue that his previous political statements reflected the views of administrations he had served, not views that he would carry to the bench. Thomas sought to underscore his impartiality by telling an incredulous Judiciary Committee that he had no view on abortion and had never discussed this controversial issue with anyone.

Bill Clinton has appointed two justices—Ruth Bader Ginsburg and Stephen Breyer—with reputations as political moderates. Of the two, the Ginsburg nomination presented a greater potential for confirmation problems. As an attorney, Ginsburg had worked for the American Civil Liberties Union, had won several important sexual discrimination cases, and had identified herself as a strong supporter of sexual equality and abortion rights. But her record as an appeals court judge suggested that she was a moderate who often voted with Republican colleagues. Justice Ginsburg had the further advantage of appealing to women's groups, who advocated more women on the federal bench, and to Jews, who had not had a representative on the Supreme Court since 1969.

Who becomes a federal judge?   What, then, are the results of the play of power surrounding judicial appointments? What types of people are most likely to survive the appointment process and take seats on the federal bench? Tables 16.1 and 16.2 (on pp. 616 and 617) list some salient characteristics of district court and appeals court judges appointed during the past several presidential administrations.

Several conclusions are readily apparent from the data presented in these tables. First, almost all appointees come from the president's party. At the district court level (see Table 16.1), 93 percent of Jimmy Carter's appointees were Democrats whereas 93 percent of Ronald Reagan's and 89 percent of George Bush's appointees were Republicans. Table 16.2 shows a similar pattern for court of appeals judges, except that Carter falls well below the 90 percent mark in appointees from his own party. In most administrations, appointees not only are affiliated with the president's party but also have been party activists.

The importance of partisan background and the highly political nature of the nomination and confirmation process mean that the winners of presidential and senatorial elections have a great deal to say about the shape of the

TABLE 16.1
CHARACTERISTICS OF FEDERAL DISTRICT COURT APPOINTEES IN RECENT ADMINISTRATIONS

	CARTER	REAGAN	BUSH
Political identification			
Democrat	92.6%	4.8%	5.4%
Republican	4.4	93.1	88.5
Independent	2.9	2.1	6.1
Past party activism	60.9	58.6	60.8
Gender			
Male	85.6	91.7	80.4
Female	14.4	8.3	19.6
Race			
White	78.7	92.4	89.2
African-American	13.9	2.1	6.8
Hispanic	6.9	4.8	4.0
Asian	0.5	0.7	—
Percent white male	68.3	84.8	72.9
Religion			
Protestant	60.4	60.3	64.2
Catholic	27.7	30.0	28.4
Jewish	11.9	9.3	7.4
Net worth			
Less than $200,000	35.8	17.6	10.1
$200,000–499,999	41.2	37.6	31.1
$500,000–999,999	18.9	21.7	26.4
$1 million or more	4.0	23.1	32.4

Source: Sheldon Goldman, "Bush's Judicial Legacy: The Final Imprint," *Judicature* 76 (April–May 1993), p. 287, Table 2. Reprinted by permission of *Judicature*, the journal of the American Judicature Society.

next generation's judicial branch. Research on the impact of the appointing president and of partisan affiliation on the voting behavior of district and appeals court judges suggests at least a modest relationship. Although results vary depending on the issues examined, judges affiliated with the Democratic Party, typically appointed by Democratic presidents, are generally more liberal in their votes and decisions than are their Republican colleagues, who are typically appointed by Republican presidents.[21] In recent years the impact of partisanship and of the appointing president has appeared to be

TABLE 16.2
CHARACTERISTICS OF FEDERAL APPEALS COURT APPOINTEES IN RECENT ADMINISTRATIONS

	CARTER	REAGAN	BUSH
Political identification			
Democrat	82.1%	—	5.4%
Republican	7.1	97.4%	89.2
Independent	10.7	1.3	5.4
Other	—	1.3	—
Past party activism	73.2	69.2	70.3
Gender			
Male	80.4	94.9	81.1
Female	19.6	5.1	18.9
Race			
White	78.6	97.2	89.2
African-American	16.1	1.3	5.4
Hispanic	3.6	1.3	5.4
Asian	1.8	—	—
Percent white male	60.7	92.3	70.3
Religion			
Protestant	60.7	55.1	59.4
Catholic	23.2	30.8	24.3
Jewish	16.1	14.1	16.3
Net worth			
Less than $200,000	33.3	15.6	5.4
$200,000–499,999	38.5	32.5	29.7
$500,000–999,999	17.9	33.8	21.6
$1 million or more	10.3	18.2	43.2

Source: Sheldon Goldman, "Bush's Judicial Legacy: The Final Imprint," *Judicature* 76 (April–May 1993), p. 293, Table 4. Reprinted by permission of *Judicature,* the journal of the American Judicature Society.

much stronger on certain issues, such as abortion, that have served as a presidential platform–based appointment criterion.[22]

An examination of the demographic and socioeconomic attributes of district and appeals court judges indicates that, far from reflecting the diversity of the American population, they are a fairly homogeneous group. Although both Jimmy Carter and George Bush sought to appoint more women to the federal bench, and Carter took steps to increase the number of African-American judges, federal judges tend to be upper-class or upper-middle-

DO WOMEN JUDGES DECIDE CASES DIFFERENTLY THAN MEN?

Prior to 1977, when President Jimmy Carter developed a plan to diversify the federal courts, only eight women had served as federal judges. Carter's commitment to diversity and the establishment of regional panels to identify women and minority judicial candidates resulted in the appointment of 11 women to the United States Courts of Appeal and 29 to the federal district courts. Subsequently, Ronald Reagan appointed 24 women to district courts and 4 to the courts of appeal, and George Bush added 29 to the district courts and 7 to the courts of appeal.[1] Reagan appointed the first female Supreme Court justice, Sandra Day O'Connor, and Bill Clinton added a second, Ruth Bader Ginsburg. By all accounts, Clinton will surpass by the end of his first term all records set by previous presidents in appointments that diversify the federal bench.

Although the number of women judges presiding in federal courts remains relatively small, research has begun to assess whether these women decide cases differently than men do. The results of this research are mixed but generally point to some differences in certain issue areas. One study comparing decisions of male and female district court judges found no significant differences in cases involving criminal procedure. In the areas of personal liberties and minority policy, however, they found women to be more conservative than their male counterparts, contrary to the researchers' expectations.[2]

More recent research on the courts of appeal suggests more readily understandable gender differences in judicial decision making. One study, for example, found differences between male and female judges in two areas studied. In employment discrimination cases, women supported plaintiffs more often than men did: Sixty-three percent of female judges' votes supported plaintiffs, as compared with 46 percent of male judges' votes. In cases involving possible illegal searches and seizures of evidence, female judges were more likely than male judges to support the claims of criminal defendants. It appeared, then, that female judges tended to sympathize with members of groups

Supreme Court justices have overwhelmingly been white men. In the history of the Court only two African Americans, Thurgood Marshall and Clarence Thomas, and two women, Sandra Day O'Connor and Ruth Bader Ginsburg, have served as associate justices. O'Connor (*left*) and Ginsburg, shown here, sit today along with Clarence Thomas on the most diverse Supreme Court in our nation's history.

precluded from participating as full members of the community.[3]

Finally, a study of Supreme Court justice Sandra Day O'Connor's judicial behavior found that she usually voted for the plaintiff in sex discrimination cases. Further, her arrival on the Court appears to have influenced her colleagues, as all but two of them have begun to vote for plaintiffs in sex discrimination cases more frequently than they did before she was appointed.[4]

[1] Sheldon Goldman, "Bush's Judicial Legacy: The Final Imprint," *Judicature* 76 (April–May 1993), pp. 282–97.

[2] Thomas G. Walker and Deborah J. Barrow, "The Diversification of the Federal Bench: Policy and Process Ramifications," *Journal of Politics* 47 (1985), pp. 596–617.

[3] Sue Davis, Susan Haire, and Donald R. Songer, "Voting Behavior and Gender on the U.S. Court of Appeals," *Judicature* 77 (November–December 1993), pp. 129–33.

[4] Karen O'Connor and Jeffrey A. Segal, "Justice Sandra Day O'Connor and the Supreme Court's Reaction to Its First Female Member," in Naomi B. Lynn, ed., *Women, Politics, and the Constitution* (New York: Haworth Press, 1990), pp. 95–104.

class white Protestant men. This pattern applies with even greater force to Supreme Court justices, who—with the exceptions of Ruth Bader Ginsburg, Sandra Day O'Connor, Thurgood Marshall, and Clarence Thomas—have been white males from privileged socioeconomic and educational backgrounds.[23] You will not find many—or even any—socialists, gay activists, African-American separatists, or advocates for the homeless sitting on the bench. Nor will you find more than a few federal judges who understand from personal experience what it means to be poor in a relatively affluent country or a person of color in a predominantly white society.

During its first year in office, the Clinton administration took steps to diversify the federal bench.[24] Indeed, of 48 nominees sent to the Senate, only 18 were white men. Clinton's nominees included 18 women, 11 African Americans, and three Hispanics, a record that led an expert in federal judicial selection to conclude that "this tremendous push toward ethnic and gender diversity could break all historic records."[25] Clinton's appointment of Martha Vazquez to a district court judgeship in New Mexico was especially noteworthy, as she became the first Mexican-American woman to sit on the federal bench.

Although women continue to be vastly underrepresented on the federal bench, their increasing numbers in recent years have raised important questions about how, if at all, gender influences judicial decision making. This question serves as the topic for "Outcomes: Do Women Judges Decide Cases Differently Than Men?"

LAWYERS

Lawyers play a central role in American politics. With important exceptions, such as court staff and the police, nearly all crucial players in the American legal process are lawyers. A large proportion of both U.S. House members and senators hold law degrees. A significant portion of Washington-based lobbyists are lawyers, as are many lobbyists in the states. Attorneys inhabit bureaucratic agencies and legislative staffs as well. Besides occupying key legal and political positions and engaging in overtly political behavior, lawyers serve an important political function as the judicial system's gatekeepers. In their daily activities, lawyers render significant decisions regarding which individuals and groups receive representation, which cases and clients receive the most zealous representation, and which issues are placed on the agendas of courts.[26]

Recruitment of lawyers The political prominence and gatekeeping role of lawyers make the question of who becomes what kind of lawyer important for understanding American politics. Politics has always shaped entry into the legal profession. In fact, the key to understanding the politics of lawyer recruitment rests in finding out what authoritative rules govern admission to the bar, who makes them, and why. These authoritative rules, which differ somewhat from state to state, require lawyers to pass a state-administered bar exam and meet certain "fitness of character" standards. In most states one cannot take the bar exam unless one has graduated from an accredited law school, and one cannot enroll in most law schools without an undergraduate degree and a means of covering the costs of tuition.

TABLE 16.3

ENROLLMENT OF WOMEN AND RACIAL MINORITIES IN FIRST YEAR OF LAW SCHOOL, 1977–1992

	1977	1982	1987	1992
Total enrollment	39,676	42,034	41,055	42,793
Women	11,928 (30%)	16,136 (38%)	17,506 (42.6%)	18,325 (42.8%)
African Americans	1,945 (.05%)	2,217 (.05%)	2,339 (.06%)	3,303 (.07%)
Mexican Americans	588 (.01%)	628 (.01%)	610 (.01%)	807 (.02%)
Puerto Ricans	138 (.003%)	171 (.004%)	178 (.004%)	202 (.004%)
Other Hispanics	257 (.006%)	520 (.01%)	750 (.02%)	1,210 (.03%)
Native Americans/ Alaskan Natives	137 (.003%)	154 (.004%)	189 (.005%)	313 (.007%)
Asian Americans/ Pacific Islanders	509 (.01%)	731 (.017%)	1,064 (.025%)	2,235 (.05%)

Source: American Bar Association, Section on Legal Education and Admission to the Bar, *A Review of Legal Education in the United States, 1992* (Chicago: American Bar Association, 1992), pp. 67–69. Copyright © 1992 by American Bar Association. Reprinted by permission.

Obviously, middle- and upper-income people have a better chance of meeting the financial costs of college and law school than do lower-income people. It is hardly surprising, then, that people from historically disadvantaged groups in American society are virtually unrepresented among the ranks of American lawyers. Only about 3 percent are African American, and even fewer are Hispanic.[27] As Table 16.3 shows, in 1992 the numbers of racial minorities—African Americans, Hispanics, Native Americans, and Asian Americans—enrolled in law school remained minuscule. But Table 16.3 also shows that the number of women entering law school has skyrocketed in recent years. Women now account for more than 40 percent of law school enrollments. However, prior to 1980 so few women practiced law that their total proportion in the profession in 1980 was only 12 percent.[28] Although the increasing numbers of women entering law school suggest that women will have an ever more significant role in the play of power, they still face many obstacles in pursuing a legal career. Some of these are described in "Participants: The Feminization of the Legal Profession."

Stratification in the legal profession   Common images of American lawyers depict a profession of solo practitioners who strike out on their own, hang out a shingle, and wait eagerly to handle any problem brought by any client

THE FEMINIZATION OF THE LEGAL PROFESSION

Participation by women in the American legal profession has soared in the past 25 years. As recently as 1970, women represented only 5 percent of the graduates of American law schools. Today women make up more than 40 percent of law school graduates.

Although women have undeniably made dramatic progress in entering this once all-male profession, many important obstacles to gender equality remain. For one thing, women lawyers work disproportionately in the least prestigious and least lucrative practices: in public interest organizations, as governmental lawyers, and in solo private practice.[1] Women now constitute one-third of the associates (lower-level attorneys seeking to become partners) in the most prestigious and lucrative practices—namely, large corporate firms. Yet women constitute less than 10 percent of the partners, or managing attorneys who share in the firms' profits.[2] Moreover, a variety of studies conducted during the 1980s found that women earned less than did men practicing the same type of law.[3]

Women also face gender bias as they do their jobs. A survey in Massachusetts found that two-thirds of women lawyers in the state reported being addressed by court personnel and male lawyers by their first name or by a term of endearment (e.g., "honey," "babe," "sweetie") in situations in which they expected to be called by a surname or title. Further, a large number of women reported being subjected to inappropriate comments about their personal appearance or comments of a sexually suggestive nature by other attorneys. More than 10 percent reported being touched in an inappropriate way by another attorney.[4] These reports were confirmed for women working in large law firms in a national survey of more than 900 women lawyers. Some 60 percent reported that they had experienced unwanted sexual attention on the job.[5]

Finally, many of those surveyed said that their career was in conflict with their personal life—a conflict that most men do not experi-

As recently as the 1960s, few women attended law school. This photo of a recent graduating class at Boston University's law school shows how times have changed. Women now constitute nearly half of the students graduating from law school nationally.

ence to the same degree. Many reported delaying marriages or serious relationships, or having to decide between marriage and work. Ninety percent believed that taking advantage of the part-time or flexible work schedules offered by some firms would slow or totally obstruct their path to partnership. Nearly half believed that simply using the officially allotted amount of maternity leave would hurt their career by creating a perception that they were less professional than their male colleagues.

---

[1] Richard L. Abel, *American Lawyers* (New York: Oxford University Press, 1989), pp. 90–99.
[2] Rita Henley Jensen, "Minorities Didn't Share in Firm Growth," *National Law Journal*, February 19, 1990, p. 1.
[3] Abel, *American Lawyers*, pp. 92–93.
[4] "Report of the Gender Bias Study of the Supreme Judicial Court, Commonwealth of Massachusetts, 1989," reproduced in Mary Jo Frug, ed., *Women and the Law* (Westbury, N.Y.: Foundation Press, 1992), pp. 2–16.
[5] The survey was conducted by the *National Law Journal* and reported in Tamar Lewin, "Women Say They Face Obstacles as Lawyers," *New York Times*, December 4, 1989.

who walks through the door. These images, shaped by popular-culture images of lawyers such as Abraham Lincoln and Perry Mason, suggest that all lawyers are essentially the same, performing similar tasks and sharing the same values, knowledge, and ability to practice law.

Research on the American legal profession paints a very different picture. Only about one attorney in three engages in solo private practice. Almost as many work for a private organization, such as a business or association; serve in government; or have retired. Increasingly, lawyers in private practice work in law firms, and the size of these firms is growing. Currently about 25 percent of all lawyers work in firms employing more than 11 attorneys.[29] Furthermore, most lawyers, regardless of where they work, specialize in only one or two areas of law.

The most significant characteristic of the legal profession is its division into distinct layers or segments. The legal profession is not homogeneous; it is highly stratified. At the top sit lawyers in the large firms that serve the needs of organizations, usually corporations and financial institutions with a continuing need for legal help on a variety of matters. Most of these attorneys come from high-status backgrounds (they tend to be white, male, Protestant, and from an affluent background) and attended prestigious undergraduate and law schools.[30] They spend little time in court; when they do, it tends to be in the more respected federal courts rather than in state courts. In the middle layer are attorneys in small and medium-size firms who serve smaller businesses and government. They are more likely to be from an "ethnic" background and to have attended public college. At the bottom are those lawyers who represent individuals, often on a single matter such as the writing of a small will, the sale of property, a divorce, an auto accident, or a minor criminal charge. They engage in solo practice or work with one or two others. They tend to come from the less prestigious law schools, including those operated for profit and offering night courses, and they are more likely to be Roman Catholic or Jewish than are attorneys in large firms. They also make much less money than do lawyers in the other strata.[31]

The overt discrimination against African Americans, Hispanics, Asian Americans, women, Jews, and Catholics that once characterized law school admissions and the hiring of attorneys by elite firms has largely disappeared, but the patterns created by these practices linger.[32] Elite firms in New York and other major cities still draw heavily from elite law schools such as Harvard, Yale, Columbia, and Stanford. In choosing from among the graduates of such schools, the partners tend to hire younger copies of themselves, perpetuating existing patterns. Some firms still have few members of minority groups,[33] and women remain underrepresented among partners in large firms.[34]

Patterns in the recruitment and stratification of attorneys are important because the structure of the bar shapes American politics and society in general. Graduates of the most prestigious law schools generally gravitate to the large and prestigious law firms serving corporate clients.[35] Can you blame them? In the mid-1990s, annual starting salaries at top firms hovered around $80,000. Moreover, the training received at such firms is generally considered to be second to none. As a result, those who have most distinguished themselves academically serve the needs of those who are able to afford them. Because they can allocate tremendous resources and a corps of well-trained, specialized attorneys to individual cases, large firms can provide exhaustive defenses against claims that their clients have violated governmental regulations, failed to pay their taxes, or manufactured defective products.

When cases affecting the interests of organizations with the money to hire specialized lawyers go against them, the firm can and does appeal the decisions vigorously. But the interests of the less well off and the less well organized receive much less vigorous and effective representation.[36] When these clients win, they must fight off expensive appeals. When they lose, they often cannot afford to appeal. No single decision is likely to alter the shape of politics dramatically, but the cumulative effects of thousands of decisions over decades are something else. The recruitment and deployment of legal talent exert a long-term, subtle but crucial effect not just on who gets what in individual decisions but on the rules, established by legal precedents, that affect how such decisions are made.

LITIGANTS AND OTHER INTERESTED PARTIES

Judges and lawyers participate directly in the play of power only if individuals, organizations, and groups perceive some harm and bring their cases to the judicial system for resolution. Crucial to the case of *Texas v. Johnson* was the fact that Gregory Johnson perceived a constitutional violation in his arrest for burning the flag, sought legal counsel, retained a very effective attorney, and pursued his case through several stages before being heard by the Supreme Court. In some important respects the mobilization of law by individual and group litigants is a significant form of political participation. Litigants using the judicial system are essentially petitioning governmental institutions (i.e., courts) for redress of grievances—a petitioning that is more direct than what is typically possible in more traditional modes of political participation, such as voting.[37]

Increasingly, parties in federal courts consist, either directly or indirectly, of interest groups.[38] Organized interests behave differently in courts than they do in legislative and executive arenas because the judicial system's rules prohibit direct, face-to-face lobbying. Therefore, groups with an interest in litigation must find alternative avenues that are consistent with the legal system's rules and norms.

Organized interests participate in the judicial play of power primarily in two ways. First, groups participate directly in litigation through the **sponsorship of test cases**. In such cases, groups seek to challenge or change existing laws by supplying litigants with attorneys and other resources necessary to carry an appeal to the Supreme Court. For example, in the early 1950s, the Legal Defense Fund of the National Association for the Advancement of Colored People (NAACP) appointed Thurgood Marshall and other attorneys to represent Linda Brown and other plaintiffs who wished to challenge the constitutionality of school segregation—a challenge that resulted in the Supreme Court's landmark decision in *Brown v. Board of Education of Topeka*.[39] Organized interests may also try to pave the way for favorable court decisions by collecting statistical data to support their legal arguments and by writing articles for legal journals. For instance, the NAACP's Legal Defense Fund published numerous articles stating its views in a wide variety of legal journals before it became involved in litigation seeking to end racial discrimination in housing.[40]

A second major avenue of interest group participation in litigation is through the writing of **amicus curiae**, or "friend of the court," **briefs**—that is, legal arguments filed with courts that support one side of the controversy.

In the case of *Texas v. Johnson*, 24 groups signed a brief supporting Gregory Johnson, whereas 6 signed a brief in favor of upholding the Texas law banning flag burning. In *Webster v. Reproductive Health Services*, a 1989 Supreme Court decision upholding some restrictions on abortion in Missouri law, an unprecedented 78 amicus curiae briefs were filed.[41] Groups file amicus briefs because it is less expensive and more cost efficient than sponsoring litigation, because they are asked by other organizations to form coalitions, and ultimately because they believe the arguments may influence the outcome.[42] Although much remains to be learned about the impact of amicus curiae briefs on legal decisions, it has been suggested that in the *Webster* case the briefs shaped the terms of the Court's debate. Justices struggled openly with arguments made in the briefs, sometimes refuting statements contained in them, modifying their own arguments because of statements contained in them, or accepting and integrating some of the points into specific written opinions.[43]

The frequent participation of interest groups in federal court litigation is illustrated in research on the Supreme Court's 1987 term.[44] Groups sponsored some 65 percent of cases decided with full opinions, and 80 percent of the Court's cases had at least one amicus brief filed by an organized or governmental interest. During this term the Court received nearly 460 amicus curiae briefs signed by more than 1,600 groups and governmental units.

The frequency of interest group participation in judicial matters does not appear to be evenly distributed among different types of organized interests. At the Supreme Court level, commercial interests—such as chambers of commerce, corporations, and trade associations—dominate the sponsorship of litigation. As Table 16.4 shows, commercial and governmental inter-

TABLE 16.4
AMICUS CURIAE PARTICIPATION IN SUPREME COURT LITIGATION, 1987 TERM

PARTICIPANT	% OF FILINGS	NO. OF FILINGS
Commercial interests	24.4%	111
Governments	24.4	111
Legal groups	12.9	59
Civil liberties groups	10.8	49
Religious groups	4.8	22
Public affairs groups	4.4	20
Women's groups	3.7	17
Health groups	3.3	15
Educational groups	3.3	15
Labor groups	3.3	15
Other groups	2.9	13
Consumers	1.5	7
Total		454

Source: Lee Epstein, "Courts and Interest Groups," in John B. Gates and Charles A. Johnson, eds., *The American Courts: A Critical Assessment* (Washington, D.C.: Congressional Quarterly Press, 1991), p. 356.

ests (e.g., state and local governments, and agencies of the U.S. government) are the most frequent filers of amicus briefs. In 1987 nearly 50 percent of all amicus briefs were filed by commercial and governmental interests. Legal organizations and civil rights and liberties groups accounted for another 25 percent, and the rest were dispersed among other types of organizations.

AGENDA SETTING BY THE SUPREME COURT

The three principal participants in the judicial play of power have differing amounts of influence on which issues reach the legal agenda at the three federal court levels. On the one hand, lawyers, litigants, and interest groups determine which cases to bring to district courts and to courts of appeal. Judges in these courts have little discretion concerning which cases to hear and which issues to resolve.[45] On the other hand, Supreme Court justices since 1925 have exercised substantial—and today exercise nearly total—control over the cases they will hear.

AVENUES OF APPEAL

Most cases arrive at the Supreme Court from courts of appeal or state courts of last resort. Cases are appealed to the Court in the form of a **writ of certiorari**—a written petition outlining the issues presented in the case and an argument for Court review. Filing fees for appeals may be waived if a petitioner is indigent and files an affidavit **in forma pauperis** ("in the manner of a pauper"). The Court receives more than 4,000 requests for review per year but selects only a relative handful (about 170) for consideration. Lower-court decisions are allowed to stand for cases that the Court chooses not to review.

Decisions to grant review occur only when four or more justices agree that a case deserves consideration—an unwritten rule known as the **rule of four**. The justices' law clerks screen all petitions that arrive at the Court and prepare summaries. The nine justices render decisions on the petitions at twice-weekly conferences, where they also vote on cases previously argued before the Court. Before a conference the chief justice circulates a "discuss list," a list of petitions considered worthy of serious consideration for review.[46]

Why does the Court choose to hear some cases and not others? Certain characteristics of cases described in certiorari petitions communicate to judges that they ought to take a closer look. Ultimately, these signs increase the likelihood of acceptance. For example, conflicts among judges in lower courts, signified by nonmajority opinions and written dissenting opinions, increase the chances that the Court will review a case.[47] The most significant sign is the involvement of the **solicitor general**—the lawyer who represents the federal government before the Supreme Court. The solicitor general is respected by Supreme Court justices as one who appeals only the most appropriate and important cases and, consequently, is "the most important person in the country, except the justices themselves, in determining which cases are heard in the Supreme Court."[48] Finally, the likelihood of review is increased when amicus curiae briefs are filed by interest groups on behalf of petitioners.[49]

PROCEDURAL REQUIREMENTS FOR SUPREME COURT REVIEW

Along with its discretion in choosing which cases it will review, the Court may use several procedural rules to determine the issues it will hear. These rules may be invoked to avoid hearing a case, even if four of nine justices have voted to grant certiorari. For example, the requirement of **standing to sue** stipulates, among other things, that a litigant must show a real and direct stake in the case's outcome and some injury from the action or law that is challenged. This rule serves to prohibit hypothetical cases or cases filed by third parties; and it forces organizations, such as the NAACP and the American Civil Liberties Union (ACLU), to identify persons directly harmed rather than simply filing a suit on the organization's behalf.

Likewise, if a case is not **ripe**—if it merely speculates that some harm will occur to someone in the future—the Court may refuse to hear the case. This rule seeks to prohibit the adjudication of issues that are premature. For instance, the Court in a 1972 case, *Laird v. Tatum*, ruled that a challenge to the Army's surveillance of civilian political activity lacked ripeness because no actual injuries had occurred.[50] Further, the Court may refuse to hear a case where an issue has become **moot** (i.e., is no longer a live issue) because a change in the challenged law or practice has dissolved the controversy. Finally, the Court has developed the **political question doctrine**, which allows it to avoid issues it believes are "political" rather than legal and, thus, are more appropriately decided by other branches of government.

The use of these rules, along with the Court's certiorari decisions, has a clear political dimension. The procedural rules and the political question doctrine are vague and can be difficult to interpret. Justices sometimes disagree in their interpretations of these rules. Such disagreements often "reflect the views of justices about the underlying merits of cases. The justices most likely to grant standing to environmental groups generally are those who are most favorable to the policy positions of those groups."[51] Different Courts examining issues at different times apply these rules in different ways. Prior to 1962, the Court considered legislative apportionment to be a "political question" better handled by legislatures. But in 1962 the Court ruled in *Baker v. Carr* that such issues fell within its jurisdiction. Had apportionment become less political over time? Of course not. The activist Warren Court reversed the judgment because it wanted to decide these issues.

In recent years the conservative Rehnquist Court has restricted standing for organizational plaintiffs in environmental cases.[52] For example, in a 1992 decision, *Lujan v. Defenders of Wildlife*, the Court dismissed an environmental group's challenge to a narrow interpretation of the Endangered Species Act by the Reagan administration's Interior Department. The interpretation limited the statute's reach to federal activities within the United States, an interpretation at odds with the Carter administration's policy requiring all federal agencies to ensure than no threatened

In *Lujan v. Defenders of Wildlife*, a challenge by environmentalists to a narrow interpretation of the Endangered Species Act by Interior Secretary Manuel Lujan (*left*) of the Reagan administration was rejected by the Rehnquist Court. In a majority opinion, Justice Antonin Scalia (*right*) contended that environmental groups had failed to show sufficient injury to merit the case's appearance before the Supreme Court.

species would be harmed by governmental activities abroad. The Court, in an opinion written by Antonin Scalia, ruled that the environmental group bringing the legal challenge had failed to show sufficient injury to meet the standing requirement.

The Court may relax procedural rules of standing, ripeness, and mootness when it wants to hear a case. For example, for years the Court considered challenges to election rules excluding certain candidates to be moot, because the cases reached it long after the election in question had taken place. But during the 1960s the Court became more receptive to such cases, arguing that the involved issues were likely to recur. A similar logic permits the Court to hear challenges to state laws restricting abortion, even though the need for an abortion is long past by the time the particular case makes its way onto the Court's docket. Thus, these rules and discretionary certiorari decisions may be viewed as resources the Court can use in shaping its agenda. Decisions on certiorari and whether or not to invoke procedural rules all serve to determine who gets what. Specifically, these decisions determine which issues are heard by the Court and which are ignored.

At times the Court uses these resources to avoid hearing cases perceived as too controversial. These cases have the potential to weaken public support for the Court and hurt its relations with other governmental branches. The Court avoided choosing sides on the issue of American involvement in Vietnam, refusing to hear several challenges to the war's constitutionality. And for years the Court refused to hear cases involving the rights of gay men and lesbians.[53]

ORAL ARGUMENTS AND PERSUASION

Oral arguments are scheduled for all cases accepted by the Supreme Court for review. In many cases the justices have already read the lawyers' legal briefs before oral argument and have come to some judgment on the issues presented in the case.

Oral arguments may be used by the lawyers to persuade uncommitted justices, although lawyers are rarely given an opportunity to argue their position fully. Instead, justices continually interrupt them, using the time to ask questions meant to persuade their colleagues to reach a particular decision. Oral argument in the *Texas v. Johnson* case may be used to illustrate how this works.[1]

Kathi Alyce Drew, an assistant district attorney from Dallas, argued the state's case. Johnson was represented by William Kunstler, a prominent private attorney known for representing political dissenters.

Drew had the lectern for the first 30 minutes. She quickly found that the justices would not permit her to make her argument uninterrupted. Only on rare occasions could she utter more than a few sentences before being abruptly stopped by a pointed question from one of the nine justices. Drew began by arguing that Texas had an overriding interest in the "preservation of the flag as a symbol of nationhood and national unity." Justice Antonin Scalia quickly broke in. "Why did the defendant's actions here destroy the symbol? His action does not make it any less a symbol." "Your Honor," Drew politely responded, "we believe that if a symbol over a period of time is ignored or abused . . . it can, in fact, lose its symbolic value." Scalia shot back, "I think not at all. It seems to me you're running a different argument, not that he's destroying its symbolic character, but that he is showing disrespect for it. . . . You don't want just a symbol, but you want a venerated symbol." This argument seemed to invoke a previous Court decision, *West Virginia Board of Education v. Barnette,* which prohibited the state from compelling public school students to salute the flag.

In the midst of this exchange, Chief Justice William Rehnquist weighed in, seeking to provide support for Drew's argument. Although not arguing with Scalia directly, a practice pro-

DECISION MAKING
BY THE SUPREME COURT

How does the Supreme Court process the cases it accepts for review? It follows three major stages: oral argument, conference deliberations, and opinion assignment and writing.

The Court typically sits for 36 weeks a year, from the first or second Monday in October until the end of June.[54] Oral arguments are usually heard for three or four days of two weeks per month. The other two weeks are reserved for legal research, consideration of cases, and writing opinions. Although the justices have reviewed written briefs on opposing sides of a case, lawyers for each party are allocated a half-hour per side to summarize their arguments. Lawyers often find that their 30 minutes are devoted primarily to receiving probing questions from the justices. Indeed, as "Resources: Oral Arguments and Persuasion" shows, the justices may use questions asked at oral arguments as resources in their efforts to persuade their colleagues to vote in a particular way.

hibited by Court custom, Rehnquist took issue with his colleague and tried to rescue Drew from the barrage of questions. "Well, in a sense," Rehnquist began, "you're arguing for a minimal form of respect for the flag." After conceding Scalia's point, Rehnquist continued by differentiating this case from the Court's decision in *West Virginia v. Barnette.* "Not that you have to take your hat off or salute when it goes by. . . . The state can't require that, but at least can't it insist that you not destroy it?" A relieved Drew responded, "Yes, Your Honor."

William Kunstler then strode to the lectern and began by quoting from the *Barnette* decision. Immediately, Rehnquist went to work on him. "Well, the facts of *West Virginia v. Barnette* were quite different from this. There the students were required to salute the flag." Kunstler and Rehnquist argued about *Barnette*'s applicability for several minutes. Kunstler then referred to several other cases in which the Supreme Court had overturned the convictions of a man who burned a flag to protest the shooting of James Meredith, a civil rights leader; a college student who had sewn a peace symbol onto a flag that flew outside his apartment window to protest American involvement in Vietnam; and a man who had sewn a small flag onto the seat of his pants. Rehnquist challenged each analogy, suggesting that these cases presented different facts and issues.[2]

It is difficult to say whether oral arguments have any effect on justices' votes, although Justice David Souter recently suggested that he had changed his mind during oral argument in 5 to 10 percent of the cases heard during his first judicial term.[3] At the very least, oral arguments sometimes identify conflicting positions among the justices, positions that may be further debated in conference.

[1] Oral arguments in *Texas v. Johnson* are included in Phillip B. Kurland and Gerhard Casper, eds., *Landmark Briefs and Arguments of the Supreme Court of the United States: Constitutional Law,* 1988 Term Supplement, Vol. 190 (Bethesda, Md.: University Publications of America, 1990), pp. 336–688.
[2] *Street v. New York,* 394 US 576 (1969); *Spence v. Washington,* 418 US 405 (1974); and *Smith v. Goguen,* 415 US 566 (1974).
[3] Henry J. Abraham, *The Judicial Process,* 6th ed. (New York: Oxford University Press, 1993), p. 191.

After hearing oral arguments, the justices meet behind closed doors in conference to deliberate and reach a decision. They meet two days a week— on Wednesday afternoons and Fridays—to select cases for review from certiorari petitions and to decide cases recently argued before them. Although it is difficult to learn what goes on in these meetings (they are so secret that even law clerks and secretaries are prohibited from attending), we do know that the chief justice presides and typically begins by summarizing the cases to be considered and arguing how they should be decided. The chief justice then yields to the others, who state their opinions in descending order of seniority.

If the chief justice votes with the majority, he assigns the opinion to himself or one of the others in the majority. Chief justices typically assign the most important constitutional cases to themselves. If the chief justice is on the minority side, the most senior justice in the majority assigns the opinion. Opinion assignment is a coveted prize, and a variety of strategic considerations go into the decision. The chief justice has been known to assign an opinion in a way that may make it more palatable to segments of the public who will likely disagree with the outcome. In the 1940s, for example, after initially assigning Felix Frankfurter, a Jew from the Northeast, to write the opinion striking down the southern practice of excluding African Americans from voting in primary elections—a decision certain to upset some in the South—Chief Justice Harlan Fiske Stone was persuaded by another member of the Court to reassign the opinion to Justice Stanley Reed, a Protestant from Kentucky.[55]

Along similar lines, decisions that are easily identified as liberal are often assigned to the most "conservative" justice possible, and conservative decisions are often assigned to "liberal" justices. For example, Chief Justice Stone assigned the Court's 1944 decision in *Korematsu v. United States*—a decision upholding the internment of more than 100,000 Japanese Americans in detention camps—to one of the court's leading liberals, Hugo Black.[56] Finally, when the Court is divided at conference, the chief justice may assign the opinion to the member of the majority whose views are closest to those in dissent, hoping to pull them into the majority and produce a unanimous decision. Most opinions go through numerous revisions as they circulate among the justices. After an opinion is completed, a final vote is taken, at which time the justices may change their votes from the conference.[57]

What explains these final votes? Attempts have been made to identify the most crucial factors in predicting how the justices will vote. One approach examines the justices' values, attitudes, and ideological predilec-

The Supreme Court's chief justice sometimes operates politically in assigning the writing of opinions to particular judges. Chief Justice Harlan Fiske Stone presided over the case of *Korematsu v. United States*, in which the Court upheld the internment of Japanese Americans during World War II. Stone assigned the opinion to Justice Hugo Black (*shown here*), a leading liberal, in the hope that Black would persuade other liberals that the decision was legitimate.

tions.[58] A second approach suggests that social backgrounds and previous experience are good predictors of votes.[59] A third approach treats the Court as a small decision-making group, focusing on the bargaining, negotiations, and persuasion that occur as the justices interact.[60] Using the personal papers and diaries of retired justices as evidence, this research finds that a great deal of give-and-take occurs among the justices, that justices may be persuaded to change their minds about a case by other justices, and that language in opinions has been changed to accommodate the concerns of justices who threaten to vote in opposition, a particularly effective maneuver in close decisions. "Strategies: The Internal Politics of Abortion Rights" (on pp. 632–33) describes an especially good example of the internal politics of judicial decision making: the Court's deliberations over whether to overturn *Roe v. Wade* in the 1989 case of *Webster v. Reproductive Health Services*. Although each of these approaches highlights an important feature of judicial decision making, none of them alone captures the great complexity of voting behavior, in large part because the strictly confidential nature of judicial conferences makes it impossible to fully reconstruct the decision-making process.

None of the decision-making approaches attributes much weight to legal factors commonly thought to play an important role in judicial decisions. Legal approaches emphasize the role of precedent—that is, previously rendered decisions of the Court—in judicial choice. Judges are thought to follow the doctrine of **stare decisis** ("stand by the decision"), moving away from previously decided principles only when it is absolutely necessary.

Although it can hardly be denied that precedent constrains judicial decisions and that about one-third of all decisions are unanimous, courts and judges desiring to move in new or different directions may use a variety of strategies. Judges may distinguish the precedent from the case at hand, arguing that the case they are considering presents very different facts or issues. This strategy was used by Justice Rehnquist in *Texas v. Johnson* as he sought to differentiate the law prohibiting flag burning from the one struck down in *West Virginia v. Barnette* compelling students to salute the flag. Because a variety of precedents may relate to the case at hand, justices may "precedent shop," choosing from among previous decisions that support their decision and distinguishing them from those that do not. Finally, the Court may overturn precedent, a strategy it avoids whenever possible for fear of the potential damage to its popular prestige.

If courts routinely overturned precedents, the public could lose faith both in the judiciary's ability to remain neutral and nonpartisan and in constitutional law's certainty and consistency. Consequently, the Court may prefer to subtly and incrementally chip away at a precedent that a majority does not like, redefining the law in a way that preserves the precedent but strips it of content and force. The Rehnquist Court has engaged in such a strategy in the area of civil rights, and some people believe that this has been the ultimate fate of the Court's 1973 abortion decision, *Roe v. Wade*.[61] By refusing to overturn the constitutional right to privacy—the underpinning of the Court's decision in *Roe*—while simultaneously upholding a variety of restrictions on abortion, the Court effectively changes the meaning of the right to privacy announced in *Roe*.

THE INTERNAL POLITICS OF ABORTION RIGHTS

Recently released papers of the late justice Thurgood Marshall provide a rare glimpse at the internal politics of the contemporary Supreme Court. One of the documents' most important insights involves the fate in the late 1980s of *Roe v. Wade,* the controversial 1973 Supreme Court decision establishing a right to privacy that protects women in their first trimester of pregnancy from laws prohibiting abortion. Throughout the 1980s, supporters of abortion rights feared that a new conservative majority on the Supreme Court would overturn *Roe.*

The case of *Webster v. Reproductive Health Services,* a challenge to a Missouri law imposing restrictions on abortion, caused an immediate stir when it was appealed to the Court. Supporters of abortion rights so feared that *Roe* might be overturned that they considered voting to deny certiorari for the case. A law clerk to Justice Marshall summarized the strategy: "For defensive reasons," supporters of *Roe* should vote to keep the case away from the conservatives. Ultimately, on January 9, 1989, Justice Harry Blackmun, the author of *Roe,* and Justice William Brennan, a strong supporter, voted to hear the case over Marshall's objection.

Marshall's handwritten tally sheet showed that initially, in conference, five justices voted to uphold the Missouri law: Chief Justice William Rehnquist and Justices Byron White, Antonin Scalia, Anthony Kennedy, and Sandra Day O'Connor. Being the senior justice in the majority, Rehnquist got to decide who would write the majority opinion. Not surprisingly, he chose himself—as is customary in highly significant cases.

In his first draft of the opinion, circulated to

On controversial issues likely to divide the Court, justices try to form voting coalitions of at least five justices. This photo shows the Court that heard arguments in the abortion case of *Webster v. Reproductive Health Services.* In this case the justices who sought to overturn *Roe v. Wade* seemed to have mobilized a majority at one point during the Court's deliberations, only to see Justice Sandra Day O'Connor decide to join four others to uphold *Roe.*

the other justices on May 25, Rehnquist upheld the Missouri law and took direct aim at *Roe.* He called *Roe* "unsound in principle and unworkable in practice." This language was designed to appeal to Justice O'Connor, considered a swing voter in abortion cases, who had used virtually the same language in a previous decision. Rehnquist offered a new test for state abortion laws: Did they "reasonably further the state's interest in protecting potential human life"?

The next day Justice Blackmun signaled his displeasure with Rehnquist's opinion by circulating a memo that read simply, "I will be writing something in this case." On May 30, Jus-

THE IMPACT OF THE SUPREME COURT

Few significant political issues fail to reach the Supreme Court's docket. Flag burning as political protest, abortion, pornography, racial discrimination in housing, affirmative action, sexual harassment in the workplace, voting rights—these are but a few of the issues that have been considered by the

tices Brennan and Marshall wrote to Rehnquist that they would wait to see Blackmun's opinion. The coalitions were beginning to form.

On June 21, Blackmun circulated the first draft of an angry and passionate dissent. "Today," wrote the author of *Roe*, "a bare majority of this Court disservices the people of this nation, and especially the millions of women who have lived and come of age in the 16 years since the decision in *Roe v. Wade*." Blackmun went on, "I rue this day. I rue the violence that has been done to the liberty and equality of women."

The next day Rehnquist circulated a two-page scheduling memo indicating that he would announce the *Webster* decision on June 29, the final day of the term. On this same day an extremely important event occurred. Justice O'Connor circulated a first draft opinion suggesting that Rehnquist might no longer have a majority. She argued that she would uphold the Missouri law but would not overturn *Roe*. Scalia then circulated his own draft of an opinion that called for a more explicit overturning of *Roe*. Rehnquist, fearful of losing O'Connor and his majority, incorporated some points from her opinion that were unrelated to the question of *Roe*. In a memo he wrote, "Sandra had indicated that she had no objection to such modest plagiarism."

Rehnquist's attempt to appeal to O'Connor did not work. In a new draft of her opinion, she referred to Rehnquist's opinion as the "plurality" opinion. Blackmun changed his dissent to refer to the plurality opinion and deleted the phrase "*Roe* no longer survives." And instead of saying, "I rue this day," he said, "I fear for the future." Rehnquist's final draft made it clear that the Court had voted only to uphold the Missouri law but not to overturn the *Roe* precedent. On July 3, when the decision was announced in open court, Blackmun read a final section that he had added to his dissent after Rehnquist failed to mobilize a majority: "For today, the women of this Nation still retain the liberty to control their destinies. But the signs are evident and very ominous, and a chill wind blows."

After the *Webster* decision the justices supporting *Roe* worked on O'Connor's vote. In *Hodgson v. Minnesota*, the Court during the following term examined a state law requiring women under the age of 18 to notify both parents before obtaining an abortion. In conference, O'Connor voted with four others to uphold that part of the law. But in a later memo to the chief justice she explained that she had been persuaded by Justice John Paul Stevens's arguments in conference and that "this leads me to change my vote." Although Brennan, Marshall, and Blackmun—the most reliable supporters of abortion rights—had some objections to the opinion written by Justice Stevens in this case, they decided for strategic reasons to sign on to maintain the 5–4 majority and forge a coalition in support of *Roe* with Justice O'Connor. Brennan, in a memo to Marshall, reasoned, "I think it is important for John [Stevens] to get as much support as possible, now that Sandra has for the first time joined us in holding invalid a law regulating abortion." In a 5–4 vote, the Court struck down the two-parent notification requirement.

Source: Benjamin Weiser and Bob Woodward, "For a Moment, Abortion Was All but Outlawed," *Washington Post National Weekly Edition*, May 31–June 6, 1993, pp. 8–9.

Court. Thus, the Supreme Court has the potential to be an effective policy-making institution, one rendering decisions that authoritatively allocate who gets what.

Although the issues considered by the Court give it the potential to exert influence in the policy-making process, the Court's status in the political system provides it with a very fragile basis for making a material impact.

Quite simply, the Supreme Court lacks important political resources. It does not control money, coercive force, or votes. Its chief resource is its legitimacy—its ability to command voluntary compliance and the political support of those who do control important political resources. If people believe that the Court has the right to decide certain questions, that its decisions are arrived at appropriately and competently, and that its decisions ought to be obeyed—important indicators of legitimacy—then its influence will likely be felt. But if the public were to believe that the Court did not have jurisdiction to decide certain questions, that its decisions were the product of illegitimate considerations, and that such decisions should be ignored rather than obeyed, its influence would rapidly erode. To exert influence, then, the Court relies on public support.

The Court also depends on other components of government. Contrary to popular conceptions, the Court is not an independent institution. Its dependence on Congress is substantial.[62] Congress, with the constitutional power to establish the Court's appellate jurisdiction, can remove nearly all the important cases heard by the Court. It also holds the Court's purse strings. And Congress may reverse specific decisions of the Court by passing new legislation or beginning the process of amending the Constitution. For example, the Constitution's Sixteenth Amendment authorizes a federal income tax, a tax held unconstitutional by the Supreme Court. As of 1992, Congress had enacted more than 120 statutes reversing, totally or in large measure, decisions of the Court.[63] In the years 1985–1990, Congress introduced 82 bills to overturn or modify 56 separate Supreme Court decisions.[64] The Court's decision in *Texas v. Johnson* is only one of many that Congress has sought to reverse.

The Court's reliance on the executive branch is substantial as well. Having no direct control over armed forces, police, or even U.S. marshals, the Court depends on the executive branch to enforce its decisions. If the Court fails to obtain the cooperation of the executive branch, it is forced to stand helplessly on the sidelines. This is illustrated in the famous, although perhaps apocryphal, comment of President Andrew Jackson on the Supreme Court's decision in *Worcester v. Georgia.*[65] The opinion, written by Chief Justice John Marshall, upheld the rights of the Cherokee Nation in a dispute with the state of Georgia and strongly implied that the president had a duty to protect the Cherokees. According to Justice Marshall's biographer, this decision prompted Jackson's angry response, "John Marshall has made his decision—now let him enforce it."[66] The Court had more luck with the executive branch in the area of school desegregation: Federal troops were used at Little Rock High School in 1957, after some delay from the Eisenhower administration, and at the University of Mississippi in 1963 to enforce school desegregation orders issued by lower federal courts, which were implementing the Supreme Court's desegregation decisions.

Finally, the Court relies on the cooperation of state and lower federal court judges for implementation of its policies. Yet compliance by lower-court judges with Supreme Court mandates is not always forthcoming.[67] For example, some federal district court judges in the South were unwilling to implement the Court's school desegregation decisions. The Court's dependence on lower courts poses a serious problem. By and large, the Court is a *passive institution*, one that must wait patiently for losing parties in lower courts to appeal cases containing important issues. Because the number of cases it may hear is limited in comparison to the number of lower-court decisions, its ability to control lower-court decisions is quite limited. Here again we see the critical significance of the Court's legitimacy.

Because the Court is so dependent on the attitudes of the public and other components of government, it must be careful to exercise some degree of self-restraint.[68] It cannot decide every controversial case or issue that presents itself. It cannot nullify every piece of legislation that contains flaws. The Court's institutional limitations, emerging from its lack of resources for use in the play of power, help to explain at least some of the discretionary decisions it makes.

FEDERAL COURTS IN THE PLAY OF POWER

Given the significance of political issues entering the courts, and the judiciary's dependence on other political institutions, what can we say about the role of courts in the broader play of power? By and large, scholars who have examined this question systematically focus on the Supreme Court because of its position at the top of the judicial hierarchy and its practical and symbolic significance.

The Court's major resource in the broader play of power is the decisions it renders. For example, the Court enters politics in a dramatic way when it exercises its power of *judicial review*—the power to nullify federal laws. This power, which is not explicitly stated anywhere in the Constitution, was established by Chief Justice John Marshall in the Court's 1803 decision in *Marbury v. Madison*.[69] In perhaps the most important decision in the Court's history, Marshall justified this potentially enormous grant of authority, and the strong judiciary it created, in a few simple statements. The Constitution, Marshall argued, is the supreme law of the land, and if other laws contradict it, they are unconstitutional. Since judges interpret law, and

Chief Justice John Marshall used the 1803 case of *Marbury v. Madison* to establish the Court's most significant power, judicial review. This power, which allows the Court to nullify federal laws, is not explicitly mentioned in the Constitution.

since the Constitution is the supreme law, federal court judges have the authority to strike down laws violating constitutional provisions.

The Court also enters the play of power directly when it overturns state laws and constitutional provisions. But the Court also legitimizes the actions of states and other governmental branches when it renders the opposite decision, affirming the constitutionality of actions brought before it. Finally, when the Court engages in **statutory construction**—interpreting the meaning of federal legislation that is often broad and vague—it may flesh out the law's content, much as do bureaucratic agencies that are asked to implement such laws.

When the Court exercises its power of judicial review, it appears to act in a way that is counter to the will of the majority. However, the Court rarely uses this power, having overturned federal laws fewer than 150 times during its history.[70] An influential study of the Court's role found that the Court rarely used its power of judicial review to protect individual rights or those of disadvantaged groups.[71] Instead, such decisions most frequently protected the interests of powerful economic groups. For example, during the 1930s the Court overturned many of President Franklin Roosevelt's New Deal programs, arguing that economic regulation and social welfare programs violated a liberty of contract. The intended beneficiaries of much of this legislation—women and child laborers, for example—were members of historically disadvantaged groups. Because the Court's decisions nullified legislative enactments so infrequently, and because those that did tended to support the values of powerful interests in society, the study's author concluded that the Court's primary role in the play of power is to confer legitimacy on the policies of the ruling political coalition. When viewed historically, then, the Court's decisions seem to agree with and, significantly, to provide the judicial "seal of constitutionality" for policy decisions of Congress and the executive branch.

This view of the Court as a democratic institution is further enhanced by research on the Court and public opinion. For example, one researcher found that three-fifths of all Supreme Court decisions he studied agreed with public opinion when polls showed a clear majority on one side of the issue.[72] This relationship held even when the Court overturned laws. In nearly half of the instances studied where the Court nullified laws, the Court's decisions reflected national public opinion as measured in polls.[73]

This is not to say that the Court never counters majority sentiment or protects individual rights and disadvantaged groups. Although it is true that the Court rarely overturns federal law, it has nullified state laws and provisions of state constitutions much more frequently.[74] It has overturned some 1,200 state actions since 1789, about 850 of these since 1870.[75] These decisions often find the Court protecting unpopular individuals and minority interests, in areas such as prayer in public schools, criminal defendants' rights, and racial discrimination. In some areas, such as abortion and racial discrimination, the Court has rendered decisions on issues that other political institutions and branches try to avoid.

Although the legitimacy of specific decisions of the Supreme Court may be questioned, by and large the Court's role in the play of power reflects dominant values and policy preferences. In entering the policy-making arena, courts are not independent forces uninfluenced by politics, imposing their will on an unsuspecting or resistant society. Instead, they are institutions that are inextricably linked to the political system. The Court's legitimacy must be cultivated and maintained for it to have a material impact on the broader play of power. Consequently, it must be cautious in confronting dominant institutions and majority sentiment.

SUMMARY

In this chapter we have explained the varied links between politics and the operations of the federal judiciary. Contrary to common beliefs about courts and law that emphasize their apolitical nature, key participants in federal courts—judges and lawyers—are recruited to their practices in processes that are political, making use of authoritative rules that advantage some and disadvantage others. The federal judiciary, like much of American government, is disproportionately white, male, and upper-class. African Americans, Asian Americans, Hispanics, and other people of color form a tiny minority of the legal profession. Discrimination against women in law school admissions has ended, but the profession as a whole remains dominated by males. It is highly stratified, with an upper crust composed largely of white Protestant men from large corporate firms who have graduated from prestigious law schools.

Interest groups are important players that sponsor test cases and file amicus curiae briefs. However, the frequency of interest group participation in federal courts is not evenly distributed. Commercial interests—corporations, financial institutions, and trade associations—dominate these processes.

The federal judiciary has three levels. District courts are the trial courts, where disputes are first heard and typically resolved once and for all. A middle level is composed of courts of appeal. The Supreme Court resides at the top of the judicial hierarchy.

The Supreme Court has total discretion over which cases it will hear. Cases are appealed to it through the writ of certiorari. The Court looks for certain cues, such as the participation of the solicitor general, to decide whether or not to accept cases. The decision is governed by a rule of four: Four justices must vote to hear a case before it is placed on the docket. Indigents may file an affidavit in forma pauperis, waiving filing fees for appeals. The Court may also use procedural rules of standing, ripeness, mootness, and the political question doctrine to shape its agenda. Cases that are accepted for review are scheduled for oral argument. After argument, the justices deliberate in secret conferences, and opinions are assigned to particular justices to be written.

Through the decisions it renders, the Court plays an important role in American politics. Because it lacks important political resources, it depends for its material impact on its legitimacy in the eyes of the public and other components of government. But through its use of judicial review and statutory construction, the Court helps to fashion policy for the nation. Although some of its most notable decisions run counter to majority sentiment, the Court primarily confers legitimacy on the policies of dominant ruling coalitions.

Here we reproduce two paragraphs from Justice David Souter's opinion in the abortion case of *Planned Parenthood of Southeastern Pennsylvania v. Casey,* 112 S.Ct. 2791 (1992). Part of the debate in this case addressed the wisdom of overturning the Supreme Court's major abortion precedent, *Roe v. Wade.* What do these passages from Justice Souter's opinion illustrate about the resources of federal courts, the Supreme Court in particular? What sense do you make of the use of capital letters in these paragraphs?

Our analysis would not be complete, however, without explaining why overruling *Roe*'s central holding would not only reach an unjustifiable result under principles of stare decisis, but would seriously weaken the Court's capacity to exercise the judicial power and to function as the Supreme Court of a nation dedicated to the rule of law. To understand why this would be so it is necessary to understand the source of this Court's authority, the conditions necessary for its preservation, and its relationship to the country's understanding of itself as a constitutional Republic.

The root of American Governmental power is revealed most clearly in the instances of the power conferred by the Constitution upon the Judiciary of the United States and specifically upon this Court. As Americans of each succeeding generation are rightly told, the Court cannot buy support for its decisions by spending money, and, except to a minor degree, it cannot independently coerce obedience to its decrees. The Court's power lies, rather, in its legitimacy, a product of substance and perception that shows itself in the people's acceptance of the Judiciary as fit to determine what the Nation's law means and to declare what it means.

Key Terms

law *(p. 605)*

precedent *(p. 606)*

diversity of citizenship *(p. 608)*

concurrent jurisdiction *(p. 608)*

forum shopping *(p. 608)*

district courts *(p. 609)*

courts of appeal *(p. 609)*

original jurisdiction *(p. 609)*

United States Supreme Court *(p. 609)*

appellate jurisdiction *(p. 609)*

senatorial courtesy *(p. 612)*

sponsorship of test cases *(p. 623)*

amicus curiae brief *(p. 623)*

writ of certiorari *(p. 625)*

in forma pauperis *(p. 625)*

rule of four *(p. 625)*

solicitor general *(p. 625)*

standing to sue *(p. 626)*

ripe case *(p. 626)*

moot issue (*p. 626*)

political question doctrine
(*p. 626*)

stare decisis (*p. 631*)

statutory construction (*p. 636*)

Recommended Readings

Auerbach, Jerold S. *Unequal Justice: Lawyers and Social Change in Modern America.* New York: Oxford University Press, 1976. A historical examination of social stratification and discrimination against minority groups in the legal profession.

Carp, Robert A., and C. K. Rowland. *Policymaking and Politics in the Federal District Courts.* Knoxville: University of Tennessee Press, 1983. A study of judicial decisions and policy making in federal trial courts.

Epstein, Cynthia Fuchs. *Women in Law,* 2nd ed. Urbana: University of Illinois Press, 1993. A study of the experiences and changing roles of women in the legal profession.

Gates, John B., and Charles A. Johnson, eds. *The American Courts: A Critical Assessment.* Washington, D.C.: Congressional Quarterly Press, 1991. A collection of essays by leading scholars that provide a good overview of social science literature on judicial processes and policy making.

Howard, J. Woodford, Jr. *Courts of Appeals in the Federal Judicial System: A Study of the Second, Fifth, and District of Columbia Circuits.* Princeton, N.J.: Princeton University Press, 1981. Examines the policy-making roles of intermediate federal courts of appeal.

Murphy, Walter. *Elements of Judicial Strategy.* Chicago: University of Chicago Press, 1964. A classic study of how Supreme Court justices interact with one another as they seek to exert influence on decisions.

Perry, H. W., Jr. *Deciding to Decide: Agenda Setting in the United States Supreme Court.* Cambridge, Mass.: Harvard University Press, 1991. Based on interviews with Supreme Court justices and law clerks, this study explores how the Court decides which cases it will hear.

Segal, Jeffrey A., and Harold J. Spaeth. *The Supreme Court and the Attitudinal Model.* Cambridge, Eng.: Cambridge University Press, 1993. A study assessing the view that the votes of Supreme Court justices on particular cases are primarily a product of their personal and political attitudes.

Vose, Clement. *Caucasians Only.* Berkeley: University of California Press, 1959. A classic study examining the role of the NAACP in the legal struggle for racial equality.

Appendixes

The Declaration

of Independence

When in the Course of human events, it becomes necessary for one people to dissolve the political bands which have connected them with another, and to assume among the Powers of the earth, the separate and equal station to which the Laws of Nature and of Nature's God entitle them, a decent respect to the opinions of mankind requires that they should declare the causes which impel them to the separation.

We hold these truths to be self-evident, that all men are created equal, that they are endowed by their Creator with certain unalienable Rights, that among these are Life, Liberty and the pursuit of Happiness. That to secure these rights, Governments are instituted among Men, deriving their just powers from the consent of the governed. That whenever any Form of Government becomes destructive of these ends, it is the Right of the People to alter or to abolish it, and to institute new Government, laying its foundation on such principles and organizing its powers in such form, as to them shall seem most likely to effect their Safety and Happiness. Prudence, indeed, will dictate that Governments long established should not be changed for light and transient causes; and accordingly all experience hath shown, that mankind are more disposed to suffer, while evils are sufferable, than to right themselves by abolishing the forms to which they are accustomed. But when a long train of abuses and usurpations, pursuing invariably the same Object evinces a design to reduce them under absolute Despotism, it is their right, it is their duty, to throw off such Government, and to provide new Guards for their future security.—Such has been the patient sufferance of these Colonies; and such is now the necessity which constrains them to alter their former Systems of Government. The history of the present King of Great Britain is a history of repeated injuries and usurpations, all having in direct object the establishment of an absolute Tyranny over these States. To prove this, let Facts be submitted to a candid world.

He has refused his Assent to Laws, the most wholesome and necessary for the public good.

He has forbidden his Governors to pass Laws of immediate and pressing importance, unless suspended in their operation till his Assent should be obtained; and when so suspended, he has utterly neglected to attend to them.

He has refused to pass other Laws for the accommodation of large districts of people, unless those people would relinquish the right of Representation in the Legislature, a right inestimable to them and formidable to tyrants only.

He has called together legislative bodies at places unusual, uncomfortable, and distant from the depository of their public Records, for the sole purpose of fatiguing them into compliance with his measures.

He has dissolved Representative Houses repeatedly for opposing with manly firmness his invasions on the rights of the people.

He has refused for a long time, after such dissolutions, to cause others to be elected; whereby the Legislative Powers, incapable of Annihilation, have returned to the People at large for their exercise; the State remaining in the mean time exposed to all the dangers of invasion from without, and convulsions within.

He has endeavoured to prevent the population of these States; for that purpose obstructing the Laws of Naturalization of Foreigners; refusing to pass others to encourage their migration higher; and raising the conditions of new Appropriations of Lands.

He has obstructed the Administration of Justice, by refusing his Assent to Laws for establishing Judiciary powers.

He has made Judges dependent on his Will alone, for the tenure of their offices, and the amount and payment of their salaries.

He has erected a multitude of New Offices, and sent hither swarms of Officers to harass our People, and eat out their substance.

He has kept among us in times of peace, Standing Armies without the Consent of our legislature.

He has affected to render the Military independent of and superior to the Civil power.

He has combined with others to subject us to a jurisdiction foreign to our constitution, and unacknowledged by our laws; giving his Assent to their acts of pretended Legislation:

For quartering large bodies of armed troops among us;

For protecting them, by a mock Trial, from punishment for any Murders which they should commit on the inhabitants of these States;

For cutting off our Trade with all parts of the world;

For imposing taxes on us without our Consent;

For depriving us in many cases, of the benefits of Trial by Jury;

For transporting us beyond Seas to be tried for pretended offences;

For abolishing the free System of English Laws in a neighbouring Province, establishing therein an Arbitrary government, and enlarging its Boundaries so as to render it at once an example and fit instrument for introducing the same absolute rule into these Colonies;

For taking away our Charters, abolishing our most valuable Laws, and altering fundamentally the Forms of our Governments;

For suspending our own Legislature, and declaring themselves invested with Power to legislate for us in all cases whatsoever.

He has abdicated Government here, by declaring us out of his Protection and waging War against us.

He has plundered our seas, ravaged our Coasts, burnt our towns, and destroyed the lives of our people.

He is at this time transporting large Armies of foreign Mercenaries to compleat the works of death, desolation and tyranny, already begun with circumstances of Cruelty & perfidy scarcely paralleled in the most barbarous ages, and totally unworthy the Head of a civilized nation.

He has constrained our fellow Citizens taken Captive on the high Seas to bear Arms against their Country, to become the executioners of their friends and Brethren, or to fall themselves by their Hands.

He has excited domestic insurrections amongst us, and has endeavoured to bring on the inhabitants of our frontiers, the merciless Indian Savages, whose known rule of warfare, is an undistinguished destruction of all ages, sexes and conditions.

In every stage of these Oppressions We have Petitioned for Redress in the most humble terms: Our repeated Petitions have been answered only by repeated injury. A Prince, whose character is thus marked by every act which may define a Tyrant, is unfit to be the ruler of a free People.

Nor have We been wanting in attention to our British brethren. We have warned them from time to time of attempts by their legislature to extend an unwarrantable jurisdiction over us. We have reminded them of the circumstances of our emigration and settlement here. We have appealed to their native justice and magnanimity, and we have conjured them by the ties of our common kindred to disavow these usurpations, which, would inevitably interrupt our connections and correspondence. They too have been deaf to the voice of justice and of consanguinity. We must, therefore, acquiesce in the necessity, which denounces our Separation, and hold them, as we hold the rest of mankind, Enemies in War, in Peace Friends.

We, therefore, the Representatives of the United States of America, in General Congress, Assembled, appealing to the Supreme Judge of the world for the rectitude of our intentions, do, in the Name, and by Authority of the good People of these Colonies, solemnly publish and declare, That these United Colonies are, and of right ought to be Free and Independent States; that they are Absolved from all Allegiance to the British Crown, and that all political connection between them and the State of Great Britain, is and ought to be totally dissolved; and that as Free and Independent States, they have full Power to levy War, conclude Peace, contract Alliances, establish Commerce, and to do all other Acts and Things which Independent States may of right do. And for the support of this Declaration, with a firm reliance on the protection of divine Providence, we mutually pledge to each other our Lives, our Fortunes and our sacred Honor.

The Constitution of

the United States

of America

WE THE PEOPLE OF THE UNITED STATES, in Order to form a more perfect Union, establish Justice, insure domestic Tranquility, provide for the common defence, promote the general Welfare, and secure the Blessings of Liberty to ourselves and our Posterity, do ordain and establish this Constitution for the United States of America.

[THREE BRANCHES OF GOVERNMENT]

[*The legislative branch*]

ARTICLE I

[*Powers vested*]
SECTION 1   All legislative Powers herein granted shall be vested in a Congress of the United States, which shall consist of a Senate and House of Representatives.

[*House of Representatives*]
SECTION 2   The House of Representatives shall be composed of Members chosen every second Year by the People of the several States, and the Electors in each State shall have the Qualifications requisite for Electors of the most numerous Branch of the State Legislature.

No Person shall be a Representative who shall not have attained to the Age of twenty-five Years, and been seven Years a Citizen of the United States, and who shall not, when elected, be an Inhabitant of that State in which he shall be chosen.

[Representatives and direct Taxes shall be apportioned among the several States which may be included within this Union, according to their respective Numbers, which shall be determined by adding to the whole Number of free Persons, including those bound to Service for a Term of Years, and excluding Indians not taxed, three fifths of all other Persons.][1]

The actual Enumeration shall be made within three Years after the first Meeting of the Congress of the United States, and within every subsequent Term of ten Years, in such Manner as they shall by Law direct. The Number of Representatives shall not exceed one for every thirty Thousand, but each State shall have at Least one Representative; and until such enumeration shall be made, the State of New Hampshire shall be entitled to chuse three, Massachusetts eight, Rhode-Island and Providence Plantations one, Connecticut five, New York six, New Jersey four, Pennsylvania eight, Delaware one, Maryland six, Virginia ten, North Carolina five, South Carolina five, and Georgia three.

When vacancies happen in the Representation from any State, the Executive Authority thereof shall issue Writs of Election to fill such Vacancies.

The House of Representatives shall chuse their Speaker and other Officers; and shall have the sole Power of Impeachment.

[*The Senate*]
SECTION 3   The Senate of the United States shall be composed of two Senators from each State, [chosen by the Legislature thereof],[2] for six Years; and each Senator shall have one Vote.

Immediately after they shall be assembled in Consequence of the first Election, they shall be divided as equally

[2]Changed by Amendment XVII.

[1]Changed by Section 2 of Amendment XIV.

as may be into three Classes. The Seats of the Senators of the first Class shall be vacated at the Expiration of the Second Year, of the second Class at the Expiration of the fourth Year, and of the third Class at the Expiration of the sixth Year, so that one-third may be chosen every second Year; [and if Vacancies happen by Resignation, or otherwise, during the Recess of the Legislature of any State, the Executive thereof may make temporary Appointments until the next Meeting of the Legislature, which shall then fill such Vacancies].[3]

No person shall be a Senator who shall not have attained to the Age of thirty Years, and been nine Years a Citizen of the United States, and who shall not, when elected, be an Inhabitant of that State for which he shall be chosen.

The Vice President of the United States shall be President of the Senate, but shall have no Vote, unless they be equally divided.

The Senate shall chuse their other Officers, and also a President pro tempore, in the absence of the Vice President, or when he shall exercise the Office of President of the United States.

The Senate shall have the sole Power to try all Impeachments. When sitting for that Purpose, they shall be on Oath or Affirmation. When the President of the United States is tried, the Chief Justice shall preside: And no Person shall be convicted without the Concurrence of two-thirds of the Members present.

Judgment in Cases of Impeachment shall not extend further than to removal from Office, and disqualification to hold and enjoy any Office of honor, Trust, or Profit under the United States: but the Party convicted shall nevertheless be liable and subject to Indictment, Trial, Judgment, and Punishment, according to Law.

[Elections]

SECTION 4   The Times, Places and Manner of holding Elections for Senators and Representatives, shall be prescribed in each State by the Legislature thereof; but the Congress may at any time by Law make or alter such Regulations, except as to the Places of chusing Senators.

The Congress shall assemble at least once in every Year, and such Meeting shall be on the first Monday in December, [unless they shall by Law appoint a different Day].[4]

[Powers, duties, procedures of both bodies]

SECTION 5   Each House shall be the Judge of the Elections, Returns, and Qualifications of its own Members, and a Majority of each shall constitute a Quorum to do Business; but a smaller Number may adjourn from day to day, and may be authorized to compel the Attendance of absent Members, in such Manner, and under such Penalties as each House may provide.

Each House may determine the Rules of its Proceedings, punish its Members for disorderly Behavior, and, with the Concurrence of two thirds, expel a Member.

Each House shall keep a Journal of its Proceedings, and from time to time publish the same, excepting such Parts as

[3]Changed by Amendment XVII.
[4]Changed by Section 2 of Amendment XX.

B-2

may in their Judgment require Secrecy; and the Yeas and Nays of the Members of either House on any question shall, at the Desire of one fifth of those Present, be entered on the Journal.

Neither House, during the Session of Congress, shall, without the Consent of the other, adjourn for more than three days, nor to any other Place than that in which the two Houses shall be sitting.

[Compensation, privileges, limits on other governmental service]

SECTION 6   The Senators and Representatives shall receive a Compensation for their Services, to be ascertained by Law, and paid out of the Treasury of the United States. They shall in all Cases, except Treason, Felony and Breach of the Peace, be privileged from Arrest during their Attendance at the Session of their respective Houses, and in going to and returning from the same; and for any Speech or Debate in either House, they shall not be questioned in any other Place.

No Senator or Representative shall, during the Time for which he was elected, be appointed to any civil Office under the Authority of the United States, which shall have been created, or the Emoluments whereof shall have been encreased during such time; and no Person holding any Office under the United States, shall be a Member of either House during his Continuance in Office.

[Origin of revenue bills; presidential approval or disapproval of legislation; overriding the veto]

SECTION 7   All Bills for raising Revenue shall originate in the House of Representatives; but the Senate may propose or concur with Amendments as on other Bills.

Every Bill which shall have passed the House of Representatives and the Senate, shall, before it become a Law, be presented to the President of the United States; if he approve he shall sign it, but if not he shall return it, with his Objections to that House in which it shall have originated, who shall enter the Objections at large on their Journal, and proceed to reconsider it. If after such Reconsideration two thirds of that House shall agree to pass the Bill, it shall be sent, together with the Objections, to the other House, by which it shall likewise be reconsidered, and if approved by two thirds of that House, it shall become a Law. But in all such Cases the Votes of both Houses shall be determined by Yeas and Nays, and the Names of the Persons voting for and against the Bill shall be entered on the Journal of each House respectively. If any Bill shall not be returned by the President within ten Days (Sundays excepted) after it shall have been presented to him, the Same shall be a Law, in like Manner as if he had signed it, unless the Congress by their Adjournment prevent its Return, in which Case it shall not be a Law.

Every Order, Resolution, or Vote to which the Concurrence of the Senate and House of Representatives may be necessary (except on a question of Adjournment) shall be presented to the President of the United States; and before the Same shall take Effect, shall be approved by him, or being disapproved by him, shall be repassed to two thirds of the Senate and House of Representatives, according to the Rules and Limitations prescribed in the Case of a Bill.

SECTION 8   The Congress shall have power

To lay and collect Taxes, Duties, Imposts and Excises, to pay the Debts and provide for the common Defence and general Welfare of the United States; but all Duties, Imposts and Excises shall be uniform throughout the United States;

To borrow money on the credit of the United States;

To regulate Commerce with foreign Nations, and among the several States, and with the Indian Tribes;

To establish an uniform Rule of Naturalization, and uniform Laws on the subject of Bankruptcies throughout the United States;

To coin Money, regulate the Value thereof, and of foreign Coin, and fix the Standard of Weights and Measures;

To provide for the Punishment of counterfeiting the Securities and current Coin of the United States;

To Establish Post Offices and post Roads;

To promote the Progress of Science and useful Arts, by securing for limited Times to Authors and Inventors the exclusive Right to their respective Writings and Discoveries;

To constitute Tribunals inferior to the Supreme Court;

To define and punish Piracies and Felonies committed on the high Seas, and Offences against the Law of Nations;

To declare War, grant Letters of Marque and Reprisal, and make Rules concerning Captures on Land and Water;

To raise and support Armies, but no Appropriation of Money to that Use shall be for a longer Term than two Years;

To provide and maintain a Navy;

To make Rules for the Government and Regulation of the land and naval Forces;

To provide for calling forth the Militia to execute the Laws of the Union, suppress Insurrections and repel Invasions;

To provide for organizing, arming, and disciplining the Militia, and for governing such Part of them as may be employed in the Service of the United States, reserving to the States respectively, the Appointment of the Officers, and the Authority of training the Militia according to the discipline prescribed by Congress;

To exercise exclusive Legislation in all Cases whatsoever, over such District (not exceeding ten Miles square) as may, by Cession of particular States, and the acceptance of Congress, become the Seat of the Government of the United States, and to exercise like Authority over all Places purchased by the Consent of the Legislature of the State in which the Same shall be, for the Erection of Forts, Magazines, Arsenals, dock-Yards, and other needful Buildings;—And

[Elastic clause]

To make all Laws which shall be necessary and proper for carrying into Execution the foregoing Powers, and all other Powers vested by this Constitution in the Government of the United States, or in any Department or Officer thereof.

[Powers denied to Congress]

SECTION 9   The Migration or Importation of Such Persons as any of the States now existing shall think proper to admit, shall not be prohibited by the Congress prior to the Year one thousand eight hundred and eight, but a tax or duty may be imposed on such Importation, not exceeding ten dollars for each Person.

The privilege of the Writ of Habeas Corpus shall not be suspended, unless when in Cases of Rebellion or Invasion the public Safety may require it.

No Bill of Attainder or ex post facto Law shall be passed.

[No capitation, or other direct, Tax shall be laid, unless in Proportion to the Census or Enumeration herein before directed to be taken.][5]

No Tax or Duty shall be laid on Articles exported from any State.

No preference shall be given by any Regulation of Commerce or Revenue to the Ports of one State over those of another: nor shall Vessels bound to, or from, one State be obliged to enter, clear, or pay Duties in another.

No money shall be drawn from the Treasury, but in Consequence of Appropriations made by Law; and a regular Statement and Account of the Receipts and Expenditures of all public Money shall be published from time to time.

No Title of Nobility shall be granted by the United States: And no Person holding any Office of Profit or Trust under them, shall, without the Consent of the Congress, accept of any present, Emolument, Office, or Title, of any kind whatever, from any King, Prince, or foreign State.

[Powers denied to states]

SECTION 10   No State shall enter into any Treaty, Alliance, or Confederation; grant Letters of Marque and Reprisal; coin Money; emit Bills of Credit; make any Thing but gold and silver Coin a Tender in Payment of Debts; pass any Bill of Attainder, ex post facto Law, or Law impairing the Obligation of Contracts, or grant any Title of Nobility.

No State shall, without the Consent of the Congress, lay any Imposts or Duties on Imports or Exports, except what may be absolutely necessary for executing its inspection Laws: and the net Produce of all Duties and Imposts, laid by any State on Imports or Exports, shall be for the Use of the Treasury of the United States; and all such Laws shall be subject to the Revision and Control of the Congress.

No State shall, without the Consent of Congress, lay any duty of Tonnage, keep Troops, or Ships of War in time of Peace, enter into any Agreement or Compact with another State, or with a foreign Power, or engage in War, unless actually invaded, or in such imminent Danger as will not admit of delay.

[The executive branch]

ARTICLE II

[Presidential term, choice by electors, qualifications, payment, succession, oath of office]

SECTION 1   The executive Power shall be vested in a President of the United States of America. He shall hold his Office during the Term of four Years, and, together with

---

[5]Changed by Amendment XVI.

the Vice President, chosen for the same Term, be elected, as follows:

Each State shall appoint, in such Manner as the Legislature thereof may direct, a Number of Electors, equal to the whole Number of Senators and Representatives to which the State may be entitled in the Congress: but no Senator or Representative, or Person holding an Office of Trust or Profit under the United States, shall be appointed an Elector.

[The Electors shall meet in their respective States, and vote by Ballot for two persons, of whom one at least shall not be an Inhabitant of the same State with themselves. And they shall make a List of all the Persons voted for, and of the Number of Votes for each; which List they shall sign and certify, and transmit sealed to the Seat of the Government of the United States, directed to the President of the Senate. The President of the Senate shall, in the Presence of the Senate and House of Representatives, open all the Certificates, and the Votes shall then be counted. The Person having the greatest Number of Votes shall be the President, if such Number be a Majority of the whole Number of Electors appointed; and if there be more than one who have such Majority, and have an equal Number of Votes, then the House of Representatives shall immediately chuse by Ballot one of them for President; and if no Person have a Majority, then from the five highest on the List the said House shall in like Manner chuse the President. But in chusing the President, the Votes shall be taken by States, the Representation from each State having one Vote; A quorum for this Purpose shall consist of a Member or Members from two-thirds of the States, and a Majority of all the States shall be necessary to a Choice. In every Case, after the Choice of the President, the Person having the greatest Number of Votes of the Electors shall be the Vice President. But if there should remain two or more who have equal Votes, the Senate shall chuse from them by Ballot the Vice President.][6]

The Congress may determine the Time of chusing the Electors, and the Day on which they shall give their Votes; which Day shall be the same throughout the United States.

No person except a natural born Citizen, or a Citizen of the United States, at the time of the Adoption of this Constitution, shall be eligible to the Office of President; neither shall any Person be eligible to that Office who shall not have attained to the Age of thirty-five Years, and been fourteen Years a Resident within the United States.

[In case of the removal of the President from Office, or of his Death, Resignation, or Inability to discharge the Powers and Duties of the said Office, the same shall devolve on the Vice President, and the Congress may by Law provide for the Case of Removal, Death, Resignation or Inability, both of the President and Vice President, declaring what Officer shall then act as President, and such Officer shall act accordingly, until the Disability be removed, or a President shall be elected.][7]

The President shall, at stated Times, receive for his Services, a Compensation, which shall neither be increased nor diminished during the Period for which he shall have been elected, and he shall not receive within that Period any other Emolument from the United States, or any of them.

Before he enter on the Execution of his Office, he shall take the following Oath or Affirmation:—"I do solemnly swear (or affirm) that I will faithfully execute the Office of President of the United States, and will to the best of my Ability, preserve, protect and defend the Constitution of the United States."

[Powers to command the military and executive departments, to grant pardons, to make treaties, to appoint governmental officers]

SECTION 2   The President shall be Commander in Chief of the Army and Navy of the United States, and of the Militia of the several States, when called into the actual Service of the United States; he may require the Opinion, in writing, of the principal Officer in each of the executive Departments, upon any subject relating to the Duties of their respective Offices, and he shall have Power to grant Reprieves and Pardons for Offenses against the United States, except in Cases of Impeachment.

He shall have Power, by and with the Advice and Consent of the Senate, to make Treaties, provided two-thirds of the Senators present concur; and he shall nominate, and by and with the Advice and Consent of the Senate, shall appoint Ambassadors, other public Ministers and Consuls, Judges of the Supreme Court, and all other Officers of the United States, whose Appointments are not herein otherwise provided for, and which shall be established by Law; but the Congress may by Law vest the Appointment of such inferior Officers, as they think proper, in the President alone, in the Courts of Law, or in the Heads of Departments.

The President shall have Power to fill up all Vacancies that may happen during the Recess of the Senate, by granting Commissions which shall expire at the End of their next Session.

[Formal duties]

SECTION 3   He shall from time to time give to the Congress Information of the State of the Union, and recommend to their Consideration such Measures as he shall judge necessary and expedient; he may, on extraordinary Occasions, convene both Houses, or either of them, and in Case of Disagreement between them, with Respect to the Time of Adjournment, he may adjourn them to such Time as he shall think proper; he shall receive Ambassadors and other public Ministers; he shall take Care that the Laws be faithfully executed, and shall Commission all the Officers of the United States.

[Conditions for removal]

SECTION 4   The President, Vice President and all civil Officers of the United States, shall be removed from Office on Impeachment for, and Conviction of, Treason, Bribery, or other high Crimes and Misdemeanors.

[6]Changed by Amendment XII.
[7]Changed by Amendment XXV.

[The judicial branch]

ARTICLE III

[Courts and judges]

SECTION 1   The judicial Power of the United States, shall be vested in one supreme Court, and in such inferior Courts as the Congress may from time to time ordain and establish. The Judges, both of the supreme and inferior Courts, shall hold their Offices during good Behaviour, and shall, at stated Times, receive for their Services a Compensation which shall not be diminished during their Continuance in Office.

[Jurisdictions and jury trials]

SECTION 2   The judicial Power shall extend to all Cases, in Law and Equity, arising under this Constitution, the Laws of the United States, and Treaties made, or which shall be made, under their Authority;—to all Cases affecting Ambassadors, other public Ministers and Consuls;—to all Cases of admiralty and maritime Jurisdiction;—to Controversies to which the United States shall be a Party;—to Controversies between two or more States;—[between a State and Citizens of another State;—][8] between Citizens of different States;—between Citizens of the same State claiming Lands under Grants of different States, [and between a State, or the Citizens thereof, and foreign States, Citizens or Subjects].[9]

In all Cases affecting Ambassadors, other public Ministers and Consuls, and those in which a State shall be Party, the supreme Court shall have original Jurisdiction. In all the other Cases before mentioned, the supreme Court shall have appellate Jurisdiction, both as to Law and Fact, with such Exceptions, and under such Regulations as the Congress shall make.

The trial of all Crimes, except in Cases of Impeachment, shall be by Jury; and such Trial shall be held in the State where the said Crimes shall have been committed; but when not committed within any State, the Trial shall be at such Place or Places as the Congress may by Law have directed.

[Treason and its punishment]

SECTION 3   Treason against the United States, shall consist only in levying War against them, or, in adhering to their Enemies, giving them Aid and Comfort. No Person shall be convicted of Treason unless on the Testimony of two Witnesses to the same overt Act, or on Confession in open Court.

The Congress shall have power to declare the Punishment of Treason, but no Attainder of Treason shall work Corruption of Blood, or Forfeiture except during the Life of the Person attainted.

[THE REST OF THE FEDERAL SYSTEM]

ARTICLE IV

[Relationships among and with states]

SECTION 1   Full Faith and Credit shall be given in each State to the public Acts, Records, and judicial Proceedings of every other State. And the Congress may by general Laws prescribe the Manner in which such Acts, Records and Proceedings shall be proved, and the Effect thereof.

[Privileges and immunities, extradition]

SECTION 2   The Citizens of each State shall be entitled to all Privileges and Immunities of Citizens in the several States.

A Person charged in any State with Treason, Felony, or other Crime, who shall flee from Justice, and be found in another State, shall on demand of the executive Authority of the State from which he fled, be delivered up, to be removed to the State having Jurisdiction of the Crime.

[No Person held to Service or Labour in one State, under the Laws thereof, escaping into another, shall, in Consequence of any Law or Regulation therein, be discharged from such Service or Labour, but shall be delivered up on Claim of the Party to whom such Service or Labour may be due.][10]

[New states]

SECTION 3   New States may be admitted by the Congress into this Union; but no new State shall be formed or erected within the Jurisdiction of any other State; nor any State be formed by the Junction of two or more States, or parts of States, without the Consent of the Legislatures of the States concerned as well as of the Congress.

The Congress shall have Power to dispose of and make all needful Rules and Regulations respecting the Territory or other Property belonging to the United States; and nothing in this Constitution shall be so construed as to Prejudice any Claims of the United States, or of any particular State.

[Obligations to states]

SECTION 4   The United States shall guarantee to every State in this Union a Republican Form of Government, and shall protect each of them against Invasion; and on Application of the Legislature, or of the Executive (when the Legislature cannot be convened) against domestic Violence.

[MECHANISM FOR CHANGE]

ARTICLE V

[Amending the Constitution]

The Congress, whenever two-thirds of both Houses shall deem it necessary, shall propose Amendments to this Constitution, or, on the Application of the Legislatures of two-thirds of the several States, shall call a Convention for proposing Amendments, which, in either Case, shall be valid to all Intents and Purposes, as part of this Constitution, when ratified by the Legislatures of three-fourths of the several States, or by Conventions in three-fourths thereof, as the one or the other Mode of Ratification may be proposed by the Congress; Provided that no Amendment which may be made prior to

---

[8]Changed by Amendment XI.
[9]Changed by Amendment XI.

[10]Changed by Amendment XIII.

the Year One thousand eight hundred and eight shall in any Manner affect the first and fourth Clauses in the Ninth Section of the first Article; and that no State, without its Consent, shall be deprived of its equal Suffrage in the Senate.

[FEDERAL SUPREMACY]

ARTICLE VI

All Debts contracted and Engagements entered into, before the Adoption of this Constitution shall be as valid against the United States under this Constitution, as under the Confederation.

This Constitution, and the Laws of the United States which shall be made in Pursuance thereof; and all Treaties made, or which shall be made, under the Authority of the United States, shall be the supreme Law of the Land; and the Judges in every State shall be bound thereby, any Thing in the Constitution or Laws of any State to the Contrary notwithstanding.

The Senators and Representatives before mentioned, and the Members of the several State Legislatures, and all executive and judicial Officers, both of the United States and of the several States, shall be bound by Oath or Affirmation, to support this Constitution; but no religious Test shall ever be required as a Qualification to any Office or public Trust under the United States.

[RATIFICATION]

ARTICLE VII

The Ratification of the Conventions of nine States shall be sufficient for the Establishment of this Constitution between the States so ratifying the Same.

Done in Convention by the Unanimous Consent of the States present the Seventeenth Day of September in the year of our Lord one thousand seven hundred and eighty seven and of the Independence of the United States of America the twelfth. In witness whereof We have hereunto subscribed our Names.

[BILL OF RIGHTS AND OTHER AMENDMENTS]

Articles in addition to, and amendment of, the Constitution of the United States of America, proposed by Congress, and ratified by the several States, pursuant to the fifth Article of the original Constitution.

AMENDMENT I [1791]

[Freedoms of religion, speech, press, assembly]
Congress shall make no law respecting an establishment of religion, or prohibiting the free exercise thereof; or abridging the freedom of speech, or of the press; or the right of the people peaceably to assemble and to petition the Government for a redress of grievances.

AMENDMENT II [1791]

[Right to bear arms]
A well regulated Militia, being necessary to the security of a free State, the right of the people to keep and bear Arms, shall not be infringed.

AMENDMENT III [1791]

[Quartering of soldiers]
No Soldier shall, in time of peace be quartered in any house, without the consent of the Owner, nor in time of war, but in a manner to be prescribed by Law.

AMENDMENT IV [1791]

[Protection against search and seizure]
The right of the people to be secure in their persons, houses, papers, and effects, against unreasonable searches and seizures, shall not be violated, and no Warrants shall issue, but upon probable cause, supported by Oath or affirmation, and particularly describing the place to be searched, and the persons or things to be seized.

AMENDMENT V [1791]

[Protection of citizens before the law]
No person shall be held to answer for a capital, or otherwise infamous crime, unless on a presentment or indictment of a Grand Jury, except in cases arising in the land or naval forces, or in the Militia, when in actual service in time of War or public danger; nor shall any person be subject for the same offence to be twice put in jeopardy of life or limb; nor shall be compelled in any criminal case to be a witness against himself, nor be deprived of life, liberty, or property, without due process of law; nor shall private property be taken for public use, without just compensation.

AMENDMENT VI [1791]

[Rights of the accused in criminal cases]
In all criminal prosecutions, the accused shall enjoy the right to a speedy and public trial, by an impartial jury of the State and district wherein the crime shall have been committed, which district shall have been previously ascertained by law, and to be informed of the nature and cause of the accusation; to be confronted with the witnesses against him; to have compulsory process for obtaining witnesses in his favor, and to have the Assistance of Counsel for his defence.

AMENDMENT VII [1791]

[Rights of complainants in civil cases]
In suits at common law, where the value in controversy shall exceed twenty dollars, the right of trial by jury shall be preserved, and no fact tried by jury, shall be otherwise reexamined in any Court of the United States, than according to the rules of the common law.

AMENDMENT VIII [1791]

[Constraints on punishments]
Excessive bail shall not be required, nor excessive fines imposed, nor cruel and unusual punishments inflicted.

AMENDMENT IX [1791]

[Rights retained by the people]
The enumeration in the Constitution, of certain rights, shall not be construed to deny or disparage others retained by the people.

AMENDMENT X [1791]

[Rights reserved to states]
The powers not delegated to the United States by the Constitution, nor prohibited by it to the States, are reserved to the States respectively, or to the people.

AMENDMENT XI [1798]

[Restraints on judicial power]
The Judicial power of the United States shall not be construed to extend to any suit in law or equity, commenced or prosecuted against one of the United States by Citizens of another State, or by Citizens or Subjects of any Foreign State.

AMENDMENT XII [1804]

[Mechanism for presidential elections]
The electors shall meet in their respective states and vote by ballot for President and Vice-President, one of whom, at least, shall not be an inhabitant of the same state with themselves; they shall name in their ballots the person voted for as President, and in distinct ballots the person voted for as Vice-President, and they shall make distinct lists of all persons voted for as President, and of all persons voted for as Vice-President, and of the number of votes for each, which lists they shall sign and certify, and transmit sealed to the seat of the government of the United States, directed to the President of the Senate;—The President of the Senate shall, in presence of the Senate and House of Representatives, open all the certificates and the votes shall then be counted;—

The person having the greatest number of votes for President, shall be the President, if such number be a majority of the whole number of Electors appointed; and if no person have such majority, then from the persons having the highest numbers not exceeding three on the list of those voted for as President, the House of Representatives shall choose immediately, by ballot, the President. But in choosing the President, the votes shall be taken by states, the representation from each state having one vote; a quorum for this purpose shall consist of a member or members from two-thirds of the states, and a majority of all the states shall be necessary to a choice. [And if the House of Representatives shall not choose a President whenever the right of choice shall devolve upon them, before the fourth day of March next following, then the Vice-President shall act as President, as in the case of the death or other constitutional disability of the President.—][11] The person having the greatest number of votes as Vice-President, shall be the Vice-President, if such number be a majority of the whole number of Electors appointed, and if no person have a majority, then from the two highest numbers on the list, the Senate shall choose the Vice-President; a quorum for the purpose shall consist of two-thirds of the whole number of Senators, and a majority of the whole number shall be necessary to a choice. But no person constitutionally ineligible to the office of President shall be eligible to that of Vice-President of the United States.

AMENDMENT XIII [1865]

[Abolishment of slavery]
SECTION 1 Neither slavery nor involuntary servitude, except as a punishment for crime whereof the party shall have been duly convicted, shall exist within the United States, or any place subject to their jurisdiction.

SECTION 2 Congress shall have power to enforce this article by appropriate legislation.

AMENDMENT XIV [1868]

[Citizens' rights and immunities, due process, equal protection]
SECTION 1 All persons born or naturalized in the United States, and subject to the jurisdiction thereof, are citizens of the United States and of the State wherein they reside. No State shall make or enforce any law which shall abridge the privileges or immunities of citizens of the United States; nor shall any State deprive any person of life, liberty, or property, without due process of law; nor deny to any person within its jurisdiction the equal protection of the laws.

[Basis of representation]
SECTION 2 Representatives shall be appointed among the several States according to their respective numbers, counting the whole number of persons in each State, exclud-

---

[11]Superseded by Section 3 of Amendment XX.

ing Indians not taxed. But when the right to vote at any election for the choice of electors for President and Vice-President of the United States, Representatives in Congress, the Executive and Judicial officers of a State, or the members of the Legislature thereof, is denied to any of the male inhabitants of such State, being twenty-one years of age, and citizens of the United States, or in any way abridged, except for participation in rebellion, or other crime, the basis of representation therein shall be reduced in the proportion which the number of such male citizens shall bear to the whole number of male citizens twenty-one years of age in such State.

[*Disqualification of Confederates for office*]
SECTION 3   No person shall be a Senator or Representative in Congress, or elector of President and Vice-President, or hold any office, civil or military, under the United States, or under any State, who, having previously taken an oath, as a member of Congress, or as an officer of the United States, or as a member of any State legislature, or as an executive or judicial officer of any State, to support the Constitution of the United States, shall have engaged in insurrection or rebellion against the same, or given aid or comfort to the enemies thereof. But Congress may by a vote of two-thirds of each House, remove such disability.

[*Public debt arising from insurrection or rebellion*]
SECTION 4   The validity of the public debt of the United States, authorized by law, including debts incurred for payment of pensions and bounties for services in suppressing insurrection or rebellion, shall not be questioned. But neither the United States nor any State shall assume or pay any debt or obligation incurred in aid of insurrection or rebellion against the United States, or any claim for the loss or emancipation of any slave; but all such debts, obligations and claims shall be held illegal and void.
SECTION 5   The Congress shall have power to enforce, by appropriate legislation, the provisions of this article.

AMENDMENT XV [1870]

[*Explicit extension of right to vote*]
SECTION 1   The right of citizens of the United States to vote shall not be denied or abridged by the United States or by any State on account of race, color, or previous condition of servitude.
SECTION 2   The Congress shall have power to enforce this article by appropriate legislation.

AMENDMENT XVI [1913]

[*Creation of income tax*]
The Congress shall have power to lay and collect taxes on incomes, from whatever source derived, without apportionment among the several States, and without regard to any census or enumeration.

AMENDMENT XVII [1913]

[*Election of senators*]
The Senate of the United States shall be composed of two Senators from each State, elected by the people thereof, for six years; and each Senator shall have one vote. The electors in each State shall have the qualifications requisite for electors of the most numerous branch of the State legislatures.

When vacancies happen in the representation of any State in the Senate, the executive authority of such State shall issue writs of election to fill such vacancies: *Provided,* That the legislature of any State may empower the executive thereof to make temporary appointments until the people fill the vacancies by election as the legislature may direct.

This amendment shall not be so construed as to affect the election or term of any Senator chosen before it becomes valid as part of the Constitution.

AMENDMENT XVIII [1919]

[*Prohibition of alcohol*]
[SECTION 1   After one year from the ratification of this article the manufacture, sale, or transportation of intoxicating liquors within, the importation thereof into, or the exportation thereof from the United States and all territory subject to the jurisdiction thereof for beverage purposes is hereby prohibited.
SECTION 2   The Congress and the several States shall have concurrent power to enforce this article by appropriate legislation.
SECTION 3   This article shall be inoperative unless it shall have been ratified as an amendment to the Constitution by the legislatures of the several States, as provided in the Constitution, within seven years from the date of the submission hereof to the States by the Congress.][12]

AMENDMENT XIX [1920]

[*Voting rights and gender*]
The right of citizens of the United States to vote shall not be denied or abridged by the United States or by any State on account of sex.

Congress shall have the power to enforce this article by appropriate legislation.

AMENDMENT XX [1933]

[*Terms of executives, assembly of Congress, presidential succession*]
SECTION 1   The terms of the President and Vice President shall end at noon on the 20th day of January, and the terms of Senators and Representatives at noon on the 3d day of January, of the years in which such terms would have ended if this article had not been ratified; and the terms of their successors shall then begin.

[12]Repealed by Amendment XXI.

SECTION 2   The Congress shall assemble at least once in every year, and such meeting shall begin at noon on the 3d day of January, unless they shall by law appoint a different day.

SECTION 3   If, at the time fixed for the beginning of the term of the President, the President elect shall have died, the Vice President elect shall become President. If a President shall not have been chosen before the time fixed for the beginning of his term, or if the President elect shall have failed to qualify, then the Vice President elect shall act as President until a President shall have qualified; and the Congress may by law provide for the case wherein neither a President elect nor a Vice President elect shall have qualified, declaring who shall then act as President, or the manner in which one who is to act shall be selected, and such person shall act accordingly until a President or Vice President shall have qualified.

SECTION 4   The Congress may by law provide for the case of the death of any of the persons from whom the House of Representatives may choose a President whenever the right of choice shall have devolved upon them, and for the case of the death of any of the persons from whom the Senate may choose a Vice President whenever the right of choice shall have devolved upon them.

SECTION 5   Sections 1 and 2 shall take effect on the 15th day of October following the ratification of this article.

SECTION 6   This article shall be inoperative unless it shall have been ratified as an amendment to the Constitution by the legislatures of three-fourths of the several States within seven years from the date of its submission.

AMENDMENT XXI [1933]

[Repealing of prohibition]
SECTION 1   The eighteenth article of amendment to the Constitution of the United States is hereby repealed.

SECTION 2   The transportation or importation into any State, Territory, or possession of the United States for delivery or use therein of intoxicating liquors, in violation of the laws thereof, is hereby prohibited.

SECTION 3   This article shall be inoperative unless it shall have been ratified as an amendment to the Constitution by conventions in the several States, as provided in the Constitution, within seven years from the date of the submission hereof to the States by the Congress.

AMENDMENT XXII [1951]

[Limits on presidential term]
SECTION 1   No person shall be elected to the office of the President more than twice, and no person who has held the office of President, or acted as President, for more than two years of a term to which some other person was elected President shall be elected to the office of the President more than once. But this Article shall not apply to any person holding the office of President when this Article was proposed by the Congress, and shall not prevent any person who may be holding the office of President, or acting as President, during the term within which the Article becomes operative

from holding the office of President or acting as President during the remainder of such term.

SECTION 2   This article shall be inoperative unless it shall have been ratified as an amendment to the Constitution by the legislatures of three-fourths of the several States within seven years from the date of its submission to the States by the Congress.

AMENDMENT XXIII [1961]

[Voting rights of District of Columbia]
SECTION 1   The District constituting the seat of Government of the United States shall appoint in such manner as the Congress may direct:

A number of electors of President and Vice President equal to the whole number of Senators and Representatives in Congress to which the District would be entitled if it were a State; but in no event more than the least populous State; they shall be in addition to those appointed by the States, but they shall be considered, for the purposes of the election of President and Vice President, to be electors appointed by a State; and they shall meet in the District and perform such duties as provided by the twelfth article of amendment.

SECTION 2   The Congress shall have power to enforce this article by appropriate legislation.

AMENDMENT XXIV [1964]

[Prohibition of poll tax]
SECTION 1   The right of citizens of the United States to vote in any primary or other election for President or Vice President, for electors for President or Vice President, or for Senator or Representative in Congress, shall not be denied or abridged by the United States or any State by reason of failure to pay any poll tax or other tax.

SECTION 2   The Congress shall have power to enforce this article by appropriate legislation.

AMENDMENT XXV [1967]

[Presidential disability and succession]
SECTION 1   In case of the removal of the President from office or his death or resignation, the Vice President shall become President.

SECTION 2   Whenever there is a vacancy in the office of the Vice President, the President shall nominate a Vice President who shall take the Office upon confirmation by a majority vote of both houses of Congress.

SECTION 3   Whenever the President transmits to the President pro tempore of the Senate and the Speaker of the House of Representatives his written declaration that he is unable to discharge the powers and duties of his office, and until he transmits to them a written declaration to the contrary, such powers and duties shall be discharged by the Vice President as Acting President.

SECTION 4 Whenever the Vice President and a majority of either the principal officers of the executive departments, or of such other body as Congress may by law provide, transmit to the President pro tempore of the Senate and the Speaker of the House of Representatives their written declaration that the President is unable to discharge the powers and duties of his office, the Vice President shall immediately assume the powers and duties of the office as Acting President.

Thereafter, when the President transmits to the President pro tempore of the Senate and the Speaker of the House of Representatives his written declaration that no inability exists, he shall resume the powers and duties of his office unless the Vice President and a majority of either the principal officers of the executive department, or of such other body as Congress may by law provide, transmit within four days to the President pro tempore of the Senate and the Speaker of the House of Representatives their written declaration that the President is unable to discharge the powers and duties of his office. Thereupon Congress shall decide the issue, assembling within 48 hours for that purpose if not in session. If the Congress, within 21 days after receipt of the latter written declaration, or, if Congress is not in session, within 21 days after Congress is required to assemble, deter-mines by two-thirds vote of both houses that the President is unable to discharge the powers and duties of his office, the Vice President shall continue to discharge the same as Acting President; otherwise, the President shall resume the powers and duties of his office.

AMENDMENT XXVI [1971]

[*Voting rights and age*]
SECTION 1 The right of citizens of the United States, who are eighteen years of age, or older, to vote shall not be denied or abridged by the United States or by any state on account of age.

SECTION 2 The Congress shall have the power to enforce this article by appropriate legislation.

AMENDMENT XXVII [1992]

[*Congressional pay raises*]
No law varying the compensation for the services of the Senators and Representatives shall take effect, until an election of Representatives shall have intervened.

From

The Federalist,

Nos. 10 and 51

FEDERALIST NO. 10 [1787]

To the People of the State of New York: Among the numerous advantages promised by a well-constructed union, none deserves to be more accurately developed than its tendency to break and control the violence of faction. The friend of popular governments, never finds himself so much alarmed for their character and fate, as when he contemplates their propensity to this dangerous vice. He will not fail, therefore, to set a due value on any plan which, without violating the principles to which he is attached, provides a proper cure for it. The instability, injustice, and confusion introduced into the public councils, have, in truth, been the mortal diseases under which popular governments have everywhere perished; as they continue to be the favourite and fruitful topics from which the adversaries to liberty derive their most specious declamations. The valuable improvements made by the American constitutions on the popular models, both ancient and modern, cannot certainly be too much admired; but it would be an unwarrantable partiality, to contend that they have as effectually obviated the danger on this side, as was wished and expected. Complaints are everywhere heard from our most considerate and virtuous citizens, equally the friends of public and private faith, and of public and personal liberty, that our governments are too unstable; that the public good is disregarded in the conflicts of rival parties; and that measures are too often decided, not according to the rules of justice, and the rights of the minor party, but by the superior force of an interested and overbearing majority. However anxiously we may wish that these complaints had no foundation, the evidence of known facts will not permit us to deny that they are in some degree true. It will be found, indeed, on a candid review of our situation, that some of the distresses under which we labour have been erroneously charged on the operation of our governments; but it will be found, at the same time, that other causes will not alone account for many of our heaviest misfortunes; and, particularly, for that prevailing and increasing distrust of public engagements, and

alarm for private rights, which are echoed from one end of the continent to the other. These must be chiefly, if not wholly, effects of the unsteadiness and injustice, with which a factious spirit has tainted our public administrations.

By a faction, I understand a number of citizens, whether amounting to a majority or minority of the whole, who are united and actuated by some common impulse of passion, or of interest, adverse to the rights of other citizens, or to the permanent and aggregate interests of the community.

There are two methods of curing the mischiefs of faction: The one, by removing its causes; the other, by controlling its effects.

There are again two methods of removing the causes of faction: The one, by destroying the liberty which is essential to its existence; the other, by giving to every citizen the same opinions, the same passions, and the same interests.

It could never be more truly said, than of the first remedy, that it was worse than the disease. Liberty is to faction what air is to fire, an ailment without which it instantly expires. But it could not be a less folly to abolish liberty, which is essential to political life, because it nourishes faction, than it would be to wish the annihilation of air, which is essential to animal life, because it imparts to fire its destructive agency.

The second expedient is as impracticable, as the first would be unwise. As long as the reason of man continues fallible, and he is at liberty to exercise it, different opinions will be formed. As long as the connection subsists between his reason and his self-love, his opinions and his passions will have a reciprocal influence on each other; and the former will be objects to which the latter will attach themselves. The diversity in the faculties of men, from which the rights of property originate, is not less an insuperable obstacle to a uniformity of interests. The protection of these faculties is the first object of government. From the protection of different and unequal faculties of acquiring property, the possession of different degrees and kinds of property immediately results; and from the influence of these on the sentiments and views of the respective proprietors, ensues a division of the society into different interests and parties.

The latent causes of action are thus sown in the nature of man; and we see them everywhere brought into different degrees of activity, according to the different circumstances of civil society. A zeal for different opinions concerning religion, concerning government, and many other points, as well of speculation as of practice; an attachment to different leaders ambitiously contending for preeminence and power; or to persons of other descriptions whose fortunes have been interesting to the human passions, have, in turn, divided mankind into parties, inflamed them with mutual animosity, and rendered them much more disposed to vex and oppress each other, than to cooperate for their common good. So strong is this propensity of mankind, to fall into mutual animosities, that where no substantial occasion presents itself, the most frivolous and fanciful distinctions have been sufficient to kindle their unfriendly passions and excite their most violent conflicts. But the most common and durable source of factions, has been the various and unequal distribution of property. Those who hold, and those who are without property, have ever formed distinct interests in society. Those who are creditors, and those who are debtors, fall under alike discrimination. A landed interest, a manufacturing interest, a mercantile interest, a moneyed interest, with many lesser interests, grow up of necessity in civilized nations, and divide them into different classes, actuated by different sentiments and views. The regulation of these various and interfering interests forms the principal task of modern legislation, and involves the spirit of the party and faction in the necessary and ordinary operations of the government.

No man is allowed to be a judge in his own cause; because his interest will certainly bias his judgment, and, not improbably, corrupt his integrity. With equal, nay, with greater reason, a body of men are unfit to be both judges and parties at the same time; yet what are many of the most important acts of legislation, but so many judicial determinations, not indeed concerning the right of single persons, but concerning the rights of large bodies of citizens? And what are the different classes of legislators, but advocates and parties to the causes which they determine? Is a law proposed concerning private debts? It is a question to which the creditors are parties on one side, and the debtors on the other. Justice ought to hold the balance between them. Yet the parties are, and must be, themselves the judges; and the most numerous party, or, in other words, the most powerful faction, must be expected to prevail. Shall domestic manufactures be encouraged, and in what degree, by restrictions on foreign manufactures? are questions which would be differently decided by the landed and the manufacturing classes; and probably by neither with a sole regard to justice and the public good. The apportionment of taxes, on the various descriptions of property, is an act which seems to require the most exact impartiality; yet there is, perhaps, no legislative act, in which greater opportunity and temptation are given to a predominant party to trample on the rules of justice. Every shilling,

with which they overburden the inferior number, is a shilling saved to their own pockets.

It is in vain to say, that enlightened statesmen will be able to adjust these clashing interests, and render them all subservient to the public good. Enlightened statesmen will not always be at the helm: nor, in many cases, can such an adjustment be made at all, without taking into view indirect and remote considerations, which will rarely prevail over the immediate interest which one party may find in disregarding the rights of another, or the good of the whole.

The inference to which we are brought is, that the *causes* of faction cannot be removed; and that relief is only to be sought in the means of controlling its *effects*.

If a faction consists of less than a majority, relief is supplied by the republican principle, which enables the majority to defeat its sinister views, by regular vote. It may clog the administration, it may convulse the society; but it will be unable to execute and mask its violence under the forms of the constitution. When a majority is included in a faction, the form of popular government, on the other hand, enables it to sacrifice to its ruling passion or interest, both the public good and the rights of other citizens. To secure the public good, and private rights, against the danger of such a faction, and at the same time to preserve the spirit and the form of popular government, is then the great object to which our inquiries are directed. Let me add, that it is the great desideratum, by which alone this form of government can be rescued from the opprobrium under which it has so long laboured, and be recommended to the esteem and adoption of mankind.

By what means is this object attainable? Evidently by one of two only. Either the existence of the same passion or interest in a majority, at the same time, must be prevented; or the majority, having such coexistent passion or interest, must be rendered, by their number and local situation, unable to concert and carry into effect schemes of oppression. If the impulse and the opportunity be suffered to coincide, we well know that neither moral nor religious motives can be relied on as an adequate control. They are not found to be such on the injustice and violence of individuals, and lose their efficacy in proportion to the number combined together; that is, in proportion as their efficacy becomes needful.

From this view of the subject, it may be concluded, that a pure democracy, by which I mean a society consisting of a small number of citizens, who assemble and administer the government in person, can admit of no cure for the mischiefs of faction. A common passion or interest will, in almost every case, be felt by a majority of the whole; a communication and concert, results from the form of government itself; and there is nothing to check the inducements to sacrifice the weaker party, or an obnoxious individual. Hence, it is, that such democracies have ever been spectacles of turbulence and contention; have ever been found incompatible with personal security, or the rights of property; and have in general been as short in their lives, as they have been violent in their deaths. Theoretic politicians, who have patronized this species of government, have erroneously supposed, that by reducing mankind to a perfect equality in their political rights, they would, at the same time, be perfectly equalized and assimilated in their possessions, their opinions, and their passions.

A republic, by which I mean a government in which the scheme of representation takes place, opens a different prospect, and promises the cure for which we are seeking. Let us examine the points in which it varies from pure democracy, and we shall comprehend both the nature of the cure and the efficacy which it must derive from the union.

The two great points of difference, between a democracy and a republic, are, first, the delegation of the government, in the latter, to a small number of citizens, elected by the rest; secondly, the greater number of citizens, and greater sphere of country, over which the latter may be extended.

The effect of the first difference is, on the one hand, to refine and enlarge the public views, by passing them through the medium of a chosen body of citizens, whose wisdom may best discern the true interest of their country, and whose patriotism and love of justice, will be least likely to sacrifice it to temporary or partial considerations. Under such a regulation, it may well happen, that the public voice, pronounced by the representatives of the people, will be more consonant to the public good, than if pronounced by the people themselves, convened for the purpose. On the other hand the effect may be inverted. Men of factious tempers, of local prejudices, or of sinister designs, may by intrigue, by corruption, or by other means, first obtain the suffrages, and then betray the interest of the people. The question resulting is, whether small or extensive republics are most favourable to the election of proper guardians of the public weal; and it is clearly decided in favour of the latter by two obvious considerations.

In the first place, it is to be remarked that, however small the republic may be, the representatives must be raised to a certain number, in order to guard against the cabals of a few; and that however large it may be, they must be limited to a certain number, in order to guard against the confusion of a multitude. Hence, the number of representatives in the two cases not being in proportion to that of the constituents, and being proportionally greatest in the small republic, it follows, that if the proportion of fit characters be not less in the large than in the small republic, the former will present a greater option, and consequently a greater probability of a fit choice.

In the next place, as each representative will be chosen by a greater number of citizens in the large than in the small republic, it will be more difficult for unworthy candidates to practise with success the vicious arts, by which elections are too often carried; and the suffrages of the people being more free, will be more likely to centre in men who possess the most attractive merit, and the most diffusive and established characters.

It must be confessed, that in this, as in most other cases, there is a mean, on both sides of which inconveniences will be found to lie. By enlarging too much the number of electors, you render the representatives too little acquainted with all their local circumstances and lesser interests; as by reducing it too much, you render him unduly attached to these, and too little fit to comprehend and pursue great and national objects. The federal constitution forms a happy combination being referred to the national, the local and particular, to the state legislatures.

The other point of difference is, the greater number of citizens, and extent of territory, which may be brought within the compass of republican, than of democratic government;

and it is this circumstance principally which renders factious combinations less to be dreaded in the former, than in the latter. The smaller the society, the fewer probably will be the distinct parties and interests composing it; the fewer the distinct parties and interests, the more frequently will a majority be found of the same party; and the smaller the number of individuals composing a majority, and the smaller the compass within which they are placed, the more easily will they concert and execute their plans of oppression. Extend the sphere, and you take in a greater variety of parties and interests; you make it less probable that a majority of the whole will have a common motive to invade the rights of other citizens; or if such a common motive exists, it will be more difficult for all who feel it to discover their own strength, and to act in unison with each other. Besides other impediments, it may be remarked, that where there is a consciousness of unjust or dishonourable purposes, communication is always checked by distrust, in proportion to the number whose concurrence is necessary.

Hence, it clearly appears, that the same advantage, which a republic has over a democracy, in controlling the effects of faction, is enjoyed by a large over a small republic,—is enjoyed by the union over the states composing it. Does this advantage consist in the substitution of representatives, whose enlightened views and virtuous sentiments render them superior to local prejudices, and to schemes of injustice? It will not be denied that the representation of the union will be most likely to possess these requisite endowments. Does it consist in the greater security afforded by a greater variety of parties, against the event of any one party being able to outnumber and oppress the rest? In an equal degree does the increased variety of parties, comprised within the union, increase the security? Does it, in fine, consist in the greater obstacles opposed to the concert and accomplishment of the secret wishes of an unjust and interested majority? Here, again, the extent of the union gives it the most palpable advantage.

The influence of factious leaders may kindle a flame within their particular states, but will be unable to spread a general conflagration through the other states; a religious sect may degenerate into a political faction in a part of the confederacy; but the variety of sects dispersed over the entire face of it, must secure the national councils against any danger from that source: a rage for paper money, for an abolition of debts, for an equal division of property, or for any other improper or wicked project, will be less apt to pervade the whole body of the union than a particular member of it; in the same proportion as such a malady is more likely to taint a particular county or district, than an entire state.

In the extent and proper structure of the union, therefore, we behold a republican remedy for the diseases most incident to republican government. And according to the degree of pleasure and pride we feel in being republicans, ought to be our zeal in cherishing the spirit, and supporting the character of federalists.

James Madison

FEDERALIST NO. 51 [1788]

To the People of the State of New York: To what expedient then shall we finally resort for maintaining in practice the necessary partition of power among the several departments, as laid down in the constitution? The only answer that can be given is, that as all these exterior provisions are found to be inadequate, the defect must be supplied, by so contriving the interior structure of the government, as that its several constituent parts may, by their mutual relations, be the means of keeping each other in their proper places. Without presuming to undertake a full development of this important idea, I will hazard a few general observations, which may perhaps place it in a clearer light, and enable us to form a more correct judgment of the principles and structure of the government planned by the convention.

In order to lay a due foundation for that separate and distinct exercise of the different powers of government, which to a certain extent, is admitted on all hands to be essential to the preservation of liberty, it is evident that each department should have a will of its own; and consequently should be so constituted, that the members of each should have as little agency as possible in the appointment of the members of the others. Were this principle rigorously adhered to, it would require that all the appointments for the supreme executive, legislative, and judiciary magistracies, should be drawn from the same fountain of authority, the people, through channels, having no communication whatever with one another. Perhaps such a plan of constructing the several departments would be less difficult in practice than it may in contemplation appear. Some difficulties however, and some additional expense, would attend the execution of it. Some deviations therefore from the principle must be admitted. In the constitution of the judiciary department in particular, it might be inexpedient to insist rigorously on the principle; first, because peculiar qualifications being essential in the members, the primary consideration ought to be to select that mode of choice, which best secures these qualifications; secondly, because the permanent tenure by which the appointments are held in that department, must soon destroy all sense of dependence on the authority conferring them.

It is equally evident that the members of each department should be as little dependent as possible on those of the others, for the emoluments annexed to their offices. Were the executive magistrate, or the judges, not independent of the legislature in this particular, their independence in every other would be merely nominal.

But the great security against a gradual concentration of the several powers in the same department, consists in giving to those who administer each department, the necessary constitutional means, and personal motives, to resist encroachments of the others. The provision for defense must in this, as in all other cases, be made commensurate to the danger of attack. Ambition must be made to counteract ambition. The interest of the man must be connected with the constitutional rights of the place. It may be a reflection on human nature, that such devices should be necessary to con-trol the abuses of government: But what is government itself but the greatest of all reflections on human nature? If men were angels, no government would be necessary. If angels were to govern men, neither external nor internal controls on government would be necessary. In framing a government which is to be administered by men over men, the great difficulty lies in this: You must first enable the government to control the governed; and in the next place, oblige it to control itself. A dependence on the people is no doubt the primary control on the government; but experience has taught mankind the necessity of auxiliary precautions.

This policy of supplying by opposite and rival interests, the defect of better motives, might be traced through the whole system of human affairs, private as well as public. We see it particularly displayed in all the subordinate distributions of power; where the constant aim is to divide and arrange the several offices in such a manner as that each may be a check on the other; that the private interest of every individual, may be a sentinel over the public rights. These inventions of prudence cannot be less requisite in the distribution of the supreme powers of the state.

But it is not possible to give to each department an equal power of self defense. In republican government the legislative authority, necessarily, predominates. The remedy for this inconveniency is, to divide the legislature into different branches; and to render them by different modes of election, and different principles of action, as little connected with each other, as the nature of their common functions, and their common dependence on the society, will admit. It may even be necessary to guard against dangerous encroachments by still further precautions. As the weight of the legislative authority requires that it should be thus divided, the weakness of the executive may require, on the other hand, that it should be fortified. An absolute negative, on the legislature, appears at first view to be the natural defense with which the executive magistrate should be armed. But perhaps it would be neither altogether safe, nor alone sufficient. On ordinary occasions, it might not be exerted with the requisite firmness; and on extraordinary occasions, it might be perfidiously abused. May not this defect of an absolute negative be supplied, by some qualified connection between this weaker department, and the weaker branch of the stronger department, by which the latter may be led to support the constitutional rights of the former, without being too much detached from the rights of its own department?

If the principles on which these observations are founded be just, as I persuade myself they are, and they be applied as a criterion, to the several state constitutions, and to the federal constitution, it will be found, that if the latter does not perfectly correspond with them, the former are infinitely less able to bear such a test.

There are moreover two considerations particularly applicable to the federal system of America, which place that system in a very interesting point of view.

First. In a single republic, all the power surrendered by the people, is submitted to the administration of a single government; and usurpations are guarded against by a division of the government into distinct and separate departments. In the compound republic of America, the power surrendered by the people, is first divided between two distinct governments, and then the portion allotted to each, subdivided among distinct and separate departments. Hence a double security arises to the rights of the people. The different governments will control each other; at the same time that each will be controlled by itself.

Second. It is of great importance in a republic, not only to guard the society against the oppression of its rulers; but to guard one part of the society against the injustice of the other part. Different interests necessarily exist in different classes of citizens. If a majority be united by a common interest, the rights of the minority will be insecure. There are but two methods of providing against this evil: The one by creating a will in the community independent of the majority, that is, of the society itself; the other by comprehending in the society so many separate descriptions of citizens, as will render an unjust combination of a majority of the whole, very improbable, if not impracticable. The first method prevails in all governments possessing an hereditary or self appointed authority. This at best is but a precarious security; because a power independent of the society may as well espouse the unjust views of the major, as the rightful interests, of the minor party, and may possibly be turned against both parties. The second method will be exemplified in the federal republic of the United States. While all authority in it will be derived from and dependent on the society, the society itself will be broken into so many parts, interests and classes of citizens, that the rights of individuals or of the minority, will be in little danger from interested combinations of the majority. In a free government, the security for civil rights must be the same as for religious rights. It consists in the one case in the multiplicity of sects. The degree of security in both cases will depend on the number of interests and sects; and this may be presumed to depend on the extent of country and number of people comprehended under the same government. This view of the subject must particularly recommend a proper federal system to all the sincere and considerate friends of republican government: Since it shows that in exact proportion as the territory of the union may be formed into more circumscribed confederacies or states, oppressive combinations of a majority will be facilitated; the best security under the republican form, for the rights of every class of citizens, will be diminished; and consequently, the stability and independence of some member of the government, the only other security, must be proportionally increased. Justice is the end of government. It is the end of civil society. It ever has been, and ever will be pursued, until it be obtained, or until liberty be lost in the pursuit. In a society under the forms of which the stronger faction can readily unite and oppress the weaker, anarchy may as truly be said to reign, as in a state of nature where the weaker individual is not secured against the violence of the stronger: And as in the latter state even the stronger individuals are prompted by the uncertainty of their condition, to submit to a government which may protect the weak as well as themselves: So in the former state, will the more powerful factions or parties be gradually induced by alike motives, to wish for a government which will protect all parties, the weaker as well as the more powerful. It can be little doubted, that if the state of Rhode Island was separated from the confederacy, and left to itself, the insecurity of rights under the popular form of government within such narrow limits, would be displayed by such reiterated oppressions of factious majorities, that some power altogether independent of the people would soon be called for by the voice of the very factions whose misrule had proved the necessity of it. In the extended republic of the United States, and among the great variety of interests, parties and sects which it embraces, a coalition of a majority of the whole society could seldom take place on any other principles than those of justice and the general good; and there being thus less danger to a minor from the will of the major party, there must be less pretext also, to provide for the security of the former, by introducing into the government a will not dependent on the latter; or in other words, a will independent of the society itself. It is no less certain than it is important, notwithstanding the contrary opinions which have been entertained, that the larger the society, provided it lie within a practicable sphere, the more duly capable it will be of self government. And happily for the *republican cause,* the practicable sphere may be carried to a very great extent, by a judicious modification and mixture of the *federal principle.*

James Madison

Presidential

Elections,

1789–1992

CANDIDATES	PARTY	ELECTORAL VOTES
1789		
George Washington	Federalist	69
John Adams	Federalist	34
Others		35
1792		
George Washington	Federalist	132
John Adams	Federalist	77
George Clinton		50
Others		5
1796		
John Adams	Federalist	71
Thomas Jefferson	Democratic-Republican	68
Thomas Pinckney	Federalist	59
Aaron Burr	Democratic-Republican	30
Others		48
1800		
Thomas Jefferson[1]	Democratic-Republican	73
Aaron Burr	Democratic-Republican	73
John Adams	Federalist	65
Charles C. Pinckney		64
1804		
Thomas Jefferson	Democratic-Republican	162
Charles C. Pinckney	Federalist	14
1808		
James Madison	Democratic-Republican	122
Charles C. Pinckney	Federalist	47
George Clinton	Independent-Republican	6
1812		
James Madison	Democratic-Republican	122
DeWitt Clinton	Federalist	89

CANDIDATES	PARTY	ELECTORAL VOTES
1816		
James Monroe	Democratic-Republican	183
Rufus King	Federalist	34
1820		
James Monroe	Democratic-Republican	231
John Quincy Adams	Independent-Republican	1
1824		
John Quincy Adams[1]	Democratic-Republican	84
Andrew Jackson	Democratic-Republican	99
Henry Clay	Democratic-Republican	37
William H. Crawford	Democratic-Republican	41
1828		
Andrew Jackson	Democratic	178
John Quincy Adams	National-Republican	83
1832		
Andrew Jackson	Democratic	219
Henry Clay	National-Republican	49
William Wirt	Anti-Masonic	7
John Floyd	National-Republican	11
1836		
Martin Van Buren	Democratic	170
William H. Harrison	Whig	73
Hugh L. White	Whig	26
Daniel Webster	Whig	14
1840		
William H. Harrison[2]	Whig	234
(John Tyler)	Whig	—
Martin Van Buren	Democratic	60

Candidates	Party	Electoral Votes
1844		
James K. Polk	Democratic	170
Henry Clay	Whig	105
James G. Birney	Liberty	—
1848		
Zachary Taylor[2]	Whig	163
(Millard Fillmore)	Whig	—
Lewis Cass	Democratic	127
Martin Van Buren	Free Soil	—
1852		
Franklin Pierce	Democratic	254
Winfield Scott	Whig	42
1856		
James Buchanan	Democratic	174
John C. Fremont	Republican	114
Millard Fillmore	American	8
1860		
Abraham Lincoln	Republican	180
Stephen A. Douglas	Democratic	12
John C. Breckinridge	Democratic	72
John Bell	Constitutional Union	39
1864		
Abraham Lincoln[2]	Republican	212
(Andrew Johnson)	Republican	—
George B. McClellan	Democratic	21
1868		
Ulysses S. Grant	Republican	214
Horatio Seymour	Democratic	80
1872		
Ulysses S. Grant	Republican	286
Horace Greeley	Democratic	66
1876		
Rutherford B. Hayes	Republican	185
Samuel J. Tilden	Democratic	184
1880		
James A. Garfield[2]	Republican	214
(Chester A. Arthur)	Republican	—
Winfield S. Hancock	Democratic	155
James B. Weaver	Greenback-Labor	—
1884		
Grover Cleveland	Democratic	219
James G. Blaine	Republican	182
Benjamin F. Butler	Greenback-Labor	—
1888		
Benjamin Harrison	Republican	233
Grover Cleveland	Democratic	168
1892		
Grover Cleveland	Democratic	277
Benjamin Harrison	Republican	145
James R. Weaver	People's	22

Candidates	Party	Electoral Votes
1896		
William McKinley	Republican	271
William J. Bryan	Democratic, Populist	176
1900		
William McKinley[2]	Republican	292
(Theodore Roosevelt)	Republican	—
William J. Bryan	Democratic, Populist	155
1904		
Theodore Roosevelt	Republican	336
Alton B. Parker	Democratic	140
Eugene V. Debs	Socialist	—
1908		
William H. Taft	Republican	321
William J. Bryan	Democratic	162
Eugene V. Debs	Socialist	—
1912		
Woodrow Wilson	Democratic	435
Theodore Roosevelt	Progressive	88
William H. Taft	Republican	8
Eugene V. Debs	Socialist	—
1916		
Woodrow Wilson	Democratic	277
Charles E. Hughes	Republican	254
1920		
Warren G. Harding[2]	Republican	404
(Calvin Coolidge)	Republican	—
James M. Cox	Democratic	127
Eugene V. Debs	Socialist	—
1924		
Calvin Coolidge	Republican	382
John W. Davis	Democratic	136
Robert M. LaFollette	Progressive	13
1928		
Herbert C. Hoover	Republican	444
Alfred E. Smith	Democratic	87
1932		
Franklin D. Roosevelt	Democratic	472
Herbert C. Hoover	Republican	59
Norman Thomas	Socialist	—
1936		
Franklin D. Roosevelt	Democratic	523
Alfred M. Landon	Republican	8
William Lemke	Union	—
1940		
Franklin D. Roosevelt	Democratic	449
Wendell L. Willkie	Republican	82
1944		
Franklin D. Roosevelt[2]	Democratic	432
(Harry S Truman)	Democratic	—
Thomas E. Dewey	Republican	99

Candidates	Party	Electoral Votes
1948		
Harry S Truman	Democratic	303
Thomas E. Dewey	Republican	189
J. Strom Thurmond	States' Rights	39
Henry A. Wallace	Progressive	—
1952		
Dwight D. Eisenhower	Republican	442
Adlai E. Stevenson	Democratic	89
1956		
Dwight D. Eisenhower	Republican	457
Adlai E. Stevenson	Democratic	73
1960		
John F. Kennedy[2]	Democratic	303
(Lyndon B. Johnson)	Democratic	—
Richard M. Nixon	Republican	219
1964		
Lyndon B. Johnson	Democratic	486
Barry M. Goldwater	Republican	52
1968		
Richard M. Nixon	Republican	301
Hubert H. Humphrey	Democratic	191
George C. Wallace	American Independent	46

Candidates	Party	Electoral Votes
1972		
Richard M. Nixon[3]	Republican	520
(Gerald R. Ford)	Republican	—
George S. McGovern	Democratic	17
1976		
Jimmy Carter	Democratic	297
Gerald R. Ford	Republican	240
1980		
Ronald Reagan	Republican	489
Jimmy Carter	Democratic	49
John Anderson	Independent	—
1984		
Ronald Reagan	Republican	525
Walter Mondale	Democratic	13
1988		
George Bush	Republican	426
Michael Dukakis	Democratic	111
1992		
Bill Clinton	Democratic	370
George Bush	Republican	168
Ross Perot	Independent	—

[1] Elected by the House of Representatives.
[2] Died while in office.
[3] Resigned from office.

Party Control

of Congress,

1901–1995

	SENATE			HOUSE			
	DEM.	**REP.**	**OTHER**	**DEM.**	**REP.**	**OTHER**	**PRESIDENT**
57th Congress, 1901–1903	31	55	4	151	197	9	McKinley
							T. Roosevelt
58th Congress, 1903–1905	33	57	—	178	208	—	T. Roosevelt
59th Congress, 1905–1907	33	57	—	136	250	—	T. Roosevelt
60th Congress, 1907–1909	31	61	—	164	222	—	T. Roosevelt
61st Congress, 1909–1911	32	61	—	172	219	—	Taft
62nd Congress, 1911–1913	41	51	—	228	161	1	Taft
63rd Congress, 1913–1915	51	44	1	291	127	17	Wilson
64th Congress, 1915–1917	56	40	—	230	196	9	Wilson
65th Congress, 1917–1919	53	42	—	216	210	6	Wilson
66th Congress, 1919–1921	47	49	—	190	240	3	Wilson
67th Congress, 1921–1923	37	59	—	131	301	1	Harding
68th Congress, 1923–1925	43	51	2	205	225	5	Coolidge
69th Congress, 1925–1927	39	56	1	183	247	4	Coolidge
70th Congress, 1927–1929	46	49	1	195	237	3	Coolidge
71st Congress, 1929–1931	39	56	1	167	267	1	Hoover
72nd Congress, 1931–1933	47	48	1	220	214	1	Hoover
73rd Congress, 1933–1935	60	35	1	319	117	5	F. Roosevelt
74th Congress, 1935–1937	69	25	2	319	103	10	F. Roosevelt
75th Congress, 1937–1939	76	16	4	331	89	13	F. Roosevelt
76th Congress, 1939–1941	69	23	4	261	164	4	F. Roosevelt
77th Congress, 1941–1943	66	28	2	268	162	5	F. Roosevelt
78th Congress, 1943–1945	58	37	1	218	208	4	F. Roosevelt
79th Congress, 1945–1947	56	38	1	242	190	2	Truman
80th Congress, 1947–1949	45	51	—	188	245	1	Truman
81st Congress, 1949–1951	54	42	—	263	171	1	Truman
82nd Congress, 1951–1953	49	47	—	234	199	1	Truman
83rd Congress, 1953–1955	47	48	1	211	221	—	Eisenhower
84th Congress, 1955–1957	48	47	1	232	203	—	Eisenhower
85th Congress, 1957–1959	49	47	—	233	200	—	Eisenhower
86th Congress, 1959–1961	65	35	—	284	153	—	Eisenhower

	SENATE			HOUSE			
	DEM.	REP.	OTHER	DEM.	REP.	OTHER	PRESIDENT
87th Congress, 1961–1963	65	35	—	263	174	—	Kennedy
88th Congress, 1963–1965	67	33	—	258	177	—	Kennedy Johnson
89th Congress, 1965–1967	68	32	—	295	140	—	Johnson
90th Congress, 1967–1969	64	36	—	247	187	—	Johnson
91st Congress, 1969–1971	57	43	—	243	192	—	Nixon
92nd Congress, 1971–1973	54	44	2	254	180	—	Nixon
93rd Congress, 1973–1975	56	42	2	239	192	1	Nixon Ford
94th Congress, 1975–1977	60	37	2	291	144	—	Ford
95th Congress, 1977–1979	61	38	1	292	143	—	Carter
96th Congress, 1979–1981	58	41	1	276	157	—	Carter
97th Congress, 1981–1983	46	53	1	243	192	—	Reagan
98th Congress, 1983–1985	45	55	—	267	168	—	Reagan
99th Congress, 1985–1987	47	53	—	252	183	—	Reagan
100th Congress, 1987–1989	54	46	—	257	178	—	Reagan
101st Congress, 1989–1991	55	45	—	262	173	—	Bush
102nd Congress, 1991–1993	56	44	—	276	167	—	Bush
103rd Congress, 1993–1995	57	43	—	258	176	1	Clinton
104th Congress, 1995–1997[1]	47	53	—	204	230	1	Clinton

[1]Numbers indicate initial composition of Congress.

Sources: Department of Commerce, Bureau of the Census, *Statistical Abstracts of the United States* (Washington, D.C.: Government Printing Office); and *Members of Congress since 1789,* 2nd ed. (Washington, D.C.: Congressional Quarterly Press, 1981), pp. 176–77.

United States

Supreme Court

Justices, 1789–1995

JUSTICE[1]	PRESIDENT	YEARS OF SERVICE
John Jay	Washington	1789–1795
John Rutledge	Washington	(1789–1791)[2]
William Cushing	Washington	1789–1810
James Wilson	Washington	1789–1798
John Blair Jr.	Washington	1789–1796
James Iredell	Washington	1790–1799
Thomas Johnson	Washington	1791–1793
William Paterson	Washington	1793–1806
John Rutledge	Washington	(1795)[2]
Samuel Chase	Washington	1796–1811
Oliver Elsworth	Washington	1796–1800
Bushrod Washington	J. Adams	1798–1829
Alfred Moore	J. Adams	1799–1804
John Marshall	J. Adams	1801–1835
William Johnson	Jefferson	1804–1834
Henry B. Livingston	Jefferson	1806–1823
Thomas Todd	Jefferson	1807–1826
Gabriel Duval	Madison	1811–1835
Joseph Story	Madison	1811–1845
Smith Thompson	Monroe	1823–1843
Robert Trimble	J. Q. Adams	1826–1828
John McLean	Jackson	1829–1861
Henry Baldwin	Jackson	1830–1844
James M. Wayne	Jackson	1835–1867
Roger B. Taney	Jackson	1836–1864
Philip P. Barbour	Jackson	1836–1841
John Catron	Jackson	1837–1865
John McKinley	Van Buren	1837–1852
Peter V. Daniel	Van Buren	1841–1860
Samuel Nelson	Tyler	1845–1872
Levi Woodbury	Polk	1846–1851
Robert C. Grier	Polk	1846–1870
Benjamin R. Curtis	Fillmore	1851–1857

JUSTICE[1]	PRESIDENT	YEARS OF SERVICE
John A. Campbell	Pierce	1853–1861
Nathan Clifford	Buchanan	1858–1881
Noah H. Swayne	Lincoln	1862–1881
Samuel F. Miller	Lincoln	1862–1890
David Davis	Lincoln	1862–1877
Stephen J. Field	Lincoln	1863–1897
Salmon P. Chase	Lincoln	1864–1873
William Strong	Grant	1870–1880
Joseph P. Bradley	Grant	1870–1892
Ward Hunt	Grant	1872–1882
Morrison R. Waite	Grant	1874–1888
John M. Harlan	Hayes	1877–1911
William B. Woods	Hayes	1880–1887
Stanley Matthews	Garfield	1881–1889
Horace Gray	Arthur	1881–1902
Samuel Blatchford	Arthur	1882–1893
Lucius Q. C. Lamar	Cleveland	1888–1893
Melville W. Fuller	Cleveland	1888–1910
David J. Brewer	Harrison	1889–1910
Henry B. Brown	Harrison	1890–1906
George Shiras Jr.	Harrison	1892–1903
Howell E. Jackson	Harrison	1893–1895
Edward D. White	Cleveland	1894–1910
Rufus W. Peckham	Cleveland	1895–1909
Joseph McKenna	McKinley	1898–1925
Oliver W. Holmes Jr.	T. Roosevelt	
William R. Day	T. Roosevelt	1903–1922
William H. Moody	T. Roosevelt	1906–1910
Horace H. Lurton	Taft	1909–1914
Charles E. Hughes	Taft	1910–1916
Edward D. White	Taft	1910–1921

Justice[1]	President	Years of Service	Justice[1]	President	Years of Service
Willis Van Devanter	Taft	1910–1937	**Earl Warren**	Eisenhower	1953–1969
Joseph R. Lamar	Taft	1910–1916	John M. Harlan	Eisenhower	1955–1971
Mahlon Pitney	Taft	1912–1922	William J. Brennan Jr.	Eisenhower	1956–1990
James C. McReynolds	Wilson	1914–1941	Charles E. Whittaker	Eisenhower	1957–1962
Louis D. Brandeis	Wilson	1916–1939	Potter Stewart	Eisenhower	1958–1981
John H. Clarke	Wilson	1916–1922	Byron R. White	Kennedy	1962–1993
William H. Taft	Harding	1921–1930	Arthur J. Goldberg	Johnson	1962–1965
George Sutherland	Harding	1922–1938	Abe Fortas	Johnson	1965–1969
Pierce Butler	Harding	1922–1939	Thurgood Marshall	Johnson	1967–1991
Edward T. Sanford	Harding	1923–1930	**Warren E. Burger**	Nixon	1969–1986
Harlan F. Stone	Coolidge	1925–1941	Harry A. Blackmun	Nixon	1970–1994
Charles E. Hughes	Hoover	1930–1941	Lewis F. Powell Jr.	Nixon	1972–1988
Owen J. Roberts	Hoover	1930–1945	William H. Rehnquist	Nixon	1972–1986
Benjamin N. Cardozo	Hoover	1932–1938	John Paul Stevens	Ford	1975–
Hugo Black	F. Roosevelt	1937–1971	Sandra Day O'Connor	Reagan	1981–
Stanley F. Reed	F. Roosevelt	1938–1957	**William H. Rehnquist**	Reagan	1986–
Felix Frankfurter	F. Roosevelt	1939–1962	Antonin Scalia	Reagan	1986–
William O. Douglas	F. Roosevelt	1939–1975	Anthony M. Kennedy	Reagan	1988–
Frank Murphy	F. Roosevelt	1940–1949	David H. Souter	Bush	1990–
James F. Byrnes	F. Roosevelt	1941–1942	Clarence Thomas	Bush	1991–
Harlan F. Stone	F. Roosevelt	1941–1946	Ruth Bader Ginsburg	Clinton	1993–
Robert H. Jackson	F. Roosevelt	1941–1954	Stephen G. Breyer	Clinton	1994–
Wiley B. Rutledge	F. Roosevelt	1943–1949			
Harold H. Burton	Truman	1945–1958			
Fred M. Vinson	Truman	1946–1953			
Tom C. Clark	Truman	1949–1967			
Sherman Minton	Truman	1949–1956			

[1]Bold type indicates chief justice.
[2]Rutledge resigned after his confirmation to become chief justice of South Carolina; in 1795 he served during a Court recess.

GLOSSARY

activists People who are seriously and passionately involved in politics by protesting, participating in campaigns, or contacting officials. (Ch. 11)

advisory committee A permanent group of advisers who help a government agency make technical decisions beyond the scope of its employees' expertise. (Ch. 15)

age cohort A group with common characteristics that is identifiable by age. (Ch. 7)

agenda A plan of content that determines the order in which items are considered and the manner in which they are formulated and presented. (Ch. 1)

agenda setting The process by which issues prominently featured in the media become the issues citizens regard as important. (Ch. 8)

agents of socialization The people and institutions imparting knowledge about politics, politicians, and policies. (Ch. 7)

amicus curiae brief "friend of the court" brief or argument filed with a court in support of one side of a controversy. (Ch. 16)

Antifederalists Those who opposed the adoption of the U.S. Constitution in the late 1780s and favored state sovereignty and decentralization. (Ch. 3)

apathetics People who do not engage in political activity, never vote, and rarely pay attention to politics. (Ch. 11)

appellate jurisdiction The authority of a court to review legal decisions rendered in other courts. (Ch. 16)

arenas of government The sphere of executive, legislative, and judicial activity in which formal governmental decisions are made in the United States. (Ch. 1)

attack ad A campaign advertisement that attempts to raise doubts and questions about a candidate's opponents. (Ch. 8)

attentive public Special public those people who know and care about a particular issue. Also known as *special public*. (Ch. 7)

attitudes Predispositions to favor or oppose, like or dislike, or approve or disapprove of that which one encounters. (Ch. 7)

authority The feature of a leader or an institution that compels other's obedience, usually because of some ascribed legitimacy. (Ch. 1)

authorized discretion The power to make a decision resulting from explicit and legitimate grants of authority. Also known as *delegated discretion*. (Ch. 15)

bicameral Legislature A legislature that consists of two separate chambers or houses. (Ch. 3)

block grant A type of financial aid provided to states and localities by the federal government for implementation of very broadly defined policies. (Ch. 4)

bureaucracy A large, formal organization that administers programs and policies, both in government and in the private sector, with a staff and functions that remain in place irrespective of changes in political leadership. (Ch. 15)

cabinet A group including the heads of the 14 executive departments, the director of the Office of Management and Budget, and the ambassador to the United Nations and reporting to and advising the president. (Ch. 14)

capitalism An economic system characterized by the private ownership and control of productive resources. (Ch. 2)

checks and balances The notion that constitutional devices can prevent any power within a nation from becoming absolute by requiring it to be balanced against, or checked by, another power within that same nation. (Ch. 3)

citizens' group An organization that seeks collective benefits and membership in which is open to anyone. (Ch. 9)

civil liberties Freedoms to which every individual has a right, such as the right to own property or to have children. (Ch. 5)

civil religion A set of beliefs and rituals subscribed to by a people about its collective political and economic life. (Ch. 2)

civil rights Protections the government is obligated to provide to all U.S. citizens; these are outlined in amendments to the Constitution. (Ch. 5)

clear and present danger rule The Supreme Court's test of whether an exorcise of the First Amendment's right of free speech should be restricted or punished, based on the ability of words to create a clear and present danger (e.g., when someone creates a panic by shouting "Fire!" in a theater. (Ch 5)

closed primary A primary election in which a voter must declare a political party affiliation and vote only that party's ballot in the primary. (Ch. 11)

cloture A motion to close debate; the process by which a filibuster can be ended in the U.S. Senate other than by unanimous consent. (Ch. 13)

coalition A temporary alliance of political actors to advance legislation, elect candidates, or achieve a common goal. (Ch. 1)

coattail effect The ability of the head of a political election ticket to help attract voters to other members of that same ticket. (Ch. 11)

collective benefits Commodities typically provided by a government that cannot, or would not, be parceled out to individuals, since no one can be excluded from their benefits. (Ch. 9)

communal democracy See *participatory democracy*.

compact theory of government A theory suggesting that the authority for national government in the United States is ultimately rooted in the states and that the Constitution is a compact between the states. (Ch. 4)

comparable worth A premise for legal action which demands that wages for women be equal to those earned by men in comparable jobs. (Ch. 6)

competitive sector The part of the economy occupied by small businesses, partnerships, and small incorporated firms and in which the conditions of the classic ideal of a true market are approximated. (Ch. 2)

conceptual framework The underlying set of concepts, beliefs, and predispositions that guides the thoughts and actions of individuals and helps them to accomplish specific tasks. (Ch. 1)

concurrent jurisdiction A legal situation in which two or more court systems have the power to deal with a problem or a case. (Ch. 16)

confederation A system of government that is dominated by regional governments that have joined together and that grant limited power to the central government. (Ch. 3)

conference committees Formal meetings between the representatives of the two houses of a legislature to reconcile differences over the provisions of a particular bill. (Ch. 13)

Congressional Budget Office (CBO) A support agency created in 1974 to provide Congress with basic budget data and with analyses of fiscal, budgetary, and policy issues, independent of the executive branch and of the Office of Management and Budget. (Ch. 13)

constituency The groups and organizations affected by an agency's actions. (Ch. 15)

constituency committees Congressional committees that deal with domestic matters, such as the Public Land and Resources Committee. (Ch. 13)

constitutional rights Provisions in the Constitution and its amendments that outline relationships between government and citizens. (Ch. 5)

contract clause The portion of Article I of the Constitution asserting that no state shall pass any "law impairing the obligation of contracts," meaning that states cannot relieve debtors of their contractual obligations, prevent creditors from foreclosing on mortgages, or declare moratoriums on debt. (Ch. 3)

cooperative federalism A model of government in which national, state, and local governments are interacting agents with intermingling activities, sharing responsibilities and working together to solve common problems. Also known as *marble-cake federalism*. (Ch. 4)

corporate sector The part of the economy occupied by major private corporations. (Ch. 2)

courts of appeal Intermediate level appellate courts that hear appeals from trial courts. (Ch. 16)

cross-cutting memberships The diversity of combinations resulting when people of varied race, religion, and ethnicity are members of the same organization or group. (Ch. 2)

cue networks Members of Congress known for their integrity, good judgment, or expert knowledge in a particular area and whose opinions are sought by their colleagues when the latter are voting on complicated issues. (Ch. 13)

defensive advantage The advantage that opponents of proposed legislation enjoy in the legislative process; legislators who oppose a bill merely need to form a majority, regardless of their disagreement on its specific provisions. (Ch. 13)

delegated discretion See *authorized discretion.*

delegation of powers The process by which Congress confers greater responsibilities on the president and the executive branch to create agencies and administer programs especially during times of national crisis. (Ch. 14)

democracy A system of government that seeks to place control of state authority in the hands of the people to ensure that policies reflect their wishes. (Ch. 2)

determinative public opinion Public opinion so strong that it becomes an irresistible force that compels decision makers to act. (Ch. 7)

developmental policies Policies in which the government plays a role in fostering economic growth. (Ch. 4)

direction of opinion Whether an opinion is favorable or unfavorable. (Ch. 7)

discretion The ability of bureaucracies to choose between significantly different courses of action without facing severe penalties for choosing one course over another. (Ch. 15)

district courts Courts of original jurisdiction in most federal cases; the only federal courts that hold trials in which juries and witnesses are used. (Ch. 16)

diversity of citizenship The situation that exists when parties involved in a lawsuit reside in different states. (Ch. 16)

due process clause The constitutional requirement that no person be deprived of life, liberty, or property without due process of law. (Ch. 5)

elastic clause See *necessary and proper clause.*

Electoral College The 538 electors who, on the first Monday after the second Wednesday in December of a presidential election year, officially elect the president and the vice president of the United States. The electors are chosen by the states and represent the total number of senators and representatives in Congress. (Ch. 10)

electoral party organization An organization that manages elections. (Ch. 10)

enumerated powers Those powers of the U.S. government specifically provided for and listed in the Constitution. (Ch. 3)

equal protection clause The constitutional requirement that a government not treat people unequally, nor give unfair or unequal treatment to a person based on that person's race or religion. (Ch. 5)

establishment clause The part of the First Amendment to the Constitution that erects a wall between church and state, asserting that "Congress shall make no law respecting an establishment of religion". (Ch. 5)

exchange A form of bartering of resources; it can consist of positive or negative inducements that affect the behavior of others. (Ch. 1)

exclusionary rule The constitutional ruling that evidence obtained through an illegal search or seizure may not be used in a criminal trial. (Ch. 5)

Executive Office of the President (EOP) The umbrella office consisting of a variety of formal agencies that assist the president in making administrative and political decisions. (Ch. 14)

executive order A rule or regulation issued by the president that, because of precedent and existing legislative authorization, has the effect of law. (Ch. 14)

exit costs The burdens that are associated with leaving an economic transaction or market. (Ch. 4)

external efficacy See *sense of external efficacy.*

federalism A system of government providing for a division of power between national and state governments. (Ch. 3)

Federalists Those who supported the U.S. Constitution before its ratification and favored a strong central government and a national commercial economy. (Ch. 3)

feel-good ad A campaign advertisement that mimics a commercial and associates the candidate with positive images having little substantive content. (Ch. 8)

filibuster An effort to prevent or postpone a vote by prolonging debate. (Ch. 13)

fiscal policy The manipulation of government finances by raising or lowering taxes or levels of spending to promote economic stability and growth. (Ch. 2)

folkways Long-standing informal rules and practices that supplemented formal rules and governed interpersonal relations among members of Congress for a number of years. (Ch. 13)

forum shopping The careful selection of the level of the court system that will be most sympathetic to a legal claim. (Ch. 16)

foundation A noneconomic organized interest that does not lobby directly or advocate positions on major policy issues, but instead indirectly affects by through funding research and demonstration projects. (Ch. 9)

franking privilege The free use of the mails by congressional incumbents for oficial business such as sending newsletters and other information to their constituents. (Ch. 12)

free coverage Media coverage not paid for by a political candidate. (Ch. 8)

free exercise clause The part of the First Amendment to the Constitution that prohibits the government from interfering with the practice of religious groups. (Ch. 5)

free rider A person who does not belong to an organized group, such as a union or a political party, but benefits from its activities. (Ch. 9)

free-market economy An economy in which free, competitive markets determine all prices. (Ch. 2)

front-loading Compression of the primary election season into a brief, concentrated period. (Ch. 12)

full faith and credit clause The clause in Article IV, Section 1 of the Constitution that requires states to honor and enforce the governmental actions of all other states. (Ch. 3)

funding constituency A group of contributors who provide the money a candidate needs in order to conduct a campaign. (Ch. 12)

general public The adults in a nation or state. (Ch. 7)

gerrymandering The process of shaping electoral districts to enhance the political advantage of one party or group. (Ch. 12)

good-faith exception The premise that evidence obtained in an illegal search or seizure may be used at a trial provided that those who obtained it acted in an honest manner, or in "good faith". (Ch. 5)

government The formal institutions and processes through which binding policy decisions are made for a society. (Ch. 1)

government corporation A bureaucratic organization run by bipartisan boards that closely resembles a private corporation. (Ch. 15)

Great Compromise A proposal accepted by the Constitutional Convention of 1787 that called for a bicameral legislature allowing equal state representation in the Senate and representation based on population in the House of Representatives. (Ch. 3)

gross domestic product (GDP) The total value of all economic activity within the geographical boundaries of a nation in a given year. (Ch. 2)

habeas corpus The right of individuals to appear in court prior to being jailed; one of the oldest protections of personal liberty, and considered to be fundamental to the due process of law. (Ch. 3)

horse-race coverage Media coverage of a political campaign that focuses on the candidates' standings in the polls rather than on the issues. (Ch. 8)

ideology A highly structured set of beliefs grounded in fundamental principles that produces highly consistent attitudes and opinions to explain politics. (Ch. 7)

in forma pauperis The Latin term for "in the manner of a pauper"; refers to an affidavit that an indigent petitioner may file in order to waive filing fees for legal appeals. (Ch. 16)

incremental model of bureaucratic decision making An approach to government decision making in which policymakers begin with the current situation, consider a limited number of changes to that situation, and test those changes by instituting them one at a time. (Ch. 15)

independent federal agency A government agency that does not fall under the organizational jurisdiction of either a federal cabinet department or the Executive Office of the President. (Ch. 15)

influence The power to affect or alter the behavior or condition of others. (Ch. 1)

inherent powers Presidential powers and resources not explicitly stated in the Constitution but rather inferred from constitutional provisions. (Ch. 14)

intensity of opinion How strongly an opinion is held. (Ch. 7)

interagency group (IG) A committee that seeks to resolve conflicts created between lateral agencies by drawing members from contending political parties. (Ch. 15)

interest group A group with a formal structure and a minimum frequency of interaction whose members share attitudes, interests, or attributes. (Ch. 9)

internal efficacy See *sense of internal efficacy.*

iron triangles Mutually supportive relationships among interest groups, government agencies, and the legislative committee or subcommittee with jurisdiction over their particular areas of concern. Also known as *subgovernments.* (Ch. 9)

issue network A group of public and private actors with shared knowledge who combine either to propose policy and enact it into law or to oppose public policy initiatives. (Ch. 9)

judicial review The power of the U.S. Supreme Court to declare actions of the president, the Congress, or other governmental agencies at any level to be invalid or unconstitutional. (Ch. 3)

lateral agency A counterpart of a subordinate agency within a department that is located elsewhere in the bureaucracy. (Ch. 15)

law The set of written and unwritten materials used by lawyers and judges in formulating arguments and rendering decisions. (Ch. 16)

layer-cake federalism A model of government in which the powers of the two levels of government are as separate and distinct as the layers in a cake. (Ch. 4)

legislative clearance The procedure by which agencies of the federal government must submit proposed legislative changes to the Office of Management and Budget. (Ch. 15)

legislative party The elected members of each political party in a city council, a state legislature, and the U.S. Congress. (Ch. 10)

levels of government The structure of American government consists of three separate divisions or levels—local, state, and national. (Ch. 1)

libel The publication of a story so false and malicious that it injures a person's reputation or character. (Ch. 5)

line-item veto The power to delete spending on specific programs or projects in a spending bill. (Ch. 13)

lobbyist A person who makes a living by receiving money to exert political influence on behalf of the individuals, organizations, or associations that hire him or her. (Ch. 9)

local factors In congressional elections, factors unique to a local district and its candidates. (Ch. 11)

logrolling The legislative practice of trading votes on unrelated issues. Legislators who engage in it are essentially promising, "You vote for my bill, I'll vote for yours." (Ch. 13)

low-tension politics The low level of conflict in American politics over issues involving religion, language, ethnicity, region, and economic conditions. (Ch. 2)

majority leader The chief strategist and floor spokesperson for the party having nominal control of either house of a legislature. (Ch. 13)

marble-cake federalism See *cooperative federalism.*

market failure The deterioration of a market system due to economic externalities, inability to produce public goods, or lack of consumer information on product quality. (Ch. 2)

material outcomes Tangible goods and services that individuals obtain or fail to obtain as the result of politics. (Ch. 1)

midterm election A U.S. congressional election that occurs in the middle of a four-year presidential term. (Ch. 11)

minority leader The leader of the minority party in a legislature. (Ch. 13)

Miranda warnings Information about the rights that a person accused or suspected of committing a crime has during an interrogation, and of which he or she must be informed prior to questioning. (Ch. 5)

modified market economy A market economy that departs significantly and frequently from the classic ideal of the free market. (Ch. 2)

monetary policy A government's formal efforts to manage the money in its economy, through control of interest rates and other devices, in order to realize specific economic goals. (Ch. 2)

moot issue An issue in a legal case that a court believes is no longer viable because a change in the challenged law or practice that dissolves the controversy. (Ch. 16)

multimember district An electoral district that elects more than one candidate to a legislature at the same time. (Ch. 10)

myth of rights A set of common beliefs about rights, law, courts, and litigation which suggests that constitutional rights are possessed by all citizens regardless of distinction. (Ch. 6)

national factors In congressional election, factors that operate irrespective of local candidates or issues. (Ch. 11)

national theory of government A theory suggesting that sovereignty is located in the entire citizenry of the United States and that the ultimate authority of the government, the Constitution, comes from a national constituency. (Ch. 4)

necessary and proper clause The portion of Article I, Section 8 of the Constitution that makes it possible for Congress to enact all "necessary and proper" laws to carry out its responsibilities. Also known as the *elastic clause. (Ch. 3)*

New Jersey Plan The proposal put before the Constitutional Convention of 1787 by William Patterson of New Jersey that called for equal representation of the states in a unicameral legislature. (Ch. 3)

nonpartisan election An election in which candidates run without indicating political party affiliation, as in a race for town council. (Ch. 10)

objectivity The fair and neutral reporting of facts or conditions as perceived by journalists, without distortion by personal bias or interpretation. (Ch. 8)

oligopoly A situation in which a handful of producers controlling a significant proportion of a market. (Ch. 2)

open primary A primary election in which a voter may vote for the nomination of any of the candidates, regardless of the voter's political party affiliation. (Ch. 11)

open-seat election An election in which incumbent legislators are not running for re-election. (Ch. 11)

opinion cluster A set of views about aspects of a general topic. (Ch. 7)

opinions Expressions of support or opposition regarding specific policies or people based on underlying attitudes. (Ch. 7)

original jurisdiction The authority of a court to hear a case initially. (Ch. 16)

oversight Congress's monitoring of executive branch activities to ensure their conformity to policies established by law. (Ch. 13)

paid coverage Media coverage paid for by a political candidate. (Ch. 8)

participatory democracy The direct involvement of individuals and groups in the decision-making processes of government. Also known as *communal democracy.* (Ch. 4)

partisan election An election in which candidates running for office formally indicate political party affiliation. (Ch. 10)

party as contester of elections The role of a political party in contesting and winning elections. (Ch. 10)

party as governing mechanism The role of a political party in organizing and controlling government and implementing a set of policies. (Ch. 10)

party as symbol The role of a political party in evoking widely shared, complex, intense feelings and images that are derived from political history, past events, and personalities. (Ch. 10)

party identification Loyalty or attachment to a political party. (Ch. 10)

party in government A political party as represented by organizations in the legislative and executive branches of government. (Ch. 10)

party in the electorate A political party as represented by organizations in the ordinary voting public. (Ch. 10)

party unity   The support that party members in a legislature are supposed to give to their party's legislative proposals. (Ch. 10)

party vote   In a legislative body, a vote in which a majority of one party votes differently in a roll-call vote from a majority of the other party. (Ch. 10)

patronage   The exercise by elected and appointed officials of the power to make partisan appointments to office and to confer contracts, honors, or other benefits on their political supporters. (Ch. 15)

permissive public opinion   Public opinion that allows wide latitude in the range of policies that can be adopted by decisionmakers without fear of significant public reaction. (Ch. 7)

persuasion   The act or process of moving others to a course of action by means of argument and logic. (Ch. 1)

picket-fence federalism   A model of government in which bureaucratic specialists, rather than elected officials, at the national, state, and local levels exercise considerable power over highly technical government programs, thus forming narrow ties that run between higher and lower levels of government like the crosspieces in a fence. (Ch. 4)

policy committees   Congressional committees (e.g., the International Relations Committee) that deal with policy matters. (Ch. 13)

political action committee (PAC)   An organization that raises money and disperses it to political candidates in the hope of gaining effective access to them if they are elected to office. (Ch. 9)

political interest group   An organized body of individuals who share political interests and policy goals and act to influence public policy. (Ch. 9)

political opinions   Explicit expressions of support or opposition about public matters. (Ch. 7)

political question doctrine   A doctrine developed by the U.S. Supreme Court that allows the Court to avoid issues it feels are "political" rather than legal in nature and therefore better decided by other branches of government. (Ch. 16)

political realignment   A shift in the interests and groups supporting a political party. (Ch. 10)

political socialization   The process by which citizens come to adopt political values, attitudes, and opinions. (Ch. 7)

political symbols   Words, gestures, objects, and images that condense and convey emotion and political meaning. (Ch. 1)

politics   The process by which a community selects rulers and empowers them to make decisions, takes action to attain common goals, and reconciles conflicts within the community. (Ch. 1)

politics of rights   The view that constitutional rights are political resources of unknown value that are often employed in political struggles. (Ch. 6)

power committees   Congressional committees (e.g., the Ways and Means Committee) that deal with matters of vital concern to virtually all members. (Ch. 13)

power of the purse   The power of Congress to control the Treasury and spending of tax dollars. (Ch. 13)

precedent   The legal principle that previous decisions made by judges influence future decisions. (Ch. 16)

presidential party   The coalition of voters and party leaders supporting an incumbent president. (Ch. 10)

priming   The process in which the media focus on some issues and not others, thereby shaping the criteria citizens use to make political judgments about important issues. (Ch. 8)

prior restraint   The power to prevent the publication or broadcast of a story. (Ch. 5)

private sector   The segment of the economy composed of partnerships, private businesses, corporations, and individuals who engage in economic exchange. (Ch. 2)

privileges and immunities clause   A clause in Article IV, Section 2 of the Constitution that compels states to guarantee citizens of other states all citizenship rights in their states and thereby promotes uniformity among states. (Ch. 3)

progressive tax   A tax policy whereby people in successively higher income brackets pay a progressively higher tax rate. (Ch. 2)

prospective voting   Voting based on evaluations of what candidates promise to do and what a voter believes they might really do. (Ch. 11)

public choice perspective   An approach to public administration and politics that is based on microeconomic theory and views the citizen as a consumer of government goods and services. (Ch. 4)

public opinion   Political opinions held by private persons that governments find it prudent to heed. (Ch. 7)

public sector The involvement of the government in economic activity. (Ch. 2)

rational-comprehensive model of bureaucratic decision making An approach to government decision making holding that bureaucratic decisions are rational and scientific and result from nonpolitical, neutral administrative procedures. (Ch. 15)

reapportionment The process of assigning a state a new number of congressional seats based on changes in the states' population as determined by a census. (Ch. 12)

redistributive policies Policies that benefit poorer citizens and are paid for by wealthier citizens. (Ch. 4)

reinforcing cleavages Patterns of social characteristics in which distinct yet predictable traits such as race, religion, ethnicity, language, and social class tend to separate people into classes. (Ch. 2)

representation The role of Congress to act in the interests of the people they represent and in a manner responsive to them. (Ch. 13)

representative democracy A form of government in which the citizens delegate decision-making authority to representatives, who are periodically elected so that they will remain accountable. (Ch. 4)

representative government A governing system in which elected or appointed representatives freely chosen by the people exercise substantial power on their behalf. (Ch. 3)

republican form of government A form of government in which the people exercise their sovereignty through elected representatives. (Ch. 3)

resources Sources of information or expertise possessed that permit influence to be exerted. (Ch. 1)

retrospective voting Voting for candidates on the basis of their past performance in political office. (Ch. 11)

revenue sharing A program that distributes federal tax revenues among state and local levels of government. (Ch. 4)

revolving door The movement of governmental employees to the influence industry. (Ch. 9)

ripe case A legal case relating to actual harm that has been done to someone. (Ch. 16)

roll-call vote The calling of the names of members of a legislature to determine their position on an issue; this method, although often replaced by electronic tallies, makes the results of a vote a matter of public record. (Ch. 13)

rule of four The Supreme Court's policy of granting a petition for a writ of certiorari if four or more justices consider a case to be worthy of review. (Ch. 16)

salience of opinion The significance, prominence, or degree of importance an opinion has for an individual. (Ch. 7)

selective benefits Benefits of membership in a group that are of direct and personal value to individuals belonging to the group. (Ch. 9)

senatorial courtesy The practice whereby senators defer to the preference of senators of the president's party in the state of a federal judicial appointment. (Ch. 16)

seniority system The custom whereby the member of Congress who has served the longest on the majority side of a committee automatically becomes its chair. (Ch. 13)

sense of external efficacy People's assessments concerning a political system's responsiveness. (Ch. 11)

sense of internal efficacy People's feelings about politics based on their evaluations of themselves. (Ch. 11)

sense of political efficacy An attitude referring to how effective people feel they can be in politics. (Ch. 11)

separation of powers The allocation of distinct powers among the three branches of government—legislative, executive, and judicial. (Ch. 3)

short-term partisan forces The short-lived party identification of candidates and issues in a presidential election. (Ch. 11)

single-issue interest group An organized body of individuals devoted to one particular political issue. (Ch. 9)

single-issue voter A voter who will support any candidate in an election who favors that voter's position on a particular issue. (Ch. 11)

single-member district An electoral district that elects only one candidate (chosen by a plurality) to a legislature. (Ch. 10)

social pluralism The diversity found in the races, religions, languages, and ethnicities of the American people. (Ch. 2)

socialism An economic system characterized by government ownership and control of productive resources. (Ch. 2)

soft money Funds given by national political parties to their state and local parties for nonfederal uses, such as voter registration drives. (Ch. 12)

solicitor general The attorney who represents the federal government before the Supreme Court. (Ch. 16)

sovereignty The establishment unchallenged claim to rule over a particular geographic area and people. (Ch. 4)

Speaker The presiding officer of the U.S. House of Representatives. The Speaker is elected by House members and holds the most powerful office in that legislative body. (Ch. 13)

special public See *attentive public.*

spectators People who stay somewhat informed about politics, vote regularly, and participate in patriotic events. (Ch. 11)

sponsorship of test cases The process whereby organized interest groups, in their quest to challenge or change existing laws, supply litigants with attorneys and other resources necessary to carry an appeal to a higher court. (Ch. 16)

stability of opinion The extent to which an opinion changes over time. (Ch. 7)

standing committees The permanent committees in each chamber of Congress that deal with bills within a specified subject area. (Ch. 13)

standing to sue A legal requirement demanding that a litigant show a real and direct stake in the outcome of a legal case and some injury from the action or law that is being challenged. (Ch. 16)

stare decisis Latin for "to stand by the decision," the phrase refers to the legal principle that once a precedent is established, all similar cases should be decided in the same manner. (Ch. 16)

statutory construction Occurs when the Supreme Court interprets the meaning of federal legislation that often is broad and vague and fills in some of the law's substantive content. (Ch. 16)

strategic politician A candidate who decides to run for office when the prospects for his party are favorable. (Ch. 11)

subgovernments See *iron triangles.*

suggestive public opinion When public opinion provides decisionmakers with knowledge of the public's preferences but leaves many of the important details undecided. (Ch. 7)

supremacy clause That portion of Article VI of the Constitution asserting that the Constitution, federal laws, and federal treaties "shall be the supreme law of the land," implying that federal law takes precedence over state law. (Ch. 3)

Supreme Court See *United States Supreme Court.*

surge and decline A theory arguing that in presidential election years, candidates from the president's party benefit from a surge of support from independents; however, these independents, being less involved in politics, do not bother to vote in the following midterm election. (Ch. 12)

swing voter A voter who has no loyalty to a political party and leans toward one candidate or another depending on policy or personality. (Ch. 12)

symbolic outcomes Intangible psychological and emotional feelings individuals experience as the result of politics. (Ch. 1)

symbolic speech Conduct with a purpose to express a political view that is constitutionally protected. (Ch. 5)

term limits Restrictions on the number of terms an elected official may serve. (Ch. 12)

theory of the impact of economic conditions A theory stating that economic performance affects elections. (Ch. 11)

think tank A private research group that explores policy questions through social science research and advocates specific proposals. (Ch. 9)

Three-Fifths Compromise An agreement reached between northern and southern delegates at the Constitutional Convention in 1787 which specified that the number of representatives in the House would be determined by a population formula whereby one slave equaled three-fifths of a person. (Ch. 3)

unauthorized discretion The power to make a decision without official sanction and without the knowledge or approval of other political players. Also known as *undelegated discretion.* (Ch. 15)

undelegated discretion   See *unauthorized discretion.*

unfunded mandate   A program or project that the federal government requires without providing funds to implement it. (Ch. 4)

unit rule   See *winner-take-all rule.*

unitary system of government   A system in which sovereignty is vested entirely in the national government. (Ch. 4)

United States Supreme Court   The highest U.S. court and the only federal court explicitly established in the Constitution. (Ch. 16)

universal adult suffrage   The granting of equal voting rights to all adult citizens. (Ch. 11)

values   The basic principles and beliefs from which attitudes flow in a society. (Ch. 2)

Virginia Plan   The proposal for abolishing the Articles of Confederation and creating a strong central government that was submitted to the Constitutional Convention in 1787 by Edmund Randolph on behalf of the Virginia delegation. (Ch. 3)

voter registration   The process whereby prospective voters are required to establish their identities prior to an election in order to be declared eligible to vote in a particular jurisdiction. (Ch. 11)

voting constituency   The individuals and groups providing votes to a particular candidate. (Ch. 12)

White House Office (WHO)   The personal office of the president of the United States, staffed by advisers close to the president. (Ch. 14)

winner-take-all rule   The requirement that state delegations to a national convention must cast all of their votes for the issue or candidate having the majority of the votes of the state delegates. Also known as the *unit rule.* (Ch. 12)

writ of certiorari   A written petition outlining the issues presented in a case and an argument for review by a higher court. (Ch. 16)

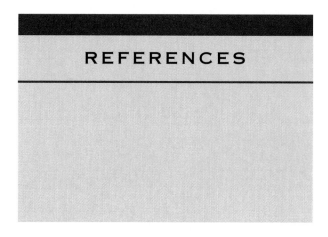

REFERENCES

Chapter 1

1. This definition draws on the classic and influential work of Harold Lasswell, *Politics: Who Gets What When How* (New York: McGraw-Hill, 1936).

2. For a discussion of this view, see Dorothy McBrinde Stetson, *Women's Rights in the U.S.A.* (Pacific Grove, Calif.: Brooks/Cole, 1991).

3. Barbara J. Nelson, "Women and Knowledge in Political Science: Texts, Histories, and Epistemologies," *Women and Politics* 9: 2(1989), p. 2.

4. For a recent example of such work, see Sue Thomas, *How Women Legislate* (New York: Oxford University Press, 1994).

5. This definition of politics and government draws on the influential definition found in David Easton's, *The Political System* (New York: Knopf, 1953). He defines politics as "the authoritative allocation of values for society."

6. See, for example, the discussion of power in Nancy Hartsock's *Money, Sex, and Power* (Boston: Northeastern University Press, 1985), pp. 224–26.

7. Charles Lindblom, *The Policy Making Process*, 2nd ed. (Englewood Cliffs, N.J.: Prentice Hall, 1980), Chapter 6.

8. George Lakoff and Mark Johnson, *Metaphors We Live By* (Chicago: University of Chicago Press, 1980), p. 3.

9. See, for example, William T. Bianco, *Trust: Representatives and Constituents* (Ann Arbor: University of Michigan Press, 1994).

10. Hedrick Smith, *The Power Game: How Washington Works* (New York: Random House, 1988).

11. This statement was made by Aneurin Bevan, a minister of health in Great Britain. *The Macmillan Dictionary of Political Quotations* (New York: Macmillan, 1993).

12. This definition draws on the seminal discussion of power, influence, and control (terms used interchangeably) by Robert Dahl. He initially defines power as follows: "A has power over B to the extent that he can get B to do something that B would not otherwise do" (pp. 202–3). Refining the concept, he notes that influence can occur when the *probability* of a particular behavior by B changes because of A. See "The Concept of Power," *Behavioral Science* 2 (July 1957), pp. 201–15.

13. This notion is best described in a classic study of American politics by E. E. Schattschneider, *The Semi-Sovereign People* (New York: Holt, 1960). Schattschneider refers to this as the *direction* of conflict.

14. Schattschneider, *The Semi-Sovereign People*, p. 74.

15. Two prominent political scientists deserve the credit for recognizing and emphasizing the symbolic dimension of politics. One is Harold Lasswell, who deeply influenced the entire postwar generation of political scientists. See, for example, his seminal *Politics: Who Gets What When How*. The other is Murray Edelman, whose *The Symbolic Uses of Politics* (Urbana: University of Illinois Press, 1964) refined and applied some of Lasswell's insights to the next generation.

16. Edelman, *The Symbolic Uses of Politics*, p. 5.

Chapter 2

1. William Greider, "The Global Sweatshop," *Rolling Stone*, June 30, 1994, pp. 43–44.

2. Kenneth M. Parzych, *A Primer to Antitrust Law and Regulatory Policy* (Lanham, Md.: University Press of America, 1987), p. 17.

3. Laurie McGinley, "Airline Rivalry Declines Sharply at Major Hubs," *Wall Street Journal*, September 20, 1989. The author notes that at the six largest airports, a single airline accounted for 75 percent or more of the departures. In general, the degree of concentration in the airline industry increased substantially after the deregulation of the industry in 1979.

4. Edward Nissan and Regina Caveny, "Relative Concentration of Sales and Assets in American Business," *Southern Economic Journal* 54 (April 1988), p. 929, Table 1. The proportion of the *Fortune* 500's total sales and assets ac-

counted for by the 100 largest among them in 1967 were 63.7 percent and 68.5 percent, respectively.

5. For a discussion of the competitive sector and its characteristics, see James O'Connor, *The Fiscal Crisis of the State* (New York: St. Martin's Press, 1973).

6. U.S. Bureau of the Census, *Statistical Abstracts of the United States: 1993*, 113th ed. (Washington, D.C.: Government Printing Office, 1993), Table 862.

7. Gross domestic product can also be described as the market value of the total output of goods and services produced by labor and property located within the United States. The figures on government revenue, expenditures, and debt in this paragraph are calculated from *Statistical Abstracts: 1993*, Table 509.

8. U.S. Bureau of the Census, *Statistical Abstracts of the United States: 1994*, 114th ed. (Washington, D.C.: Government Printing Office, 1994), Table 493.

9. Frank Levy, *Dollars and Dreams: The Changing American Income Distribution* (New York: Russell Sage, 1987), p. 14, Table 2.1.

10. For a discussion of some of these problems, see Edward N. Wolff, ed., *International Comparisons of the Distribution of Household Wealth* (New York: Oxford University Press, 1987), especially Chapter 1.

11. James D. Smith, "Recent Trends in the Distribution of Wealth in the U.S.: Data, Research Problems, and Prospects," in Wolff, *International Comparisons of the Distribution of Household Wealth*, Chapter 4. The 1981 study reported by Smith was conducted by Marvin Schwartz and published as "Trends in Personal Wealth, 1976–1981" in *Statistics of Income Bulletin* (Summer 1983).

12. U.S. House of Representatives, Committee on Ways and Means, *Background Material and Data on Programs Within the Jurisdiction of the Committee on Ways and Means*, 102nd Congress, 2nd Session, May 15, 1992, Table 2, p. 1567.

13. U.S. Bureau of the Census, *Statistical Abstracts of the United States: 1991*, 111th ed. (Washington, D.C.: Government Printing Office, 1991), calculated from Table 729. The table gives median income in constant 1991 dollars. For men with income, the 1970 median income was $21,996; for women, it was $7,477. In 1991, the equivalent figures were $20,469 and $10,476.

14. *Statistical Abstracts: 1993*, Table 731.

15. *Statistical Abstracts: 1994*, Table 619.

16. These figures are all from Kevin Phillips, *The Politics of Rich and Poor* (New York: Random House, 1990), pp. 202–3.

17. *Statistical Abstracts: 1993*, Table 713.

18. *Statistical Abstracts: 1994*, Table 712.

19. *Statistical Abstracts: 1993*, Table 717.

20. *Statistical Abstracts: 1993*, Table 737.

21. *Statistical Abstracts: 1993*, Table 738.

22. *Statistical Abstracts: 1993*, Table 18.

23. *Statistical Abstracts: 1993*, Table 20. These figures are for the "middle series" estimates. Recognizing how inexact such projections are, the Census Bureau also presents a "lowest" and "highest" estimate series.

24. "Catholics in America: A Church in Transition," *New York Times*, May 29, 1994, p. A21. The survey classified 0.6 percent of respondents as being of "other religions"; 2.3 percent refused to state any religious preference.

25. Gerald M. Pomper, "The Presidential Election," in Gerald M. Pomper, ed., *The Election of 1992* (Chatham, N.J.: Chatham House, 1993), p. 139. White born-again Christians, 17 percent of those voting, gave Bush a solid 61 percent of their votes; other white Protestants gave Bush 46 percent of their votes.

26. For an intriguing discussion of regional differences, see Joel Garreau, *The Nine Nations of North America* (Boston: Houghton Mifflin, 1981).

27. See, for example, Jonathan Kozol, *Savage Inequalities* (New York: Crown, 1991).

28. Jeffery Schmalz, "Ethnic Split Fuels Miami Campaign," *New York Times*, August 29, 1989, p. A12.

29. From a transcript of the speech printed in the *New York Times*, September 7, 1971.

30. "Excerpts from President Clinton's State of the Union Message," *New York Times*, January 26, 1994, p. A15.

31. Herbert McClosky and John Zaller, *The American Ethos* (Cambridge, Mass.: Harvard University Press, 1984), p. 1.

32. For a thorough treatment of the notion of equality in American history and politics, see Sidney Verba and Gary R. Orren, *Equality in America: The View from the Top* (Cambridge, Mass.: Harvard University Press, 1985).

33. Phillips, *The Politics of Rich and Poor*, Chapter 2.

34. McClosky and Zaller's *The American Ethos* provides an excellent summary of the evidence supporting the assertion that Americans truly do believe in these values.

35. Albert R. Hunt, "How to Put a Democrat in the White House," *Wall Street Journal*, August 3, 1989, p. A10.

36. Robert Bellah, *Habits of the Heart: Individualism and Commitment in American Life* (Berkeley: University of California Press, 1985).

37. James Monroe, *The Democratic Wish* (New York: Basic, 1990).

38. McClosky and Zaller, *American Ethos*, p. 38.

39. McClosky and Zaller, *American Ethos*, pp. 24–25.

40. McClosky and Zaller, *American Ethos*, p. 135.

Chapter 3

1. This portrait of the social and political structure of colonial America is drawn from Charles M. Andrews, *The Colonial Background of the American Revolution* (New Haven: Yale University Press, 1924); Jackson Turner Main, *The Social Structure of Revolutionary America* (Princeton: Princeton University Press, 1965); Merrill Jensen, *The Articles of Confederation: An Interpretation of the Social Constitutional History of the American Revolution* (Madison: University of Wisconsin Press, 1940); and Gordon S. Wood, *The Radicalism of the American Revolution* (New York: Knopf, 1991).

2. See Wood, *The Radicalism of the American Revolution*.

3. This narrative draws on the account in Alfred A. Kelly, Winfred A. Harbison, and Herman Belz, *The American Constitution: Its Origins and Development* (New York: Norton, 1983).

4. See Linda K. Kerber, *Women of the Republic: Intellect and Ideology in Revolutionary America* (Chapel Hill: University of North Carolina Press, 1980); and Mary Beth Norton, *Liberty's Daughters: The Revolutionary Experience of American Women* (Boston: Little, Brown, 1980).

5. Lerone Bennett Jr., *Before the Mayflower: A History of Black America* (New York: Penguin, 1982), p. 60.

6. Gordon S. Wood, *The Creation of the American Republic, 1776–1787* (New York: Norton, 1969), p. 32.

7. Bernard Bailyn, *The Ideological Origins of the American Revolution* (Cambridge, Mass.: Harvard University Press, 1967), p. 118.

8. Joan Hoff, *Law, Gender, and Injustice: A Legal History of U.S. Women* (New York: New York University Press, 1991), p. 67.

9. Bailyn, *The Ideological Origins of the American Revolution*, p. 120.

10. Bailyn, *The Ideological Origins of the American Revolution*, p. 20.

11. The classic source on the Declaration is Carl L. Becker, *The Declaration of Independence: A Study in the History of Political Ideas* (New York: Vintage, 1922).

12. Clinton Rossiter, *Seedtime of the Republic: The Origin of the American Tradition of Political Liberty* (New York: Harcourt, Brace & World, 1953), p. 344.

13. On the Iroquois Confederacy, see Bruce E. Johansen, *Forgotten Founders: How the American Indian Helped Shape Democracy* (Boston: Harvard Commons Press, 1982); Vine Deloria Jr. and Clifford M. Lytle, *American Indians, American Justice* (Austin: University of Texas Press, 1983); Elizabeth Tooker, "The United States Constitution and the Iroquois League," *Ethnohistory* 35 (Fall 1988), pp. 305–36; and Donald A. Grinde Jr. and Bruce E. Johansen, *Exemplar of Liberty: Native Americans and the Evolution of Democracy* (Los Angeles: American Indian Studies Center, 1991).

14. Jensen, *The Articles of Confederation*; Wood, *The Creation of the American Republic*.

15. Kelly, Harbison, and Belz, *The American Constitution*, p. 86.

16. Jensen, *The Articles of Confederation*; Wood, *The Creation of the American Republic*.

17. On Shays's Rebellion, see David P. Szatmary, *Shays's Rebellion: The Making of an Agrarian Insurrection* (Amherst, Mass.: University of Massachusetts Press, 1980).

18. Wood, *The Creation of the American Republic*, p. 414.

19. Wood, *The Creation of the American Republic*, p. 413.

20. Robert A. McGuire and Robert L. Ohsfeldt, "An Economic Model of Voting Behavior over Specific Issues at the Constitutional Convention of 1787," *Journal of Economic History* 46 (1986), pp. 79–111.

21. Calvin C. Jillson, "Constitution-Making: Alignment and Realignment in the Federal Convention of 1787," *American Political Science Review* 75 (1981), pp. 598–612.

22. Jackson Turner Main, *The Antifederalists: Critics of the Constitution, 1781–1788* (Chapel Hill: University of North Carolina Press, 1961), p. 105.

23. Jensen, *The Articles of Confederation*, p. 4.

24. Wood, *The Creation of the American Republic*, p. 513.

25. Charles A. Beard, *An Economic Interpretation of the Constitution of the United States* (New York: Macmillan, 1913).

26. See Robert E. Brown, *Charles Beard and the Constitution* (Princeton: Princeton University Press, 1956); and Forrest McDonald, *We the People* (Chicago: University of Chicago Press, 1958).

27. John F. Manley, "Class and Pluralism in America: The Constitution Reconsidered," in John F. Manley and Kenneth M. Dolbeare, eds., *The Case against the Constitution: From the Antifederalists to the Present* (Armonk, N.Y.: Sharpe, 1987).

28. Donald L. Robinson, *Slavery in the Structure of American Politics, 1765–1820* (New York: Harcourt Brace Jovanovich, 1971). Also see John Hope Franklin, *From Slavery to Freedom: A History of Negro Americans*, 4th ed. (New York: Knopf, 1974).

29. Winthrop Jordan, *White over Black: American Attitudes toward the Negro, 1550–1812* (Baltimore: Penguin, 1969), p. 323.

30. Donald G. Nieman, *Promises to Keep: African-Americans and the Constitutional Order, 1776 to the Present* (New York: Oxford University Press, 1991), p. 11.

31. Nieman, *Promises to Keep*, p. 10.

32. Nieman, *Promises to Keep*, p. 12.

33. Robinson, *Slavery in the Structure of American Politics*, p. 71.

34. John P. Roche, "The Founding Fathers: A Reform Caucus in Action," *American Political Science Review* 55 (1961), pp. 799–816.

35. Richard E. Neustadt, *Presidential Power* (New York: Wiley, 1960), p. 33.

36. On the views of the Antifederalists, see Herbert J. Storing, *What the Antifederalists Were For* (Chicago: University of Chicago Press, 1981); and Main, *The Antifederalists*.

37. Jacob E. Cooke, ed., *The Federalist* (Middletown, Conn.: Wesleyan University Press, 1961).

38. Roche, "The Founding Fathers."

39. Quoted in Leonard W. Levy, "The Bill of Rights," in Leonard W. Levy, ed., *Essays on the Making of the Constitution*, 2nd ed. (New York: Oxford University Press, 1987), p. 285.

40. Michael Wines, "House Approves Bill to Mandate Balanced Budget," *New York Times*, January 27, 1995, p. 1; Michael Wines, "Senate Rejects Amendment on Balancing the Budget," *New York Times*, March 3, 1995, p. 1.

41. On the Equal Rights Amendment, see Jane Mansbridge, *Why We Lost the ERA* (Chicago: University of Chicago Press, 1986).

42. See Stuart A. Scheingold, *The Politics of Rights: Lawyers, Public Policy, and Political Change* (New Haven: Yale University Press, 1974); and Mary Ann Glendon, *Rights Talk: The Impoverishment of Political Discourse* (New York: Free Press, 1991).

43. Jordan, *White over Black*, p. 339.

44. Quoted in Jordan, *White over Black*, p. 339.

Chapter 4

1. See David Broder, "Feds Can't Dismiss Gun Law's Burden," *Ann Arbor News*, June 1, 1994, p. A11.

2. Martin Diamond, "What the Framers Meant by Federalism," in Lawrence J. O'Toole, ed., *American Intergovernmental Relations*, 2nd ed. (Washington, D.C.: Congressional Quarterly Press, 1993).

3. William Riker, *Federalism: Origin, Operation, Significance* (Boston: Little, Brown, 1964).

4. Figures are from Barbara J. Nelson and Kathryn A. Carver, "Many Voices but Few Vehicles: The Consequences for Women of Weak Political Infrastructure in the United States," in Barbara J. Nelson and Najma Chowdhury, eds., *Women and Politics Worldwide* (New Haven: Yale University Press, 1994), p. 740.

5. This discussion relies heavily on Samuel H. Beer, *To Make a Nation: The Rediscovery of American Federalism* (Cambridge, Mass.: Harvard University Press, 1993).

6. Arthur M. Schlesinger Jr., *The Disuniting of America: Reflections on a Multicultural Society* (New York: Norton, 1993).

7. For a good example of this position, see Stanley Fish, *There's No Such Thing as Free Speech* (New York: Oxford University Press, 1993).

8. Ronald Takaki, *A Different Mirror: A History of Multicultural America* (New York: Back Bay Books, 1993), p. 4.

9. Quoted in Takaki, *A Different Mirror*, p. 4.

10. Beer, *To Make a Nation*.

11. Beer, *To Make a Nation*, p. 16.

12. Beer, *To Make a Nation*, p. 2.

13. Jennifer Nedelsky, *Private Property and the Limits of American Constitutionalism: The Madisonian Framework and Its Legacy* (Chicago: University of Chicago Press, 1990).

14. For a good summary of how this debate has developed in the twentieth century, see Ronald Steele, *Walter Lippmann and the American Century* (New York: Vintage, 1981); Robert Westbrook, *John Dewey and American Democracy* (New York: Columbia University Press, 1992).

15. Paula Baker, "The Domestication of Politics: Women and American Political Society, 1780–1920," in Ellen Carol DuBois and Vicki L. Ruiz, eds., *Unequal Sisters: A Multicultural Reader in U.S. Women's History* (New York: Routledge, 1990), p. 66.

16. Takaki, *A Different Mirror*.

17. Dirk Johnson, "Economies Come to Life on Indian Reservations," *New York Times*, July 3, 1994, p. A1.

18. Johnson, "Economies Come to Life," p. A1.

19. Robert D. Bullard, "Environmental Justice for All," in Robert D. Bullard, ed., *Unequal Protection: Environmental Justice and Communities of Color* (San Francisco: Sierra Club Books, 1994), pp. 16–17.

20. This distinction is clearly laid out in Theodore Lowi and Benjamin Ginsberg, *American Government* (New York: Norton, 1993).

21. Morton Grodzins, "The Federal System," in *Goals for Americans: The Report of the President's Commission on National Goals* (Englewood Cliffs, N.J.: Prentice Hall, 1960), pp. 365–66.

22. The information in this paragraph is from James D. Savage, *Balanced Budgets and American Politics* (Ithaca, N.Y.: Cornell University Press, 1988), Chapter 3.

23. For a candid discussion of these issues during the Reagan administration, see William Greider, *The Education of David Stockman and Other Americans* (New York: Dutton, 1986). Stockman served as Ronald Reagan's first director of the Office of Management and Budget.

24. Figures are from Laurence J. O'Toole Jr., "American Intergovernmental Relations: An Overview," in Laurence J. O'Toole Jr., ed., *American Intergovernmental Relations*, 2nd ed. (Washington, D.C.: Congressional Quarterly Press, 1993), p. 9.

25. Quoted in Russell L. Hanson, "Intergovernmental Relations," in Virginia Gray, Herbert Jacob, and Robert B. Albritton, eds., *Politics in the American States* (Glenview, Ill.: Scott, Foresman/Little, Brown Higher Education, 1990), p. 54.

26. Hanson, "Intergovernmental Relations," p. 54.

27. Rochelle L. Stanfield, "Thanks a Lot for Nothing, Washington," *National Journal*, March 26, 1994, p. 726.

28. Quoted in Margaret Kriz, "Cleaner Than Clean?" *National Journal*, April 23, 1994, p. 726.

29. For a fuller discussion of Superfund, the federal program aimed at cleaning up leaking hazardous waste sites, see Bruce A. Williams and Albert R. Matheny, *Democracy, Dialogue, and Environmental Disputes: The Contested Languages of Social Regulation* (New Haven: Yale University Press, 1995).

30. "States Look to Kentucky's Experiment in Education," *Ann Arbor News*, January 28, 1993, p. A1.

31. See Robert L. Bish, *The Political Economy of Metropolitan Areas* (Chicago: Markham Books, 1971); Charles Tiebout, "A Pure Theory of Local Expenditures," *Journal of Political Economy* 4 (October 1956); Vincent Ostrom, *The Intellectual Crisis in Public Administration* (University: University of Alabama Press, 1965).

32. For example, they have opposed reforms aimed at combining localities into larger metropolitan governments. See Ostrum, *Intellectual Crisis*; Bish, *Political Economy*.

33. Nelson and Carver, "Many Voices but Few Vehicles," p. 745.

34. This section is drawn from Bruce A. Williams, "Regulation and Economic Development," in Virginia Gray, Herbert Jacob, and Robert B. Albritton, eds., *Politics in the American States* (Glenview, Ill.: Scott, Foresman/Little, Brown Higher Education, 1990).

35. Nancy S. Lind, "Economic Development and Diamond-Star Motors: Intergovernmental Competition and Cooperation," in Ernest J. Yanarella and William C. Green, eds., *The Politics of Industrial Recruitment* (New York: Greenwood Press, 1989).

36. Robert Perrucci and Madhavi Patel, "Local Media Images of Indiana and Kentucky's Recruitment of Subaru-Isuzu and Toyota," in Ernest J. Yanarella and William C. Green, eds., *The Politics of Industrial Recruitment* (New York: Greenwood Press, 1989) p. 22.

37. Although only Georgia and Tennessee bid for the Nissan factory in 1980, over 30 states bid for the Toyota plant in 1985, and 14 bid for the Fuji-Isuzu factory in 1986.

38. In a study of the factors that determine the location of Japanese investments in the United States, Mamoru Yoshida interviewed Japanese executives and found that they ranked tax incentives as only the seventh most important consideration and "other state incentives" as tenth. The first six factors were quality of labor, proximity to markets, nonunionized labor force, cost of land, cost of labor, and quality of life. Cited in Brinton H. Milward and Heidi Newman, "State Incentive Packages and Industrial Location Decisions," in Ernest J. Yanarella and William C. Green, eds., *The Politics of Industrial Recruitment* (New York: Greenwood Press, 1989).

39. Bruce A. Williams and Albert R. Matheny, "Testing Theories of Social Regulation," *Journal of Politics* (May 1984).

40. "Many Harried S&Ls, Irked by States' Rules, Seek Federal Charters," *Wall Street Journal*, August 13, 1981, p. 1.

41. Robert Sherrill, "The Looting Decade," *Nation* (November 23, 1983).

42. For an excellent history of the early civil rights movement, see Taylor Branch's award-winning *Parting the Waters* (New York: Simon and Schuster, 1988).

43. The discussion of *Katzenbach v. McClung* draws heavily on Lowi and Ginsberg, *American Government*.

44. Abigail Thernstrom, *Whose Votes Count?* (Cambridge, Mass.: Harvard University Press, 1987).

45. William H. Riker, "Federalism," in O'Toole, ed., *American Intergovernmental Relations*, p. 98.

46. John J. Harrigan, *Empty Dreams, Empty Pockets: Class and Bias in American Politics* (New York: Macmillan, 1993), p. 63.

Chapter 5

1. Information on Summerton is drawn from Peter Applebome, "Legacy of a Southern Town: Schools Split on Racial Lines," *New York Times*, May 21, 1991, p. A18; and Joe Klein, "The Legacy of Summerton," *Newsweek*, May 16, 1994, pp. 26–31.

2. 163 US 537 (1896).

3. 347 US 483 (1954).

4. On the *Brown* decision, see Richard Kluger, *Simple Justice* (New York: Knopf, 1975); and Alfred H. Kelly, "The School Desegregation Case," in John A. Garraty, ed., *Quarrels That Have Shaped the Constitution* (New York: Harper & Row, 1962). On the NAACP, its lawyers, and the litigation strategy leading to *Brown*, see Mark V. Tushnet, *The NAACP's Legal Strategy against Segregated Education, 1925–1950* (Chapel Hill: University of North Carolina Press, 1987).

5. See Kenneth J. Meier and Joseph Stewart Jr., *The Politics of Hispanic Education* (Albany: State University of New York Press, 1991). The citation for the *Keyes* decision is 413 US 921 (1973).

6. Meier and Stewart, *The Politics of Hispanic Education*, p. 70.

7. See, for example, *Schenck v. United States*, 249 US 47 (1919); *Debs v. United States*, 249 US 211 (1919); *Abrams v. United States*, 250 US 616 (1919).

8. 249 US 47 (1919).

9. See Alexander Meiklejohn, *Political Freedom: The Constitutional Powers of the People* (New York: Oxford University Press, 1960); Thomas I. Emerson, *Toward a General Theory of the First Amendment* (New York: Random House, 1966); and Mark A. Graber, *Transforming Free Speech* (Berkeley: University of California Press, 1991).

10. See, for example, *Gitlow v. New York*, 268 US 653 (1925).

11. See *Bradenburg v. Ohio*, 395 US 444 (1969).

12. See *Chaplinsky v. New Hampshire*, 315 US 568 (1942).

13. *Terminiello v. Chicago*, 337 US 1 (1949).

14. Concurring opinion in *Jacobellis v. Ohio*, 378 US 184. This quotation is discussed in Henry J. Abraham and Barbara A. Perry, *Freedom and the Court*, 6th ed. (New York: Oxford University Press, 1994), p. 204.

15. 354 US 476 (1957).

16. 413 US 15 (1973).

17. See, for example, *Near v. Minnesota*, 283 US 697 (1931), and *New York Times Company v. United States*, 403 US 713 (1971).

18. Important decisions discussing libel include *New York Times Company v. Sullivan*, 376 US 254 (1964); *Gertz v. Robert Welch, Inc.*, 418 US 323 (1974); and *Time, Inc. v. Firestone*, 424 US 448 (1976).

19. 391 US 367 (1968); 491 US 397 (1989).

20. For example, see *Clark v. Community for Creative Non-Violence*, 468 US 288 (1984).

21. See *Lloyd Corporation v. Tanner*, 407 US 551 (1972).

22. See the discussion in John C. Domino, *Civil Rights and Liberties: Toward the 21st Century* (New York: HarperCollins, 1994), pp. 105–28.

23. These rules were drawn from previous cases and pulled together in *Lemon v. Kurtzman*, 403 US 602 (1971).

24. *Engel v. Vitale*, 370 US 421 (1962).

25. *Abington School District v. Schempp*, 374 US 203 (1963).

26. *Wallace v. Jaffree*, 472 US 38 (1985).

27. *West Virginia State Board of Education v. Barnette*, 319 US 624 (1943).

28. *Employment Division v. Smith*, 110 S.Ct. 1595 (1990).

29. 367 US 643 (1961).

30. 384 US 436 (1966).

31. 372 US 335 (1963).

32. 468 US 897 (1984).

33. 467 US 649 (1984).

34. 496 US 292 (1990).

35. Quoted in Abraham and Perry, *Freedom and the Court*, p. 132.

36. 408 US 238 (1972).

37. 428 US 513 (1976).

38. 478 US 1019 (1987).

39. 381 US 479 (1965).

40. 410 US 113 (1973).

41. See, for example, *Maher v. Roe*, 432 US 464 (1977); and *Harris v. McRae*, 448 US 297 (1980).

42. See, for example, *Planned Parenthood of Central Missouri v. Danforth*, 428 US 552 (1976); *Akron v. Akron Center for Reproductive Health*, 462 US 416 (1983); and *Thornburgh v. American College of Obstetricians and Gynecologists*, 476 US 747 (1986).

43. 492 US 490 (1989).

44. David Johnston, "Marshall Papers Reveal Court behind the Scenes," *New York Times*, May 24, 1993, p. A10.

45. 112 S.Ct. 2791 (1992).

46. 478 US 186 (1986).

47. See *The Slaughterhouse Cases*, 16 Wall. 36 (1873).

48. See *Barron v. Baltimore*, 7 Pet. 243 (1833).

49. See *Cherokee Nation v. Georgia*, 30 US 1 (1831); and *Worcester v. Georgia*, 31 US 515 (1832). These cases and others are discussed in Vine Deloria Jr. and Clifford M. Lytle, *American Indians, American Justice* (Austin: University of Texas Press, 1983); and Stephen L. Pevar, *The Rights of Indians and Tribes* (Carbondale: Southern Illinois University Press, 1992).

50. *Talton v. Mayes*, 163 US 376 (1896).

51. See Deloria and Lytle, *American Indians*, pp. 126–30; Pevar, *The Rights of Indians and Tribes*, Chapter 14.

52. Deloria and Lytle, *American Indians*, p. 120.

53. Deloria and Lytle, *American Indians*, p. 128.

54. 323 US 214 (1944).

55. 323 US 214, p. 216.

56. *Swann v. Charlotte-Mecklenburg Board of Education*, 402 US 1 (1971).

57. For example, see *Regents of the University of California v. Bakke*, 438 US 265 (1978); and *City of Richmond v. J. A. Croson*, 488 US 469 (1989). On affirmative action programs for women, see *Johnson v. Santa Clara Transportation Agency*, 480 US 616 (1987).

58. See *Reed v. Reed*, 404 US 71 (1971); and *Craig v. Boren*, 429 US 190 (1976).

59. *Rostker v. Goldberg*, 453 US 57 (1981).

60. *Craig v. Boren*, 429 US 190 (1976).

Chapter 6

1. Alfred H. Kelly, "The School Desegregation Case," in John A. Garraty, ed., *Quarrels That Have Shaped the Constitution* (New York: Harper & Row, 1962), p. 266.

2. *Brown v. Board of Education*, 349 US 295 (1955).

3. On southern resistance to school desegregation, see G. Theodore Mitau, *Decade of Decision: The Supreme Court and the Constitutional Revolution 1954–1964* (New York: Scribner's, 1967), Chapter 2; J. Harvie Wilkinson, *From Brown to Bakke: The Supreme Court and School Integration 1954–1978* (New York: Oxford University Press, 1979), Chapter 5.

4. Joe Klein, "The Legacy of Summerton," *Newsweek*, May 16, 1994, p. 27.

5. Gerald N. Rosenberg, *The Hollow Hope: Can Courts Bring about Social Change?* (Chicago: University of Chicago Press, 1991), p. 78.

6. Mitau, *Decade of Decision*, p. 63.

7. Mitau, *Decade of Decision*, p. 63.

8. Charles A. Johnson and Bradley C. Canon, *Judicial Policies: Implementation and Impact* (Washington, D.C.: Congressional Quarterly Press, 1984), p. 256.

9. Randall Kennedy, "Martin Luther King's Constitution: A Legal History of the Montgomery Bus Boycott," *Yale Law Journal* 98 (April 1989), p. 1000.

10. Mitau, *Decade of Decision*, pp. 68–69.

11. See *Swann v. Charlotte-Mecklenburg Board of Education*, 402 US 1 (1971).

12. See Kenneth J. Meier and Joseph Stewart Jr., *The Politics of Hispanic Education* (Albany: State University of New York Press, 1991).

13. On considering Mexican Americans as a minority group for purposes of desegregation, see *Keyes v. School District No. 1, Denver, Colorado*, 413 US 921 (1973). The lower federal court decisions that applied this reasoning to Puerto Ricans are discussed in Meier and Stewart, *The Politics of Hispanic Education*, p. 70.

14. William Celis III, "Study Finds Rising Concentration of Black and Hispanic Students," *New York Times*, December 14, 1993, p. A1; William Celis III, "40 Years after *Brown*, Segregation Persists," *New York Times*, May 18, 1994, p. A1.

15. Celis, "Study Finds Rising Concentration"; and Celis, "40 Years after *Brown*."

16. For a view that the way in which legal arguments are framed is crucial for explaining outcomes in courts, see Lee Epstein and Joseph F. Kobylka, *The Supreme Court and Legal Change: Abortion and the Death Penalty* (Chapel Hill: University of North Carolina Press, 1992).

17. Robert W. Bennett, "The Burger Court and the Poor," in Vincent Blasi, ed., *The Burger Court: The Counter-Revolution That Wasn't* (New Haven: Yale University Press, 1983), p. 61.

18. Stuart A. Scheingold, *The Politics of Rights: Lawyers, Public Policy, and Political Change* (New Haven: Yale University Press, 1974), pp. 5–6.

19. Scheingold, *The Politics of Rights*, p. 6.

20. Rhonda Copelon, "Beyond the Liberal Idea of Privacy: Toward a Positive Right of Autonomy," in Michael W. McCann and Gerald L. Houseman, eds., *Judging the Constitution: Critical Essays on Judicial Lawmaking* (Glenview, Ill.: Scott, Foresman, 1989), p. 293. See also Mary Becker, "The Politics of Women's Wrongs and the Bill of Rights: A Bicentennial Perspective," *University of Chicago Law Review* 59 (1992), pp. 453–518.

21. Michael W. McCann, "Equal Protection for Social Inequality: Race and Class in Constitutional Ideology," in

Michael W. McCann and Gerald L. Houseman, eds., *Judging the Constitution: Critical Essays on Judicial Lawmaking* (Glenview, Ill.: Scott, Foresman, 1989).

22. Alan Freeman, "Antidiscrimination Law: The View from 1989," in David Kairys, ed., *The Politics of Law*, rev. ed. (New York: Pantheon, 1990).

23. McCann, "Equal Protection for Social Inequality," p. 233.

24. The study, referred to as the Civil Litigation Research Project, is described in "Special Issue on Dispute Processing and Civil Litigation," *Law & Society Review* 15 (1980–1981).

25. See Richard E. Miller and Austin Sarat, "Grievances, Claims, and Disputes: Assessing the Adversary Culture," *Law & Society Review* 15 (1980–1981), pp. 525–66.

26. See Harry Stumpf, *Community Politics and Legal Services* (Beverly Hills, Calif.: Sage, 1975); Earl Johnson Jr., *Justice and Reform: The Formative Years of the American Legal Services Program* (New Brunswick, N.J.: Transaction Books, 1978).

27. Mark Kessler, *Legal Services for the Poor: A Comparative and Contemporary Analysis of Interorganizational Politics* (New York: Greenwood Press, 1987); Mark Kessler, "Legal Mobilization for Social Reform: Power and the Politics of Agenda Setting," *Law & Society Review* 24 (1990), pp. 121–43.

28. Kristin Bumiller, *The Civil Rights Society: The Social Construction of Victims* (Baltimore: Johns Hopkins University Press, 1988).

29. In her research, Bumiller used data collected by the Civil Litigation Research Project. The data are described in "Special Issue on Dispute Processing and Civil Litigation," *Law & Society Review* 15 (1980–1981). Bumiller supplemented the data by interviewing 18 people who had experienced discrimination.

30. Robert A. Dahl, "Decision-Making in a Democracy: The Role of the Supreme Court as a National Policy-Maker," *Journal of Public Law* 6 (1957), pp. 275–95. A revised version of the argument appears in Robert A. Dahl, *Pluralist Democracy in the United States: Conflict and Consent* (Chicago: Rand-McNally, 1967), pp. 154–70.

31. Dahl's conclusions were criticized in Jonathan Casper, "The Supreme Court and National Policy-Making," *American Political Science Review* 70 (1976), pp. 50–63. Casper emphasizes the times when the Court's decisions overturned the policies of state legislatures and protected minority rights.

32. Dahl, *Pluralist Democracy*, p. 156.

33. Gerald N. Rosenberg, "Judicial Independence and the Reality of Political Power," *Review of Politics* 54 (1992), pp. 369–98.

34. Rosenberg, "Judicial Independence," p. 391.

35. Rosenberg, "Judicial Independence," pp. 391–92.

36. Rosenberg, "Judicial Independence," p. 398.

37. For example, Donald L. Horowitz, *The Courts and Social Policy* (Washington, D.C.: Brookings Institution, 1977); Joel F. Handler, *Social Movements and the Legal System: A Theory of Law Reform and Social Change* (New York: Academic Press, 1978); Gerald N. Rosenberg, *The Hollow Hope*.

38. Quoted in Scheingold, *The Politics of Rights*, pp. 85–86.

39. Rosenberg, *The Hollow Hope*, pp. 339, 341.

40. Scheingold, *The Politics of Rights*, pp. 6–7.

41. On these types of psychological effects, see Elizabeth M. Schneider, "The Dialectics of Rights and Politics: Perspectives from the Women's Movement," *New York University Law Review* 61 (1986), pp. 589–652; Patricia J. Williams, "Alchemical Notes: Reconstructing Ideals from Deconstructed Rights," *Harvard Civil Rights–Civil Liberties Law Review* 22 (1987), pp. 401–33; and Kimberle Williams Crenshaw, "Race, Reform, and Retrenchment: Transformation and Legitimation in Antidiscrimination Law," *Harvard Law Review* 101 (1988), pp. 1331–87.

42. Williams, "Alchemical Notes," p. 416.

43. Richard Delgado, "The Ethereal Scholar: Does Critical Legal Studies Have What Minorities Want?" *Harvard Civil Rights–Civil Liberties Law Review* 22 (1987), pp. 301–22.

44. Constance Baker Motley, quoted in the documentary film *Eyes on the Prize*, "Awakening 1954–1956," prod. and dir. Judith Vecchione, PBS Home Video, 1987.

45. Bernice Reagon, quoted in Michael W. McCann, "Reform Litigation on Trial," *Law & Social Inquiry* 17 (Fall 1992), pp. 722–23, n. 8.

46. Jo Ann Robinson, quoted in the documentary film *Eyes on the Prize*, "Awakening 1954–1956." The Supreme Court ruled that segregation in Montgomery city buses was unconstitutional in *Gayle v. Browder*, 352 US 903 (1957).

47. E. E. Schattschneider, *The Semi-Sovereign People: A Realist's View of Democracy in America* (New York: Holt, Rinehart & Winston, 1960).

48. On the use of rights and litigation for publicity purposes, see Handler, *Social Movements and the Legal System*, pp. 214–22.

49. This case is discussed in Joel F. Handler, with George Edgar and Russell F. Settle, "Public Interest Law and Employment Discrimination," in Burton Weisbrod, Joel F. Handler, and Neil Komesar, eds., *Public Interest Law: An Economic and Institutional Analysis* (Berkeley: University of California Press, 1978), pp. 275–76.

50. See Michael W. McCann, "Legal Mobilization and Social Reform Movements: Notes on Theory and Its Application," in Austin Sarat and Susan S. Silbey, eds., *Studies in Law, Politics, and Society*, Vol. 11 (Greenwich, Conn.: JAI Press, 1991), pp. 225–54. The most extended treatment of this research is Michael W. McCann, *Rights at Work: Pay Equity Reform and the Politics of Legal Mobilization* (Chicago: University of Chicago Press, 1994).

51. *County of Washington v. Gunther*, 452 US 161 (1981); *AFSCME v. State of Washington*, 578 F.Supp. 846 (D.Wash.1983).

52. McCann, "Legal Mobilization and Social Reform Movements," p. 236.

53. See Schneider, "The Dialectics of Rights and Politics."

54. 88 Wash.2d 221 (1977).

55. Schneider, "The Dialectics of Rights and Politics," p. 606.

56. Schneider, "The Dialectics of Rights and Politics," p. 608.

57. Schneider, "The Dialectics of Rights and Politics," p. 608, n. 87.

58. McCann, "Legal Mobilization and Social Reform Movements."

59. McCann, "Legal Mobilization and Social Reform Movements," p. 236.

60. For a view that *Brown* had little, if any, effect on political mobilization, see Rosenberg, *The Hollow Hope.* For an excellent critique of Rosenberg's argument, see McCann, "Reform Litigation on Trial."

61. Quoted in McCann, "Reform Litigation on Trial," p. 722.

62. Aldon Morris, *The Origins of the Civil Rights Movement* (New York: Free Press, 1984), p. 34, quoted in McCann, "Reform Litigation on Trial," p. 736.

63. McCann, "Reform Litigation on Trial," p. 723, n. 9.

64. On this point, see Stuart Scheingold, "Constitutional Rights and Social Change: Civil Rights in Perspective," in Michael W. McCann and Gerald L. Houseman, eds., *Judging the Constitution: Critical Essays on Judicial Lawmaking* (Glenview, Ill.: Scott, Foresman, 1989), pp. 73–91.

65. McCann, "Reform Litigation on Trial," p. 739.

66. This analysis is drawn from Scheingold, "Constitutional Rights and Social Change."

67. Scheingold, "Constitutional Rights and Social Change," p. 84.

68. Scheingold, "Constitutional Rights and Social Change," p. 84.

69. Rosenberg, *The Hollow Hope.*

70. Scheingold, *The Politics of Rights,* p. 74.

71. Mary Ann Glendon, *Rights Talk: The Impoverishment of Political Debate* (New York: Free Press, 1991).

Chapter 7

1. Barry Sussman, *What Americans Really Think* (New York: Pantheon, 1988), p. 4.

2. Sussman, *What Americans Really Think,* p. 4.

3. For discussions of the definition of attitudes and opinions and the connection between them, see Michael Corbett, *American Public Opinion: Trends, Processes, and Patterns* (New York: Longman, 1991), p. 21; and Robert S. Erikson, Norman R. Luttbeg, and Kent L. Tedin, *American Public Opinion: Its Origins, Content, and Impact,* 4th ed. (New York: Macmillan, 1991), pp. 44–45.

4. For one description of the civil rights movement, see Francis Fox Piven and Richard Cloward, *Poor People's Movements: Why They Succeed; How They Fail* (New York: Vintage, 1979), Chapter 4.

5. See the discussion of opinion clusters in Corbett, *American Public Opinion,* pp. 24–28.

6. V. O. Key, *Public Opinion and American Democracy* (New York: Knopf, 1961), p. 14.

7. See George Gerbner, Larry Gross, Michael Morgan, and Nancy Signorielli, "Growing Up with Television: The Cultivation Perspective," in Jennings Bryant and Dolf Zillman, eds., *Media Effects: Advances in Theory and Research* (Hillsdale, N.J.: Lawrence Erlbaum, 1994).

8. For a discussion of the measurement and meaning of political trust, see Paul R. Abramson, *Political Attitudes in America* (San Francisco: Freeman, 1983), Chapter 11.

9. Abramson, *Political Attitudes in America,* Chapter 11.

10. Seymour Martin Lipset and William Schneider, *The Confidence Gap: Business, Labor, and Government in the Public Mind,* rev. ed. (Baltimore: Johns Hopkins University Press, 1987), pp. 48–49.

11. Robin Toner, "Health Impasse Sours Voters, New Poll Finds," *New York Times,* September 13, 1994, p. A1.

12. William G. Mayer, *The Changing American Mind: How and Why American Public Opinion Changed between 1960 and 1988* (Ann Arbor: University of Michigan Press, 1992), p. 366.

13. Mayer, *The Changing American Mind,* p. 366.

14. Angie Cannon, "Poll Portrays an Angry and Disenchanted Public," *Philadelphia Inquirer,* September 21, 1994, p. A1.

15. Mayer, *The Changing American Mind,* p. 373. The 1993 figures are calculated from *The General Social Survey, 1972–1993, Codebook* (Chicago: University of Chicago, National Opinion Research Center, 1993), p. 191.

16. Mayer, *The Changing American Mind,* p. 394.

17. Mayer, *The Changing American Mind,* p. 394.

18. Tom W. Smith, "General Social Survey Topic Report No. 19," University of Chicago, National Opinion Research Center, December 1990.

19. George Gallup Jr., *The Gallup Poll: Public Opinion, 1993* (Wilmington, Del.: Scholarly Resources, 1994), pp. 271–72 and p. 251.

20. See James R. Kluegel and Eliot R. Smith, *Beliefs about Inequality: Americans' Views of What Is and What Ought to Be* (New York: Aldine De Gruyter, 1986), especially pp. 295–96.

21. Corbett, *American Public Opinion,* p. 173.

22. Thomas R. Dye, "Legitimacy, Governments, and Markets," Foreword to Thomas R. Dye, ed., *The Political Legitimacy of Markets and Governments* (Greenwich, Conn.: JAI Press, 1989), pp. 6–7.

23. Christopher Jencks, "Economic Inequality and Political Legitimacy," in Thomas R. Dye, ed., *The Political Legitimacy of Markets and Governments* (Greenwich, Conn.: JAI Press, 1989), p. 56. The poll was conducted by the National Opinion Research Center.

24. Erikson et al., *American Public Opinion,* p. 60.

25. This finding came from a 1987 poll summarized by Corbett, *American Public Opinion,* p. 8.

26. Corbett, *American Public Opinion,* p. 8.

27. Erikson et al., *American Public Opinion,* p. 44.

28. Erikson et al., *American Public Opinion,* pp. 42–43.

29. Erikson et al., *American Public Opinion,* p. 43.

30. William H. Flanigan and Nancy H. Zingale, *Political Behavior of the American Electorate,* 5th ed. (Boston: Allyn and Bacon, 1983), p. 102.

31. Toner, "Health Impasse Sours Voters, New Poll Finds," p. A1.

32. The study, conducted by Philip Converse, is described briefly in Erikson et al., *American Public Opinion,* pp. 46–47.

33. Sussman, *What Americans Really Think,* pp. 42–43.

34. James A. Stimson, Michael B. MacKuen, and Robert S. Erikson, "Opinion and Policy: A Global View," *PS: Political Science & Politics* 27 (March 1994), pp. 29–35. For an earlier outline of this argument, see James A. Stimson, *Public Opinion in America: Moods, Cycles, and Swings* (Boulder: Westview, 1991).

35. For a comprehensive treatment of these differences and their causes, see Lee Sigelman and Susan Welch, *Black Americans' Views of Racial Inequality: The Dream Deferred* (Cambridge, Eng.: Cambridge University Press, 1991).

36. Kluegel and Smith, *Beliefs about Inequality,* p. 190.

37. Sigelman and Welch, *Black Americans' Views of Racial Inequality,* p. 91.

38. Kluegel and Smith, *Beliefs about Inequality,* p. 202.

39. Sigelman and Welch, *Black Americans' Views of Racial Inequality,* p. 129.

40. The poll results reported here are found in Kluegel and Smith, *Beliefs about Inequality,* pp. 222–23, Table 8.2.

41. Barbara Hinkson Craig and David M. O'Brien, *Abortion and American Politics* (Chatham, N.J.: Chatham House, 1993), p. 259.

42. The poll results reported in this paragraph are taken from Erikson et al., *American Public Opinion,* p. 199, Table 7.11.

43. Rodolfo O. de la Garza, Louis DeSipio, F. Chris Garcia, John Garcia, and Angelo Flacon, *Latino Voices: Mexican, Puerto Rican, and Cuban Perspectives on American Politics* (Boulder: Westview, 1992), p. 110, Table 7.36.

44. Paul R. Abramson, *Political Attitudes in America: Formation and Change* (San Francisco: W. H. Freeman, 1983), p. 57, Table 4.3.

45. Erikson et al., *American Public Opinion,* p. 74, Table 3.8.

46. Erikson et al., *American Public Opinion,* p. 84, Table 4.3.

47. Erikson et al., *American Public Opinion,* p. 85, Table 4.4.

48. See the discussion in Erikson et al., *American Public Opinion,* pp. 76–79.

49. For a summary of this incident through July 1981, see Bill Keller, "Democrats Run with Social Security Issue," *Congressional Quarterly Weekly Report* (August 1, 1981), pp. 1379–80.

50. Stimson, MacKuen, and Erikson, "Opinion and Policy: A Global View," p. 29.

Chapter 8

1. Patrick J. Sloyan, "Gulf War Missile Failures Power Rocket Research," *Ann Arbor News,* May 23, 1993, p. A1.

2. See John McArthur, *Second Front* (Berkeley: University of California Press, 1993); and Douglas Kellner, *The Persian Gulf TV War* (Boulder: Westview, 1993).

3. For an excellent discussion of the origins and negative implications of the "marketplace of ideas," see Benjamin Ginsberg, *The Captive Public: How Mass Opinion Promotes State Power* (New York: Basic, 1986).

4. These figures on media outlets, ownership, and concentration are all taken from Ben H. Bagdikian, *The Media Monopoly,* 4th ed. (Boston: Beacon, 1992).

5. Figures are from the Minority Telecommunications Development Program, National Telecommunications and Information Administration, U.S. Department of Commerce, "Compilation by State of Minority-Owned Commercial Broadcast Stations," November 1992.

6. Susan Faludi, *Backlash* (New York: Crown, 1991), p. 373.

7. The company is Gannett Company, Inc., the largest newspaper chain in the country. Figures are taken from Maurine H. Beasley and Sheila J. Gibbons, *Taking Their Place: A Documentary History of Women and Journalism* (Washington, D.C.: American University Press, 1993), p. 268.

8. Michael Hoyt, "When the Walls Come Tumbling Down," *Washington Journalism Review,* March/April 1990.

9. Faludi, *Backlash.* See also Naomi Wolf, *The Beauty Myth* (New York: Anchor, 1991).

10. Doris A. Graber, *Mass Media and American Politics* (Washington, D.C.: Congressional Quarterly Press, 1993), p. 49.

11. Daniel C. Hallin, *The "Uncensored War": The Media and Vietnam* (Berkeley: University of California Press, 1986), pp. 22–23.

12. See, for example, Herbert Gans, *Deciding What's News* (New York: Pantheon, 1979).

13. A survey of journalists revealed that more than 90 percent of publishers, editors, and staff agreed that examining willingness to expose wrongdoing was one of the most important ways of evaluating media companies. Reported in David L. Protess, Fay Lomax Cook, et al., *The Journalism of Outrage* (New York: Guilford, 1991).

14. Robert S. Lichter, Stanley Rothman, and Linda S. Lichter, *The Media Elite* (Bethesda, Md.: Adler and Adler, 1986).

15. See, for example, Noam Chomsky and Edward S. Herman, *Manufacturing Consent: The Political Economy of the Mass Media* (New York: Pantheon, 1988).

16. Michael Kelly, "A History of Gergenism," *New York Times,* June 2, 1993, p. A10.

17. *TV Guide* 41: 14 (April 3–9, 1993), p. 3.

18. Beasley and Gibbons, *Taking Their Place,* p. 307.

19. Cornelius F. Foote Jr., "Minority, Total Newsroom Employment Shows Slow Growth, 1994 Survey Says," *ASNE Bulletin* (April/May 1994), p. 20.

20. Faludi, *Backlash,* p. 374.

21. Sue A. Lafky, "The Progress of Women and People of Color in the U.S. Journalistic Workforce," in Pamela J. Creedon, ed., *Women in Mass Communication* (Newbury Park, Calif.: Sage, 1993).

22. Beasley and Gibbons, *Taking Their Place*, p. 286.

23. All figures are from Beasley and Gibbons, *Taking Their Place*, p. 307.

24. Carlin Romano, "The Grisly Truth about Bare Facts," in Robert Karl Manoff and Michael Schudson, eds., *Reading the News* (New York: Pantheon, 1986), p. 44.

25. Leon V. Sigal, "Sources Make the News," in Robert Karl Manoff and Michael Schudson, eds. *Reading the News* (New York: Pantheon, 1986).

26. Mercedes Lynn de Uriarte, "A Minority Perspective," in *The Next Step: Toward Diversity in the Newspaper Business* (Reston, Va.: American Newspaper Publishers Association Foundation, 1991), p. 19, cited in Beasley and Gibbons, *Taking Their Place*, p. 309.

27. David Broder, *Behind the Front Page* (New York: Touchstone, 1987), p. 215.

28. Broder, *Behind the Front Page*, p. 202.

29. David Caute, *The Great Fear: The Anti-Communist Purge under Truman and Eisenhower* (New York: Simon and Schuster, 1979).

30. This discussion is based on Anthony Lewis, "The Press and the 1988 Campaign," *New York Review of Books*, January 25, 1989.

31. Gary Orren, "Introduction," in Martin Linsky, *Impact: How the Press Affects Federal Policymaking* (New York: Norton, 1986).

32. Martin Linsky, *Impact: How the Press Affects Federal Policymaking* (New York: Norton, 1986).

33. Richard Campbell, *60 Minutes and the News: A Mythology for Middle America* (Champaign: University of Illinois Press, 1991).

34. Robert Entman, "Blacks in the News: Television, Modern Racism and Cultural Change," *Journalism Quarterly* 69: 2 (Summer 1992), pp. 341–61.

35. David Sear, "Symbolic Racism," in Phyllis Katz and Dalmas Taylor, eds., *Eliminating Racism* (New York: Plenum, 1985).

36. Entman, "Blacks in the News," p. 355.

37. Joseph Klapper, *The Effects of Mass Communication* (New York: Free Press, 1960).

38. Shanto Iyengar and Donald Kinder, *News That Matters* (Chicago: University of Chicago Press, 1987); and Shanto Iyengar, *Is Anyone Responsible?* (Chicago: University of Chicago Press, 1991).

39. Iyengar, *Is Anyone Responsible?*

40. Jan A. Krosnick and Donald R. Kinder, "Altering the Foundations of Support for the President through Priming," *American Political Science Review* 84: 2 (1990), pp. 497–511.

41. Russell W. Neuman, Marion R. Just, and Ann N. Crigler, *Common Knowledge: News and the Construction of Political Meaning* (Chicago: University of Chicago Press, 1992).

42. Here we are borrowing heavily from the work of media scholar Robert Entman. He argues that media coverage of politics in America is not the result of a neutral marketplace of ideas. Instead, news coverage is the outcome of economic and political marketplaces. Robert Entman, *Democracy without Citizens* (New York: Oxford University Press, 1989).

43. Entman, *Democracy without Citizens*, Chapter 1.

44. For an excellent first-person account of these pressures, see Ed Joyce, *Prime Time, Bad Times* (New York: Anchor, 1988).

45. The statistics on cable television and VCRs are from Doris Graber, *Mass Media and American Politics*, 4th ed. (Washington, D.C.: Congressional Quarterly Press, 1993), p. 415.

46. Jon Katz, "Say Good Night, Dan," *Rolling Stone*, June 27, 1991, p. 82.

47. The following is drawn from Michael X. Delli Carpini and Bruce A. Williams, "Fictional and Non-Fictional Television Celebrate Earth Day: or, Politics Is Comedy Plus Pretense," *Cultural Studies* 8: 1 (January 1994).

48. See John Fiske, *Television Culture* (London: Methuen, 1987), Chapter 15.

49. For example, a CBS docudrama on Oliver North was rushed onto the air so that its broadcast would coincide with the actual trial of North. The trial, the docudrama, and people's responses to both become subjects of stories on the nightly news.

50. Edwin Diamond, *Good News, Bad News* (Cambridge, Mass.: MIT Press, 1978).

51. Mark Hertsgaard, *On Bended Knee* (New York: Schocken, 1989).

52. See especially Kathleen Hall Jamieson, *Dirty Politics* (New York: Oxford University Press, 1992).

53. Jamieson, *Dirty Politics*, Chapter 7.

54. Broder, *Behind the Front Page*, p. 181.

55. Broder, *Behind the Front Page*, pp. 181–82.

56. Jamieson, *Dirty Politics*, p. 5.

57. Joe McGinnis, *The Selling of the President 1968* (New York: Trident, 1968).

58. Mark Crispin Miller, *Boxed-In: The Culture of Television* (Evanston, Ill.: Northwestern University Press, 1988), p. 103.

59. Michael J. Robinson and Margaret A. Sheehan, *Over the Wire and on TV* (New York: Russell Sage Foundation, 1983).

60. Daniel Hallin, *We Keep America on Top of the World: Television and the Public Sphere* (New York: Routledge, 1994), p. 145.

61. Thomas Patterson, "The Press and Candidate Images," *International Journal of Public Opinion Research* 1: 2 (1989), pp. 123–35.

62. Broder, *Behind the Front Page*.

63. Entman, *Democracy without Citizens*, p. 55.

64. Daniel C. Hallin, *"The Uncensored War."*

65. Entman, *Democracy without Citizens*, p. 61.

66. W. Lance Bennett, *News: The Politics of Illusion* (New York: Longman, 1988),

Chapter 9

1. For a summary of the industry's troubles, from which this discussion is drawn, see Margaret E. Kriz, "Saving the S&Ls," *National Journal* 21: 2 (January 14, 1989), pp. 60–66.

2. Brooks Jackson and Paulette Thomas, "Waning Power: As S&L Crisis Grows, U.S. Savings League Loses Lobbying Clout," *Wall Street Journal,* March 7, 1989, p. A1.

3. Jackson and Thomas, "Waning Power."

4. Jackson and Thomas, "Waning Power."

5. Jackson and Thomas, "Waning Power."

6. For a discussion of the relationship between social movements and associated political interest groups, see Joyce Gelb and Ethel Klein, *Women's Movements: Organizing for Change* (Washington, D.C.: American Political Science Association, 1988).

7. For a summary of the nature of "organized interests" that fall outside the traditional definition of interest groups, see Allan J. Cigler, "Interest Groups: A Subfield in Search of an Identity," in William Crotty, ed., *Political Science: Looking to the Future* (Evanston, Ill.: Northwestern University Press, 1991), p. 103.

8. "Ex–Housing Aide Tells Jury of Pressure for Grants," *New York Times,* October 7, 1993, p. A1.

9. Many public interest groups circumvent this rule by forming tax-exempt aligned organizations, such as those devoted to research and education, and encouraging donations to them. See Ronald G. Shaiko, "More Bang for the Buck: The New Era of Full Service Public Organizations," in Allan J. Cigler and Burdette A. Loomis, eds., *Interest Group Politics,* 3rd ed. (Washington, D.C.: Congressional Quarterly Press, 1991), pp. 109–30.

10. The following discussion draws on the excellent summary by Allan J. Cigler, "Interest Groups: A Subfield in Search of an Identity," in Crotty, ed., *Political Science: Looking to the Future,* Vol. 4, pp. 99–135.

11. Jack L. Walker, "The Origins and Maintenance of Interest Groups in America," *American Political Science Review* 77: 2 (1983), pp. 390–406.

12. Walker, "Origins and Maintenance," p. 394.

13. The study, by Scholzman and Tierney, is summarized in Cigler, "Interest Groups," p. 103.

14. Cigler, "Interest Groups," p. 103, citing Kay Schlozman and John Tierney, *Organized Interests and American Democracy* (New York: Harper & Row, 1986).

15. Mary Lyndon Shanley and Shelby Lewis, *Women's Rights, Feminism, and Politics in the United States* (Washington, D.C.: American Political Science Association, 1988), pp. 3–4.

16. Gelb and Klein, *Women's Movements,* p. vi.

17. See Gelb and Klein, *Women's Movements,* for a brief overview of each of these categories of groups.

18. Edward Zigler and Susan Muendrew, *Head Start: The Inside Story of America's Most Successful Educational Experiment* (New York: HarperCollins, 1992).

19. Jon Wiener, "The Olin Money Tree: Dollars for Neocon Scholars," *Nation,* January 1, 1990, pp. 11–13. Wiener reported that Olin gave $3.6 million to Allan Bloom, author of the best-selling 1987 book *The Closing of the American Mind,* to establish a center at the University of Chicago.

20. *The Foundation Directory* (New York: Foundation Center, 1993), p. vii.

21. Robert H. Salisbury, John P. Heinz, Edward O. Laumann, and Robert L. Nelson, "Who Works with Whom? Interest Group Alliances and Opposition," *American Political Science Review* 81 (December 1987), p. 1220.

22. Cigler, "Interest Groups," p. 102.

23. Elizabeth Newlin Carney, "Liquor Groups Fear Another Round—On Them," *National Journal,* October 2, 1993, p. 2379.

24. Jill Abramson and Paulette Thomas, "Lobbyists for S&L Industry Hound Congress, Hoping to Influence Bush's Bailout Legislation," *Wall Street Journal,* March 29, 1989, p. A16.

25. Nathaniel C. Nash, "Rift Runs Deep in Savings Bill Battle," *New York Times,* April 17, 1989, p. D1. Much to the League's embarrassment, several S&L heads withdrew their institutions from the League and denounced its lobbying efforts.

26. For a good summary of this position, see Jeffrey M. Berry, *The Interest Group Society,* 2nd ed. (Glenview, Ill.: Scott, Foresman, 1989).

27. Charles E. Lindblom, *Politics and Markets: The World's Political-Economic Systems* (New York: Basic, 1977), Chapter 13. See also E. E. Schattschneider, *The Semi-Sovereign People* (New York: Holt, 1975).

28. For a description and analysis of this effort, see Richard L. Berke, "Expensive Lobbying Pays Off for Rifle Association," *New York Times,* September 22, 1988, p. A32.

29. Berke, "Expensive Lobbying."

30. Charles Mohr, "Gun Control Plan Rejected in House," *New York Times,* September 16, 1989, p. A1.

31. Joel Brinkley, "Cultivating the Grass Roots to Reap Legislative Benefits," *New York Times,* November 1, 1993, p. A1.

32. Glenn Rifkin, "I.B.M. Urges 110,000 Workers to Help Defeat Health-Care Bills," *New York Times,* August 19, 1994, p. A1.

33. See Brinkley, "Cultivating the Grass Roots," for a description of these techniques.

34. Gelb and Klein, *Women's Movements,* p. 8.

35. Cigler, "Interest Groups," p. 118.

36. The examples in this paragraph are reported in Abramson and Thomas, "Lobbyists for S&L Industry."

37. Douglas Jehl, "Lobbying Rules for Ex-Officials at Issue Again," *New York Times,* December 8, 1993, p. A1.

38. Tom Webb, "Former Regulator Blames Lobbyists and Congress for S&L Problems," *Philadelphia Inquirer,* February 6, 1989.

39. Brooks Jackson, "As Thrift Industry's Troubles and Losses Mounted, Its PACs' Donations to Key Congressmen Surged," *Wall Street Journal,* February 7, 1989, p. A22.

40. "At a Glance: Politics," *National Journal,* September 19, 1993, p. 2257.

41. Elizabeth Drew, "A Reporter at Large (Lobbyist)," *New Yorker,* January 9, 1978, p. 41. Quoted in David J. Vogler, *The Politics of Congress,* 4th ed. (Boston: Allyn and Bacon, 1983), p. 260.

42. Berke, "Expensive Lobbying."

43. Shanley and Lewis, *Women's Rights, Feminism, and*

Politics in the United States, p. 2. For a brief summary of the women's suffrage movement, see pp. 1–3 and 7–9.

44. For a history of the National Welfare Rights Organization, see Guida West, *The National Welfare Rights Movement: The Social Protest of Poor Women* (New York: Praeger, 1981).

45. Ronald J. Hrebenar and Clive S. Thomas edited a series of books on interest groups that demonstrate their growth in the states. See, for example, *Interest Group Politics in the Northeastern States* (University Park: Pennsylvania State University Press, 1993). In this volume, political scientist Douglas I. Hodgkin, for example, shows how Maine's politics is no longer dominated by a "big three" of utilities, timber, and textile and shoe manufacturers. Many new groups have formed to provide a more diverse political environment there. See Douglas I. Hodgkin, "Maine: From the Big Three to Diversity," in Hrebenar and Thomas, eds., *Interest Group Politics in the Northeastern States.*

46. The executive director of the League of Conservation Voters, for example, wrote to 1,000 of the group's largest contributors about the disaster. The National Resources Defense Council planned to mail half a million letters seeking funds based on the disaster. E. J. Dionne Jr., "Big Oil Spill Leaves Its Mark on Politics of Environment," *New York Times*, April 3, 1989, p. A1.

47. Cigler, "Interest Groups," p. 121.

48. Robert H. Salisbury, "The Paradox of Interest Groups in Washington—More Groups, Less Clout," in Anthony King, ed., *The New American Political System*, 2nd ed. (Washington, D.C.: AEI Press, 1990), pp. 209–10.

49. Salisbury, Heinz, Laumann, and Nelson, "Who Works with Whom? Interest Group Alliances and Opposition," pp. 1226–28.

50. Cigler, "Interest Groups," p. 125.

Chapter 10

1. For a description of this battle, see *Congressional Quarterly Almanac 1993* (Washington, D.C.: Congressional Quarterly Press, 1994), pp. 170–81.

2. See David S. Cloud, "Defection of House Leaders Reflects Deeper Concerns," *Congressional Quarterly Weekly Report*, September 11, 1993, pp. 2373–75; and Jon Healey and Thomas H. Moor, "Clinton Forms New Coalition to Win NAFTA's Approval," *Congressional Quarterly Weekly Report*, November 20, 1993, pp. 3181–83.

3. Katherine Tate, *From Protest to Politics: The New Black Voters in American Elections* (New York: Russell Sage, 1993), Chapter 3.

4. Quoted in Larry Sabato, *The Party's Just Begun: Shaping Political Parties for America's Future* (Glenview, Ill.: Scott, Foresman, 1987), p. 32.

5. For a good description of this classification, see John F. Bibby, *Politics, Parties, and Elections in America* (Chicago: Nelson-Hall, 1987), Chapter 2, pp. 20–34.

6. Joel H. Silbey, "The Rise and Fall of American Political Parties, 1790–1990," Chapter 1 in L. Sandy Maisel, ed., *The Parties Respond: Changes in the American Party System* (Boulder: Westview, 1990). The following discussion is drawn from Silbey.

7. Silbey, "The Rise and Fall of American Political Parties," p. 9.

8. Cornelius P. Cotter, James Gibson, John F. Bibby, and Robert J. Huckshorn, *Party Organizations in American Politics* (Pittsburgh: University of Pittsburgh Press, 1984), p. 42.

9. Harold W. Stanley and Richard G. Niemi, *Vital Statistics on American Politics*, 4th ed. (Washington, D.C.: Congressional Quarterly Press, 1994), p. 149.

10. Stanley and Niemi, *Vital Statistics*, p. 90.

11. For a history of these parties, see Hanes Walton Jr., *Black Political Parties: An Historical and Political Analysis* (New York: Free Press, 1972).

12. For a description of the Populists and their effects, see the classic work on political parties by V. O. Key, *Politics, Parties, and Pressure Groups*, 4th ed. (New York: Thomas Crowell, 1958), pp. 282–86.

13. For discussions of factors affecting the number of parties, see Maurice Duverger, *Political Parties* (New York: Wiley, 1951); and Clinton Rossiter, *Parties and Politics in America* (Ithaca, N.Y.: Cornell University Press, 1960).

14. Nelson W. Polsby and Aaron Wildavsky, *Presidential Elections*, 6th ed. (New York: Scribner's, 1984), p. 42.

15. Paul Allan Beck and Frank J. Sorauf, *Party Politics in America*, 7th ed. (New York: HarperCollins, 1992), p. 70.

16. See Anthony Gierzynski, *Legislative Party Campaign Committees in the American States* (Lexington: University Press of Kentucky, 1992).

17. For an entertaining and penetrating description of Mayor Richard J. Daley's reign, see Mike Royko, *Boss* (New York: Dutton, 1971).

18. For a discussion of party organizations, see Cornelius P. Cotter et al., *Party Organizations in American Politics* (New York: Praeger, 1984), especially Chapter 3.

19. For a good summary of state party organizations, see Beck and Sorauf, *Party Politics in America*, pp. 86–91; and John F. Bibby et al., "Parties in State Politics," in Virginia Gray et al., eds., *Politics in the American States: A Comparative Analysis* (Glenview, Ill.: Scott, Foresman, 1992), pp. 101–11.

20. Cotter et al., *Party Organizations in American Politics*, Chapter 2.

21. John F. Bibby, "Party Organization at the State Level," in L. Sandy Maisel, ed., *The Parties Respond: Changes in the American Party System* (Boulder: Westview, 1990), pp. 36–38.

22. Beck and Sorauf, *Party Politics in America*, p. 95.

23. Beck and Sorauf, *Party Politics in America*, p. 92.

24. Austin Ranny and Willmore Kendall, *Democracy and the American Party System* (New York: Harcourt, 1956), p. 508.

25. Beck and Sorauf, *Party Politics in America*, pp. 90–91.

26. Frank J. Sorauf, *Money in American Elections* (Glenview, Ill.: Scott, Foresman/Little, Brown, 1988), p. 55.

27. Federal Election Commission, "1994 Congressional Fundraising Climbs to New High," press release, April 28, 1995, p. 45.

28. Beck and Sorauf, *Party Politics in America*, p. 187, Table 7.3. In 1952, 86 percent of strong Democrats and 85 percent of strong Republicans voted a straight ticket. By 1984 the rate had fallen to 69 percent and 59 percent, respectively.

29. John F. Bibby et al., "Parties in State Politics," in Virginia Gray et al., eds., *Politics in the American States: A Comparative Analysis* (Glenview, Ill.: Scott, Foresman, 1992), p. 92.

30. A. James Reichley, *The Life of the Parties: A History of American Political Parties* (New York: Free Press, 1992), pp. 371–72.

31. Barbara Sinclair, "The Congressional Party: Evolving Organizational, Agenda-Setting, and Policy Roles," in L. Sandy Maisel, ed., *The Parties Respond: Changes in the American Party System* (Boulder: Westview, 1990), pp. 246–47.

32. Herbert McClosky, Paul J. Hoffman, and Rosemary O'Hara, "Issue Conflict and Consensus in Political Parties," *American Political Science Review* 54 (1960), pp. 405–27. The study also compared the views of rank-and-file voters; it found that they held very similar views on a wide range of issues.

33. For a brief portrait of Democratic delegates, see "The Democratic Delegates: Who Are They?" *New York Times*, July 13, 1992, p. B6.

34. Sabato, *The Party's Just Begun*, p. 15. Sabato notes that the Democratic platform pledged to make ratification of the Equal Rights Amendment (ERA) a top priority, proposed tough restraints on handguns, and criticized the "Star Wars" program. The Republicans did not support ratification of the ERA, supported the "right to keep and bear arms," and enthusiastically supported the development of Star Wars.

35. Michael B. Berkman, *The State Roots of National Politics: Congress and the Tax Agenda, 1978–1986* (Pittsburgh: University of Pittsburgh Press, 1993), p. 83, Figure 11.

36. The figures on party support in this paragraph come from Stanley and Niemi, *Vital Statistics*, p. 277.

37. "When Congress Had to Choose, It Voted to Back Clinton," *Congressional Quarterly Weekly Report*, December 18, 1993, p. 3427.

38. Benjamin Ginsberg, "Elections and Public Policy," *American Political Science Review* 70 (1976), pp. 41–49.

39. Ginsberg, "Elections and Public Policy," p. 43.

40. For a summary of this transformation, see Edward G. Carmines and James A. Stimson, *Issue Evolution: Race and the Transformation of American Politics* (Princeton: Princeton University Press, 1989), especially Chapters 8 and 9.

41. Stanley and Niemi, *Vital Statistics*, pp. 402–3.

42. Susan J. Carroll, *Women as Candidates in American Politics* (Bloomington: Indiana University Press, 1985), Chapter 3.

43. Barbara C. Burrell, "Party Decline, Party Transformation, and Gender Politics: The USA," in Joni Lovenduski and Pippa Norris, eds., *Gender and Party Politics* (London: Sage, 1993), p. 295.

Chapter 11

1. Glen Burkins, "S. Africans Undeterred by Long Lines," *Philadelphia Inquirer*, April 28, 1994, p. A1.

2. Lester W. Milbrath and M. L. Goel, *Political Participation: How and Why Do People Get Involved in Politics?* 2nd ed. (Chicago: Rand-McNally, 1977).

3. Andrew Kohut, *The Vocal Minority in American Politics* (Washington, D.C.: Times Mirror Center for The People & The Press, 1993), p. 25.

4. Sidney Verba and Norman H. Nie, *Participation in America: Political Democracy and Social Equality* (New York: Harper & Row, 1972). See especially Chapter 4, "The Modes of Participation: An Empirical Analysis," pp. 56–81.

5. Kohut, *The Vocal Minority*, p. 26.

6. M. Margaret Conway, *Political Participation in the United States*, 2nd ed. (Washington, D.C.: Congressional Quarterly Press, 1991), p. 8, Table I.2.

7. William H. Flanigan and Nancy H. Zingale, *Political Behavior of the American Electorate*, 5th ed. (Boston: Allyn and Bacon, 1983), p. 176.

8. Conway, *Political Participation*, p. 10, Table I.3.

9. See, for example, Gabriel A. Almond and Sidney Verba, *The Civic Culture* (Princeton, N.J.: Princeton University Press, 1963); or Milbrath and Goel, *Political Participation*, pp. 22–23, Table 1.2.

10. For a description of these events, see Frances Fox Piven and Richard A. Cloward, *Poor People's Movements: Why They Succeed, How They Fail* (New York: Pantheon, 1977), Chapters 2 and 3.

11. Harold W. Stanley and Richard G. Niemi, *Vital Statistics on American Politics*, 4th ed. (Washington, D.C.: Congressional Quarterly Press, 1994), p. 54, Table 2.2.

12. Lawrence Bobo and Franklin D. Gilliam Jr., "Race, Sociopolitical Participation, and Black Empowerment," in Richard G. Niemi and Herbert F. Weisberg, eds., *Controversies in Voting Behavior*, 3rd ed. (Washington, D.C.: Congressional Quarterly Press, 1993), p. 43, Table 4.1.

13. Katherine Tate, *From Protest to Politics: The New Black Voters in American Elections* (New York: Russell Sage, 1993), p. 5, Table 1.1. These figures are from 1984.

14. Bobo and Gilliam, "Race, Sociopolitical Participation, and Black Empowerment," p. 44.

15. John A. Garcia and Carlos H. Arce, "Political Orientations and Behaviors of Chicanos: Trying to Make Sense out of Attitudes and Participation," in F. Chris Garcia, ed., *Latinos and the Political System* (Notre Dame, Ind.: University of Notre Dame Press, 1988), p. 128.

16. Virginia Sapiro, *Women, Political Action, and Political Participation* (Washington, D.C.: American Political Science Association, 1988), pp. 10, 17.

17. Stanley and Niemi, *Vital Statistics*, p. 87, Table 3.1.

18. For a report of recently published research confirming this pattern, see Sidney Verba, Kay Lehman Scholzman, Henry Brady, and Norman H. Nie, "Citizen Activity: Who Participates? What Do They Say?" *American Political Science Review* 87 (1993), p. 306, Figure 1.

19. For an excellent full discussion of political efficacy, see Paul Abramson, *Political Attitudes in America* (San Francisco: Freeman, 1983), Chapters 8–10.

REFERENCES

20. Michael M. Gant and Norman R. Luttbeg, *American Electoral Behavior: 1952–1988* (Itasca, Ill.: F. E. Peacock, 1991), based on data in Table 3.3. These figures are from the 1988 election.

21. Milbrath and Goel, *Political Participation*, p. 64.

22. Verba et al., "Citizen Activity," p. 306, Figure 1.

23. Kohut, *The Vocal Minority*, pp. 8–9.

24. Raymond E. Wolfinger and Steven J. Rosenstone, *Who Votes* (New Haven: Yale University Press, 1980), p. 1.

25. Ruy A. Teixeira, *The Disappearing American Voter* (Washington, D.C.: Brookings Institution, 1992), p. 1.

26. Teixeira, *The Disappearing American Voter*, p. 8.

27. "A Correction: Voter Turnout in 20 Primaries So Far," *New York Times*, April 15, 1992, p. A24.

28. Teixeira, *The Disappearing American Voter*, pp. 13–18.

29. Teixeira, *The Disappearing American Voter*, p. 57.

30. See Chandler Davidson and Bernard Grofman, eds., *The Quiet Revolution in the South: The Impact of the Voting Rights Act, 1965–1990* (Princeton, N.J.: Princeton University Press, 1994).

31. Neal R. Pierce and Jerry Hagstrom, "The Hispanic Community: A Growing Force to Be Reckoned With," in F. Chris Garcia, ed., *Latinos and the Political System* (Notre Dame, Ind.: University of Notre Dame Press, 1988), p. 14.

32. Don Edwards, "The Voting Rights Act of 1965, As Amended," in Lorn S. Foster, ed., *The Voting Rights Act: Consequences and Implications* (New York: Praeger, 1985), pp. 5–6.

33. Gant and Luttbeg, *American Electoral Behavior*, p. 89.

34. Michael Wines, "Senate Approves Bill to Protect Abortion Clinics," *New York Times*, May 13, 1994, p. A1.

35. Gant and Luttbeg, *American Electoral Behavior*, p. 40. The reported voting rates are somewhat inflated because more people say they voted than actually did. The point remains, however, that strong identifiers vote at higher rates.

36. Stanley and Niemi, *Vital Statistics*, pp. 140–41, Table 4.4.

37. Flanigan and Zingale, *Political Behavior of the American Electorate*, p. 49, Table 3.3.

38. The classic formulation of this approach, which was used to explain voting in 1952 and 1956, can be found in Angus Campbell, Philip E. Converse, Warren E. Miller, and Donald E. Stokes, *The American Voter* (New York: Wiley, 1960), especially Chapters 3 and 4.

39. Campbell et al., *The American Voter*, p. 74.

40. John H. Kessel, *Presidential Campaign Politics*, 3rd ed. (Chicago: Dorsey Press, 1988), p. 269, Table 9.2.

41. Gerald M. Pomper et al., "The Presidential Election," in *The Election of 1992: Reports and Interpretations* (Chatham, N.J.: Chatham House, 1993), p. 146, Table 5.3.

42. Edward R. Tufte, *Political Control of the Economy* (Princeton, N.J.: Princeton University Press, 1978), p. 9.

43. Tufte, *Political Control of the Economy*, p. 137.

44. Quoted by Jonathan Rauch, "Election-Day Economy," *National Journal*, October 24, 1987, p. 2661.

45. "The Economy and the Candidates," *New York Times*, September 14, 1988, p. A29.

46. Tufte, *Political Control of the Economy*, p. 128, Table 5.7.

47. Tufte, *Political Control of the Economy*, pp. 122–23.

48. For a widely cited and influential statement of this argument, see Morris P. Fiorina, *Retrospective Voting in American National Elections* (New Haven: Yale University Press, 1981).

49. Michael B. MacKuen, Robert S. Erickson, and James A. Stimson, "Peasants or Bankers? The American Electorate and the U.S. Economy," *American Political Science Review* 86 (1992), pp. 597–611.

50. Michael S. Lewis-Beck, *Economics and Elections: The Major Western Democracies* (Ann Arbor: University of Michigan Press, 1988), p. 132.

51. William G. Mayer, "The New Hampshire Primary," in Gary R. Orren and Nelson W. Polsby, eds., *Media and Momentum* (Chatham, N.J.: Chatham House, 1987), pp. 26–29.

52. Larry M. Bartels, *Presidential Primaries and the Dynamics of Public Choice* (Princeton, N.J.: Princeton University Press, 1988), p. 147, Table 7.1.

53. For a dissenting view on the unrepresentativeness of primary voters, see John G. Geer, *Nominating Presidents: An Evaluation of Voters and Primaries* (New York: Greenwood Press, 1989), Chapter 2.

54. Pat Dunham, *Electoral Behavior in the United States* (Englewood Cliffs, N.J.: Prentice Hall, 1991), p. 175.

55. For a good overview of this research and its problems, see Richard G. Niemi and Herbert F. Weisberg, *Controversies in Voting Behavior*, 3rd ed. (Washington, D.C.: Congressional Quarterly Press, 1993), Chapter 13: "What Determines Congressional and State-Level Voting?"

56. See Alan I. Abramowitz and Jeffrey A. Segal, *Senate Elections* (Ann Arbor: University of Michigan Press, 1992), p. 37, Table 2.7; and Dunham, *Electoral Behavior*, pp. 188–89.

57. Robert S. Erikson and Gerald C. Wright, "Voters, Candidates, and Issues in Congressional Elections," in Lawrence C. Dodd and Bruce I. Oppenheimer, eds., *Congress Reconsidered*, 5th ed. (Washington, D.C.: Congressional Quarterly Press, 1993), p. 97.

58. Stanley and Niemi, *Vital Statistics*, p. 147, Table 4.8.

59. Stanley and Niemi, *Vital Statistics*, pp. 206–7, Table 7.5.

60. Everett Carll Ladd, ed., *America at the Polls: 1994* (Storrs, Conn.: Roper Center, 1995), p. 1.

61. Stanley and Niemi, *Vital Statistics*, p. 142, Table 4.5. These figures underestimate the role of party identification. They lump "independent leaners" with stronger identifiers; approximately 7 percent of the total vote came from independents.

62. The classic formulation of this theory appeared in Edward R. Tufte, "Determinants of the Outcomes of Midterm Congressional Elections," *American Political Science Review* 69 (1975), pp. 812–26.

63. Niemi and Weisberg, *Controversies in Voting Behavior*, p. 209.

64. Gary C. Jacobson and Samuel Kernell introduced this explanation in their book *Strategy and Choice in Congressional Elections* (New Haven: Yale University Press, 1983). See also Gary C. Jacobson, *The Electoral Origins of Divided Government* (Boulder: Westview, 1990), Chapter 4.

65. Niemi and Weisberg, in *Controversies in Voting Behavior*, p. 215, cite four separate studies that find a relationship between the quality of challengers and their share of the vote.

66. Gary C. Jacobson, *The Politics of Congressional Elections*, 2nd ed. (New Haven: Yale University Press, 1980), p. 110.

67. Based on Table 2.9 in Abramowitz and Segal, *Senate Elections*, p. 39.

68. Niemi and Weisberg, *Controversies in Voting Behavior*, p. 214.

69. For a statement of this argument with supporting evidence, see Benjamin Ginsberg, *The Consequences of Consent: Elections, Citizen Control and Popular Acquiescence* (Reading, Mass.: Addison-Wesley, 1982), p. 178.

70. Jan E. Leighley and Jonathan Nagler, "Socioeconomic Class Bias in Turnout, 1964–1988: The Voters Remain the Same," *American Political Science Review* 86 (1992), pp. 725–36.

Chapter 12

1. *Congressional Quarterly Weekly Report*, July 18, 1992, p. 2131.

2. Information on the backgrounds of presidents is drawn from Richard A. Watson and Norman C. Thomas, *The Politics of the Presidency*, 2nd ed. (Washington, D.C.: Congressional Quarterly Press, 1988), pp. 119–33.

3. Richard A. Watson, *The Presidential Contest*, 2nd ed. (New York: Wiley, 1984), Appendix C, lists all "major candidates" between 1932 and 1980. No African Americans, women, Hispanics, Jews, or members of other traditionally disadvantaged groups appear on the list.

4. For a selected list of women who have run for president, see Marjorie P. K. Weister and Jean S. Arbeiter, *Womanlist* (New York: Atheneum, 1981), pp. 364–65.

5. For a discussion of obstacles to women's seeking the presidency, see Bella Abzug, *Gender Gap* (Boston: Houghton Mifflin, 1984), pp. 192–97. Susan J. Carroll, *Women as Candidates in American Politics* (Bloomington: Indiana University Press, 1985), provides a more general treatment of obstacles to women's winning political office.

6. Dianna Gordon, "Republican Women Make Gains," *State Legislatures*, February 1995, p. 15. Gordon reports that the overall total actually dropped from 1,547 to 1,533 after the 1994 elections.

7. Abzug, *Gender Gap*, p. 194.

8. Warren Weaver Jr., "Schroeder, Assailing 'the System,' Decides Not to Run for President," *New York Times*, September 29, 1987, p. A1.

9. George Gallup Jr., *The Gallup Poll: Public Opinion 1983* (Wilmington, Del.: Scholarly Resources, 1984), p. 70.

10. Federal Election Commission, "1994 Congressional Fundraising Climbs to New High," press release, April 28, 1995, p. 2.

11. B. Drummond Ayres Jr., "California Primary Sets a Lavish Tone, Lifting a Long Shot," *New York Times*, June 7, 1994, p. A1.

12. Frank J. Sorauf, *Inside Campaign Finance: Myths and Realities* (New Haven: Yale University Press, 1992), Chapter 2.

13. Sorauf, *Inside Campaign Finance*, p. 150.

14. F. Christopher Arterton, "Campaign '92: Strategies and Tactics of the Candidates," in Gerald M. Pomper et al., eds., *The Election of 1992: Reports and Interpretations* (Chatham, N.J.: Chatham House, 1993), p. 83.

15. David J. Jefferson, "Keating of American Continental Corp. Comes Out Fighting," *Wall Street Journal*, April 18, 1989.

16. Federal Election Commission, "FEC Reports on Political Party Activity for 1993–94," press release, April 13, 1995, pp. 6–7.

17. For a discussion of leadership PACs, see Ross K. Baker, *The New Fat Cats: Members of Congress as Political Benefactors* (New York: Priority Press, 1989).

18. Anthony Lewis of the *New York Times* wrote that Roger Ailes, George Bush's media consultant, "played a key part in persuading Mr. Bush to pick Dan Quayle, a former Ailes client, as a running mate." See "A Corrupted Process," *New York Times*, October 13, 1988, p. A23. For a general discussion of the role of consultants in candidate recruitment, see Larry Sabato, "How Political Consultants Choose Candidates," in Peter J. Woll, ed., *Behind the Scenes in American Government*, 7th ed. (Glenview, Ill.: Scott, Foresman, 1989), Chapter 12.

19. Larry J. Sabato, "Political Influence, the News Media, and Campaign Consultants," *PS: Political Science & Politics* 22: 1 (March 1989), p. 16.

20. James A. Barnes, "Courting California," *National Journal*, April 17, 1993, p. 954.

21. "Mr. Perot's Ballot Blues," *New York Times*, April 27, 1992, p. A16.

22. James A. Barnes, "Back in the Picture," *National Journal*, October 2, 1993, p. 2398.

23. Half of the large cities in the United States elected all council members at large in the 1980s. See Richard L. Engstrom and Michael D. McDonald, "The Effect of At-Large versus District Elections on Racial Representation in U.S. Municipalities," in Bernard Grofman and Arend Lijphart, eds., *Electoral Laws and Their Political Consequences* (New York: Agathon Press, 1986), Chapter 13, p. 204.

24. See *Baker v. Carr*, 369 US 186 (1962).

25. Peter Woll, *Congress* (Boston: Little, Brown, 1985), p. 227.

26. Arterton, "Campaign '92," p. 84.

27. Harold W. Stanley and Richard G. Niemi, *Vital Statistics on American Politics*, 4th ed. (Washington, D.C.: Congressional Quarterly Press, 1994), p. 264, Table 8.5.

REFERENCES

28. For a discussion of these committees and their effects, see Anthony Corrado, *Creative Campaigning: PACs and the Presidential Selection Process* (Boulder: Westview, 1992).

29. For a description of these campaigns, see Kenneth D. Wald, "Ministering to the Nation: The Campaigns of Jesse Jackson and Pat Robertson," in Emmett H. Buell Jr. and Lee Sigelman, eds., *Nominating the President* (Knoxville: University of Tennessee Press, 1991), Chapter 5.

30. Data on 1992 presidential prenomination campaign finance come from Stanley and Niemi, *Vital Statistics*, p. 264, Table 8.5.

31. Stanley and Niemi, *Vital Statistics*, p. 264, Table 8.5.

32. According to Clyde Wilcox, "Financing the 1988 Prenomination Campaigns," in Emmett H. Buell Jr. and Lee Sigelman, eds., *Nominating the President* (Knoxville: University of Tennessee Press, 1991), p. 96, just 1 percent of presidential primary campaign funds came from PACs in 1988.

33. Richard L. Berke, "Dole Takes First Real Steps toward '96 Presidential Race," *New York Times*, June 15, 1994, p. A1.

34. Quoted by Jonathan Rauch, "Election-Day Economy," *National Journal*, October 24, 1987, p. 2661.

35. These figures are drawn from Herbert B. Asher, *Presidential Elections and American Politics: Voters, Candidates, and Campaigns since 1952*, 5th ed. (Belmont, Calif.: Wadsworth, 1992), p. 299, Table 10.1.

36. The discussion that follows draws on the analysis of Arterton, "Campaign '92," pp. 81–82.

37. Arterton, "Campaign '92," p. 87, quotes Clinton's campaign manager as saying that visits to states produced a two- to three-point rise in polls.

38. Arterton, "Campaign '92," pp. 87–88.

39. James Ceaser and Andrew Busch, *Upside Down and Inside Out: The 1992 Elections and American Politics* (Lanham, Md.: Littlefield & Adams, 1993), p. 161.

40. Attributed to an "unnamed campaign official" in a *Washington Post* article cited by Arterton, "Campaign '92," p. 80.

41. Paul J. Quirk and Jon K. Dalager, "The Election: A 'New Democrat,'" in Michael Nelson, ed., *The Elections of 1992* (Washington, D.C.: Congressional Quarterly Press, 1993), p. 63.

42. Stephen J. Wayne, *The Road to the White House: The Politics of Presidential Elections* (New York: St. Martin's Press, 1988), pp. 61–62.

43. This discussion is based on Stephen J. Wayne, *The Road to the White House 1992: The Politics of Presidential Elections* (New York: St. Martin's Press, 1992), pp. 176–77.

44. For a discussion of the new political technology, see Benjamin Ginsberg, *The Captive Public: How Mass Opinion Promotes State Power* (New York: Basic, 1986), Chapter 5, pp. 160–66; and Joel L. Swerdlow, ed., *Media Technology and the Vote: A Source Book* (Boulder: Westview, 1988).

45. Russell W. Getter and James Emerson Titus, "Voter Registration Tapes: Mining for New Votes, New Voters, and New Money," in Larry Sabato, ed., *Campaigns and Elections: A Reader in Modern American Politics* (Glenview, Ill.: Scott, Foresman, 1989), p. 85. One can obtain the phone numbers of approximately 75 percent of registered voters for about six cents each.

46. Quoted by Arterton, "Campaign '92," p. 90. The reporter was the *Washington Post*'s Howard Kurtz.

47. See Frank Sorauf, *Money in American Elections* (Boston: Little, Brown, 1988), Chapter 2, for a history of campaign finance.

48. Sorauf, *Money in American Elections*, p. 61.

49. Sorauf, *Money in American Elections*, p. 58.

50. Figures for 1978 are from Sorauf, *Money in American Elections*, p. 55. Figures for 1994 are from Federal Election Commission, "1994 Congressional Fundraising Climbs to New High," p. 2.

51. Calculated from data from Federal Election Commission, "PAC Activity in 1994 Remains at 1992 Levels," press release, March 31, 1995, p. 2.

52. Calculated from data from Federal Election Commission, "1994 Congressional Fundraising Climbs to New High," p. 2.

53. National Committee for an Effective Congress, "Election Update," January 1995, p. 4.

54. Marjorie Randon Hershey, "The Congressional Elections," in Gerald M. Pomper et al., eds., *The Election of 1992: Reports and Interpretations* (Chatham, N.J.: Chatham House, 1993), pp. 173–74.

55. Data on turnout are from Everett Carll Ladd, ed., *America at the Polls: 1994* (Storrs, Conn.: Roper Center for Public Opinion Research, 1995), p. 15.

56. For a good summary of these efforts, see Richard G. Niemi and Herbert F. Weisberg, *Controversies in Voting Behavior*, 3rd ed. (Washington, D.C.: Congressional Quarterly Press, 1993), Chapter 13.

57. George Pettinico, "1994 Vote: Where Was Public Thinking at Vote Time?" in Everett Carll Ladd, ed., *America at the Polls: 1994* (Storrs, Conn.: Roper Center for Public Opinion Research, 1995), p. 36.

58. Ladd, ed., *America at the Polls: 1994*, p. 27.

59. "Has Gingrich Gone Too Far with GOPAC?," *National Journal*, October 1, 1994, p. 2272. In the article, a GOPAC fundraising letter is cited that claimed GOPAC helped elect 41 of the 48 new Republican members of Congress in 1992.

60. Michael J. Malbin, "1994 Vote: The Money Story," in Ladd, ed., *America at the Polls: 1994*, p. 127.

61. Ladd, ed., *America at the Polls: 1994*, p. 71.

62. For a statement of this argument, see Benjamin Ginsberg, *The Consequences of Consent: Elections, Citizen Control and Popular Acquiescence* (Reading, Mass.: Addison-Wesley, 1982), p. 178.

63. E. E. Schattschneider, *The Semi-Sovereign People* (Hinsdale, Ill.: Dryden Press, 1975), p. 138.

64. For a succinct, clear summary of the many factors that make voting a blunt instrument for citizens to use in shaping policy, see Charles Lindblom, *The Policy Making Process*, 2nd ed. (Englewood Cliffs, N.J.: Prentice Hall, 1980), Chapter 12, "The Imprecision of Voting."

65. Edward R. Tufte, *Political Control of the Economy* (Princeton, N.J.: Princeton University Press, 1978), p. 21.

66. Tufte, *Political Control of the Economy*, p. 22.

67. Tufte, *Political Control of the Economy*, p. 137.

68. For an excellent discussion of the occasions when the agenda of politics shifts, including elections, see John W. Kingdon, *Agendas, Alternatives, and Public Policies* (Boston: Little, Brown, 1984), especially Chapter 8.

Chapter 13

1. *The Budget of the United States Government, FY 1996* (Washington, D.C.: Government Printing Office, 1995), Historical Documents, p. 187. This total does not include employees of the General Accounting Office and the Library of Congress who do not work directly for Congress.

2. *Congressional Quarterly Weekly Report*, November 12, 1988, pp. 3293–95.

3. Alissa J. Rubin, "1994 Elections Are Looking Like the 'Off Year' of the Woman," *National Journal*, October 15, 1994, p. 2972.

4. Richard F. Fenno Jr., *Home Style* (Boston: Little, Brown, 1978). For a good summary of Fenno's work, see Morris P. Fiorina and David W. Rohde, "Richard Fenno's Research Agenda and the Study of Congress," in Fiorina and Rohde, eds., *Home Style and Washington Work: Studies of Congressional Politics* (Ann Arbor: University of Michigan Press, 1989), pp. 1–15.

5. Fenno, *Home Style*, p. 168.

6. Harrison W. Fox Jr. and Susan Webb Hammond, *Congressional Staffs: The Invisible Force in American Law Making* (New York: Free Press, 1977), p. 143.

7. Jeffrey E. Birnbaum, "Many Ex-Aides to Congressional Tax Writers Now Lobby for Companies to Undo '86 Reforms," *Wall Street Journal*, August 15, 1989, p. A16.

8. Louis Jacobson, "Too Hard a Hill to Climb?" *National Journal*, June 4, 1994, p. 1321.

9. Louis Jacobson, "In Charge at the Committees," *National Journal*, January 21, 1995, p. 182.

10. Jacobson, "Too Hard a Hill," p. 1321.

11. Richard Fenno Jr., *Congressmen in Committees* (Boston: Little, Brown, 1973).

12. Lawrence C. Dodd and Bruce I. Oppenheimer, "Maintaining Order in the House: The Struggle for Institutional Equilibrium," in Dodd and Oppenheimer, eds., *Congress Reconsidered*, 5th ed. (Washington, D.C.: Congressional Quarterly Press, 1993), pp. 51–53.

13. Jill Abramson and Paulette Thomas, "Lobbyists for S&L Industry Hound Congress, Hoping to Influence Bush's Bailout Legislation," *Wall Street Journal*, March 29, 1989, p. A16.

14. For a discussion of these changes, see William J. Keefe, *Congress and the American People*, 3rd ed. (Englewood Cliffs, N.J.: Prentice Hall, 1988), p. 107.

15. Jeff Shear, "Pain's the Game," *National Journal*, January 14, 1995, p. 108.

16. Adam Clymer, "House Democrats Face Painful Choices," *New York Times*, December 10, 1994, p. A30.

17. Norman Ornstein et al., *Vital Statistics on Congress: 1991–1992*, p. 111, Table 4.2.

18. For a discussion of the growth in subcommittee power, see David J. Vogler, *The Politics of Congress*, 4th ed. (Boston: Allyn and Bacon, 1983), pp. 151–53.

19. "Rules Changes Open the Process . . . but Strengthen the Reins of Power," *Congressional Quarterly Weekly Report*, January 7, 1995, pp. 14–15.

20. Keefe, *Congress and the American People*, p. 95.

21. For a description of these committees, see Steven S. Smith, *Call to Order: Floor Politics in the House and Senate* (Washington, D.C.: Brookings Institution, 1989), Chapter 7.

22. For a detailed discussion of conference committees, see Kenneth A. Shepsle and Barry R. Weingast, "Penultimate Power: Conference Committees and the Legislative Process," in Morris P. Fiorina and David W. Rohde, eds., *Home Style and Washington Work: Studies of Congressional Politics* (Ann Arbor: University of Michigan Press, 1989), pp. 199–217.

23. For a discussion of these and other powers of the Speaker, see Vogler, *Politics of Congress*, pp. 117–18; or Peter J. Woll, *Congress* (Boston: Little, Brown, 1985), pp. 154–56.

24. Quoted in Vogler, *Politics of Congress*, p. 202.

25. For a summary of folkways, see David J. Vogler, *The Politics of Congress*, 5th ed. (Boston: Allyn and Bacon, 1988), pp. 203–10.

26. Donald S. Matthews, *U.S. Senators and Their World* (New York: Vintage, 1960), is the classic study of Senate folkways.

27. For a description of one such episode, see Katharine Q. Seelye, "In Attack on Gingrich, Democrats Use His Tactics and Reap Chaos," *New York Times*, January 19, 1995, p. A1.

28. "Excerpts from Byrd and Dole Remarks on Balanced-Budget Amendment," *New York Times*, March 1, 1995, p. B16.

29. Eric M. Uslander, *The Decline of Comity in Congress* (Ann Arbor: University of Michigan Press, 1993), p. 21.

30. One student of Congress, Steven S. Smith, wrote an entire book—*Call to Order: Floor Politics in the House and Senate* (Washington, D.C.: Brookings Institution, 1989)—on the politics of floor action in the House and Senate.

31. Richard L. Hall, "Participation, Abdication, and Representation in Congressional Committees," in Lawrence C. Dodd and Bruce I. Oppenheimer, eds., *Congress Reconsidered*, 5th ed. (Washington, D.C.: Congressional Quarterly Press, 1993), p. 162. The discussion of behavior in committees draws heavily on Hall's excellent treatment.

32. Ornstein et al., *Vital Statistics on Congress: 1991–1992*, p. 113, Table 4.4.

33. In 1993, an average of 96 percent of House members participated in roll-call votes; in the Senate, 97.6 percent did. See *Congressional Quarterly Almanac: 1993* (Washington, D.C.: Congressional Quarterly Press, 1994), pp. 31–32.

34. *Congressional Quarterly Almanac: 1993*, p. 31-C.

35. Ann Cooper, "House Use of Suspensions Grows Drastically," *Congressional Quarterly Weekly Report*, September

30, 1978, p. 2693, quoted by Vogler, *The Politics of Congress*, p. 109. Vogler notes (p. 108) that in 1979–80 Congress took more than 2,300 recorded votes.

36. Donald Matthews and James Stimson, *Yeas and Nays: Normal Decision-Making in the U.S. House of Representatives* (New York: Wiley, 1974), p. 94.

37. Julie Rovner, "Catastrophic-Coverage Law Narrowly Survives Test," *Congressional Quarterly Weekly Report*, June 10, 1989, p. 1402.

38. Vogler, *Politics of Congress*, 4th ed., pp. 114–15.

39. John W. Kingdon, *Congressmen's Voting Decisions* (New York: Harper & Row, 1981), p. 42.

40. Lewis Anthony Dexter, "The Representative and His District," in Nelson Polsby and Robert Peabody, eds., *New Perspectives on the House of Representatives*, 2nd ed. (Chicago: Rand-McNally, 1969), pp. 3–29.

41. For a summary of research supporting this conclusion, see Frank Sorauf, *Money in American Elections* (Boston: Little, Brown, 1988), pp. 307–17.

42. Richard L. Hall and Frank W. Wayman, "Buying Time: Moneyed Interests and the Mobilization of Bias in Congressional Committees," *American Political Science Review* 84 (1990), pp. 797–820.

43. These data are drawn from Ornstein et al., *Vital Statistics on Congress: 1991–1992*, p. 151, Table 6.1; p. 153, Table 6.2; p. 156, Table 6.4.

44. "Clinton Keeps Southern Wing on His Team in 1993," *Congressional Quarterly Almanac: 1993*, p. 22-C.

45. Hanna F. Pitkin, *The Concept of Representation* (Berkeley: University of California Press, 1963), Chapter 6.

46. Dexter, "The Representative and His District," p. 6.

47. "Senate Debaters Call Canal Treaty Unpopular Issue," *Boston Globe*, February 10, 1978, p. 5, quoted in Vogler, *Politics of Congress*, 4th ed., pp. 76–77.

48. The delegate/trustee/politico distinction was introduced by John C. Wahlke, Heinz Eulau, William Buchanan, and Leroy Ferguson in *The Legislative System: Explorations in Legislative Behavior* (New York: Wiley, 1962).

49. For an insightful discussion of trust, see William Bianco, *Trust* (Ann Arbor: University of Michigan Press, 1994).

50. Maureen Dowd, "Senate Approves a 4-Star Rank for Admiral in Tailhook Affair," *New York Times*, April 20, 1994, p. A1.

51. Quoted in Aaron Wildavsky, *The Politics of the Budgetary Process*, 4th ed. (Boston: Little, Brown, 1984), p. 99.

Chapter 14

1. Neil A. Lewis, "Facing Opposition, Clinton Abandons Rights Nominee," *New York Times*, June 4, 1993, p. A1.

2. Richard L. Berke, "President Backs a Gay Compromise," *New York Times*, May 28, 1993, p. A1.

3. David E. Rosenbaum, "Snip, Snip, Snip, Clinton's Budget Takes Shape," *New York Times*, May 23, 1993, p. D1.

4. Robert Pear, "Clinton, in Compromise, to Drop Parts of Childhood Vaccine Plan," *New York Times*, May 5, 1993, p. A1.

5. Maureen Dowd, "Clinton's Reversals," *New York Times*, June 4, 1993, p. A16.

6. Richard L. Berke, "President Braces the Middle Class for Tax Increases," *New York Times*, February 16, 1993, p. A1.

7. Michael Duffy, "That Sinking Feeling," *Time*, June 7, 1993, p. 24.

8. Gwen Ifill, "Clinton Sees Need to Focus His Goals and Sharpen Staff," *New York Times*, May 5, 1993, p. A1.

9. Gwen Ifill, "Ex-Reagan Adviser Appointed to Post in the White House," *New York Times*, May 30, 1993, p. A1.

10. President's Committee on Administrative Management, *Report with Special Studies* (Washington, D.C.: Government Printing Office, 1937), p. 5. The commission was headed by Louis Brownlow and is often referred to as the Brownlow Commission.

11. For histories of the development and growth of the presidential office, see John Hart, *The Presidential Branch* (New York: Pergamon, 1987); and John P. Burke, *The Institutional Presidency* (Baltimore: Johns Hopkins University Press, 1992).

12. These figures are drawn from John P. Burke, "The Institutional Presidency," in Michael Nelson, ed., *The Presidency and the Political System* (Washington, D.C.: Congressional Quarterly Press, 1995), pp. 381–407.

13. Philip J. Hilts, "Into the Maelstrom," *New York Times*, June 27, 1993, p. A23.

14. Quoted in Charles O. Jones, *The Presidency in a Separated System* (Washington, D.C.: Brookings Institution, 1994), p. 59.

15. See Ifill, "Clinton Sees Need to Focus," p. A1; R. W. Apple Jr., "Clinton's Refocusing," *New York Times*, May 6, 1993, p. A22; Ifill, "Ex-Reagan Adviser Appointed to Post in the White House," p. A1.

16. H. R. Haldeman, *The Ends of Power* (New York: New York Times Books, 1978), pp. 58–59, quoted in Burke, *The Institutional Presidency*, p. 41.

17. Hugh Heclo, *A Government of Strangers: Executive Politics in Washington* (Washington, D.C.: Brookings Institution, 1977), p. 104.

18. Jones, *The Presidency in a Separated System*, p. 63.

19. Robert Pear, "Two in Bush's Cabinet Attack Democrats on Health Care," *New York Times*, January 10, 1992, p. A12.

20. Sam Howe Verhovek, "Selling Clinton Program, the Cabinet Pledges Better Times, but Not Soon," *New York Times*, February 19, 1993, p. A16.

21. Both quotations are from Richard A. Watson and Norman C. Thomas, *The Politics of the Presidency*, 2nd ed. (Washington, D.C.: Congressional Quarterly Press, 1988), p. 223.

22. Paul Light, *Vice Presidential Power: Advice and Influence in the White House* (Baltimore: John Hopkins University Press, 1984).

23. Some of the various roles performed by Gore are discussed in Bob Woodward, *The Agenda: Inside the Clinton White House* (New York: Simon and Schuster, 1994).

24. On Jackson's "kitchen cabinet," see Sidney M. Milkis and Michael Nelson, *The American Presidency: Origin and*

Development, 1776–1990 (Washington, D.C.: Congressional Quarterly Press, 1990), p. 125.

25. See Burke, *The Institutional Presidency*, p. 370.

26. See Robert Pear, "Economic Advisors Caution President on Medical Costs," *New York Times*, May 22, 1993, p. A1; Robert Pear, "A White House Fight," *New York Times*, May 23, 1993, p. A20.

27. Burke, *The Institutional Presidency*, p. 37.

28. *Youngstown Sheet and Tube Co. v. Sawyer*, 343 US 579 (1952).

29. Richard E. Neustadt, *Presidential Power: The Politics of Leadership, with Reflections on Johnson and Nixon* (New York: Wiley, 1976).

30. Neustadt, *Presidential Power*, Chapter 3.

31. Neustadt, *Presidential Power*, pp. 102–3.

32. This story and others discussed here regarding personal contacts between Clinton and members of Congress are taken from Richard L. Berke, "Courting Congress Nonstop, Clinton Looks for an Alliance," *New York Times*, March 8, 1993, p. A1.

33. See Clifford Krauss, "When the President Calls, Mavericks Run for Cover," *New York Times*, May 29, 1993, p. A1.

34. Douglas Jehl, "Scramble in the Capital for Today's Trade Pact Vote," *New York Times*, November 17, 1993, p. A1.

35. Berke, "Courting Congress Nonstop," p. A1.

36. Rosenbaum, "Snip, Snip, Snip, Clinton's Budget Takes Shape," p. D1.

37. Michael Wines, "Clinton Now Seems Prepared to Drop Any Wide Fuel Tax," *New York Times*, June 10, 1993, p. A1.

38. Woodward, *The Agenda*, p. 182.

39. Neustadt, *Presidential Power*.

40. David E. Rosenbaum, "Clinton's Plan for Economy May Hinge on His Popularity," *New York Times*, April 29, 1993, p. A1.

41. Kristen Renwick Monroe, *Presidential Popularity and the Economy* (New York: Praeger, 1984).

42. Richard L. Berke, "Poll Shows Raid on Iraq Buoyed Clinton's Popularity," *New York Times*, June 29, 1993, p. A7.

43. Richard A. Brody, *Assessing the President: The Media, Elite Opinion, and Public Support* (Stanford, Calif.: Stanford University Press, 1991).

44. On "priming" by the news media, see Shanto Iyengar and Donald R. Kinder, *News That Matters: Television and American Opinion* (Chicago: University of Chicago Press, 1987).

45. Quoted in Hedrick Smith, *The Power Game: How Washington Works* (New York: Random House, 1988), pp. 405–6.

46. Smith, *The Power Game*, p. 405.

47. See George C. Edwards III, *At the Margins: Presidential Leadership of Congress* (New Haven: Yale University Press, 1989); and Jon R. Bond and Richard Fleisher, *The President in the Legislative Arena* (Chicago: University of Chicago Press, 1990).

48. Smith, *The Power Game*, p. 339.

49. Woodward, *The Agenda*, p. 297.

50. Woodward, *The Agenda*, p. 217.

51. Samuel Kernell, *Going Public: New Strategies of Presidential Leadership*, 2nd ed. (Washington, D.C.: Congressional Quarterly Press, 1993).

52. Jeffrey K. Tulis, *The Rhetorical Presidency* (Princeton, N.J.: Princeton University Press, 1987), p. 64.

53. Tulis, *The Rhetorical Presidency*, pp. 138–39.

54. Kernell, *Going Public*, p. 99.

55. Truman and Carter are quoted in Richard P. Nathan, *The Administrative Presidency* (New York: Wiley, 1983), p. 2.

56. Joel D. Aberbach and Bert A. Rockman, "Clashing Beliefs within the Executive Branch: The Nixon Administration Bureaucracy," *American Political Science Review* 70 (1976), pp. 456–68.

57. Richard D. Cole and David A. Caputo, "Presidential Control of the Senior Civil Service: Assessing the Strategies of the Nixon Years," *American Political Science Review* 73 (1979), pp. 399–413.

58. Nathan, *The Administrative Presidency*, Chapter 7.

59. See James W. Davis, *The President as Party Leader* (New York: Greenwood Press, 1992).

60. Adam Clymer, "Win or Lose, Vetoes by Bush Frustrate Democrats," *New York Times*, September 29, 1993, p. A22.

61. Clymer, "Win or Lose."

62. Bruce Miroff, "The Presidency and the Public: Leadership as Spectacle," in Michael Nelson, ed., *The Presidency and the Political System*, 2nd ed. (Washington, D.C.: Congressional Quarterly Press, 1988), pp. 271–92.

63. See, for example, Paul B. Sheatsley and Jacob J. Feldman, "A National Survey on Public Reactions and Behavior," in Bradley S. Greenberg and Edwin B. Parker, eds., *The Kennedy Assassination and the American Public* (Palo Alto, Calif.: Stanford University Press, 1965).

Chapter 15

1. The following case is drawn from Graham Allison, *Essence of Decision: Explaining the Cuban Missile Crisis* (New York: Little, Brown, 1971).

2. Allison, *Essence of Decision*, p. 110.

3. Max Weber, "On Bureaucracy," in *From Max Weber* (New York: Oxford University Press, 1946).

4. Paul Craig Roberts, *The Supply-Side Revolution: An Insider's Account of Policymaking in Washington* (Cambridge, Mass.: Harvard University Press, 1984), p. 254.

5. For a more formal treatment of this issue, see William Niskanen, *Bureaucracy and Representative Government* (New York: Aldine, 1971). After writing this book, Niskanen served as a budget official in the Reagan and Bush administrations.

6. George E. Berkley, *The Craft of Public Administration*, 4th ed. (Boston: Allyn and Bacon, 1984), p. 88.

7. See Kenneth J. Meier, *Politics and Bureaucracy: Policymaking in the Fourth Branch of Government*, 2nd ed.

(Monterey, Calif.: Brooks/Cole, 1987), pp. 32–39. For a complete description of the civil service system up to 1958, see Paul P. Van Riper, *History of the United States Civil Service* (Evanston, Ill.: Row, Patterson, 1958).

8. Hugh Heclo, *A Government of Strangers: Executive Politics in Washington* (Washington, D.C.: Brookings Institution, 1977).

9. Meier, *Politics and the Bureaucracy*, Chapter 2.

10. These figures come from James Q. Wilson, "The Rise of the Bureaucratic State," in Frederick S. Lane, ed., *Current Issues in Public Administration*, 2nd ed. (New York: St. Martin's Press, 1982), pp. 55–74.

11. One source estimates that in 1880 there were 100,000 homeless children in New York City alone. See Otto L. Bettman, *The Good Old Days: They Were Terrible* (New York: Random House, 1974), p. 44.

12. See Robert S. Friedman, "Representation in Regulatory Decision Making: Scientific, Industrial, and Consumer Inputs to the FDA," *Public Administration Review* 38: 3 (May–June 1978), pp. 205–14.

13. For an excellent discussion of the nature of bureaucratic expertise on which our treatment is based, see Frances E. Rourke, *Bureaucracy, Politics, and Public Policy*, 2nd ed. (Boston: Little, Brown, 1976), especially Chapter 1.

14. Herbert Kaufman, "Fear of Bureaucracy: A Raging Pandemic," *Public Administration Review* 41 (January–February 1981), p. 4.

15. Richard Nixon to John Ehrlichman, presidential transcripts published in *Washington Star-News*, July 20, 1974, and quoted by Thomas E. Cronin, *The State of the Presidency*, 2nd ed. (Boston: Little, Brown, 1980), p. 223.

16. Charles Lindblom, in *Public Policy-Making* (Englewood Cliffs, N.J.: Prentice Hall, 1983), calls this "scientific analysis" and distinguishes it from "partisan analysis." His discussion in Chapter 5 is an excellent summary of the role of analysis in bureaucratic politics.

17. Eugene Lewis, *American Politics in a Bureaucratic Age* (Cambridge, Mass.: Winthrop, 1978).

18. John Berbers, "How 43 Agencies Keep Surviving the Reagan Ax," *New York Times*, September 9, 1986, p. A20. The person quoted was Robert D. Reischauer, a senior fellow at the Brookings Institution in Washington, D.C.

19. Alexander M. Haig Jr., *Caveat: Realism, Reagan, and Foreign Policy* (New York: Macmillan, 1984), p. 60.

20. For a discussion of the relationship between the Agriculture Department and the Farm Bureau Federation, see Grant McConnell, *Private Power and American Democracy* (New York: Vintage, 1966), pp. 230–43.

21. See Philip Selznick, *TVA and the Grass Roots: A Study in the Sociology of Formal Organizations* (Berkeley: University of California Press, 1949).

22. Rourke, *Bureaucracy, Politics, and Public Policy*, p. 54.

23. Roberts, *The Supply-Side Revolution*, p. 135.

24. Charles Lindblom, "The Science of Muddling Through," *Public Administration Review* 19 (Spring 1959).

25. David Stockman, *The Triumph of Politics* (Boston: G. K. Hall, 1987), p. 154.

26. For an insightful discussion of the concept of policy windows and how the changing of administrations opens them, see John W. Kingdon, *Agendas, Alternatives, and Public Policies* (Boston: Little, Brown, 1984).

27. Martin Linsky, *Impact: How the Press Affects Federal Policymaking* (New York: Norton, 1986), pp. 185–87.

28. Haig, *Caveat*, p. 53.

29. Haig, *Caveat*, p. 54.

30. Haig, *Caveat*, p. 76.

Chapter 16

1. Unless otherwise noted, this narrative draws from *Texas v. Johnson*, 491 US 397 (1989); Phillip B. Kurland and Gerhard Casper, eds., *Landmark Briefs and Arguments of the Supreme Court of the United States: Constitutional Law*, 1988 Term Supplement, Vol. 190 (Bethesda, Md.: University Publications of America, 1990), pp. 336–688; and Daniel H. Pollitt, "The Flag Burning Controversy: A Chronology," *North Carolina Law Review* 70 (1992), pp. 553–614.

2. 491 US 397 (1989).

3. The briefs are included in Kurland and Casper, *Landmark Briefs*.

4. This discussion is drawn from David G. Savage, *Turning Right: The Making of the Rehnquist Court* (New York: Wiley, 1992), pp. 259–60.

5. Pollitt, "The Flag Burning Controversy," p. 557.

6. Marcia Coyle, "Rally 'Round the Flag?" *National Law Journal*, May 14, 1990, p. 1.

7. Pollitt, "The Flag Burning Controversy," p. 591.

8. Robin Toner, "Patriotism and Politics Mix in Reaction to Flag Burning," *New York Times*, June 12, 1990, p. B6.

9. Robert A. Carp and Ronald Stidham, *Judicial Process in America*, 2nd ed. (Washington, D.C.: Congressional Quarterly Press, 1993), p. 127.

10. 1 Wheaton 304 (1816).

11. Carp and Stidham, *Judicial Process*, pp. 44, 121.

12. Carp and Stidham, *Judicial Process*, pp. 36, 124.

13. Carp and Stidham, *Judicial Process*, p. 127.

14. On Native American courts, see Vine Deloria Jr. and Clifford M. Lytle, *American Indians, American Justice* (Austin: University of Texas Press, 1983); and Stephen L. Pevar, *The Rights of Indians and Tribes* (Carbondale: Southern Illinois University Press, 1992).

15. Pevar, *The Rights of Indians and Tribes*, p. 97.

16. For a discussion of the development and early years of the committee's work, see Joel B. Grossman, *Lawyers and Judges: The ABA and the Politics of Judicial Selection* (New York: Wiley, 1965). Also see "The Committee on Federal Judiciary: What It Is and How It Works," *ABA Journal* 63 (June 1977), pp. 803–7. As part of a negotiated settlement of concerns expressed by the Bush administration, the ABA agreed to eliminate its top rating, "exceptionally well qualified." See Sheldon Goldman, "The Bush Imprint on the Judiciary: Carrying on a Tradition," *Judicature* 74 (April–May 1991), pp. 294–306.

17. Neil D. McFeeley, *Appointment of Judges* (Austin: University of Texas Press, 1987), p. 16. The brief description here only touches the surface of what can be a very complex process. What happens when two senators from the president's party disagree over a nomination? What happens when a senator chooses not to try to dominate the process, or when the administration insists on its candidate and the senators insist on theirs? These situations do occur, and the outcomes vary.

18. Lawrence Baum, *The Supreme Court*, 4th ed. (Washington, D.C.: Congressional Quarterly Press, 1992), p. 39.

19. For example, on Clinton's role in the nomination of Stephen Breyer, see Neil A. Lewis, "Clinton Nears a Court Choice as Lobbying Becomes Intense," *New York Times*, May 10, 1994, p. A21; Gwen Ifill, "Clinton Again Puts Off Decision on Nominee for Court," *New York Times*, May 11, 1994, p. A16.

20. Baum, *The Supreme Court*, p. 47.

21. On district courts, see Robert A. Carp and C. K. Rowland, *Policymaking and Politics in the Federal District Courts* (Knoxville: University of Tennessee Press, 1983). On courts of appeal, see Sheldon Goldman, "Voting Behavior on the United States Courts of Appeals Revisited," *American Political Science Review* 69 (1975), pp. 491–506; and J. Woodford Howard Jr., *Courts of Appeals in the Federal Judicial System: A Study of the Second, Fifth, and District of Columbia Circuits* (Princeton: Princeton University Press, 1981). On the conservative posture of George Bush's appointees to district and appeals courts, see Robert A. Carp, Donald Songer, C. K. Rowland, Ronald Stidham, and Lisa Richey-Tracy, "The Voting Behavior of Judges Appointed by President Bush," *Judicature* 76 (April–May 1993), pp. 298–302.

22. Steve Alumbaugh and C. K. Rowland, "The Links between Platform-Based Appointment Criteria and Trial Judges' Abortion Judgments," *Judicature* 74 (October–November 1990), pp. 153–62.

23. On the attributes of Supreme Court justices, see John Schmidhauser, *Judges and Justices: The Federal Appellate Judiciary* (Boston: Little, Brown, 1979). Schmidhauser concludes that throughout its history "only a handful of the members of the Supreme Court were of essentially humble origins" (p. 49).

24. See Ruth Marcus, "Building Diversity in Lifetime Positions: Women and Minorities Dominate Initial Selections for the Federal Bench," *Washington Post*, January 16, 1994, p. A27; and Naftali Bendavid, "Adding Diversity to the Bench," *New Jersey Law Journal*, January 3, 1994, p. 4.

25. Sheldon Goldman, quoted in Bendavid, "Adding Diversity to the Bench."

26. For an examination of the gatekeeping role of lawyers for the poor, see Mark Kessler, *Legal Services for the Poor: A Comparative and Contemporary Analysis of Interorganizational Politics* (New York: Greenwood Press, 1987); and Mark Kessler, "The Politics of Legal Representation: The Influence of Local Politics on the Behavior of Poverty Lawyers," *Law & Policy* 8 (April 1986), pp. 149–67.

27. On African Americans in the legal profession, see Geraldine R. Segal, *Blacks in the Law: Philadelphia and the Nation* (Philadelphia: University of Pennsylvania Press, 1983).

28. Cynthia Fuchs Epstein, *Women in Law* (New York: Basic, 1981), p. 4.

29. The most complete statistics on American lawyers can be found in Barbara A. Curran et al., *The Lawyer Statistical Report: A Statistical Profile of the U.S. Legal Profession in the 1980s* (Chicago: American Bar Foundation, 1985) and Curran's *Supplement to the Lawyer Statistical Report* (Chicago: American Bar Foundation, 1986). The most useful recent sociological study of the local bar is John P. Heinz and Edward O. Laumann, *Chicago Lawyers: The Social Structure of the Bar* (New York: Russell Sage Foundation and American Bar Foundation, 1982).

30. See Heinz and Laumann, *Chicago Lawyers.*

31. For example, a study by the *National Law Journal* reported that in 1991 the average income of attorneys practicing alone was $89,000. For attorneys in small firms in the Northeast, the average income was $97,000. Partners in large firms, who share in the firm's profits, made anywhere from $156,000 (at Armstrong, Teasdale, Schlafly & Davis in St. Louis) to $637,000 (at Debevoise & Plimpton in New York City). See "What Lawyers Earn," *National Law Journal*, April 27, 1992, pp. S1–S8.

32. For an impassioned description of the history of discrimination in the legal profession, see Jerold S. Auerbach, *Unequal Justice: Lawyers and Social Change in Modern America* (New York: Oxford University Press, 1976), especially Chapter 4, "Cleansing the Bar."

33. See Robert L. Nelson, *Partners with Power: Social Transformation of the Large Law Firm* (Berkeley: University of California Press, 1988). In his study of large firms in Chicago, Nelson reports that only 1.3 percent of lawyers were black, Hispanic, or Asian American. But the numbers of Jews and Catholics in Chicago firms had increased throughout the late 1970s and 1980s. In a national survey of large firms, the *National Law Journal* reported that only 2 percent of associates and fewer than 1 percent of partners were black, with even smaller numbers of Hispanics and Asians. See Rita Henley Jensen, "Minorities Didn't Share in Firm Growth," *National Law Journal*, February 19, 1990, p. 1.

34. See Jensen, "Minorities Didn't Share in Firm Growth."

35. A study of eight top law schools by the *National Law Journal* found that nearly 65 percent of their graduates between 1983 and 1987 took jobs in law firms. Nearly 20 percent clerked for a judge. The remaining graduates either worked for the government, worked as house counsel for a corporation, or took an uncategorized position. Only 2.1 percent—or 243 of 11,000 graduates—took jobs with public interest organizations, such as legal services for the poor, consumer groups, foundations, or advocacy organizations such as the NAACP, the National Organization for Women, or the American Civil Liberties Union. See David A. Kaplan, "Out of 11,000, 243 Went into Public Interest," *National Law Journal*, August 8, 1988, p. 1. See also Robert Stover, *Making It and Breaking It: The Fate of Public Interest Commitment during Law School* (Urbana: University of Illinois

REFERENCES

Press, 1989); and Robert Granfield, *Making Elite Lawyers: Visions of Law at Harvard and Beyond* (New York: Routledge, 1992).

36. See Kessler, *Legal Services for the Poor;* and Marc Galanter, "Why the Haves Come Out Ahead: Speculations on the Limits of Legal Change," *Law & Society Review* 9 (1974), pp. 95–160.

37. Frances Kahn Zemans, "Legal Mobilization: The Neglected Role of the Law in the Political System," *American Political Science Review* 77 (1983), pp. 690–703. For a view that legal mobilization takes place on an uneven playing field, constraining the types of issues that disadvantaged groups may bring to courts, see Mark Kessler, "Legal Mobilization for Social Reform: Power and the Politics of Agenda Setting," *Law & Society Review* 24 (1990), pp. 121–43.

38. For a brief review of interest groups as participants in the American legal system, see Lee Epstein, "Courts and Interest Groups," in John B. Gates and Charles A. Johnson, eds., *The American Courts: A Critical Assessment* (Washington, D.C.: Congressional Quarterly Press, 1991), pp. 335–71.

39. 347 US 483 (1954).

40. Clement Vose, *Caucasians Only* (Berkeley: University of California Press, 1959).

41. See Susan Behuniak-Long, "Friendly Fire: Amici Curiae and *Webster v. Reproductive Health Services,*" *Judicature* 74 (February–March 1991), pp. 261–70.

42. Epstein, "Courts and Interest Groups," pp. 345–49.

43. Behuniak-Long, "Friendly Fire."

44. Epstein, "Courts and Interest Groups," pp. 354–57.

45. On agenda setting in federal district courts, see Joel B. Grossman and Austin Sarat, "Litigation in the Federal Courts: A Comparative Perspective," *Law & Society Review* 9 (1975), pp. 321–46; and Joel B. Grossman, Austin Sarat, Herbert M. Kritzer, Stephen McDougal, Kristin Bumiller, and Richard Miller, "Dimensions of Institutional Participation: Who Uses the Courts, and How?" *Journal of Politics* 44 (1982), pp. 86–114.

46. For an excellent description of the procedures used in the certiorari process, see H. W. Perry Jr., *Deciding to Decide: Agenda Setting in the United States Supreme Court* (Cambridge, Mass.: Harvard University Press, 1991).

47. Perry, *Deciding to Decide,* pp. 135–36.

48. Perry, *Deciding to Decide,* p. 129.

49. Gregory A. Caldeira and John R. Wright, "Organized Interests and Agenda Setting in the U.S. Supreme Court," *American Political Science Review* 82 (1988), pp. 1109–28.

50. This example is mentioned in Stephen L. Wasby, *The Supreme Court in the Federal Judicial System,* 3rd ed. (Chicago: Nelson-Hall, 1988), p. 177.

51. Baum, *The Supreme Court,* p. 103.

52. See Linda Greenhouse, "Court Limits Legal Standing in Suits," *New York Times,* June 6, 1992, p. A12.

53. Perry, *Deciding to Decide,* p. 257.

54. For an excellent extended treatment of the Court's decision-making process, see Henry J. Abraham, *The Judicial Process,* 6th ed. (New York: Oxford University Press, 1993), pp. 188–231.

55. The case was *Smith v. Allwright,* 321 US 649 (1944). The story is told in Abraham, *The Judicial Process,* p. 209. It is adapted from Alpheus T. Mason, *Harlan Fiske Stone: Pillar of the Law* (New York: Viking, 1956), pp. 614–15.

56. Abraham, *The Judicial Process,* p. 210.

57. For an excellent treatment of the fluid nature of justices' decision making, see J. Woodford Howard Jr., "On the Fluidity of Judicial Choice," *American Political Science Review* 62 (1968), pp. 43–57.

58. Classic studies include C. Herman Pritchett, *The Roosevelt Court: A Study in Judicial Politics and Values 1937–1947* (New York: Macmillan, 1948); and Glendon Schubert, *The Judicial Mind* (Evanston, Ill.: Northwestern University Press, 1965). A recent example of this approach is Jeffrey A. Segal and Harold J. Spaeth, *The Supreme Court and the Attitudinal Model* (Cambridge, Eng.: Cambridge University Press, 1993).

59. For examples, see S. Sidney Ulmer, "Social Background as an Indicator to the Votes of Supreme Court Justices in Criminal Cases: 1947–1956 Terms," *American Journal of Political Science* 19 (1973), pp. 622–30; and C. Neal Tate, "Personal Attribute Models of the Voting Behavior of United States Supreme Court Justices: Liberalism in Civil Liberty and Economic Decisions, 1946–1978," *American Political Science Review* 75 (1981), pp. 355–67.

60. See Walter Murphy, *Elements of Judicial Strategy* (Chicago: University of Chicago Press, 1964).

61. Linda Greenhouse, "Court Could Overturn *Roe* in an Unexplicit Way," *New York Times,* January 24, 1992, p. A12.

62. See Walter Murphy, *Congress and the Court: A Case Study in the American Political Process* (Chicago: University of Chicago Press, 1962); Edward Keynes and Randall K. Miller, *The Court vs. Congress: Prayer, Busing, and Abortion* (Durham, N.C.: Duke University Press, 1989).

63. Abraham, *The Judicial Process,* p. 328.

64. Abraham, *The Judicial Process,* p. 328, n. 74.

65. 6 Peters 515 (1832).

66. Albert J. Beveridge, *The Life of John Marshall,* Vol. 4 (Boston: Houghton Mifflin, 1919), p. 551, discussed in Abraham, *The Judicial Process,* p. 333.

67. See Walter Murphy, "Lower Court Checks on Supreme Court Power," *American Political Science Review* 53 (1959), pp. 1017–31.

68. For a view that the institutional limitations of the Supreme Court make it a poor avenue for litigation seeking significant social change, see Gerald N. Rosenberg, *The Hollow Hope: Can Courts Bring About Social Change?* (Chicago: University of Chicago Press, 1991).

69. 1 Cranch 137 (1803).

70. Abraham, *The Judicial Process,* p. 273, Table 8.

71. Robert A. Dahl, "Decision-Making in a Democracy: The Supreme Court as a National Policy-Maker," *Journal of Public Law* 6 (1957), pp. 279–95.

72. Thomas R. Marshall, *Public Opinion and the Supreme Court* (Boston: Unwin Hyman, 1989).

73. Marshall, *Public Opinion,* p. 192.

74. See Jonathan Casper, "The Supreme Court and National Policy Making," *American Political Science Review* 70 (1976), pp. 50–63.

75. Abraham, *The Judicial Process*, p. 272.

Illustration credits continued from p. iv

INDEX TO REFERENCES

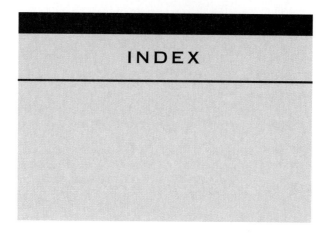

INDEX

Throughout this index, the lowercase letters *c, t,* and *f* indicate *captions, tables,* and *figures,* respectively.